The ecology of algae

The ecology of algae

F. E. ROUND

Professor of Botany, University of Bristol

Cambridge University Press

CAMBRIDGE

LONDON   NEW YORK   NEW ROCHELLE

MELBOURNE   SYDNEY

CAMBRIDGE UNIVERSITY PRESS
Cambridge, New York, Melbourne, Madrid, Cape Town, Singapore, São Paulo, Delhi

Cambridge University Press
The Edinburgh Building, Cambridge CB2 8RU, UK

Published in the United States of America by Cambridge University Press, New York

www.cambridge.org
Information on this title: www.cambridge.org/9780521269063

First published 1981
First paperback edition 1984
Reprinted 1985
Re-issued in this digitally printed version 2009

A catalogue record for this publication is available from the British Library

Library of Congress Cataloguing in Publication data
Round, Frank Eric.
The ecology of algae.
Includes index.
1. Algae–Ecology. I. Title.
QK565.R76 589'.35   79–50886

ISBN 978-0-521-22583-0 hardback
ISBN 978-0-521-26906-3 paperback

CONTENTS

ACKNOWLEDGEMENTS

I am exceedingly thankful for fascinating discussions with and for the help and hospitality of many phycologists, too numerous to mention by name. My research students have continuously provided stimulation and critical comments over a wide range of algal problems. In particular I thank Dr J. W. G. Lund, FRS, CBE, who allowed me to use his collection of reprints, Professor M. Doty, Professor B. Wommersley, Dr C. Happey-Wood and Dr Brian Moss, for ideas and discussion, and Dr P. Broady for the illustrations of algae. The librarians of the Marine Biological Association, the Freshwater Biological Association and the British Museum (Natural History) eased my task of searching the literature. I am particularly indebted to Dr H. Lund for supplying me with so many excellent photographs from which to choose just a few. The idea for the book grew out of the experience of teaching a summer school at the Lake Itasca Field Station of the University of Minnesota and I am grateful to Dr A. J. Brook for inviting me to such an idyllic limnological setting. My colleague Dr R. Campbell has kindly read the entire text and made many suggestions for improvement; I thank him and everyone else who has assisted but I alone accept responsibility for the statements.

Finally I thank my family who have tolerated algal literature scattered around the house and Mrs P. Calrow and Mrs J. B. Hancock who have typed and retyped the many drafts.

January 1979   *F. E. Round*

INTRODUCTION

This book is a first attempt to present a unified account of algal ecology without the artificial division into freshwater and marine communities. The variation in salinity is continuous from rainwater, through freshwaters, the ocean and to hypersaline land-locked waters and similar basic principles apply to the study of all waters.

Ecology is the study of organisms in their natural environment and therefore the 'organisms' and 'environments' are of equal concern to the ecologist. However, both 'organisms' and 'environments' have intrinsic properties and these, though not directly the concern of the ecologists, nevertheless require understanding. For example, environmental factors include the anomalous properties of water, the solubility of gases and solids in water, the penetration of light, and temperature–density relationships, while organic factors include intrinsic growth rates, life cycles, dispersal mechanisms of species, etc. Some of these facets are to a varying degree independent of one another (e.g. the input of solar radiation, current effects, time of nuclear division, selection of certain metabolic pathways) and can be discussed in relative isolation, though it is well to realise that there is some inter-reaction between most, if not all, properties of the environment and of the organism. Other aspects are so interlocked that discussion of one in isolation is virtually meaningless, for example the availability of nutrients such as nitrogen, phosphorus and silicon is intricately bound up with the rate of cycling by organisms, and the rates of change of populations are inextricably bound up with the fluctuations of the physico-chemical factors. In the aquatic environment, much more so than on the land, the growth of the organisms themselves has a rapid and dramatic effect on the environment and whilst seasonal climatic changes mould the environment and are often intensively studied the effects of the organisms are less studied but often more dramatic. Inevitably some habitats have been given greater consideration – some deserve it – but equally the treatment is a reflection of the unequal effort in the field and I hope that I shall succeed in bringing neglected aspects to the consideration of the reader. On the land the angiosperms of the grasslands, forests, etc. provide the fixed carbon for utilisation by all other organisms but in the aquatic environment the algae almost alone, provide the complex organic molecules for the vast range of animals in the water. It is perhaps worth emphasising that the ocean waters form, by far, the world's largest habitat. As work on this book progressed I began to realise what a daunting task I had undertaken since the range of algal ecology is comparable to that of land ecology.

Clearly the literature is too vast to cover in detail but one of my aims is to lead the reader into the relevant literature by illustrating points with *quoted* examples rather than generalised statements. Most examples have been taken from recent accounts and to keep the literature list within reasonable limits I have not referred to much of the earlier work. This does not imply any criticism of the early workers, and students should certainly consult the historical data for there is much of value to be discovered and much modern work is merely repetitive; however, to have included the references would have doubled the number. No doubt I have failed to refer to some important papers and to many which readers regard as important, but a choice simply had to be made. Works which merely list species are often valuable for phytogeographical studies but I have quoted them only if they have some special value. Individual lakes, sites, etc. are quoted whenever possible. Although I realise they may have little relevance to many readers, they are nevertheless essential to ecologists since they enable future workers to return to sites to check earlier data, continuity, pollution effects, etc. It is as important to an ecologist to be able to return to a site as it is to a taxonomist or experimentalist to be able to obtain a type species or authenticated culture. A further aim is to pin-point the diverse aspects of algal ecology, references to many of which are buried in obscure journals, or journals outside the normal range consulted by botanists. This necessitates referring to topics which have been investigated only in a cursory manner and the numerous mentions in the text to the paucity of data are added purposely to highlight the need for intensive studies. No apologies are given, for many fields are utterly neglected and desperately require attention.

Throughout the reading of the literature, an obvious split between the floristic and experimental approaches is revealed: *both* are absolutely essential parts of ecology, but rare are the publications which attempt to combine them. Those workers who disapprove of lists of species are denying the reality of the complexity of algal communities which *must* be described somewhere. Poore (1962) summed this up neatly: 'the exact description and characterisation of the community thus becomes essential to all who work in the science of vegetation; for it is important that the ecological findings should be related to communities which are well described. Not to do so is tantamount to carrying out a physical experiment without stating the conditions'. Equally, those who do not welcome data based on chlorophyll estimations, ^{14}C uptake studies etc., are rejecting valuable even if simplified approaches to complex situations.

Both approaches are *essential* elements of ecology and neither takes precedence but data from the former have simply been accumulating over a longer period of time. There are unfortunate comments in the literature that the organismal–community approach is difficult and to be side-stepped wherever possible. It is not possible to do this and I can best quote Margalef & Estrada (1971): 'Identification of species is tedious and time consuming, but it brings high quality information – provided the identifications are reliable. *No short cuts exist* for this and of course nothing, no information – is given freely' and Nicholls & Dillon (1978) 'direct microscopic determination of biomass appears to provide the best estimate of phytoplankton biomass'. Nevertheless

I have in the succeeding chapters cut down to a minimum the listing of species, though I realise this will be criticised by those who would like to have complete accounts of communities. The full data can however often be found in the quoted papers and one must remember that the large number of species in most associations is such that to quote them would make a most unwieldy account. I have therefore attempted to strike a balance and point out other facets of the biology of the species, associations, etc.

Ecologists are concerned with many levels of organisation although many workers confine themselves to one level and few publications attempt an overall view. This is not surprising since the levels of organisation involve single species, communities of many species, interacting communities and the ecosystem itself. At the first level there has been only slight study of genetic variability but this has been sufficient to show that such variability is considerable and important.

Some of the work in ecology is moving towards biochemical functioning within natural ecosystems rather than in the unnatural conditions in laboratories and indeed little ecological significance can be attached to many laboratory-based experiments until similar effects have been demonstrated in nature. The biochemical aspect is also compounded by interactions between algae and other organisms. Another approach, which is in its infancy and too often has to rely on extremely crude data, is that of mathematical modelling of the ecosystems. This relies on the skill with which the data are obtained from the traditional approaches and is only as good as the initial observations.

I have attempted to indicate areas where studies are needed but I am all too conscious that someone, somewhere, may have completed studies and it is I who have regrettably missed the appropriate publications; in all such instances I apologise and I shall be delighted to hear of the results.

The aquatic environment is not uniform but contains within it numerous spatial niches and these are each colonised by very well-defined communities of algae – the open water by the phytoplankton, the sediments by the epipelon, plant surfaces by the epiphyton, rock surfaces by the epilithon, etc. There is slight interplay between these communities but they have a high degree of independence and need to be discussed separately. The interactions with other communities and between species within the community are as yet largely undocumented.

Though the subject of this book is algal ecology this, like higher plant ecology, cannot be isolated from interaction with the animal world. The algae live in a solution of animal waste products of varying concentration, are grazed directly, or indirectly following bacterial decay, and in many habitats are increasingly influenced by the single most active animal – man. Algal ecology has, however, one fortunate feature which is absent from higher plant–animal ecology and that is the almost cosmopolitan distribution of most freshwater species, of oceanic phytoplankton and even of some seaweeds (though here the greatest degree of geographical isolation occurs). The genera therefore tend to be familiar to workers in different continents even if some of the species are different. Equally, however, algae are not ubiquitous throughout each habitat-type and each water body has its own peculiar selection of species.

As mentioned above there are two equally important sides to ecology and it is not always easy or desirable to separate them. One aspect is the detailed analysis of communities, the kinds and numbers of organisms and seasonal changes within each discrete community, and the second is the functional aspect of the individual species and of the communities as a whole. For the study of both aspects it is necessary to define quite clearly the individual, often restricted, habitats in which algae live and then to analyse the physico-chemical background of these habitats. The approach to community structure involves the use of precise sampling techniques and is dependent upon the continuing development of basic and all-important taxonomic revisions of groups of algae.* Here much remains to be achieved in spite of 150 or more years of study. Wherever the problems of identification are too

great, effort can always be diverted from a synecological to an autecological approach in which a single, more readily identified, species is selected for study. The functional aspect is dependent upon developments in physiology and biochemistry. It requires the same attention to habitats but also to sophisticated techniques for detecting the operation of processes and for measuring the rate of such processes under natural conditions, for it is here in nature that the rates of carbon fixation, nitrogen fixation, carbohydrate secretion etc., are of prime significance.

Ideally it is possible to describe the conditions under which every alga lives, how its functioning is affected by and how it interacts with the habitat. Equally, the precise habitat of each alga is theoretically definable and this is easier since a relatively small number of habitats are occupied by a large number of species. However I find it difficult to define *some* of the rather diffuse habitats which algae occupy; it is essential for comparative purposes to be able to identify habitats precisely at numerous sites and many more detailed studies of microhabitats need to be undertaken. Few autecological studies of individual algae have been attempted (examples are *Ascophyllum nodosum* (Baardseth, 1970), *Laminaria hyperborea* (Kain, 1971)) and many more need tackling along the lines of the excellent Biological Flora accounts of the angiosperms published by the British Ecological Society; but even here only a fraction of the species has been tackled. The problems with algae are more acute than with angiosperms but they are not insurmountable and present a challenge to phycologists. A similar series of accounts of algae is slowly being published by the British Phycological Society, but there is a massive amount of work to complete throughout the world.

There is a direct interaction resulting from man's waste products but even more important are the indirect and often unconscious effects of his economic activities; these threaten to be even more disastrous than his natural activity. The study of ecology is thus in the broadest sense the study of interaction – give and take, stimulus and response, stimulation and feed-back. Since quantitative field and laboratory observations can be made, a mathematical base is being gradually defined and from the generalities of this base a new systems ecology is evolving.

---

* As far as has been possible, names of European marine algae have been brought into line with the 3rd edition of Parke & Dixon (1976) *Checklist of British Marine Algae.* A few lists compiled by European workers have not been altered since there is some slight doubt about the application of the British names to the more distant material. The Cyanophyta have been quoted as in authors' papers and not altered since I believe that there are important ecological variants which should not be submerged in a few taxa. This does not imply that revisions are unnecessary, they are necessary, but must be based more on ecological–cultural studies.

Algal ecology thus involves a continuum of web-like interactions between the environment, the algae themselves, the animals and man's industrial disturbance, but readers and publishers expect chapters and not a continuum of discussion. As previously suggested some sections of the subject can be isolated, e.g. the intrinsic physico-chemical environment (Chapter 1), the littoral benthos (Chapter 3), etc., but compartmentalisation into chapters inevitably tends to isolate the various aspects. One must always stress that nothing is operating in isolation in ecology and even where the data are sparse, e.g. in the all-important aspect of algal–animal relationships, their importance is immense and bears no relationship to the number of papers published. A further important aspect of algal ecology is the breakdown of algae by fungi and bacteria and the re-cycling of their components, but regrettably little can be written on this aspect because the literature is almost devoid of studies. Algae are the primary carbon fixing organisms in the aquatic environment and are thus an indispensable link between solar radiation, the complex solution of chemicals in the waters, *all* aquatic animals and eventually man, whose existence is dependent on oxygen evolved in plant photosynthesis; some 50–90 % of this oxygen is estimated to come from algal growth. The original oxygen of the earth's atmosphere is presumed all to have been derived from algal photosynthesis (Cloud, 1968) and it plays a vital role in geochemical evolution of the lithosphere, hydrosphere and atmosphere (Cloud *et al.*, 1965).

Two branches of science, limnology and oceanography, have developed and have, unfortunately, to some extent grown apart. Limnology is concerned with the freshwaters on the land surface, and oceanography with saline water, though the most saline water occurs in inland lakes and the salinity of the seas varies from 'freshwater' (e.g. the melting ice of the polar regions and the upper regions of enclosed seas such as the northern part of the Baltic) to the somewhat more saline tropical evaporation basins (e.g. the Red Sea, Gulf of California, etc.). The distinctions within these compartments of science are thus merely of degree. Few if any organisms are present in both fresh and saline waters but the same *kinds* of organisms with *similar* life forms are present, as is obvious if one compares the forms of epiphytic or epipelic algae in marine and freshwater habitats

(Fig. 2.1*a*). Many factors operate in a similar manner in the sea and in lakes, e.g. light penetration and temperature stratification, and can be discussed together. Techniques of sampling and of measurement are essentially similar and the distinction between the two is grossly distorted by the fact that sampling in the Atlantic Ocean or in the Pacific Ocean requires a large expensive ship whilst a rowing boat is adequate on the much studied Linsley Pond in Connecticut.

Another important habitat in which algae grow is the so-called sub-aerial habitat: they occur on the surfaces of plants, especially in the tropics but commonly on tree bark throughout the world and on soils of all types. Algal floras exist in such unlikely habitats as moist rock surfaces, even within rocks in arid zones, and on the surfaces of permanent snow fields. These floras tend to be confined to a relatively small number of highly characteristic species often with an almost world-wide distribution.

I see no virtue whatsoever in separating off a section of so-called applied algal ecology but considerable studies have been undertaken on man-made systems utilising water, e.g. sewage disposal and water supply installations, where the algae are beneficial or detrimental to the system. Thus algae in reasonable amounts provide oxygen, take up nutrients and *clean* the water but in too large amounts produce filtration problems (e.g. in waterworks) or decay and utilise excessive amounts of oxygen (e.g. in the deep waters of lakes such as Lake Erie and in relatively enclosed oceanic basins) with subsequent detrimental effects. Other systems have been evolved to utilise algal growth for food (especially in the Pacific region), for extraction of organic compounds, and for production of Crustacea and fish and all of these benefit from and contribute to the study of ecology. Their problems are intricately bound up with basic limnological and oceanographic studies and cannot be studied in isolation or without consideration of the natural systems.

The sequence of topics is deliberately chosen to deal first with a brief survey of the physico-chemical background (Chapter 1), followed by a general discussion of the habitats and communities (Chapter 2). Then habitats exposed first to rainwater as it falls, through increasing salinity and complexity due to percolation, flow and interaction of the water with sediments, etc. are discussed in more detail. Thus,

Chapter 3 deals with the flora of bare rock surfaces on mountains subjected almost entirely to rainwater, then with that of the rock surfaces of streams, rivers and lakes and finally with the flora of the rocky sea shore. A similar but geographically restricted habitat involves the free-living species of the algal community of coral reefs and this is dealt with in Chapter 4. The breakdown of the rock and its intermixture with organic matter to give soil, and underwater sediments provides the next habitat group on the land surface, in rivers, lakes and shallow seas, all richly colonised by algae (Chapter 5). At some point a discussion of snow and ice algae had to be incorporated and since these both grow on 'sediments' they are described at the end of Chapter 5. Into this chapter I also fit the algae of hot springs since these algae are growing on and forming a thick mat over the sediments of the outflow streams. In the sediments (and on the rocks) macroscopic algae, other plants and animals grow and these are colonised by attached algae (Chapter 5). The water which has percolated, seeped and drifted past all these communities then collects in rivers, lakes and the open ocean and here the surface layers are colonised by the phytoplankton (Chapter 7). In the chapters devoted to habitats I have made no attempt to present a standardised account but have treated each in the way I felt to be most suitable. Algal floras are the appropriate place to find illustrations of all the algae mentioned. I have, however, included sufficient illustrations to show the diversity of form within each habitat. On these, no attempt has been made to draw the genera to scale but I have indicated the range in the captions.

The remaining chapters deal with certain general and inter-linked aspects – dispersal and phytogeography (Chapter 8), symbiosis–parasitism–grazing (Chapter 9), seasonal succession (Chapter 10), energy flow and cycling of nutrients (Chapter 11), sedimentation of algal remains (Chapter 12), palaeoecology (Chapter 13), and finally eutrophication and pollution (Chapter 14). I have not eliminated all mention of the latter aspects from the earlier chapters, especially when dealing with the less well-studied communities, where it seemed more appropriate to treat all aspects together.

I The physical and chemical characteristics of the environment

Slightly over 70% of the surface of this planet is submerged under water and much lies at great depths beneath the water. In fact 84% of the ocean has a depth of more than 2000 m (Fig. 1.1). It has been calculated that if the earth were smoothed out it would be covered by water to a depth of about 2440 m. There are two theories to explain the origin of the water on the earth's surface. The first assumes that heating of the material which accumulated to form the earth produced a metallic iron core accompanied by release of vapour and salts. This so-called 'degassing' process is presumed to have occurred primarily during the first 500 000 000 years of the earth's existence. The second theory only differs in that this 'degassing' process is presumed to continue throughout geological time so that the gaseous material released by volcanoes and hot springs has steadily accumulated. Whatever the process or processes giving rise to the aqueous media there can be no doubt that suitable, but changing,

Fig. 1.1. The fractions of the earth's and of the oceans' surface areas covered by each depth of water. From Russell-Hunter, 1970.

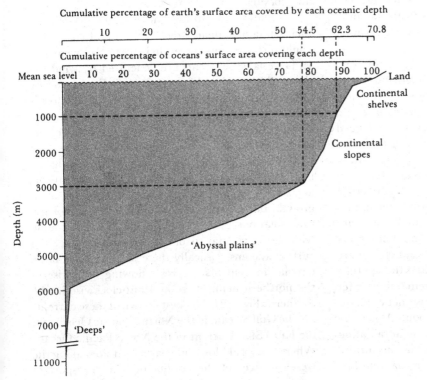

conditions for algal growth have existed for a considerable part of geological time. Algal ecology is mainly concerned with the present-day situation, but some aspects require an appreciation of past climates, movement of land masses, etc. On the other side of the coin the residues from algal growth are contained in sediments and rocks and are of considerable importance in geological and geochemical studies.

Whilst the oceans are ancient features, lakes and rivers are relatively recent and the majority of the world's lakes are situated in temperate to subarctic regions. They have only existed since the last retreat of the Pleistocene ice sheet from their sites. Most of these lakes, from the smallest to the sea-like Laurentian Lakes, are the result of water accumulation in hollows deepened by the movement of the ice sheet and of glaciers over the surface of the land. They therefore vary in age from recently formed lakes abutting the present ice sheet to those more temperate lakes formed between 12000 and 15000 years ago. Lakes in tropical regions have a different origin, for here the water accumulates in basins formed by tectonic activity, e.g. the rift valley lakes of East Africa and the crater lakes located in extinct volcanic cones. There are many more minor agencies involved in lake formation and for a full and fascinating account the student should consult Hutchinson (1957).

Freshwaters form only about 1 % of the total global area of water, so the total contribution to organic production from this source is relatively small. There are however large areas of bog and swamp, flood plains, etc. which are waterlogged (often termed wet lands, e.g. the Sud along the River Nile) and these are good habitats for the growth of algae though, as yet, little investigated.

Only a relatively small number of angiosperms, pteridophytes and bryophytes, form associations within the aquatic environment and they are abundant only in shallower waters and along some temperate and tropical coasts. Algae of course are intimately mixed into these associations though they do not dominate the scene. In the remaining water masses the algae alone of photosynthetic plants colonise the medium and fix carbon. Algae are however dependent upon the influx of solar radiation into the water and hence only grow in the surface veneer which is penetrated by this radiation. The vast mass of water lies below this veneer, which in depth is less than that represented by the wavy line of Fig. 1.1; the mean depth of the ocean is 3800 m whilst the depth to which photosynthetically usable light penetrates is less than 200 m even in the clearest water.

It is impossible to consider algal ecology without an appreciation of the physico-chemical environment and the remainder of this chapter gives a brief outline, for the greater the understanding of these aspects the more perceptive will be the algal ecology.

Circulation of water

No water mass is stationary and therefore aquatic algae are subjected to a flow of water. This relative movement between algae and water results in exposure of the algae to fresh media and a continual removal of extra-cellular products and only in very special circumstances is it reduced to a minimum, e.g. in brine cells in ice, on the surface of soil, in cavities within rock or in epiphytic situations such as within moss clumps. Even the water of the smallest ponds is moved by wind stress on the surface and by convection currents caused by diurnal heating and cooling.

Inflow and outflow move water masses and their contained algae laterally through freshwaters, whilst turbulent mixing results in vertical movements. In large lakes, currents are generated by wind, by the earth's rotational forces (Coriolis Force) and by inflow–outflow movements. Oceanic circulation is even more complex for not only is there a system of surface currents but also a slower circulation of the deeper layers. The surface currents move algae from place to place and the circulation of the deep layers is in part responsible for the supply of nutrients for algal growth. The sun provides energy for algal growth, but is additionally linked to algal ecology, since solar energy generates the winds which couple with the surface layers of the ocean to produce the current systems. Basically these currents form large circular movements ('gyres') flowing clockwise in the northern hemisphere and anticlockwise in the southern (Fig. 1.2). The best known of these currents are the Gulf Stream in the North Atlantic Ocean and the Kuro Shio Current in the North Pacific Ocean. Where the wind blows off coasts, as it does along the western edges of the continents, the transport of surface water seaward results in upwelling of deep, and therefore nutrient-rich, waters and has a dramatic effect on plankton growth (*see* Chapter 7).

In recent years undercurrents have been found in the tropics flowing eastward along the equator; the Pacific undercurrent, also known as the Cromwell Current, is symmetrical about the equator, 300 km wide and has been followed some 3500 miles from 150° W to the Galapagos Islands. The upper boundary of this current is in places no deeper than 30 m and at its maximum velocity it moves at 150 cm s^{-1} or three times as fast as the South Equatorial Current flowing above it in the opposite direction (Knauss, 1965). Thus in this region phytoplankton assemblages can be mixing and moving in different directions within the upper 100 m, and this feature requires detailed study.

The deeper water can often be divided into four layers (upper, intermediate, deep, and bottom water) each characterised by different densities, dependent upon salinity and temperature. Owing to this relationship the deep circulation is often referred to as a thermohaline system. Water differing in salinity and/or temperature, and hence in density, arises in several ways, e.g. inflows of river water, evaporation in the tropics, formation of ice in polar regions. Two examples are often quoted; one is North Atlantic deep water, formed when highly saline Gulf Stream water flowing northward, meets cold, less saline Arctic water resulting in mixed dense water which sinks and drifts southward (Fig. 1.3), and the other example is Antarctic bottom water, which forms when sea ice accumulates as relatively salt-free ice crystals and leaves behind cold brine. This heavy brine sinks and forms the densest known bottom water which drifts northward and can be detected in the bottom of the North Atlantic Ocean (Figs. 1.3, 7.26). In general, most of the deep water is formed by sinking of surface water in the polar regions and to counteract this there is a general slow rise of deep water over the rest of the ocean. Such slow movement of bottom water may transport sedimented algae great distances from their original sites of initial sedimentation, e.g. diatoms from the Antarctic into South Atlantic sites.

At the surface, tongues of water of lower salinity outflowing from massive drainage basins, such as that of the Amazon, can be detected for hundreds of miles out to sea. This system alone is reckoned to discharge some 18% of the total river output of the world and the resulting dilution was thought to lead to a decrease in surface fertility over a million square miles (Ryther, Menzel & Corwin, 1967). More recent work (Cadée, 1975) has shown

Fig. 1.2. The major surface current systems in the oceans of the world.

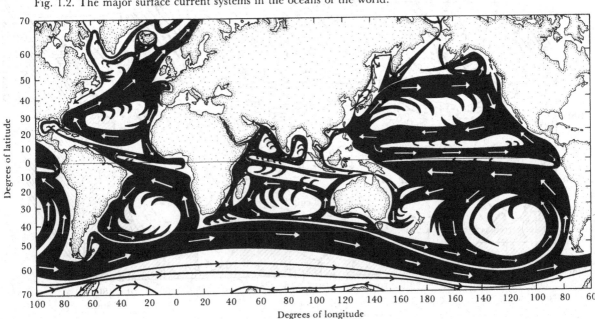

that outflow of Amazon water does not decrease the nutrient content to the extent reported by Ryther *et al.* and productivity in the offshore waters is not particularly lower than the ocean average. Indeed from the numerous reports of optimal algal growth at salinities of 20–24‰ (e.g., Provasoli, 1965) one might expect greater production in the diluted water. In fact mixing of the Amazon water with water of the countercurrent system actually maintains the growth of the inshore species (Hulburt & Corwin, 1969). The immensity of some of these current systems can perhaps be illustrated by the statement of Pratt (1966*a*): 'The Gulf Stream carries a volume of water north through the Straits of Florida that is more than 70 times the combined flow of all the land rivers of the world'.

Some current systems alter their direction of flow at different times of the year and cause dramatic changes in phytoplankton production and migration of fish, e.g. along the Guinea Coast of Africa. If these currents change irregularly they can seriously alter fish production. The El Niño current which flows in a southerly direction along the west coast of South America brings warm water in contact with the cold Humboldt current flowing northward. The southerly current forms a wedge of nutrient-poor water in the coastal region and in turn depresses phytoplankton production: if this wedge becomes so wide that it spreads outward over the deep offshore water it drives the anchovy out to sea with disastrous effects on the coastal fisheries.

Water motion is a much neglected physical factor which operates on all algae. There are hardly any observations on the relationship between the form of the plankton and the degree of motion of the water. Schöne (1970) found that increasing sea force reduced the length of filamentous diatoms. He thought that the downward motion of air bubbles was responsible and in experiments using streams of air bubbles found that the chain length of *Skeletonema* could be reduced. In addition, the vigorous movement increased the number of cells in the culture, although the cell division of another alga (*Chaetoceros*) was reduced by excessive water movement. Turbulent seas also cause fragmentation of the colonies of the blue-green alga *Trichodesmium* (*Oscillatoria*) and this allows oxygen to enter the central colourless cells where nitrogen fixation normally occurs. The

Fig. 1.3. The surface water and deep-water circulation in the Atlantic Ocean. From Defant, 1961.

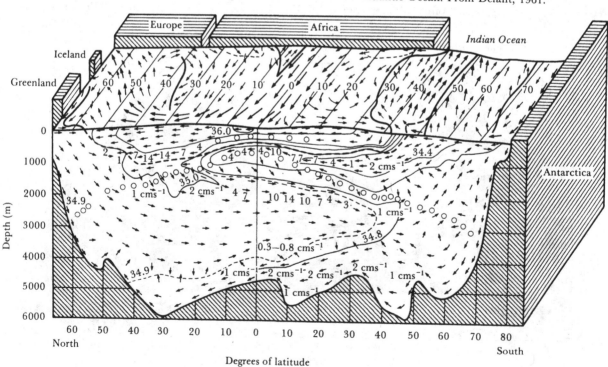

Fig. 1.4. Some examples of different tidal patterns. *a, b, c*, predicted tides at three different localities, 18 June–18 July 1951; *d*, levels at San Francisco; *e*, predicted range of tides at Holyhead during 1959 showing the alternation of neap and spring tides, the annual variation in the range of neap and spring tides and the diurnal inequalities; *f*, data from *e* for 20–31 July 1959 enlarged on a day-to-day basis. MHWS, mean high water spring tide; MHWN, mean high water neap tide; MLWN, mean low water neap tide; MLWS, mean low water spring tide. *a, b, c, d* from Doty, 1957; *e, f* from Lewis, 1972.

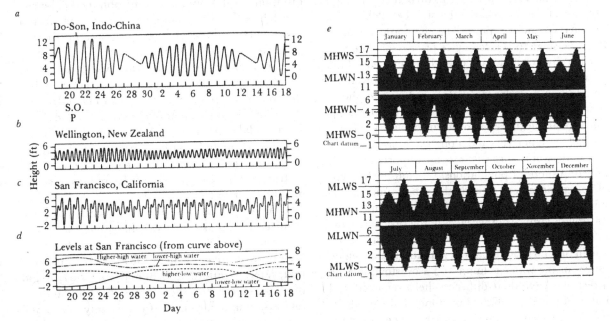

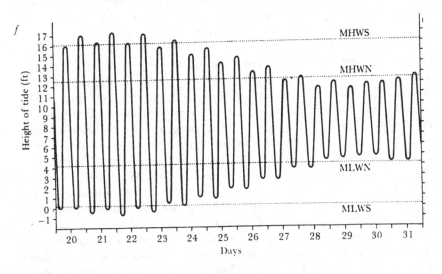

subsequent reduction in growth due to lack of nitrogen keeps the population small and only when calm seas prevail do the colonies remain intact, grow rapidly and form water blooms (Carpenter & Price, 1976). The division rate of *Peridinium* in Lake Kinneret is much reduced when turbulence is high because of wind stress on the surface (Pollingher, personal communication). The effect appears to be on some stage in the cell cycle prior to the onset of nuclear division, since once this has commenced the formation of daughter cells is in no way prevented. In addition turbulence does not reduce the photosynthesis rate of *Peridinium* and hence a simple nutritional effect seems to be ruled out. Motion is most obvious in rivers and along wave-beaten coasts but underwater surge is also an important factor affecting subtidal populations. Obviously this type of water movement moulds the substratum, determines the type of alga which will settle and influences its growth form. Another facet of the problem is the influence of water movement on ion diffusion rates and in turn on growth rates. A planktonic alga which moves in the water column and a crustose alga on a reef will have their diffusion shells removed and influx of nutrients and efflux of metabolic products will be affected to varying degrees. Few attempts to discuss or measure the rates have been made although Munk & Riley (1952) pointed out the importance of the problem in a classic paper. Conover (1968) did however show that increase in standing crop occurs as current velocity increases up to a certain point, after which there is a decrease.

An even less studied aspect is that of measurement of the magnitude of the effects in nature; Jones & Demetropoulos (1968) measured wave force on rocky shores with a spring dynamometer and found drag readings up to 1 kg cm^{-2}. They showed that this factor can be quantified, though the measurements involve embedding apparatus into the rocks. The giant kelps of the Californian coast resist the drag imposed by the waves by having an extensible stipe; *Nereocystis* stipes can extend up to 38 % (Koehl & Wainwright, 1977). The work needed to break these stipes is comparable to that needed to break wood, bone or cast iron.

Doty (1971) recognised the importance of water movement on diffusion rates of tropical reef algae and adapted methods used for studying current velocities to the reef flat environment. The technique involves dissolution from Plaster of Paris (calcium sulphate) blocks cemented to plastic cards placed in the natural habitat and held in position by masking tape. The amount of calcium sulphate lost can be compared to that lost in *still* water and an index of diffusion (or dissolution) can be obtained. The rise and fall of the tide produces water movement in a plane varying from vertical to almost horizontal. The extent of this flow varies considerably from place to place and the force of the water varies according to wind conditions. Tidal patterns are complex but the common types are illustrated in Fig. 1.4.

There is movement of the water up and down shores in lakes but it is much less than along coasts; it is caused by inflow–outflow patterns, wind-induced oscillation (seiches, *see* p. 25) and changes in barometric pressure, e.g. a change of 10 mm produces an oscillation of 13 cm.

In rivers the rate of flow has been shown to influence the uptake of ^{32}P by *Oedogonium*, the uptake increasing in current equivalents up to 0.4 m s^{-1} (Whitford & Schumacher, 1961). It is interesting that, later, Whitford & Schumacher (1964) found that lake species also showed a slight stimulation of respiration in a current, but it was only very small compared to that exhibited by the river forms. On the other hand both the lake and river species showed enhanced nutrient uptake in a current.

Properties of water

Excluding fluids of organic origin, water and mercury are the only naturally occurring liquids. Extrapolation of the melting and boiling points of the dihydrides of oxygen, sulphur, selenium and tellurium would suggest that water should freeze at − 150 °C and boil at 80 °C and since it does not, it is obviously an unusual fluid. Water has the highest heat capacity (except for liquid ammonia), latent heat of fusion (except for liquid ammonia), latent heat of evaporation (of all substances), surface tension (other than mercury) and dielectric constant and it dissolves more substances than any other liquid. It is thus not only an anomalous but also a highly specialised compound. The structure of water is complex and it is still under investigation by physicists, but its unusual properties are related to the tendency of water molecules to associate by means of hydrogen bridges, so that water consists of continually changing, branched chains of imperfect

Table 1.1. *Composition of various waters. Data from Gorham (1955), Hutchinson (1957), Livingston (1963), Culkin (1965) and Visser & Villeneuve (1975): dashes indicate that values were not given*

	Seawater (g kg^{-1} at a salinity of 35‰)	Rainwater (mg l^{-1})	Minimum concentration of Wisconsin lakes (mg l^{-1})	Riverwater (mg l^{-1})	Larger East African Rift lakes (mg l^{-1})	East African saline lakes[a] (mg l^{-1})
Chloride	19.353	0.2–12.6[b]	0.1	7.8	29.9	16 385
Sodium	10.77	0.2–7.5[b]	0.13	6.3	105	18 261
Sulphate	2.712	1.1–9.6[c]	0.75	11.2	39.5	2175
Magnesium	1.294	Nil–0.8[b]	0.5	4.1	33.7	0.5–6.9
Calcium	0.413	0.1–10[d]	0.13	15.0	139	0–22
Potassium	0.387	0.03–0.7	0.25	2.3	62	854
Bicarbonate	0.142	Nil–2.8	—	58.4	—	—
Bromide	0.067	0.03	—	—	—	—
Strontium	0.008	—	—	—	—	—
Boron	0.004	0.01	—	—	—	—
Fluoride	0.001	—	—	—	—	—

[a] From Visser & Villeneuve with calcium and magnesium values added from Degens, Okada, Honjo & Hathaway (1972).

[b] High values found particularly in coastal areas and derived from sea spray.

[c] High values associated with soot and industrial effluent.

[d] High values often over continents and derived from terrestrial dust.

oxygen tetrahedra linked by hydrogen atoms. This tetrahedral lattice structure exists in ice and in water at a temperature near freezing, but as the temperature increases and the association decreases, the increased thermal agitation results in a looser packing; the outcome of these two processes is that the close random packing of the molecules increases and a slight contraction occurs up to the temperature of maximum density (3.94 °C). This explains why ice is less dense than water. The temperature of maximum density of seawater is decreased by increasing salinity so that, at salinities above 24.7‰, it is below the freezing point of freshwater. Although the quasi-crystalline state results in two hydrogen atoms at a distance of 0.99 Å and two at 1.77 Å from a single oxygen atom, a few oxygen atoms are also associated with three or one close hydrogen atom(s) giving hydroxonium (H_3O^+) and hydroxyl (OH^-) ions. In pure water the concentration of these will be equal and at 25 °C is 10^{-7} moles l^{-1}, corresponding to a pH of 7. References to and further details of the properties of water can be found in Hutchinson (1957): whilst they may not directly concern most algal ecologists it is worth remembering that the growth medium for natural algal populations is based on a

very remarkable liquid, the formula of which, written as H_2O, gives no hint of its complexities, and that on a weight basis even seawater is 96–97% water.

Dissolved salts

Pure water probably does not exist in any natural habitat on earth; even if it did, no alga would be able to colonise it owing to the absence of dissolved salts. The cycling of water via evaporation to the atmosphere and its subsequent return in rain results in a dilute solution (*see* Table 1.1) which is then distributed in a most unequal manner back to the surface waters either directly or after percolating through soil.

Sea salt is carried up into the atmosphere in the form of fine particles which were once assumed to provide the nuclei around which cloud forms. However, later workers consider nitrous acid or ammonia to be more likely nuclei. These compounds are formed mainly by photochemical reactions which take place on dust particles in the atmosphere and give rise to ammonia some of which is then oxidised to nitrate. Yet other workers claim that combustion products, volatile plant products, etc. act as nuclei.

As rain forms, the salts which have been carried up into the atmosphere are dissolved, and analyses show that rainwater from oceanic regions is a very dilute seawater but over continents the proportions of ions can vary considerably from those in seawater; for example the calcium content can increase owing to the occurrence of calcareous dust particles in the atmosphere. Junge & Werby (1958) recorded 0.3–0.5 mg l^{-1} of calcium in rainwater over the east and west coasts of the United States but as much as 3.0 mg l^{-1} over central Texas, whilst chloride and sodium contents showed a reverse trend. Thus the ratio of ions in what might be considered the 'mother' solution for algal growth is variable and it is then further altered by percolation through chemically diverse soils. In some areas of the world, dry fallout of salts may be appreciable and exceed that derived from rainfall, adding yet another source of chemicals to the waters.

Atmospheric pollution from industrial sources is increasing in importance as a source of ions to natural waters; in some regions it affects the composition and productivity of the algal floras. A good example of this is the distribution of sulphur which is of ecological importance in two forms, as sulphuric acid formed in the atmosphere and subsequently deposited into lakes (the deposition back into the sea has little biological effect since it is neutralised by the excess salts) and as hydrogen sulphide produced in anaerobic basins and in many shallow bodies of water which contain excessive amounts of organic matter. A major source of sulphur in the atmosphere is man-made emission of sulphur dioxide which amounts to around 100×10^6 tons of sulphate per year. Of this 93% is produced in the northern hemisphere and it has been predicted that the figure may rise to 175×10^6 tons by A.D. 2000 (Kellogg *et al.*, 1972). In addition, this sulphur in the atmosphere is supplemented by sulphur derived from volcanic activity, from that carried up by sea spray and from hydrogen sulphide formed by biological activity on the land. It has been calculated that sulphur is deposited at the rate of 2.7 tons km^{-2} in the United States and 2.2 tons km^{-2} in Europe. However this deposition is not evenly distributed and greater concentrations occur windward of producing areas; thus it has been shown that Scandinavia receives sulphur produced in Great Britain and this is altering the acidity of its lakes and rivers with serious con-

sequences for the biological balance. Man is now contributing to the atmosphere about half as much sulphur as nature and by 2000 A.D. man-made emission will possibly equal that from natural sources. But in industrial regions man's effect is already swamping that of nature. Around such sites, there is in addition a fallout of heavy metals which must be affecting algal populations, if only to the extent of being absorbed in amounts excess to normal requirements, and these elements then pass to the next level in the food chain.

Only in a few habitats does the dilute solution which is rain provide almost all the salts required by algae, e.g. in rock hollows in some mountains and in bog pools, although even here the water is enriched by solution, however slight, from dust or sediment; a similar effect is the chemical enrichment of snow fields by wind-blown dust. Nevertheless, Hutchinson (1957) considers that certain dilute lake waters differ only very slightly from rainwater and the minimum concentrations of Wisconsin lakes given in Table 1.1 bear this out. Recently Parker & Wachtel (1972) have recorded small amounts of vitamins (B_{12}, biotin and niacin) in rainwater falling over St Louis. They believe that they are derived from wind-borne pollen and spores. The total organic content in this rainwater was found to be 8 mg l^{-1}, which is about 10^6 times the sum of the concentration of vitamins found in the rain, but is similar to or in excess of that found in some small lakes in the region. A recent study of rain over Lake George, New York State, has shown that during 3 h precipitation the pH of the rain changed from 3.1 to 5.2 and the sulphate content fell from 3.11 to 0.20 mg l^{-1}, showing the pronounced washing out of particulate matter from the air (Stensland, 1976).

Percolation of rain water through soils and into streams and lakes yields solutions of increasing concentration, the exact composition of which varies according to geology, soil type, vegetation, evaporation, etc. During this solution process, four cations (calcium, magnesium, sodium and potassium) and three anions (bicarbonate, sulphate and chloride) are increased greatly so that they make up the bulk of ions in freshwater. In addition there is usually a variable amount of undissociated silicic acid present. Gorham (1961) considers that five principal environmental factors – climate, geology, topography, biota and time – interact to determine the ionic concentration of rain, soil solutions, lake and river waters. These variables all contrive to make the composition of each body of water unique and much more variable than that of normal oceanic water which is given in Table 1.1. Additionally, evaporation alters the concentration of salts in all waters, causing, for example, the high salinities of relatively closed tropical seas such as the Red Sea (40–41‰). However, even more extreme salinities occur in some inland basins, particularly those without outflows, e.g. some East African lakes (Table 1.1).

The annual fluctuations in concentration of some major ions are only slight and can therefore be used to characterise freshwaters, whilst others, such as nitrate, silicate, and phosphate, and some minor ions, such as iron, copper and zinc, fluctuate greatly and either the range, or at least the mean value, must be quoted if they are used to characterise a particular water. The ability of algae to colonise a body of water may be related to the absolute levels of certain important ions whilst the growth of populations is more likely to be related to the changing levels of other less conservative ions, e.g. the diatom *Melosira distans* occurs only in lakes with low concentrations of nutrients (i.e. soft water) whilst *M. granulata* occurs in lakes with high concentrations (i.e. hard water) but both only grow when silicates are in sufficient supply and growth ceases when they are used up. Ammonium nitrogen is the major source of nitrogen for euglenoid algae and provides a good example of nutrient control of algal occurrence since ammonium nitrogen is not common in well aerated habitats but is common in organically enriched waters. There is evidence that the more stable the chemical environment the smaller the species complement, for example Patrick (1967) reports that in chemically stable freshwater springs about 60 species of diatom were present whereas in an unstable river system the number is likely to be 250. The marine algae tolerate high salinity but such conditions are not necessarily the optimum for their growth, for example, it has been shown that many marine phytoplankton species grow better under experimental conditions at lower salinities than those found over much of the ocean surface (*see* p. 293).

The total amount of any particular ion may bear little relationship to that available to algae, for example iron tends to be complexed with organic molecules or to be in a colloidal state and thus it is only the so-called reactive iron which is available and it is necessary to use methods of chemical analysis which measure the state of the element *assimilated* by the algae. The common use of ethylene diamine tetra-acetate (EDTA) and NTA for complexing metal ions in artificial culture media, illustrates the use of chelating substances to allow the supply of nutrients in a controlled manner. Soil extract is commonly used to perform a similar function in cultural studies where the exact chemical composition is not an important consideration.

The ability of water to contain large amounts of substances in solution as electrolytes plus a small quantity as non-electrolytes means that all the elements are present in some form or other. A valuable source of methods for their determination is given by Strickland & Parsons (1968). Whilst the main constituents listed above give the overall chemical stamp to the solution, many others occur, some of which are of greater biological importance than the dominants, such as nitrogen (as NO_3^-, NO_2^-, NH_4^+), phosphate ($H_2PO_4^-$ and HPO_4^-), silicate ($SiOH_4$), trace elements, and organic complexes. It is important to remember that the amount of the biologically utilised nutrients measured in the water at any one time is not a measure of the total available nor if in minimal quantity is it indicative of a limiting nutrient since it is the *rate* of cycling of the elements which is of prime importance. Almost all values for dissolved substances in waters are for subsurface water and are perfectly adequate for reference to the contained populations, but there is a surface film which is beginning to attract attention. In this film the concentrations of the surfactant compounds and also of some other organic and inorganic molecules are often in excess of those in waters below the surface.

Salinity is the feature usually associated in peoples' minds with oceans and it is defined as the weight in grams of the dissolved inorganic matter in one kilogram of water after all bromide and iodide have been replaced by the equivalent amount of chloride and all carbonate converted to oxide. This can be expressed as grams (milligrams) per kilogram and although rarely quoted can be used also for fresh and inland saline waters where it is simply regarded as the concentration of all the ionic constituents.

Determination of the salinity of seawater is too complex for routine analysis. Since there is a fairly constant relationship between chloride content and salinity, it is easier to measure the halide concentration by reaction with silver nitrate and, assuming that it is all chloride, salinity $S‰ = 1.805 + 0.03 × Cl‰$. Nowadays, conductivity measurements are used and 'salinometers' can be towed behind ships and the results transmitted to the ship so that continuous readings can be obtained.

Seawater must be regarded as a dynamic chemical system* in which the dissolved substances have had thousands of years to reach equilibrium. The residence times, i.e. the time taken for a given amount of introduced element to become incorporated into the sediments varies from $2.6 × 10^8$ years for sodium, to 100 years for aluminium and to 250 years for the biologically recycled silica in the sea. Although the evidence is slight it seems that the basic composition of seawater has been virtually constant for very long periods and hence the algal floras have not been subjected to changes in the chemical environment such as have occurred, for example, in

glaciated lake basins over the last 12 000 years. The state of equilibrium reached by most elements (so-called conservative elements) in seawater is a result of slow physico-chemical processes. Other elements are more directly involved in biochemical cycling through the algae and other organisms. This cycling is also accompanied by fractionation of the elements from one layer of water to another. Phosphorus and silicate are good examples. They are rapidly cycled through organisms in the upper layer and hence the amounts in the water are small, while lower down both are regenerated from sinking dead organisms and faecal matter; and this, combined with relatively low utilisation at these depths, results in higher and relatively uniform concentrations throughout the deeper water masses (Fig. 1.5, 1.6). Great increases in nutrients also occur in deep stabilised lakes, for example in some of the East African lakes the deep waters are anoxic (Table 1.2)

One striking feature of seawater is that, regardless of the absolute concentrations of solids, the ratio between the elements is almost constant wherever the water is sampled. In addition the range of salinity (or chlorinity) in the seas, with the exception of certain evaporation basins and those grossly affected by freshwater, is not great, for example it fluctuates between 33‰ and 37‰ over 97% of the ocean surface. The average chlorinity value is 19‰. Although these values do not appear to be very divergent there is ample evidence that they are effective in influencing the occurrence of some planktonic algae. The surface waters of tropical seas can be rapidly diluted by monsoon rains and this is thought to have important consequences for phytoplankton. Although less precise and exhibiting greater ranges the ratios of the major elements in certain types of freshwater vary only slightly with increasing salinity, with a tendency for the alkali content to fall and those of magnesium and, particularly, calcium to rise.

Salinities greatly in excess of 35‰ are only found in relatively enclosed seas and in inland saline lakes, e.g. Great Salt Lake, Utah (203.4‰), Dead Sea (226‰ given by some authorities and 300‰ by others, e.g. Begin, Ehrlich & Nathan, 1974). These waters are characterised by an increase in chloride over other anions (they are termed chloride lakes) whilst other lakes have greatly increased sulphate or

Fig. 1.5. The vertical distribution of phosphate (P), oxygen (O_2) and temperature (T) at three stations in the North-east Sargasso Sea (February–March 1932). From Seiwell (1935) in Armstrong (1966).

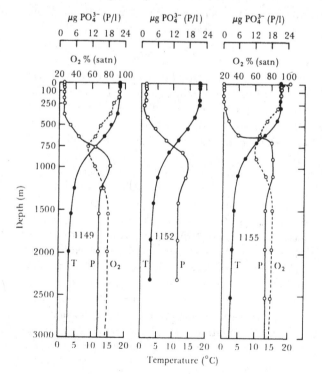

* Freshwaters are also dynamic chemical systems but are subject to more rapid fluctuations.

Table 1.2. *The chemical composition of water in Lake Kivu: maximum depth, 485 m; anoxic from 70 m. From Degens et al. 1972*

Depth (m)	Calcium (mg l^{-1})	Magnesium (mg l^{-1})	Potassium (mg l^{-1})	Sodium (mg l^{-1})	Sulphate (mg l^{-1})	Phosphorus (μg atoms l^{-1})	Silicon (μg atoms l^{-1})	Carbon dioxide (mmole kg^{-1})	Ammonia (μg l^{-1})
Surface	4.8	87	92	121	23	0.8	231	12	18
100	64	147	145	192	25	18	424	29	487
200	83	182	178	244	166	22	470	24	1314
350	110	394	315	465	214	53	1050	133	5400
440	112	417	338	487	220	54	1226	171	7100

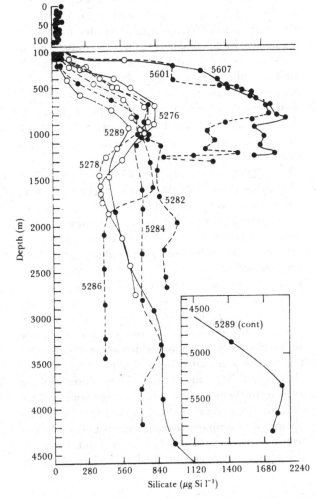

Fig. 1.6. The distribution of silicate with depth at stations in the Antilles Arc region. 5601, 5607, Cariaco Trench; 5282, 5284, Venezuelan Basin; 5276, 5278, Atlantic Ocean, east of Windward Islands; 5289, Puerto Rico Trench; 5286, Jungfern Passage region. From Richards (1958) in Armstrong (1966).

carbonate concentrations. At the lower end of the scale there is great variation, but the mean salinity of river water is usually taken as 0.1‰ and that of water from igneous regions only 0.05‰. Some enclosed seas have lowered salinities, e.g. the Black Sea where the surface waters have a salinity of 18‰ and in other enclosed areas the innermost regions can be freshwater (e.g. the Baltic) whilst similar areas in the tropics with little inflow can be extra-saline.

Hypersaline waters have not been studied intensively but in general the algal flora is impoverished and certain groups do not appear at all, for example there are no diatom species reported for the Dead Sea but in a saline pond near Eilat a few species occur at 140‰ and at 100‰ *Anomoeoneis* and *Nitzschia* species are recorded by Begin *et al.* who consider that the upper limit for diatoms is approximately 150‰. At this salinity *Amphora coffeaeformis* occurs abundantly (Ehrlich, 1975) but it is a very widespread diatom throughout a great salinity range. In fact Ehrlich comments that there are no diatom species *confined* to hypersaline waters but merely a few species which extend from the lower ranges into the upper. In Great Salt Lake *Dunaliella*, a naked flagellate, is dominant in the phytoplankton; in spite of the salinity the average carbon fixation for the year amounts to 145 g C m^{-2} which is considerably more than in many lakes but nowhere near the amount for some eutrophic lakes (*see* p. 469; Stephens & Gillespie, 1976).

The basic chemical features control the occurrence of algal species to such an extent that few if any freshwater species penetrate into marine waters and vice versa. There is much greater interchange of species between lakes and rivers, as might be expected from the closer chemical compositions of the waters. Coupled with the relative constancy of chemical composition of seawater is the fact that the phytoplankton content of the sea is also remarkably constant and a relatively small number of *common* species, perhaps 100–200, mainly from three phyla (Bacillariophyta, Prymnesiophyta and Dinophyta) are involved. The diatoms and dinoflagellates are the conspicuous species in samples collected with nets but the smaller Prymnesiophyta (flagellates with organic scales or with organic scales plus calcareous plates – Coccolithophorids) are often much more important, both numerically and in biomass. Freshwater phytoplankton on the other hand encompasses a much greater number of common species (500 or more) belonging to Cyanophyta, Chlorophyta, Bacillariophyta, Dinophyta, Chrysophyta, Xanthophyta and Euglenophyta. Inland saline waters continue the trend from seawater in that the number of species is drastically reduced until only a single phylum is dominant, for example the Cyanophyta in the phytoplankton of some African saline lakes or a single species, e.g. *Dunaliella* in the Dead Sea. Although there is little information on the correlation between water chemistry and the distribution of species it seems certain that each species is linked in nature to a particular chemical environment. In addition some may tolerate a wide range of concentrations and others a narrow range. Many algae accumulate excessive amounts of elements, in the cytoplasm, vacuole, and complexed to wall and mucilage material, so that amounts, especially of minor metals, rarely relate directly to the requirements of the alga. However in the natural habitats the absolute levels of individual elements may be of importance in determining the occurrence of species. The tolerance of species to varying chemical conditions must be determined in nature and then tested experimentally but few such studies have been undertaken and overall little is known of the relationships between the presence or absence of algae and the chemistry of the waters. Superimposed on the chemical limitation of species is the physical, e.g. some tropical seas and temperate seas have almost identical chemical status but support quite different algal floras, a finding which is probably related to differences in temperature and irradiance.

Organic components

Early work on algae assumed that photoautotrophic organisms merely required inorganic nutrients, and only in the last three decades has it been adequately shown that organic compounds, in particular vitamins, are absolute requirements for many algae. Algae with absolute requirements for organic compounds are termed auxotrophic. The amount of dissolved organic matter in natural water is small and variable (up to 50 mg l^{-1}) but Menzel (1974) considers a reasonable average for the ocean as a whole to be 0.5 mg l^{-1}. Equally large variations are recorded in freshwaters, e.g. Saunders (1972*a*) reported that organic detritus in Lake Michigan ranged from 1.3 to 16.9 times as much as the phytoplankton biomass. Measurements of cellulose in the water column suggest that there is more detrital cellulose than that found in the algae in the column and it is not degraded until it reaches the sediments. The organic fraction is made up of a large number of compounds derived from secretion, excretion, conversion of organic matter from dead organisms, inflow of detritus, etc. The main components are carbohydrates, organic acids, fatty acids, amino acids, vitamins, and plant hormones. The micro-organisms in the water contribute to the actual process of aggregation of organic matter into flocs, on and within which organisms also live (Paerl, 1973). Extraction and characterisation of such small amounts of organic matter from the 50–35 000 mg l^{-1} of salts in water is not easy. The vitamins are probably the most important to the algae and they are produced by a wide variety of bacteria present in both freshwater and seawater. Some algae can however synthesise their own vitamins and also secrete them for use by other algae (*see* p. 340). Organic compounds are not all advantageous to algal growth and some may inhibit algae or, by a process of auto-inhibition, react back on the alga which produced the substance (*see* p. 317). Another example of the effect of organic compounds is the action of specific hormones which are responsible for the control of sexuality in some algae (e.g. in *Volvox* (Darden, 1970; Starr, 1970)). These hormones are known to be released into cultures and induce sexuality in vegetative colonies and they may also operate in a similar manner in nature. The filtrates from bacterial growth (presumably containing organic molecules) can be used to restore the leafy morphology of some marine algae, e.g. *Monostroma*, which is lost if the alga is cultured in bacteria-free conditions (Provasoli, 1971). Presumably such organic compounds are present normally in nature and only experiments of this kind reveal the importance of the compounds.

It is convenient to aportion organic matter in water into particular organic carbon (POC) and dissolved organic carbon (DOC). The latter is often much more abundant in both freshwater (Wetzel, Rich, Miller & Allen, 1972) and marine environments and consists of a vast variety of organic compounds. Of the POC, up to 70% may be detrital and the remainder in the live organisms and Menzel (1974) estimates that it is present at an average concentration of 10 mg C l^{-1} for the ocean as a whole. The term detritus is often applied exclusively to the remains of organisms but it really should include inorganic material as well, and the terminological problem can be overcome by the use of adjectives such as organic, inorganic, etc. Thus although to the microscopist the algae appear to float in relatively clean water this is clearly not so. Algae float in a mass of organic material, a feature which is evident when one considers that all algae secrete some organic material, all animals excrete organic material and all organisms die and add to the organic material. A portion of this organic matter ultimately becomes incorporated into the sediments after cycling through many organisms (*see* Chapter 11).

In marl lakes, partially labile organic compounds are ultimately lost to the sediments because particulate and colloidal calcium carbonate occurs in quantity forming aggregates on the surface of which organic compounds can be strongly adsorbed (Wetzel & Allen, 1971). Some of the extracellular organic products released into the water are bound in this way before they can be utilised by bacteria or other organisms. In marl lakes the rate of heterotrophic activity in the water is therefore reduced below that of either eutrophic or oligotrophic lakes. This is yet another example of the complexity of the physico-chemical environment and its interaction with the algae.

Mineral components

Although they have received little attention, there are always mineral particles in the water column. These can form the major component in some waters and according to Lisitzin (1962) the mineral material exceeds the biogenic by a factor of 2 to 5.

Dissolved gases

Oxygen. The amount of oxygen dissolved in water depends on the partial pressure in the gas phase, temperature and salinity; increases in temperature and salinity both decrease the solubility. Thus freshwaters absorb more oxygen than the sea and the freshwater lying in the coldest regions absorbs the most. The oxygen content of seawater at 15 °C and under atmospheric pressure is approximately 5.86 ml l^{-1}, and that of freshwater is 6.85 ml l^{-1}. Oxygen is an exceptional element in that it is the only one constantly produced and released into the water by algae and since equilibration with the atmosphere is a relatively slow process many situations exist where supersaturation with oxygen occurs. Below the euphotic (illuminated, *see* p. 313) zone in all waters there is a net consumption of oxygen since animal and bacterial metabolisms consume oxygen; replenishment of oxygen is achieved by downward circulation. Oxygen becomes depleted from the lower waters in enclosed seas (e.g. the Black Sea), in many fjords, and during thermal stratification in lakes rich in plankton, where the rate of supply of organic matter to the lower waters is greater than the oxygen supply. Remarkable changes in concentration can occur over very short distances in a water column, e.g. in the small Abbot's Pool near Bristol a range from greater than 100 to 10% over a distance of 3.5 m is regularly recorded in the summer (Fig. 1.7). Some basins are completely anoxic, e.g. the Black Sea at depths below 125–250 m, certain of the deep oceanic trenches and Lake Tanganyika below about 80–140 m (the figure varies seasonally). In others, and particularly in temperate eutrophic lakes, the anoxia disappears when oxygen-rich surface water is mixed down into the lower water; this is always associated with the breakdown of a thermocline (p. 23) or halocline (p. 26). Under anoxic conditions the surface layer of sediments becomes reduced and results in the release of nutrients into the deoxygenated water, thus unlocking some of the store of nutrients contained in the sediments (Fig. 1.7). An oxidised surface layer prevents this escape, although subsurface sediments are always reducing. Oxygen concentration is one of the easily measured variables and a study of its distribution in a water body gives a large amount of direct and indirect information, e.g. on bacterial activity, photosynthesis, stratification, availability of nutrients, etc. However there is also a direct effect on algal growth though it has rarely been investigated; thus Gessner & Pannier (1958) found that respiration of both freshwater and marine algae decreased as oxygen saturation of the water decreased. As might be expected supersaturation up to 300% increased the respiration. The morphology of two algae, *Hydrodictyon reticulatum* and *Ulva lactuca*, grown under differing oxygen saturations, was investigated by Pannier (1957–8) who found cell size was affected. In *Hydrodictyon reticulatum* it was reduced by both supersaturation and undersaturation and in *Ulva lactuca* it was reduced in supersaturated solution but increased at 25% saturation. It must be remem-

bered that these and other experiments involving relatively unnatural concentrations of soluble gases, salts, unusual pressures, temperature, etc., while interesting in themselves, may not be very relevant to the natural situation.

Enormous diurnal changes in oxygen saturation due to algal photosynthesis and respiration are quoted in Hutchinson (1957); thus in Clearwater Lake, Minnesota a dense population of *Hydrodictyon* raised the surface water to 248% saturation at 16.00 h but the level was then lowered to 27% saturation at 05.00 h. Lowering of the oxygen concentration by decay of algae in undisturbed lakes in hot weather can cause mass mortalities of fish. Inhibition of photosynthesis occurs under some circumstances when water becomes supersaturated with oxygen (Warburg effect) especially at low carbon levels. These conditions may occur when algae are enclosed in bottles during photosynthesis

Fig. 1.7. The distribution of temperature, oxygen, nitrate, ammonia, phosphate and silica down the vertical profile of Abbot's Pool from February 1967 to June 1968. The overturn is marked by an arrow. From Happey (1970).

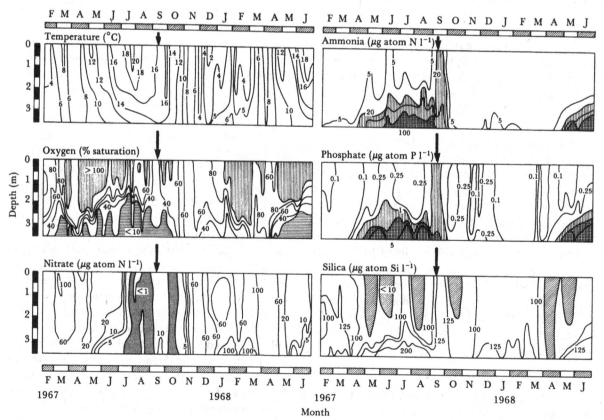

Month

Table 1.3. *Proportions of carbon dioxide,*
bicarbonate and carbonate ions in water at various
pH values. From Hutchinson, 1957

pH	Total free carbon dioxide	Bicarbonate ions	Carbonate ions
4	0.996	0.004	1.25×10^{-9}
5	0.962	0.038	1.20×10^{-7}
6	0.725	0.275	0.91×10^{-5}
7	0.208	0.792	2.6×10^{-4}
8	0.025	0.972	3.2×10^{-3}
9	0.003	0.966	0.031
10	0.000 (2)	0.757	0.243

experiments, particularly during long exposures in saline waters. Perhaps this feature should also be investigated in surface waters when low rates of photosynthesis are recorded.

Nitrogen. Values varying from 94 to 109% saturation have been detected in seawater, where saturation less than 100% may be the result of nitrogen fixation (uptake) by Cyanophyta. Supersaturation above these figures has been recorded in lake waters and may be derived from underground water flowing into the lake. Nitrogen is a gas which is rarely measured since it is of little direct biological importance except when nitrogen-fixing species are present.

Carbon dioxide. Carbon dioxide is some 200 times more soluble in water than oxygen and forms, in association with the carbonate/bicarbonate system, the single most important source of carbon for algal photosynthesis. Carbon is of course taken up in other forms to varying degrees by algae but, by and large, the vast mass of algae are dependent on carbon dioxide as a source, though the uptake through the cell membrane may be in the form of bicarbonate. The amount of carbon dioxide in air varies somewhat but is around 0.033%, although there is considerable evidence that the average values have been rising over this century due to combustion, etc. and values over 0.04% have been recorded. Deep ocean water contains carbon dioxide at about 50 ml l^{-1} a value which exceeds that of gaseous carbon dioxide in the air by a factor of about 60, whereas oxygen and rare

gases are present in water only as a fraction of that present in the atmosphere. On solution two reactions can occur

$$CO_2 + H_2O \rightleftharpoons H_2CO_3, \qquad (1.1)$$

$$CO_2 + OH^- \rightleftharpoons HCO_3^-. \qquad (1.2)$$

The first reaction is the only important one below pH 8, and the second above pH 10. Since carbonic acid is strongly dissociated the proportion present in the water is very small

$$H_2CO_3 \rightleftharpoons H^+ + HCO_3^-, \qquad (1.3)$$

$$HCO_3^- \rightleftharpoons H^+ + CO_3^{2-}. \qquad (1.4)$$

Table 1.3 gives the molecular proportions of total free carbon dioxide, bicarbonate ions and carbonate ions calculated from the apparent dissociation constants in dilute solution at 15 °C. From this table it is clear that below pH 5 only total free carbon dioxide is of any quantitative importance, between pH 7 and pH 9 bicarbonate is most important and above pH 9.5 carbonate becomes important. At high pH values calcium carbonate starts to precipitate and in freshwaters above pH 9 the main cations in solution are magnesium, sodium and sometimes appreciable amounts of potassium. In seawater at pH 8.3 the bicarbonate level is 18×10^{-4} g.mole l^{-1} and carbonate is 3.5×10^{-4} g mole l^{-1} and hence algal growth is dependent on these two sources and never on carbon dioxide. Carbon dioxide concentration varies only slightly in seawater compared with the very considerable variations in freshwater. Effects of high carbon dioxide concentrations are difficult to study owing to the pH changes involved but Paasche (1964) showed that photosynthesis and coccolith formation decreased under high carbon dioxide concentrations. pH values ranging from 1.7 to 12.0 have been recorded in freshwater (Hutchinson, 1967). pH values below 4.0 are associated with volcanic lakes, bog pools or acid polluted waters whilst those above 9.0 (other than temporary increases as a result of removal of carbon dioxide by photosynthesis) are due to carbonates of sodium and magnesium. Most lakes have pH values between 6.0 and 8.0 whilst that of seawater varies between 7.8 and 8.3.

The occurrence of carbon dioxide, bicarbonate or carbonate as the predominant form is correlated with the distribution of certain species in freshwater; there is evidence that some can only use carbon dioxide, whereas others use bicarbonate and a few use carbonate, although carbonate can be toxic to other species. Moss (1973c) has shown that species from acid (oligotrophic) lakes can be grown in water from alkaline lakes (eutrophic) if the pH is reduced whilst if the pH of the acid water is increased growth of oligotrophic species is reduced. A recent excellent paper on some of the complexities of the removal of carbon dioxide from water is that of Talling (1976). McLean (1978) has some interesting comments on the impact of the oceanic uptake of carbon dioxide in controlling carbon dioxide to oxygen ratios through geological time.

Solar radiation and temperature

The sun is the only noteworthy source of heat for the earth, which traps merely a two billionth part of the solar output, an amount equivalent to 23 trillion horsepower. It has been calculated that the earth receives every minute as much energy as man uses in a year. This minute percentage of the sun's output affects many of the physical phenomena and all the biological phenomena on the earth's surface.

The elliptical orbit of the earth around the sun, taking $365\frac{1}{4}$ days for one cycle, only affects the solar radiation received by the earth to the extent of about 3%. As the earth follows this ellipse it rotates once every 24 h but its axis of rotation is not perpendicular to the plane of the elliptic except during the equinoxes (at which time the sun's rays are perpendicular to a point on the equator). The axis of rotation tilts 23° from the perpendicular, so part of the year the north pole is tilted towards the sun and part of the year away from the sun. The tilting is responsible for our seasons and to calculate the amount of solar radiation received at any one point on earth both the latitude and the angle of tilting must be considered. The complexities affecting the amount of solar radiation received at any point on the earth's surface are reflected in the algal populations which change from season to season though, as might be expected, the changes tend to be least in tropical regions.

The amount of solar radiation and hence of heating varies enormously from latitude to latitude. At the equator the variation is slight with peaks in March and September (i.e. at the spring and autumn equinoxes); this variation is simply due to the tilting mentioned above. Moving north or south the two peaks soon merge into a single June peak at

30° latitude, whilst at 60° the amplitude has become enormous, with a tenfold difference between June and December, and nearer the poles there is almost no solar radiation during the winter months but the amount received in June actually exceeds that at the equator in June. These radiation influxes are reflected in the water temperature and differing amounts of energy available for photosynthesis at the different latitudes. There is a high input and slight annual fluctuation in the tropics, greater fluctuations at temperate latitudes with cold winters and warm summers, and again only slight fluctuations in polar waters with a single short heating period.

Solar radiation is absorbed by water and the water is thus heated; this heating and the distribution of the heat down the water column are among the most fundamental features of the aquatic environment and have far-reaching implications. Equally important for the algae is the availability of the radiation for photosynthesis. The depth to which the photosynthetically usable radiation penetrates is very limited and this limits the amount of primary fixation of carbon. It was a consideration of these features which led Riley, Stommel & Bumpus (1949) to write in one of the first and very important keystone papers 'radiation is the most fundamental ecological factor in the marine environment'. It is, however, one of the most complicated since it is a 'complex of variables which cannot be expressed by a single numerical value as can temperature' (Talling, 1971). The complex effects of radiation are due to the changes in spectral composition with depth, diurnal variation, etc. In some waters, penetration may be merely a few centimetres whilst in very clear oceanic water it may be 120 m or more. It has been estimated that in a hypothetical ocean water, free of particulate and dissolved organic matter, the blue wavelengths (475 nm) would be reduced to 1% of the surface value at 160–165 m and the green (525 nm) to 1% of the surface value at about 90–95 m (quoted in Steeman-Nielsen, 1974). In fact, in the cleanest ocean water, 1% penetrates to approx 120 m, e.g. in the Sargasso Sea (Steemann-Nielsen, 1962). The illuminated zone is known as the euphotic zone* and in this zone almost the total

---

* Strictly, the euphotic zone is that part of the water column with its base at the point where 1% of the photosynthetically available radiation (approx. 400–700 nm) reaches.

photosynthesis takes place. The dark region below is often termed the dysphotic zone. A rough measurement is often obtained by lowering a white disc (Secchi disc) overboard and noting the depth at which it just disappears from view. The depth of the euphotic zone is then generally estimated as three times the Secchi disc depth. The amount of downwelling radiation varies greatly with the wavelength and the amount of pigment and particulate matter in the water. In general the attenuation with depth follows a straight line on a semi-log plot and a curve on a linear plot (Fig. 1.8). The differences in the slopes of the semi-log plots are due to continuous variations in the colour and the content of particles in the water as populations change. The attenuation of the various wavelengths* is very variable but in general the red wavelengths penetrate least and the blue and green deepest (Fig. 1.8). This situation is however reversed in some situations where there is dense plankton growth or other substances in the water which absorb strongly in the blue and green regions. Thus not only is the intensity of irradiance different at each depth but so is the spectral composition and therefore an alga in shallow water may be receiving a mixture of blue, green and red light but one at depth may be living in a virtually monochromatic environment (*circa* 500 nm in the sea according to Eppley & Strickland, 1968). Photosynthesis is a quantum reaction and clearly the number of quanta available varies greatly with depth. However, cells circulate in most waters and

so they live in a constantly changing light climate and this makes estimation of photosynthesis a complex matter.

The radiant energy which is absorbed by water is converted into another form of energy, heat, and this also is stratified by the passage of light through the water, being greatest at the surface. If the water were perfectly still, a graph of temperature plotted against depth would have a similar form to the plots of irradiance with depth (Fig. 1.8). However, cooling results from surface-water evaporation and back-radiation during night-time whilst wind stress on the water surface sets the water in turbulent motion and mixes the warm water downwards. The temperature–depth graphs regularly plotted by limnologists and oceanographers tend therefore to have the form illustrated in Fig. 1.5. The waters are described as stratified, the upper warm mass is termed the epilimnion in lakes and the mixed layer, surface layer, or trophosphere in the sea. The region of rapid temperature decrease is known as the thermocline or metalimnion and the lower cool mass of water the hypolimnion in lakes and the deep layer or stratosphere in the sea. The region where the density changes rapidly over a short distance in the sea is termed the pycnocline. Since no mass of water is completely still, there is circulation of an upper mass

---

* Measured either as broad bands, with coloured filters on selenium photocells, or as narrower band widths, with a spectroradiometer in conjunction with a monochromator.

Fig. 1.8. The light extinction in Lake Tahoe, California–Nevada on 27 August 1960 plotted, *a*, on a linear and, *b*, on a log scale. From Goldman (1967). *c*. The light penetration into the sea off California measured in the red, blue, and green spectral regions and for the total energy (T). From Talling (1960).

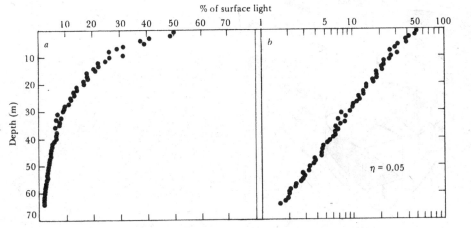
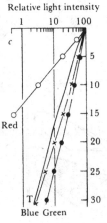

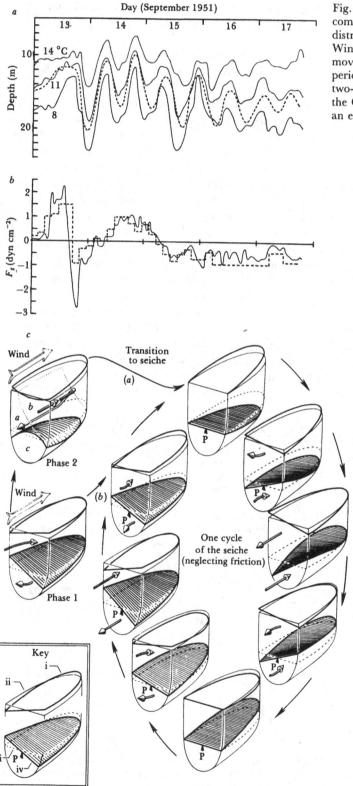

a

Day (September 1951)

Fig. 1.9. *a*. Observed (unbroken lines) and computed (broken lines) isotherm–depth distribution in the thermocline region in Windermere North basin showing seiche movements. *b*. The wind stress over the same period at a nearby site. *c*. An internal seiche in a two-layered lake model, deflected to the right by the Coriolis Force. From Mortimer (1974), where an excellent discussion of the phenomena is given.

of less dense, warm water, lying on a cooler mass of dense, slower circulating water; in the ocean depths this circulation is very reduced. In most temperate lakes the upper warm mass of water is mixed into the lower cool water in the autumn, a phenomenon known as the overturn and the result of rapid surface cooling and often of increased wind stress (Fig. 1.7). Shallow tropical lakes tend not to have such pronounced thermoclines and some circulate completely each day. Diurnal changes in temperature in tropical lakes are often greater than seasonal variations in temperate lakes. Thus in Lake George, on the equator in Uganda, the diurnal fluctuation can be 10 °C *each* day. In this lake a thermocline forms during day-time, isolating a shallow epilimnion, and an overturn follows each night. In the sea there is no thermocline in arctic and antarctic seas, but a permanent thermocline occurs in tropical oceans at a depth varying between 250 and 750 m, e.g. at 400 m in the Sargasso Sea with a seasonal thermocline at 100 m in summer (Ryther, 1963). In cooler seas a summer thermocline often forms at 30–35 m but even here a deep thermocline may also form. Since the stability of the water mass is high in the thermocline and the coefficient of vertical eddy diffusion* is small, mixing in the vertical plane is much greater above than below, so that the thermocline forms a barrier between the two water masses. This thermal stratification is of the greatest importance when considering the ecology of phytoplankton since it determines the depth through which the organisms will circulate, e.g. 10 m in summer as opposed to 60 m in winter in a lake such as Lake Windermere. The algae will thus spend a much greater time in the photic zone in summer than in winter. Seiches develop in stratified water when the wind moves warm surface water in the direction of the wind. This sets up an oscillation of the water which can be measured by recording temperature down the depth profile (Fig. 1.9). In some basins the operation of the Coriolis Force sets up a deflection of the internal wave and this is also illustrated as movement of a plane of water in Fig. 1.9. These movements will affect the position of plankton pop-

ulations in and on either side of the thermocline. Species trapped below the thermocline tend to sink and die unless they can swim sufficiently to reach the upper levels; many phytoplankton species are however non-motile. In shallow seas the layer beneath the thermocline is less static than in freshwaters since there are subsurface currents and also tidal current movements and hence there may be two circulating systems even during the period of temperature stratification.

A most drastic side effect of thermal stratification is that owing to reduced downward mixing of water any replenishment of nutrients from the deeper water is greatly restricted. Imagine a layer of water lying on top of thick syrup; if one blows on the surface a little of the syrup is stirred up into the water and this is exactly parallel to the situation in thermally stratified waters. Coupled with the relative isolation of the upper water mass, there is an increased rate of uptake of nutrients in the upper mixed zone; the cells are also in a more favourable position to utilise solar energy and they divide rapidly. This effectively reduces certain nutrients to such a low level that they become limiting to algal growth, yet below the thermocline a rich pool of nutrients lies almost unavailable until stratification breaks down. In the deep oceans such a situation is a permanent feature, with partial circulation at periods of destratification although wind stress on the surface never supplies sufficient energy to stir up the lowermost nutrient-rich water. The small amounts which are drawn up into the uppermost layers in the tropics are rapidly utilised by the organisms and are insufficient for luxuriant growth; hence the 'desert-like' conditions of these waters. In regions where the lower water upwells for various hydrographic reasons, usually currents flowing against the land or wind-induced surface water flow away from the land, nutrients come to the surface and the result is enormous algal production (p. 474).

After the breakdown of stratification (overturn), in the autumn in temperate lakes, the whole turbulently mixed water mass continues to cool down to 4 °C and even below; since the water below 4 °C is less dense it floats on the surface and under calm conditions or extremely low air temperatures the surface layer cools rapidly and ice forms. This effectively cuts off the water mass from the wind stress and the circulation of the lake slows down. An

---

* The eddy diffusion coefficient (A) is calculated from a formula incorporating temperature changes with time and depth, $\dfrac{dT}{dt} = \dfrac{A\,dT}{dz^2}$ where T is temperature, t time and z depth below the surface.

inverse stratification is set up, with water just beneath the ice at a temperature just above zero and increasing with depth to 4 °C or whatever temperature the mass of lake water reached when ice formed. Heating of the lower water then occurs, since solar radiation entering through the ice will heat the water; there is also a slight transference of heat from the sediments to the bottom water. Light penetrating the ice will be available for algal photosynthesis and there are many records of populations growing beneath the ice (*see* Chapter 7); these tend to be composed of species which are smaller and often more motile than the species which grew prior to the ice formation. Lack of turbulent mixing results in the sedimentation of many species, especially the heavy forms which were growing in the mixed water prior to ice cover (Fig. 10.17). Only when a layer of snow is deposited on the ice, is light penetration dramatically reduced. Sea ice forms at −1.91 °C where the salinity is 35.00‰ and at higher temperatures as the salinity decreases. Pure ice crystals separate out and as these fuse to form a matrix, brine collects in cells between the crystals. As the temperature decreases more ice forms and the brine cells decrease in size but increase in salt concentration and crystals may even appear in the cells. Within these brine cells, algae continue to live and even to divide, whilst on the under-surface of the ice mass other species may live in great quantity; diatoms are common, often giving the ice a brown colour (Chapter 5). Since the areas over which sea ice forms are great and the depth of ice may also be great this algal flora is of considerable extent and importance.

Another kind of stratification occurs in some, especially deep, lakes and fjords where a layer forms at the bottom which is chemically stabilised (like a layer of sugary tea at the bottom of an unstirred cup). This stratification can be brought about in many ways, such as accumulation of decaying organic matter, entry of basal springs, deoxygenation and the build up of hydrogen sulphide-rich water, the isolation of a saline basin by uplift of the land and the maintenance of a pool of seawater at the bottom of a freshwater lake. It is probable that all deep tropical lakes are of this type. Such waters are usually maintained in a permanently stratified state since the downward mixing by wind-induced turbulence is insufficient to stir the density stabilised bottom waters. These waters are termed meromictic and the region of rapid change in chemical status is known as the chemocline (or halocline if the lower layer is saline). The stable water below the chemocline is called the monimolimnion and the water above, which can stratify thermally as in other lakes, is called the mixolimnion.

Summary

The origin of water and water bodies is mentioned briefly and is followed by a discussion of circulation patterns and the physical effect of water movement on algae. Common tidal patterns are illustrated. The anomalous properties of water are shown to be related to the complex structure of water. The origin of the dissolved salts, the importance of their concentration, balance, residence times, and vertical distribution in the water column are all shown to affect the composition of algal floras. Organic molecules are also abundant in all waters and affect algal populations. Dissolved gases, especially oxygen and carbon dioxide, have interesting distributions in aquatic systems and a study of these is an essential part of algal ecology. The carbon dioxide–bicarbonate–carbonate equilibration in waters is a complex study and essential details are given. The understanding of algal ecology is intimately bound up with the effects of the penetration of solar radiation into water and the subsequent heating and the action of wind stress on the surface causing mixing of the waters. These important effects are also coupled to circulation of waters and hence to concentration of chemicals and organisms down the water column. The annual changes occurring at different latitudes greatly affect the algal populations. This chapter stresses that the physico-chemical environment is intimately linked with the occurrence, growth and decay of algae and their ecology can often only be interpreted when the non-living environmental factors are studied simultaneously.

2 Habitats and communities of algae

Algae are always thought of as aquatic organisms, a concept reinforced by the abundant, but by no means universal, formation of swimming gametes and zoospores. They are indeed the dominant photo-autotrophic organisms in the aquatic environment taken as a whole; angiosperms, bryophytes and pteridophytes (*sensu lato*) only achieve prominence in small bodies of freshwater and angiosperms only in some shallow coastal waters. But much of the land surface is also wet enough to support considerable algal growth; this growth is associated with surfaces and there are few surfaces which, if examined microscopically, yield no algal cells, for example in deserts even the surface cracks of stones, the roofs, and sides of buildings support some coccoid or short filamentous algae. Intermittent precipitation or moisture derived from dew is sufficient to enable motile stages of the life cycle to function. It is worth remembering that algal growth is usually possible, and frequently abundant, wherever vegetation is obvious to the naked eye, though the microscopic algae themselves are only visible when growing in aggregate. Equally when no vegetation is visible, populations of microscopic algae are usually present, for example in the middle of the ocean or on ice floes.

This chapter is concerned with a review of the places in which algae grow and with the gross communities therein, their classification and inter-relationships. The details of species and their biology follows in Chapters 3–7.

Habitats

The habitats of algae need not detain us for long for they are mostly part of the physiography of the earth and can be described from a purely geographical standpoint without reference to the algae themselves. One must accept the constraint that most algae are autotrophic and therefore habitats

not exposed to solar radiation are unlikely to support algae.* Algae can maintain themselves at extremely low light intensities, but under these circumstances endogenous organic substances are probably utilised and growth as such does not occur, for example, in deep-water populations in high latitudes during winter.

Terrestrial habitats include rock faces and rock debris, and these are particularly well colonised when they are moistened by percolating water.† Artificial habitats are provided by concrete surfaces of buildings, paths, cooling towers of power stations and other industrial plants. A classic 'terrestrial' habitat is that of the tree trunks where '*Pleurococcus*' grows ('*Pleurococcus*' is usually a collection of minute unicells, aggregates or weakly filamentous algal species belonging to several genera and taxonomically most confused) whilst in moist tropical regions the algal habitats extend on to the leaf surfaces as well. Almost all soils provide an important habitat for microscopic algae. Permanent ice and snow are also colonised.

---

* Heterotrophy is proven for some algae, and undoubtedly occurs in nature, though its extent has not yet been ascertained and its contribution to algal growth as a whole is likely to be slight. It may however be of importance in certain specialised circumstances. Phagotrophy also occurs amongst colourless algae but again there are almost no data on its ecological significance. Practically all algal ecological studies are concerned only with the pigmented species.

† Petersen (1935) termed these 'aerial' habitats and kept them as a group separated from 'terrestrial', which he confined to the soil. I see little need to make such a distinction since there is a range of moisture conditions in both. There may however be greater merit in separating off the habitats which occur on objects above the soil, litter or water surface and terming them subaerial as Schlichting (1975) does. These are subject to greater desiccation and less contamination from soil solutions or organisms.

a

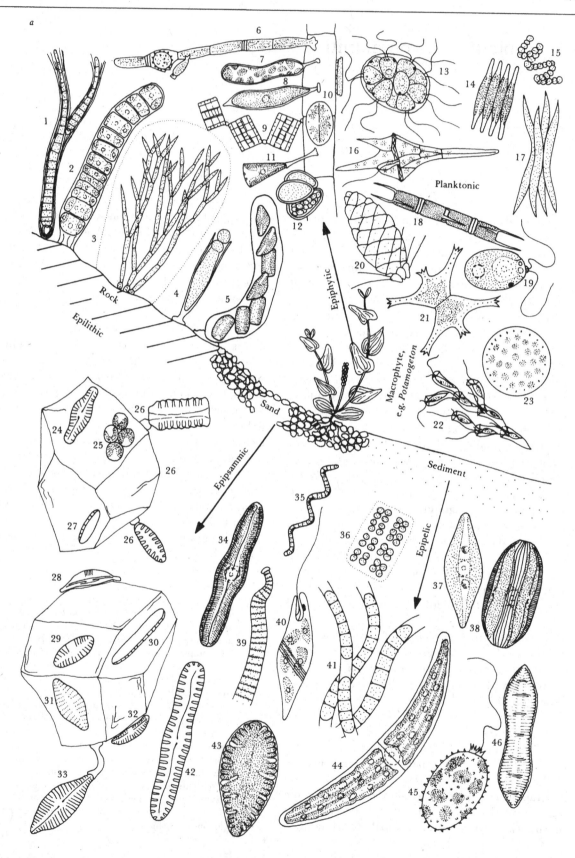

Planktonic

Epiphytic

Rock

Epilithic

Sand

Epipsammic

Macrophyte, e.g. *Potamogeton*

Sediment

Epipelic

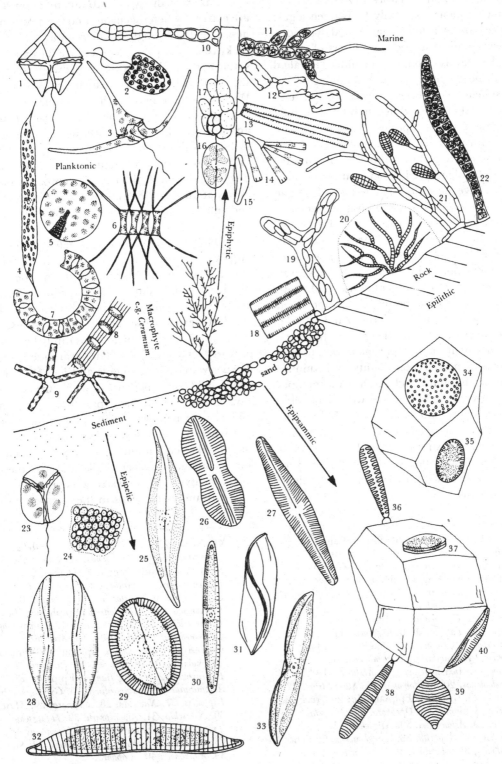

Fig. 2.1: for legend see page 30.

Many mosses and liverworts present a favourable habitat for algae, especially blue-green algae, whilst lichens are of course algal–fungal symbiotic systems. Sewage disposal beds and filters of waterworks are abundantly colonised, and indeed the algal growths can be essential to the functioning of the cleansing systems. Wherever water collects in temporary pools a slightly modified habitat is formed in which some of the soil species and opportunistic flagellates, with rapid life cycles, proliferate. A few terrestrial animals (e.g. sloths) provide a habitat for some specialised algae but I imagine many possible animal sites have never been investigated. Algae have even been found growing on human teeth.

Aquatic habitats are similar in both freshwater and marine situations, the open water habitat being the most extensive of all. Marine ice floes form suitable sites for colonisation by algae. Along shores, the rock surfaces provide another habitat, as do both the intermittently wet and permanently submerged silt and sand. Certain types of intertidal and subtidal sands are, however, almost devoid of algae due to the grinding movement of the particles. Submerged microscopic and macroscopic plants and animals provide further habitats, and the hosts sometimes seem to be overwhelmed by the weight of algal growth; surprisingly little is known of the possible detrimental or beneficial effects of such growth. Any suitable artificial substratum placed in the water affords further algal habitats, and the 'fouling' of these substrata by algae is a considerable problem, e.g. on ships' hulls.

Many of the gross habitats mentioned above are extremely heterogeneous and for most work it is desirable to distinguish microhabitats; on many rocky coasts, for example, these are the exposed rock surfaces, the rock pools where the water stands, and the rock gulleys down which the water drains. Within the rock pool further microhabitats can be distinguished, for example the intermittently wet and dry rim, the permanently wet bottom, and the surfaces of the various animals and plants growing in the pool. In many studies the microhabitats involved are not distinguished; comparison of data from station to station is then a very difficult matter and many subtleties of algal ecology are not revealed.

The habitats associated with flowing water are largely confined to the channels draining the land, and the stream–river complex of environments is relatively little studied. Unidirectional flow does occur at sea through some passages between islands, etc., but there is no special flora associated with these regions and the effect of flow has rarely been studied (but *see* Conover, 1968). The term lotic is used for flowing water environments and lentic for those of ponds and lakes. There are some features of the lotic environment which have parallels in the marine sphere, e.g. where water flows past artificial objects, and where there is flow across rocky shores, small 'stream-like' communities develop in the channels and in the outflows from rock pools. At the upper end of the river–stream system there is often a spring where groundwater seeps to the surface and this is

---

Fig. 2.1. Examples of the life forms of algal growths in various habitats. *a*. Freshwater. These are all microscopic genera, a few of which may form filaments or crusts visible to the naked eye. Epilithic: 1. *Calothrix*, 2. *Ulothrix*, 3. *Chaetophora*, 4. *Chamaesiphon*, 5. *Cymbella* (in mucilage). Epiphytic: 6. *Oedogonium*, 7. *Ophiocytium*, 8. *Characium*, 9. *Tabellaria*, 10. *Cocconeis*, 11. *Gomphonema*, 12. *Dermocarpa*. Planktonic: 13. *Pandorina*, 14. *Fragilaria*, 15. *Anabaena*, 16. *Ceratium*, 17. *Ankistrodesmus*, 18. *Melosira*, 19. *Chlamydomonas*, 20. *Mallomonas*, 21. *Staurastrum*, 22. *Dinobryon*, 23, *Cyclotella*. Epipsammic: 24. *Navicula*, 25. Green Unicells, 26. *Opephora*, 27. *Nitzschia*, 28. *Amphora*, 29. *Achnanthes*, 30. *Nitzschia*, 31. *Achnanthes*, 32. *Cymbella*, 33. *Gomphonema*. Epipelic: 34. *Caloneis*, 35. *Spirulina*, 36. *Merismopedia*, 37. *Navicula*, 38. *Amphora*, 39. *Oscillatoria*, 40, *Euglena*,

41. *Phormidium*, 42. *Pinnularia*, 43. *Surirella*, 44. *Closterium*, 45. *Trachelomonas*, 46. *Cymatopleura*. *b*. Marine. Planktonic: 1. *Peridinium*, 2. *Hymenomonas*, 3. *Ceratium*, 4. *Rhizosolenia*, 5. *Coscinodiscus*, 6. *Chaetoceros*, 7. *Eucampia*, 8. *Sceletonema*, 9. *Thalassionema*. Epiphytic: 10. *Erythrotrichia*, 11. *Achrochaetium*, 12. *Grammatophora*, 13. *Synedra*, 14. *Licmophora*, 15. *Amphora*, 16. *Cocconeis*, 17. *Pleurocapsa*. Epilithic: 18. *Fragilaria*, 19. *Navicula* (in mucilage), 20. *Rivularia*, 21. *Ectocarpus*, 22. Young *Bangia*. Epipelic: 23. *Amphidinium*, 24. *Holopedium*, 25. *Gyrosigma*, 26. *Diploneis*, 27. *Navicula*, 28. *Amphora*, 29. *Mastogloia*, 30. *Nitzschia*, 31. *Amphiprora*, 32. *Hantzschia*, 33. *Tropidoneis*. Episammic: 34. *Coscinodiscus*, 35. *Cocconeis*, 36. *Opephora*, 37. *Cocconeis*, 38. *Synedra*, 39. *Raphoneis*, 40. *Amphora*.

a very special niche within the lotic environment. Most of these springs are low-temperature habitats but some provide the only natural sites for high temperature (thermophilic) species, e.g. the hot springs in Iceland, New Zealand, North America. An excellent account of river ecology in general is provided by Hynes (1970).

Habitats are fortunately rather few compared to the number of species, and it is striking how similar the flora is in similar habitats in many parts of the world. The only one showing pronounced differences is that of the rocky coast, where dispersal problems coupled with geographic isolation of shores result in quite different assemblages of algae in the various phytogeographic regions (*see* Chapter 8), though even then the life forms are similar. Elsewhere, also, almost identical life forms tend to dominate within each habitat, for example the planktonic flora or the epiphytic flora of marine and freshwater habitats show remarkable similarities (*see* Fig. 2.1*a*, *b*). In contrast to much of higher plant ecology, in which *stands* of vegetation can be recognised, selected, sampled, and studied, in algal ecology the *habitats* are recognised, selected, sampled, and studied. The sampling of many algal habitats can be likened to sampling the terrestrial vegetation by means of a grab lowered through clouds from a helicopter, at least it tends to be random within the limits of the patchiness of algal vegetation. Only on the rocky shore can stands of vegetation be consciously selected for study.

Communities

Much of the work in algal ecology has been concerned with describing communities of algae; if one believes, as I do, in the concept of communities as collections of species living together and recurring in spatially separated habitats, albeit as open systems, then it is as *essential* to identify and describe these communities as it is to identify and describe the constituent species. It has been said that classification is essential for conceptual thinking and certainly to clarify concepts in ecology it is helpful to have classes, just as it is in taxonomy. Problems frequently arise in algal ecology from the lack of identification of communities; to quote Fager (1963) on this subject 'one cannot study something until it can be fairly objectively identified and described'. However the identification and description of the community

is *no more nor less* important than its functional aspect, and, in fact, structure and function form an interacting system which is only separated for the convenience of scientific studies. The alternative and virtually unworkable concept is that of a continuum of species throughout the habitats. Of course, to some degree there is such a continuum but only a very few species form the links. It is necessary to clarify the terminology which has been used to describe communities. This is dealt with in a comprehensive manner by Hutchinson (1967) for freshwaters and wherever possible the same terminology should be applied to the marine ecosystem. *Community* itself is a wide term with various meanings, but essentially it is used to denote a collection of species living together and usually linked to a particular habitat, e.g. the phytoplankton, or the organisms of the intertidal zone. The more cumbersome term, *recurrent floristic assemblage* has been used as a synonym for community. Within this loose concept, one can distinguish *populations*, which are collections of individuals all of *one* species living within a defined area or volume. A collection of populations is termed an *assemblage*; this is merely a collective term and is best used when there has been no attempt to define dominance. When, however, characteristic or dominant species can be listed the term association comes into use; it can be defined as a segment of the biocoenosis.* The *association* is 'an assemblage of species that recurs under comparable ecological conditions in different places' (Hutchinson, 1967); it is the keystone of ecological studies, since it forms a recognisable entity, which can be described, and within which the species are in intimate interaction. As writers on terrestrial ecology have emphasised, the association, though known by its dominants, includes all the species within this unit, e.g., the Wärming definition is 'a community of definite (floristic) composition, within a formation' (see below for a definition of formation).

The study of communities is a topic which has engaged many land plant ecologists but has been

* A biocoenosis is the totality of organisms living in a biotope, which itself is defined as any segment of the environment which has convenient but arbitrary upper and lower limits, characterised by dominant or characteristic species. It is virtually synonymous with community and plant biocoenoses are sometimes termed phytocoenoses whilst the term phycocoenosis has been applied to algal communities (e.g. Pérterfi, 1972).

Table 2.1. *Classification of algal life forms based on mode of 'overwintering'. Based on Feldmann, 1937*

Annual algae

Algae found throughout the year; one or several generations per year; spores or zygotes germinating immediately.

EPHEMEROPHYCEAE

(e.g. *Enteromorpha*)

Algae confined to one period of the year.

   Present during the other period in a microscopic
      vegetative state

ECLIPSIOPHYCEAE

(e.g. many Phaeophyceae)

Present during the other period as a resting stage

HYPNOPHYCEAE

(e.g. Desmids)[a]

Perennial algae

Complete frond perennial.

   Frond erect   PHANEROPHYCEAE

(e.g. *Fucus*)

   Frond a crust   CHAMAEPHYCEAE

(e.g. *Lithophyllum*)

Only part of the frond persisting for several years.

   Part of erect frond persisting

HEMIPHANEROPHYCEAE

(e.g. *Cystoseira*)

   Only the basal portion persisting

HEMICRYPTOPHYCEAE

(e.g. *Cladostephus*)

---

[a] It is not known exactly how long desmids persist in nature, dividing vegetatively and without undergoing conjugation leading to the formation of zygotes.

relatively ignored by phycologists, though much of the philosophy and technology is common to both. Throughout this century, and in isolated instances before, two approaches have been applied. The first is concerned with classification of communities, and several methods have been evolved, only two of which are of immediate concern to phycologists, the *physiognomic* and the *phytosociological*. The physiognomic method deals with the growth forms of plants; it is exemplified by Wärming (1909) in his 'Oecology of Plants', and involves large units, such as tropical rain forest, savanna, cold desert, etc., which have been termed *formations* (*see* also Beveridge & Chapman (1950) for definitions of this and other terms in relation to rocky shore ecology). Rarely have algae been introduced into such schemes though

Rübel (1936) did have two special formations, Aquarrantia (equivalent to phytoplankton) and Solerrantia (equivalent to phytoedaphon, soil microorganisms). Clearly, one can distinguish formations in algal ecology, e.g., the phytoplankton formation, the rocky coast formation, the sediment microfloral formation (possibly this should be thought of as a submerged extension of Rübel's phytoedaphon, *see* Chapter 5) but these are perhaps the only ones which are dominated by algal life forms and exhibit similarity of intrinsic habitat features, though climate, which is such a feature of higher plant formations, is not important for these algal examples. Beveridge & Chapman (1950) use the term formation and apply it to rocky coasts but obviously believe that several formations are present, e.g., the laminarian zone. This concept was developed by den Hartog (1955, 1959) who recognised 7(8) formations along the Dutch coast which, from his description, are clearly based on life forms, e.g. *Hildenbrandia*, *Cladophora*, *Fucus*, *Laminaria*, etc. It seems to me that these can hardly be equated with the formations of the land ecologists. Another physiognomic approach, developed by Raunkier in his 'The life forms of plants and statistical plant geography' (1934), was based on life forms and involves the position and nature of the perennating organs. Attempts have been made to carry this system over into algal ecology but with little apparent following and few phycologists use it. The commonest example of this approach is the system devised by Feldmann (1937) and given in Table 2.1. It is unfortunate that the 'classes' of this scheme have been given endings which imply actual classes of algae and I suggest the ending -phycon may be more appropriate. Earlier attempts to classify algal forms recognised groups such as 'Crustida', 'Corallida' and 'Silvida' (Gislen, 1930) but these have not acquired any following.

The *phytosociological* method has its exponents *par excellence* in the Braun-Blanquet school of terrestrial ecologists but it is also being applied by phycologists, especially those working in the Mediterranean (e.g. Boudouresque, 1971a, b) and has also been used very successfully on coral reefs by van den Hoek, Cortel-Breeman & Wanders (1975). It is an approach involving rather formal, rigid methods for classifying communities and it deserves much greater consideration as a method in algal ecology and certainly much more detailed testing by phycologists. I realise that it has its drawbacks but selection of homogenous stands is only a problem on rocky coast vegetation; in other situations the habitat itself has to be selected. Numerical methods also have many problems (*see* discussions in higher plant ecological literature) and as yet are not greatly used by phycologists. A fairly simple example of an association table for a single station on a rocky coast is given in Table 3.9. A true phytosociological approach involves many such tables combined to yield a very complex mass of data; one such example is given (Table 2.2). It might perhaps be advisable to produce individual tables, e.g., one for the uppermost shore stations, one for the mid-shore and one for the lower shore zone, etc. The method depends on the concept that plant communities can be recognised by their full (floristic) composition. Within these communities some species are more sensitive than others to environmental factors; these are termed diagnostic species, and can be used to organise communities into a hierachical classification with the association as the basic unit. The association again emerges as the fundamental unit, but here it is an abstraction obtained from comparison of lists, although now with character species (*Charakterart*) which possess 'fidelity' to the association. These are not necessarily important in terms of biomass, yet they will occur in all stands of the association (samples of a suitable size to characterise the association). Examples of this approach in algal ecology would be subtidal stands containing a *Cystoseira* species, as in Table 2.2, or intertidal stands each containing *Ascophyllum nodosum*, or samples of plankton each dominated by *Skeletonema costatum* as the character species. The methods of studying the stands are formalised, e.g., as to degree of cover, abundance, etc., but this process is familiar to algal ecologists since, because of the microscopic nature of many of the organisms in the stands, the quantitative analyses tend to be more precise than many used by terrestrial ecologists. The associations can be grouped into units of a higher order (*alliances*) or on the other hand divided into *sub-associations* and *facies* on the occurrence of different subgroups of species in the association. The methodology is not difficult, but numerous stands have to be studied so that an association table can be built up, and the associations, sub-associations, etc. can be determined. A fairly rigid scheme is needed in order that results from different workers may be strictly comparable. The most detailed study based on this system is that of Boudouresque (1971a) though others (e.g., Kornas & Medwecka-Kornas, 1949; Den Hartog, 1959; Russell, 1972, 1973) have used it in a less detailed manner. It would be interesting to apply the method to the phytoplankton in a range of waters, for example, those dominated by the clearly recognisable *Asterionella formosa*, which tends to occur in the spring, to determine the relationship between this alga and the other components of the plankton during a segment of the year. Further studies would be needed at other times of the year so that finally comparison of the phytoplankton of lakes with this '*Charakterart*' could be achieved. The interesting

Table 2.2. *An example of an association table compiled after the manner of Braun-Blanquet from eight sites described in the table*

Species grouping according to their affinities and in relation to detail of the underwater habitats, e.g. the *Cystoseiretum strictae* occurs in well illuminated, rough water. The numbers in the columns indicate abundance and habitat, e, epiphyte; i, endophyte; j, young plant; l, weak plant. The final two columns give the overall abundance/dominance in a percentage estimate and in five classes of increasing presence from I to V. From Boudouresque, 1969, whose papers (Boudouresque, 1971a, b) should be consulted for further study

Number of site	292	245	251	282	259	258	260	238	Mean global abundance (%)	Abundance
Date	17 i.69	10 ii.67	10 ii.67	30 v.68	15 vi.67	16 ix.67	16 ix.67	28 xii.67		
Depth (cm)	20	20	10	20	15	15	20	20		
Slope (in degrees)	5	10	10	5	5	5	10	5		
Aspect	NE	NE	S	NE	NE	S	NE	NE		
Cover	100%	100%	100%	100%	100%	100%	98%	100%		
Cystoseiretum strictae (photophilic, rough-water species)										
Cystoseira mediterranea Sauv.	3.2	3.2	4.4	5.5	5.5	3.3	3.4	4.5	56.2	V
Gelidium latifolium (Grev.)	2.3	1.4	2.1	1.5j	.	2.3	3.4	1.4	10.6	V
Laurencia pinnatifida (Gmel.)	1.2j	1.1	2.2	2.3	1.2	.	1.2	2.2	6.9	V
Ceramium rubrum (Huds.)	1.2e	2.4e	1.3e	2.3e	1.1e	1.3e	+e<	+e	5.0	V
Gelidium spathulatum (Kütz.)	1.2	.	+	.	1.3	1.3	1.3	1.3	1.6	IV
Boergeseniella fruticulosa (Wulf.)	+e<	2.3e	.	.	.	1.4e	+e	.	2.2	III
Herponema valiantei (Born.)	.	.	.	1.3i	1.4i	.	.	.	0.6	II
Dilophus fasciola (Roth) var. *repens* (J. Ag.)	.	.	1.5	.	.	+j	.	.	0.3	II
Sphacelaria hystrix Suhr.	.	.	.	+ij	.	+i	1.3i	.	0.3	II
Feldmannia paradoxa (Mont.)	.	.	.	.	1.5e	+e	.	.	0.3	II
Cystoseiretalia (photophilic species, sensu lata)										
Crouania attenuata (Bonnem.)	1.2e	1.3e	1.4e	+e	+e	1.4e	+e	2.4e	3.2	V
Sphacelaria cirrosa (Roth.)	1.1e	+ej	1.4e	.	+ej	+ej	+e	+ej	0.4	V
Laurencia obtusa (Huds.)	.	.	1.2	.	1.3	1.3	2.4e	+	2.8	IV
Amphiroa rigida Lamour.	1.3	.	.	1.3e	1.1	.	1.1	1.1	1.3	IV
Jania rubens (L.)	+ej	1.4e	+e	+e<	.	+e	.	.	0.4	IV
Lithophyllum incrustans Phil.	.	.	1.3	1.5	2.3	1.3	.	.	2.8	III
Lithoderma adriaticum Hauck	.	.	.	.	1.4	1.4	1.3	.	0.9	II
Feldmannia lebelii (Aresch.)	.	.	.	.	1.3e	.	.	.	0.3	I

Species	1	2	3	4	5	6	7	8	%	
Cystoseiretum crinitae (photophilic, calm-water species)										
Herposiphonia tenella (C. Ag.)	+e	+e	+e	.	+e	+e	.	+e	0.1	IV
Halopteris scoparia (L.)	.	1.1	1.1	.	1.1	.	.	1.1	0.9	II
Cystoseira fimbriata (Desf.)	.	1.1	1.1	.	+	.	.	1.2	0.6	II
Jania corniculata (L.)	+e	+e	+e	+e	.	.	.	.	≈0	II
Liebmannia leveillei J. Ag.	.	.	.	+e	.	+	.	.	≈0	I
Rhodymenietalia (sciaphilic species, sensu lata)										
Peyssonnelia polymorpha (Zan.)	2.4	1.4	1.3e	1.3	1.4	1.3	1.4	2.3	5.6	V
Mesophyllum lichenoides (Ellis)	2.3	(+)	1.5	1.5	.	1.1	1.4	2.3	4.7	IV
Apoglossum ruscifolium (Turn.)	.	.	.	.	.	.	+	+e	0.3	II
Rhodymenia ardissonei Feldm.	1.3e	.	.	.	.	.	.	.	≈0	I
Amphiroa beauvoisii Lamour.	+e	.	.	.	+	.	+	+	≈0	I
Petroglosso-plocamietum (sciaphilic, rough-water species)										
Caulacanthus (?) rayssiae Feldm.	+e	+e	.	+	.	.	+e	.	≈0	II
Myriogramme minuta Kylin	+e <	1.2e	.	.	.	.	.	.	≈0	II
Callithamnion tetragonum (With.)	.	1.5	.	.	.	.	.	.	0.3	I
Gymnothamnion elegans (Schousb.)	1.5	.	.	.	.	.	1.3	.	0.3	I
Cladophora coelothrix Kütz.	.	.	.	.	.	.	.	1.3	0.3	I
Ceramium echionotum J. Ag.	+e	+e	.	.	.	.	.	.	0	I
Rhodophyllis divaricata (Stack.)	+e	+e	.	.	.	.	.	.	0	I
Acrochaetietalia (mediolittoral species)										
Gastroclonium clavatum (Roth.)	+ej	+	1.2	+	1.1e	1.1e	+j	.	1.0	V
Ceramium ciliatum (Ellis)	.	1.5	1.4e	1.3e	+	+	1.3e	.	1.6	IV
Porphyra leucosticta Thur.	1.3ej	+ej	+ej	.	1.2e	.	.	.	0.3	II
Chaetomorpha capillaris (Boerg.) var. crispa (Schousb.)	.	+	.	.	+	+e	.	.	0.3	II
Cladophora dalmatica Kütz.	.	.	.	.	+	.	.	.	≈0	I
Ralfsia verrucosa (Aresch.)	.	.	.	+	.	.	.	.	0.3	I
Mesospora mediterranea ? Feld.	.	.	.	.	.	+	.	.	≈0	I
Cladophora laetevirens (Dillw.)	.	.	+	.	+	.	.	.	≈0	I
Gelidium crinale (Turn.)	.	.	.	.	.	.	.	.	≈0	I
Neogoniolithon notarisii (Duf.)	.	.	.	+	.	.	.	.	≈0	I
Chondria boryana (De Not.)	.	.	+	.	.	.	.	.	≈0	I
Callithamnion granulatum (Ducl.)	+e	.	.	.	.	.	.	.	≈0	I

Table 2.2. (*cont.*)

	292	245	251	282	259	258	260	238	Mean global abundance (%)	Abundance
Number of site	292	245	251	282	259	258	260	238		
Date	17.i.69	10.ii.67	10.ii.67	30.v.68	15.vi.67	16.ix.67	16.ix.67	28.xii.67		
Depth (cm)	20	20	10	20	15	15	20	20		
Slope (in degrees)	5	10	10	5	5	5	10	5		
Aspect	NE	NE	S	NE	NE	S	NE	NE		
Cover	100%	100%	100%	100%	100%	100%	98%	100%		
Ulvetalia (species adapted to high sulphur and nitrogen)										
Ulva rigida? C. Ag.	1.1	1.5e	+	1.4	~2.4	1.2e	2.2	1.4	5.3	V
Gigartina acicularis Lam.	1.3e	1.3e	1.3e	+e	+e	+e	+e	+e	1.0	V
Colpomenia sinuosa (Mert.)	+	.	.	.	.	.	1.2e	+	0.3	II
Species with wide or universal distribution										
Corallina mediterranea Aresch.	3.4	4.5	3.3	3.4	2.4	3.3	2.5	2.4	32.2	V
Cyanophyceae (undetermined)	.	+e	+e	+e	+e	2.1e	2.1e	+e	3.8	V
Diatoma (undetermined)	+e	1.1e	+e	.	.	.	+e	+e	0.4	IV
Melobesiae (undetermined)	+	+e	1.3e	1.3e	+e	.	1.1e	1.1e	1.3	V
Aglaozonia parvula? (Grev.)	1.3	1.3	2.4	1.3	1.4	+	+	2.4	5.0	V
Ceramium diaphanum (Lightf.)	+e	1.4e	2.4e	1.1e	.	2.3e	.	+e	4.4	IV
Dermatolithon sp.	1.4e	.	1.4e	.	.	2.4e	1.4e	1.1e	3.1	IV
Ceramium gracillimum (Griff.)	2.4e	.	.	+e	+e	.	.	1.4e	2.2	IV
Licmophora sp.	1.3e	+e	+e	+e	+e	.	.	+e	0.7	IV
Cladophora sp.	1.2e	.	+	1.2	+j	.	+j	.	0.7	IV
Falkenbergia rufolanosa (Harv.)	1.4e	.	+e	+e <	+e <	.	.	+e	0.4	IV
Lomentaria clavellosa (Turn.)	+e	+e	+e	+e	+e	+e	.	.	0.1	IV
Dermatolithon pustulatum var. *corallinae* Crouan	.	1.3e	1.2e	.	.	+e	.	.	0.6	III
Acrochaetium sp.	.	.	+ej	+ej	.	+e	+ej	.	≈0	III
Ceramium tenuissimum (Lyngb.)	.	2.4e	1.5e	.	.	.	.	.	2.2	II
Aglaozonia melanoidea (Schous.)	.	2.4	.	.	.	1.4	.	.	2.2	II
Melobesia farinosa Lamour.	.	.	.	.	.	+e	1.2e	+e	0.3	II
Pringsheimiella scutata? (Reinke)	.	.	.	1.4e	.	.	.	.	0.3	II
Dasya sp.	.	+e	.	+e	+e	+e	+ej	.	≈0	II
Polysiphonia sp.	+e	.	+e	.	.	.	.	+e	≈0	II

relationships which might emerge can only be guessed at. Certainly we shall not know the value of this methodological approach, or what limitations or new ideas await us, until someone undertakes such an analysis. Equally valid would be a similar study of algal communities in waters selected on characteristic chemical (habitat) features, e.g., marl lakes or acid bog lakes.

Another more recent development involves the study of communities along an environmental gradient and is hence termed *gradient analysis*. Arrangement of the communities along the gradient is the process known as *ordination*. The ordination may be *direct* e.g., when the series of samples is arranged along a temperature gradient or *indirect* when the samples are arranged along directions of change in communities. Gradient analyses of the direct type are exemplified by transects taken down rocky shores or, at sea, through a series of stations differing in temperature. If such data are obtained in sufficient detail to plot each species or association a series of Gaussian curves expresses the data (Fig. 2.2). It is interesting that very similar curves also result from a seasonal analysis of many microscopic algal communities. This is only to be expected, since in effect it is a gradient analysis linked to the gradient of annual climatic variables (*see* Figs. in Chapter 10). Such analyses are essential to plot the range of factors under which an association can exist. Samples are to some extent randomised but strict randomisation seems to be of little value and fixed intervals are used, e.g., along transects on shores or at sea. Very detailed data along such lines have been collected by the use of continuous plankton recorders towed behind ships on regular routes in the North Atlantic. The combination of years of data shows clear species boundaries and when all the data are synthesised will show distinct associations in spite of intergrading at the boundaries (*see* Fig. 8.12).

If association data are plotted against two axes representing different environmental gradients, then a mosaic type representation is produced, comparable in many ways to the time–depth plots of species (e.g. Fig. 10.15) or the Ca–Mg–Na plots of the growth of individual species (e.g. Fig. 7.35). These approaches deserve greater study and application by phycologists.

The technique of arranging the associations in relation to one or more environmental gradients (*ordination*) is well documented, and many procedures have been devised for terrestrial ecology; they can be used in algal ecology. Indeed the arrangement in Table 3.9 forms a primary matrix and the species can be rearranged subjectively, or by using similarity coefficients, recurrent group analysis, etc. (Whittaker, 1973; Allen & Skagen, 1973).

There are, of course, two views about the discreteness of communities on the land; the phytosociologically minded tend to the concept of discreteness, as did the approach of workers such as Du Rietz in Sweden, whilst the other school of thought tends to think of continuous variation (though each recognises a degree of gradation). Obviously in algal ecology there is a degree of continuous variation at sea, but discreteness in a particular lake. Each lake will have its own discrete population and the problem resolves itself into the degree of grouping of lakes which can be achieved. At sea there are probably also discrete populations, such as those revealed from the data of the continuous plankton recorder, but if they are sampled at their edges, clines can be distinguished.

One thing is certain: aquatic vegetation is exceedingly complex and anything which enables phycologists to classify it into broad units will help understanding. Nevertheless one must always remember that these broad (phytosociological) units can, and must, be split into smaller 'bits' for even more intensive study. The approach is however the opposite of the one starting with the assumption that no two are alike and all must be treated as individuals, which is merely unworkable and a sure way to chaos.

A danger in gradient analysis is that the physical parameter might assume too great an importance when in fact it is merely one variable among

Fig. 2.2. The appearance of species or associations along an environmental gradient.

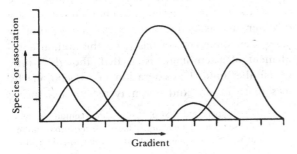

many, and indeed, some parameters may not grade along the same direction; but it is a valuable approach providing that the overall complexity is not forgotten. Interplay between associations occurs to a lesser degree and usually via their physico-chemical influence on habitats, e.g., shading of a sediment-dwelling association by a rich phytoplankton association or grazing by species characteristic of one association at an interface between associations. An example of these factors acting together can be found in the association of algae growing epiphytically on the underside of a floating angiosperm leaf which is bathed in water in which planktonic associations grow. The phytoplankton may deplete the water of nutrients, or modify its composition by secretion, whilst planktonic animals can graze the epiphytes; thus an interaction is set up between two associations, across a boundary between them since it is impossible for two associations to occupy the same spatial niche. The dead remains of one or more associations can however drift into the space occupied by other associations, e.g., the association of algae living on sediments is continually contaminated by dead (and more rarely live) plankton and epiphyton descending from above. Under *no* circumstances should the contaminant species be ascribed to the sediment association and methods must always be employed to exclude such species, otherwise totally misleading species lists (and hence associations) are compiled. Conversely, after storms, species of the epipelon and epiphyton are washed into the open water to contaminate the plankton. Some impressive lists in the literature are meaningless owing to the inclusion of such accidental species. It is particularly difficult to separate such species in the analysis of associations dominated by diatoms, especially when the estimation of numbers is made merely on material cleaned by boiling acids.

Associations are usually designated by the name of the dominant/subdominants, e.g., the intertidal *Pelvetia canaliculata* association of rocky shores in Europe, the intertidal *Hormosira banksii* association of Australasian shores, the subtidal *Macrocystis pyrifera* association off the Californian coast. These associations are relatively easy to define though much work is still needed to clarify many aspects of their biology.

The naming of regions of the shore and subtidal zone after dominant plants is fraught with dangers,

for example, the '*Laminarian*' zone of European workers is occupied by *Sargassum/Turbinaria* in the Red Sea (Simonsen, 1968) and by other large brown algae elsewhere.

I also incline to the view of Poore* (1955) that mere naming of communities by the dominant is not satisfactory unless the name has been chosen after a thorough sociological study; hence the *Pelvetietum* of European coasts would be much more valuable as an abstraction if one could produce association tables compiled from many sites to justify the naming, as can be done for the *Cystoseiretum strictae* of the Mediterranean. The endings -etum, etc., are adopted and used as in higher plant ecology (*see* Boudouresque, 1968, 1969, 1971a, b).

In these associations, and many others of the rocky intertidal and subtidal habitats, the dominance of macroscopic species throughout the year makes recognition of the associations relatively easy. Even in these apparently well-known habitats detailed comparative studies of the associations from different regions are still lacking, and rarely have they been fully described, especially with regard to the smaller species of the associations, whilst only rarely is the associated epiflora studied. Many of these associations, dominated by macroscopic species, can be likened to a land forest association; the conspicuous alga is comparable to the tree layer, the lower growing forms to the shrub layer, and the creeping species to the herb layer and in addition all have their associated epiflora. Clearly, these associations are not simple, yet have a degree of stability in that many of the species are perennial and occupy large areas with relatively easily recognised dominants.

It is, however, much more difficult to apply the association concept to the phytoplankton since the dominance changes throughout the year and dominance itself is not so easily determined, e.g., it could be based on either cell number or cell volume. If one wishes to use the association to characterise a body of water such as a lake, and contrast it with another lake, then some system based on the annual complement of assemblages is needed, though it may be cumbersome. Consideration of associations by seasons is simpler and one may recognise a spring

* These papers and that of Poore (1962) contain many valuable concepts which are based on terrestrial studies but which are equally applicable to the aquatic field.

Asterionella formosa, a summer *Pandorina morum*, an autumn *Aphanizomenon flos-aquae* and a winter *Peridinium willei* association in many lakes. Equally in the marine environment one often finds a *Skeletonema costatum* association in spring, e.g., in the Menai Straits (Jones & Spencer, 1970) and off the coast of Rhode Island (Pratt, 1966*b*). Microscopic communities growing on sediments, etc., are not yet sufficiently documented to allow definition of associations.

Associations are intimately linked to spatial (physical) aspects of the environment and it is convenient to distinguish these spatial entities or habitats. As noted above (p. 27), habitats can be defined irrespective of the associations growing in them and are relatively few in number, though hundreds of associations are found within each. They are conveniently divided into benthic, i.e., essentially associated with the bottom of a water body (now by extension to other submerged surfaces) and open water or planktonic* in which the algae drift passively in the water, or move through the water; intrinsic movement of planktonic algae is usually slight relative to the turbulent water movement induced currents. The phytoplankton is the algal community occupying this spatial niche and will be discussed in detail in Chapter 7. The algal component of the benthos is often termed the phytobenthos.

Plankton

The phytoplankton may be subdivided. The *euplankton* is the permanent community of the open water, and is by far the most important of the floating communities. Accidental plankters, caught up in water currents or washed into the habitat are termed *pseudoplankton* (*tychoplankton*). The prefixes heleo-, limno-, potamo- have been employed for pond, lake, and river plankton respectively. The only other floating community which has received any attention is that at the water–air interface. Here a special community of micro-organisms develops and is termed the *neuston*. In fact this comprises two groups of organisms, one living on the upper surface of the water film (*epineuston*) and one attached to

---

* The term planktonic is used frequently for both the spatial and organismal concept but strictly speaking it should refer to the organisms. Various adjectival forms are in use, e.g., planktic, benthonic (*see* Rodhe (1974), Hutchinson (1974).

the under surface (*hyponeuston*). There is yet another, and almost completely uninvestigated, community floating in water and that is the one which develops on suspended particulate matter, usually organic; the term for this non-living particulate matter is *tripton* or *abioseston*. Tentatively, I suggest that this kind of plankton might be termed *detritiplankton* (*see* also p. 246). If it is necessary to refer to the living plankton and tripton combined, then the term *seston* is used. In a sense, the classification of the neuston and the community associated with the tripton, within the plankton, is arbitrary, since these two communities are surface living assemblages and the component species are morphologically similar to the other algae growing at solid–liquid interfaces.

Benthos

There have been many schemes for subdividing the benthos. Some terms are widely used and others less common, whilst some of the terminology evolved for describing animal communities is not necessary for algae since it is based on size range of organisms. In oceanographic and animal ecological studies the region where plant growth is possible is often termed the *phytal* zone in contrast to the deeper, dark *profundal* zone. The terms *trophogenic* and *tropholytic* are also used, implying an upper zone of production and a lower zone of breakdown; they are also applied to the upper and lower water masses.

The benthic environment in freshwaters comprises the zone of the shore which is occasionally wetted by waves or spray, or is periodically exposed, and the sediments below low water level.

The terminology based on the word littoral has been extended to freshwaters; it has not been subject to such great variation of meaning as when applied to marine benthos. The proper terms for the three zones are supralittoral, eulittoral and infralittoral (sublittoral) but they are rarely used in limnology.

There is little work on the algal flora of the zone above normal water level in lakes but undoubtedly certain algae are only abundant in this situation. The extent of the benthic environment for *algal growth* is confined to the bottom which is sufficiently illuminated for photosynthetic purposes; this may extend only centimetres in turbid water, or to many metres depth in clear waters. The region below the limit of light penetration is still the benthic environment but is not populated by photosynthetic plants.

In marine habitats a complex system of classification has been developed for the zone adjacent to the land. It is complicated by consideration of the belts of algae and animals which live within it, but a purely physical classification can be adopted as in Rigg & Miller (1949), Hesse, Allee & Schmid (1951), Clarke (1965), Odum (1971), and Round, (1971a, 1973a) with a simple division into three regions. These are the uppermost exposed region subjected only to occasional storm waves or spray and termed the supratidal, the region between tides termed the intertidal* and the region below low tide termed the subtidal. These terms are virtually synonymous with supralittoral, mediolittoral and infralittoral as used by French workers, especially in the study of Mediterranean shores, and by Spanish workers (e.g. Seoane-Camba, 1969). The intertidal can be divided on many shores into upper, mid and lower zones with the upper grading to varying degrees into the supratidal, and the low more distinctly continuing into the subtidal. These zones are used in Ricketts, Calvin & Hedgepeth's stimulating book *Between Pacific Tides* and I see no reason to enter into the controversies of naming and delimiting extra zones where for most instances the simple terms referring to the all-important tidal cover regimes are sufficient. Obviously highs and lows of tides vary daily, and from place to place, but time and time again, three distinct intertidal zones are recognised, although the boundaries may be somewhat blurred, or occupied by characteristic associations (for examples *see* p. 90). The exact vertical extent of these zones can be

* The term intertidal is also used for the time span between successive high tides but it should always be obvious from the context whether the spatial or temporal concept is implied. An increasing number of papers are appearing using these simple terms (e.g. Doty, 1967; Mann, 1972b, c, d), rather than those based on littoral. Indeed Stephenson & Stephenson (1972) comment on the fundamentally different senses applied to the word littoral and therefore only use it with prefixes 'whereas if we use the term "tidal zone" or "intertidal" for the strip of coast between extreme high water and extreme low water, there is no ambiguity'. A suggestion that the term 'littoral' be yet again redefined from a biological standpoint was made by Lewis (1961) even though he writes 'how thoroughly, if *confusingly* "littoral" is now rooted in thought and literature' (my italics). There is a table in Hedgepeth (1963) which amply illustrates the confusion over the term littoral but, in spite of this, he argued for its retention.

plotted from long-term measurements coupled with data on tidal cover, and points down the shore can be designated according to the length of time of tidal submersion or of emersion; this is not a linear function and especially on shores with complex tidal patterns there are points close together on the shore which are exposed for very different lengths of time (Fig. 3.7). These time elements should be worked out for any particular shore under investigation so that the broad intertidal zone can be sub-divided according to the particular local tidal and physiographic factors. The physical subdivisions *may* then be found to relate to algal distributions (*see* Chapter 3) but it must always be borne in mind that the tidal heights, or mean tide levels, are *not* causative agents in algal distribution but merely a framework of convenient and essential reference points on the shore. Other factors are also involved in the distribution of organisms in this zone, e.g., type of substratum, angle of slope, degree of insolation, degree of wave action, degree and type of grazing, making it one of the most complex of algal environments. However Southward (1958) writes 'the rise and fall of the water line produced by the tides or waves, or both, must be regarded as the primary cause of the observed patterns of zonation'; a coincidence between tide levels and zonation clearly recognised also by Hedgepeth (1963). An objection sometimes raised to systems based on tidal data is that on some shores, e.g. in relatively enclosed seas (Mediterranean, Baltic) or in many tropical regions, the rise and fall of the tide is small, but this is merely an extreme situation resulting in a very narrow intertidal zone, or merely the upper and lower edge of such a zone, if comparisons with other shores are made (*see* Coleman & Stephenson, 1966). It is certain however that even in almost tideless habitats (some vertical movement occurs in all waters) both supratidal and subtidal habitats can still be distinguished. The common terms supralittoral, littoral, sublittoral, and many others such as epilittoral, prelittoral, eulittoral, mesolittoral, midlittoral, infralittoral, have been used and many arguments, both terminological and semantic, blur and confuse their application to the zones at the edge of the sea. The belts and associations of algae (and animals) are nonetheless the real biological expression of all the factors operating in this complex region and the belts can be named by dominants. I acknowledge that such belts may not occupy exactly similar positions on different shores, but they do occupy the same general position relative to tidal heights and are more easily fitted to these than to zones defined from biological horizons.

As in freshwaters, the marine benthic habitats colonised by algae extend down below low tide level to a depth to which photosynthetically usable light penetrates. This subtidal has also been subdivided, for example Feldmann (1937) recognised an upper (étage infralittoral superieur) and lower (étage infra-littoral inferieur) region based on the distribution of sun and shade forms (*photophilic* and *sciaphilic* species). These appear above or below the 5–10 m depth in the Mediterranean Sea. Neushul (1965) and Shepherd & Womersley (1971) recognised three zones, upper, mid and lower subtidal, though ad-mitting that to some extent there is a continuum of species in the subtidal. The situation regarding any subdivision of the subtidal is similar to that of the intertidal in that it is possible to find some species throughout the zone but, nevertheless, certain species tend to be confined, and it is these which must be used to define the zones and their very existence gives credence to the schemes of zonation. Data derived from dredging must be treated very cautiously in attempting subdivisions in this zone. However, diving may not be possible to the lower limit of algal growth and a lower zone is only valid if its *lowermost* limit has been charted. At some sites this has not been possible owing to the limitations of Scuba diving.

The terminology applicable to the communities of algae living in the supratidal, intertidal and subtidal zones is compounded from terms referring to the mode of growth of the algae (p. 32) or the type of substratum on which the communities grow. There has been a regrettable tendency to confuse concepts which are really basically physical–geographical with the biological associations. Most shore zonation studies would benefit if the physical features were measured first and then the biological associations studied. In some instances the latter may fit into subdivisions based on physical conditions but some may not, indicating that further physical or biological parameters require study. Furthermore, additional experimental evidence is required before the factors controlling the distribution of species or associations can be determined exactly.

A further series of associations can be recognised in the rock pools of the supratidal and intertidal zones. Some of the pools are closed and when the tide recedes they remain filled, others outflow over sills and merely drain more slowly than the surrounding shore. Clearly these are special habitats, which should be studied independently of the surrounding rock faces since the factors operating on the algal associations are less related to drying and wetting and more to changes in concentration of dissolved substances, temperature, and pH.

The plants of the intertidal region are consid-ered by some authors to be a climax development, or *biome*, developed under given climatic or physio-graphic features exactly as a coniferous forest forms a biome. Beveridge & Chapman (1950) wrote 'The biome of the rocky seashore represents a physio-graphic climax since it is dependent upon the tide rather than the climate'. Again the emphasis is on the tide as a controlling factor.

The organisms (and communities) growing in the benthos have been classified on the growth form of the algae and the following summary gives the main forms.

Rhizobenthos comprises the vegetation rooted (by means of rhizoids) in the sediment, e.g. the beds of *Chara* in a lake or the underwater swards of *Halimeda*, *Caulerpa* etc. in tropical seas. Some workers have distinguished associations 'rooted' in silt (*rhizo-pelon*) or in sand (*rhizopsammon*).

Haptobenthos is adnate to solid surfaces and is conveniently divided into *epiphyton* (epiflora) grow-ing on other plants, e.g., the coating of *Oedogonium* and other species on reeds, of *Ectocarpus* on *Fucus* species; *epilithon** growing on rock surfaces, e.g. *Fucus* on the intertidal rocks, *Rivularia* on stones in streams, *Stigonema* on wet rocks exposed on the land, green coccoid algae on chalk fragments; *epipsammon* grow-ing on sand grains in both freshwater and marine sites (the total community living in sand is sometimes referred to as *psammon*, e.g. in Sassuchin, Kabanov & Nieswestnova, 1927); *epizoon* (epifauna) growing on animals. An additional term is necessary for organisms growing on artificial surfaces, e.g. ship hulls, piers etc; the term *fouling* has been used, but is not entirely acceptable though it is the only usable term at the moment. The terms, periphyton,

---

* Similar terminology is used by Friedrich (1969) in discussing animal communities and by Peres & Ricard (1957) in their excellent discussion of communities in the Mediterranean.

phycoperiphyton and the even more superfluous combination epiphytic periphyton (Wetzel & Westlake, 1969) have crept into the literature as portmanteaux terms to include algae growing on objects placed in the water, and unfortunately have been loosely applied to epiphytic, epipelic and epilithic communities, merely adding confusion, and worse still, imprecision to the subject. I strongly urge that the term periphyton and its variants be dropped or at most be confined to growths on artificial substrata.

Herpobenthos is the community living on, or moving through, sediments (strictly the term refers to organisms creeping in sediments; *herpon* is used by Margalef, 1960) and it can be subdivided into *epipelon* (Round (1956) and Naumann (1925), who used the term pelophytic) living on the surface of the deposit (mud or sand), *endopelon*, living and moving within muddy sediments and *endopsammon* living within sandy sediments. Since the surface sediments are always being moved by water currents, and especially by animals, these communities occupy the sediment to a variable but generally fairly shallow depth.

Endobenthos is the community living and often boring into solid substrata. The commonest example occurs on rock and this is termed *endolithon*, e.g., Cyanophyta boring into calcareous rock. In a sense the endopelic and endopsammic communities of algae are also endobenthic and could perhaps equally be classified here. Investigations are now also revealing algal species associated with the internal mucilage of other algae and these form a genuine *endophyton*, e.g. *Navicula endophytica* in the conceptacles of *Fucus* (Hasle, 1968). The free-living algae, e.g. diatoms and *Chlorochytrium* living in mucilage tubes, have been termed *endotubular* (Cox, 1975*b*).

A further and relatively uninvestigated community of algae is found loosely associated with the epiphyton but since the species lack attachment organs they can be easily washed out from amongst the epiphyton. This has been termed the *metaphyton* (Behre, 1956) and is found in both freshwater and marine habitats. It is an important and in some lakes a quantitatively quite extensive community which deserves much greater study. Ilmavirta, Jones & Kairesalo (1977) also recognised the significance of this community and rightly suggest it should be treated as a separate entity but unfortunately they

termed it 'pseudoplankton' though recognising its derivation from the loosely attached community around *Equisetum* plants (*see* also footnote p. 232).

A community with some characteristics of the metaphyton occurs lying loose on the sediments. The algae are unattached species moved at random by local water movements. It has received little study though it is extremely common; the name *plocon* has been applied to it (e.g. in Margalef, 1960) and, earlier, Naumann (1925) coined the very clumsy adjectival term *epipythmetic*. Kairesalo (1976) has compared this assemblage with the epilithon of Lake Pääjärvi, but terms it 'littoral plankton' a term which I believe should be avoided.

The algae of the soil form a further community which has many characteristics in common with the epipelon; it has been termed the *phytoedaphon* by some workers (e.g. Francé, 1913) to bring the naming in line with plankton, etc. Algae of snow fields and of ice floes have been termed *cryophyton* but they do not easily fit into the above scheme.

Clearly one should distinguish between the descriptive terms for habitats and those for communities but make combinations when necessary, e.g. epiphyton can be found on supratidal, intertidal and subtidal plants and these should be distinguished since they may be quite distinct assemblages. A vast amount of work is needed on the associations of all aquatic habitats in order to describe them adequately and relate their species composition to the environmental parameters. When a situation is found where the number of species is reduced the recognition becomes easier, e.g., the epiphyton of the under surface of *Lemna* leaves is often dominated by a single diatom *Achnanthes hungarica* with *Cocconeis placentula* subdominant and the associated species number less than 50 (Round, unpublished). Several other relatively simple communities exist, e.g., the community of Chlorophyta on tree bark, the community of algae on permanent snow fields, the community on chalk fragments, the community on whale skins. Simplicity is added to these since there is little, if any, seasonal variation such as it common in planktonic, epipelic and epiphytic communities. Variations of community structure in time are surprisingly inadequately described except for the freshwater plankton (p. 422), and to a lesser extent for the algae of the intertidal zone (p. 449). If one assumes that the plankton of a lake is a single association then its delineation is complex since several dominants and subdominants succeed one another during the year and must be quoted in its description (p. 272). It cannot therefore be described from infrequent sampling as can a land community and the analysis of dominance should extend (preferably) over a period of years since it will then be possible to distinguish the species which re-appear each year from those which are casuals. If only a few analyses of a community have been completed within a defined season, the time of year or months should be quoted, e.g. a spring *Asterionella formosa* association or an autumn *Aphanizomenon flos-aquae* association. It is misleading to do otherwise since it is *rare and perhaps unknown* for a single species to be abundant throughout an annual cycle in such communities. In some habitats the seasonal components may be sequential assemblages in a plant succession leading to a climax algal vegetation, e.g., on a rocky coast where wave action has removed the flora, or on a newly exposed soil surface.

A major problem in the study of algal communities is classification of the algae themselves for the number involved in some communities is extremely large and each must be identified accurately. To give but one example, Skuja (1948) estimated that 440 species occur in the phytoplankton of Lake Erken. Identification is becoming easier as monographs on classes and genera appear, but most common genera still require detailed study. The problem can be side-stepped in at least two ways: either a single species can be selected and studied in detail, or an attempt can be made to measure the biomass of the community irrespective of its specific composition, e.g. from chlorophyll *a* estimations (*see* p. 252). The former, or autecological approach, has yielded some most impressive data, whilst the latter approach is often combined with rate studies, involving either ^{14}C uptake or oxygen output. Carbon fixation studies have been widely adopted, resulting in some generalised patterns of biomass distribution, related in some works to environmental parameters. However this approach has to be combined with detailed studies of the species composition of the communities before anything other than overall approximations can be made; examples are the work of Stull, Amezaga & Goldman (1973) and Watt (1971) (p. 320) where the contribution of individual species has been considered.

Algal communities, as listed above, are simply defined as groups of species living together, which almost always means groups of species living in clearly defined habitats. The community/habitat can be recognised on a purely subjective basis or can be made objective by application of a mathematical approach. However all objective approaches are based on sampling and are therefore affected by the technique, site, size, frequency of sampling and by seasonal variations and other variables. Closeness of communities is a further complicating factor, e.g., comparison of the planktonic community sampled 20 miles from the shore in a deep lake would show few, or almost no, species representative of the epipelic or epiphytic communities which are often almost in contact with one another and bathed by the water containing the planktonic community. In this situation the three communities can be separated by sampling techniques, or simply by statistical techniques since although the plankton bathes the other communities, cell counts of the epiphytic or epipelic communities usually show a tremendous dominance of epiphytic or epipelic species over planktonic. In a very small pond near Bristol the coefficient of similarity between the diatom component of two epipelic stations was 74.65 whilst between these and the plankton it was only 0.61 and 0.57, between these and the epiphytic only 2.78 and 2.14; and the coefficient of similarity between the planktonic and epiphytic species was 0.99 (Eaton, 1967). The total number of diatom species used in this study was 138, but analysis of the complete communities would require the identification of 500 or more organisms; few attempts have been made to compare communities of algae in this way, but one which does (Hodgkiss & Tai, 1976) found a greater degree of similarity between the communities in a Hong Kong reservoir. It is a simpler matter to compare associations, e.g., those on *Lemna* leaves in different ponds, or in the marine intertidal on different coasts.

Climax communities

The concept of climax communities is much more difficult to apply to algal vegetation owing to the rather rapid seasonal changes which take place and it needs therefore to be based on long-term work which integrates the seasonal variations in the flora (cf. the plankton recorder data p. 265). The climax in the Clements sense as applied to terrestrial vegetation is determined by climate and this concept could easily apply to communities of algae. Fig. 8.2 gives examples of the phytoplankton community of each sector of the North Atlantic Ocean in which one element is plotted on each diagram. Between these sectors or zones, in each of which a distinct community of phytoplankton develops, are transitional zones which in the view of Venrick (1971) should be considered as discrete biogeographical regions: she terms these 'non-conservative' areas, whereas the major oceanic zones, in which the species associations appear relatively permanent, are termed 'conservative areas'. Perusal of the literature certainly indicates large oceanic areas of relative stability and therefore of climax vegetation. Confidence that such communities form climax communities is increased when one compares the flora in an area at the turn of the century and again now, and finds the same species in abundance (e.g. the dinoflagellates, p. 380). The ecologists' concern with succession impinges on this view, but in many habitats succession is a relatively slow, almost imperceptible progression, and does not detract from the idea of climax vegetation.

Niche

The term niche is used in a spatial sense to describe the place in which a species lives, or in an abstract sense to designate the requirements of the species abstracted from the spatially extended habitat; these two may be thought of as the habitat centred and the organism centred approach. Which definition is better or more correct raises much argument but so long as the context is clear, both can be used. Every alga occupies a well-defined spatial niche and in general only one abstract niche is occupied by any alga, e.g. *Asterionella formosa* only occurs suspended in the open water of a lake, and only occurs when a given set of requirements is met. It is probable that the abstract niches for species do not overlap, in other words no species has precisely the same set of requirements. A definition is given by Maguire (1973): 'The niche is the genetically (evolutionarily) determined capacity (range of tolerance) and pattern of biological response of an individual, a species population or the whole species to environmental conditions'.

Hutchinson (1967) defines the ecological niche of a species as a set of points in an abstract Cartesian space, in which the co-ordinates are various environ-

mental factors. However, as Levandowsky (1972) points out, these factors are variable and mixed, and he distinguishes five aspects of an environmental variable. 1. An aspect which is not 'used up', e.g., temperature and this he terms a 'linking' variable, since all the species in a habitat are adapted to this aspect (or are tolerant of a wide range, i.e., are eurythermal). 2. An aspect that can be 'used up' and hence limit population size. Some, such as nutrients, are used and need to be replenished; others, such as space, are occupied but later become available again. This aspect results in competition and Levandowsky terms it a 'separating' aspect. 3. An aspect involving biotic interaction, e.g., predation, which he terms a 'rarifying' aspect. If predators are specialists, they may only affect certain species; predation reaches an extreme case with specialised parasites. Rarifying aspects are also coupled with population size, since there is a feedback between prey and predator. 4. A 'reinforcing' aspect in which an organism is favoured by, or dependent on, the mere presence of another; and the reverse, a 'negative reinforcing' aspect where the organism is inhibited by the presence of another. There are many examples of these aspects amongst algae. 5. A 'variational' aspect, i.e., the variability or regularity of recurrence of an environmental variable (i.e. of one of the previous four aspects).

Maturity

There have been many attempts and diverse approaches applied to measuring the characteristics of ecosystems; and one of these has been concerned with assessing the maturity of the system. Undoubtedly some are more mature than others in the sense that they have existed for a longer period of time, e.g., African rift lakes as opposed to glacially formed lakes arising since the commencement of retreat of the last Ice Age. But evidence from newly formed natural or artificial lakes, or from newly erupted islands such as Surtsey, show that colonisation and stabilisation is rapid and since all systems are dynamic one might not expect to find evidence of maturity. Apart from the time factor, which is probably only meaningful during formative periods, what measurements can be used to assess maturity? Three attributes are often proposed (Margalef, 1963; Odum, 1963): (a) production–biomass ratios (low in mature systems); (b) species diversity (low in mature systems); (c) pigment diversity. Margalef suggests that the latter can be measured adequately by determining the optical density of acetone extracts at 430 and 665 nm, though this hardly gives a measure of the numerous pigments which occur in algae, and there are thin layer chromatographic techniques which are easy and give more detail (Jeffrey, 1974) and which should be used if pigments are to indicate diversity (Hallegraeff, 1977). Winner (1972) points out that there are in fact few data which can be used for evaluation of the concept of maturity; he analysed the above three attributes in five very diverse lakes in Colorado. Mature systems are assumed to be characterised by a low productivity–biomass ratio, yet in Winner's study the more productive lakes appear the most mature, a finding which is contrary to the current theory. Again if species diversity is measured the two most productive lakes had the greatest diversity which is exactly the opposite of Margalef's prediction. Also, on the basis of pigment ratios, all these quite diverse lakes would classify as mature (as one might expect in lakes which have been stable for several thousand years) and Winner concludes that this ratio has no consistent relationship to variations in ecosystem structure and function; a similar result was reached by Mathis (1972) and Skeen (1975) for both aquatic and terrestrial vegetation. Interestingly, however, Winner found that the data could be used to rank the lakes in a eutrophication series, utilizing primary production (per m^3) of the trophogenic zone, chlorophyll a content, seston content and Secchi disc transparency. It is clear from this work that conclusions concerning maturity based on single samples, such as used in many of the data presented by Margalef, are not valid; indeed Winner (1969) found that species diversity measured at weekly intervals over a year, in a single pond, varied almost as much as did the data from 33 lakes reported by Margalef.

In a study of waters in Holland, Hallegraeff (1976) also found no increase in diversity of pigments from eutrophic to oligotrophic. Nor could Hallegraeff (1977) find any evidence for Margalef's suggestion that there is an increase in the pigment ratio as the season progresses. Any attempt to formulate general ecological theories and mathematical models can only be successful if data are collected over long periods and take into account the reality of the complexities, e.g., species diversity, pigment values, productivity values all change seasonally. Measures of diversity require a detailed knowledge of the taxonomy of algae and of community structure, so as to avoid inclusion of casual species. There is a considerable literature on 'diversity' and 'diversity indices' related to terrestrial communities and the techniques are relevant and indeed have been frequently used by phycologists. Considerable doubt is cast upon the value of such indices by some ecologists, 'Diversity *per se* does not exist' (Hurlburt, 1971) and as suggested by Dickman (1968a) an index based on relative productivity may be more ecologically significant, whilst an index combining abundance of species and relative biomass is more valuable than species abundance. 'Stability' is also a feature which has been discussed but whilst many algal communities appear very stable there is a considerable problem when attempts are made to quantify the concept (Margalef, 1969).

Summary

The range of algal habitats is far greater than the immediately obvious streams, ponds, lakes and ocean. Algae occur on buildings, woodwork, terrestrial rocks, tree trunks, permanent snow, on the under surface of Arctic and Antarctic ice, on the sediments of hot springs etc. Temporary rain pools are often densely populated by blue-greens, diatoms, euglenoids and chlorophycean flagellates. A few terrestrial animals and plant surfaces, especially in moist tropical regions, support algal growths, and soil surfaces are very rich sites for algal growth. Within the truly aquatic habitats there is a range of habitats supporting similar life forms whether the water is fresh or saline. These habitats are the rock surfaces on which *epilithic* algae grow, plant surfaces supporting *epiphytic* algae, sediments, such as sand with its *epipsammic* flora and silt–mud with its *epipelic* flora. Algae which penetrate into rock are known as *endolithic* and those which penetrate sediments are *endopelic*. Amongst the firmly attached epiphyton a looser non-attached community known as the *metaphyton* is found; it often lives in the mass of mucilage secreted by the epiphytes. Classification of these into *rhizobenthos* (rooted by rhizoids), *haptobenthos* (attached to solid surfaces), *herpobenthos* (moving on sediments) and *endobenthos* (within substrata) is dealt with. The floating communities of algae form the *plankton. Euplankton (true plankton), pseudoplankton* (accidental species) and *neuston* (confined to the upper water–air interface; *epineuston*, associated with the upper surface, and *hyponeuston*, associated with the under surface) are also distinguished. *Tripton* is the non-living particulate matter and *seston* the combined non-living and planktonic particles of the water. A further designation, *detritiplankton*, is suggested for forms living on floating particulate matter.

Communities, populations, assemblages, associations are defined and mention is made of the classification of algae into *life forms*. The ecological approaches based on *phytosociological* methods and through *gradient*

analysis and *ordination* are briefly surveyed and some examples given. *Climax* communities, *niche*, *maturity* and *diversity* have, apart from diversity, been only rarely discussed in relation to algal ecology.

This chapter provides a framework for the consideration of the spatial occurrence of algae and methods for their analysis. The precise definition of habitats and discrete communities is essential to the study of algal ecology but much work still remains to be done in refining the techniques of analysis and describing the communities of precisely defined spatial niches.

3 Phytobenthos: epilithon, epipsammon, endolithon, endopsammon, artificial surfaces

This chapter deals with the phytobenthos associated with rock and sand surfaces and therefore includes the very conspicuous flora of rocky coasts. Also, since the forms growing on artificial substrata (permanent structures or those placed temporarily in the water for experimental purposes) develop a flora which is essentially the same as that colonising the non-living 'lithos', they are dealt with here.

Epilithon

Associations of algae occur on the surface of exposed rock on the land, especially in mountainous regions, on rocks in flowing water, along the margins of lakes (Fig. 3.6) and, most conspicuously, forming belts of vegetation on rocks along tidal coasts (Figs. 3.53, 3.54). Only the last habitat has been worked extensively and then only with respect to the macroscopic algae,* which are so conspicuous in it. In temperate zones this flora is very well developed in the intertidal region, especially where there is a considerable tidal rise and fall and not too excessive desiccation.

In colder zones ice scour, and in the tropics excessive heating and desiccation, often prevent the development of an extensive vegetation.

Sub-aerial rocks

The epilithic algae of bare rock surfaces have received very little attention apart from the classic works of Jaag (1945), on rocks in the Swiss Alps, and of Golubić (1967) on calcareous rocks in Yugoslavia, yet all rock surfaces in moist climates probably have some algal colonists, and even in the tropics the

---

* The terms macrolithophytic (macroepilithic) and microlithophytic (microepilithic) have been used occasionally to distinguish the gross size classes of algae. Wood (1965, 1967) used the term epontic for microorganisms living on surfaces but the term is superfluous.

---

crevices of the surface rock support unicellular algae. In this habitat fungi also occur, and it is here that primitive lichen-like associations can be found. Not only is the habitat subject to extremes of wetting and desiccation, but also to temperatures which can be several degrees higher than the air temperature. Angle and aspect affect temperature of the rock surface. That the habitat is subjected to great extremes is shown by the fact that on summer days the surface can reach 23 °C above the surrounding air, the daily variation can be as much as 34 °C, and the annual variation 73 °C (Jaag, 1945). Even in the tropics, extremes from 20 °C at 6.00 a.m. to 50 °C at mid-day have been recorded (Zehnder, 1953). The algae often form coloured stripes ('Tintenstrichen') down the rock faces. The flora of these sites is dominated by Cyanophyta (compare the situation in the uppermost zone of rocky shores) and Table 3.1 is compiled from the extensive survey by Jaag; observations by the present author in various regions amply confirm the widespread occurrence of such communities dominated by Cyanophyta (*Gloeocapsa, Scytonema, Stigonema* (Fig. 3.52)), though there are sites where almost pure masses of diatoms occur (e.g., of *Frustulia rhomboides* var. *saxonica*). There are few data on the tropical rock flora but Zehnder (1953) found *Stigonema* and *Gloeocapsa* in Cameroun; there does not seem to be a special tropical flora. Schorler (1914) recognised five sub-associations of diatoms in European habitats – *Fragilarietum virescentis, Pinnularietum borealis, Pinnularietum appendiculatae, Frustulietum saxonicae* and *Melosiretum roeseanae* – and *Eunotia praerupta* is also common at some sites. Schade (1923) added another sub-association in which *Chromulina rosanoffi* was the dominant. The other alga which commonly forms mucilaginous masses on rock is the desmid *Mesotaenium*. Strøm (1920) described three different communities in Norway – the reddish-black crusts

Table 3.1. *The relative frequency of sub-aerial epilithic species in the Alps. Adapted from Jaag, 1945*

	Number of occurrences		Number of occurrences
Cyanophyceae			
Gloeocapsa sanguinea	331	*Gloeocapsa itzigsohnii*	5
Gloeocapsa kützingiana	133	*Aphanocapsa grevillei*	5
Scytonema myochrous	105	*Chroococcus helveticus*	5
Stigonema minutum	72	*Chroococcus minutus*	5
Calothrix parietina	63	*Clastidium setigerum*	5
Nostoc microscopicum	62	*Phormidium lividum*	5
Synechococcus aeruginosus	37	*Rivularia biasolettiana*	5
Gloeocapsa nigrescens	26	*Synechococcus major*	5
Gloeocapsa fusco-lutea	25	**Chlorophyceae**	
Dichothrix orsiniana	24	*Trentepohlia aurea*	36
Gloeocapsa shuttleworthiana	14	*Protococcus viridus*	21
Schizothrix heufleri	13	*Haematococcus pluvialis*	12
Chroococcus turgidus	13	*Trentepohlia jolithus*	12
Phormidium favosum	12	*Stichococcus bacillaris*	10
Dichothrix gypsophila	10	*Hormotila mucigena*	8
Chroococcus turicensis	10	*Coccomyxa thallosa*	7
Chlorogloea microcystoides	10	*Chlorococcum humicolum*	6
Chroococcus tenax	9	*Trentepohlia umbrina*	5
Chamaesiphon polonicus	8	**Zygnemaphyceae**	
Gloeocapsa dermochroa	8	*Mesotaenium macrococcum*	18
Microcoleus vaginatus	8	var. *micrococcum*	
Desmonema wrangelii	8	*Cylindrocystis brebissonii*	9
Tolypothrix byssoidea	8	*Zygnema cylindricum*	6
Tolypothrix epilithica	8	**Bacillariophyceae**	
Nostoc sphaericum	7	*Tabellaria flocculosa*	20
Microcoleus paludosus	6	*Melosira roeseana*	7
Gloeocapsa atrata	6		

dominated by *Gloeocapsa*, the brown, cushion-like, jelly masses of *Scytonema/Stigonema* and the light-green, hanging, jelly-like masses of Zygnemaphyceae/diatoms. These associations certainly need careful analysis of data from many sites. In Finland, Cedercreutz (1941) found some 37 species in the rock face habitat, but considered that the Chroococcaceae were the pioneers in the habitat. It is, however, clear that the habitat is colonised by only a very small collection of species capable of survival in this very specialised niche. The flora does of course become much enriched in regions where soil water flows down the rock face.

Green powdery layers of algae, comparable to those on tree bark, are also encountered, usually on the more dry rock surfaces. The species are similar, even perhaps identical to the '*Pleurococcus*'* of tree bark and, as in that habitat, occur together with

Stichococcus bacillaris and *Hormidium* spp. Orange felts are often seen growing on rocks and belong to the genus *Trentepohlia* which like '*Pleurococcus*' also grows on trees (*see* p. 214). P. Broady (personal communication) found 34 algal species on rock surfaces on the antarctic Signy Island; thirteen of these were found only in the rocks, while the others occurred also on soils. 43 out of 51 sites examined had fewer than four algae present, a large number of sites having only *Prasiococcus calcarius* and *Desmococcus vulgaris* growing as a green powder on the rock.

* '*Pleurococcus*' is merely a collective term for a group of algae (*Desmococcus*, *Apatococcus*, etc.) which require detailed taxonomic study. The recent excellent work on the classification of the 'difficult' genus *Chlorella*, from morphological (Fott & Nováková, 1969) and from physiological/biochemical standpoints (Kessler, 1974 and earlier publications), points the way to more definitive studies of '*Pleurococcus*'.

Where there was any animal presence the 'Hormidium' stage of *Prasiola crispa* occurred and where the rock face was enriched by water seeping down it, a number of Cyanophyta (*Oscillatoria*, *Phormidium*) and diatoms (*Melosira* sp., *Diatoma*, *Achnanthes*, *coarctata* var. *elliptica*, *Navicula mutica*, *Pinnularia globiceps* and *Nitzschia palea*) occurred. On marble, the Cyanophyta were dominant, especially *Gloeocapsa gelatinosa* and *Nostoc* sp., but diatoms and Chlorophyta were almost completely absent. Very similar assemblages were recorded growing on rocks and stones, in Iceland by Petersen (1928) and by Strøm (1920) in Norway and it is clear that the flora is similar the world over.

An interesting aspect of the blue-green algae living on calcareous rock faces, e.g. *Gloeocapsa sanguinea*, *G. kutzingiana* is their ability to deposit crystals of calcium carbonate in their sheaths. This feature does not occur on acidic rock.

Solution of rock by algal growth is probably rare or unknown on acidic rocks; on the other hand Jaag does quote the example of *Gloeocapsa kützingiana* producing small pits in the surface when growing on calcite crystals and on marble in quarries. These algae are, however, often in loose association with fungal hyphae and it is not clear to what extent it is fungal or algal action which produces pitting of the rock surface. More frequently algae penetrate the fine cracks in the surface of the rock and grow endolithically.

Some algae occur more frequently on acid rock surfaces, e.g. *Stigonema minutum*, desmid species (32 spp. in Jaag (1945)), *Eunotia* spp., *Pinnularia borealis*, whilst on alkaline surfaces *Aphanocapsa* spp. *Schizothrix* spp., only one desmid species and *Melosira roeseana* are common.

Bare rock surfaces within caves where sufficient light penetrates are also colonised by algae, and again Cyanophyta tend to be dominant; Golubić (1967) found grey felt masses of *Hapalosiphon intricatus* with an undergrowth of *Aphanocapsa* in the darkest situations in a cave in Yugoslavia, Table 3.2*a*. On the rock face outside this cave *Scytonema myochrous* was dominant and in an intermediate position *Aphanocapsa* and *Chroococcus* were dominent; only the Cyanophyta and one or two other algae are recorded and the various 'statuses'* of the species show how complex the forms are in this vegetation. Interestingly, the dominant species in Table 3.2*c* are those

also found most frequently by Jaag in the Alps.

On old church walls and grave stones an epilithic flora is very common and in many sites it is probably contributing to the breakdown of the stone. Golubić (1967) records *Gloeocapsa sanguinea*, *G. kützingiana*, *G. compacta* and *Scytonema myochrous* on such dryish walls. Where more water percolates over the surface, *Stichococcus*, *Gloeocystis*, *Coccomyxa* and *Chlorococcus* occur; these powdery green growths are universal on buildings of all kinds, but are most obvious on the more ancient monuments, where time and the breakdown of the surface stone has yielded a favourable habitat. Such growths are causing concern to conservationists in places such as Italy and Greece. Stone surfaces which are not periodically inundated with soil water tend to support only Cyanophyta and Chlorophyta and it is noticeable that diatoms are absent, e.g. in all but one site investigated by Schlichting (1975). This may simply reflect the lack of soluble silica in the moisture available to the algae.

When they are lying on soils, fragments of rock, volcanic glass, crystals of calcite, etc., which transmit light can develop a flora (epilithic) on their sides or undersurface ('Fensterflora'), especially in deserts. Bone and shell fragments also often support a similar algal flora. Cameron & Blank (1966) give a detailed list and bibliography of algae in such sites; they are much more common than is generally believed, especially in desert and semi-desert areas. Equally, however, they occur in temperate sites but then often on the upper surface of stones; they are conspicuous on chalk and bone fragments since the green–blue–green algal colour is obvious against the creamy-white background but I suspect they are equally abundant but less obvious on other rock fragments. On quartz and chalcedony, Cameron & Blank record *Nostoc muscorum*, *Schizothrix calcicola*, *Protococcus grevillei*, *Phytoconis* (= *Protococcus*) *viridis* as rich growths, mainly on the undersurface but also in some cavities of the upper surface. Quartz stones have been shown to have a lower temperature than

---

* Some workers reduce the number of taxa in the Cyanophyta to a mere handful. This may have merit and is certainly a simple solution to the problem of recording forms in nature but ecologically I suspect it is totally misleading and the recognition of varieties and forms is necessary for comprehensive floristic surveys.

Table 3.2. *Algal associations inside a cave in Yugoslavia: a, in the darkest zone; b, in the lighted zone near the entrance; c, on the cliff outside the cave. Nine stands taken from each site. From Golubić* (1967)

	1	2	3	4	5	6	7	8	9
Zone *a*									
Aphanocapsa muscicola	.	.	+	2	+	.	.	.	.
Aphanocapsa grevillei	.	.	+	1	.	1	+	.	.
Gloeocapsa kützingiana									
st. simplex	.	.	.	.	3	.	.	.	.
st. rupestris, simplex	.	.	.	.	1	.	.	.	.
st. nannocytosus	.	.	.	+	1	.	.	.	.
Gloeocapsa compacta									
st. simplex	4	3	.	.	.	.	.	.	.
st. lam. coloratus	+	+	.	.	.	.	.	.	.
st. nannocytosus	1	.	.	.	.	.	.	.	.
Gloeocapsa biformis									
st. punctatus	2	2	.	.	.	.	.	.	1
st. nannocytosus	1	.	.	.	.	.	.	.	+
Xenococcus kerneri	.	.	.	.	+	1	3	.	.
Hapalosiphon intricatus	+	4	+	3	2	5	4	+	+
Scytonema myochrous	.	.	.	.	.	.	.	5	4
Schizothrix delicatissima	.	.	3	+	.	.	.	.	.
Pleurococcus viridis	1	.	+	.	.	.	.	.	.
Zone *b*									
Aphanocapsa grevillei	5	3	2	3	2	3	.	.	.
Chroococcus turgidus	2	.	1	.	.	+	.	.	2
var. spelaeus	.	3	4	2	3	2	+	.	.
Gloeocapsa sanguinea									
st. simplex	+	+	.	1	1	1	.	2	.
st. lam. col. alpinus	.	.	.	.	.	.	.	+	.
Gloeocapsa kützingiana									
st. rupestris, simplex	.	.	.	.	.	1	.	+	.
Gloeocapsa compacta	.	.	.	.	.	.	.	1	.
Gloeocapsa biformis									
st. punctatus	+	.	1	.	.	.	.	.	.
Gloeothece rupestris	.	.	.	.	3	.	.	.	.
Scytonema myochrous	.	.	.	.	1	+	.	.	5
Nostoc microscopicum	1	3	.	.	3	1	.	.	.
Nostoc macrosporum	1	.	.	.	.	.	.	.	.
Schizothrix sp.	.	1	2	1	3	.	5	.	.
Trentepohlia aurea	.	.	.	.	.	.	.	5	.
Zone *c*									
Chroococcus turgidus	.	.	.	1	+	.	2	.	.
Chroococcus westii	.	.	.	.	.	+	.	.	.
Gloeocapsa sanguinea									
st. lam. coloratus	.	.	.	.	.	.	.	+	+
st. lam. col. alpinus	+	+	+	3	3	1	3	.	.
st. lam. col. alp. magma	.	.	.	+	.	.	.	.	.
st. perdurans	.	.	.	+	+	.	.	.	.
Gloeocapsa kützingiana									
st. lam. coloratus	.	.	.	2	2	.	.	.	.
st. rupestris	.	.	.	1	+	.	+	.	.
st. perdurans	.	.	.	.	.	.	3	.	.
Gloeocapsa compacta	.	.	.	1	1	+	2	+	1
Gloeocapsa biformis									
st. dermochrous	.	.	.	1	+	+	.	.	.
Chlorogloea microcystoides	.	.	.	.	.	.	.	5	4
Scytonema myochrous	5	4	5	+	2	.	.	.	.
st. petalonema	.	.	.	5	1	.	.	.	.
Calothrix parietina	.	.	.	.	.	+	3	.	.
Schizothrix lardacea	1	1	+	.	5	.	.	.	.
Schizothrix affinis	+	2	1	.	.	.	2	.	.

the surrounding soil and in addition the undersides will form moisture traps for the algal growth. Experiments show that the rate of desiccation around such stones is lower than that for the adjacent soil. The majority of algal species associated with the surface of stones in deserts seem to be identical to those found in the soils themselves, though coccoid Cyanophyta (*Anacystis, Coccochloris*) and coccoid green algae especially *Troschiscia* are either more common on or occur only associated with the stones. Some of the early records were made by workers studying soil algae (e.g. Tchan & Beadle, 1955; Durrell, 1962) and it is not always clear exactly which community is involved; in some instances the presence of the translucent stones may simply yield a moist habitat for soil algae whereas those studied by Friedmann, Lipkin & Ocampo-Paus (1967) live actively on and in the stones.

Friedmann *et al.* (1967) report work from the central Negev desert which showed that in one year, whilst rainfall was merely 18 mm, the amount of dewfall was 22 mm and, more significantly, there were 146 dew nights as opposed to ten rainy days. Dew is probably the single most significant source of moisture for desert algae. These workers investigated the flora on the lower surfaces of stones; they term these the *hypolithic* algae, and the associations growing in fissures open to the surface the *chasmolithic*, but admit that many of the same algae occur in both sites. They also list some of the same species from the desert soil. Their lists are realistic in containing numerous forms which could not be identified to species, e.g., forms of *Bracteococcus, Chlorosarcinopsis, Hormidium, Stichococcus, Trebouxia, Trochiscia* amongst the Chlorophyta and *Aphanocapsa, Aphanothece, Borzia, Chroococcus, Chroococcidiopsis* (?), *Gloeocapsa, Lyngbya, Myxosarcina, Nostoc, Plectonema, Schizothrix* and *Scytonema* amongst the Cyanophyta, together with new species such as *Chlorosarcinopsis negevensis, Friedmannia israeliensis* and *Radiosphaera negevensis* and the more frequently recorded *Microcoleus chthonoplastes, Schizothrix calcicola, Scytonema ocellatum* and *Tolypothrix byssoidea.*

Running water

The running water (lentic) habitats tend to form a more constant environment than those of the lotic; temperature fluctuates less and, perhaps more important, the flow removes extra-cellular products

and keeps a constant supply of nutrients available. Whitford (1960) found that current speed must exceed 15 cm s^{-1} to produce a sweeping effect and displace the water in the diffusion shells extending out to 0.25 mm from the plant surface. He also found that the faster the water flowed the greater was the oxygen concentration. There is a tendency for certain genera, e.g. *Lemanea*, to occur in fast-flowing reaches (Round, unpublished). Equally, stability of the habitat is as important a factor in streams as on rocky shores and it is not surprising that Douglas (1958) found the effect of flooding to be greater in the removal of the flora from small stones than from stable rock surfaces.

The actual patterns of flow over surfaces in a river bed are quite complex and undoubtedly influence algal colonisation. The formation of a boundary layer zone where flow is greatly reduced is a common feature (Fig. 3.1); the innermost region of this layer is where unicellular algae and spores settle and here there is even a sublayer, 10–100 μm thick, where the velocity of flow is extremely low and frictional forces are important.

The pattern of movement is largely determined by the frictional and pressure forces and an expression known as the Reynolds number (Re) can be defined to indicate the kind of flow past an object.

$$Re = \frac{\rho VL}{\mu}, \qquad (3.1)$$

where ρ is the density of water, V is the velocity of the water, L is the length of the object, and μ is the viscosity of water.

Since the density of water is 1 g/ml and viscosity is 0.1 the equation can be approximated:

$$Re = 100 . VL \qquad (3.2)$$

If Re is small (< 1) frictional forces mainly apply and if large (> 1000) pressure is predominant; with a small Reynolds number there will be laminar flow and with a large one turbulent flow.

There are three, or possibly four, zones available for colonising on a stone such as that illustrated in Fig. 3.1. These are the 'nose' pointing into the water, the upper surface boundary layer, the turbulent water of the wake and possibly the shaded underside where it is not buried. The flora described by Fritsch (1929) which consists of crusts a few cells thick (*Hildenbrandia, Lithoderma, Chamaesiphon* etc.) probably occurs within the boundary layer and does not raise any filaments above it. Gessner (1955) found that objects placed in a flowing stream had maximal growth of *Cocconeis* on the region facing the flow and in the wake, and minimal growth along the sides.

The abundant microscopic epilithic species are the forms which also attach to glass slides. Butcher (1949) used the number of such cells per mm² to differentiate eutrophic and oligotrophic rivers, the former having up to 10^5 cells per mm² and the latter 10^3–5×10^3 per mm². These algae growing on artificial substrata 'often show considerable differences from substrata occurring naturally in a stream or river' (Whitton, 1975). Glass slides certainly collect mainly attached species and Hohn (1964) found that less than 5 % of the cells were planktonic. A much more appropriate sampling procedure is given by Saunders & Eaton (1976) and Fig. 3.2 shows how such a technique can yield interesting data on cover and dry weight.

The epilithon of running waters has received very little attention: macroscopic growths of species such as *Cladophora glomerata*, *Lemanea fluviatilis*, *Batrachospermum boryanum*, *Chaetophora incrassata* and *Ulothrix* are common on rock(s) in many fast flowing waters – slower flow tends to result in silt deposition on the stones and this eliminates the epilithon. The normally marine alga, *Enteromorpha*, also occurs here and appears to be spreading rapidly in rivers as the 'salinity' increases. There appears to be a degree of latitudinal separation of these filamentous forms, for example in regions such as northern Scandinavia the rocks are coated with species of *Zygnema*, *Vaucheria*, *Lemanea*, *Chaetophora* (*Spirogyra* and *Mougeotia* are also intermixed), whilst throughout temperate zones *Cladophora glomerata* is the most abundant (a *Clado-*

phoretum glomerata is recognised by some workers, e.g., Roll, 1938) with *Chaetophora incrassata* secondary. In tropical regions the red alga *Compsopogon* seems to be the common genus, though as yet this habitat has received almost no attention in warm climates. Although tangled masses of green algae are richly developed in some clear northern streams, e.g., in Finland (Round, 1959a) a recent survey of streams on Baffin Island revealed an epilithic river flora dominated by Cyanophyta (*Aphanocapsa pulchra*, *Oscillatoria tenuis*, *Schizothrix meulleri*, *Lyngbya* spp. *Stigonema mamillosum*) making up 90 % of the biomass during the short summer season (Moore, 1974a). Only immediately after the break-up of the ice were diatoms abundant (*Tabellaria flocculosa*, *Eunotia tenella* and *Achnanthes marginulata*). In this arctic habitat the rock was almost devoid of algae when first exposed but populations developed rapidly and were hardly grazed. The population reverted to a low level before ice re-covered the streams, presumably from loss by sloughing off the rock. Some species of algae tend to grow in the slower flowing parts and in pool-like reaches in late winter and early spring and then disappear as the temperature rises; they then begin to grow in the regions of more rapid flow but some, e.g., *Chaetophora incrassata*, disappear even from the rapids when temperature rises above 10 °C (Whitford, 1960). Dillard (1969) gives the upper temperature for this alga as 20 °C and at this temperature it is replaced by a summer *Spirogyra/Oedogonium* association. Temperature exerts a direct effect on the lentic environment and algae such as *Hydrurus foetidus* and *Lemanea* occur only in winter in lowland streams but can be found throughout the year in the cool headwaters; *Hydrurus* is very common in alpine streams and rivers.

In mountain streams, Cyanophyta (*Oncobyrsa* and *Chamaesiphon* spp.) form black bead-like structures attached to the rocks and the diatom *Didymosphenia geminatum* produces whitish-grey pustules which may grow many centimetres in extent; the greyness is caused by the aggregation of the thick mucilage stalks of this diatom and in the pustules *Achnanthes* spp. grow epiphytically on the stalks. In such situations, large amounts of mucilage are formed and in this mucilage many desmids (e.g. *Cosmarium*, *Euastrum*) and diatoms (e.g. *Eunotia*, *Frustulia*, *Cymbella*, *Achnanthes*) grow. Margalef (1960) reviewed data on rivers and recognised three associations starting with

Fig. 3.1. Water motion around a large rock in water moving at 1 m s⁻¹. There is an outer region of transient flow, a thin boundary current layer and a region of separated flow or wake (w). From Neushul, 1972a.

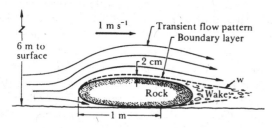

a *Hydrurus foetidus/Ceratoneis arcus* ('Ceratoneto–Hydruretium rivulare') in the uppermost fast-flowing reaches of European rivers, passing into the 'Diatometo–Meridionetum rivulare' (*Diatoma hiemale* var. *mesodon*, *Meridion circulare*, *Achnanthes lanceolata*, *Cymbella ventricosa*, *Gomphonema parvulum*, *G. angustatum*, *Cymbella sinuata* accompanied by more widespread species such as *Cocconeis placentula*, *Synedra ulna*, *Nitzschia linearis* and *Navicula cryptocephala*) and then into a 'Melosiretum rivulare' (with *Melosira varians*, *Synedra ulna*, *Achnanthes lanceolata*, *Cocconeis placentula*, *Rhoicosphenia curvata*, *Navicula gracilis*, *Diatoma vulgare*, *Meridion circulare*, *Staurastrum punctulatum*, *Cymbella ventricosa*, *Navicula cryptocephala*, *Fragilaria virescens*, *Gomphonema parvulum*, *Amphora ovalis*, etc.). This latter collection of algae is very common in British rivers and in the lowlands can

extend almost up to the headwaters (Round, unpublished). Margalef classed the above three associations in a group which includes all those algae with a morphology extending out from the stones but some of these algae also grow close to the stone surface. They are however distinguished from two other associations which are adpressed to the stones; an upper-reach association in which *Oncobyrsa*, *Chamaesiphon*, *Hildenbrandia* or combinations of these are dominants and a lower-reach association with *Phormidium* and *Hydrocoleus* dominant. It is desirable to distinguish the closely 'adpressed' and the 'pedunculate' type of growth but perhaps undesirable to maintain them as separate associations. Margalef then proceeded to an interesting discussion of further variants dominated by *Melosira arenaria*, *Amphipleura pellucida*, *Cladophora glomerata*, *Tribonema*, *Eunotia*,

Fig. 3.2. *a*, the percentage cover of the freshwater red alga *Lemanea fluviatilis* and, *b*, the dry weight of the alga in a rocky stream. From Saunders & Eaton, 1976.

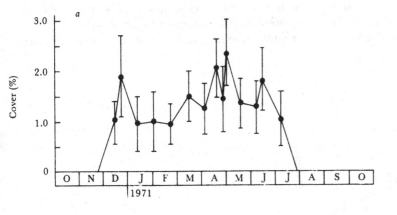

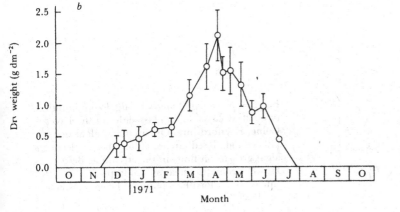

Diploneis etc., and it is clear that a whole series of associations can be recognised, although I feel that much more extensive phytosociological studies are needed to clarify these – the study of algae in rivers has hardly begun!

Seasonal changes have been followed with a chlorophyll extraction technique on algae growing on flints (siliceous nodules) in the Bere Stream in Dorset (Westlake *et al.*, 1972) (Fig. 3.3). There is a large spring maximum and a smaller mid-summer growth. The species which are common in the spring are *Syndra ulna*, *Diatoma vulgare*, *Gomphonema olivaceum*, *Navicula avenacea*, *N. gracilis*, *Nitzschia* spp. and *Fragilaria* spp. This community contains a mixture of attached and motile species and it seems likely that epipelic *Navicula* and *Nitzschia* species mingle with the attached *Synedra*, *Diatoma*, *Gomphonema* etc. and are not swept away since the surface film is not subjected to too intense a current. These motile elements perhaps form a sub-community rather similar to the metaphyton developed amongst the epiphyton of lakes (*see* p. 232). In summer the flints tend to be covered with a lime-encrusted Cyano-phyceae community composed of *Phormidium incrust-atum* and/or *Homeothrix crustacea* with some *Pleurocapsa*

and *Chamaesiphon*, followed in late summer by patchy growths of *Cladophora* and *Vaucheria* (Marker, 1976*a*). This is similar to the lime encrusted community described by Fritsch. *Rivularia haematites* also grows around small stones in rivers, and the precipitation of calcium carbonate gradually enlarges the stones so that in places extensive beds of 'growing' and 'eroding' pebbles are formed, e.g., in the Rhine at Stein-am-Rhine (Jaag, 1938); these pebbles are known from geological strata where they are termed oncolites. There are few data to compare with the studies on the Bere Stream but a study of a Montana stream also showed a peak of cell numbers in February; although there were few data, the summer months appear to be unfavourable for growth which recommenced in autumn (Gumtow, 1955). Measurements of algal photosynthesis in the Bere Stream are given in Table 3.3. and, compared with optimum phytoplankton figures, they show that the photosynthesis per unit chlorophyll is lower in this epilithic community, where, in addition, there tends to be an inverse relationship between biomass and photosynthesis per unit of chlorophyll. The extraction of chlorophyll from the algae on flints is rather variable, for whilst diatoms were completely extracted after

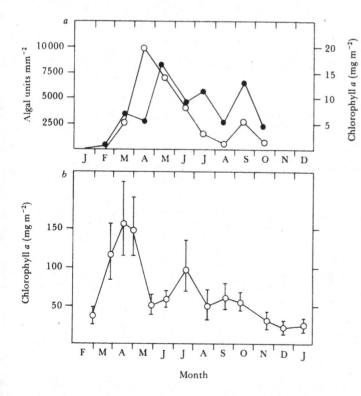

Fig. 3.3. Seasonal changes in algal communities in chalk streams. *a*, on glass slides in the River Colne, Hertfordshire. Open circles, algal units (1954) and closed circles, chlorophyll *a* (1955); *b*, algal growth on flints in the stream at Bere Heath, geometric mean of chlorophyll *a* (uncorrected for phaeophytin). From Westlake *et al.*, 1972.

Table 3.3. *Biomass measured as chlorophyll a* (mg m^{-2}) *on flints in the Bere Stream and the gross rate of photosynthesis* (mg O$_2$ mg chlorophyll a^{-1} h^{-1}) *compared with a typical phytoplankton rate. From Westlake et al., 1972*

Conditions	Biomass (chlorophyll a, mg m^{-2})	Gross photosynthetic rate (oxygen output, mg mg chlorophyll a^{-1} h^{-1})
Winter, dull	< 50	< 1.0
Summer, bright	< 50	3–5
April, ⎰ dull	> 200	1–2
diatom growth ⎱ bright	> 200	2–3
Phytoplankton, optimum (Strickland, 1960)		c. 10

24 h in acetone, the green alga *Ulvella frequens* was very much less extracted. Normal grinding procedures cannot be readily used on these epilithic algae (Marker, 1972). Methanol extractions are preferable under these conditions, but problems arise with the estimation of phaeophytin in methanol and since this degradation pigment is likely to be present in considerable quantity in epilithic situations, Marker devised a technique for transference of the two pigments from methanol to acetone. In the latter solvent the phaeophytin can be estimated by spectrophotometry. In epilithic situations, dead and dying algal cells and colonies are likely to remain attached for a considerable time and increase greatly the amount of phaeophytin which will be extracted with the chlorophyll. Marker (1976a) found up to 30% phaeophytin from November to January but less than 20% during the remainder of the year and he also found a distinct diurnal variation in photosynthesis both in spring and summer with peaks around the 12.00–14.00 h period (Marker, 1976b). Dead diatoms remaining attached in such communities are also a considerable problem when methods involving cell counting are employed. This problem may be overcome by using fluorescence microscopy to show up the live cells by the red emission (J. G. Jones, 1974), a technique which was used on stones from streams with large numbers of *Cocconeis placentula* and gave about five times as many cells as estimates obtained from scraping the algae off the stones.

In streams and rivers flowing through limestone country, the river bed often consists of dissected pavements of rock on which algae are extremely well developed. However the brownish colour is not always due to diatom growth but to iron compounds in the mucilage of blue-green algae, particularly species of *Rivularia*, *Schizothrix* and *Gloeocapsa*; Hughes & Whitton (1972) record the almost scarlet appearance of a stream due to growth of *Schizothrix lardacea*. They also record large growths of stalked species of the diatom *Cymbella*; apparently this is a primary coloniser on freshly exposed limestone.

In a Californian stream a *Nostoc* species grows on stones mainly in early summer and fixes nitrogen with a maximum rate during the day, lowest in the evening but then picking up during the night (Horne, 1975; Horne & Carmiggelt, 1975).

Many streams commence as distinct springs and in calcareous regions the springs are often forming deposits of calcium carbonate (tufa). This deposit is accumulating around algal growths in which Cyanophyta are very common and the desmid *Oocardium stratum* also occurs. The flora is not so different from that of flowing water and, indeed, can be considered as the initial stage in colonisation of such waters. Springs deserve much greater study since they are ideal natural laboratories, where the temperature and chemical composition are relatively constant whilst day length and illumination vary. Springs flowing out over rock or tufa support an epilithic flora and Table 3.4a illustrates the mainly Cyanophyta flora of such a spring (Golubić, 1967) whilst Table 3.4b presents the flora in the outflowing streams. In both, the red alga *Pseudochantransia* is prominent. The flora is quite different to that

Table 3.4. *The epilithic flora of some calcareous springs (a) and streams (b).*
From Golubić, 1967

	1	2	3	4	5	6	7	8	9
Springs, *a*									
Xenococcus kerneri	·	·	3	2	·	·	·	3	·
Xenococcus rivularis	·	·	·	·	1	·	·	·	·
Hydrococcus caesatii	·	+	·	·	·	·	·	·	·
Chamaesiphon confervicolus	·	·	1	·	·	·	·	·	·
Chamaesiphon incrustans	·	·	2	2	·	·	·	·	·
Chamaesiphon polonicus	2	2	·	·	·	·	2	·	·
Homoeothrix juliana	·	1	·	·	·	·	·	·	3
Homoeothrix varians	·	·	2	3	·	·	1	2	2
Homoeothrix caespitosa	·	·	·	·	·	3	·	·	·
Homoeothrix crustacea	·	·	·	·	·	2	·	·	·
Calothrix parietina	2	2	·	·	·	·	·	·	·
Phormidium autumnale	·	·	·	·	·	3	·	·	·
Lyngbya aerugineo-coerulea	·	·	·	·	·	·	·	·	3
Gongrosira incrustans	·	·	·	·	·	·	3	2	·
Meridion circulare	1	+	·	·	·	·	·	·	·
Cocconeis placentula	·	·	·	·	1	·	·	·	·
Epithemia muelleri	·	·	·	·	·	·	2	·	·
Pseudochantransia chalybea	·	·	·	3	·	+	·	·	·
Pseudochantransia pygmaea	·	·	3	·	4	3	2	·	3
Batrachospermum moniliforme	3	·	·	·	·	·	·	·	·
Streams, *b*									
Hydrococcus caesatii	·	+	·	·	·	·	+	·	·
Chamaesiphon incrustans	·	·	·	·	·	·	·	1	·
Homoeothrix caespitosa	2	4	1	3	2	·	+	3	2
Homoeothrix crustacea	3	·	2	2	2	+	+	2	3
Rivularia biasolettiana	1	·	·	·	·	·	1	·	·
Phormidium autumnale	·	·	·	·	·	·	3	·	·
Lyngbya aerugineo-coerulea	1	·	·	·	·	·	·	·	·
Schizothrix fasciculata	·	·	·	·	3	·	·	·	·
Gongrosira incrustans	1	3	+	·	3	5	4	3	4
Chaetophora elegans	·	·	·	1	·	·	·	·	·
Oocardium stratum	·	·	·	·	·	·	·	·	2
Epithemia muelleri	·	·	·	·	·	2	+	·	·
Cocconeis placentula	·	·	·	1	·	2	·	·	·
Diatomeae	3	2	2	2	·	·	·	·	·
Pseudochantransia pygmaea	2	4	5	3	1	·	·	1	1

of the epilithon on aerial rocks (cf. Tables 3.1, 3.2).

The influence of rock type needs careful study since Parker, Samsel & Prescott (1973) found that *Monostroma quaternarium* only occurred on iron-rich rocks, *Hydrurus* favoured limestone and sandstone, whereas *Batrachospermum* showed no apparent substratum preference in the streams in Glacier National Park, Montana.

Few experimental studies have been made on river epilithon, but the idea that increased rate of flow is advantageous to algal metabolism has been generally accepted. Rider & Wagner (1972) report the occurrence of *Batrachospermum* species in the faster flowing central areas of streams, and it appears that this genus is difficult to culture except in artificial stream systems (Whitford & Schumacher, 1964). *B. moniliforme* was found to withstand high light intensities and grew well in a clear-water stream, whereas

Fig. 3.4. The effect of current speed on algal biomass (dry wt) and on the uptake of radioactive phosphate by *Mastigocladus laminosus*. Triangles, fast flow; circles, slow flow; squares, phosphate concentration. From Sperling & Grunewald, 1969.

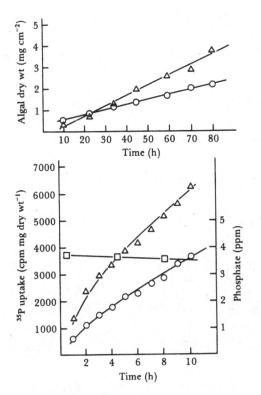

B. vagum growing in dark waters draining from bogs was found to disintegrate at high light intensity (Rider & Wagner, 1972). *B. sirodotii* is also highly sensitive to light according to Dillard (1969). Sperling & Grunewald (1969) designed an interesting flow system in tanks and grew the stream-living, thermophilic alga *Mastigocladus laminosus* on styrofoam strips; they measured its growth, and uptake of radioactive phosphorus, at two different current speeds (Fig. 3.4), and found enhanced uptake at the higher speed. Experimental streams with a bed of gravel and rocks develop a rich algal flora and McIntire (1968) found that the flora varied according to the rate of flow and illumination. Fig. 3.5 shows the effect of two different light intensities and three different rates of flow; the low rate of flow was clearly not favourable for any of these three organisms, the *Oscillatoria* grew at the high light intensities, and *Tribonema* at high light and moderate flow, whilst the diatom (*Nitzschia*) grew at both light intensities. In a very interesting study in experimental channels with slow, medium and fast flow, Zimmerman (1961) found he could reproduce the conditions of oxygenated, slightly alkaline, cool summer mountain streams and obtain growths of *Hydrurus foetidus* and *Meridion circulare* (cf. the first association of Margalef, p. 53). He also found that a palmelloid stage of *Hormidium subtile* grew in the channels; thus, there are problems involving growth forms of green algae in these habitats.

Lakes

Stones and rock surfaces along the margins of lakes, even when apparently barren to the naked eye, always have a mucilaginous coating of Cyanophyta, diatoms, desmids, etc., which becomes obvious when one tries to walk over the stones. Quantitative methods for assessing biomass on stones are not easy to devise: most techniques have involved brushing the algae off the stones but damage and loss may be considerable.

In Kootenay Lake (British Columbia) cell counts of the epilithon (mainly diatoms) ranged from 1×10^5 to 9×10^6 cells cm^{-2}. Cell counts increased with depth from 1×10^6 cm^{-2} near the surface to 3×10^6 cm^{-2} at 5 m and at all depths numbers were highest in spring and lowest in summer (Ennis, 1977). Stripping off the algae by application of a collodion film has been suggested

Table 3.5. *Epilithic algae at Lake Vrana: a, at 400–60 cm, above mean water level; b, at approximately 30–50 cm above mean water level (columns 1–5) and at mean water level, columns 6–9; c, below water level in depths of 30–80 cm; d, at 12 m deep (columns 1–6) and 43 m deep (columns 7–9). Data from Golubić, 1967*

	1	2	3	4	5	6	7	8	9
Level *a*									
Gloeocapsa sanguinea									
st. *lam. col. alpinus*	·	·	·	·	·	+	1	1	+
st. *col. alp. magma*	+	3	3	2	+	·	1	1	·
st. *perdurans*	·	·	2	·	·	·	·	·	·
Gloeocapsa kützingiana									
st. *lam. coloratus*	2	1	·	2	+	+	2	1	+
st. *rupestris*	2	+	1	3	+	+	2	+	·
Gloeocapsa compacta									
st. *simplex*	2	1	·	·	·	·	·	·	·
st. *lam. coloratus*	·	·	3	2	+	+	·	·	·
st. *lam. col. magma*	2	·	·	3	·	·	·	·	·
st. *perdurans*	·	1	·	·	·	·	·	·	·
Gloeocapsa biformis									
st. *punctatus*	·	1	·	·	·	·	·	·	·
st. *dermochrous*	1	1	+	·	·	·	·	·	·
Stigonema mamillosum	·	·	·	·	·	·	+	1	·
Scytonema myochrous	·	·	1	·	+	1	3	4	1
st. *petalonema*	·	·	·	·	+	·	·	·	·
Calothrix parietina	2	3	2	+	1	1	3	2	2
Nostoc sphaericum	2	+	+	·	·	·	·	·	·
Nostoc microscopicum	·	·	+	·	·	·	·	·	2
Nostoc minutum	·	+	·	·	·	·	·	·	·
Schizothrix affinis	2	1	·	·	·	·	·	·	·
Schizothrix delicatissima	·	+	·	·	·	·	·	·	·
Schizothrix arenaria	·	·	2	·	4	3	2	+	3
Schizothrix penicillata	·	·	·	·	4	3	·	·	3
Microcoleus vaginatus	·	·	·	·	·	1	·	·	·
Level *b*									
Gloeocapsa sanguinea	·	·	·	+	+	·	·	·	·
Gloeocapsa kützingiana	·	2	·	·	·	·	·	·	·
st. *rupestris*	·	2	·	·	·	·	·	+	+
Gloeothece confluens	·	·	·	·	·	+	+	·	·
Stigonema mamillosum	·	3	·	2	1	3	2	·	·
Scytonema myochrous	5	+	4	1	2	4	3	·	·
Tolypothrix penicillata	·	·	·	·	·	·	·	5	5
Calothrix parietina	·	·	·	+	1	+	+	·	·
Dichothrix gypsophila	·	·	·	5	4	·	·	·	·
Dichothrix compacta	·	·	·	·	·	·	·	1	+
Rivularia biasolettiana	·	·	·	·	·	3	2	·	·
Nostoc microscopicum	·	·	·	+	·	·	·	·	·
Schizothrix affinis	·	3	·	·	·	·	·	·	·
Schizothrix lardacea	·	·	4	·	·	·	·	·	·
Schizothrix delicatissima	·	·	·	+	+	·	·	·	·
Schizothrix arenaria	·	·	·	·	·	3	2	·	·
Schizothrix lacustris	·	·	·	·	·	·	+	·	·

Table 3.5. (*cont.*)

	1	2	3	4	5	6	7	8	9
Level *c*									
Aphanothece clathrata	·	3	2	·	·	·	·	·	·
Aphanothece nidulans	·	1	1	·	·	·	·	·	·
Aphanothece castagnei	·	1	2	·	·	·	·	·	·
Aphanothece microsporum	·	+	·	·	·	·	·	4	3
Stigonema mamillosum	2	2	·	·	·	1	2	·	·
Scytonema myochrous	5	4	4	4	1	5	3	·	·
Calothrix parietina	·	+	2	·	·	·	·	·	·
Dichothrix gypsophila	·	·	·	·	·	·	·	+	2
Rivularia biasolettiana	·	·	·	3	2	·	·	·	·
Schizothrix delicatissima	·	·	·	·	·	1	2	·	·
Schizothrix lacustris	·	·	·	·	·	·	·	4	4
Bulbochaete sp.	·	·	·	·	·	+	+	·	·
Spirogyra and *Zygnema* spp.	·	·	·	2	5	2	3	·	·
Level *d*									
Aphanocapsa anodontae	·	·	·	·	+	·	·	·	·
Aphanocapsa muscicola	·	·	·	·	3	2	·	·	·
Aphanocapsa pulchra	·	·	·	·	1	2	·	·	·
Aphanothece castagnei	·	·	·	·	+	3	·	·	·
Aphanothece saxicola	·	·	·	·	·	·	4	3	·
Chroococcus turgidus	·	·	+	+	·	·	·	·	·
Pseudoncobyrsa lacustris	1	1	·	·	·	·	·	·	·
Xenococcus kerneri	·	·	·	·	·	·	2	2	·
Chamaesiphon cylindricus	2	1	·	·	·	·	·	·	·
Clastidium setigerum	·	+	·	·	·	·	·	·	·
Tolypothrix penicillata	·	·	3	4	·	·	·	·	·
Calothrix fusca	·	·	2	2	·	·	·	·	·
Schizothrix delicatissima	·	·	·	·	3	2	·	·	·
Schizothrix calcicola	·	·	·	·	·	·	·	·	4
Schizothrix lacustris	·	·	1	2	·	·	·	·	·
Bulbochaete sp.	1	1	+	+	·	·	·	·	·
Spirogyra and *Zygnema* spp.	4	4	1	1	·	·	·	·	·
Diatomeae	·	·	·	·	2	+	·	2	·
Moss protonema	·	1	3	2	2	4	2	4	3

(Margalef, 1948) but not used to any extent. The technique of J. G. Jones (1974) might be applicable provided growth is not too dense but once filamentous green algae have developed some method of washing the algae off the stones would have to be employed. A brush attached to the piston of a syringe was used to removed lake epilithon by Stockner & Armstrong (1971); 60–70 % of their samples from a northwestern Ontario lake were diatoms. In some acid lakes a thick layer of mucilage formed by *Frustulia rhomboides* can coat the surface of the stones. In this mucilage,

species of *Nitzschia*, *Chlamydomonas*, etc. can live and the *Frustulia* cells are also buried in the gel. Macroscopic growths of *Batrachospermum*, *Stigeoclonium*, *Draparnaldia* etc., also occur in such acid situations. Cyanophyta similar to those listed above in running water apparently also colonise the rocks in lakes on Baffin Island (Moore, 1974c).

One investigation (Golubić, 1967), admittedly concentrating mainly on Cyanophyta, has compared the flora on rock faces above water level (Table 3.5a), near water level (Table 3.5b), at 30–80 cm depth (Table 3.5*c*), and at 12 m and 43 m depth (Table 3.5*d*). From these data one can see clearly that the aerial site is quite comparable to those investigated on high mountain rock surfaces and totally unrelated to lake situations, whilst the underwater sites have a greater preponderance of *Aphanothece, Rivularia, Dichothrix, Tolypothrix* and the green algae *Bulbochaete, Spirogyra* and *Zygnema*. Kann (1941) investigated the precipitation of calcium carbonate around Cyanophyceae in lakes and she found a loose deposit associated with *Schizothrix lacustris* and a hard stone-like coating where *Rivularia haematites* was growing.

Another example of zonation is given in Kann (1959) and shows that on a steep rock face rising out of the Traunsee there is a sequence of macroscopic species from the spray wetted stone, where algae

typical of sub-aerial habitats occur, to the permanently submerged communities (Fig. 3.6*a, b*). The occurrence of the red alga *Bangia* is interesting, and according to Kann it has been recorded in many other similar habitats; there is a marine equivalent which also occurs on rocks in the splash zone. The movement of water above or below the mean lake level is seasonal and the algae living below this point are not permanently submerged, e.g. between − 50 cm and − 25 cm the rock can be dry from 68 to 150 days per year. On blocks of stone in the Traunsee a somewhat similar flora occurs (Fig. 3.6*b*); similar associations of filamentous epilithic algae occur in many lakes and the habitat is worthy of greater study. Some examples of freshwater epilithic algae are illustrated in Fig. 3.6*c*. Only rarely has the succession of species during an annual cycle

Fig. 3.5. The occurrence of three algae in experimental streams with varying flow rates and illumination conditions. From McIntire, 1968.

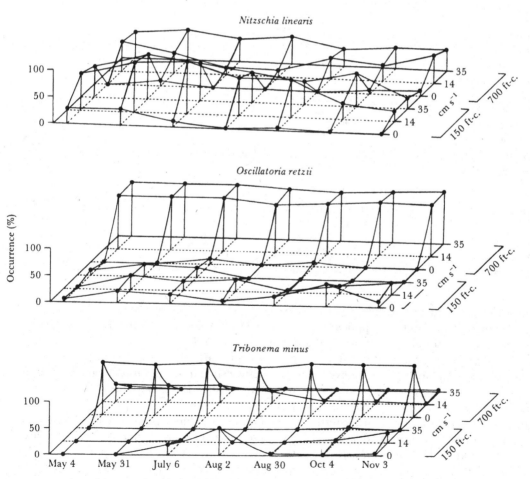

Nitzschia linearis

Oscillatoria retzii

Tribonema minus

Fig. 3.6. The distribution of epilithic algae in the Traunsee; *a*, on a steep slope above and below mean water level and, *b*, on a flatter shore with boulders. From Kann, 1959, *c*. (*p*. 62), illustrations of freshwater epilithic algae which live on exposed or submerged rocks. A, *Gloeocapsa*, B, *Chroococcus*, C, *Aphanocapsa*, D, *Chamaesiphon*, E, *Oncobyrsa*, F, *Scytonema*, G, *Stigonema*, H, *Hapalosiphon*, I, *Homeothrix*, J, *Tolypothrix*, L, *Meridion*, M, *Melosira*, N, *Didymosphenia*, O, *Frustulia*, P, *Hydrurus*, Q, *Oocardium*, R, *Mesotaenium*, S, *Gongrosira*, T, *Trentepohlia*, U, *Batrachospermum*, V, *Lemanea*, W, *Compsopogon*, X, *Cladophora*. Except for U–X, these are microscopic forms which, however, often aggregate to form mucilaginous crusts or tufts.

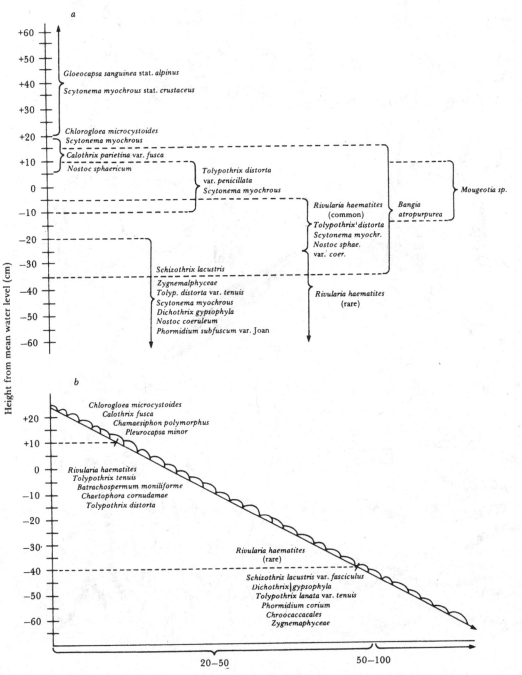

c

been studied, e.g. Mattern (1970) working on the Bodensee (Lake Constance) found a spring diatom growth (*Diatoma vulgare, Fragilaria capucina, Cymbella ventricosa, Gomphonema olivaceum* and *Synedra ulna*) intermingled with *Ulothrix zonata*; in summer filamentous green algae (*Spirogyra, Oedogonium, Stigeoclonium, Cladophora, Bulbochaete*) overgrow and kill the diatoms; in mid-summer *Cladophora/Oedogonium* becomes dominant and later *Hydrodictyon* occurs; in winter *Chaetophora incrassata* replaces the other green algae. *Cladophora* growing epilithically has become a major nuisance in some of the Laurentian Great Lakes where it grows in filaments up to 25 cm long; on decay it causes problems for water supply undertakings and it is quite unpleasant along recreational shores. Herbst (1969) found that it had a double cycle of growth, rapid in June–July then ceasing and recommencing growth in late August–September and finally declining abruptly. A similar pattern was recorded by Chudyba (1965, 1968) in Polish lakes and it is suggested that the pause might be caused either by zoospore formation or by subtle reactions to temperature.

Few studies of annual production of lake epilithon have been made; in a very cold water lake, Schindler, Frost & Schmidt (1973) report a low value of 5.18–5.19 g m^{-2} of carbon in two lakes and they also comment on the fact that the surface samples were always lower than those from 1 m and that a steady fall follows to 4–6 m. This is similar to that recorded for epipelic communities (*see* p. 176). Kairesalo (1976) also found a very low rate of carbon production (0.5 mg m^{-2} h^{-1}) in a Finnish lake.

Sea coasts

The rocky shore has been a favourite haunt of naturalists and student classes, for here the complexities of ecology *seem* to be reduced by the presence of parallel belts of organisms. It is a habitat in which sessile animals are sometimes as conspicuous as plants and, as a result, is one of the few habitats where both plants and animals are regularly recorded in ecological surveys. Some schemes of zonation utilise both plants and animals, but either can be used alone, and since the complexities of combining them are enormous the task will not be attempted here.* This is not to imply that the ecology of plants

* For European workers the book by Lewis (1972) presents an integrated account.

and animals is not interlinked, it is, as in all other habitats, but the subject becomes unwieldy if both are considered. So much has been written about the shore zonation that many biologists consider it has been overdone, but others, myself included, incline to the view that the real study has not yet started; only very elementary descriptive details are available for most shores, little experimental work has been undertaken and the all-important aspect of the dynamic relationships between the species is only now beginning to be resolved. The rocky shore is at first sight one of the most inhospitable environments, with great changes in temperature, repeated desiccation and wetting, salinity changes, grazing from two communities, and subject to the full force of the storm. Yet in temperate zones, where all these factors apply, there is a profusion of growth, with layered vegetation, abundance of epiphytes, and a teeming mass of interacting animals. On many shores, and especially those in temperate regions, the layering of the vegetation is striking, with a canopy of large algae, an understorey of small species, which exist only under the protection of the canopy (Connel, 1972; Dayton, 1975*b*), and an even less conspicuous community of species encrusting the rock surface.

Many accounts of coastal algal ecology are centred on the coast as a region and workers have therefore had the difficult task of distinguishing between a wide variety of habitats and algal life forms. Obviously degree of wave action and the nature of the substratum combine to give very different habitats and vegetation types. Extreme, moderate, or even slight wave action is associated with rocky and sandy shores but more rarely with silted shores. The largest changes in flora occur where sufficient silt collects to eliminate the algae which require an exposed rock surface for attachment, i.e., the epilithic species. This accumulation of silt does not occur on sloping surfaces in silted regions so that here a proportion of the epilithic species persist, but the number is reduced, since some are intolerant of sediment in the water which bathes the plant during high tide. The silted areas encourage a different life-form based on a branching rhizomatous growth and will be dealt with in the chapter on epipelic algae (p. 155). It is usually necessary to distinguish between the sites where the epilithic flora is subjected to considerable wave action (exposed shores), those with moderate wave action and those protected from all but the slightest wave disturbances (protected shores). Slight differences in height of colonisation also occur on rocky shores facing north compared to those facing south, e.g. Castenholz (1963) found that diatoms on pilings always extended highest on the north side.

The purely descriptive studies are essential, especially in little worked regions of the world, but for well documented areas further study by such means is unlikely to unravel the causative factors, interactions, etc. In spite of the concentration of effort on the shore hardly any floristic surveys have been made which attempt to be comprehensive. Well populated rocky shores of north Europe cannot be described adequately in terms of *Pelvetia canaliculata/Fucus spiralis/Fucus vesiculosus/Ascophyllum nodosum/Fucus serratus/Laminaria digitata* belts with perhaps a few more species added to each. Illustrations are given in Fig. 3.53. This is not to say that these belts should not be given such names based on the dominants, as is the practice in land-plant ecology. However, descriptions need at least to be comprehensive of the species present in each belt of the shore and the usually neglected microscopic and epiphytic floras (*see* p. 221) are as much a part of these belts as are the bryophytes and lichens of tropical forests. For example on many rocky shores there is a well developed, but seasonal (winter–spring in Europe) growth of *Navicula ramossisima* and *Berkleya rutilans* (Cox, 1975*a, b*), visible as brown pustules or strands due to the growth of the cells in mucilage tubes; these can be mistaken for 'ectocarpoid' growth unless checked under the microscope. Cox (1977) studied the penetration of these tube-dwelling diatoms up the estuary of the River Severn and found that *Nitzschia angularis* died out first followed by *Navicula mollis, N. incerta* and *N. comoides* whilst *Berkleya rutilans* reached almost to the freshwater zone where other tube-forming diatoms grow but do not penetrate seawards. Less obvious growths of unicellular Cyanophyta, coccoid greens, and diatoms also occur coating the rock surface. One of the most important factors operating on these algae, which are almost invisible to the naked eye, is grazing and competition for space by animals; again these are often identified in the descriptive literature but rarely is their influence studied. A detailed study of this micro-epilithic flora is long overdue, and there are many problems of quantitative sampling of irregular rock surfaces which still need to be overcome.

A further important distinction must be made on rocky shores between the flora of the draining rock face/boulder habitat, which supports one set of algae, and the rock pools in which water stands during tidal submersion. In addition there are areas of shore where during the intertidal period, outflow of drainage water produces stream-like conditions and in these channels certain species grow in excess of others. There is a degree of mixing of the three floras but many are confined, or more abundant, in one or the other habitat.

Zonation. Before discussing the algae of the marine epilithon it is necessary to digress a little further into zonation of the shore. Stephenson & Stephenson (1949) are credited with the popularisation of the idea of universal belts* present on all rocky shores of the world, though others before them had implied the same (*see* the excellent discussion in Doty, 1957). These belts are occupied by similar life forms on most coasts of the world, although only a few species are worldwide in distribution. The belts are the result of interaction between settling and growth of the species and the operation of a complex set of environmental, grazing and competition factors.

As pointed out in several places in this book, wherever a gradient exists in nature there is a zonation of algae and indeed Gause & Witt (1935) concluded, from a theoretical study, that a zonation will occur in a mixed population under the influence of a gradient and that competition factors alone can explain this without recourse to physiological thresholds. It is abundantly clear that the physical environmental factors play a prominent part, especially on the upper shore, but that, throughout the gradient, biological interaction is a prime factor (Chapman, 1973; *see* examples in Chapter 9, this volume). Unfortunately features of biological stress have rarely been considered by the main workers on shore zonation who have often been more intent on fitting schemes into the Stephenson Universal Zonation or modifications of it. Only by combining the different viewpoints will zonation be understood. There is an intrinsic biological basis (differing growth rate, etc.) for zonation along an obvious physical gradient on shores. The gradient results in physical stress and this reinforces the zonation

---

* Various terms have been used for these 'parallel' vegetation units but belt seems to convey best the concept of laterally extended vegetation.

Table 3.6. *The vertical distribution of the more conspicuous indicator algae along central and northern Pacific coasts of the United States. Corresponding tide heights (in feet above or below mean lower low water for San Francisco) are given below the table. These heights correspond to tide breaks. From Doty, 1946*

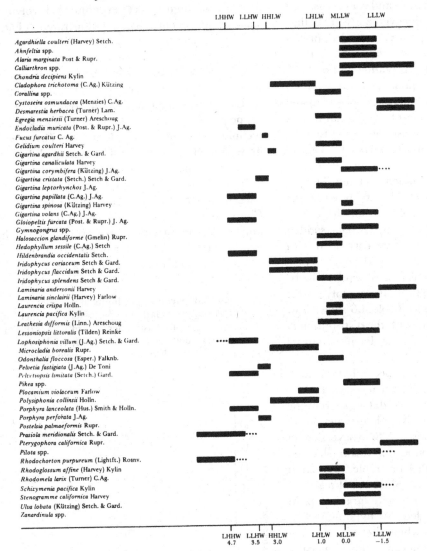

	LHHW	LLHW	HHLW		LHLW	MLLW	LLLW

Agardhiella coulteri (Harvey) Setch.
Ahnfeltia spp.
Alaria marginata Post & Rupr.
Calliarthron spp.
Chondria decipiens Kylin
Cladophora trichotoma (C.Ag.) Kützing
Corallina spp.
Cystoseira osmundacea (Menzies) C.Ag.
Desmarestia herbacea (Turner) Lam.
Egregia menziesii (Turner) Areschoug
Endocladia muricata (Post. & Rupr.) J.Ag.
Fucus furcatus C. Ag.
Gelidium coulteri Harvey
Gigartina agardhii Setch. & Gard.
Gigartina canaliculata Harvey
Gigartina corymbifera (Kützing) J.Ag.
Gigartina cristata (Setch.) Setch & Gard.
Gigartina leptorhynchos J.Ag.
Gigartina papillata (C.Ag.) J.Ag.
Gigartina spinosa (Kützing) Harvey
Gigartina volans (C.Ag.) J.Ag.
Gloiopeltis furcata (Post. & Rupr.) J. Ag.
Gymnogongrus spp.
Halosaccion glandiforme (Gmelin) Rupr.
Hedophyllum sessile (C.Ag.) Setch
Hildenbrandia occidentalis Setch.
Iridophycus coriaceum Setch & Gard.
Iridophycus flaccidum Setch. & Gard.
Iridophycus splendens Setch & Gard.
Laminaria andersonii Harvey
Laminaria sinclairii (Harvey) Farlow
Laurencia crispa Holln.
Laurencia pacifica Kylin
Leathesia difformis (Linn.) Areschoug
Lessoniopsis littoralis (Tilden) Reinke
Lophosiphonia villum (J.Ag.) Setch. & Gard.
Microcladia borealis Rupr.
Odonthalia floccosa (Esper.) Falknb.
Pelvetia fastigiata (J.Ag.) De Toni
Pelvetiopsis limitata (Setch.) Gard.
Pikea spp.
Plocamium violaceum Farlow
Polysiphonia collinsii Holln.
Porphyra lanceolata (Hus.) Smith & Holln.
Porphyra perforata J.Ag.
Postelsia palmaeformis Rupr.
Prasiola meridionalis Setch. & Gard.
Pterygophora californica Rupr.
Pilota spp.
Rhodochorton purpureum (Lightft.) Rosnv.
Rhodoglossum affine (Harvey) Kylin
Rhodomela larix (Turner) C.Ag.
Schizymenia pacifica Kylin
Stenogramme californica Harvey
Ulva lobata (Kützing) Setch. & Gard.
Zanardinula spp.

	LHHW	LLHW	HHLW		LHLW	MLLW	LLLW
	4.7	3.5	3.0		1.0	0.0	−1.5

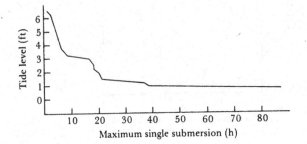

Fig. 3.7. Length of maximum single submersion correlated with tide levels. Sudden increases in length of maximum single submergence show clearly near levels of 3.5, 3.0 and 1.0 ft above mean lower low water. From Doty, 1946.

especially since different stresses occur at points along the gradient; to simplify, there is desiccation at the top, wave shock in the middle, and low light intensity in the subtidal. Since the algae are growing on a slope of varying inclination and roughness, part of which is covered and uncovered by tidal flow every day, some reference points are needed, preferably based on the operation of what is obviously a major variable, i.e. the tide. As pointed out by Doty (1946), on a plane surface equally exposed to the sea as many as ten distinct zones have been observed and in almost every one a dominant species can form a subzone of its own. On many shores, if the times of submersion* are plotted against tidal heights, certain regions of sudden increase in time of submersion (tide breaks) can be determined and this leads to the stepped form of Fig. 3.7 which is caused by the complexities of tidal movement when the tide is of the semi-diurnal type (Fig. 1.4). Such sudden changes in time of submersion are correlated with the breaks between the major algal zones. These so-called critical levels were first recognised to be important by Colman (1933) but it was Doty (1946) who developed the idea and Table 3.6 shows how these critical levels correlate with the distribution of some common species along the Pacific coast of North America.

Nevertheless experimental proof that the times of emersion or submersion are critical is scanty. Doty & Archer (1950) subjected intertidal algae from the Cape Cod region to progressively longer periods of submersion in seawater raised 5 °C above that of the natural habitat, and found that death occurred when the time was doubled or trebled over that which produced slight injury. Since the 'breaks' in some situations tend to occur where large jumps in time of exposure are recorded this hypothesis does seem viable, but in regions where there are no such abrupt jumps in times of emersion and submersion the results may not be applicable. Another attempt to study the effect of time of emersion and submersion was used by Townsend & Lawson (1972); they allowed *Enteromorpha* swarmers to settle on a glass plate, which was then lowered and raised vertically

* The terms submersion and emersion are used (as in Den Hartog, 1968) for the periods of tidal cover and uncover respectively since the term 'exposed' is used in the literature for shores open to the full force of wave action.

into seawater on a machine which could simulate different times of emersion and submersion. They found that a definite and sharp upper limit to the *Enteromorpha* growth was obtained after a few days and this was higher in the experiments involving the shortest time of emersion, as is illustrated in Fig. 3.8. Since the exact height can be modified by changes in the frequency of submersion–emersion, even though in their experiments the proportion of time in air and in water were kept constant, Townsend & Lawson considered that the results gave some support to the tide factor hypothesis. Using a similar technique, Allender (1977) found that *Padina japonica* grew best when permanently submerged, but again a definite upper limit could be detected which was reduced by high temperature; such a temperature effect is obviously important with a tropical alga. Chapman (1942) considered that critical levels probably occur on British shores around mean high-water, extreme high-water and extreme low-water neap tides, and Evans (1957) came to the same

Fig. 3.8. Mean upper limits of zones of *Enteromorpha* relative to periodicity of emersion: the range of measurements from which the mean was calculated are indicated: *A*, 1.5 h cycle; *B*, 3 h cycle; *C*, 6 h cycle; *D*, 12 h cycle. From Townsend & Lawson, 1972.

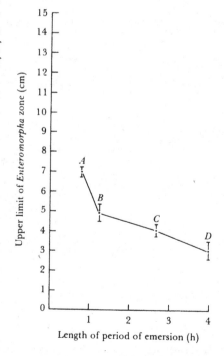

conclusion concerning these levels on French Atlantic shores.

In spite of complexities of tidal flow, (see Chapter 1) it is desirable to have some universal system of reference points on the shore so that comparative descriptions can be prepared. It must be stressed that the reference points are purely topographical, and may be used for description of vegetation though they are not in themselves indications of controlling factors, a confusion which is rife in the literature. Often the tidal data are not measured, and indeed there are obvious problems associated with measurement, yet frequently the flora is referred to a *position* related to tides on the shore (e.g., lower intertidal). The reference points can only be the ones given in Fig. 3.9 and, to quote Doty (1957), 'Tidal datum points are used as reference points from which elevations are measured physically'. They are also used in this manner by Den Hartog (1968). Measurements relative to a low tide datum point are used extensively in descriptions of the subtidal (see p. 96). Obviously, in general discussion, such detail is not essential; hence the terms proposed on p. 40 will be used in the following account. These terms are supratidal* (it is almost impossible to lay down strict limits for this belt but in general it lies above high-water spring

Fig. 3.9. An example of the complexity of tidal heights which can be encountered on a shore. Doty, 1957. Other simpler cycles are illustrated in Fig. 1.4.

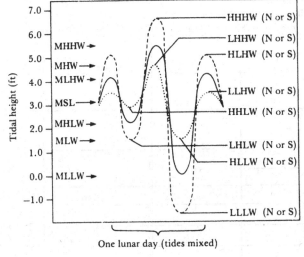

One lunar day (tides mixed)

Maximum ------ Mean tide ——— Minimum ············

tides), subtidal (below low-water spring tides) and intertidal (the belt between high-water and low-water spring tides) into which even the complexities of mixed tides can be fitted (Fig. 1.4). Intertidal is also the term used by Doty (1973a) for the intermittently covered region on coral reefs, and Kurogi (1972) uses intertidal and subtidal with upper and lower divisions to describe the sites of occurrence of *Porphyra* species in Japan. The same terms are used extensively by geologists and geographers working along coasts. Lewis (1972) rejects the use of physical reference points on the basis that the totality of the physical environment is unmeasurable, the tidal points are only aspects of the physical environment, and they do not necessarily coincide with biological levels. One cannot disagree with this view but I merely argue that one set of physical measurements (hopefully measured over as long a period as possible and with the awareness of the complicated tidal patterns which can occur, e.g., diurnal, semidiurnal, mixed, etc.) is better than none, and whenever possible should be determined, though actual zonation of the

* These correspond to the subdivisions of Den Hartog (1968) who however used supralittoral (spray–splash zone), eulittoral (intertidal, between mean low-water spring tides and either mean high-water or mean high-water spring tides) and sublittoral (always covered except perhaps its uppermost limit which is occasionally uncovered by very low tide), or equally the supralittoral–littoral–sublittoral of the extremely detailed survey of the Oslo fjord (Sundene, 1953). The flora is continuous from the uppermost supratidal point of marine algal vegetation to the lowermost point in the subtidal, where light limits growth. Schemes of zonation should not stop at low tide level simply because this is a physically measurable point, unless it is also a point where dramatic breaks in the vegetation occur, which is not usually so, at least on temperate shores. A few truly terrestrial plants (lichens) are characteristic of the supratidal zone and occur nowhere else on the shore. This adds further evidence for the concept of a discrete biological entity above the zone of tidal influence and below the truly terrestrial zone (see Den Hartog, 1968, for an excellent detailed discussion). French workers use an exactly comparable system, supralittoral, meso- (or medio-) littoral and infralittoral (e.g., Pérès & Ricard, 1955) and Dellow (1950) also used zones based on -tidal in New Zealand. Several workers mix the terms and use intertidal for the intermittently uncovered and sublittoral for permanently covered (e.g., Adey, 1968) which seems unnecessarily confusing. The distinction of a deep, subtidal region of algal growth termed either the elittoral or circalittoral also has no basis in detailed sociological studies.

shore can be described in terms of algal belts. Thus I am also inclined to use biological entities for the delimitation of belts, but without the rigidity imposed by recognising three zones as does Lewis (littoral fringe–eulittoral–sublittoral based on the lichen–barnacle–fucoid–laminarian associations) or five zones as in Stephenson & Stephenson (1949) and again in Southward's (1958) excellent discussion. A further complication is that a fucoid zone in the northern hemisphere is intertidal (however *see* p. 78) but subtidal in the southern hemisphere and therefore such extremely general concepts based on algal groups should be avoided.

Chapman (1942) was one of the first workers to distinguish between what he called 'presence and absence factors'. They are so obvious that they are generally forgotten but, to take a simple example, if the necessary substratum is absent from a shore, then the species colonising such a substratum will be absent. So lack of any algal species on a shore should not be taken to imply that the other 'causal' factors e.g., period of submersion, temperature, etc. are necessarily unfavourable. Chapman also pointed out that perfectly good examples of zonation can be found on Caribbean shores where there is only a nine-inch tide and that amplitude itself is unimportant.

Spore production and initial colonisation. To start at the beginning: the spores of the species arrive on the shore, yet little is known of their dispersal patterns. Undoubtedly most are produced on the shore itself, but some must be transported greater distances from other shores, together with fragments of thallus (*see* p. 361 where the geographical spread of species is discussed). As Neushul (1972a) points out, macroscopic, benthic algae have a 'planktonic' spore stage in their life-history, during which time 'planktonic' factors affect them, whilst during later periods the plants are subjected to differing kinds of water motion; this is illustrated in Fig. 3.10 based on *Macrocystis* but it applies to many other coastal algae. The spores settling on the various substrata are of two types, those arising from asexual processes, and those from sexual reproduction and Tables 6.2 and 3.7 give an indication of the time of formation of these spores for some common shore plants in Britain, and for a North American shore; the North American data show that the Chlorophyta produce swarmers in summer–early autumn, which is also the time when many Rhodophyta produce tetraspores or carpospores. Some Phaeophyta (*Laminaria*) produce their zoospores (meiospores) during the months between September and early April with a peak in January (Kain, 1975a). Clearly it is important to

Fig. 3.10. The four water-motion regions (orders of magnitude of water velocities are given at the head of each column) in a *Macrocystis* forest together with the developmental stages (or life forms) of the algae in the community. From Neushul, 1972a.

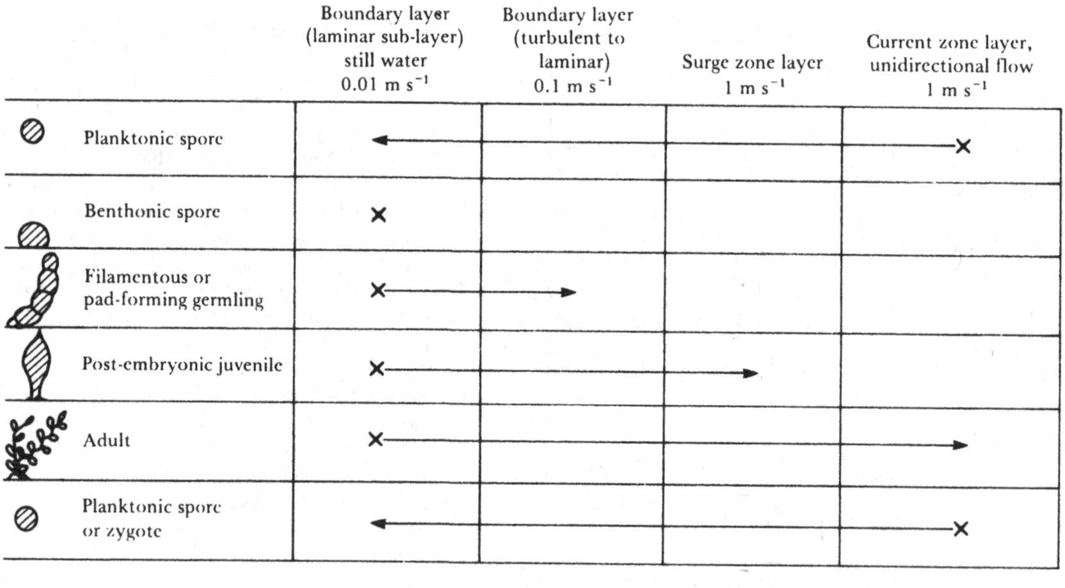

	Boundary layer (laminar sub-layer) still water 0.01 m s^{-1}	Boundary layer (turbulent to laminar) 0.1 m s^{-1}	Surge zone layer 1 m s^{-1}	Current zone layer, unidirectional flow 1 m s^{-1}
Planktonic spore	←————			————×
Benthonic spore	×			
Filamentous or pad-forming germling	×————	—→		
Post-embryonic juvenile	×————		—→	
Adult	×————			—→
Planktonic spore or zygote	←————			————×

Table 3.7. *Phenology of reproductive structures of some lower Chesapeake Bay benthic algae. From Zaneveld & Barnes, 1965*

Season	Winter			Spring				Summer			Autumn		
Month of the year	12	1	2	3	4	5	6	7	8	9	10	11
Chlorophyta												
Enteromorpha intestinalis	+	+	+	+	+	+	s	s	s	s	s	+
Enteromorpha minima	+	+	+	–	–	–	s	s	s	–	s	+
Enteromorpha linza	+	+	+	+	+	+	s	s	s	s	s	+
Ulva lactuca	+	+	+	+	+	+	s	s	s	s	+	+
Cladophora flexuosa	–	+	+	–	–	s	s	s	s	–	s	–
Cladophora gracilis	–	+	+	–	–	–	s	g	s	s	s	–
Bryopsis hypnoides	–	–	–	–	–	–	+	g	g	+	–	–
Bryopsis plumosa	–	+	g	+	+	–	g	g	+	+	–	g
Phaeophyta												
Punctaria plantaginea	–	–	–	–	+	+	u, p	'u, p	u, p	+	–	–
Petalonia fascia	–	+	+	p	p	p	+	+	+	–	–	–
Scytosiphon lomentaria	–	+	+	+	+	+	+	+	+	–	–	–
Rhodophyta												
Bangia fuscopurpurea	+	+	m	–	–	m	m	m	–	–	–	+
Porphyra leucosticta	+	+	+	+	c, a	–	+	–	–	–	+	+
Porphyra umbilicalis	–	+	+	+	c, a	+	+	+	t	+	+	+
Gelidium crinale	–	–	–	+	–	–	t	t	t	+	–	–
Gracilaria verrucosa	+, *	+, *	t, c	+	–	+	t, c	t, c	+	+	t	t
Gracilaria foliifera	–	–	l	–	–	–	t, c	–	–	+	+	c
Agardhiella tenera	+	+	+	+	+	+	t, c	t, c	c	+	+	+
Champia parvula	–	+, *	–	±	+	+	+	+	t	–	±	±
Callithamnion byssoides	–	–	–	+	–	–	+	t	t	–	–	–
Ceramium fastigiatum	–	–	–	–	+	+	t	l	t	–	–	–
Ceramium strictum	–	–	–	–	+	+	t	l	t	–	–	–
Ceramium diaphanum	–	–	–	+	–	–	c	l	l	+	–	l
Ceramium rubrum	+	+	t, c	+	+	+	+	t, c	+	+	+	t
Grinnellia americana	+	+	l	+	–	+	t, c	+	t	+	+	c
Dasya pedicellata	–	+	t, c, a	+	–	–	t, c	t	t	–	–	–
Polysiphonia harveyi	–	±	t, a	–	–	+	t, c	t, c, a	t, c	+	+	t
Polysiphonia denudata	–	–	–	–	–	+	+	t, c	t	l	+	–
Polysiphonia nigrescens	–	–	t, c, a	±	+	+	c	c, a	c	–	–	t

+, present; –, absent; ±, dormant; a, spermatangia; c, cystocarps or, in *Porphyra*, carpogonia; g, gametes; m, monospores; p, plurilocular reproductive structures; s, swarmers; t, tetrasporangia; u, unilocular reproductive structures; *, washed ashore.

have such phenological information on phenomena controlled primarily by genetic processes, in order to understand fully the ecology of a species but unfortunately it is not often available in any detail. In addition, spore production is obviously also affected by environmental factors, e.g., Smith (1967) reported that, in the subtidal, the deeper the alga grew the longer its sporulation period. It is generally accepted that algae produce large numbers of spores but how large is rarely appreciated. Kain (1975) estimated at least 7300 sporangia per cm² of *Laminaria hyperborea* frond surface, which would give 3 300 000 zoospores per mm² of rock surface at Port Erin. If they all settled these would form a layer 70 spores deep on the rock surface! Comparably large numbers of sexual cells are formed, e.g., Knight & Parke (1950) estimated that *Fucus serratus* and *F. vesiculosus* could release over a million eggs in a season from each of the largest plants they measured.

A few studies have been made on the factors affecting sporulation, e.g. in *Rhodochorton purpureum*, photoregimes are important and tetrasporangia (this plant is only known in the tetrasporophytic phase)

are formed during short days, but there are complicated temperature-dependent responses in different clones, e.g., a clone from Chile showed light-break effects* (i.e., inhibition of sporulation) at 15 °C but not at 10 °C, whereas a Californian strain shows the effect at 10 °C, but less so at 15 °C, and an Alaskan clone showed little or no effect of light breaks at 10 °C, and no sporulation at all at 15 °C (West, 1972). Salinity also affected sporulation of *Rhodochorton*, with the somewhat expected result that clones from the uppermost intertidal sporulated better in lower salinities than did a clone from the mid-intertidal and these responses remained constant over long periods in culture, which suggests a genetically stable character. In nature, West found that sporulation was confined to the winter period, in day lengths less than 12 h. Tetrasporangia formation in *Achrochaetium pectinatum*, another related red alga, occurs only with day lengths less than 10 h but a light break in the middle of the dark period is *not* inhibiting; induction is therefore not strictly a photoperiodic effect. Temperature control of sporulation was reported for *Ectocarpus siliculosus* by Müller (1962) who found unilocular sporangia at 13 °C, plurilocular at 19 °C and both at 16 °C. At 16 °C the ratio of the two could be varied by altering the light intensity, salinity or day length.

---

* Interruption of the long dark period with a light break causing inhibition of the phenomena is a criterion for genuine photoperiodic response.

Fig. 3.11. The percentage germination of carpospores of *Chondrus crispus* (*a*) and *Gigartina stellata* (*c*) at various temperatures and salinities and 440 ft-c.; and the growth of *Chondrus crispus* (*b*) and *Gigartina stellata* (*d*) at various temperatures and salinities and 440 ft-c. From Burns & Mathieson, 1972.

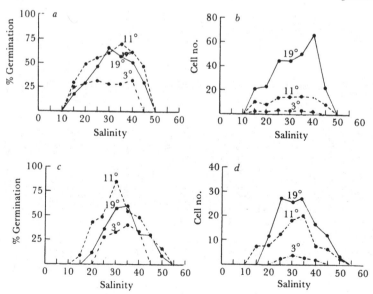

Burns & Mathieson (1972) showed that carpospores of *Chondrus crispus* and *Gigartina stellata* germinate over wide ranges of temperature and salinity (Fig. 3.11), but it seems that the latter has a slightly more restricted range, with a reduction in germination at higher salinities, such as it might experience at times in rock pools. Optimal temperatures are probably higher than those experienced by the plants at their collection site.

Spore release is not a continuous process, e.g., in *Nitophyllum punctatum* Sagromsky (1961) found that more spores were released in the light than in the dark; red light and continuous light reduced release. It would be interesting to know how this type of control operates on genera in the polar winter and summer. Boudouresque (1971a) comments that some of the populations living under high-light conditions (photophilic) in the Mediterranean seem to be short-lived and their reproduction is strongly in-

hibited, but it is not clear whether this is an effect on spore release or formation.

Occurrence of a species does not imply a normal reproductive capacity, for example in estuaries species often exist but do not reproduce normally; Russell (1971) gives many interesting details from the River Mersey, amongst which is the occurrence of *Pilayella littoralis* with both unilocular and plurilocular sporangia, but without meiosis in the unilocular, so that the population is diploid. Species near the end of their geographic range are also often sterile (Dixon & Richardson (1970) and p. 361) or triploid parasporangial plants with apomictic life cycles; e.g. *Plumaria elegans* in Nova Scotia (Whittick, 1977). Obviously climatic conditions vary during the maturation periods of asexual and sexual phases of the plants and also during spore germination and initial growth of the new plants and conditions for each phase must be studied. The very fact that some plants of an epilithic association are male, others female and yet others sporophytic, and that these may occupy different belts on the shore, is often ignored by ecologists and therefore we have few data on the distribution of the gametophytes and sporophytes on the shore. Liddle (1971) reported that the sporophytes and gametophytes of the

Fig. 3.12. Photographic records of individual algal spores falling through still water: *a*, *Cryptopleura violacea* (diam. 55 μm, timed through 2.85 s); *b*, *Agardhiella tenera* (38.1 μm, 2.55 s); *c*, *Gelidium robustum*; *d*, *Myriogramme spectabilis* (45.1 μm, 4.40 s); *e*, *Callophyllis flabellulata* (17.3 μm, 25 s). From Coon *et al.* in Neushul, 1972a.

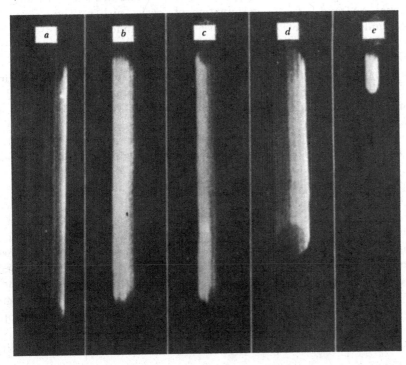

Carribbean *Padina sanctae-crucis* do not occur mixed together.

Spores of some species aggregate in slime masses and this will affect their sinking rate and distribution. Only recently have any measurements been made of rate of fall of algal spores. Fig. 3.12 shows some data from Coon, Neushul & Charters (1972); *Cryptopleura violacea* carpospores would take three and a half days to settle through the water column (circa 30 m) of a *Macrocystis* bed, assuming still-water conditions, which of course do not exist in nature. From the site of production on the parent plant, however, the spores would only take 10 min to reach the sediment; a more relevant figure. Whilst such data show, as might be expected, that the rates of fall vary from species to species, it is still necessary to study in some manner what actually happens in the turbulent water in the natural habitat. Spores of red algae are often embedded in large mucilage spheres, e.g., Boney (1975) found that the mucilage sheath occupies 52–88% of the combined spore and mucilage volume; the sheath may affect the rate of sinking but it may also be protective. Boney found that newly released spores sank more slowly than those 12 h old.

Starting with the germinating stage, it is frequently observed that numerous young plants are aggregated in, and around, mature stands of a species and these are almost certainly the products of the mature plants in the stand. Yet the element of chance dispersal is everywhere evident on the shore. Any spore which has not attached soon after release will be transported up or down the shore by tidal flow and so, by sheer chance, almost all species establish themselves at any point on the shore (*see* also evidence from cleared strips of shore, p. 118). Most isolated plants, however, are subjected to unfavourable conditions, e.g., of exposure, grazing etc., and do not survive, except perhaps as random plants with a tenuous existence. Such plants can be found on any shore and their inclusion in lists of species for any particular belt is confusing, unless accompanied by qualifying statements; equally they confuse the student when included in quadrats, line transects, etc. unless the data are subjected to statistical analysis. Establishment of sporelings is likely to be a critical phase in the life cycle of any species and local year-to-year variations, especially of annual species, may depend on conditions during this early growth period. Similarly the year-to-year fluctuations can be influenced by the size and vitality of the previous year's population through its effects on the inoculum size at the spore settling stage. Thus, at each level on the shore conditions must be suitable *both* for germination and growth. Assuming that the sporeling stage is the most susceptible, there is little point in looking for factors in, for example, the upper intertidal preventing the vegetative growth of a lower intertidal plant, when in fact it may be merely the spore stage which is susceptible to desiccation. A simple example is provided by *Himanthalia elongata*, the spores of which will not germinate when salinity is halved (Moss *et al.*, 1973) and only a few germinate in the dark. However every alga requires individual study since Moss found that *Halidrys* germlings can grow in the dark for several weeks. Another possibility is that spores of such high-level forms as *Pelvetia* may require organic factors which are present in the decaying material usually found in a belt on the upper shore. Spore attachment in the uppermost zones may require either a desiccating environment or a frequent wetting by freshwater. As Baker (1909) pointed out, the species growing highest on the shore do not grow in rock pools whereas the lower growing species often do. She concluded that time of submersion was critical, and was led to investigate the germination of species such as *Fucus serratus* and *Ascophyllum nodosum* under various submersion–emersion regimes. It was found that the fertilised eggs of these species would not germinate when exposed to one hour in water and 11 h dry; the upper shore species could germinate under such conditions but growth was slow. It is unlikely that spore germination and sporeling growth of upper intertidal plants is affected to a similar degree by desiccation and factors must be sought controlling both stages. For example zygotes of *Pelvetia* secrete a firm wall within 24 h of release and many collect in the channelled thallus which may act in a protective manner against desiccation (Moss, 1974). *Ascophyllum nodosum* secretes a substance which firmly binds the zygote to the substratum (Moss, 1975) and which may bind only with certain surfaces. The mature holdfast of *Laminaria digitata* gives no clue to the fact that the meristoderm produces rhizoids which fill every crevice of the rock surface and secrete copious mucilage to bind to the substratum (Tovey & Moss, 1978). Competition for space and susceptibility to grazing must also be considered at this very early stage but little is known about these factors.

The actual site of spore attachment on the rock surface is in the boundary zone, where current effects are minimal in the subtidal, but possibly extreme in the exposed wave-beaten intertidal. The flow in the boundary layer is laminar if the Reynolds number is small (*see* p. 51), but turbulent if it is large. Nevertheless the actual surface is a region where frictional effects dominate and velocity may be very low. The larger the spore, the greater the surface area exposed above the area of minimal flow and so, theoretically, the smaller the spore, the less its attachment is affected by water flow. Charters, Neushul & Coon (1973) found that the firmness of attachment of spores increased with time for *Agardhiella* and *Gracilariopsis* but decreased with time for *Cryptopleura*. When *Gracilariopsis* spores are once attached they resist removal by sheer forces 100 times their weight. An important factor in this region of attachment is silting, and a moderate water flow may help to prevent accumulation of particles; certainly a richer flora tends to be found on unsilted rock surfaces. Grazing pressure, competition for space, nutrients, and light are all undoubtedly extreme in this attachment region, but there is no reason to suppose that spores of microscopic algae are any less able than other organisms to compete with bacteria and unicellular algae, the initial colonisers of the rock surfaces. The presence of microscopic algae in a slime layer on the rock may indeed be an advantage, offering many more sites for colonisation by spores of macrophytic algae. In fact it is probably rare, if not impossible, to find a bare rock face or indeed any uncolonised space in the epilithon and spores of many macroscopic fucoids, laminarians etc., often commence growth as epiphytes and then overgrow the other algae attached to the rock. Nevertheless, Hruby & Norton (1979) found the greatest species diversity under a canopy of *Ascophyllum*. In this habitat, speed is probably a vital factor and Christie *et al.* (1970) found that *Enteromorpha* zoospores adhere within minutes of contact by means of a mucopolysaccharide. Adhesion might be expected to be greater on rough surfaces but Linskens (1966) found that some algae prefer smooth surfaces and indeed one alga may favour smooth for its gametes and rough for its zygotes, e.g. *Acetabularia*. As might be expected, spores of intertidal algae must be unaffected by high light intensity and, even when spores of Rhodophyta have lost their phycoerythrin, growth continues, whereas spores of subtidal species die if this pigment is lost (Jones & Dent, 1971).

The reproductive units released by Rhodophyta are non-motile, whereas many of those from Chlorophyta and Phaeophyta are motile, flagellate swarmers. There is equally little information on the longevity, dispersal, settling etc., of the motile cells and about the conditions affecting release of gametes, etc. As early as 1910 Baker showed that *Ascophyllum nodosum* needed exposure to air for the expulsion of its gametes though, surprisingly, species growing higher on the shore could expel their gametes when submerged. More gametes were expelled in freshwater than in seawater and so rain on the shore may have a favourable effect at the time of gamete release and not, as might have been expected, a destructive effect. Jones & Babb (1965) found that under experimental conditions swarmers of *Enteromorpha intestinalis* can remain motile for up to eight days in suitable light and temperature regimes, whereas zoospores of *Ulva reticulata* are motile for only two hours first in a positively phototactic state and then later ceasing motility (Kemp, 1974). The swarmer release occurs at dawn, but can be delayed by keeping *Ulva reticulata* plants in the dark; the gametes can be delayed only until mid-afternoon, whereas the zoospores can be delayed until the evening. Release of swarmers in these plants is remarkably fast, occurring within five minutes of transference to suitable light conditions, and is not affected by lunar periodicity. The release of swarmers of other closely related algae is known, however, to be linked to lunar periodicity, e.g., *Enteromorpha* releases its swarmers three to five days before the high tide of each lunar period (Christie & Evans, 1962) whereas *Ulva lobata* (Fig. 3.13) releases zoospores towards the end of the spring tide and the gametes early in the cycle of spring tides; the gametes are released between 3 and 6 a.m. and zoospores between 6 and 9 a.m. (Smith, 1947).

In *Dictyota dichotoma* growing on North Atlantic shores there is a periodic release of eggs from the sori on the female plants during summer only; the release of sperm must be more or less synchronised but it does not appear to have been investigated whilst there is *no* rhythmic release of tetraspores from the tetraspore plant. The time of release relative to the tidal cycle varies from place to place, e.g. it occurs

2–3 days after spring tides in Wales, 5–6 days after spring tides at Plymouth, but in Beaufort, North Carolina it occurs only once per month and then 4–7 days after full moon. On the other hand in the Mediterranean and in Jamaica, egg release is a prolonged process which tends to obscure the rhythm (Hoyt, 1927). Hoyt also reported that about 80% of the gametangia discharged on one day and 60–70% in a single hour of the day, just after dawn. How many other rhythmic phenomenon may occur in the almost totally uninvestigated marine algae of most shores? The *Dictyota* system has lent itself to experimental study and it is now known that the rhythm is a complex of a diurnal and a lunar response. The trigger seems to be full moonlight (approx. 0.3 lux) since in the laboratory a short interruption given at any time during the dark phase of plants kept in alternate light:dark conditions is sufficient to induce egg release some ten days later and then repeatedly every 16–17 days after this (Müller, 1962). The 16–17 day rhythm is not quite what is expected since a semi-lunar rhythm would occur every 14.5 days; but it is well known that these 'circadian' rhythms vary slightly under laboratory conditions. Indeed, Vielhaben (1963) followed up this work and showed that the period is lengthened if the experimental studies are conducted in a

light:dark regime of 14.25:10.25 (i.e. a 24.5 h day) and shortened in a light:dark regime of 13.5:9.5 (i.e. a 23 h day). In fact the lengthening produces a 16–17 day and the shortening an 11–12 day fruiting rhythm. The precision and short period of gamete release means that a great number of eggs are available for fertilisation at one time and presumably this synchronisation is advantageous under the ebb and flow conditions of a rocky shore or in the turbulence of the subtidal zone.

Vielhaben (1963) extended her work to investigate the photosynthetic rhythm of *Dictyota* and produced some unexplained and striking data. After setting the rhythm with a period of 'moonlight' during a dark cycle and then measuring the photosynthetic capacity of the thalli on subsequent days there is a very distinct peak in capacity six hours into the succeeding light period and on subsequent days this peak occurs 0.7 h later (strikingly similar to the 0.8 h shift in the lunar period).

Under experimental conditions *Enteromorpha* gametes and zoospores behaved differently, in that the gametes move to regions of high light intensity at the air–water interface, whilst the zoospores settled in lower light regions (Christie & Evans, 1962). Both temperature and salinity affect this settlement as shown in Fig. 3.14; settlement is an

Fig. 3.13. The daily tidal range at Pt Cypress, Monterey Peninsula, the days on which gametophytic plants of *Ulva lobata* liberated gametes (*g*) and on which sporophytic plants liberated zoospores (*s*) at tidal heights of 0 ft and 1 ft. From Smith, 1947.

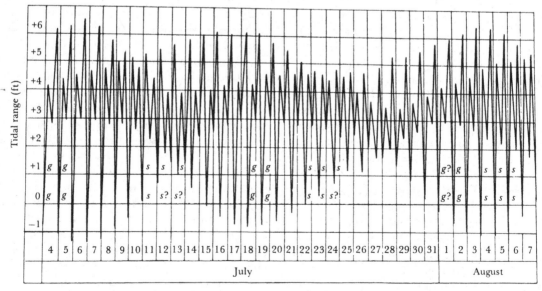

active process due to the rapid spinning motion of the swarmer as its anterior (flagellate) end comes into contact with the substratum. Brown algal swarmers, e.g. those of *Chorda filum* (South & Burrows, 1967) also undergo a similar movement although the flagella are inserted laterally. Kain (1964) deduced from laboratory studies that spores of *Laminaria hyperborea* could remain in a non-motile, planktonic state for some days and could even undergo the first stages of gametophyte development whilst still planktonic. Most zoospores of this alga lost motility within one day of release.

Enteromorpha is often growing very abundantly where freshwater flows on to the shore and spore settlement was poor at normal seawater salinities compared with the range 17–25‰ (Christie & Shaw, 1968).

Swarmers of *Ectocarpus confervoides* tend to settle in grooves rather than on flat surfaces, e.g., Müller (1964) found that, in experimental studies, the swarmers chose the grooves in the substrata, presum-

ably where water movement would not wash them off the surfaces. The grooves may provide a rough surface for attachment; Moss *et al.* (1973) found that *Himanthalia* spores did not attach as firmly to glass as to washed rock and, more significantly perhaps, the growth rate was greater on rock than glass. The fate of spores and factors affecting their development in nature have been followed in an indirect manner by staining the spore walls with a fluorescent dye which does not affect their physiology (Cole, 1964). Spores of *Laminaria* species can be stained and allowed to settle on glass slides which can subsequently be placed into the natural habitat and the development of the germlings followed (Hsiao & Druehl, 1973). These workers followed the production of spores by *L. saccharina* in Burrand Inlet, British Columbia; spore production was greatest between May and June and again between October and December, but the gametophytes growing from these spores produced gametes throughout the year, although the shortest time taken for gametogenesis was 10–14 days (antheridia) and 12–16 days (oogonia) during March to May. During November to January these times were increased to 15–18 and 20–26 days respectively. Hsiao & Druehl believe that this finding is simply a function of total light energy available. Young sporophytes were, however, found only during February–March and September–October on natural substrata, though they could be found on the glass slides at any time of the year; Hsiao & Druehl suggest that light intensity is too great in the summer period. However, this may only apply at shallow stations and it would be interesting to check development in the lower subtidal. Kain (1964) found that gametophytes of *L. hyperborea* could certainly survive at very low light intensities (*see* also comments on p. 544) and that survival was patchy at high temperatures. Lüning & Neushul (1978) have confirmed this low light requirement for Californian *Laminaria* spp. but have also shown that 2–3 times higher light intensity is necessary to induce fertility.

Longevity of spores has rarely been studied; however the monospores of *Audouinella* (*Acrochaetium*) have been shown to remain viable some 3–4 months in the dark (White & Boney, 1969). These spores also undergo changes in shape during the first few hours after release; such activity may be important in enabling the cells to penetrate into cavities on the

Fig. 3.14. The effect of *a*, light, *b*, temperature and, *c*, salinity on the settlement of *Enteromorpha* zoospores. From Christie & Shaw, 1968.

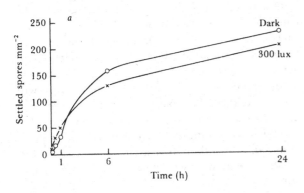

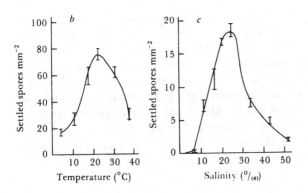

host substrata (the particular observations were on endophytic spores). The first cell division of the spore occurs after 3–4 days in culture. Spores of *L. hyperborea* remained viable for up to 40 days at 17 °C and 60 days at 5 °C. In this respect it may be an added advantage that this alga produces its sporangia during the cold months of the year (October to the beginning of April) thus allowing a long period for initial attachment and growth.

Sporeling and crustose stage. Assuming spore germination has been successful on the surface, or more likely in small crevices of the rock, the early stages of growth, either as a short filament anchored by a basal cell, or as a layer of cells 'glued' to the rock surface, will be highly susceptible to the grazing activities of myriads of animal colonists. I suspect that the life form of the early stages has a considerable bearing upon survival and some aspects of shore zonation may be related to this. The amazing feature is that any alga establishes at all when, in addition, one considers the force with which waves break on rocky shores. At this sporeling stage variations in salinity can be damaging, e.g., Yamauchi (1973) found that *Porphyra* sporelings grow normally at 17–18‰ chlorine but were abnormal at 19–22‰ chlorine.

Larger marine algae appear to be relatively indifferent to the chemical composition of the substratum (Den Hartog, 1972) but are affected by physical factors such as roughness and hardness. Den Hartog comments on the importance of porosity in retaining water, so that on some limestones the algae can colonise to a higher level on the shore than when growing on granite and still higher than on basalt (Den Hartog, 1968). In fact a whole association can be missing, e.g., on basalt the upper Cyanophyta–*Fucus spiralis* community can be absent and the lower zonation displaced downwards some 1–2 m compared to the position on concrete dykes (Nienhuis, 1969). The stability of the substrata is obviously important; moving pebbles cannot maintain an epilithic flora yet similar sized pebbles in a protected habitat will be colonised by genera such as *Enteromorpha*, *Ectocarpus* and even *Fucus*. Rocks such as chalk tend to crumble and form unstable habitats, though it is surprising how rich some of these shores are when properly surveyed, see for

instance Tittley & Price's (1978) survey of the Eastern English Channel shores.

Physiological studies on sporelings have received little attention, though they are important for an understanding of the ecology. Boney & Corner (1962, 1963) showed that optimal growth of sporelings of the red algae *Plumaria elegans* and *Antithamnion plumula* was doubled when some of the green wavelengths were removed from the light source and this suggests that, in nature, conditions would not be optimal for growth. Sundene (1973) measured growth rates of *Ascophyllum* and *Fucus* species and found that germlings of the latter grew faster and would therefore outcompete *Ascophyllum* especially on ice-scoured shores. Under cultural conditions species of *Monostroma* and *Ulva* have been shown to require organic compounds before the blades can form, but such effects have not been investigated in nature.

Little is known about the physiological effects of surfaces on spore attachment and growth. In a very neat work, Luther (1976) showed by exposing blocks of natural rock with different granular surfaces that *Enteromorpha* settled on granule sizes < 0.5 mm, *Porphyra* on granules > 0.5 mm and red algal colonisation increased with increasing granule size; he concluded that surface structure is more important than hardness. Another interesting study used monolayers of particles of different sizes cemented to acrylic discs (Harlin & Lindbergh, 1977); there were no differences in initial settlement but later growth was affected with few sporelings on the smooth plate and increasing amounts on the larger particles. *Corallina* had its optimal development on 0.1–0.5 mm granules, and although it attached to the smooth plates, development was limited to the basal system only.

In habitats where inorganic' or organic materials are in very dilute concentrations they may be adsorbed and concentrated on the surfaces. Work on bacteria (quoted in Zobell, 1972) has shown this to be important in allowing organisms to develop in media otherwise too dilute for growth. Such surface phenomena should also be studied in relation to algal colonisation. Many algae produce extracellular mucilage in which clay particles become attached and the influence of these particles on nutrient exchanges requires investigation, since such deposits are very common around the algae. Although not studied in relation to algae, clay particles have been found to stimulate bacterial respiration, nutrient uptake and enzyme reaction rates.

Whilst it is not obvious to collectors who tear the algae off the rocks, a majority of species tend to form basal pseudoparenchymatous crusts. This is so even for those which grow up initially as filaments, e.g. *Enteromorpha*. Once a crust has become established it is common for several plants to grow from the surface, e.g., this is easily seen on *Corallina*, and so grazing at this crust stage may only destroy part of the germling. Experimental removal of animals from areas of rock, followed by enclosure of the algae in cages, has shown how drastically grazing affects this community. Equally the destruction of the animals (limpets and other grazing molluscs) by the combined effects of oil and detergent after the *Torrey Canyon* accident revealed how much more widely the algae (e.g. *Enteromorpha*, *Ulva* and *Porphyra*) became established, and how richly they grew. 'An unusual carpet of vivid green covered almost the whole of the shore from about half tide level downwards. Nothing like this has been seen here before.' (Smith, 1968.) These rich growths were present less than four months after the oil was washed ashore. The crusts when once established form perennial structures from which new plants grow by purely vegetative means. Of course they get damaged, eroded, eaten and often overgrown by other algae, and by animal growths, but still often function effectively. No study of rocky shore zonation is complete without observations of these encrusting growths; often they can only be studied after chipping off fragments of rock, with a geological hammer used with discretion so as not to damage vital habitats. As Dixon & Richardson (1970) point out 'careful examination in the field is needed to discriminate between the product of spore germination and a new frond formed by regeneration from a minute permanent fragment'. Unfortunately many of the encrusting brown and red algae are extremely difficult to identify, though some progress is now being made. In this field one is very aware of the dependence of algal ecology upon advances in systematics based on wide ranging herbarium–culture–field studies.* The crustose brown algae of the Ectocarpales are particularly difficult (Ravanko, 1970*b*) and many of the characters used to separate

them in keys are untenable, for example, Ravanko considers that most of the species composing the *Ectocarpus* complex are only developmental stages of a few species. In addition, each species has several developmental stages (prostages, i.e., sexual, plethysmothalli, asexual, dwarf forms, etc.) each of which is subject to variation under the influence of environmental factors, especially the substratum and degree of wave action. In a classic study of the rocky shore algae in the Öregrund (part of the Baltic Sea), Waern (1952) concluded that the epilithic brown encrusting algae belonged to three different taxa (*Petroderma maculiforme*, *Lithodesmium subextensum* and *L. rosenvingii*) but recently Ravanko (1970*b*) has shown that these are all forms of *L. subextensum*.

The belts of algae. The growth of the sporelings on the shore soon results in roughly parallel tiers of vegetation, although in some regions of contorted topography and varied direction of wave action the vegetation may not be quite so clearly zoned. On simple physical grounds, three zones (supra-, inter-, and subtidal) would be expected with exposure to air and freshwater the dominant factor in the uppermost, alternating emersion and submersion in the middle and permanent submersion and low light intensity in the lowermost zone. These and other associated physico-chemical factors ought to produce three rather uniform zones, but uniformity is not a feature of the shore and from the uppermost limit of the supratidal to the lowermost of the subtidal a varying number of belts (often nine) can be distinguished. The causative agent in their production is almost certainly biological competition and grazing, a factor regrettably neglected and not even mentioned in Stephenson & Stephenson (1949) and only cursorily discussed in many other papers. But the random growth on cleared shores followed by the gradual development of belts is clearly produced by the operation of biological factors except perhaps on the uppermost shore (and on tropical shores) where desiccation is more important.

In some literature the shore is regarded as a

* An excellent attempt to clarify the British Phaeophyta by means of a numerical taxonomic survey has now been published by Russell & Fletcher (1975). It has thrown up some new problems and indicated areas which require detailed study.

transitional zone between the terrestrial and marine environments; in terrestrial ecology this would be an ecotone and in such ecotones mixing of species from either side is to be expected. But this hardly occurs on the shore, and no land or freshwater plants extend into the intertidal zone.* A few lichens, a moss (*Schistidium maritimum*) and a fern (*Asplenium marinum*) occur in the supratidal and occasionally the lichens penetrate slightly downwards into the intertidal; these plants may be classed as terrestrial components of the upper shore, and they and some algae then form a discrete collection of plants faithful to the supratidal. All the algal species occurring in the shore region are essentially marine, i.e. they are never found in terrestrial environments and those of the supratidal are an *upwardly* extending element from the intertidal. Further evidence for this concept is perhaps provided by Ramus, Lemons & Zimmerman (1977) who found that the normally intertidal *Fucus vesiculosus* and *Ascophyllum nodosum* actually had higher photosynthetic rates when cultured at 4 m depth than at the surface. Brackish water elements occurring along the shore are also mostly of marine origin, e.g. *Enteromorpha* spp. The only common genus which extends from brackish to freshwater conditions is *Vaucheria* (*see* p. 172), whilst a very small number of species actually extend from supratidal to freshwater (e.g. *Bangia fuscopurpurea* and *Blidingia minima*). Only at the heads of inlets, where sheltered conditions allow a slight mingling of algae and salt-marsh plants, is there any real contact between terrestrial angiosperms and intertidal algae. In fact some shore algae can and do grow *permanently* submerged below sea level in areas where tidal movements are slight and they therefore do not *need* a period of emersion in the air. Table 3.8 lists the species capable of such submergence. If we look at shore zonation beginning from this premise, i.e. that the organisms are marine (Den Hartog, 1968), then all the shore associations are upward extensions of essentially subtidal organisms. If we then heed Doty's (1957) words, and seek out the simplest

situation, we might start with a submerged zonation e.g. in the relatively enclosed, almost tideless mid-Baltic.† In such protected situations, wave action is slight, yet nevertheless there is a supratidal belt (not merely a splash zone) with lichens and some algae, and a subtidal series of belts. The first of these subtidal belts is dominated by *Cladophora glomerata* and normally extends a metre or so outwards from the shore to a depth of 0.1–0.2 m; associated species are *Enteromorpha* sp., *Pilayella littoralis* and *Ceramium tenuicorne*. In some places *Ceramium* or *Pilayella* may take over as dominants, whilst an undergrowth on the rocks is formed by the crustose species, *Hildenbrandia prototypus*, *Lithoderma fatiscens* and *Rivularia atra* (Ravanko, 1972). Below this is a *Fucus vesiculosus* community usually starting at about 0.5 m depth, extending to a depth of 1–8 m and forming a belt extending out to about 6 m from the shore. The vegetation below the *Fucus* was generally composed of angiosperms, *Potamogeton* spp., which occur in this only slightly saline water, with *Pilayella littoralis* and *Ceramium tenuicorne* growing epiphytically here; it is interesting that in one part of the shore these species are epilithic and in another epiphytic. This is not an isolated occurrence of intertidal species in the subtidal but one which has been reported from many sites where freshwater inflow results in lowered salinity and which tend, of course, to be related to ice and snow melt, so that the records tend to be from arctic–boreal situations. The phenomenon has been termed 'brackish water submergence'. These belts are independent of emersion–submersion cycles and their vertical extension is more likely to be governed by light penetration rather than any other factor; the salinity was 5–6‰. In the southern Baltic Sea the communities growing at similar salinities have been investigated using plant sociological methods (Kornas & Medwecka-Kornas, 1949); a Fuceto–Furcellarietum (*Fucus vesiculosus*, *Furcellaria fastigiata*) occurs and nearby is found a community more frequently associated with fresh–brackish waters, Chareto–Tolypelletum (*Chara aspera* and *Tolypella*

---

* The only organisms of the intertidal zone which have been derived from the terrestrial sphere are some Arthropoda, Rotifera and Oligochaeta.

† In such areas, changes in barometric pressure cause quite appreciable water-level fluctuations, e.g., about 13 cm for a 10 mm change in barometric pressure. The species occurring in the lowest 'terrestrial' zone

(geolittoral) designated by Scandinavian workers (e.g. Du Rietz, 1950; Waern, 1952) are identical to those in the supratidal elsewhere. The uppermost 'marine' belt, hydrolittoral of Scandinavian workers is uncovered in the spring (covered with ice in winter) and submerged in summer. This summer subtidal state merely submerses what would normally be an intertidal flora.

Table 3.8. *Species confined to the littoral complex but varying in relation to sea water flooding. Modified from Den Hartog, 1968*

a. Species of algae occurring on the upper shore and not exhibiting submergence in brackish water

Supratidal (Den Hartog's supralittoral) Intertidal (Upper) (Den Hartog's eulittoral)

Chlorophyceae
  Blidingia marginata
  Monostroma groenlandicum
  Prasiola stipitata
  Rosenvingiella constricta
  R. polyrhiza
  Urospora hartzei
  U. penicilliformis
Phaeophyceae
  Waerniella lucifuga
Rhodophyceae
  Bangia fuscopurpurea

Cyanophyceae
  Calothrix scopulorum
  Entophysalis deusta
  Lyngbya confervoides
  L. majuscula
  L. semiplena
  Microcoleus tenerrimus
  Plectonema battersii
  Schizothrix calcicola
  Symploca atlantica

Chlorophyceae
  Blidingia minima
  Capsosiphon fulvescens
  Monostroma oxyspermum
Rhodophyceae
  Bostrychia scorpioides
  Catenella repens
  Porphyra elongata

b. Species which require occasional submergence in undiluted seawater to survive

Intertidal

Phaeophyceae
  Fucus spiralis
  F. virsoides
  Mesospora macrocarpa
  Nemoderma tingitanum
  Pelvetia canaliculata
  Petalonia zosterifolia
  Bifurcaria bifurcata
  Cystoseira myriophylloides
  Elachista scutulata
  Herponema velutinum
  Himanthalia elongata
Rhodophyceae
  Porphyra linearis
  Callithamnion arbuscula
  C. tetricum

  Caulacanthus ustulatus
  Ceramium deslonchampsii
  C. shuttleworthianum
  Erythrotrichia welwitschii
  Gelidium spinulosum
  Lithophyllum tortuosum
  Nemalion helminthoides
  Polysiphonia lanosa
  Porphyra leucosticta
  P. purpurea
  P. umbilicalis
  Rissoella verruculosa
Cyanophyceae
  Rivularia mesenterica
  R. bullata

c. Species which are capable of living completely submerged

Intertidal

Chlorophyceae
  Monostroma grevillei
  Tellamia contorta
Phaeophyceae
  Ascophyllum nodosum
  Elachista fucicola
  Fucus distichus
  F. serratus
  F. vesiculosus
  Fucus vesiculosus

  Pilayella littoralis
  Ralfsia verrucosa
  Spongonema tomentosa
Rhodophyceae
  Callithamnion scopulorum
  Dumontia incrassata
  Gigartina stellata
  Plumaria elegans

Fig. 3.15. *a*, shallow-water (0.5 m) plant of
Laminaria digitata and, *b*, a plant of the ecad *L.
digitata* f. *cucullata* developed under calm-water
conditions. From Svendsen & Kain, 1971.

nidifica). Later studies revealed further subtleties and the communities were split,. the Chareto–Tolypelletum into a Characetum balticae and a Tolypelletum nidificae and the Fuceto–Furcellarietum into two facies, a Fuceto–Furcellarietum facies with *Potamogeton*, and a Zostero–Furcellarietum with *Zostera* (Kornas, Pancer & Brzyski, 1960). Low salinity and shelter confine *Laminaria digitata* to the subtidal in Loch Etive (Powell, 1972) and here the plant is morphologically modified; the lamina becomes cape-like and brittle and the stipe and haptera poorly developed. The same modifications occur in *Saccorhiza* growing in calm water in Loch Ine and this effect is I believe more related to lack of water movement than lowered salinity (author's observations). These are phenotypic modifications and the cape-like forms have been termed *cucullata* varieties. Fig. 3.15 shows a normal plant of *L. digitata* and a *cucullata* variant (Svendsen & Kain, 1971). Another modification under reduced salinity is the increase in vesiculation in *Fucus vesiculosus* (Jordan & Vadas, 1972). Where there is still no tide, but strong winds cause the water level to fluctuate, a much wider 'supratidal' belt begins to appear (cf. Du Rietz, 1925, 1932; Levring, 1940; Waern, 1952) and in extreme cases it ought to be possible to detect the beginning of an 'intertidal' belt.

Information is required on algal distribution in tideless, but fully saline, situations. Whilst the algal ecologist is dealing with belts of algae on the shore, it must be noted that on many shores belts occur in which algae are absent, e.g. there is frequently a zone above *Pelvetia* in which no algae occur and below the *Pelvetia* there may be belts occupied only by animals (plus, as yet, relatively uninvestigated microscopic algae). The reasons for the absence of algae from such belts should also be the concern of the algal ecologist. That there is a zonation on the shore and not merely a continuum is clearly shown in Table 3.9 (Russell, 1973). This is an excellent example of how to study a rocky shore and shows a distinct upper shore community (i.e. supratidal) and a less distinct lower intertidal which is really the upward extension of the subtidal flora. The method of recording in quadrats and determination of 'minimal area' (Fig. 3.16) is similar to that used in higher plant ecology, e.g., by estimation of cover, with classes *1–5* representing cover of 0–20%, 21–40%, etc. (Taniguti, 1962; Russell, 1973). Fig. 3.17 shows comparison of quadrats by Sorensen, Gleason-cover and Gleason-frequency indices and the latter in particular emphasises the three zones. Fig. 3.17 also shows an example of cluster analysis of these data and illustrates the difficulty of distinguishing zones by this method. Even here it is possible to discern upper (with *F. spiralis*), mid (with *F. vesiculosus*), and lower (with *F. serratus*) intertidal belts albeit with some overlap. The recognition of three belts within the intertidal is widespread, e.g., Womersley & Edmonds (1952) and King (1973) also divide Australasian shores in this way.

Many more such detailed analyses are needed, followed by a synthesis and comparison of the data from each zone and of the many finer subdivisions which can be recognised from species presence. For example associations can be characterised by *Fucus spiralis*, *F. vesiculosus*, *Cladophora rupestris*, etc., and, despite some species overlap obvious from the matrices of Fig. 3.17, they are worthy of detailed analysis and comparison at different sites. Russell (1973) showed that the upper shore community was very distinct; it contains species characteristic of the supratidal (*see* list of Den Hartog, p. 79) and confirms his view that it is 'a separate biological entity'). It is the lower boundary of this upper shore community which corresponds to the 'litus' line used by North European phycologists and defined by Sjöstedt (1928) as 'the ecological boundary line between land and sea'. The litus line has been used as a base line to describe rocky shore zonations in Scandinavia and later research shows the validity of the concept.

The individual regions of the shore will now be considered in greater detail.

The supratidal region. The supratidal belt on an exposed coast is determined by spray and splash from wave action and in its lowermost part by occasional inundation by high waves. There is, therefore, the necessary saline 'climate' for the essentially marine species which can extend upwards, in some instances many metres above the zone normally reached by the tides e.g., in the Faroes and other Northern Islands. The greater the 'storminess' of the sea the higher the zones are 'raised'. Operating also on this supratidal zone are two important factors which undoubtedly lead to impoverishment of the flora since few marine algae can withstand their effects.

Table 3.9. Table of species and abundance in ten quadrat samples from five sites at Port St Mary, Isle of Man. Abundance expressed as percentage frequency (F) and cover (C). × denotes uncertain abundance. From Russell, 1973

Quadrats	1		2		3		4		5		6		7		8		9		10	
	F	C	F	C	F	C	F	C	F	C	F	C	F	C	F	C	F	C	F	C
Verrucaria maura	×	×	×	×	—	—	—	—	—	—	—	—	—	—	—	—	—	—	—	—
Porphyra linearis	88	1	72	1	—	—	—	—	—	—	—	—	—	—	—	—	—	—	—	—
Prasiola stipitata	92	2	88	1	—	—	—	—	—	—	—	—	—	—	—	—	—	—	—	—
Entophysalis deusta	92	2	96	2	—	—	—	—	—	—	—	—	—	—	—	—	—	—	—	—
Lyngbya lutea	92	1	100	1	—	—	—	—	—	—	—	—	—	—	—	—	—	—	—	—
Lyngbya semiplena	64	1	24	1	—	—	—	—	—	—	—	—	—	—	—	—	—	—	—	—
Chthamalus stellatus	16	1	20	1	—	—	—	—	—	—	—	—	—	—	—	—	—	—	—	—
Bangia fuscopurpurea	4	1	—	—	—	—	—	—	—	—	—	—	—	—	—	—	—	—	—	—
Fucus spiralis	—	—	—	—	100	4	80	1	56	2	36	1	—	—	—	—	—	—	—	—
Balanus balanoides	—	—	—	—	100	2	100	4	100	3	100	4	100	5	84	3	—	—	—	—
Patella spp.	—	—	—	—	8	1	8	1	36	1	40	1	32	2	24	1	16	1	8	1
Rivularia atra	—	—	—	—	—	—	40	1	—	—	—	—	—	—	—	—	—	—	—	—
Littorina saxatilis	—	—	—	—	—	—	—	—	8	1	—	—	—	—	—	—	—	—	—	—
Fucus vesiculosus	—	—	—	—	—	—	—	—	56	2	24	1	88	5	88	3	—	—	—	—
Elachista fucicola	—	—	—	—	—	—	—	—	12	1	24	1	—	—	—	—	—	—	—	—
Spongonema tomentosum	—	—	—	—	—	—	—	—	32	1	—	—	16	1	52	1	—	—	—	—
Urospora speciosa	—	—	—	—	—	—	—	—	16	1	—	—	4	1	4	1	—	—	—	—
Enteromorpha compressa	—	—	—	—	—	—	—	—	4	1	8	1	4	1	4	1	—	—	—	—
'Lithothamnia'	—	—	—	—	—	—	—	—	24	1	—	—	—	—	32	1	100	5	100	5
Corallina officinalis	—	—	—	—	—	—	—	—	—	—	16	1	—	—	24	1	96	1	76	1
Ulva lactuca	—	—	—	—	—	—	—	—	—	—	4	1	32	1	20	1	12	1	4	1
Ectocarpus siliculosus	—	—	—	—	—	—	—	—	—	—	—	—	8	1	4	1	—	—	—	—
Littorina littoralis	—	—	—	—	—	—	—	—	—	—	—	—	8	1	—	—	—	—	—	—
Cladophora rupestris	—	—	—	—	—	—	—	—	—	—	—	—	4	1	4	1	8	1	40	1
Rhodymenia palmata	—	—	—	—	—	—	—	—	—	—	—	—	36	1	16	1	68	1	56	1
Actinia equina	—	—	—	—	—	—	—	—	—	—	—	—	—	—	12	1	—	—	—	—
Nucella lapillus	—	—	—	—	—	—	—	—	—	—	—	—	—	—	12	1	—	—	—	—
Cystoclonium purpureum	—	—	—	—	—	—	—	—	—	—	—	—	—	—	8	1	—	—	—	—

Acrosiphonia arcta	–	–	–	–	–	–	–	–	–	–	–	–	–	–	4	1	8	1	–	–
Gigartina stellata	–	–	–	–	–	–	–	–	–	–	–	–	–	–	4	1	–	–	–	–
Ceramium rubrum	–	–	–	–	–	–	–	–	–	–	–	–	–	–	16	1	–	–	–	–
Chaetomorpha sp.	–	–	–	–	–	–	–	–	–	–	–	–	–	–	4	1	–	–	–	–
Callithamnion arbuscula	–	–	–	–	–	–	–	–	–	–	–	–	–	–	4	1	–	–	4	1
Himanthalia elongata	–	–	–	–	–	–	–	–	–	–	–	–	–	–	24	1	84	1	48	1
Laurencia pinnatifida	–	–	–	–	–	–	–	–	–	–	–	–	–	–	24	1	12	1	4	1
Gelidium latifolium	–	–	–	–	–	–	–	–	–	–	–	–	–	–	4	1	4	1	4	1
Fucus serratus	–	–	–	–	–	–	–	–	–	–	–	–	–	–	–	–	80	1	64	1
Lomentaria articulata	–	–	–	–	–	–	–	–	–	–	–	–	–	–	–	–	20	1	20	1
Halichondria panicea	–	–	–	–	–	–	–	–	–	–	–	–	–	–	–	–	48	1	44	1
Chondrus crispus	–	–	–	–	–	–	–	–	–	–	–	–	–	–	–	–	4	1	–	–
Membranoptera alata	–	–	–	–	–	–	–	–	–	–	–	–	–	–	–	–	8	1	8	1
Laminaria digitata	–	–	–	–	–	–	–	–	–	–	–	–	–	–	–	–	100	5	100	5
Ectocarpus fasciculatus	–	–	–	–	–	–	–	–	–	–	–	–	–	–	–	–	80	1	52	1
Laminariocolax tomentosoides	–	–	–	–	–	–	–	–	–	–	–	–	–	–	–	–	76	1	44	1
Patina pellucida	–	–	–	–	–	–	–	–	–	–	–	–	–	–	–	–	–	–	8	1
Plumaria elegans	–	–	–	–	–	–	–	–	–	–	–	–	–	–	–	–	–	–	8	1
Rhodochorton floridulum	–	–	–	–	–	–	–	–	–	–	–	–	–	–	–	–	–	–	4	1
Cladostephus spongiosus	–	–	–	–	–	–	–	–	–	–	–	–	–	–	–	–	–	–	4	1

These are desiccation from sun, heat and wind and large variations in surface salinity: increases in surface salinity occur as water evaporates, and at other times there is greatly decreased surface salinity when, for example, rain, snow and dew settle on the algae. These terrestrial factors, superimposed on the marine factors operating from below, create one of the most severe habitats one can imagine, a feature reflected in the small number of species in the associations. This is in fact a brackish environment, i.e., one with fluctuating salinity–emersion–submersion features, comparable to a salt marsh, or the edge of a dune system; in the latter habitats species also occur in parallel belts and the associations are species-poor. A facet which requires further investigation is the grazing of the supratidal zone, for under certain conditions intertidal animals may invade from below and at other times terrestrial animals invade from above. Some of the species growing in the supratidal belt do not penetrate below the water surface in any environment and one must conclude that they have an absolute requirement for exposure to air (Den Hartog, 1968). Table 3.8 is based on Den Hartog's paper, which is one of the most interesting modern discussions of the border between land and sea.

In very protected regions, e.g., in the inner parts of inlets or on the leeward side of islands in places where wave action is minimal, marine algae such as *Bostrychia* grow in a turf-like manner amongst the salt marsh flora (e.g., amongst *Armeria maritima* and *Festuca rubra*).

Fig. 3.16. 'Minimal area' graphs of species present in quadrat samples; *a.* from the upper shore community dominated by *Prasiola stipitata* and *b*, a mid-shore community dominated by *Fucus vesiculosus*. From Russell, 1972.

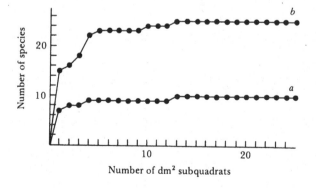

In the supratidal (and also below) the blue-green alga *Calothrix scopulorum* has been shown to fix nitrogen under field conditions (Stewart, 1965, 1967). Fixation was greatest in spring and autumn but reduced in summer when the alga became desiccated and peeled off the rock, and again in winter when low temperatures reduced fixation. Jones (1974) estimated the amount of nitrogen fixed to be some 2.5 g m^{-2} yr^{-1} in areas uniformly covered by algae and showed that the fixed nitrogen was rapidly made available to other algae.

The supratidal is also conspicuous as a detrital zone, with material drifting in from above and below, decaying and providing a rich source of organic matter. In some regions it is a place where sea birds congregate and their faeces enrich the habitat so that algae such as *Prasiola* grow; interestingly, this grows also on rocks projecting from lakes when they are visited by birds.

The supratidal habitat seems to support a similar flora throughout many latitudes e.g., at Amchitka Island between the Bering Sea and Pacific Ocean the very common *Prasiola–Rhodochorton–Porphyra–Ulothrix* communities occur one below the other (Lebednik, Weinmann & Norris, 1971). At a similar position at Hilbre Island off the mid-west coast of England, Russell (1971) recorded *Prasiola stipitata, Blidingia marginata, B. minima, Ulothrix* and *Bangia fusco-purpurea*.

On calcareous rocks the supratidal often has an extensive community of endolithic Cyanophyta (Gilet, 1954; Golubić, 1969). This community is well developed in the region splashed by waves on some tropical coasts e.g., the coral limestones of Curacao are perforated by the *Hyella*-phase of *Entophysalis deusta*, whilst *Gloeocapsa*-like and *Hormathonema*-like plants occur abundantly on the surface rock (van den Hoek, 1969).

The intertidal region. Whereas tidal movement itself was of little consequence to the supratidal associations it is obviously or prime importance in the intertidal zone and on purely theoretical grounds one would expect belts of organisms to relate in some way to length of time of emersion–submersion. Confusion has arisen in the literature where it is stated either that the tide controls or does not control zonation: in almost all instances it is clear that authors do not regard the tidal flow, but the

Fig. 3.17. *a*, similarity matrices of the whole quadrat samples using coefficients of Sörensen (Q_s), Gleason cover (Q_{Gc}), and Gleason frequency (Q_{Gf}) on the data in Table 3.9. *b*, a cluster analysis of the data based on highest χ^2 values of species presence in 250 subquadrats. Cluster a corresponds to the supratidal (littoral fringe in Russell), cluster b to the intertidal and cluster c to the species of the subtidal which penetrate upwards into the lower intertidal. Horizontal double lines indicate negative correlations at the appropriate χ^2 levels. Key to species: 1, *Porphyra linearis*; 2, *Prasiola stipitata*; 3, *Lyngbya lutea*; 4, *Entophysalis deusta*; 5, *Fucus spiralis*; 6, *Balanus balanoides*; 7, *Patella*; 8, *Fucus vesiculosus*; 9, *Lithothamnia*; 10, *Corallina officinalis*; 11, *Himanthalia elongata*; 12, *Laminaria digitata*; 13, *Fucus serratus*. From Russell, 1973.

a

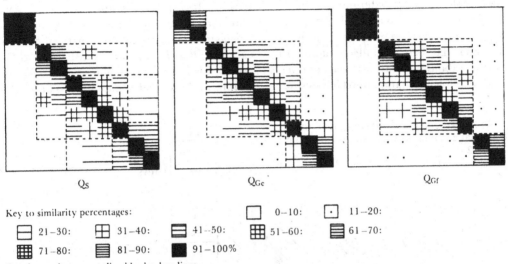

Key to similarity percentages:

0–10: 11–20: 21–30: 31–40: 41–50: 51–60: 61–70: 71–80: 81–90: 91–100%

Zonal groupings are outlined by broken lines.

b

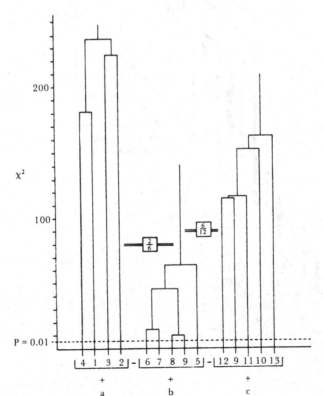

covering–uncovering by seawater as the controlling factor. Since some of the algae of the intertidal zone can live completely submerged in brackish waters (e.g. the Baltic Sea) they have no absolute requirement for emersion and the zonation on the rocky shore is related to a complex of factors, only one of which is their ability to withstand varying periods of exposure to air. Other algae, however (Table 3.8), neither grow completely out of water nor completely submerged. These can be considered the prime intertidal species and they tend to occur in the upper part of the intertidal belt. Detailed studies are needed of the metabolism of various kinds of intertidal algae. The simplest situation in which to study

them would be an evenly sloping shore which was so protected that wave action was at a minimum, e.g., at the head of a long sea loch protected from excessive wind by hills. Lewis (1972) studied such a site at Clachan Sound, Argyll (Fig. 3.18). Here the lowermost lichen (*Verrucaria*) belt is submerged by mean spring tides and in the intertidal there is a characteristic zonation of *Pelvetia canaliculata*, *Fucus spiralis*, *Ascophyllum nodosum*, with *Fucus vesiculosus* appearing in the lower intertidal, together with *Fucus serratus* in its 'normal' position on the shore. With the exception that *F. vesiculosus* is rather low at Clachan Sound this is a very common sequence on northern European shores and apparently has little to do with

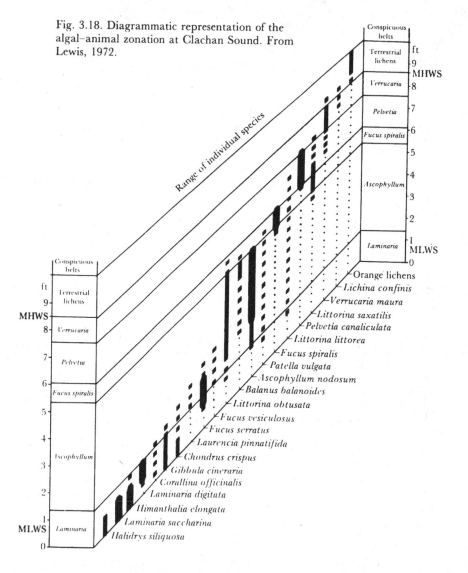

Fig. 3.18. Diagrammatic representation of the algal–animal zonation at Clachan Sound. From Lewis, 1972.

Table 3.10. *Development of* Fucus vesiculosus *germlings on various substrata, 15 °C 12 h light and dark alternating. Measurements (in mm) taken after 46 days in Petri dishes. From Barnes & Topinka, 1969*

Algal growth	*Balanus balanoides* Shell	*Balanus balanoides* Dish	*Mercenaria mercenaria* Shell	*Mercenaria mercenaria* Dish	Marble Marble	Marble Dish	Rock Rock	Rock Dish	Dish alone
Mean length of thallus	2.1	2.1	4.0	—	1.1	1.3	1.8	1.5	1.1
Mean number of rhizoids/germling	32.2	31.0	35.9	—	19.3	23.0	14.8	26.0	9.8
Mean total rhizoid length of germlings	166.3	176.3	171.2	—	59.0	78.3	33.2	86.7	25.6
Mean length of rhizoid	5.16	5.68	4.77	—	3.06	3.40	2.24	3.33	2.61
Mean no. thalli/germling	1.47	1.50	7.71	—	1.08	1.33	1.47	1.67	1.0

wave action; arguments suggesting such action as a *primary* factor in the zonation are not valid. A good account with lists of all the common species for a European shore is given by Gibb (1938) whilst rather similar zonations are described from Nova Scotia with distinct *Fucus* and *Ascophyllum* zones (Mann, 1972c) and from Massachusetts (Lamb & Zimmermann, 1964) with *Calothrix scopulorum* → *Porphyra linearis*–*Bangia fuscopurpurea* → *Fucus vesciculosus* → *Ascophyllum nodosum* → *Chondrus crispus* → *Gigartina stellata* → *Fucus edentatus*. Obviously on exposed shores the forces of wave action must operate on the flora. These forces superimpose a further stress on the primary emersion–submersion factor the effect of which can be most easily detected in long, relatively enclosed fjords (*see* for example the data in Sundene, 1953). Intensity of wave action varies according to the topography of the intertidal zone, and often the most affected zones are devoid of macroscopic algae. Evans (1947) considered that the wave-shock effect was greatest in the mid-intertidal yet this is just the position where the apparently susceptible *Ascophyllum* grows. It is apparent that wave force damages some of the frondose algae either directly, or after the plants have been weakened by grazing animals. However, in these belts, epilithic animals are often abundant and grazing and competition from the animals is also an excluding factor for the algae. As Barnes & Topinka (1969) showed, algae are more easily removed from animal shells than from rock surfaces and, therefore, in rough water one would expect an animal dominated zone. These workers used a spring balance attached just above the holdfast

and measured the force required at right angles to pull *Fucus* off rocks and barnacles. The plants attached to the shells always came away with pieces of shell, which had been penetrated by rhizoids and weathered. In experiments using marble, they found that the holdfasts would slightly etch the rock: this may however be caused by bacteria growing on the carbohydrate secreted by the alga. Table 3.10 shows that growth of germlings is better on shells than on rock and certainly it is common to find shells, such as limpets, densely coated with algae in nature. This work complements that of Burrows & Lodge (1950), who found that on cleared shores new *Fucus* plants were lost because they attached to barnacles rather than rocks. It would be interesting to determine any 'symbiotic' relationship between the shells or nutrients from the animals and the growth of the young germlings, especially since the growth on dishes without any shell or rock is clearly much less than the growth on dishes containing other substrata also.

Such differential removal also affects the age structure of the communities, those on rock being composed of longer living, older, heavier, and larger plants. This feature had been demonstrated earlier by Burrows & Lodge (1950) who found that the *Fucus* plants growing on barnacles had an annual cycle, but on rock surfaces, 2–3 year old plants were present.

On shores strewn with loose boulders, and protected from excessive wave action, a very rich intertidal flora can develop with the boulders covered by the same species as on stable rock. On similar shores exposed to more intense wave action, the upper and mid zone can be prevented from developing and only the lower *Fucus serratus* zone is present, giving a truncated zonation. On smaller stones only *Fucus vesiculosus* seems able to attach and the typical mid-intertidal *Ascophyllum* is absent.

In the upper region of protected shores (e.g., in protected lochs in Scotland (Gibb, 1950; see also Brinkhuis, 1976) the slight tidal movement allows the formation of loose-lying aggregates of Fucoids, e.g. *Ascophyllum nodosum* f. *mackaii*, etc. Desiccation of these populations results in a considerable decrease in photosynthesis and probably limits their upper limit (Brinkhuis, Tempel & Jones, 1976). The rolling movement of the water results in almost ball-like thalli in some situations; similar balls of *Cladophora*, *Bryopsis* etc. are sometimes found in fully submerged habitats (*see* also p. 185) and *Ascophyllum nodosum* f. *mackaii* can also occur fully submerged. According to Powell (1963) all the British fucoids can grow as loose lying aggregates. These are merely environmental forms of species, which elsewhere grow in a more fan-like manner, and are a further illustration of the effect of environment on growth form (Brink-

huis & Jones, 1976). This environmental control of form is emphasised by the finding of a gradient of morphology between the species and the ecad of *Ascophyllum nodosum* on North American shores (Chock & Mathieson, 1976). Free-living ecads of the branching filamentous *Ectocarpus* are also recorded and shown to be forms of attached taxa (Russell, 1967*b*).

One of the most extensive studies of the biology of intertidal algae is that of Knight & Parke (1950), devoted to *Fucus vesiculosus* and *F. serratus*, which is a model of what is required for other species; Knight & Parke show the variation in form on different shores and Fig. 3.19 is a plot of the length–dichotomy relationships from this paper. Growth rates were shown to vary from station to station, in the case of *F. serratus* increasing from 0.49 cm week^{-1}, on the Devon coast, to 0.68 cm week^{-1}, on the Isle of Man, and to 0.85 cm week^{-1} on the Argyll coast: shelter from rough water enhances growth rate. Such rates compare well with data from *F. spiralis* in North America where an average 1.2 cm month^{-1} has been recorded (Niemeck & Mathieson, 1976). Another interesting aspect which the work of Knight & Parke deals clearly with is the little appreciated 'defoliation' which occurs after fruiting and involves necrosis of the tissues. Although these two common plants occur close to one another on shores they have alternating reproductive and vegetative periods and a complexity which is best summed up in the concluding paragraph of Knight & Parke: 'It is interesting to reflect that in the course of a long evolution in the sea, such relatively "simple" marine plants as these two species of *Fucus* have developed growth rhythms, habits of branching, controlled defoliation after fruiting and apical control of remote parts of subtending tissue, in a degree comparable to that shown by land plants with an entirely different history.'

The intertidal flora of temperate–cool shores tends to be very dense (e.g., along the maritime provinces of Canada and New England) whilst further north the flora is sparse owing to the scouring action of ice, and permanent communities occur only in protected rock niches where only the dwarf species can survive, e.g., *Rhodochorton purpureum*, *Hildenbrandia prototypus* and *Ralfsia fungiformis*. Wilce (1959) is one of the few phycologists who have spent long periods in inhospitable northern habitats and have provided

Fig. 3.19. Length of thallus and number of dichotomies in *Fucus vesiculosus*: A. From Church Reef, Wembury; B. Port Erin, Isle of Man; C. Sgeir Bhuidhe off the west coast of the island of Luing, Argyll. From Knight & Parke, 1950.

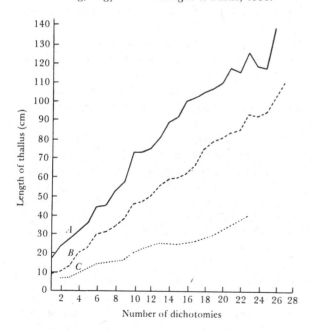

invaluable data on the flora; others may be deterred by his quoting of Kjellmann (1883): 'The vegetation of the Arctic Sea lacks variety, not only in form, but in colour. The general tone is gloomy; the dark brown colour of the Laminariaceae is the prevailing one'. In the summer, however, a population of small, annual species covers the rocky intertidal and adds some colour (*Prasiola crispa, Pilayella littoralis* and *Spongomorpha arcta*). On rocks on mud flats and in protected bays a sparse flora of larger species such as *Ascophyllum nodosum, Fucus vesiculosus, F. distichus*, subsp. *evenescens* can occur. At the highest levels, on moderately exposed shores, a band of *Prasiola crispa* occurs and mixed in with it are some Cyanophyta which also tend to form a belt below the *Prasiola*. Unlike the situation further south the mid zone of moderately exposed shores does not support *Ascophyllum nodosum* but in the mid and lower intertidal a late summer sward of *Pilayella, Spongomorpha, Chordaria* and *Scytosiphon* develops. On even more wave-swept Arctic coasts no algal vegetation occurs, except in the cracks of the rock. In addition to the ice scour, cold and lack of light, this flora also has to withstand the almost freshwater conditions during the summer ice melt (Ellis & Wilce, 1961).

At a similar latitude in the Pacific, Lebednik *et al.* (1971) found an intertidal zonation with an extreme upper belt of *Fucus distichus*, an upper-mid *Hedophyllum sessile–Halosaccion glandiforme* and a lower belt of *Alaria crispa*. This is an interesting example since the shore is very flat and yet there is a clear zonation of species (Fig. 3.20*a*). Another example of a three-part zonation from the north Pacific is given in Fig. 3.20*b*, and shows the variation superimposed on this pattern at three sites in Japan (Chihara & Yoshizaki, 1972). The upward penetration of the *Laminaria* zone and the different forms present in each zone are clearly shown at each site.

The very common three-part intertidal zonation of temperate shores is replaced by a two-part system in the Mediterranean, e.g., between Nice and the Italian border, Gilet (1954) found an upper zone (upper mesolittoral) with *Rivularia atra* and *Nemalion helminthoides* and a lower zone of coralline algae, *Neogoniolithon notarisii* and *Tenarea tortuosa*. Pérès & Ricard (1957) make a similar subdivision but in the upper zone record an uppermost endolithic Cyanophyceae zone and below this, belts of *Bangia fuscopurpurea, Porphyra leucosticta* and *Risoella verrucu-* *losa*. In the lower zone they record all the species Gilet partitioned between his two zones. This compaction of the intertidal belts seems to increase still further in the tropics, e.g., along the shore of Curacao (van den Hoek, 1969) there are just two zones, an upper of Chlorophyceae (various subdivisions of this can be discerned e.g. *Enteromorpha compressa* type, *Cladophoropsis membranacea–Cladophora latevirens*, and even one dominated by Phaeophyta, *Ectocarpus breviarticulatus, 'Giffordia duchassaingiana*) and a lower Rhodophycean turf community (three subdivisions: *Laurencia papillosa–Hydroclathrus clathratus–Padina gymnospora, Laurencia papillosa–Gelidiella acerosa, Laurencia papillosa–Ceramium tenerrimum*). Similar low growing communities are common on tropical shores and contrast markedly with the largely 'fucoid' dominated flora of temperate zones.

Along many shores there are sites which, although affording rocky substrata for the attachment of epilithic algae, are nevertheless basically sandy shores where the sand is moved on to or off the rock. Such sites are frequent along the Pacific coast of North America, where the epilithic flora can at certain times of the year be completely covered by drifting sand. Some algae survive months of such burial and recommence growth when currents wash away the sand. Few studies have been made on the ecology of such forms but Markham (1973) showed that *Laminaria sinclairii* grew best in the lowermost zone of the intertidal on the Pacific Coast, where it was subjected to the greatest surf action and sand burial. Burial starts in April and continues throughout the summer until autumn gales again remove the sand. Growth however is greatest in early summer, prior to burial of the holdfasts and stipe; the lamina usually remains above the surface of the sand. The blade growth during the spring and early summer is new material which arises from the meristematic zone of the split stipe in January. In the autumn, after uncovering, there is little growth and around November the blades begin to be lost by erosion of disintegrating tissue. It is interesting that this occurs also in laboratory cultures and is linked with internal disorganisation of tissue, the total result of which is akin to leaf fall in deciduous trees. Sori develop twice a year, on new blades in the spring and on old in the autumn, but although spores from both fruiting periods produce gametophytes, those from the old blades do not usually lead to sporophytes.

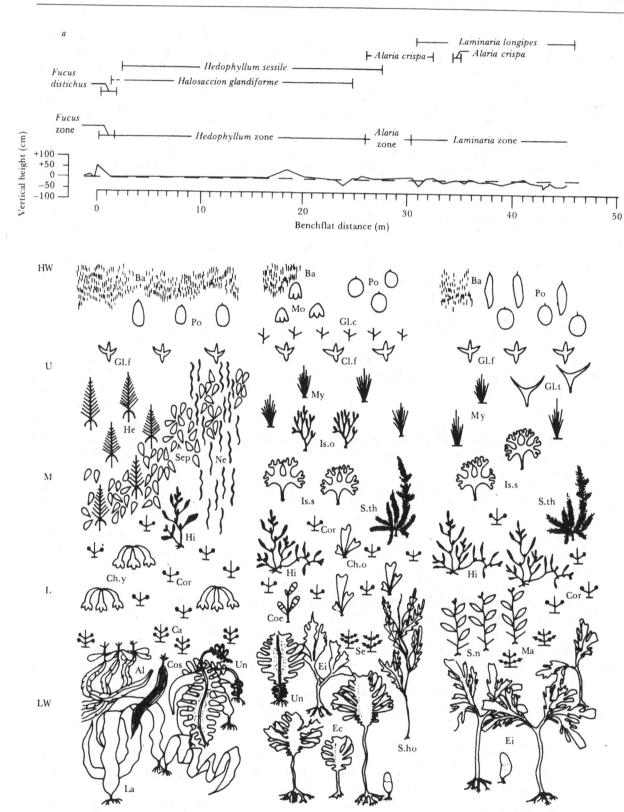

Markham records a number of other, smaller, algae growing on the same rocks and it would be instructive to study their annual cycles also. Daly & Mathieson (1977) noted the interesting points that in areas of sand movement the species diversity was low, the species tended to be tough and wiry (e.g., *Ahnfeltia, Gymnogongrus, Gigartina, Phyllorphora, Sphacelaria* and *Chaetomorpha*), vegetative regeneration was common and the species were often those with incomplete alternation of generations.

Growth studies of large Laminarians must now take into account that translocation of photosynthate can occur (Schmitz, Lüning & Willenbrink, 1972) and that this takes place in the basipetal direction with accumulation in the meristematic zone and growing hapteron. Mannitol (53%), amino acids (45%) and malate (2%) were the main translocated substances and these are of course the major products of photosynthesis in the brown algae. Schmitz & Lobban (1976) showed that this activity is widespread in at least 13 genera of Laminariales and that it is important for the supply of photosynthate from old blades to developing and/or regenerating new blades. Steinbiss & Schmitz (1973) in an elegant series of experiments proved that the transport only occurred in the trumpet hyphae and that the young hyphae were most active.

Australasian shores also tend to exhibit a very sparse flora in the upper intertidal zone where only *Bostrychia* seems to penetrate. The mid-intertidal is still only sparsely vegetated with *Rivularia* spp. and *Splachnidium rugosum* and a little *Hormosira* extending from below. The lower intertidal is the only rich zone and here the characteristic Australasian Fucoid *Hormosira banksii* is the most striking species on many shores (*see* Fig. 3.54 for this species and for illustrations of some other tropical forms), together with an algal 'turf' which varies somewhat in composition according to latitude but contains *Corallina, Jania, Laurencia, Centroceras, Gigartina, Wrangelia, Hypnea,*

Spyridia, Bryopsis, Padina, Leathesia, etc. (for details and further references *see* Womersley, 1948; Dellow, 1950; Cribb, 1954, 1973; Batham, 1956, 1958; Womersley & Edmonds, 1958; King, 1973). *Hormosira* can form loose-lying populations in mangrove swamps (Bergquist, 1960*b*) just as *Ascophyllum* does on northern salt marshes. In South Africa the lowest intertidal has a vegetation of encrusting corallines forming a basal layer and growing upwards one finds *Champia lumbricalis, Gigartina striata* and *G. radula* with *Champia* occupying the seaward zone. Above this, *Iridaea capensis* grows together with *Caulacanthus ustulatus* and *Splachnidium rugosum* (Isaac, 1937, 1949). Desiccation of the shore in the mid-intertidal often leaves a zone devoid of algae except in the protected cracks, but as Isaac points out it is difficult to equate the bareness of the zone with tidal conditions, since it is covered at each tide. Above the bare zone, a belt occupied by *Chaetangium saccatum* and *C. ornatum* occurs, whilst the uppermost zone, as on many shores, is dominated by a *Porphyra* species (*P. capensis*). At some sites there is an additional zone dominated by *Bifurcaria brassicaeformis* between the *Ecklonia* and the *Iridaea* associations and a *Gelidium pristoides* association above the *Iridaea*. Isaac distinguished the associations of algae in each belt and named them from the dominants.

An interesting aspect of intertidal zonation has been revealed by Lawson (1957) in a study of Ghanaian algae. He has shown that on the coast of Ghana which has semi-diurnal tides the lower low waters occur mainly in the day and the higher low waters during the night when the sun is south of the equator (September–March), whilst after the equinox, when the sun is in the northern hemisphere, the position is reversed and this reversal affects the vertical extent of the belts of algae on the lower part of the shore. It was shown (Fig. 3.21) that *Hypnea* 'migrates' up the shore during the months of the northern summer and 'retreats' down the shore

Fig. 3.20. *a*, the zonation on a very flat shore at Amchitka Island (51° 25′ N). From Lebednik *et al.,* 1971; *b*, the distribution of intertidal algae on three shores Rikuchu Kaigan, Izu and Tsushima in Japan. Ba, *Bangia fuscopurpurea*; Po, *Porphyra* spp., Mo, *Monostroma nitidum*; Gl.c, *Gloiopeltis complanata*; Gl.f, *Gloiopeltis furcata*; Gl. t, *Gloiopeltis tenax*; He, *Heterochordaria abietina*; My, *Myelophycus simplex*; Is.o, *Ishige okamurai*; Is.s, *Ishige sinicola*; Ne, *Nemalion vermiculare*; Sep, *Septifer* sp.; Hi, *Hizikia fusiforme*; S.th, *Sargassum thunbergii*; Ch.y, *Chondrus yendoi*; Ch.o, *Chondrus ocellatus*; Cor, *Corallina pilulifera*; Coe, *Coeloseria pacifica*; Ca, *Calliarthron yessoense*; Se, *Serraticardia maxima*; My, *Marginosporum crassissima*; S.ho, *Sargassum horneri*; Al, *Alaria crassifolia*; Cos, *Costaria costata*; Un, *Undaria pinnatifida*; S.n, *Sargassum nigrifolium*; Ei, *Eisenia bicyclis*; Ec, *Ecklonia cava*; La, *Laminaria* spp. From Chihara & Yoskizaki, 1970.

during the northern winter. This is related to desiccation, since from March–September the lower low water occurs during night-time whilst from September onwards it occurs during day-time. At this latitude, drying during low tides is fairly intense but is obviously less when low tides occur at night, hence the movements of *Hypnea* up and down the shore. Table 3.11 gives a generalised scheme for rocky shores off Ghana, for comparison with other tropical sites.

This seasonal rise and fall of belts of algae is not confined to tropical coasts, but occurs also in Europe, e.g., Den Hartog (1968) reports that the *Bangia–Urospora*, *Blidingia minima* and *Enteromorpha–Porphyra* associations have an upward shift in winter and a downward shift in summer, both upper and lower boundaries being affected. It is clear that on tropical shores desiccation plays a major role in determining the upper limit of many species and the same effect has been assumed with very little experimental evidence on temperate shores; data of Hatton (1938) show that increased humidity allowed greater Fucoid growth and Schonbeck & Norton (1978) have shown that drying at neap tides prunes back the upper shore Fucoids. The damage can be seen 21–28 days after a drying neap tide and the higher the temperature the greater the effect. They also found that neither rain nor frost had any effect.

Taking the view that almost all the organisms in the intertidal zone require intermittent submergence in seawater (*see* Table 3.8 for the few exceptions), how much does this contribute to the existence of the discrete belts which clearly occur in the intertidal zone? The uppermost belt in which *Pelvetia canaliculata* and *Fucus spiralis* tend to be the common organisms is one of lengthy periods of emersion and prolonged drying of the thalli. Early experiments (Baker, 1909, 1910), in which species were grown under conditions of varying periods of submersion every 12 h, showed that *Fucus spiralis* grew best with six hours' submersion every 12 h but survived with only one hour's submersion in 12 h whilst *Ascophyllum nodosum* grew best with one or six hours' in 12 h but not with 11 h submersion out of 12 h. She found that the species which resist desiccation grow slowly, and vice versa, and commented that the quicker growing species may supercede others on the lower shore, i.e. competition is critical, whilst the most important factor in the upper zone was resistance to desiccation. It is interesting that *F. spiralis*, although it germinates when covered by water for only one hour in 12 h, divides much less frequently than under six hours' submersion; this is further evidence that these upper shore species are really submerged forms which have penetrated into a marginal environment. It is now known that *F. spiralis* can photosynthesise at the same rate whether in or out of the water (Kremer & Schmitz, 1973) whereas *F. vesiculosus* and *F. serratus* photosynthesise out of water at only 75 % and 50 % respectively of their submerged rates and the even lower growing *Laminaria saccharina* cannot photosynthesise at all when exposed. An early report

Fig. 3.21. Relationship between seasonal tidal changes and the zonation and quantity of *Hypnea musciformis* at a station on the coast of Ghana. Continuous line is the tidal curve of the heights of the lowest of the daytime low waters in each month. The broken line represents the seasonally fluctuating upper limit of *Hypnea* and the dotted line the percentage cover of *Hypnea*. From Lawson, 1957.

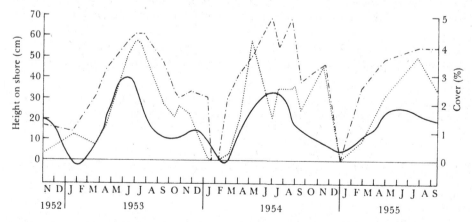

Table 3.11. *A generalised distribution of algae on Ghana coasts. Adapted from G. W. Lawson & D. M. John (personal communication)*

Supratidal
  Bostrychia spp.
  Lophosiphonia septabunda
  Rhizoclonium
  Entophysalis deusta

  Bare or occasional
    blue-green algae, filamentous
    brown algae or
    Centroceros

  Basispora africana
  Ralfsia expansa

Intertidal

Algal turf in rough water	Algal turf in calmer water
Chaetomorpha antenninia	*Bryopsis*
	Caulerpa
'*Lithothamnia*'	*Colpomenia*
Taenioma	*Hypnea*
Laurencia	*Gracilaria*
Gelidium	*Amphiroa*
Herposiphonia	*Corallina*
Centroceras	*Jania*

Subtidal
  Dictyopteris delicatula
  Sargassum vulgare

Sargassum filipendula	
Dictyota	*Champia*
Spathoglossum	*Halymenia*
Hypnea	*Caulerpa*
Valdivia	*Dictyuris*
Botryocladia	*Corynomorpha*

by Fischer (1928) that *Pelvetia* and *Fucus* die when continually submerged requires re-investigation.

The evidence points to a complex of factors operating to establish the belts in the intertidal; in the absence of grazing the rates of establishment and rates of growth would ultimately lead to the formation of belts and grazing probably modifies the composition of the belts. Vertical thinning of belts can occur when factors operate to depress the upper limit of a species, or when factors operate to raise the lower limit; some factor(s) can operate simultaneously on both boundaries.

Experiments conducted by Zaneveld (1937)

showed that specimens of *Fucus spiralis* var. *platycarpus* took 48 h to reach their minimum water content, by which time they had lost 76.2 % of their fresh weight. Lower growing fucoids, e.g., *F. serratus* lost almost as much water in 18 h exposure, and in general it seems that the lower an alga grows on the shore the faster it looses water. One might expect that the algae occurring higher up the shore would be less susceptible to water loss than those lower down, but apparently this is not so since Dorgelo (1976) has shown that *Pelvetia canaliculata* has the greatest water loss (other than *Laminaria saccharina*). The rate of water loss is supposedly connected with cell wall thickness: much of the water is lost from the walls of the fucoids growing in the upper region, but since approximately the same amount of water is lost from the lower growing, thinner walled species this water must come from the cell contents and perhaps is why they do not survive desiccation. The amount of water lost during desiccation varies according to the salinity of the seawater in which plants have been growing and in hypersaline conditions least will be lost. Fig. 3.22 shows the rate of loss from *Enteromorpha* thalli. Submergence in seawater is not essential for all intertidal species, many of which can grow in water of salinity down to 5–6‰ (*see* p. 78) but then apparently only in a subtidal position. On the other hand, species of *Ectocarpus*, *Fucus vesiculosus* and *Enteromorpha* can survive in freshwater streams running across the shore where they spend lengthy alternating periods in fresh and fully saline water. Such distributions are common, with *Ectocarpus* penetrating as high as the upper intertidal, *Fucus* above this and *Enteromorpha* even higher: it would be interesting to investigate the metabolism of such algae at full freshwater and full saline submersion. *Enteromorpha* has penetrated to many inland river sites and must therefore have adapted to low salinities.

An intriguing aspect of the distribution of algae in belts on the shore is recorded by De Silva & Burrows (1973), who showed that unbranched *Enteromorpha* plants of the *E. intestinalis–compressa* complex occur on the upper shore and exclusively branched plants occur on the lower shore (Fig. 3.23), even in rock pools. They experimented with the branched and unbranched populations and suggest that the two species are separating out from a single species and, although there is a high degree of intersterility, it is not yet complete. They also suggest that the tide is acting as an isolating mechanism, since as the tide rises, the gametes of the lower shore plants fuse with one another (fusion of gametes is rapid, at least under laboratory conditions) and so on up the shore. Although the settling position of the zygotes is not known it seems likely that most settle in the vicinity of the parents, otherwise the distribution in Fig. 3.23 would not occur.

On a broader scale, the discovery by Müller (1976) that there is a genetic barrier between otherwise morphologically similar *Ectocarpus siliculosus* populations in Europe and North America is of great interest. He found that, although the initial step of sexual attraction of male and female gametes occurs between gametes of both populations, the final stage of interaction of the male flagellum with the female gamete cell surface and subsequent fusion is blocked.

There have been very few studies of the factors involved in the formation of male and female gametangia or tetrasporangia. West (1972) showed that temperature and light regimes affect cultured *Rhodochorton purpureum* populations in a rather complex manner, in that Californian, Washington and Alaskan clones sporulate only when grown in short daylengths, whereas a clone from Chile sporulated in all daylengths. Vegetative growth of all was greatest in long daylengths and sporulation is also greatest at the higher light intensities. Low light greatly reduces sporulation. Spermatia and gametangia of the Californian strain develop richly in short-day conditions but will also develop in long days. However there is not a true photoperiodic response (*see* also p. 70). West showed that at temperatures above or below optimum, sporulation is strikingly inhibited and this is almost certainly one of the factors involved in the often reported sterility of red algal populations at the limit of their distribution (Dixon, 1965). Few other red algae have been investigated in any detail but the concensus seems to be that the general induction of sexual reproductive structures is not under photoperiodic control (Dixon & Richardson, 1970). There are many problems in the life histories of *Porphyra** and *Bangia* but at least the production of 'fertile cell rows' in the *Conchocelis* phase giving rise to the leafy phase (*Porphyra tenera*) has been shown to be associated with short-day conditions (a light break in the dark period being inhibiting) and a phytochrome system is operating (Dring, 1967*a*, *b*). Under long-day conditions monospores were formed which perpetuated the *Conchocelis* phase. Equally complex is the situation in *Bangia* which is summarised in Fig. 3.24. Suppression of sexual reproduction, but a high incidence of asexual reproduction, appears to be common in estuarine strains of marine algae (Russell, 1971) and this

* This alga has been investigated intensively, especially by Japanese workers, since it is cultivated on a large scale as a food plant.

Fig. 3.23. Percentages of plants of *Enteromorpha intestinalis–compressa* showing different types of branching, *a*, on the rock surface and, *b*, in rock pools. Stippled columns, plants with branches resembling main frond; horizontally hatched columns, plants with branches differing from main frond; vertically hatched columns, unbranched plants. l.f, uppermost region of shore; u.eu, upper intertidal; m.eu, mid-intertidal; l.eu, lower intertidal; s.l, subtidal. From Silva & Burrows, 1973.

Fig. 3.22. Desiccation curves obtained in the laboratory for *Enteromorpha flexuosa* grown in various strengths of seawater; values expressed as a percentage of total initial water content. From Townsend & Lawson, 1972.

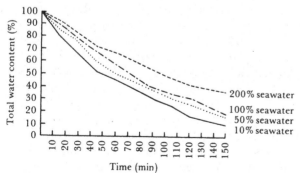

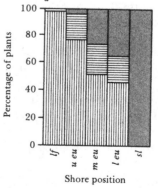
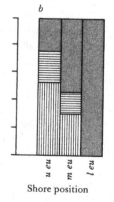

feature may be genetically controlled, as are other features of estuarine strains, such as tolerance of low salinity.

How many more complexities are there in the study of *Porphyra*? Kurogi (1972) distinguishes 31 species for Japan, of which six are subtidal in distribution, and the remainder intertidal, whilst many are also geographically separated. Conclusions from experiments on photoperiodic responses cannot however be immediately translated to natural populations since the red and far-red wavelengths penetrate only short distances into seawater and therefore phytochrome-mediated responses need to be checked

under natural conditions. There is a need to study the effect of blue wavelengths since these are also to some extent implicated in day-length effects (Dring, 1970).

The production of plurilocular sporangia on *Ectocarpus confervoides* and the release and germination of the spores is stimulated by high light intensities (Boalch, 1961) whilst in *Ectocarpus siliculosus* the ratio of unilocular to plurilocular sporangia (Table 3.12) can be changed by varying light intensity and day length whilst maintaining the temperature at 16 °C (Müller, 1962). At 13 °C only unilocular sporangia form and at 19 °C only plurilocular.

Fig. 3.24. Interrelationship of the various phases in the life history of *Bangia fuscopurpurea* together with the photoregimes necessary for the formation of the reproductive structures. From Dixon & Richardson, 1970.

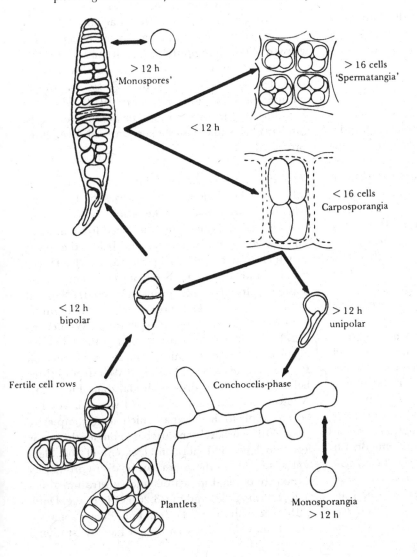

Table 3.12. *Percentages of unilocular and plurilocular sporangia on* Ectocarpus siliculosus *grown at 16 °C at various light intervals and day lengths*

Light intensity (lux)	3500	750	3500	3500
Day length (h)	14	14	9	24
Percentage of unilocular sporangia	75	94	96	16
Percentage of plurilocular sporangia	25	6	4	84

The effect of intertidal stress on an alga has rarely been studied; where the same species grows in both the intertidal and the subtidal zones an ideal situation exists. Liddle (1975) found that intertidal stress inhibited cell expansion of the brown alga *Padina sanctae-crucis* and the plants were therefore smaller than those growing submerged, although the growth rates were similar.

Rock pools. On many shores there is a distinct zonation of rock pool epilithic floras, from the supratidal region seaward. The uppermost pools in the supratidal zone usually have a coating of Cyanophyta, diatoms or coccoid Chlorophyta, but to the casual observer they appear barren. The water in these pools is brackish or hypersaline, rather than saline and it is here that a planktonic flora is also found (*see* p. 291). A brown coating on the rock is often caused by brown tinted sheaths of Cyanophyta, rather than by diatoms or other brown pigmented algae. This coating of microscopic forms on the bottom of the pools is present right down the shore but it is the macroscopic species which catch the eye even in the next lowest pools where *Enteromorpha* tends to grow. Stones are often green with a coating of coccoid Chlorophyta of the 'chlorelloid' type or of colonies of *Prasinocladus*. Pools in the upper intertidal can be quite rich with a coating of *Lithothamnia* giving the rock a pinkish-white appearance and on this, and especially on limpets, strands of *Scytosiphon lomentaria* are common. Also abundant at least on some British shores, are growths of *Corallina officinalis, Cladophora rupestris, Polysiphonia, Pterocladia, Ceramium, Laurencia, Ectocarpus*, etc. In the mid and lower intertidal all these species occur in addition to the fucoids *Cystoseira, Bifurcaria, Halidrys* (in the south-west) and even

Himanthalia elongata. The *Corallina* tends to become a substratum for the growth of the lichen-like *Mesophyllum lichenoides* and *Sedum*-like plants of *Gastroclonium ovatum*, 'bushes' of *Sphacelaria*, leaf-like growths of *Petalonia* and even small laminarians.

Particularly in the upper intertidal pools a brown felt-like growth of diatoms living in mucilage tubes is common, especially in spring. Further down the shore these diatoms often occur also on the *Corallina* in the rock pools.

Baker (1909) commented on the fact that some of the species growing on rocks high on the shore do not tolerate submersion in rock pools, whereas some of the lower intertidal forms can live under permanent submersion in rock pools; many such interesting 'experimental systems' can be found in nature!

Rock pools are often sites of accumulation for plant detritus which must have some effect on the growth of the community though it may be difficult to demonstrate in nature. A clue to this may be provided by experiments adding *Zostera* detritus to cultures of *Ulva* discs (Harrison, 1978) where he found a 50 % increase in growth over the controls. Similar effects may also be encountered in shallow, protected seas.

The subtidal region. The subtidal zone, like the intertidal, can be divided into a series of belts but, whereas in the intertidal the whole extent can be investigated and therefore subdivided from an overall view, the subtidal may extend below the depth to which SCUBA diving apparatus is safe. Hence division into upper, mid and lower is not always feasible, although the terms are used. Neushul (1967) suggested that a three-zone pattern might be a universal feature, but Shepherd & Womersley (1971) found that in South Australia three belts of algae could be distinguished in the mid-zone off Pearson Island (Fig. 3.34) and off West Island three belts occurred in the upper subtidal zone (Fig. 3.25). A lower zone can only be defined when survey has been possible to the point at which algae completely cease. Other subdivisions are based on depths below low-tide level and horizontal extent of the algal zones; sometimes the zones can be distinguished by a mixture of algal indicators and substratum zones (e.g., Smith (1967) and Fig. 3.26). Others, e.g. Doty, Gilbert & Abbott (1974), could not detect zonation but it may be obscured somewhat by studying

Fig. 3.25. The three major subtidal zones, together with a subdivision of the upper subtidal zone into three subzones. West Island, South Australia. From Shepherd & Womersley, 1971.

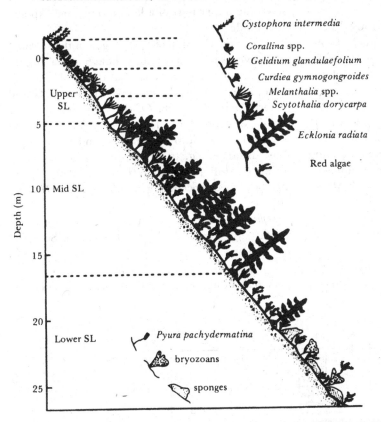

Cystophora intermedia

Corallina spp.

Gelidium glandulaefolium

Curdiea gymnogongroides

Melanthalia spp.

Scytothalia dorycarpa

Ecklonia radiata

Red algae

Pyura pachydermatina

bryozoans

sponges

Fig. 3.26. Subtidal profile showing four zones (*1, 2, 3, 4*). Vertical series of figures refer to the species occurring in each zone. 1, Blue-green mat; 2, *Porphyra*; 3, *Ulva*; 4, *Enteromorpha*; 5, *Pelvetia*; 7, *Fucus vesiculosus*; 10, *Himanthalia*; 11, *Polysiphonia nigrescens*; 12, *Chondrus*; 13, *Alaria*; 14, *Laminaria digitata*; 15, *Laminaria hyperborea*; 16, *Corallina*; 18, *Plocamium*; 19, *Heterosiphonia*; 20, *Delesseria*; 22, *Callophyllis*; 23, *Cryptopleura*; 24, *Rhodymenia*, *Membranoptera, Phycodrys*; 25, *Chorda, Ulva, Enteromorpha, Tilopteris, Halopteris, Aglazonia (Cutleria)*; 26, *Dictyota*; 27, *Dictyopteris, Halopteris*; 28, *Naccaria, Halarachnion, Taonia*; 29, *Halidrys*; 30, *Laminaria saccharina*. From Smith, 1967.

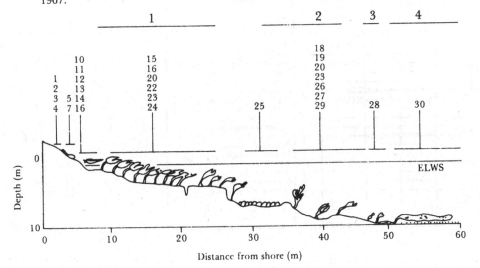

dredged material; they do comment that some of the 101 species reported off Hawaii seem to be deep-water species.

In one of the rare studies of the subtidal belt in which the algae were removed from quadrats by aqualung divers and lifted to the surface by an airlift system, analysed in the laboratory and the data submitted to numerical analysis, three distinct zones could be characterised (Prentice & Kain, 1976). At a site off the Isle of Man, the zone from 0–4.9 m below lowest astronomical tide was designated the *Laminaria* zone (*L. hyperborea*, *Odonthalia dentata* and *Delesseria sanguinea* in quantity and presence alone of *Phycodrys rubens*, *Plocamium cartilagineum* and *Membranoptera alata*), from 5–7.9 m below lowest astronomical tide occurred a *Saccorhiza* zone (*S. polyschides* and *Desmarestia aculeata*) and from 8–11 m below

lowest astronomical tide an *Echinus* zone with few algae owing to the grazing by this animal. Fig. 3.29 shows the upper part of a laminarian zone exposed at low tide.

The distribution of many algae in the subtidal is however well circumscribed though actual depths of penetration, etc., may vary at different sites. An example from Heligoland is given in Fig. 3.27 and pictorially in Fig. 3.28. Of the large macroscopic rock-face algae of the lower intertidal only *Fucus serratus* descends into the subtidal and then only a short distance, e.g. to 1 m below mean low water spring tide off Heligoland (Lüning, 1970). Other intertidal species occurring in rock pools, e.g. *Halidrys siliquosa*, *Polyides rotundus*, *Furcellaria fastigiata*, *Corallina officinalis* also descend, but generally not to a great depth (e.g., 4 m off Heligoland, 10 m off Scotland

Fig. 3.27. The vertical distribution of algae in the subtidal zone off Heligoland. The data are expressed quantitatively only for the abundant algae (thickened lines, scale at upper left of Table). Dense vegetation of *Laminaria* at 1–4 m. From Lüning, 1970.

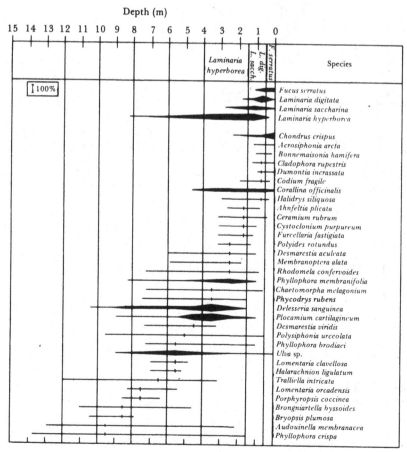

Fig. 3.28. Pictorial representation of the algal
distribution in the subtidal zone off Heligoland.
From Lüning, 1970.

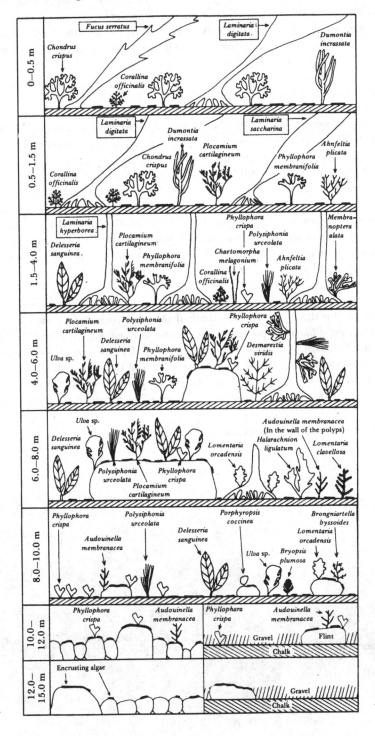

according to McAllister, Norton & Conway, 1967 or 12 m according to Norton & Milburn, 1972). Lüning (1970) found that *Lomentaria clavellosa, L. orcadensis* and *Halarachnion ligulatum* occurred only between 4.5 and 8.5 m below mean low water spring tide. He also found that there was a distinction between the species occurring in the 0–4 m depth interval in regions where the large laminarians were absent (e.g., here he found *Rhodomela confervoides, Phyllophora membranifolia, P. pseudoceranoides, P. truncata, Chaetomorpha melagonium, Plocamium cartilagineum, Polysiphonia urceolata* and *Ulva* sp.) but at the same depth in the shade under the laminarians he found *Phycodrys rubens, Delesseria sanguinea, Phyllophora truncata, Trailliella intricata, Audouinella membranacea* and *Phyllophora crispa*. These species also grow in the deeper waters and obviously invade the *Laminaria* beds because these afford shade; they also extend up into the lower intertidal but only in the most shaded places, e.g. under overhangs of rock pools. Off the eastern North American coast in Nova Scotia, Edelstein, Craigie & McLachlan (1969) report a belt of *Alaria esculenta* at low tide level and below this three belts, *Laminaria–Desmarestia* to 15 m, *Agarum–Ptilota* 10–30 m and, from 30 to 40 m, a *Phyllophora–Polysiphonia* association. Off Massachusetts there is a similar upper subtidal of *Laminaria* spp. and a lower zone of *Agarum* (Lamb & Zimmermann, 1964).

In the Mediterranean the upper subtidal is characterised by the occurrence of species of *Cystoseira* (*see* Table 2.2) and there are no large beds comparable to the shallow water laminarian beds of temperate zones. On sandy substrata a comparable community develops composed of the angiosperms *Posidonia* and *Cymodocea*.

It is rarely reported exactly which species are confined to the various zones of the shore, but such information is of great interest: e.g., in a survey off Argyll, Scotland, some 40 species were found to be confined to the intertidal, 36 species common to this and the subtidal and 85 species confined to the subtidal (Norton & Milburn, 1972). Off the Scilly Isles, Norton (1968) recorded the disappearance of *Laminaria ochroleuca* at 27–30 m and below this depth approximately 12 species of smaller red and brown algae occurred but some of these are common species occurring also at shallower depths and even in the lower intertidal, e.g. *Dictyota dichotoma* and *Cryptopleura ramosa*.

In subtidal habitats, as in intertidal, the algae may be attached to stable rock or to loose stones or even be loose on the sea bed, e.g. in Port Erin bay, loose stones seaward of the *Laminaria hyperborea* zone tend to support a flora of mainly small species, e.g. *Sphacelaria radicans, S. plumula, Polysiphonia nigresens, P. urceolata, Delesseria, Plocamium, Cryptopleura, Chaeto-*

Fig. 3.29. The upper subtidal *Laminaria* forest exposed at extreme low water spring tides.

morpha, *Acrochaetium, Cladophora* (*glaucescens*) *sericea, Phycodrys, Membranoptera, Desmarestia ligulata, Brongniartella, Halarachnion, Taonia, Dictyota, Sporochnus, Sciania, Sphondylothamnion,* and *Laminaria saccharina.* Loose lying subtidal populations of healthy plants occur in some favourable sites; presumably these sites are determined by the pattern of bottom currents and in Port Erin Bay Burrows (1958) found a population composed of *Laminaria digitata, L. saccharina, L. hyperborea,* and *Saccorhiza polyschides* and many smaller plants of which eleven species were only found in this loose lying community (*Arthrocladia villosa, Bonnemaisonia asparagoides, B. hamifera, Chaetopteris plumosa, Cutleria multifida, Dictyopteris membranacea, Dudresnaya verticillata, Halarachnion ligulatum, Naccaria wiggii, Seirospora griffithsiana,* and *Sporochnus pedunculatus*) and these were confined to the deepest populations (below 8 m). Many of these algae form attachment organs on contact with each other and often grow in a most diffuse manner. Loose lying collections of 'lithothamnia' also occur and are known as 'maerl'; only two species, *Lithothamnium calcareum* and *L. coralloides,* are recognised off the French coast (Cabioch, 1966).

Smith (1967) noted the separation in the subtidal of haploid and diploid plants of *Brongniartella;* the tetrasporic at about 6–7 m whilst at 10 m the plants were all cystocarpic. The same generations growing in the subtidal and intertidal can show distinct morphological differences, e.g., *Padina sancrae-cruci* growing in Puerto Rico produced larger plants with conspicuously larger cells in the subtidal populations (Liddle, 1975).

Physiological studies of cystocarpic and tetrasporophytic subtidal algae (Fig. 3.30) show that the two forms of the same genus may have slightly different responses to light and temperature (Mathieson & Norall, 1975) and certainly even small variations would result in spatial separation as one form outgrew the other.

Of the species almost confined to the subtidal, one of the most intensively studied is the massive kelp, *Macrocystis,* which is commercially harvested and processed to yield alginic acid. *Macrocystis pyrifera* occurs along open coasts off California, usually only in water of at least 5–10 m in depth but if the coast is very protected from wave action it can occur even up into the lower intertidal (North, 1971);* this genus is only common in cold waters e.g., the Falkland Islands (where it is also harvested) and parts of the South American coast. Off California it grows in the cold water flowing southward down the coast and off Baja California only in isolated regions where cold water upwells. The inshore fringes of Californian beds are frequently found to support only young plants, and here wave surge is regarded as the limiting factor; in this position *Egregia laevigata* with its tougher flatter stipes replaces the *Macrocystis.* Wave surge in shallow water is illustrated in Fig. 3.31

* The volume edited by North contains a comprehensive account of *Macrocystis* up to 1971.

Fig. 3.30. The net photosynthesis of tetrasporic and cystocarpic *Ptilota serrata* at different light intensities and 5 °C, sampled by SCUBA at a depth of 40 ft. From Mathieson & Norall, 1975.

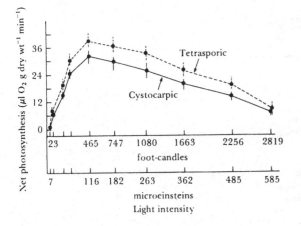

Fig. 3.31. An example of water movement in the surge zone at a site off Santa Cruz Island, California. A 5 m excursion of water was measured in the surge zone at a depth of 6 m, when a 2 m wave moved past at a velocity of 8 m s^{-1} with a wavelength of 120 m and a wave period of 15 s to produce the orbital velocities (arrows) indicated. Near the bottom, these flattened to produce the back-and-forth motion in the surge zone. From Neushul, 1972.

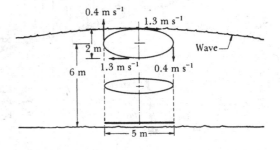

which shows how the orbital movement of the water associated with waves is translated near the bottom into a back-and-forth motion which moves the bases of the plants backwards and forwards and exerts a stress quite unlike that in areas of laminar flow. Many macroscopic algae, eg., *Pterygophora*, *Eisenia*, *Laminaria*, *Agarum*, *Desmarestia*, and *Dictyoneuropsis* are associated with this surge zone in Californian waters; some do not project their thalli up beyond this turbulent water layer, e.g., *Laminaria*, *Agarum*, and *Dictyoneuropsis*, whereas others grow up into the current zone. Within the bed at La Jolla, *Laminaria farlowii* is one of the few large algae growing exclusively associated with the *Macrocystis* (Dawson, Neushul & Wildmann, 1960). The outer fringes are most likely to be determined by the light which is sufficient to depths of 15–20 m in turbid coastal waters and 20–30 m in clear waters, though North (1971) observed attached *Macrocystis* as deep as 40.3 m. Aerial photographs show a rather even, outer border suggesting the operation of a single factor here whilst the inner border is diffuse and controlled by a complex of factors. *Macrocystis* is frequently found attached to rocky surfaces on exposed coasts but it can also grow in loose sandy substrata where the sporelings commence growth on worm tubes (Dawson *et al.*, 1960). The massive *Macrocystis* plant cannot support itself in still water, where the fronds growing upwards from the extensive holdfast will hang vertically, hence a further prerequisite for occurrence is a fairly strong current, sufficient to buoy the plants up (the air-filled pneumatocytes at the base of the blades are not sufficient). In the outer zones of the beds, the plants tend to be intermingled with the Elk Kelp (*Pterygophora*) which in some places forms a fairly prominent community seaward of the main *Macrocystis pyrifera* beds, e.g., off Anacapta Island, whilst on the seaward edge of the coastal *M. angustifolia* beds there is a zone of *Pelagophycus porra* (Clarke & Neushul, 1967). Fig. 3.32 shows three variations of species distribution across kelp beds.

Whilst the upper parts of the *Macrocystis* stream

Fig. 3.32. Diagrammatic cross sections of typical kelp beds for three geographic regions, showing differences in species composition. From North, 1971.

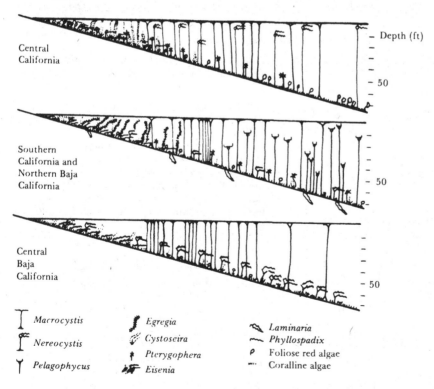

out into the current the lower parts grow in the surge zone and the dwarf *Macrocystis* gametophytes occupy the boundary layer, along with some 44 other species (excluding coralline and unicellular algae) which are less than 1 cm high (Dawson & Neushul, 1960). The diversity of species then decreases such that in the surge zone (in the 1–10 cm zone above the substratum) there are about 23 species, whilst only 13 species extend into the current zone above, and all these bear floats. Many of the algae in the subtidal produce their spores at the base of the lamina, e.g., many laminarians, including *Macrocystis*, and this is of course close to the site of germination. Since male and female plants need to grow relatively close together this is an advantage. Plants such as *Cystoseira* and *Halidrys*, growing in the current zone off California, form their gametes in this zone but as Neushul (1972a) points out they presumably extrude them on to the surface where they are fertilised and it is the zygotes which are then distributed. Neushul, Clarke & Brown (1967) measured the force on *Macrocystis* stipes in moderate wave surge as 8.1 kg and comment on the extensible nature of the fronds, which operate in a spring-like manner to dampen down the force exerted on the holdfast. *Macrocystis* is a large plant, but the depth at which it grows is no indication of the size of the plant since it streams out into the water and plants in 10 m depth usually have a maximum length of 23–28 m and those in 23 m are usually 50–60 m in length (North, 1971). Growth rates of fronds of up to 50 cm day^{-1} have been reported and of 6 cm day^{-1} for the stipe (Scagel, 1947), described by Clendenning (1960) as the fastest of any known plant. The maximum life span of the fronds is about six months, which at this upper rate of growth would yield a frond of about 90 m and suggest that the order of magnitude of the growth rate is almost certainly correct. *M. pyrifera* off the South American coast achieves a density of one to two plants a square metre (Barrales & Lobban, 1975). This is low compared with the beds off California where the plants have an average 15–30 stipes per plant (North, 1968).

The species growing under the *Macrocystis* form a 'turf'* and Neushul & Dahl (1967) found that whilst it was difficult to study this community underwater, a similar community could be reproduced on shells, rocks, etc., in the laboratory. Aleem (1973) gives the composition of the layer (Table 3.13) which grows beneath a 'shrub' layer of *Egregia laevigata*, *Eisenea aborea*, and *Pterygophora californica* (Fig. 3.33).

In the southern hemisphere, the zonation in South Africa has been described by Isaac (1937, 1949). Here, the subtidal is also characterised by large laminarians, *Ecklonia maxima* (*buccinalis*)† *Macrocystis integrifolia*, and *Laminaria pallida*, *E. maxima* being the dominant.

On southern Australian shores, Shepherd & Womersley (1970, 1971) found a distinct zonation in the subtidal and, as in Womersley's previous studies, made the important distinction between rough water and sheltered shores and demonstrated the interplay between light penetration and degree of water movement due to surge on the bottom. The light penetration at the actual attachment surface will depend to a varying degree upon the amount of sediment present and the degree of wave surge. Fig. 3.34 gives an example of the subtidal zonation at Pearson Island. The upper subtidal often has a characteristic flora of large brown algae, such as *Carpophyllum*, *Cystophora*, *Durvillea*, *Xiphophora*, and *Sargassum* and some of these species extend up slightly into the lower intertidal; especially further south, around Antarctica, those which do occur are not confined to the permanently submerged region. Similarly in New Zealand, three zones were recognised by Bergquist (1960a) with a shallow, brown algal zone (*Carpophyllum*) a red algal mid-zone (*Pterocladia*, *Vidalia*, *Melanthalia*) and lowermost, another brown algal zone of *Ecklonia radiata*.

The arctic subtidal has been studied by Wilce (1959) who found enormous populations of *Halosaccion ramentaceum*, *Rhodymenia palmata* and at times *Monostroma fuscum* in the upper belt whilst, below this, beds of *Laminaria longicruris* occur with intermixed *Agarum cribosum* and *Saccorhiza*. Beneath the latter two species he found a single storey undergrowth of *Phyllophora* and *Kallymenia*. Underneath the *Laminaria* on the other hand two understories occurred, the larger being composed of kelps (*Agarum*, *Alaria grandifolia*, *Laminaria* spp. and *Saccorhiza dermatodea*), and beneath these a collection of species

---

* Neushul & Dahl (1967) used the term 'parvisilvosa' to describe this morphological assemblage. Such terms have not been widely applied to algal communities.

† The name derived from a study of the later literature is used and Isaac's original identification is in parentheses.

Table 3.13. *The composition and biomass of plants and animals at four different depths in the* Macrocystis *region off California. From Aleem, 1973*

Density[a] and biomass of *Phyllospadix* community (0.25 m² in 6.5 m depth)

Taxa	g 0.25 m⁻²
Plants	
Phyllospadix scouleri[b] (∞) (cover 100%)	751
Porphyra naiadum (∞)	148
Gracilariopsis sp. (1)	1.5
Ectocarpus granulosus (8)	2
Melobesia mediocris (∞)	1
Kelp sporlings (8)	2
	905.5
Animals	
Isopods (2)	0.5
Small gastropods on leaves (10)	2
Bryozoa (*Membranipora*) on rhizomes and leaves (∞)	1
Caprella sp. (4)	0.5
	4

Biomass/m²	
Plants	3622 g m⁻²
Animals	16 g m⁻²
Total	3638 g m⁻²

Density and biomass of the Coralline community in 8 m depth

Taxa		g 0.25 m⁻²
Plants		
Corallina chilensis *C. gracilis* } (88), cover 80%		170
Bossea orbigniana (2), cover < 5%		3
Articulated corallines[c]		173
Acrosorium uncinatum *Callophyllis marginifructa* } Epiphytes *Plocamium pacificum* } on corallines		3
Dictyota binghami (3)		37
Dictyopteris zonarioides (3)		23.5
Colpomenia sinuosa (3)		3.5
Rhodymenia lobata (2)		15
Prionitis filiformis (1)		7
Gelidium purpurascens (7)		166
Cystoseira osmundacea (1)		63
Laurencia sp. (3)		9
Flora		500
Animals		
Sponges		13.5
Hydroids		2.5
Bryozoa		36
Top snails (3)		13
Spider crab (1) Ophiuroidae (5) Hermit crab (2) Nemertine (1) Dove snails (10) }		4.5
Fauna		69.5
Total plants and animals		569.5 g 0.25 m⁻²

Biomass/m² (average of 2 quadrants)	
Plants	2728 g m⁻²
Animals	408 g m⁻²
Total	3136 g m⁻²

Density and biomass of the Coralline community in 15 m depth

Taxa	g 0.25 m⁻²
Plants	
Corallina chilensis (32) *C. gracilis* (3) } (cover 20–30%)	187
Bossea orbigniana and *B. gardneri* (30) } (cover *Calliarthron cheilosporioides* (5) } 50%)	69
Lithothrix aspergillum (6)	53
Articulated corallines	309
Acrosorium uncinatum (∞)	2
Herposiphonia pygmea (∞)	1
Polyneura latissima (1)	0.5
Nienburgia andersoniana (5)	2
Plocamium pacificum (∞)	16
Cryptonemia angustata (?) (1)	1
Prionitis australis (2)	10
Cystoseira osmundacea (1)	40
Rhodymenia attenuata (4)	2
Gelidium coulteri (1)	21
Rhodymenia pacifica (20)	18
Flora	422.5
Animals	
Sponges	1
Hydroids	0.5
Bryozoa	19
Polychaetes	4.5
Gastropods	1
Crustacea	1
Sea cucumber (1)	88
Fauna	115

Biomass/m² (average of 2 quadrants)	
Plants	1450 g m⁻²
Animals	396 g m⁻²
Total	1846 g m⁻²

Density and biomass of the Coralline community in 21 m depth

Taxa	g 0.25 m⁻²
Plants	
Corallina chilensis (31) *C. gracilis* (1) } (cover 10–15%)	47 1
Bossea orbigniana (32) *B. gardneri* (3) } (cover 25–30%)	35.5 21
Articulated corallines	104.5
Rhodymenia californica (45) (cover 40%)	22
R. pacifica (2)	1
Nienburgia andersoniana (19)	7.5
Callophyllis marginifructa (10)	2.5
Acrosorium uncinatum (∞)	2.5
Herposiphonia pygmea (∞)	1.5
Cryptopleura sp. (2)	2.5
Prionitis australis (3)	6
Kelp sporlings	1.5
Flora	151.5

Table 3.13 (*cont.*)

Density and biomass of the Coralline community in 21 m depth

Animals[d]	
Sponges (orange)	2
Bryozoa	6
Polychaetes (free living) (2)	0.3
Tube worms (8)	26
Gastropods (4)	0.5
Phollad (1)	1
Crustacea	2.5
Ophiuroidae (5)	1
Gorgonia (1)	55
Fauna	94.3
Biomass/m²	
Plants	606 g m^{-2}
Animals	377.2 g m^{-2}
Total	983.2 gm^{-2}

[a] Number in parentheses after each taxon denotes the number of individuals, fronds, tufts or colonies according to the habit of the organism involved.

[b] Weight of *Phyllospadix* includes also that of other epiphytes such as *Callithamnion californicum* and *Ceramium pacificum*.

[c] Non-articulated corallines such as *Lithothamnion* are excluded; these grow on pebbles and stones.

[d] Large Asteroidae, holothurians and abalone are more common here but not included because of their scattered distribution.

Fig. 3.33. Cross-section of the *Macrocystis* bed from which the data presented in Table 3.13 was collected. From Aleem, 1973.

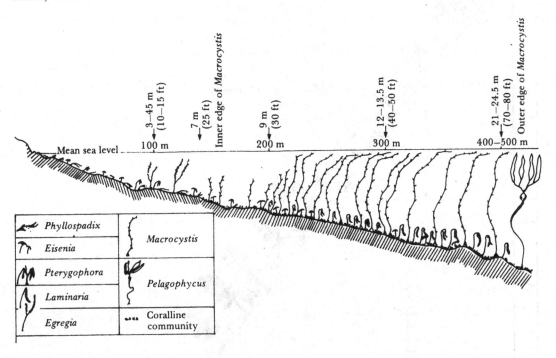

some of which are epiphytic and others epilithic (e.g. *Monostroma fuscum, Chaetomorpha melagonium, Spongomorpha arcta, Pilayella littoralis, Chaetopteris plumosa, Lithoderma extensum, Desmarestia aculeata, Litosiphon filiforme, Rhodymenia palmata, Antithamnion, Ptilota serrata, Membranoptera alata, M. denticula, Dilsea integra, Lithothamnium* spp., *Euthora cristata, Kallymenia schmitzii, Rhodophyllis dichotomum, Ahnfeltia plicata, Phyllophora interrupta, Pantoneura baerii, Phycodrys rubens, Odonthalia dentata*, and *Polysiphonia arctica*). Many of these species are also present further south. In still deeper water, *Laminaria solidangula, L. nigripes*, and *Alaria grandifolia* occur. On rough coasts the subtidal often has extensive beds of *Lithothamnium* spp. and small plants of *Alaria, Agarum*, and *Laminaria*. As in temperate zones there is a seasonal development of some species, e.g., some of the biennials shed their branches in winter and recommence growth in summer after the break-up of the ice. Others commence growth in late autumn, continue slow growth under the ice and then fruit in spring (e.g. *Chaetomorpha melagonium* and *Dilsea integra*). In the antarctic the most extensive flora was likewise found to be almost confined to the subtidal (Neushul, 1965) and here *Desmarestia menziesii, D. anceps*, and *Adenocystis utricularis* are common in protected bays. In more exposed waters these are joined by *Ascoseira mirabilis, Phaeuris antarcticus, Plocamium secundata, Myriogramme*

Fig. 3.34. A profile of the algal distribution at a sheltered site off Pearson Island at the eastern end of the Great Australian Bight showing three belts of algae in mid-region of the subtidal. From Shepherd & Womersley, 1971.

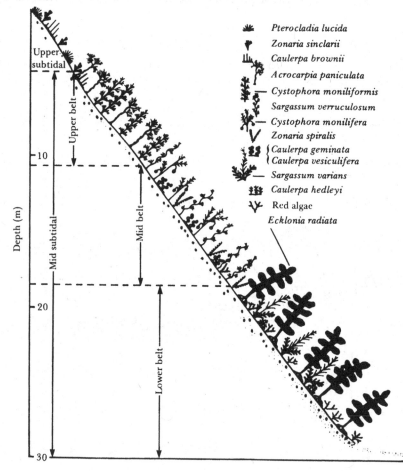

Pterocladia lucida

Zonaria sinclarii

Caulerpa brownii

Acrocarpia paniculata

Cystophora moniliformis

Sargassum verruculosum

Cystophora monilifera

Zonaria spiralis

Caulerpa geminata
Caulerpa vesiculifera

Sargassum varians

Caulerpa hedleyi

Red algae

Ecklonia radiata

mangii, Curdiea recivitzae, Ballia callitricha, Plumariopsis eatoni, Iridaea obovata, Leptosomia simplex, Porphyra sp. and *Gigartina* sp. *Phyllogigas** replaced *Desmarestia* at one site. Comparison of this brief list shows no species common to the arctic and the antarctic and in fact the degree of endemism in the southern hemisphere is very high.

Since light climate, water movement, turbidity etc., are factors which will vary with depth, it is to be expected that plants will show some morpho-logical–physiological modification according to their position in the subtidal. Plants at greater depth are less well illuminated and therefore etiolation might be expected; in fact Clendenning (1963) found that internodal lengths of *Macrocystis pyrifera* increased with depth, though North (1971) found no statis-tically significant differences in growth rate of young

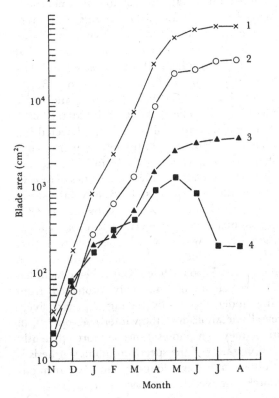

Fig. 3.35. The increase in area of blades of *Constantinea subulifera* at four different depths during a nine-month period. 1, plants just below mean lower low water, 2, about 6 m, 3, about 10 m and 4, about 14 m below low water mark. Each point is the average of five plants except for the last two points at 14 m depth where only one plant survived. From Neushul & Powell, 1964.

fronds at 8 m versus ones at 23 m. Very few experi-mental studies relating algal growth to depth have been made. Neushul & Powell (1964) devised a neat system of growing plants on small carts, which could be moved up and down rails embedded down an incline in the subtidal, measured the growth of *Constantinea* over a period of months, and showed clearly that the growth was slower at the lower depths (Fig. 3.35). The importance of this type of experiment is that it is done within the natural environment.

The maximum depth for most epilithic species is unknown, but Doty (1967) mentions crustose coralline algae occurring to depths of at least 100 m off Hawaii. Plankton can photosynthesise down to greater depths but many more data are needed on the occurrence of deep populations of epilithic species. In polar waters algae have been reported at depths of 200 m but in most temperate waters the limit seems to be around 15–40 m; Norton & Mil-burn (1972), for example, found no algae below 36 m off Argyll. On the other hand in the Mediter-ranean some populations do not even appear until this depth is reached, e.g., *Laminaria ochroleuca* in the Straits of Messina starts at 40 m depth (Drew, 1974).

Most subtidal studies are done along coasts, but occasionally they are carried out offshore, when rock ridges, etc., come close to the surface and provide interesting clear water sites for investigation. One such site, 18 km from the New England coast of North America, revealed an almost exclusive red algal community (*Ptilota serrata*) from 29 to 37 m deep and below this, down to 45 m, a *Lithothamnium glaciale* association. Below 45 m, an epizooic–endo-zooic diatom flora in sponges is reported (Sears & Cooper, 1978).

In a study from a bathyscaphe in the Medi-terranean, large algae were found down to 95 m depth (Fred, 1972). Off Corsica, *Laminaria rodriguezii* grew in a belt of vegetation between 75 and 95 m in depth and on hard substrata at 95 m the chlorophytes *Palmophyllum crassum* and *Udotea petiolata* were found amongst the red alga *Pseudolithophyllum expansum*. At these depths the temperature is around half that of the surface waters in summer (24 °C). The old concept that green algae are shallow-water species, brown are intermediate and red are deep-water forms finds little support when the subtidal is con-

* See note on p. 357.

sidered; Drew (1969*a*) found, for example, brown algae (*Dictyota dichotoma, Dictyopteris membranacea, Sargassum vulgare,* and *Padina pavonia*) in the upper 15 m in the Mediterranean, and below 15 m, to 75 m in depth, *Udotea petiolata* and *Halimeda tuna* were dominant and only grew at lesser depths when shaded by the brown algae. At such depths red and brown algae also occur e.g., off Malta *Vidalia volubilis* and a *Sargassum* sp. are reported (Larkum, Drew & Crossett, 1967); the light climate is quite adequate for all these groups and the reasons for the preponderance of Bryopsidophyceae at depths in some regions must be sought in other factors. Shepherd & Womersley (1971) comment on a similar deep penetration of green algae off South Australia and the fact is that it is a common phenomenon in regions with clear waters which allow great penetration of the blue wavelengths. They point out that the degree of wave surge must also be considered; it is perhaps not widely appreciated that wave surge reaches to about half the depth of the wave length. So movement of water from the wave surge effect can reach considerable depths and Shepherd & Womersley record ripple marks down to 70 m. The number of epilithic sites is probably small at great depths since flat surfaces will tend to be silt or sand covered and only outstanding rocks or steep slopes will be available for colonisation. Depth penetration of algae ought to vary with latitude since the total amount of light will be very small at depths in high latitudes, yet it is from just such latitudes that some of the deepest living algae have been reported. Wilce (1967) who studied the flora off Labrador and Greenland, found penetration of the flora down to 100 m, though more usually 50–55 m is the limit. At such latitudes the algae are living in almost complete darkness for many months of the year and Wilce suggested that some heterotrophic growth might occur. Macroscopic marine algae can certainly absorb organic acids and glucose but apparently only the organic acids are respired (Bidwell & Ghosh, 1963). Later, Drew (1969*b*) showed that, at least in some marine algae, exogenous sugars are readily leached out since they are only absorbed into the extracellular material. He did also show that some, albeit intertidal, genera do have metabolic pathways for conversion of hexose sugars to mannitol, so obviously studies of deep living algae may reveal other pathways. It must be admitted that wherever heterotrophy has been suspected and then tested for, e.g., in intertidal sands (Munro & Brock, 1968), ice (Horner & Alexander, 1972), and arctic lakes (Rodhe, Hobbie & Wright, 1966) the main heterotrophic activity which has been detected has been bacterial. However, at the low temperature of subtidal sites, respiration is likely to be slight and most large brown algae contain quantities of reserve polysaccharides which must form survival rations over such periods. Fig. 3.36, from Kanwisher (1966), shows a slightly lowered respiration rate for an arctic *Laminaria* during winter, which contrasts with that of a *Fucus* sp. which has a similar rate both in Labrador and Woods Hole; Kanwisher concluded that arctic algae are 'partially protected from starving under the ice by having a progressively decreased metabolism as the winter goes on'. If, in addition, they can photosynthesise more efficiently in the blue range of the spectrum, photosynthesis is probably spread over a longer period and it might be more effective than expected from a simple

Fig. 3.36. The effect of temperature on the respiration of *Fucus* and *Laminaria* in Labrador and at Woods Hole, Massachusetts in summer and winter. From Kanwisher, 1966.

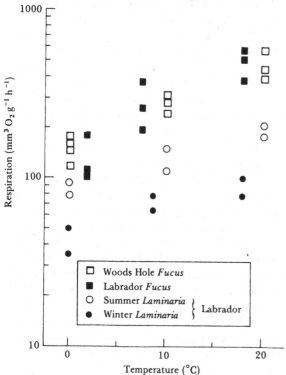

consideration of the light regime (cf., the adaptation of algae to blue light, p. 353). Algae kept in the dark are known to shut down certain intermediary metabolic pathways, e.g., in *Melosira nummuloides* there is no transfer of isotopic carbon from alanine to glucose, whilst the conversion of glutamic acid or arginine to proline and of threonine to isoleucine proceed very slowly in the dark (Hellebust, 1972). It is of course necessary to investigate the pathways in macroscopic algae but similar mechanisms may well occur.

In slightly less extreme conditions off Nova Scotia, growth of *Laminaria* and *Agarum* is fairly rapid in winter at temperatures around 0 °C and under comparatively low light conditions (Mann, 1972*d*). So possibly in the arctic, also, growth continues for longer than might be expected during the period of declining light intensity. Most recently a very valuable piece of evidence indicating how these deep living algae may survive and grow in the dark is provided by evidence that in *Laminaria hyperborea* there exists a dark carbon dioxide fixation system which is mainly localised in the growing zone (Willenbrink, Rangoni-Kübbeler & Tersky, 1975). This dark fixation mechanism also continues to function in the light and in young blades as much as 25% of light-assimilated carbon is fixed by this alternative pathway. Other brown algae have also been shown to possess the dark fixation enzymes (Akagawa, Ikawa & Nisizawa, 1972*a, b*) and some of these live also in fairly shallow, well-illuminated water.

Darkness also affects the exchange of ions, either through an effect on permeability or through active transport systems, e.g. *Ulva lactuca* and *Valonia macrophysa* lose potassium against sodium in the dark (Scott & Hayward, 1955); if this proceeds in polar darkness great changes in internal ionic composition must occur. In the arctic the habitat is extremely stable during periods of ice cover when wind-induced wave surge is minimised and hence losses from this source are cut to a low level. Loss by thinning of temperate subtidal populations during rough weather is considerable, though there are few, if any, data on its extent. However, casual observations on exposed shores is sufficient to reveal masses of seaweeds washed ashore. Drew (1974) found that the plants of *Laminaria ochroleuca* at depths below 50 m in the Straits of Messina were larger than shallower growing plants from England and Spain and he attributed this to the less damaging effects of the unidirectional currents. In addition losses from grazing are often considerable in the subtidal and indeed the associations are often modified considerably by this agent. One can only conclude that in polar waters grazing during the winter months, and perhaps even during summer, is extremely light, and this combined with all the other factors may allow plants to grow slowly over many seasons, for how else could the large plants reported for these regions survive? Indeed it needs phycologists as bold as Wilce and Neushul to dive in this cold water and estimate the age distribution of populations.

Fig. 3.37. The relative abundance of *Clathromorphum circumscriptum* (*a*) and *Phymatolithon rugulosum* (*b*) at north, south, east and west stations off Iceland. From Adey, 1968.

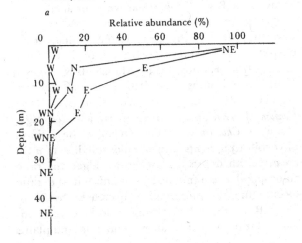
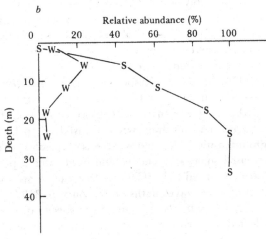

Pressure effects with depth have rarely been considered, whilst other experimental studies have used such unnatural pressures that little of ecological value can be discerned from the results. Pressure increases about one atmosphere for every 10 m depth and some effect, even if slight, might be attributable to this factor. One of the rare studies of morphological variation with depth did show that the thickness of the air bladders of *Ascophyllum* increases with depth (Damant, 1937), but this is not normally a subtidal plant and studies are needed on a range of species.

There are probably few, if any, subtidal species which occur evenly at all depths but very few measurements have been made. One example is the work of Adey (1968) who plotted the abundance of coralline algae around Iceland and Fig. 3.37 shows two contrasting distributions with depth.

Light effects

Deep water seaweeds exposed to intense light are easily damaged particularly by short wavelengths, because in such conditions photo-reduction of pigments and depression of photosynthesis occur. Subtidal species can be killed by a two-hour exposure to direct sunlight (Biebl, 1952), although such exposure to unnatural conditions adds little to the understanding of algal ecology. At high-light intensities, there is, according to Hellebust (1972), a reduction in both the light and dark reactions of photosynthesis; rapid loss of ribulose 1, 5 diphosphate carboxylase occurred during the latter part of exposure. On the other hand, intertidal species are relatively little damaged. The high light intensities of the tropics induce a yellow–green coloration in Rhodophyta; this is probably a side effect of intensity on the balance of pigments. Similar but not such intense yellowing of thalli can be observed even in Scottish sea lochs where the populations are exposed to the air for long periods of time.

Growth of calcified species, e.g., of *Acetabularia*, in low light intensity tends to reduce the calcification and the algae are then much more susceptible to damage from bleaching when transferred to high light intensities. The photosynthesis of *Acetabularia* is increased some 5–6 fold in blue light after a pretreatment in red light (Clauss, 1972); compare the effect of these wavelengths on diatoms, p. 353.

Day length also has an effect; species of *Ulva* collected on northern coasts were found to tolerate continuous illumination, whereas forms from the Mediterranean did not (Føyn, 1955).

Low light or darkness has variable effects on the viability of algae; for example, Boalch (1961) maintained *Ectocarpus confervoides* in a viable state in the dark for 150 days in spite of the fact that it did not grow on organic substrates. Collections of healthy algae from as deep as 100 m in the arctic and antarctic are further evidence of this viability. Epilithic algae extend into caves along rocky coasts and in such habitats the effects of low light intensity can be studied. Dellow & Cassie (1955) report *Nodularia*, *Calothrix*, *Rhodochorton*, *Hildenbrandia* and *Enteromorpha* growing in light intensities as low as 5 lux. Of these, *Rhodochorton* seemed to tolerate the greatest shade. Underwater caves are common along some Mediterranean coasts and in one of these Ernst (1959) showed that the *Cytoseira* community disappears first, then *Dictyopteris*, followed by *Udotea* and *Peysonnelia* and finally, in the almost total darkness, only a few melobesioid red algae survive; in fact the zonation parallels that of growth of these genera at depths in the subtidal.

There is slight evidence of 'chromatic adaptation' in algae but the old text-book concept of greens uppermost, brown algae mid and red algae at lowest depths cannot be substantiated, e.g., Crossett *et al.*, (1965), working down a vertical subtidal rock face in Malta, found brown algae (mainly *Padina pavonia*, *Dictyota dichotoma* and *Cystoseira* spp.) near the surface and green algae only became abundant at 15 m, increasing in abundance right down to 60 m. Red algae were only common between 30 and 45 m. Work by Gilmartin (1960) at Eniwetok showed that many of the Bryopsidophyceae were confined to the deeper water in the lagoon and the red algae were never important components of the biomass at the deep stations. Most recently Ramus, Beale & Mauzerall (1976); Ramus, Beale, Mauzerall & Howerd (1976) and Ramus *et al.* (1977) have shown that pigment content increases with depth in green (*Ulva*, *Codium*), brown (*Fucus*, *Ascophyllum*) and red algae (*Porphyra*, *Chondrus*). In the red algae the ratio of phycobilin pigments and of chlorophyll b to a increased with depth. In the brown algae the ratio chlorophyll c:a remained constant but the ratio fucoxanthin to chlorophyll a decreased by 20 to 30%. Ramus and his colleagues do however comment that the intertidal forms behave as land plants

Table 3.14. *The internal temperatures of* Fucus *receptacles collected at different times in Halifax Co. Nova Scotia. From Bird & McLachlan, 1974*

Date	Sea temperature (°C)	Air temperature (°C)	Species	Receptacles temperature (°C)
1970				
4 June	6	13.5	F. edentatus	14–16.5
			F. vesiculosus	17–24
29 June	8	10.5	F. spiralis	15–20
			F. vesiculosus	15.5–21
8 December	5.5	−9	F. edentatus	−6–2
21 December	3	−7	F. distichus subsp. distichus	−3.5–−3
1971[a]				
8 January	2	−8	F. edentatus	−6–−4.5
			F. distichus subsp. distichus	−3.5–−3
4 February[a]	−1.5	−16.5	F. edentatus	−10–−5
8 July	8	24	F. spiralis	18.5–20.5
			F. vesiculosus	18–20.5
21 July	12	24	F. spiralis	23.5–26.5
			F. vesiculosus	23–24.5
27 August	12	18	F. spiralis	18.5–25

[a] Exposed intertidal zone coated with ice.

in that they increase the pigment content but do not change the ratios.

Temperature effects

Cold hardiness is, as might be expected, best developed in temperate–boreal species and least in subtidal tropical species, although many cold-hardy species are sensitive to heat. The cold hardiness of marine algae varies seasonally, e.g., summer plants of *Fucus vesiculosus* can withstand −30 °C whereas winter plants can withstand −45 °C (Parker, 1960), whilst, in May, *Fucus* from the low-tide level was 15 °C more sensitive to cold than were plants from the upper intertidal. Parker also reports the interesting situation that the growing tips of the plants were hardiest in winter–early spring and least hardy in late summer. Whether or not this is the result of steady adaptation or is due to alterations in reserve products, etc., is not known. Even in cool to cold seas the species occurring in the subtidal zone are more susceptible to cold than are those in the intertidal zone. Freezing of the water in the cells of intertidal algae leads to greatly increased salt concentration

and frost plasmolysis, but not necessarily to death; Kanwisher (1957) found that up to 80 % of the water in the thallus can be frozen and that *Fucus* plants frozen into sea ice at temperatures down to −40 °C can photosynthesise immediately upon thawing. When the same species grows in both temperate and tropical seas it has a greater cold tolerance in the cooler waters, and such a feature is further evidence for the adaptability of species to temperature (Biebl, 1962). Bird & McLachlan (1974) used a hypodermic thermister probe to measure the internal temperature of *Fucus* receptacles, and at different times of the year found a range from −10 °C to 26 °C (Table 3.14). Zygotes of *F. edentatus* were found to be affected more by low temperature (−15 °C), when taken from water at 15 °C than from water just above 0 °C, showing a seasonal cold hardiness effect similar to that of thalli. Fig. 3.38 shows that cold hardiness varies from fertilisation to germination and clearly the most critical times are at fertilisation and from 4 to 24 h afterwards, whereas when germination is underway the resistance of the plants is much greater.

Heat hardiness is least developed in tropical, subtidal species whereas the temperate *Fucus vesiculosus* has been shown to resist a tissue temperature of 54 °C and tissue temperatures between 30 and 40 °C are regularly recorded (Schram, 1968). Air drying, such as occurs along coasts during tidal exposure, enhances the resistance of algae to high temperature.

It has been known from early studies that, as temperature is lowered, the rate of respiration decreases more rapidly than that of photosynthesis, yet another feature which would aid the growth of algae at low temperatures in the polar regions. Again, as might be expected, algae from almost identical habitats have different responses to temperature; these are linked with salinity effects as Schwenke (1959) showed with *Delesseria* and *Phycodrys* but he reports that it is difficult to adapt these subtidal algae artificially to different temperature regimes. Fig. 3.39a shows that, especially for *Ptilota serrata*, the photosynthetic behaviour of shallow and deep subtidal populations varies quite considerably. There is a slight seasonal variation in heat tolerance, though the differences are much less than those for cold tolerance, e.g., about 1 °C difference between February and August for *Fucus vesiculosus*, *F. serratus*, and *Ascophyllum nodosum* at around 39–42 °C (Feldmann & Lutova, 1963).

There are very few modern studies of the

metabolic functioning of epilithic algae; Kanwisher (1966) studied both the respiration and photosynthesis of some intertidal species at different temperature. Fig. 3.40 shows that respiration increases logarithmically with temperature as in higher plants. These graphs show that *Ulva*, *Enteromorpha*, and *Ceramium* respire at the same rates in winter and summer; *Fucus vesiculosus* shows the same respiration rates at summer temperatures up to 30 °C whilst above this temperature respiration decreases and the decrease in winter takes place at 20–25 °C. In *Chondrus crispus* however there is a shift to lower oxygen consumption in summer and *Ascophyllum nodosum* has a summer Q_{10} of 1.5 which rises in winter to 2.0, so that both these species show a degree of

Fig. 3.39. *a*, the net photosynthesis of shallow (−20 ft) and deep (−80 ft) subtidal plants of *Ptilota serrata* and *Phyllophora truncata* grown at 5 °C. *b*, the net photosynthesis of winter and summer plants of *Phycodrys rubens* at −40 ft, at various temperatures and 254 ft-c. From Mathieson & Norall, 1975.

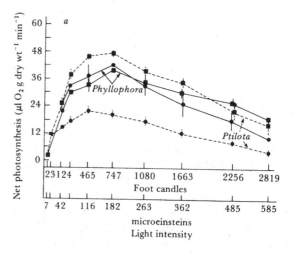

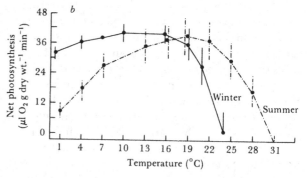

Fig. 3.38. Cold-hardiness of *Fucus serratus* zygotes from the point of fertilisation to germination. *a*, frozen to −5 °C; *b*, frozen to −10 °C; *c*, frozen to −15 °C; *d*, kept at −15 °C for 2 h. From Bird & McLachlan, 1974.

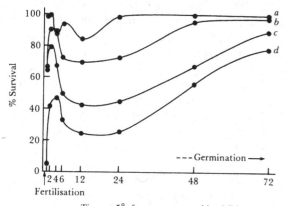

seasonal adaptation to temperature. *Phycodrys* also exhibits slightly different patterns of photosynthesis in summer and winter plants (Fig. 3.39*b*), the winter plants functioning more efficiently at low temperature and less efficiently at high temperature whilst the reverse is true of the summer plants. Photosynthesis of the species studied by Kanwisher showed a maximum rate about 20 times that of respiration (Fig. 3.41) all the rates lying around the 20:1 line

on this figure. This suggests that the plants respire only 10% of their production, a highly efficient situation. The figure also shows that the thin *Ulva* with its cells all exposed to the medium has a greater photosynthetic capacity than the massive 'tissue' based *Ascophyllum*; the days on this graph are the doubling times estimated for the algae and, as would be expected, the largest form with much non-photosynthetic tissue is the slowest; such rates are of course unlikely in nature since shading etc. will prevent full light saturation. Fig. 3.42 shows that the photosynthetic curves for two of the algae are similar to those obtained for populations of unicells and that *Chondrus* is light saturated at about half the intensity which saturates *Fucus*, a feature which is correlated with its habitat beneath the larger Fucoids. In another series of studies, Burns & Mathieson (1972) found that *Chondrus crispus* grew more rapidly than *Gigartina stellata* under various light and temperature regimes (Figs. 3.43, 3.44) which would suggest that the former has a competitive advantage, though the samples for their experiments were collected from subtidal and intertidal stations respectively and may therefore be adapted to different conditions. In many places the two species grow alongside one another and it would be interesting to compare their growth under such conditions, though one can almost certainly assume that no two genera have identical responses. Growth at varying salinities (Fig. 3.11) showed that *Chondrus* was a plant more tolerant of a range of salinity, a feature which correlates with its extension into estuaries on the New Hampshire coast whereas *Gigartina* thrives only on the open coast.

Clearly it is necessary to test competition between marine algae, not only by growing the

Fig. 3.40. The effect of temperature on the respiration of some seaweeds in summer and winter. *a*, *Ulva lactuca*; *b*, *Chondrus crispus*; *c*, *Enteromorpha linza*; *d*, *Ceramium rubrum*; *e*, *Ascophyllum nodosum*; *f*, *Fucus vesiculosus*. From Kanwisher, 1966.

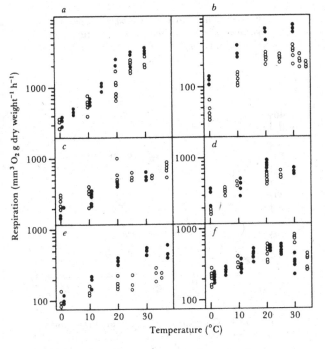

Fig. 3.41. The relationship between photosynthesis and respiration of some seaweeds. From Kanwisher, 1966.

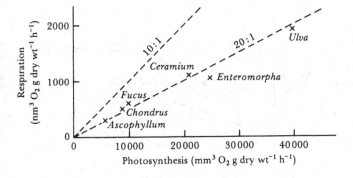

individuals in unialgal culture but also in combination with ecologically associated species. Russell & Fielding (1974) did just this by growing three species, *Ulothrix flacca*, *Ectocarpus siliculosus* and *Erythrotrichia carnea* in all possible combinations, at different temperatures, light intensities and salinities, and then measuring the yields of each species by microscopic checking of frequency of each alga, finally relating these to the total volume of harvest. Table 3.15 gives their data and shows, for example, that *Ectocarpus* completely outperformed *Ulothrix* at 15 °C and 2000 lux but the latter grew quite well with *Erythrotrichia* under the same conditions. They quite rightly point out that extrapolation from these laboratory experiments to nature should be approached with caution but comment that the emergence of *Ulothrix* as a vigorous competitor under low temperature, bright light and reduced salinity conditions is consistent with its winter–spring occurrence and its position on the upper shore where fresh water seeps across. Less strong light and higher temperatures favour *Erythrotrichia* and again this is in accordance with its occurrence in submerged communities. The ability of

Ectocarpus to compete well under most experimental conditions is in accord with its widespread distribution on British shores. In *Ectocarpus*, salinity tolerance seems to be a genetically determined attribute since even after acclimatisation at 34‰ for four years some low salinity, estuarine strains continued to perform better at lower salinities (Russell & Bolton, 1975).

Seasonal variations in net photosynthesis have been shown for *Phyllophora truncata* and *Ptilota serrata* both of which have lower light optima in winter than in spring (Mathieson & Norall, 1975). These workers also showed that deep water plants of *Ptilota* have lower optima and reduced net photosynthesis compared with shallow subtidal plants. Similar effects were noted with regard to temperature;

Fig. 3.42. The relationship between photosynthesis and illumination at two different temperatures for *Chondrus* (*a*) and *Fucus vesiculosus* (*b*). From Kanwisher, 1960.

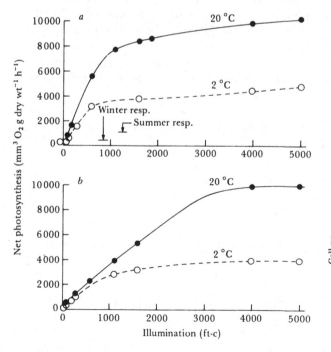

Fig. 3.43. The growth of *Chondrus crispus* and *Gigartina stellata* at different temperatures, 440 ft-c. and a salinity of 30‰. From Burns & Mathieson, 1972.

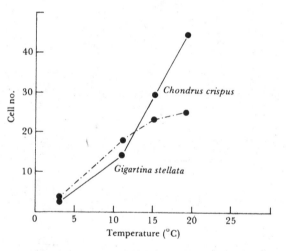

Fig. 3.44. The growth of *Chondrus crispus* and *Gigartina stellata* at different light intensities, 15 °C and a salinity of 30‰. From Burns & Mathieson, 1972.

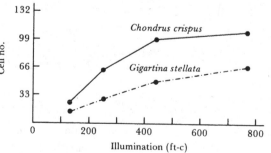

Table 3.15. *Yields of* Ulothrix (Ul), Ectocarpus (Ec), *and* Erythrotrichia (Er) *grown in mixed cultures at different light intensities, temperatures, and salinities*

1, 2, 3, inoculum ratios of 1:3, 2:2 and 3:1 respectively of the genus and its competitor (Co). The figures in bold type are those showing significant differences ($P \leqslant 0.05$). From Russell & Fielding, 1974

Species (Inoculum ratio in mixtures)

	Ulothrix							Ectocarpus							Erythrotrichia						
10 °C	1		2		3		Co	1		2		3		Co	1		2		3		Co
2000 lx	12	13	26	24	37	36	Ec	49	47	68	67	72	70	Er	2	1	5	5	7	7	Ec
	25	26	31	34	35	36	Er	12	12	24	26	37	38	Ul	0	0	7	6	9	9	Ul
1000 lx	4	4	15	14	25	23	Ec	38	37	36	39	39	39	Er	1	1	2	3	2	3	Ec
	15	15	31	27	32	34	Er	13	15	22	22	27	30	Ul	17	16	10	10	30	28	Ul
300 lx	4	6	9	8	10	10	Ec	36	35	33	33	40	42	Er	2	3	4	5	4	6	Ec
	5	5	9	9	14	15	Er	11	9	25	28	38	34	Ul	27	25	23	25	23	23	Ul
15 °C																					
2000 lx	0	0	0	0	8	8	Ec	27	26	34	33	35	34	Er	5	5	4	5	8	9	Ec
	17	20	25	23	27	23	Er	33	31	39	39	38	42	Ul	0	0	0	0	4	5	Ul
1000 lx	2	1	5	6	8	9	Ec	11	11	17	19	26	31	Er	17	17	24	26	31	32	Ec
	8	8	10	12	16	17	Er	29	31	27	30	38	35	Ul	18	16	20	21	26	27	Ul
300 lx	2	2	7	8	10	10	Ec	3	1	8	9	17	16	Er	16	16	14	14	13	13	Ec
	4	5	8	9	8	10	Er	11	13	16	14	19	19	Ul	13	13	10	9	14	12	Ul
20 °C																					
2000 lx	5	5	14	15	16	16	Ec	21	22	29	30	27	33	Er	4	4	10	9	4	5	Ec
	7	5	12	9	19	21	Er	32	34	48	43	46	44	Ul	16	16	28	26	30	26	Ul
1000 lx	0	0	4	5	8	7	Ec	13	15	34	32	30	31	Er	2	3	7	8	19	21	Ec
	4	2	2	1	1	2	Er	18	21	19	20	30	30	Ul	28	25	36	34	38	40	Ul
300 lx	3	4	5	6	9	9	Ec	8	9	18	18	25	26	Er	6	7	13	11	12	12	Ec
	3	4	5	3	4	5	Er	7	9	10	9	12	13	Ul	6	7	8	8	10	10	Ul
S 22‰	5	5	10	9	10	10	Ec	6	7	10	8	13	15	Er	8	9	17	19	20	22	Ec
	3	3	5	6	10	12	Er	3	4	10	12	10	12	Ul	10	11	18	16	20	23	Ul
S 12‰	7	8	9	10	12	12	Ec	2	4	4	6	8	10	Er	2	2	5	6	7	9	Ec
	9	9	9	10	16	15	Er	2	3	5	6	9	9	Ul	1	1	5	6	10	10	Ul
S 1‰	1	1	2	3	5	5	Ec	3	4	5	5	7	6	Er	0	0	0	0	0	0	Ec
	0	0	2	2	4	4	Er	2	2	4	5	6	6	Ul	0	0	0	0	0	0	Ul

higher summer optima were observed. These and other data such as those from studies by Biebl, McLachlan, Kanwisher and other workers point to a considerable capacity for adaptation to ambient conditions.

Relationships between growth rates and temperature are also correlated, presumably under genetic control, to the sites of occurrence of species. *Fucus distichus*, a cold-water form, has its highest growth rate at 9 °C whereas *F. vesiculosus* and *F. virsoides* increase their growth rates as temperature increases (Munda & Lüning, 1977a). In another experiment (Munda & Lüning, 1977b) sporophytes of *Alaria* from Iceland grew well in winter but less

well in summer in the sea off Heligoland and they deteriorated in August temperatures. Lüning & Neushul (1978) show similar temperature optima effects for southern and central Californian laminarians, the former peaking at 17 °C and the latter at 12 °C.

Field measurements

Measurements of biomass of epilithic algae are not easy. Standing crop measurements can be made by harvesting individual species, weighing (wet, dry or ash weight) and expressing as g/m². Alternatively, plants can be harvested, homogenised and chlorophyll *a* contents determined. In all cases problems

Fig. 3.45. Standing crops of some benthic algae
and angiosperms in relation to varying current
regimes. Modified from Conover, 1968.

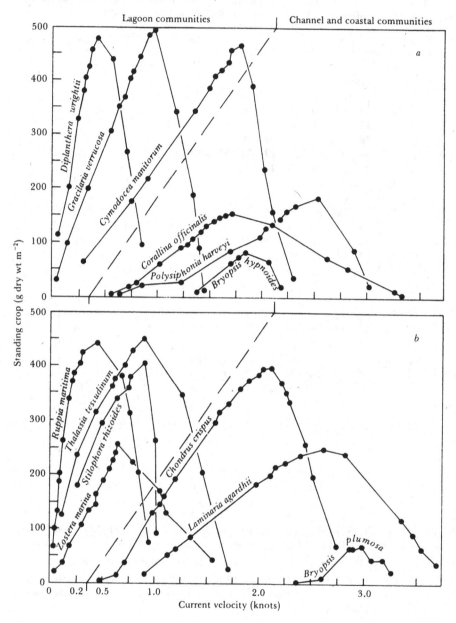

Table 3.16. *Comparison of photosynthesis in air and in seawater at 15 °C and 43000 lux for algae, the vertical distribution of which is shown in Fig. 3.46. From Johnson et al., 1974*

Species	Photosynthesis in air[a]		Photosynthesis in seawater		Ratio air/ water rate
	CO_2 $(mg\ dm^{-2}\ h^{-1})$	CO_2 $(mg\ gdw^{-1}\ h^{-1})$	CO_2[b] $(mg\ dm^{-2}\ h^{-1})$	CO_2 $(mg\ gdw^{-1}\ h^{-1})$	
Endocladia muricata	na[c]	3.4 ± 0.21	na[c]	2.1 ± 1.0	1.62
Porphyra perforata	9.4 ± 0.50	17.7 ± 0.95	3.3 ± 0.57	6.3 ± 1.1	2.84
Fucus distichus	11.2 ± 0.35	5.8 ± 0.19	1.7 ± 0.49	0.9 ± 0.13	6.59
Iridaea flaccida	4.1 ± 0.20	5.2 ± 0.25	1.4 ± 0.22	1.7 ± 0.35	2.94
Ulva expansa	3.7 ± 1.0	12.2 ± 3.0	5.6 ± 0.12	16.7 ± 0.12	0.66
Prionitis lanceolata	2.7 ± 0.10	1.1 ± 0.04	3.04 ± 0.02	1.2 ± 0.1	0.84

[a] Maximum rate recorded during desiccation.

[b] 1 mg CO_2 uptake is equivalent to 0.507 ml O_2 evolution during photosynthesis.

[c] na, not available.

$\pm$, one standard deviation of duplicate measurements.

may arise from difficulties of sampling whole plants, for example removal of basal attachment organs. Off Nova Scotia, 80% of the algal biomass is comprised of *Laminaria* and *Agarum* and subtidal standing biomass can reach 30 kg/m², the average value being 4 kg/m² (Mann, 1973). Mann comments that the *Laminaria* blade is like a 'moving belt of tissue growing from the base and eroding at the tips at approximately the same rate and the length could be renewed up to 5 times in the course of a year'. On the other hand Brinkhuis (1977) found that *Ascophyllum nodosum* and *Fucus vesiculosus* turn over their biomass twice per year on the Atlantic coast of North America. There are only scattered comments in the literature concerning the age structure of

Fig. 3.46. The vertical distribution of some algae on the California coast at Pacific Grove relative to tidal height and mean yearly exposure. From Johnson *et al.*, 1974.

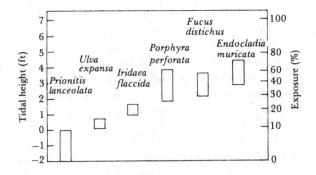

populations and of individual plants. Kain (1971) found that the maximum age of *Laminaria hyperborea* plants in the Isle of Man was ten years and thought that some Norwegian plants could be as much as 18 years old. Large plants are not necessarily old plants and in some places the mortality rates are so high that plants do not survive for any length of time. *Egregia laevigata*, for example, does not survive for more than eight months at some sites and grows essentially as an annual opportunist (Black, 1974).

There are surprisingly few data on the relationship between standing crops and local habitat conditions; one factor which is obviously important is rate of water flow over plants and Conover (1968) showed that each plant (including aquatic angiosperms) had a definite optimum related to current speed and that lagoon and channel–coastal communities could be separated (Fig. 3.45). Under conditions of increased wave action the fronds of *Fucus vesiculosus* tend to lose their vesicles (Burrows & Lodge, 1951; Jordan & Vadas, 1972). The commercially important, tropical genus *Eucheuma* grows faster in regions of high water flow and also in shallower regions where light intensity is higher (Doty, 1971). Doty points out that in tropical regions the importance of tides is more related to current movement than to the uncovering of the shore where, in any case, the biomass tends to be considerably reduced.

Another important factor rarely investigated is

the effect of exposure on photosynthetic rate. One interesting series of experiments (Fig. 3.46; Table 3.16) showed that *Ulva expansa* and *Prionites lanceolata* from the lower intertidal have a reduced rate in air but *Iridaea flaccida, Fucus distichus, Porphyra perforata* and *Endocladia muricata* assimilate 1.6–6.6 times faster in air than in water (Johnson, Gigon, Gulmon & Mooney, 1974). Rate of release of dissolved organic carbon is likewise rather neglected for these large algae but Brylinsky (1977) found fairly low rates, e.g., 1.1–3.8% released in the light, rates which are much lower than those reported by earlier workers on phytoplankton (but *see* also p. 316).

Colonisation and succession. Natural opportunities for studying this aspect of epilithic algal ecology are not often presented and artificial clearing of shores has to be attempted. Recently, volcanic eruptions have yielded sites which have been studied, one in the tropics on Hawaii (Doty, 1967) and the other in arctic waters off Iceland with the appearance of the island of Surtsey (Jónsson, 1970). The first alga to appear on the Hawaiian lava flows was a species of *Enteromorpha*; microscopic species were not investigated but no doubt they were present even earlier since Doty comments on the coating of diatom epiphytes on the mature *Enteromorpha*. The *Enteromorpha* was then joined by *Ectocarpus breviarticulatus* and about six months after the flow cooled a blue-green algal coating appeared above the *Ectocarpus*. Fig. 3.47 illustrates the sequence of changes. Of the common belt-forming species in Hawaii, *Ahnfeltia* appeared in the second year but *Ralfsia pangoensis* was delayed for four and a half years. On newly exposed surfaces on mature shores the original pioneers recur and this no doubt accounts for their inclusion in lists of so-called climax communities. Comparison of Fig. 3.47 with the zonation on mature shores in Hawaii (Fig. 3.48) shows that the process of stabilisation is slow and Doty comments that since the zonation on 100-year-old lava flows differs from that on prehistoric flows the time taken to reach stability in this tropical region may be as long as that for some terrestrial vegetation; this is the only paper I know which suggests such long-term changes. Lawson & John (1977), on the other hand, consider that as little as six years is required before bare areas on the coast of Ghana (with a similar climate to Hawaii) return to a state indistinguishable from the neighbouring undisturbed areas.

On a temperate coast where areas were cleared *Gigartina papilata* appeared in the first year and in the second year *Fucus distichus* began growth. This alga took two years to dominate the canopy by which time *Gigartina* was reduced to an encrusting system (Dayton, 1971).

The first colonists on the lava of Surtsey were found a few months after the island arose from the sea in 1963 (Jónsson, 1970) and consisted of bacteria and a few diatoms (*Navicula mollis, Nitzschia biloba* var. *minor*). These organisms were then obliterated by a new lava flow and a year later (1965) another sample yielded other diatoms (*Synedra affinis* var. *parva, Thalassionema nitzschiodes* and *Licmophora gracilis* var. *anglica*), which suggests that the primary colonists are simply opportunistic epithilic species which happen to alight on the rock and there are no specific primary colonists. Castenholz (1963) also found that diatoms occurred on cleared intertidal rock surfaces within 2 to 3 weeks. At the same time as the diatoms were noted on Surtsey the green alga *Urospora penicilliformis* was also detected and seen to spread during the succeeding months. By the time of the next visit, a year later, the whole of the rocky littoral had a growth of algae, including *Urospora, Ulothrix, Enteromorpha*, numerous species of diatoms, *Porphyra umbilicalis* in pockets in the rock around the upper intertidal and even an isolated plant of *Alaria esculenta* in the lower intertidal. Greater diversity occurred in the rock pools, where brown algae were dominant (*Petalonia zosterifolia, Scytosiphon lomentaria, Pilayella littoralis* and *Ectocarpus confervoides*), one green alga (*Enteromorpha*) occurred but, perhaps significantly, no red algae; does the complicated red algal life cycle result in a slower dispersal and colonisation rate? In later years, further species were added but again few red algae invaded except in the subtidal. This sequence of green → brown → red algae fits also with clearance experiments, e.g., those of Northcraft, 1948. According to Jónsson the subtidal colonisation was slower than that of the intertidal and again the earliest colonists were tube-forming diatoms. By 1968, *Alaria esculenta* had formed a subtidal belt (from 3 to 19 m in depth) around the whole island and in this forest occasional individuals of smaller red, brown and green algae appeared.

The recolonisation of artificially cleared rock surfaces has been studied by a number of workers and all the results tend to show settling of spores and germination over a wider range (e.g. Hatton, 1938;

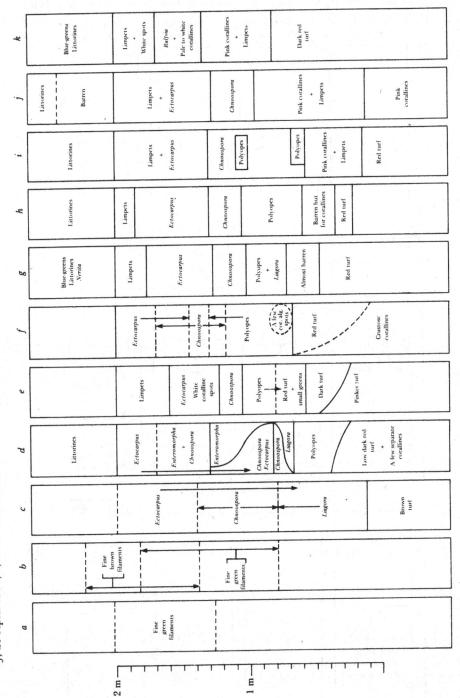

Fig. 3.47. Population changes with time on a vertical 1955 lava flow running into the sea. *a, c,* at Kaulleau on 21 June 1955 and 15 August 1955 respectively. Other readings at Kehora: *b,* 30 December 1958; *d,* 21 December 1955; *e,* 24 March 1950; *f,* 16 May 1956; *g,* 14 July 1956; *h,* 18 August 1956; *i,* 10 November 1956; *j,* 20 April 1957; *k,* 30 December 1958. From Doty, 1967.

Fig. 3.48. The vertical distribution patterns of marine algae and animals on mature Hawaiian shores on the islands of Maui and Hawaii. The distance in metres at the top of each strip indicates the distance between the two horizontal dark lines across each strip and thus provides a scale for each strip. From Doty, 1967.

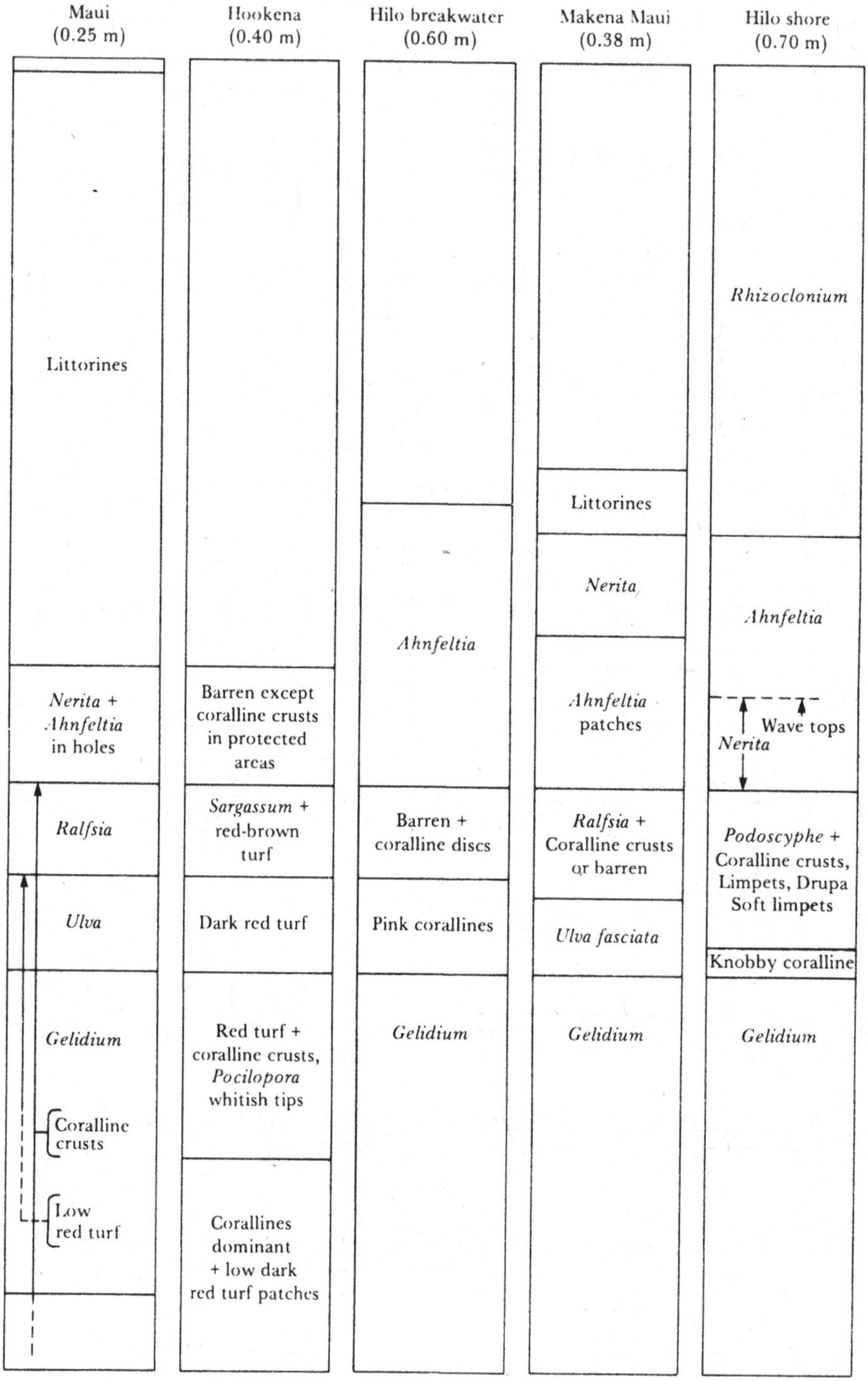

Maui (0.25 m)	Hookena (0.40 m)	Hilo breakwater (0.60 m)	Makena Maui (0.38 m)	Hilo shore (0.70 m)
Littorines				*Rhizoclonium*
			Littorines	
			Nerita	
		Ahnfeltia		*Ahnfeltia*
Nerita +Ahnfeltia in holes	Barren except coralline crusts in protected areas		*Ahnfeltia* patches	Wave tops *Nerita*
Ralfsia	*Sargassum +* red-brown turf	Barren + coralline discs	*Ralfsia +* Coralline crusts or barren	*Podoscyphe +* Coralline crusts, Limpets, Drupa Soft limpets
Ulva	Dark red turf	Pink corallines	*Ulva fasciata*	Knobby coralline
Gelidium	Red turf + coralline crusts, *Pocilopora* whitish tips	*Gelidium*	*Gelidium*	*Gelidium*
Coralline crusts				
Low red turf	Corallines dominant + low dark red turf patches			

Northcraft, 1948) than that occupied by the same species in the climax community. The latter is thus the result of the interaction of biological factors such as competition, grazing, etc. and physical ones, like exposure, slope, etc. and is not due to absence of germlings or unsuitable conditions for germination. Some algae seem to require an artificially cleared area on which to establish. *Postelsia*, for example, grows well on areas of rock which have been battered by logs; it can also clear areas by first developing on other algae and animals and then tearing them off the rock since they do not give a sufficiently stable substratum for such a large alga which favours a wave-swept shore (Dayton, 1973). Obviously the primary colonisers will be species derived from those shedding spores at the time of clearance. For example, Fahey (1953) found that transects cleared in summer and autumn were repopulated by *Entermorpha*, *Polysiphonia* or *Calothrix* at higher levels whereas on winter-cleared or ice-scoured rocks the first visible algae were the brown genera *Chordaria* and *Scytosiphon*. Northcraft (1948) and Emerson & Zedler (1978) on the other hand found little difference in the primary colonisers whatever season the rocks were cleared. These first colonisers tend to be 'fast growing' species, also termed 'opportunistic' by some workers, 'ephemerals' by others and 'fugitive' by Dayton (1975b); the climax species tend, on the other hand, to be slow growing and long lived. Burrows & Lodge (1951) made the interesting observation that on a cleared shore in the Isle of Man numerous unidentifiable *Fucus* hybrids appeared. There have been suggestions that the succession of species in recolonisation experiments is 'community controlled' but Connel (1972) thinks that its orderliness is a consequence of differences in life-history characteristics between the opportunistic species and the dominants and that community control comes only when the dominants are established, e.g., when the canopy species are well grown they then control the understorey.

Few workers have cleared areas in the subtidal zone but where clearance has been achieved, e.g., by Kain (1975b) very interesting data have resulted; for instance, removal of the *Laminaria hyperborea* cover at Port Erin in November, February or June enables *Saccorhiza polyschides* to grow but *L. hyperborea* re-achieved dominance in the second year and by the third year a virgin forest biomass had re-established.

Clearance in August, however, excluded *Saccorhiza*, and on blocks repeatedly cleared the biomass appearing in the winter was dominated by Rhodophyta, in spring by Phaeophyta and in late summer by Chlorophyta. It is interesting to compare this sequence with the phenological data in Table 10.2 and note the differing times of reproduction for the Chlorophyta, Phaeophyta and Rhodophyta. The time taken to recolonise cleared surfaces varies greatly but the minimum for large seaweeds seems to be about nine months, and it can be anything from 8 to 30 years for *Ascophyllum nodosum* (Baardseth, 1970). Obviously such communities need to be conserved rather carefully unless the habitat is to be damaged for many years. *Ascophyllum* seems to be easily removed from some shores by visitors walking over the rocks and so its recovery time is a matter of some concern.

Recolonisation is dependent upon a source of propagules yet sometimes one finds that a coastal species is sterile, or only very rarely fertile, or merely tetrasporic, e.g., *Gelidium* species around the British Isles. These perennate by over-wintering of complete plants or of the rhizoidal portions only (Dixon, 1965) but how do they manage to spread to cleared areas? Since they are being maintained as vegetative stands, exchange of genetic material, even on a local scale, is unlikely, and such exchange is even less likely over distances. However all records of these large epilithic algae occurring as only a single phase need very careful study, since recently an encrusting tetrasporophyte has been found in the life history of the very common *Ahnfeltia plicata* (Farnham & Fletcher, 1976) and an even more abundant alga *Gigartina stellata* has been shown to have a *Petrocelis*-like encrusting stage in Pacific material; such a stage ought to be looked for on European shores since it has also been found in culture from Canadian Atlantic sites (Chen, Edelstein & McLachlan, 1974). The carpospores have been shown in culture to give rise directly to the gametophyte without intercalation of a tetrasporophyte (apomictic life cycle). In laboratory experiments, photoperiods of 16 h white light have been found to prolong the crustose phase of *Scytosiphon*, whilst 8 h days allow the erect thallus to form (Dring & Lüning, 1975); how such mechanisms might operate in nature has, however, not been investigated.

Fragmentation is probably fairly common and

may assist recolonisation, though there do not seem to be many special adaptations for such propagation. One exception is the occurrence of propagules in *Sphacelaria* and another is *Codium fragile* subsp. *tomentosoides* growing in New England which has recently been shown to form special swellings along the fronds in winter; the swellings are found especially near the base of dichotomies, and result in fragmentation of practically the whole thallus, only the base being left to regenerate in the next year (Fralick & Mathieson, 1972).

Epipsammon

The flora attached to sand grains is fairly extensive in both freshwater and seawater but has, as yet, been studied less than almost any other algal community. Observation under the microscope usually reveals either quite clean looking grains or grains with a film of mucilage in which bacteria and organic particles are prolific. Small coccoid Cyanophyta and Chlorophyta often occupy the hollows of the grains but the most conspicuous components are diatoms, either adnate to the surface or raised on very short mucilage stalks (Fig 3.49). In marine beach sands the organic material is almost all associated with the sand grain surfaces and only about 5% of detrital material is in the interstitial water (McIntyre, Munroe & Steele, 1970). The sand, especially in marine sites, is regularly mixed and some algae must spend only short periods in the light, yet no trace of heterography has been detected; for example Munroe & Brock (1969) used ^{3}H-acetate and ^{3}H-glucose and found that all the label was taken up by the bacteria and none by the diatoms. However, Moss (1977b) found that the epipsammic population was much more resistant to darkness and anaerobiosis than was the epipelic.

Although attached, the diatoms must have at least short free-living periods in order to colonise new grains; this feature of free-living periods is one which probably applies to all attached communities but which has received almost no attention. Harper (1969) found that the epipsammic diatoms from a freshwater site showed a very low level of movement, e.g., *Amphora ovalis* v. *pediculus* moved an average of 1.7 μm/s whilst the epipelic *Navicula oblonga* averaged 11 μm/s. Fig. 3.50 shows that the epipsammic diatoms are moving relatively little during the light period compared with the epipelic species but never-

theless the rate of movement is sufficient for them to reach the surface. The decrease in the number at the surface around mid-day is a phenomenon which is often observed in such diurnal studies but which remains unexplained. The force attaching algae to surfaces has rarely been measured but Harper & Harper (1967) found that it is essential for diatoms to attach to a surface even when moving. They found that *Amphora ovalis* with all four raphe slits in contact with the surface was able to move whilst exerting adhesions of over 400 millidynes. The diatoms attached to sand grains were found to have very strong adhesion to the surface. Unpublished observations by the author have revealed attachment of sand grains to the valve corners of diatoms in the region of the ocelli, and the formation of considerable aggregates. The importance of this attachment coupled with mucilage formation has, I believe, been underestimated as a stabilisation factor in in-shore sediments. Holland, Zingmark & Dean (1974) show that only benthic algae, which secrete mucilage, are effective stabilisers.

Some sands are quite devoid of algae but it is not clear what factors prevent colonisation; sand in standing water is readily colonised but few records have been obtained from flowing water, though Moore (1977) reports an epipsammic flora in a eutrophic stream in southern England. Movement of grains may be responsible and needs to be measured at the actual sediment surface since the velocity of flow just above the sediment varies between $\frac{1}{3}$ and $\frac{1}{6}$ of that of the overlying water (Marshall, 1967). The angularity of the grains may also be important and it is most likely that a physical factor is involved in the cleaning of the grains; Meadows & Anderson (1968) considered that abrasion keeps the convex surfaces free but grazing also should be considered.

In freshwater, the flora is dominated by species of *Fragilaria*, *Opephora*, *Achnanthes* and *Amphora*, and in the marine habitat by *Opephora*, *Plagiogramna*, *Dimerogramma*, *Cymatosira*, *Raphoneis* and *Cocconeis*. All these are non-motile or only slowly motile genera and diatoms in general are dominant. A few cells of motile pennate species occur which probably use the raphe system more as an attachment organelle than as a motility mechanism; such genera are *Nitzschia*, *Cymbella*, *Amphora* and some small *Navicula* species. From marine sand, Meadows & Anderson (1968) recorded the Cyanophyta, *Merismopedia*, *Microcystis*,

Fig. 3.49. Two examples of sand grains with
epipsammic diatoms. *a*, from a freshwater site
with *Fragilaria* attached by mucilage pads: *b*,
from a marine site with *Synedra* cells attached by
their apices.

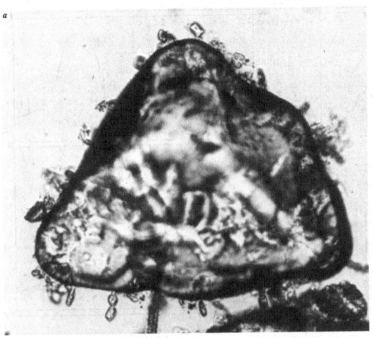

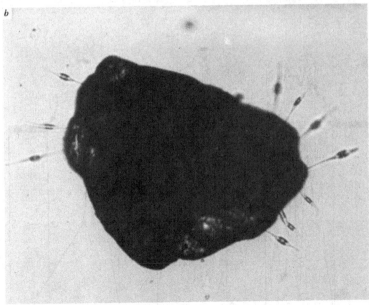

Fig. 3.50. *a*, *b*; the activity of *Amphora ovalis* and the remainder of the epipsammic diatoms during a light period. In *a* is shown the light intensity and the percentage of moving diatoms, both epipsammic (solid outline) and epipelic (dashed outline): *c*, *d*, migration rhythm of epipsammic diatoms during short-day, *c*, 3 November 1966 and long-day, *d*, 2–4 June 1967, conditions. Cell counts represent numbers which had moved from the sand on to coverglasses placed on the sand at each sampling time. From Harper, 1969.

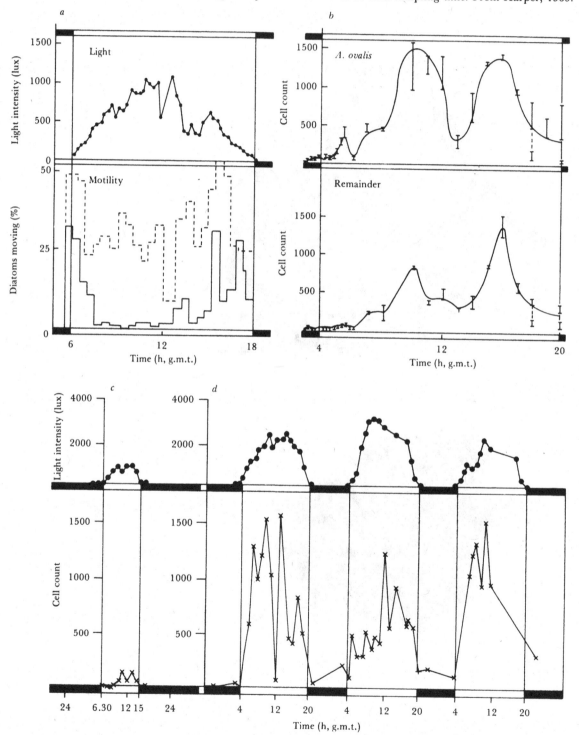

Anabaena, and possibly *Lyngbya*. These workers estimated that some of the colonies on the grains have up to 150 cells and that in general the colonies tended to occur in the hollows of the grains.

Davies (1971) used the term phytopsammon for the algal populations within sands in general, hydropsammon for those on underwater sand and eupsammon for that of the beach. He found some species confined to one or the other but the technique used collected many species which I would regard as epipelic.

Endolithon

Algae living in the fissures, and spreading laterally in rocks, are known as endolithic algae. In temperate regions, where the rocks are moist, the endolithic component has not been investigated and perhaps the habitat is not well colonised, since the surface itself is a sufficiently favourable habitat for growth. On the other hand, in arid zones the surface of the rock is less favourable for growth and few algae are known to colonise this habitat. Instead one finds the algal growth confined to the interior of the rock. Many of the algae can apparently survive at least up to ten years in storage (Friedmann, 1971), which is not surprising since they can presumably withstand intense desiccation in nature. According to Friedmann, all the algae he found belonged to the blue-green algal genus *Gloeocapsa* even though the samples were taken from both sandstone and limestone (the Negev desert in Israel and Death Valley, California). He found the algae in the interstices just beneath the crystalline rock surface layer; it is unlikely that development could occur in more amorphous rocks. The algae grow in the spaces between the particles of rock, and the gelatinous sheaths of the algae are pigmented as is common amongst blue-greens in subaerial environments. The rock in which they live is relatively transparent and so photosynthesis is no problem and moisture is supplied by night-time dew which accumulates in considerable amounts even in these desert habitats.

Friedmann *et al.* (1967) and Friedmann (1971) illustrate endolithic algae living as a subsurface layer in desert rock. They distinguish them from the chasmolithophytes in that they do not appear to have a connection to the outer rock surface though Friedmann admits that penetration must be through fissures. The chasmophytic are in fact simply epilithic

forms living in crevices. Once inside the rock the endolithic algae proliferate and penetrate in a direction horizontal to the surface. In sandstone they occupy the spaces between the insoluble crystals and may even dissolve some of the cementing matrix a few millimetres below the surface; such growths are not confined to desert regions, but can occur anywhere where suitable rock fragments are exposed along shore lines or inland (Round, unpublished). Friedmann *et al.* reported that the endolithic algae were mainly small chroococcoid Cyanophyta with occasional unicellular green algae but they could not culture nor identify them. My experience with similar endolithic green algae is that they are difficult to grow in culture; this is perhaps not unexpected since they do not live in an aqueous medium in nature but in a semi-dry state. As Friedmann *et al.* comment, these organisms have little available nitrogen except from the atmosphere and 'may represent. . . photoautotrophic organisms which live under the most extreme ecological conditions on the earths surface'. In this context it is therefore very interesting that Odintsova (1941) reported that chasmolithic algae in the Pamir desert can fix atmospheric nitrogen.

Another aspect of endolithic algae is their destructive action on the rock. This has been studied mainly by geologists and obviously it is most important in limestone. Purdy & Kornicker (1958) report that algae weaken the surface layers of limestone which then flake off and algae are 'among the most important agents of destruction of coast limestones'. Similar algae (Cyanophyta and Chlorophyta) also bore into mollusc shells and, after decay of the filaments, the cavities are filled with micritic aragonite (Bathurst, 1966).

Along the shores of the Adriatic Sea the calcareous rocks have a rich blue-green algal flora both in and on the surface and it seems that similar species are involved within the fissures and on the surface. Ercegović (1934) was able to distinguish four vertical zones in the region around Split where it seems that this flora replaces to some extent the normal marine flora. Golubić (1969) considers that this type of community is common to most coasts of the world and he distinguished between those algae which occupy existing spaces and those which actively penetrate the rock; some of the epilithic species also have penetrating 'rhizoidal' branches. Van den

Hoek *et al.* (1975) found perforating algae common on the dead coral and limestone platform off Curacao and recorded *Plectonema terebrans, Mastigocoleus testarum, Hyella caespitosa, Phaeophila dendroides, Ostreobium quekettii, O. constrictum* as well as Rhodophycean filaments, which are very common. They also comment that decalcified crustose coralline algae also always contained boring algae.

Endopsammon

A flora living beneath the surface of sand in lakes and in an estuarine situation has been investigated by the author (Round, 1979 and unpublished); I suspect that this community is fairly widely distributed but this suspicion needs confirmation. In a freshwater loch on the Isle of Skye a community of Cyanophyta (*Chroococcus, Aphanocapsa, Oscillatoria, Phormidium, Calothrix*) and the diatom *Frustula vulgaris* secreted mucilage and bound the sand together in a bluish-green layer about 0.5 mm beneath the surface. The surface layer of sand had a few of the same species in it but the sand was loose and had a very sparse epipelic flora moving on and through it.

A rather similar flora, binding the particles together, occurs in the sand of waterworks filter beds. At times this gelatinous mass forms an almost impenetrable barrier to water flow and the filter bed has to be renewed. Epipelic algae also colonise the sand surface, but it is the species which glue together the sand grains which are troublesome. It is common to find metabolically formed calcium carbonate crystals amongst the mucilage forming species (cf. its formation in *Chaetophora* mucilage) and when the bed is dried the mass of mucilage, carbonate and sand grains hardens and twice as much sand has to be removed, washed and even crushed to recover the sand for future use (Metropolitan Water Board Reports 43, 44). Adding algicides (e.g., copper sulphate) to the reservoir water often removes the large diatoms and hence improves filtration but often only slightly delays the growth of the mucilage forming species so that the biological control is intricate and only achieved by detailed study and understanding of the total system. Since slow sand filters need cleaning about seven times a year (this will of course vary according to type of water supply, etc., but is a figure given by the Metropolitan Water Board for some of their filters) reducing algal growth will reduce costs. Shading of filters has been successfully used experimentally to limit the growth but it must not be so complete as to prevent oxygenation of the water which is an important attribute of algal growth.

A similar endopsammic community in a marine environment was found by the author at Barnstaple Harbour on Cape Cod (Massachusetts). Here the sand was not bound by mucilage to the same extent as in the freshwater site, and the endopsammic species tended to be adpressed to the sand grains. The dominant species was the diatom *Amphora cymbifera*, and the lowermost layer of sand was coloured brown from the rich growth of this species. Above this layer was a colourless layer of sand, whilst at the surface was a dense flora of migrating *Hantzschia virgata* var. *intermedia* (Round, 1979).

A single paper by Williams (1965) revealed details of an intriguing endopsammic community of *Nitzschia* living in mucilage tubes. The tubes appear to be aligned vertically in the sediments and Williams photographed the surface with the diatoms half extended from these tubes: apparently they can snap back inside when disturbed, a kind of a movement which is unknown in any other diatom.

Algae are involved in cementing sand grains by carbonate deposition in freshwater lakes and also along marine beaches with the formation of beach rock (*see* p. 495), but the exact mechanism has not been studied. Cribb (1973) reported that the subsurface sand on cays (islands) in the Great Barrier Reefs was slightly compacted by Cyanophyta, mainly *Nostoc* species.

Artificial surfaces

Artificial substrata include all the structures placed in the aquatic–terrestrial environments on which algae grow such as ships' hulls, pilings, glass slides placed in waters for experiments and even the sides of buildings which in all climates, but especially in the moist tropics, are colonised by algae. Many of the sites develop very rich algal floras which are similar, but not identical, to those of natural solid, non-living substrata.

Cultivation of subtidal seaweeds for food is undertaken in Japan, and here the ability to grow on artificial substrata has been used in the successful experimental growing of *Laminaria* (Kombu) on artificial (concrete) blocks set into sandy sediments.

Glass slides, plastic plates, etc.

The use of microscope slides to study the growth of algae on easily handled substrata goes back to studies by Hentschel (according to Lund & Talling, 1957). It is a very convenient method of assessing what first colonises, and following the initial succession of species on a solid substratum, but it must be borne in mind that all it does is give a selection of species from various *natural* communities which happen to fall on the slides and then grow there. The results show a depressingly similar collection of species, mainly diatoms, which are clearly derived in the main from epiphytic and epilithic communities, and the unfortunate term periphyton has often been applied to this artificial collection (*see* p. 41). There are some unexpected features when the algal flora on glass slides is compared with the flora on the plastic used to suspend the slides in the water; for example, Hohn & Hellerman (1963) found that the polystyrene had a larger and more varied population at 3 °C but above 16 °C the populations were comparable. Bursche (1962) showed that bricks placed in a stream had a larger population than on natural rocks. As might be expected, slides held horizontally tend to yield a much greater flora on the upper than on the lower surface and also much greater than on vertically placed slides. The latter are nevertheless probably best, since they avoid to some extent the accumulation of sedimented material. I know of no comparative study on the reason

Fig. 3.51. *a.* Seasonal changes in cell numbers of epiphytic diatoms on *Elodea* and on slides in Abbot's Pool. *b.* Seasonal changes in cell numbers of *Achnanthes lanceolata* on *Fontinalis* and on slides in Langford Spring. The continuous line represents changes in cell numbers, and the fine, broken line the percentage of *Achnanthes* in the community. From Tippett, 1970.

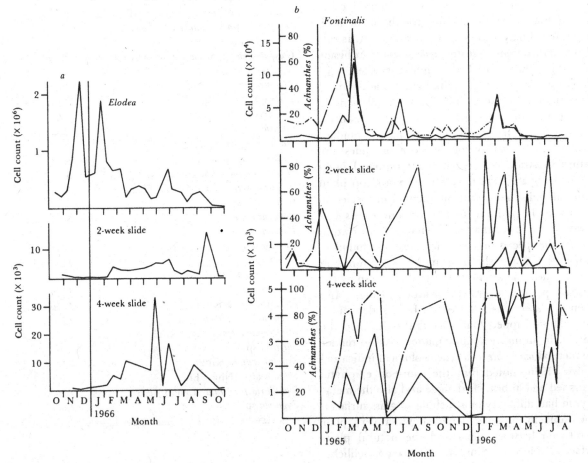

Table 3.17. *Numbers of epiphytic diatom species at each sampling site. From Brown, 1976*

Surface examined	No. of species of diatom found	
	Abbot's Pool	Langford Spring
Glass slide		
2-week immersion	34	18
4-week immersion	33	11
Natural plant		
Elodea	64	—
Cladophora	54	—
Potamogeton	69	—
Fontinalis	—	58
Chiloscyphus	—	57

Table 3.18. *The range of algae collected on vertical glass slides in Elk Lake. P, attached to plant surfaces (i.e. epiphytic); R, attached to rock surfaces (i.e. epilithic); S, living on sediment (i.e. epipelic); and PL, plankton. Adapted from Brown & Austin (1973) with the author's assessment of the habitats*

Achnanthes minutissima Kütz.	P
Amphora ovalis Kütz.	S
Ankistrodesmus sp.	PL
Asterionella formosa Hassall	PL
Cladophora glomerata (L.) Kütz.	R
Cocconeis placentula Ehr.	P; R
Cocconeis placentula var. *lineata* (Ehr.) V.H.	P; R
Coleochaete orbicularis Pringsheim	P
Cymatopleura solea (Breb.) W.Sm.	S
Cymbella sp.	S; R; P
Epithemia sorex Kütz.	P; R
Fragilaria crotonensis Kitton	PL
Fragilaria virescens Ralfs	S
Gloeotrichia echinulata (J.E. Sm.) P. Richt.	PL
Gomphonema acuminatum var. *coronatum* (Ehr.) Baben.	P
Gomphonema olivaceum (Lyngb.) Kütz.	P; R
Gyrosigma sp.	S
Melosira italica (Ehr.) Kütz.	PL
Melosira varians Ag.	P; R; Loose
Microcystis aeruginosa Kütz.	PL
Navicula sp.	S
Navicula cryptocephala Kütz.	S
Navicula pupula Kütz.	S
Nitzschia tryblionella Hantzsch.	S
Nitzschia vermicularis (Kütz.) Grun.	S
Oscillatoria sp.	S
Pinnularia sp.	S
Pinnularia gibba Ehr.	S
Pinnularia viridis (Nitz.) Ehr.	S
Quadrigula sp.	PL
Selenastrum Westii G.M. Sm.	PL
Spirogyra sp.	Loose
Spirogyra crassa Kütz.	Loose
Stephanodiscus niagarae Ehr.	PL
Stigeoclonium tenue (Ag.) Kütz.	P
Surirella sp.	S
Synedra radians Kütz.	PL
Synedra ulna (Nitz.) Ehr.	P
Tabellaria fenestrata (Lyngb.) Kütz.	PL
Zygnema sp.	Loose

for the variation of flora between vertical and horizontal; obviously the accumulation of inorganic and organic matter may stimulate growth on the horizontal but it may also eventually smother it. Is rate of growth better on these unsilted surfaces? Adhesion to surfaces has only been slightly studied, e.g., Harper & Harper (1967) found greater adhesion to glass than to plastic, and stronger adhesion to vertical than to horizontal surfaces, presumably a selective effect based on the strength of holdfast produced. Comparison of the flora on slides and natural substrata has rarely been attempted. Tippett (1970) compared the flora on slides and on plant material and showed that the number of species on slides kept in position for two or four weeks was always much lower than on the natural substratum though the species on the slides were also found as epiphytes. Fig. 3.51 shows the misleading impression gained when one attempts to use slides to show the seasonal pattern of growth of attached algae and when one species is followed on both slides and a natural host. Tippett concluded that slides affect the diatoms in an unpredictable manner and so render the method *unreliable* as a true ecological indicator! Brown (1976) noted that the common epiphytic branched and unbranched algae and also the metaphyton had difficulty in attaching to glass surfaces. The table in Brown's paper shows a very poor correlation between glass and the natural plant surface (Table 3.17). Similarly, Foerster & Schlicht-

Fig. 3.52. Production rate on glass (as ash-free dry wt, thick line), species composition (thin line) and temperature (solid dots) for two years in Falls Lake. From Castenholz, 1960.

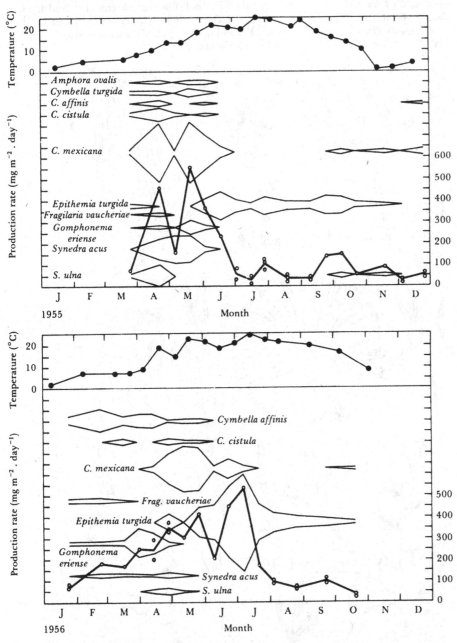

Fig. 3.53. Marine algae, mainly from northern shores. A, *Fucus*, B, *Pelvetia*, C, *Ascophyllum*, D, *Laminaria*, E, *Alaria*, F, *Chorda*, G, *Saccorhiza*, H, *Himanthalia*, I, *Scytosiphon*, J, *Punctaria*, K, *Halidrys*, L, *Ectocarpus*, M, *Enteromorpha*, N, *Ulva*, O, *Corallina*, P, *Hildenbrandia*, Q, *Acetabularia*, R, *Rhodochorton*, S, *Chondrus*, T, *Agardhiella*, U, *Ceramium*, V, *Gigartina*, W, *Nitophyllum*, X, *Furcellaria*, Y, *Phyllophora*, Z, *Phycodrys*, AA, *Delesseria*, BB, *Kallymenia*, CC, *Rivularia*, DD, *Lithoderma*, EE, *Cladophora*, FF, *Pylaiella*, GG, *Cryptopleura*. With the exception of L, R, U, EE and FF (small filamentous growths), and Q, CC and DD, which grow as crusts on the rocks, the remaining examples are variously sized macroscopic forms.

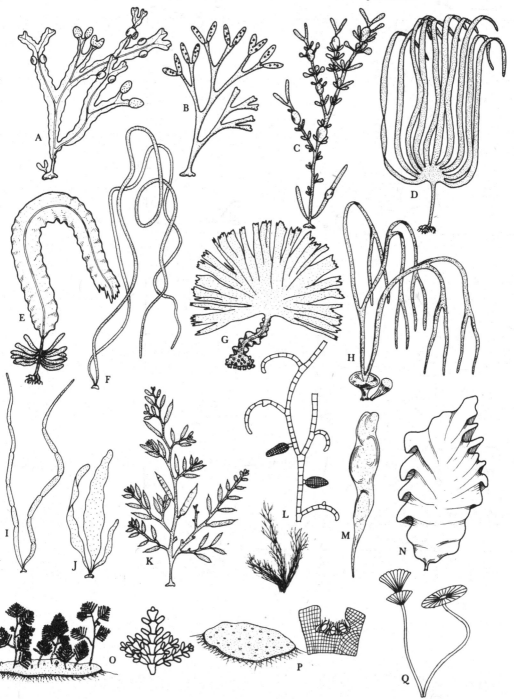

Fig. 3.53 (*cont.*)

Fig. 3.54. Epilithic marine algae, mainly from shores in tropical regions and the southern hemisphere. A, *Gelidium*, B, *Costaria costatum*, C, *Pelagophycus*, D, *Postelsia*, E, *Halosaccion*, F, *Nereocystis lutkeana*, G, *Pterocladia*, H, *Ecklonia radiata*, I, *Durvillea*, J, *Vidalia*, K, *Gigartina*, L, *Carpophyllum*, M, *Iridophycus*, N, *Splachnidium*, O, *Wrangelia*, P, *Bostrychia*, Q, *Laurencia*, R, *Hypnea*, S, *Spyridia*, T, *Desmarestia*, U, *Laminaria*, V, *Egregia*, W, *Macrocystis*, X, *Centroceras*, Y, *Pterygophora*, Z, *Agarum*, AA, *Hormosira*, BB, *Eisenia*, CC, *Dictyoneuron*. These are all macroscopic genera some of which grow to several metres in length.

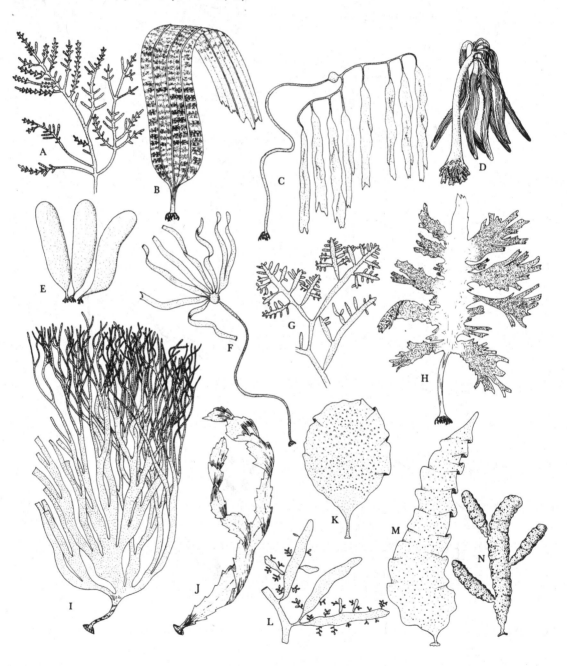

Fig. 3.54 (*cont.*)

ing (1965) in a similar study concluded that 'the artificial barren surface gave a false indication of the true productivity and indicated only some of the significant genera present in the ecosystem'. Glass slides tend to favour the growth of small species e.g. *Achnanthes minutissima* (Silver, 1977). Wetzel (1975) thought that the use of such devices in running waters needed extreme care since the growth on the slides and on natural substrates deviates considerably. Some reports in the literature that species growing on glass slides are representative of the whole algal flora have generally been disproved, though undoubtedly if left in the water long enough glass slides will eventually filter out cells of most communities and Table 3.18 shows what a mixed assemblage can appear on slides (*see also* Silver, 1977). Weber & Raschke (1970) compared the plankton and glass-slide flora and came to the conclusion that 'the composition of the communities was very different throughout the entire exposure period'. This is not surprising since the method is virtually useless to detect what actually grows in the water column. Castenholz (1960) found that the species colonising glass slides in the lakes of the Lower Grand Coulee were epilithic species; not all the epilithon were represented and so this is not a good technique even for studying epilithon. A difference in timing of the annual maximum in succeeding years and a difference in species composition was detected (Fig. 3.52).

Even after two weeks there is considerable division of the cells which have attached to the slides and numbers will merely reflect the incidence of colonisation and rates of division on an unnatural surface. According to Weber & Raschke (1970) the number of cells reaches average densities of 15 000/m² in 32 days in the Ohio River at Cincinnati but Tippett's (1970) results in pools and streams give figures an order of magnitude lower. A species such as *Tabellaria flocculosa* var. *flocculosa*, which produces chains of cells extending outwards from the site of attachment is therefore a species which could form a much larger population than could be achieved by adnate species, in fact this species yielded populations only up to 4500 cells/mm² on *Schoenoplectus lacustris* (Knudsen, 1957); this makes the figure of 15 000 seem excessive since on the whole the diatoms on glass slides adhere to the surface rather than form chains extending outwards. Figures as high as 15 000

are only possible if no larger cells or colonies occur and the remainder are very small forms such as *Achnanthes affinis*.

The slides are normally removed from the water, and either extracted to obtain chlorophyll data, or cleaned on one side and the other flooded with a mounting medium, a cover-glass placed on the surface, and the algae identified and counted as in the work quoted above. Recently Neushul (1972*b*) has devised a technique for underwater microscopy using SCUBA techniques and an incident light dipping-cone microscope sealed into a special container. Use of this would enable early stages of colonisation and growth to be followed without removal of the slides. At the same time, Staley (1971) adapted a microscope to study glass slides placed in a lake and was able to watch the increase in cell size during the day and the subsequent division during the night which neatly confirms from nature the cycle which has been reported many times in laboratory studies.

Summary

This chapter deals with the algae growing on or associated with rock, sand, and artificial surfaces, especially those used experimentally. Examples of the algae discussed are illustrated in Figs. 3.53 and 3.54. The simplest epilithic situation is that of terrestrial rock on which rain falls or over which surface water percolates and the flora of these situations is dealt with first. It is an important flora in many parts of the world but has received little attention. Even less well-known is the algal flora growing on and in the lower surfaces of stones in deserts where species of green and blue-green algae are found.

The algal flora of rock and stones in running water is often massively developed since it is in a habitat where nutrients are continually supplied and waste products removed. Here rich growths of filamentous green algae, pustules of blue-green algae and brown films of diatoms are common. Only a few macroscopic species have colonised this habitat. Seasonal changes of biomass are marked and experimental studies have shown the importance of flow rate and light regimes.

Slippery growths of algae are all too familiar to bathers in lakes and here life forms and kinds of algae are found which are similar to those in running waters. Occasionally some, such as *Cladophora*, can grow to such an extent that they become an amenity and health hazard. This flora varies with depth but many detailed studies are required before generalisations can be made.

A major and well-studied epilithon occurs on rocky coasts especially in temperate zones. Though this is conspicuous it is probably no more important on a world-wide basis than the totality of the very numerous freshwater rocky niches. Writings on zonation and terminology of rocky shores have featured prominently in the literature; these are discussed and it is suggested that the ever increasing use of the simple, widely understood, *supratidal*, *intertidal* and *subtidal* divisions is preferred to the cumbersome systems based on the word littoral. The importance of *tide breaks* and lengths of exposure to air or seawater are discussed. The microscopic species of the marine epilithon have been grossly neglected. The macrophytic species reproduce mainly by means of spores (fragmentation and vegetative regeneration also occur) and their formation and initial attachment and growth are discussed with mention of experimental work on formation and attachment, germination, rate of fall, etc.

The parallel belts of algae are dealt with in some detail with more detailed discussion of the flora in the supratidal, intertidal and subtidal and the various factors affecting growth in these regions. Examples are taken from sites ranging from arctic and antarctic regions, through temperate, to tropical waters, and special mention is made of the much studied *Macrocystis*. The problems of depth zonation, maximum depth of penetration, growth in almost dark, polar waters, light and temperature effects, and problems of field measurements are dealt with.

Opportunities to study colonisation of new areas and succession are often afforded for this community and some examples from artificial clearing, lava flows and the appearance of the island Surtsey are given.

The algal flora of sand grains consists mainly of diatoms and a few coccoid blue-green algae. Some move rather slowly from grain to grain. The flora is extremely rich and some species of diatoms are confined to this habitat.

Endolithic algae of deserts and rocky coasts are mentioned and mainly blue-green and a few filamentous green algae are involved.

A flora exists beneath the surface of sand in some lakes and in supratidal and intertidal situations where the algae appear to bind the sand into a stable gel-like layer.

Glass slides have been used widely for the study of colonisation and growth in certain habitats. Some of the results and problems are discussed; it is pointed out that the use of glass slides does not offer a suitable technique to study the natural flora since the surfaces are selective and no useful data can be obtained on the composition of the communities of the various microhabitats.

4 Coral reefs – the free-living algal flora

Coral reefs provide a highly specialised habitat in which attached algae are abundant, and in many places dominant, organisms. The name, however, conjures up a picture of a substratum covered with coelenterates, mainly of the class Scleractinia, of which the reef building species (or hermatypic corals) produce the substantial stoney calcareous skeletons much beloved by underwater photographers.*

Womersley & Bailey (1969) suggest that since algae and not corals are dominant and basic in this type of reef formation, especially in the surf zone, the misleading name coral reef should be abandoned. They suggest that the term biotic reef be used to distinguish this type, which is dependent upon photosynthesis, from reefs which are formed by sedimentation, land and sea level changes, etc. Earlier workers even suggested they should be termed 'algal reefs'. The attraction of the beautiful and spectacular animals of reefs has obscured the fact that the rock-like algal growths are of major importance and usually one has to turn to the geological writings to get a balanced account of the contribution of the various organisms to reef development. The contributions of algae and corals vary considerably even on windward and leeward shores of atolls; there is, too, a geographical variation, at least in the Pacific, where algae are more abundant in the south-east and coelenterates predominate in the western region (Doty, 1973a). In a reef off Curaçao, Wanders (1976a) calculated that under 1 m² of sea surface there was 5.5 m² of substrate surface and this was occupied by 2.1 m² of coral tissue, 2.1 m² of coralline algae and about 0.8 m² of fleshy algae. Only now, 125 years after Darwin saw these reefs and wrote with such perception about them, is any attempt

being made to investigate the algae experimentally. Reefs are the only known biological systems giving rise to massive *in situ* geological formations. Many of the coelenterates themselves contain abundant symbiotic algal cells which contribute greatly to the economy of the reefs and this aspect is dealt with in the chapter on symbiosis (Chapter 9). The habitat is essentially an extension of the coastal epilithic system, since the algae living on the reef are growing on hard substrata or forming concretions over hard rubble. It is such an extensive and important habitat that it warrants separate treatment though many of the free-living algae also occur on non-reef coasts. Some of the reef algae grow in the form of 'rock-hard' concretions to which other species may attach as though to rock itself. A further similarity is the development of topographic belts parallel to the coast, each belt supporting a characteristic flora and easily recognised from aerial photographs (Tracey, Landd & Hoffmeister, 1948). The belts are not always so easy to distinguish when transects are made across reefs.

There is a very extensive literature on formation and kinds of reef (*see* Stoddart, 1969; Maxwell, 1972), but it is clear that a simple subdivision can be made into *structural reefs*, where corals are actively contributing to the topographic development of the reef, and *coral communities* growing on substrata other than their own remains (Wainwright, 1965). It is doubtful, however, whether such distinctions are important for the algal ecology of reefs. Reefs are tropical communities, rarely found outside the 30° N or S latitudes and not developing in waters with an annual minimum temperature below 18 °C. Best growth probably occurs between 24 and 30 °C but occurrence has been recorded up to 40 °C. Structural reefs are essentially of two types – coral atolls, which are volcanic structures undergoing subsidence and

---

* A very valuable series of books on the biology and geology of coral reefs has been edited by O. A. Jones & R. Endean (Academic Press).

capped with algal–coral growths which keep pace with changing sea levels, and fringing (barrier) reefs which are elongate algal–coral growths running parallel to the shore, and likewise keeping pace with changing sea levels. Earth scientists make many more distinctions, but it is unlikely that these have great biological significance. There is considerable evidence (*see* also Chapter 13) for extensive land–sea level changes in the Pleistocene period. Sea level has certainly been lower, probably some 400 ft in the last few hundred thousand years, and it has probably been higher within the last few thousand years, e.g. during the post-glacial thermal maximum, when the sea surface was some 6 ft higher than at present; reefs have thus been eroded downwards during the last 3000–4000 years. Growth of coral reefs is only possible when the algae and the corals, with their symbiotic algae, are in the illuminated surface zone of the sea and if subsidence is too rapid development will be impeded; presumably it would have to be a fairly catastrophic drop caused by tectonic movements to cause the 'death' of a reef. Numerous sites certainly do occur where the basal rock (usually basalt) and its detrital coral capping are found some distance below the sea surface; such submerged 'islands' are known as guyots and are common in the Pacific. Drilling through Eniwetok, one of the atolls in the Pacific, has revealed a depth of nearly 5000 ft of coral capping on the basalt base (Ladd *et al.*, 1953, Ladd, 1961). This must have formed close to the sea surface, so subsidence and sea level changes have been very large indeed. In fact at various times since the Miocene, Eniwetok has been an island with tropical vegetation and even with tropical rain forest. This theory of subsidence* was first proposed by Darwin in his classic work *Geological observations on coral reefs, volcanic islands and on South America* (1851) and few new ideas have been added since.

The algal ridge

In studying the literature on reefs one finds considerable terminological difficulty since there is no uniform nomenclature for the zones. It would

help if a fairly standard series of topographical terms could be used; these could be combined with organismal names when necessary. A tentative scheme is given in Fig. 4.1 in terms which have been employed by various workers and with possible combinations.

Darwin commented on the frequent occurrence of a raised rim on the outer edge of a reef and on this raised rim 'where the waves beat the surface is covered with a Nullipora'. Later authors called this 'the Lithothamnion rim' because of its likeness to the rhodophyte *Lithothamnion*.† This is now more correctly termed the 'algal ridge' and it is certainly not composed of *Lithothamnium* but, in the Indian and Pacific Oceans, of *Porolithon onkodes* (Setchell, 1928; Womersley & Bailey, 1969; Doty, 1973a, b). Contrary to some earlier reports there is an algal ridge system in the Caribbean where reef accretion is estimated at 9–15 m per 1000 yr; 3–5 m per 1000 yr has been reported for the Pacific but more recent figures are of the same order as that reported for the Caribbean (Adey, 1978). The algal ridge is obviously an extreme habitat and as such would be expected to support only a small number of species; Foslie (1907) added *P. craspedium*, *P. gardineri* and *Neogoniolithon frutescens* to *P. onkodes* as characteristic species of this habitat and few have been added since. On this ridge, which may be 30–40 yd in width and 2 ft above the low tide level (Marshall, 1931), there are very few corals, these being developed only in the calmer water landward of the algal ridge (on the Barrier Reefs) and seaward of this in deeper water. These algal ridges occur on the seaward edge of reefs around atolls (Tracey *et al.*, 1948) and on the seaward edge of barrier reefs. Marshall (1931) classed all these as 'rough water reefs'. Between the outer 'barrier' reef and the mainland of Australia there are many other reefs; these are protected from the main force of the waves and have no algal ridge, and on such reefs algae apparently play a less prominent part. For these landward Australian reefs, Cribb (1973) terms the outer part, which is exposed at low water, the rubble crest (*see* p. 149).

Encrusting, calcareous algae are involved in

---

* It is less easy to invoke only 'subsidence' in the development of barrier and fringing reefs. Undoubtedly changes in land–sea level are involved but the situation is more complex than on atolls. Again however such variations are of minor importance for the algal ecology.

† *Lithothamnium* is the correct spelling according to Adey & MacIntyre (1973). The term 'lithothamnia' is common in the literature and is very useful as a general term denoting encrusting calcareous red algae provided it is remembered that it does not refer to a particular genus. A valuable review of crustose coralline algae is that of Littler (1972).

Fig. 4.1. *a*, diagrammatic representation of part of the intertidal reef (Capricorn and Bunker groups in the Great Barrier Reefs), not to scale and with vertical heights greatly exaggerated. From Cribb, 1973. *b*, *c*, general transects across reefs in the Solomon Islands, *b*, on a coast subject to strong wave action. *c*, under calmer conditions. From Womersley & Bailey, 1969. *d*, an attempted generalised subdivision of the regions on tropical reef systems. Not all the regions are necessarily found on all reefs. Inter-island reefs do not have the land mass shown on the left of the figure.

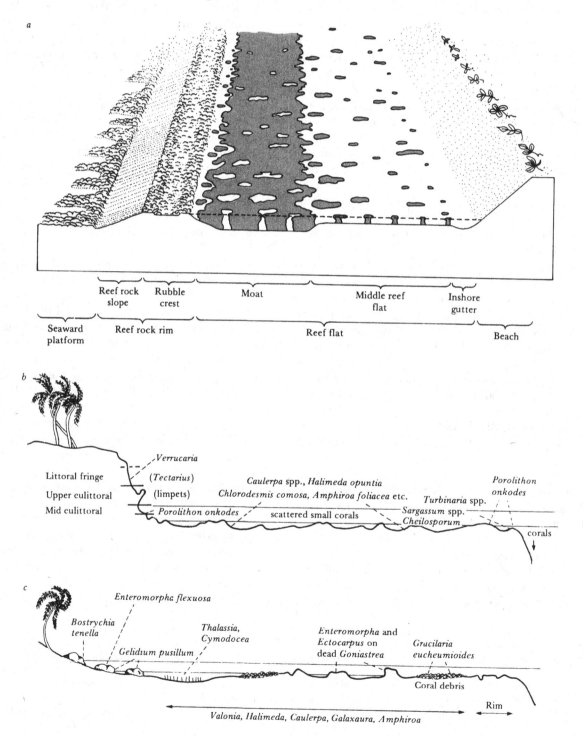

Fig. 4.1 (*cont.*)

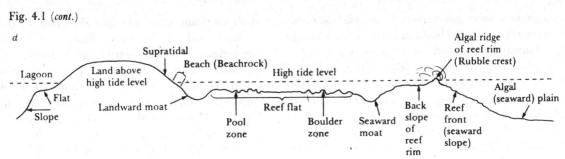

d

Lagoon

Land above
high tide level

Supratidal

Beach (Beachrock)

High tide level

Algal ridge
of reef rim
(Rubble crest)

Flat

Slope

Landward moat

Pool
zone

Reef flat

Boulder
zone

Seaward
moat

Back
slope
of
reef
rim

Reef
front
(seaward
slope)

Algal
(seaward) plain

Fig. 4.2. The distribution on the algal ridge at Waikiki of *Porolithon onkodes* (dotted line), *P. gardineri* (dashed line) and *Lithophyllum kotschyanum* (continuous line). *a*, a profile across the reef off the Natatorium at Waikiki, *b*, % cover; *c*, % density; *d*, % frequency. From Littler & Doty, 1975.

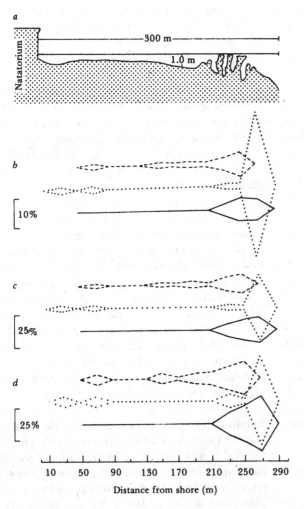

a

Natatorium

300 m

1.0 m

b

10%

c

25%

d

25%

10   50   90   130   170   210   250   290

Distance from shore (m)

two physical processes on reefs. Forms such as *Porolithon* steadily grow and build calcareous 'rock' on the basal substratum and others (*Hydrolithon, Sporolithon*) overgrow and bind together the detrital fragments into almost equally hard 'rock'. This cementing action of the coralline algae was recognised by Setchell (1930) who wrote 'By far the most, and one may say the only effective cementation is due to certain crustaceous corallines or nullipores'. Even earlier, Foslie (1907) and Howe (1912) had clearly pointed out that crustose coralline algae were the dominant organisms on many tropical reefs.

Marshall (1931) gives an impressive account of the wave action on the algal ridge, 'a single wave may throw a weight of 40,000 tons of water on to a strip of reef no more than 20 yards long, and this heavy mass has a velocity of 40 miles per hour. This mighty onslaught is repeated five times every minute'. In all the reefs examined the force of the water falls on *algal* rather than coelenterate *coral* growth. The algal ridge itself may be almost 30 yards or more in width and relatively flat but with buttresses running out to sea. Between the buttresses deep channels run out to sea. Doty & Morrison (1954) distinguish incurrent and excurrent areas on such algal ridges, where, except at high tide, the water flows inwards across the flat incurrent areas and then moves laterally along the reef and flows out to sea via excurrent channels.* The algae of the ridge extend down the reef front, 3–4 ft on the Australian barrier reef according to Marshall. In Hawaii this region is colonised by *Porolithon onkodes* which extends to depths of 6 m depending on the turbidity of the water. It seems to reach its thickest growth on the brightly lit seaward reef front (Fig. 4.2) and gradually thins out to a veneer a millimetre or so thick before ending abruptly 5–6 m below the ridge. Extending slightly landwards from the main *Porolithon* growth is a region in which *Porolithon gardineri* and *Lithophyllum kotschyarum* are developed (Fig. 4.2; Littler & Doty, 1975). In the lower seaward *Porolithon* region, other corallines occur (*Hydrolithon reinboldii* and *Sporolithon*

---

* Inter-island reefs such as the one studied by Odum & Odum (1955) at Eniwetok differ in that much of the water pours across the reef and into the lagoon and does not flow in two directions. Could this be the explanation for such high productivity on this type of reef (*see* p. 152)? The surge channels and buttresses are equally characteristic of this reef type and so are not necessarily associated with flow *onto* and *off* the reef.

erythraeum) and seaward of the *Porolithon* the bottom is dominated by *Tenarea tessellatum* and *Hydrolithon breviclavium*. Not all seaward ridges are so dominated by calcareous algae, e.g., the algal ridge at Eniwetok has much less *Porolithon* but a well developed fleshy algal mat (*Dictyosphaeria, Zonaria, Ceramium, Dictyota, Caulerpa*; Odum & Odum, 1955). On the algal ridge burrowing and rasping animals occur, and the detritus is moved off the reef towards the shore or down the excurrent channels.

The algal ridge in some reef systems is composed of great lumps of calcareous material formed by the crustose coralline algae. Tracey *et al.* (1948) termed these 'algal bosses' and showed how they extended and finally joined up to form a structure resembling mine works of the 'room and pillar' type. Blocks of calcium carbonate thrown up on to the reef also tend to become cemented into the pavement by overgrowth of coralline algae. The excurrent areas ('surge channels') may also become roofed over by algal growth which may continue until the inner openings are small and 'blowholes' are formed from which water shoots into the air after each wave. This supply of water round the 'blowhole' maintains a good upward growth of algae around the holes, which on Bikini atoll can be up to 10–30 ft in diameter and 1–3 ft above the level of the reef flat. In calm water the algal ridge is not so well developed, and the 'reef rim' is a region of coral growth with some crustose algae cementing the surface material, e.g., at sites on the Solomon Islands (Womersley & Bailey, 1969). On reefs in these islands with moderate wave action a mixture of *Lithophyllum moluccense, Neogoniolithon myriocarpum* and *Porolithon onkodes* is found but as in other Pacific islands, *Porolithon* becomes the dominant form under heavy wave action whilst *Neogoniolithon* is the most important cementing species and binds the coral debris into rock-like masses. It is not unusual to find frondose algae on the part of Pacific reefs subject to moderate wave action and on those off Mahe, Seychelles, frondose algae are reported to obscure the algal ridge (Lewis, 1969). Similar algae are involved in building the reef at St Croix, West Indies with *Neogoniolithon megacarpum* and *Porolithon pachydermum* dominant in water less than 1–2 m deep and exposed to considerable wave action (Adey & Vassar, 1978). As the algal ridge builds up to sea level, *Lithophyllum congestum* takes over as the dominant algal component.

The reef flat

The sloping ridge (back slope) on Bikini atoll is coated with crustose coralline algae, and corals only occupy 5–10 % of the surface area (Tracey *et al.* 1948). Landward of this for a distance of 25 ft to several hundred feet is a zone of rich coral growth covering 50 % or more of the area; under and between these corals the crustose algae are still abundant. Inshore of the algal ridge in the more protected reef flat waters, frondose algae can establish, especially *Sargassum* in Hawaii, and the shade cast by these eventually eliminates the *Porolithon* which is replaced by the encrusting *Sporolithon* which outgrows *Porolithon* in conditions of low light intensity. Littler (1973*a*) found that on inshore regions of the reef at Waikiki, *Hydrolithon* was the primary limestone former with an unknown melobesioid alga and *Sporolithon* subsidiary, the distribution of which is given in Fig. 4.3. Doty & Morrison (1954) suggested that *Porolithon onkodes* could develop rapidly and withstand intense illumintion and a greater range of temperature and desiccation than other coralline algae which are usually found in cavities or shady places. As far as light is concerned, Littler (1973*b*) and Littler & Doty (1975) showed that *P. onkodes* is light

saturated at 2000 ft-c. and even at 100 ft-c. is above the compensation point. Yet at natural intensities as high as 12 000 ft-c. it does not suffer any photodestruction. The effect of light intensity on the productivity of *Porolithon* is shown in the two graphs (Fig. 4.4), revealing the much greater carbon fixation at 2000 lumens ft^{-2} compared with 450 lumens ft^{-2}. *Sporolithon erythraeum*, on the other hand, is light saturated at 1000 ft-c. whilst its compensation point is at 50 ft-c. and in nature removal of the shading *Sargassum* cover leads to its death in 3–4 days. Very few detailed studies have been completed on the distribution of frondose algae on the reef flat and the only ones giving quantitative data are those of Doty (1973*a*) from which Fig. 4.5 is reproduced and Santilices (1977) from which Fig. 4.6 presents a selection of data. Clearly there are many problems of zonation in this region; the frondose algae, which are more richly developed on the outer reef flat, belong to the Phaeophyta and Rhodophyta whilst all the siphonaceous Chlorophyta are better represented in the inner zone, e.g. *Halimeda* and the angiosperm *Thalassia* are common in this region in the Bahamas (Bathurst, 1971). Gilmartin (1960) gives a comprehensive list of the species in the deep water of the reef

Fig. 4.3. The distribution of three widely distributed calcareous red algae on the reef at Waikiki. *a*, an unknown Melobesioid sp.; *b*, *Hydrolithon reinboldii*; *c*, *Sporolithon erythraeum*. The profile across the reef, % cover, % density and % frequency are given for each alga, as in Fig. 4.2. From Littler, 1973*b*.

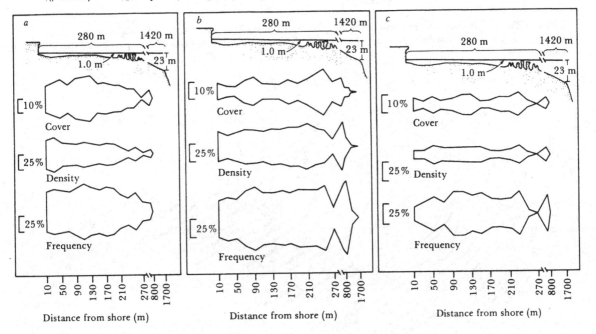

Fig. 4.4. The productivity of *Porolithon onkodes* in a strong current, at 27 °C, *a*, at 450 lm ft^{-2} and, *b*, at 2000 lm ft^{-2}. Measurement by oxygen electrode (open circles) or pH electrode (closed circles). To the left of the vertical dotted line samples were in darkness. From Littler & Doty, 1975.

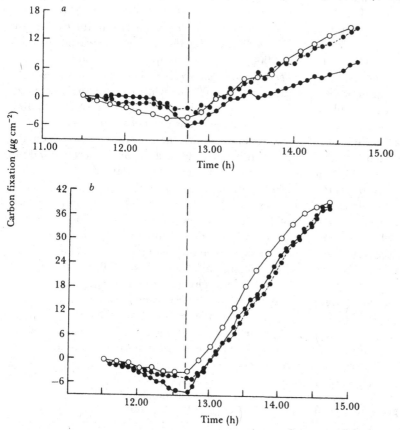

Fig. 4.5. The distribution of the non-calcified algae from the shore to just beyond the algal ridge at Waikiki. Each curve shows the distribution as a percentage of the average standing crop of that particular species. From Doty, 1973*a*.

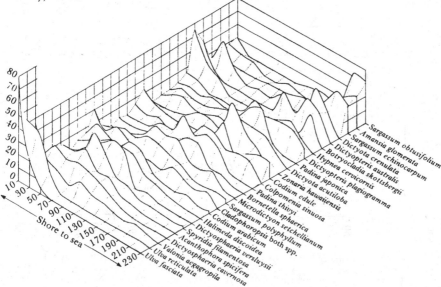

Fig. 4.6. Seasonal and horizontal changes in biomass of frondose algae across a coral reef on the north coast of Oahu, Hawaii. Distance in block diagrams refers to distance from seaward edge of the reef. *a, Sargassum echinocarpum; b, Codium edule; c, Turbinaria ornata; d, Acanthophora spicifera.* Examples selected from Santilices, 1977.

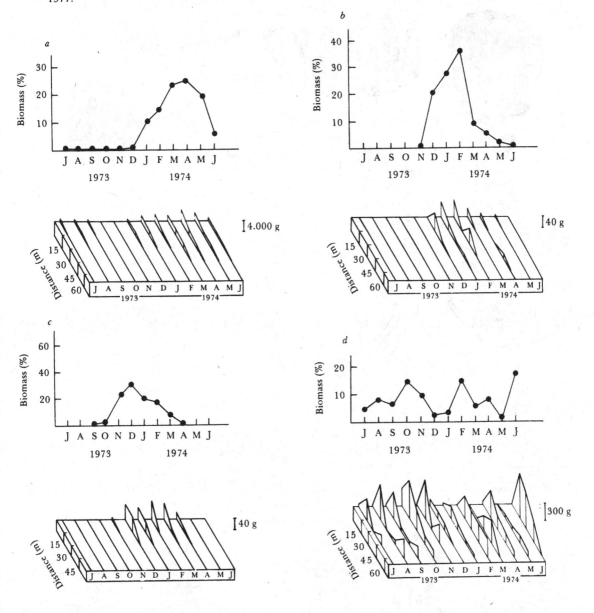

Fig. 4.7. Some alge found along tropical shores, in reef lagoons and similar habitats. A, *Dictyosphaeria*, B, *Caulerpa*, C, *Udotea*, D, *Padina*, E, *Dictyota* (also temperate in distribution), F, *Boodlea*, G, *Rhipocephalus*, H, *Penicillus*, I, *Chnoospora*, J, *Dictyopteris*, K, *Hydroclathrus*, L, *Jania*, M, *Gelidiella*, N, *Lithothamnium*, O, *Lithophylum*, P, *Sphacelaria*.

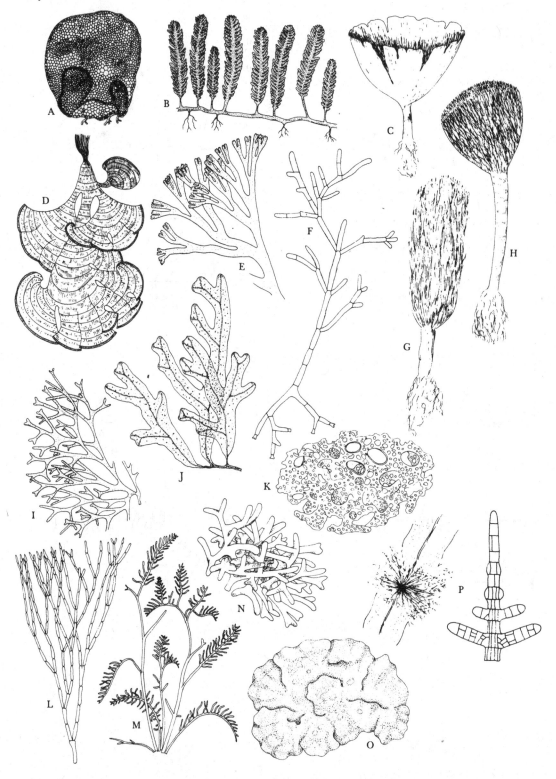

Table 4.1. *Nitrogen fixation rates of various communities at Eniwetok Atoll, January–February 1974. Values for nitrogen fixed are given for the substrate surface areas as means ±s.e. Number of replications in parentheses. From Wiebe et al., 1975*

| | Mean rates (10^{-9} mole h^{-1} cm^{-2}) | |
| | Full daylight | |
Location	intensity	Dark
Intertidal reef flat *Calothrix* community	55 ± 12.1 (11)	1.9 ± 0.14 (3)
Upper intertidal reef beach *Calothrix* community	34 ± 6.0 (7)	
Random inter-island coral community and reef rubble community samples	24 ± 6.0 (11)	0.27 ± 0.039 (2)
Outer reef slope	4.5 ± 1.2 (4)	

flat at Eniwetok where he recorded some 95 species, and he was one of the first to point out that Chlorophyta are proportionally less reduced in deep water than Phaeophyta or Rhodophyta, for only nine species of brown algae occur, of which only *Dictyopteris* and *Dictyota* are significant whilst the red algae are never important as far as biomass is concerned. Van den Hoek *et al.* (1975) found 142 fleshy algae on the reefs around Curaçao with an average 54 species per 25 m^2 but they were relatively unobtrusive owing to grazing to a turf. Caging areas allowed the fleshy algae to grow and as in Hawaii these can then outshade the corallines. Santilices (1977) found at least a doubling of algal biomass between the low period in July–September and the high in February. Other distributions of algae across reef areas are given by Womersley & Bailey (1969); there is a basic similarity in the distribution of species between these reefs and those in Hawaii and elsewhere, but the fine detail varies according to geographical distribution of organisms, varied physical chemical regimes, storm action and other factors.

In lagoons of coral atolls, meadows of algae such as *Tydemannia*, *Caulerpa*, *Halimeda* and even *Acetabularia* have been reported though they are not usually evenly developed. More frequently the algae grow in patches around corals. *Sargassum* is a common species in both channel and lagoon areas. The calcareous remains of *Halimeda* often form extensive sediments in lagoons. The lagoons are really large shallow tide pools where water motion is reduced and coralline red algae are less abundant than

the rhizomatous, siphonaceous, Bryopsidophyceae. Fig. 4.7 illustrates some examples of macroscopic algae found in lagoons (but also along other sandy shores).

On reefs and in lagoons there is a considerable amount of dead coral and coral rubble and this is colonised by algae. Filaments of blue-green algae and red algae such as *Herposiphonia tenella* penetrate into the coral skeleton; the Cyanophyta tend to accumulate in the surface layers but the red algae penetrate throughout (I. Price, personal communication). The surface itself becomes encrusted with calcified red algae and the whole system takes on the nature of a porous 'rock' substratum.

Diatoms and Cyanophyta are also common components of an epipelic and epilithic flora in the lagoons and in these reef habitats, especially on isolated atolls, Cyanophyta could be exceedingly important sources of organic nitrogen via their fixation of atmospheric nitrogen. In such areas nitrogen cannot be supplied from extensive land drainage but must come from any populations of nitrogen fixing organisms in the supra- or subtidal, from the sparse terrestrial soils or from upwelling water. Recently Wiebe, Johannes & Webb (1975) have shown that at least on the inter-island reef at Eniwetok, *Calothrix crustacea* is abundant on the reef flat and as an epiphyte. It fixes nitrogen to a level equal to that recorded for the richest nitrogen-fixing sources in marine or terrestrial environments. Table 4.1 shows that this fixation occurs throughout the reef system, even on the exposed outer reef slope.

Wiebe *et al.* found that the *Calothrix* fixed nitrogen over a wide range of salinities (3–45‰) and so would be especially effective over the wide range which might be expected on reefs which at different times are subjected to dilution, for example by rain and to concentration by evaporation. Fixation ceased at 24 °C (which is below the sea temperature at Eniwetok) but it doubled between 27 °C and 36 °C and at 39 °C it became erratic and then ceased. These figures suggest that a high temperature strain of the alga is involved.

It has been reported that *Porolithon* species overgrow and kill corals by dissolving calcareous material (Ranson, 1955) but these observations need careful rechecking though van den Hoek (1969) reported *Porolithon pachydermum* overgrowing corals at Curaçao. Also there is an interaction between *Porolithon* and the frondose algae since surf removes the fleshy species and this then allows *Porolithon* to grow in the better illuminated sites. The coralline algae, even the rock-hard *Porolithon*, are grazed by a mixture of limpets, sea urchins, molluscs, boring annelids and other animals (Littler & Doty, 1975; Wanders, 1976a). During the day fish graze over the surface and at night sea urchins. The live corals themselves have no epibiotic algal flora on the actively feeding and growing regions, since a layer of mucilage and the activity of the polyps keep the surface clean, but the basal parts often support algal species, e.g., the coral-mimicing red algae, *Eucheuma arnoldii* and *Gelidiopsis intricata* (Kraft, 1972) in the Philippines. This 'coral-base' habitat is obviously one which would repay detailed study. In the Philippines the biomass of algae amongst the corals was low (0.65–2.0 kg m^{-2}) and consisted of 15–35 algal species compared with around 4.6 kg m^{-2} and 62 species on basalt rock in areas sampled by Kraft. The fleshy, algal flora of coral reefs is also intensively grazed by fish and Dawson, Aleem & Halstead (1955) found that half the fish examined at Palmyra Island were herbivorous and that more species of algae could be found in the guts of the reef fishes than by human collecting.

Littler & Doty (1975) experimented with artificial substrata in the form of glass panels exposed on the outer reef slope; their object was to check growth in various light regimes and the effects of grazing in a habitat which is difficult to sample directly. The underside of the panels became coated with *Porolithon*

onkodes and *Peyssonelia* spp., and if they were left on the outer reef slope, or on the crest of the ridge, they continued to support just the *Porolithon*. However, if the panels were transferred to the reef flat the upper surface became covered with a growth of *Jania*, *Sphacelaria* and *Padina* spp.; in this position these species are not grazed and prevent light from reaching the underlying *Porolithon* which consequently died out. Experiments in which the panels covered with frondose algae were transferred on to the seaward reef slope, resulted in rapid removal of the soft algae within *one* day by grazing fish. This is further proof that grazing which allows light to reach the surface of the reef permits *Porolithon* to grow in deeper water on the slope. On the ridge, it is wave action which removes the competitors and allows the *Porolithon* to thrive. The *Porolithon* community is considered to be held in a sub-climax condition by the combination of the force of wave action and grazing since in the absence of these factors the corallines would be replaced by frondose species, as indeed they seem to be in certain sites (e.g. around the Seychelles).

The algal plain

The region *seaward* of reefs is termed the algal plain by Dahl (1973) and off Puerto Rico he found this was an area of wave disturbed coarse sand lying over a clay-like deposit with species of Bryopsidophyceae rooted in the deposit. Stability was provided by sponges which had spread on the bottom and partially engulfed the bases of the algae. As can be seen from Table 4.2 the density and diversity of the algae increased away from the reef; closer still to the reef there were even fewer algae. In this region the macrophytic flora consists mainly of Bryopsidophyceae and Rhodophyceae with once again only *Dictyopteris* and *Dictyota* representing the Phaeophyceae. Off the Bahamas there is a rocky, Pleistocene limestone pavement running from about 9 m to 50 m depth and covered by lime sand (partly derived from algae) and amongst the corals occur large plants of *Halimeda*, *Udotea*, *Penicillus*, *Rhipocephalus* and the non-calcified *Laurencia* (Bathurst, 1971).

Off Hawaii this region has received a little attention from the study of dredged material but Littler (1973b) has made the first detailed study by diving off Oahu. He found that between 8 m and 28 m depth, 38.9 % of the bottom is covered with

Table 4.2. *The density and relative abundance of algae living on the algal plain of a reef off Puerto Rico. From Dahl, 1973*

Description	20 m from reef Density (plants/m²)	20 m from reef Relative abundance (%)	40 m from reef Density (plants/m⁻²)	40 m from reef Relative abundance (%)
Chlorophyta		41		37
Halimeda spp.	28	36	38	18
Udotea spp.	2	3	14	6
Caulerpa spp.	1	1	13	6
Bryopsis	—	—	5	2
Anadyomene stellata	—	—	9	4
Valonia ventricosa	—	—	1	1
Rhodophyta		39		47
Agardhiella tenera	2	3	32	15
Laurencia spp.	8	10	19	9
Gracilaria and other large reds	2	2	10	5
Acanthophora spicifera	5	6	7	3
Dictyurus occidentalis	—	—	6	3
Filamentous reds	13	17	26	12
Phaeophyta		19		10
Dictyota spp.	14	18	18	9
Dictyopteris spp.	1	1	3	1
Cyanophyta	2	2	13	6
Diatoms	—	—	2	1
Total	77		216	

coralline algae and the remaining sea bed is mostly composed of dead reef, sand and rubble. The most abundant species were *Hydrolithon breviclavium* followed by *H. reinboldii* and *Tenarea tessellatum*. Fig. 4.8 shows the distribution of the algae off Waikiki where clearly there is a zonation with depth. Littler was also able to distinguish communities of algae which characteristically occurred together; e.g., between 8 m and 12 m *Sporolithon erythraeum–Dictyopteris* was found with *D. plagiogramma* in the shoreward belt and *D. australis* in the seaward. This community extends parallel to the shore in a belt around the island. Seaward of this (in depths between 13 m and 20 m) is a belt of *Hydrolithon breviclavium–H. reinboldii* and attached to the corals, usually beneath a layer of silt and sand, Littler found *Lyngbya, Halymenia, Amansia, Padina, Neomeris, Herposiphonia, Halimeda, Dictyota* and *Peyssonelia*. At 21 m depth, a pavement-like growth of *H. breviclavium* occurred and just below this is a belt

of coral (*Porites compressa*). Seaward of these at depths below 28 m there is a community of *Tenarea tessellatum–H. breviclavium*.

Littler (1971) developed a photographimetric technique to estimate cover, relative density and frequency of algae on these reefs. He comments that quantitative measurements are virtually absent from the literature since techniques for estimating melobesioid communities have been lacking. He photographed the area, first with the non-crustose seaweeds present, and then with these cleared away to reveal the melobesioids. The cover of each species was then determined from the prints by use of a planimeter. The results compared well with data from line transects, but were considered more suitable since the estimations can be done at leisure in the laboratory. Crustose corallines cover 39% of the Waikiki fringing reef and exceed all other organisms as the dominant builders and consolidators of reef material. Littler considered that the greater contribution by coralline algae at Waikiki compared with other sites around the island is due to increase in sewage pollution. He estimates that the fringing reef Melobesioideae contributes carbon at the rate of 5.7 g m^{-2} day^{-1} (Littler 1973c). This is a high rate of production, since Talling *et al.* (1973) suggest that few rich phytoplankton assemblages achieve more carbon than 4 g m^{-2} day^{-1}, though levels of 10 g can be produced in artificial mass-culture systems. However such a high rate of production on reefs is consistent with the views, many times expressed, that reefs are amongst the most productive communities in the world.

The Great Barrier Reef system

The most extensive system of reefs in the world is the 2000 km long Great Barrier Reef off the Australian coast stretching from Lady Elliott Islands at latitude 24° 5′ S, northward to the Torres Strait, and no account of reefs would be complete without consideration of this system; however, remarkably few algal studies have been made in the area. The most detailed investigations are those of Cribb (1965, 1966, 1973) who has studied reefs with a rubble crest and not those in which a massive algal ridge has been developed (Cribb, 1973). In fact, Ladd (1971) comments that this system appears to be unique in being built primarily from corals and since it figures largely in descriptions of reefs no doubt this is where the misconception about reefs has arisen. The rubble crest is usually the highest part of the reef and also the site where the surf breaks: shorewards of the crest the water is calmer and lies to a depth of 10–80 cm at low water spring tides. The sandy beach is generally not colonised by algae except at its base where *Entophysalis deusta* coats the sand and *Enteromorpha clathrata* and *Monostroma* species occur on rock fragments buried in the sand. Where this sand is compacted into beach rock in the intertidal region, e.g., at Heron Island and Wilson Island, Cribb (1973) distinguishes three belts of algae. There is an upper *Entophysalis deusta* which forms a film over the beach rock, with filaments penetrating into the rock; some *Calothrix crustacea* also occurs intermingled in this belt. Below this a middle belt of mixed Cyanophyta forms a closely adherent 'skin' up to 1.5 mm thick on the beach rock; it is pink coloured and the common algae are *Schizothrix arenaria*, *S. tenerrima*, *Microcoleus lyngbyaceous*, *Calothrix crustacea*, *Kyrtuthrix maculans*, *Entophysalis deusta* and *E. conferta* and Cribb (1966) first termed it the *Kyrtuthrix* zone: fine scratches over this 'skin' are made by grazing fish. The lowermost belt is characterised by the red alga *Gelidiella bornetii* and in winter *Enteromorpha clathrata* occurs. Numerous small filamentous forms also intermingle here with the *Gelidiella*. These communities are really part of the intertidal system of sandy, beach rock shores and can be found independently of any offshore reef formation. Cribb divides the reef flat into inshore gutter, middle reef flat and moat. The gutter,

Fig. 4.8. The standing stocks of major crustose forms in the deep water seaward of the algal ridge, at Waikiki. In each histogram cover is given by the left column, relative density by the middle column and frequency in the right column. Station 1 is at about 6 m depth, 2, 3, 4, between 10 and 15 m and 5, 6, 7 between 20 and 30 m. *a*, Coelenterate corals; *b*, *Hydrolithon breviclavium*; *c*, *Hydrolithon reinboldii*; *d*, *Tenarea tessellatum*; *e*, Melobesioid 'C'; *f*, *Sporolithon erythraeum*. From Littler, 1973*b*.

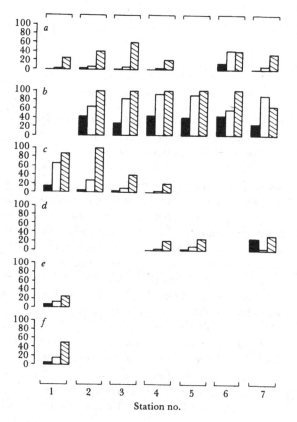

when present, takes drainage water from the reef flat at low tide and scouring keeps it relatively free of algae; but pieces of *Gelidiella*, *Enteromorpha* and stunted *Padina gymnospora*, *Penicillus sibogae*, *Champia parvula* and *Jania adherens* may occur. On the reef around Tutuila, American Samoa, a similar turf of low-growing species (*Jania*, *Polysiphonia*, *Ceramium*, *Hypnea*, *Cladophora*, *Laurencia*, *Gelidium*) occurs in the inshore 'moat' (Dahl, 1971). The middle reef flat tends to be sandy with irregular patches of coral, much of which is dead, and supports a growth of epiphytic algae. In summer there is a seasonal development of *Sargassum* species which can attain quite a size, and of *Turbinaria* which is often the first alga to be exposed by the falling tide. The tide never recedes enough to expose the whole reef flat to the air, only the upper branches of some of the organisms. In winter and spring the dominant algal cover is of other Phaeophyceae amongst which *Padina gymnospora*, *Dictyota bartayresii*, *Chnoospora implexa*, *Hydroclathrus clathratus* and *Pocockiella variegata* are the most common. Patches of the siphonaceous algae *Caulerpa racemosa*, *Boodlea composita*, *Chlorodesmus fastigiata*, *Dictyosphaerum versluysii*, *Halimeda discoidea*, *H. tuna* and *H. opuntia* occur. A complex mixture of algae often grows as a turf over the dead coral fragments in the middle reef amongst which are the green algae *Enteromorpha clathrata*, *Cladophora crystallina*, *Rhipidodesmis caespitosa*, *Udotea javanensis*, *Acetabularia moebii*, *A. clavata*, the brown algae *Ectocarpus mitchellae*, *E. indicus*, *E. irregularis*, *Sphacelaria furcigera*, *S. tribuloides*, *S. novae-hollandiae* and the red algae, *Gelidiopsis intricata*, *Ceramium gracillimum* var. *byssoideum*, *Tolypiocladia glomerata* and blue-green algae such as *Lyngbya majuscula*, *L. semiplena*, *Calothrix crustacea*, and *Fremyella grisea*. The sandy areas are relatively poor in macroscopic algae, only *Caulerpa cupressoides*, *Halimeda cylindracea*, and *H. macroloba* being recorded. Some algal species seem to be restricted to the shaded areas beneath the coral; the commonest species here are the egg-shaped *Valonia ventricosa*, encrusting *Peyssonelia* spp. *Lithothamnia* and crusts of *Lithophyllum simulans*, *Amansia glomerata*, and smaller algae such as *Microdictyon obscurum*, *Cladophora* sp., and *Hypoglossum* sp. Dead coral masses moved by storms and deposited on the reef flat and projecting out of the water are often covered by brown crusts of *Ralfsia expansa*. On some reefs (e.g., that at Lady Musgrave Island) the reef flat deepens so that a lagoon is formed. The moat area is one where the coral growth is vigorous and the patches form large platforms; fleshy algae are poorly represented but 'Lithothamnia' coat many of the surfaces with a pink 'rock-like' layer whilst some surfaces are coated with dull red crusts of *Peyssonelia*.

The reef rim on the Great Barrier Reefs may be 35–100 m wide and may rise almost vertically from the reef flat. It is a region pounded by the surf and washed by water rushing into and out from the reef flat as the tides rise and fall. The rubble crest, being emergent at low tide, supports a flora similar to that of the beach rock with *Entophysalis deusta* and *Gelidiella bornetii* recurring here. However, a distinctive alga, *Yamadaella cenomyce*, almost restricted to this habitat, occupies the position which on other reefs is colonised by the rock-like *Porolithon*. Seaward of this rubble crest is a zone where a dense algal turf grows and binds sand; the main species here are *Laurencia* spp. (*L. flexilis*, *L. obtusa* var. *obtusa*, *L. obtusa* var. *snackeyi*, and *L. perforata*) together with a mat-like growth of *Boodlea composita*, *Sphacelaria novae-hollandiae*, *Ceramium* spp., *Gelidiella acerosa*, *Griffithsia tenuis*, *Hypnea* spp., *Jania adhaerans* and *Microcoleus lyngbyaceus*. Other surfaces are covered by 'Lithothamnia') and *Peyssonelia*.

The seaward part of the rim is cut on its outer edge into a spur and groove system as in the reefs of Pacific atolls and is one of the richest areas of both coral and encrusting coralline algal growth. The latter may be locally predominant. Several species of 'Lithothamnia' are involved in forming this rock-like platform, but taxonomic studies of these are still needed. Fleshy algae are often absent from this area of the reef, though occasional plants of *Halimeda discoides*, *Turbinaria ornata* and *Amphiroa crassa* occur. *Plocamium hammatum* occurs suspended from the roofs of overhangs. In the most wave-beaten areas, fleshy algae tend to be sparse but *Chlorodesmis fastigiata*, *C. major* and *Bryopsis indica* occur.

The Great Barrier Reef system is relatively young compared with many of those of the Pacific since its development began only some 12 000–15 000 years ago during the post-Pleistocene advance of the sea over the Queensland continental shelf (Maxwell, 1972).

Cribb (1973) estimates the number of algal species on the Great Barrier Reef to be around 230 (not including unicellular algae); this compares with admittedly rough estimates from other workers which range from 145 to 320 species. The great majority of reef algae are widely distributed in the Indo-Pacific region and endemism in any one area is apparently slight, in marked contrast with the situation in southern Australia where there are no reefs but over 1000 littoral species and approximately 32% endemic species (Womersley, 1959). The Barrier Reef are relatively poor in red algae with only approximately twice the number of red as of green algae (Cribb, 1973), again in contrast with southern Australia where the factor is eight.

Material from cores and sediments of reefs is rich in remains of 'Lithothamnia' and *Halimeda*; Maxwell (1968) recorded 17–40% of the former and 10–30% of the latter making up the surface sediments in some parts of the Great Barrier Reef. Maxwell (1972) quotes a typical geological analysis of surface reef sediment as coral 28%, coralline algae 30%, *Halimeda* 30% and Foraminifera 10%. Goreau (1963) pointed out the importance of the sediment derived from the smaller calcareous algae growing on reefs. In Jamaica over 70% of the total calcium carbonate contained in the reef system is 'in the form of fine, unconsolidated sand deposited in beds over large areas adjacent to the living reef frame'. The small calcareous algae contribute a great amount of this material, followed by Foraminifera, molluscs and echinoderms whilst the large 'lithothamnioid' algae contribute very little detritus.

Futher experimental studies

Very low levels of nitrogen are generally recorded in reef waters yet there is a vigorous carbon metabolism on the reefs. Crossland & Barnes (1976) have shown, by the acetylene reduction technique, that the corals *Acropora acuminata* and *Goniastrea australensis* have organisms associated with the skeleton which fix nitrogen. Although some acetylene reduction occurs in the dark (presumably through bacterial activity) it is considerably increased in the light and since filamentous Cyanophyta are common in the skeletons these are almost certainly responsible and add to the fixation by the free-living flora investigated by Wiebe *et al.* (1975). Another possibility, however, is that it is caused by increased bacterial activity as a result of secretion of organic matter by photosynthetic algae. Crossland & Barnes consider that the organisms in the skeletons of the corals may contribute significantly to the nitrogen economy of the reef ecosystem. Their figures show much greater acetylene reduction in the lower part of the coral and an increase at the base when macroalgae (presumably epizoic) are present. Since most macroalgae have some associated filamentous blue-green algae it is reasonable to assume that these are responsible and Capone (1977) and Capone, Taylor & Taylor (1977) have demonstrated high rates of fixation associated with *Microdictyon* and *Dictyota*. Fixation also occurs in the lagoon sediments (Sournia, 1976a) where filamentous Cyanophyta are common. Webb & Wiebe (1978) have shown that nitrification at a site on the Barrier Reef increases the nitrate content of the water and that this nitrate can be removed by the corals. Very few other algae are recorded associated with corals and it would be surprising if blue-greens were the only ones present: in a recent report the diatoms *Campylodiscus*, *Podocystis* and *Triceratium* were recorded on corals on the Florida Keys and different species (*Amphora* and *Diploneis*) on the coral sand (Miller, Montgomery & Collier, 1977).

Very few studies have been made on growth or carbon fixation rates of reef algae. The rates of fixation by individual algae are extremely varied, e.g., in Jamaica, Goreau (1963) found that some algae photosynthesised 30 times faster than the slowest. Wanders (1976a) did not find quite such a variation, he measured net productivities, with oxygen outputs of $0.7–5.2$ g m^{-2} 24 h^{-1}. The dense algal vegetation growing on the Curaçao reef was about as high (0.064 mg O_2 cm^{-2} h^{-1}) as that of the corals themselves and a sparse vegetation was about half this. The red algae *Liagora*, *Galaxaura*, *Jania* and *Amphiroa* fixed carbon within the range $20–90$ μg mg N^{-1}.h^{-1} whilst the calcified Bryopsidophyceae were in the range $6–50$ μg mg N^{-1}.h^{-1}. The massive 'Lithothamnia', however, were extremely slow with rates less than 5 μg mg N^{-1}.h^{-1}. Similar studies on the incorporation of ^{45}Ca showed higher rates of calcification for the red algae (calcium above 500 μg mg N^{-1}.h^{-1}) compared to the Bryopsidophyceae (calcium below 500 μg mg N^{-1}.h^{-1}). These rates expressed on the nitrogen content of the algae are very difficult to compare with other published data. Rates for individual algae expressed on a frond area basis by Littler & Doty (1975) gave carbon values 2.2 g m^{-2} for *Porolithon onkodes* and

2.4 g m^{-2} for *P. gardineri*. These are quite high figures for calcified algae which have usually been considered slow growing.

Böhm & Goreau (1973) confirmed earlier work which had demonstrated that calcium exchange occurs in *Halimeda* in the dark and showed that outflow is even greater in the dark than in the light. They found that whilst net calcium accretion is 3.6 ± 1.1 mg g^{-1} min^{-1}, the calcium flux is much higher, about 311 ± 5 mg g^{-1} min^{-1}. Calcium exchange in dead *Halimeda* is 3.5 times lower than in live, which is further evidence that calcium deposition is a metabolic process. In Goreau's study, the amount of ^{45}Ca deposited in the dark was surprisingly high, in some algae even higher than in the light, a fact which does not suggest a very high degree of correlation with photosynthesis. Stark, Almodóvar & Krauss (1969) reported a diurnal rhythm of calcification in *Halimeda* with a peak at 8 a.m. and a low at 8 p.m. It may be of some significance that in genera such as *Halimeda*, and presumably in other Bryopsidophyceae, the calcification takes place outside the cells but isolated from the seawater by the aggregation of the outer swollen utricles (Borowitska, Larkum & Nockolds, 1976). In these algae, as in *Chara*, the initiation of calcification starts near the cell wall where a more favourable environment is created by the excretion of hydroxyl ions, at least in *Chara* (Lucas & Smith, 1973). Precipitation then occurs around the crystal nuclei and under such conditions one might expect that overall calcification is not closely connected with photosynthesis.

An average community calcification rate of 4000 g CaCO$_3$ m^{-2} yr^{-1} was found for the reef at Eniwetok and the figure was approximately the same across both the mixed algal coral and algal turf transects (Smith, 1973). Such a rate of calcium carbonate deposition is sufficient for a 3 mm yr^{-1} upward growth of the reef, but as Smith comments, there has been little change in height or sea level over the last 4000 yr* and he assumes that almost all the production is lost to the lagoon by wave transport. Adey & Vassar (1978) report similar accretion rates (1–5.2 mm yr^{-1}) on reefs in the West Indies. Overall growth rates of coral algae have rarely been measured but figures from the literature quoted in Littler (1972) suggest increases of a few millimetres to 2 cm

* Sea level has in fact been dropping gradually during this time.

per year in the diameter of encrusting forms and increases in thickness of 0.2–0.5 mm yr^{-1}. Growth on PVC panels can be much greater, e.g., 0.9–2.3 mm month^{-1} (Adey & Vassar, 1978). More recent estimates, however, suggest that the rates are not so different from those of other algae, e.g., Littler (1973c) found a net carbon contribution of 5.7 g m^{-2} day^{-1} on Hawaiian reefs. This is directly comparable to figures for the reef at Eniwetok which has been recently re-investigated (Smith & Marsh, 1973), giving gross carbon values of 0.05 g m^{-2} h^{-1} for coral–algal communities, and 0.97 g m^{-2} h^{-1} for algal turfs. Net production is given as 0.25 and 0.72 g m^{-2} h^{-1} for this reef by Smith (1973). The classic work of Odum & Odum (1955) has values of 0.80 and that by Sargent & Austin (1949) of 0.33, so in spite of later criticism the earlier results are in line with the more recent. The very considerable algal biomass relative to the animals recorded on the Eniwetok reef is given in Fig. 4.9 and though some of the animal groups may have been underestimated there is still a very conspicuous 'pyramid' of biomass. Smith & Marsh confirmed Sargent & Austin's view that in the mixed algal turf there was a markedly positive production (P:R = 1.9). Figures for a reef in the Indian ocean are also comparable at a gross carbon production of 3.84 g m^{-2} day^{-1} (Qasim, Bhattathiri & Devassy, 1972), and Wanders' (1976a) annual figure of 5650 g m^{-2} (gross) and 2330 g m^{-2} (net) for reefs in Curaçao is of the same order. The latter figures are derived from the algal and coral components only and do not take into account the animals as does the work on other reefs which give figures approximately 10% of those from the latter two regions. This reef at Curaçao has a net productivity which is estimated as being approximately 25–65 times as high as the plankton in the Caribbean and Wanders suggests that one factor contributing to this high figure is the continual flow of water across the reef; increased flow rate certainly increases algal metabolism (*see* also p. 57). But he also points out that it is no higher than that of *Sargassum* beds off Curaçao and his admittedly rough estimate of the dry weight of the primary producers on the Curaçao reef is about 1400–2800 g which is directly comparable to that of many temperate seaweed beds. Johannes *et al.* (1972) also found production by the algae of the reef flat at Eniwetok to be somewhat greater than that of the coral community.

Fig. 4.9. 'Pyramids' of biomass estimated from dry weight of organisms on various quadrats on the reef at Eniwetok. In each pyramid the bottom layer comprises the producers, the middle layer the herbivores and the upper layer the carnivores. From Odum & Odum, 1955.

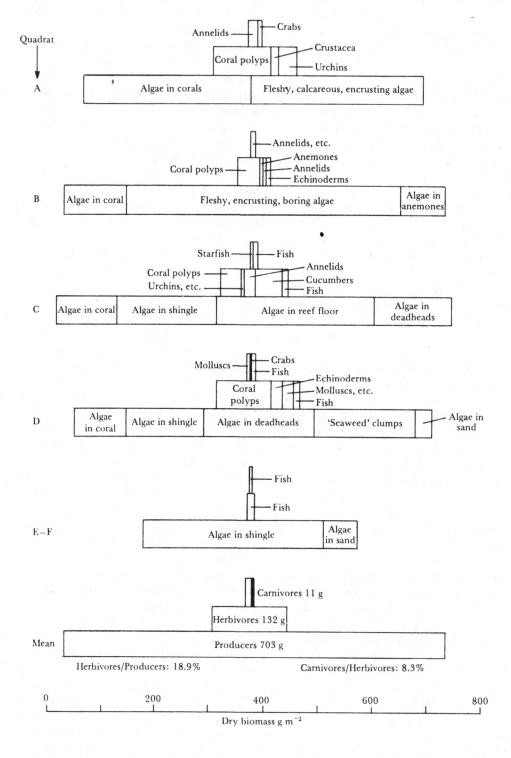

Clearly the algae of the intermittently exposed parts of reefs have to withstand considerable temperature changes as well as the stress of water movement. Very little experimental work has been done on temperature resistance of reef algae, though Schwartz & Almodóvar (1971) found that those normally exposed on the reef flat could withstand higher temperatures than those permanently submerged in the lagoon, and surprisingly, that the red algae showed greater resistance than the green algae. It is not clear, however, whether the normally submerged algae could withstand high temperatures whilst still submerged and lack of resistance under moist conditions would not be unexpected.

Algal cup reefs and 'trottoir'

Since coralline algae are a conspicuous building and binding feature of reefs, it might be expected that under certain favourable conditions they may themselves build 'algal reefs' without any association with hermatypic corals. The benthic coralline algae of temperate–boreal waters may indeed do so, but I know of no studies of these, other than comments in the literature about extensive growths over large areas of the shallow sea bed. In the tropics however there are reports of such algal activity giving rise to 'algal cup reefs' (e.g. off Bermuda, Yucatan, Brazil, Bahamas); these were previously known as 'boilers' or 'breakers' from the appearance of the water surface but Adey (1978) regards them simply as algal ridges. Earlier workers thought that they were 'eroded blocks of Pleistocene rock veneered with reef building organisms' but they are now known to be entirely formed from the growth of a mixture of coralline algae and associated invertebrates, mainly the hydrozoan *Millepora*. The algae principally concerned are species of *Neogoniolithon*, *Peyssonelia* and *Mesophyllum synthrophicum*. The vermetid gastropods *Dendropoma* and *Vermetus* also occur and become overgrown by the algae and *Millepora* which ultimately bury the gastropods in the reef. Carbon dating of these reefs has shown that in one example off Bermuda the material at 3.5 cm depth in the reef had an age of 3190 ± 210 yr B.P. and, from various samples, rates of growth ranging from 1 cm 8.5 yr^{-1} to 1 cm 39 yr^{-1}, were calculated (Ginsberg & Schroeder, 1973). Similar formations were described by Boyd, Korniker & Rezak (1963) near Cozumel Island, Mexico and their photographs show roughly

circular microatolls about 150 yd off the coast. They are circular, approximately 12–25 ft across, with a raised rim a few inches wide, extending a few inches above sea level. A lagoon is formed inside this rim in which luxuriant non-calcified algae, such as *Dictyota*, and some calcified algae, such as *Halimeda*, grow. The rims and sides are covered by coralline red algae including *Goniolithon soluble*, *Archaeolithothamium episporum*, *Lithothamnium sejunctum*, *Epilithon membranaceum*, *Lithophyllum* spp. and *Lithoporella*.* Beneath the algae some corals also occur.

The reefs just described are very similar in structure and in contributing organisms to the intertidal crust-forming growths in the Mediterranean described by Pérès & Ricard (1957) and known as 'trottoir'. These are formed by species of *Lithophyllum* or *Tenarea* growing on wave beaten rocks in the intertidal.

Coralline algae in general have been poorly studied, I suspect because of the apparent difficulties of identification. These difficulties are still present but the situation is easing, especially with the publication of papers such as that by Adey & Adey (1973), in which references to earlier papers can be found. Ecologically they have considerable importance since so many rock faces, rock pools and subtidal sites in cold and temperate waters are richly coated with pink, rock hard 'corallines' whilst the communities considered above extend their importance into the subtropics and tropics. Few marine algal growth forms are so widely dispersed; it is worth remembering that this ability to coat rock surfaces is shown not only by the *Porolithon*, *Lithothamnia*, etc. but also by the upright jointed *Corallina*, *Jania*, *Lithothrix*, etc., which also have a basal 'rock-like' crust. Yet, in 1973a, Littler was bound to state that 'there are no previous assessments of crustose coralline standing stocks from tropical deep water (below 3 m)'.

Equally there is little information on the standing stocks of the cold and temperate corallines and one has to rely on statements scattered in the literature that the subtidal in cold waters is covered by 'Lithothamnia', and the indirect evidence of massive

---

* Steneck & Adey (1976) report that *Lithophyllum congestum* is the primary ridge builder on these atolls and elsewhere in the Caribbean. But at Curaçao *Porolithon pachydermum* is the most abundant crustose coralline alga (Wanders, 1976a).

subtidal production reflected in the development of 'beaches' built up of coralline fragments. In Ireland and France this material is used as lime by farmers.

Summary

Apart from the symbiotic algae in the corals themselves (Chapter 9) most reefs and the adjacent lagoon and sea bed are rich in algal growths. In many situations and particularly where the force of the wave break prevents coral growth the reef is coated with a layer of calcareous algae which forms the surface of the reef. Live corals are not colonised by algae but damaged corals and dead fragments are incorporated into the reef by the cementing growth of crustose coralline algae. Although less spectacular than the corals themselves the algae are the most important components of reef growth.

Coral reefs are amongst the most productive carbon fixing systems in the world and this is partly due to the immense increase in surface available for colonisation under any square metre of water surface coupled with the continual flow of water across the reef. Reefs are difficult to generalise about, but in many an algal ridge can be discerned with *back* and *front* (seaward) slopes. Between the back slope and the land there may be *moats* and a *reef flat*. Seaward of the algal ridge and front slope an *algal plain* can be distinguished. Lagoons associated with reef systems have algal floras similar to those of other tropical sandy–rubble shores.

The flora of the various zones is discussed and the zonation across some reefs quoted as examples, together with experimental studies on growth rates, light intensity and grazing. Most studies distinguish between the crustose calcareous red algae which are rock-like in consistency and appearance and coat any available surface and the fleshy uncalcified flora which in composition is similar to that found on other tropical rocky shores.

Blue-green algae are often abundant around the bases of corals (and penetrating the coral when dead) and in calm parts of the reef flat, etc. They have recently been shown to fix atmospheric nitrogen and they may play an important role in these normally nutrient-poor situations.

Perhaps the most important reef systems in the world is that of the Great Barrier Reef off Australia and a brief account of the flora at some sites is given. This is such an extensive system that it is surprising how little the algal flora has been studied. Rates of carbon fixation, calcium fixation and overall productivity of reefs are given.

Coralline algae, both crustose and articulate, are common in waters outside the range of coral growth and although little studied they do form massive communities such as the 'trottoir' of the Mediterranean coasts and the algal cup reefs of the Caribbean.

5 Phytobenthos: rhizobenthos, epipelon, endopelon, stromatolites, soil algae, hot springs, snow and ice algae

The classification of the communities living on, or creeping in or on soft sediments is dealt with in chapter 2. Since stable sediments can only accumulate where water movement is slight these communities live in relatively quiet waters and do not have to adapt to excessive water flow; they still occur in streams and rivers, but at the interface between sediment and water where the disturbance is slight. The algal rhizobenthos, as its name implies, is rooted in the sediment but extends, usually as macroscopic growths, into the open water and thus has a life style similar to that of rooted angiosperms. The rooted community is absent from unstable coarse sandy sediments, e.g., where waves disturb the sea shore or in fast flowing rivers. Likewise the epipelon is best developed on relatively stable sediments where it forms a layer of cells, colonies and filaments on the surface and mixes into the upper millimetre or so of sediment. It is often visible as a green, brown or blackish colouring on the surface of the deposits. All sediments tend to be moved by the activity of benthic animals, resulting in some downward transport of algae, hence all the *major* components of the epipelon are actively motile and positively phototactic, so that they can move rapidly back to the surface after disturbance and burial. A few species even occur associated with the surf beaten sediments along sea shores where they are mixed into the water during tidal cover and can only grow in a truly epipelic manner whilst the tide is out. The endopelon is less motile, tends to live beneath the surface in mucilaginous aggregations of algae and sediment, and in some respects has affinities with the attached communities. Stromatolites are aggregations of sediment, often formed into rock-hard structures by algal secretion of mucilage and the resultant binding of sand particles; the algal species involved are basically epipelic but the particular environmental conditions allow the formation of the stromatolites. The algae of soils, of the sediments over which thermal waters flow and those in and on ice and snow are all treated in this chapter since they are all associated with inorganic 'sediments' of varying degrees of stability and when the floras are examined they are found to have life-forms similar to those of the epipelon.

Rhizobenthos

In freshwater the only algae that can be truly classed as rhizobenthic are the Charophyta which spread extensive rhizoidal anchoring systems through the sediments and raise very complex photosynthetic 'stems' into the overlying water. Remarkably little is known of the ecology of these plants, which frequently form very extensive underwater meadows. The majority of species occur in still water and the few (e.g., *Nitella opaca*, *N. flexilis*, and *Chara braunii*) which extend into rivers only occur in the slower flowing regions. Langangen (1974), in a study of Norwegian charophytes, also found some confined to brackish waters and others to waters which he classified as *Chara* lakes, *Potamogeton* lakes or *Lobelia* lakes (Table 5.1). *Chara* and *Potamogeton* lakes tend to be found in alkaline regions and *Lobelia* lakes tend to be oligotrophic and to occur in base-poor regions. This distinction into *Potamogeton* and *Lobelia* lakes has been made by other north European workers, but it is not a very satisfactory distinction unless details of chemistry and other features are recorded. The charophytes are often developed in dense belts between growths of *Lobelia* and *Isoetes*.

The *Chara* lakes are also often poor in nutrients and the charophytes themselves are usually encrusted with lime, giving them a whitish-grey appearance. Only in the *Lobelia* lakes and slightly saline habitats is the carbonate encrustation weak. The calcium carbonate sometimes occurs in bands down the

Table 5.1. *The number of finds of Charophyta in south Norwegian waters of different type. From Langangen, 1974*

| Species | Brackish water | Freshwater | | | | Rivers | No. of finds |
		Chara lakes	Potamogeton lakes	Lobelia lakes	Other lakes		
Nitella opaca	2	—	7	9	3	3	24
N. flexilis	1	—	—	2	—	3	6
N. translucens	—	—	—	1	—	—	1
N. mucronata	—	—	—	1	—	—	1
N. gracilis	—	—	—	1	—	—	1
N. confervacea	1	—	1	—	—	—	2
Tolypella nidifica	4	—	—	—	—	—	4
Lamprothamnium papulosum	2	—	—	—	—	—	2
Chara braunii	1	—	—	1	—	1	3
Ch. canescens	4	—	—	—	—	—	4
Ch. tomentosa	—	8	2	—	—	—	10
Ch. baltica	3	—	—	—	—	—	3
Ch. aculeolata	—	8	3	—	2	—	13
Ch. contraria	—	15	6	—	2	—	23
Ch. vulgaris	—	1	3	—	—	—	4
Ch. rudis	—	17	3	—	—	—	20
Ch. hispida	—	1	—	—	—	—	1
Ch. aspera	2	14	5	—	—	—	21
Ch. strigosa	—	10	1	—	2	—	13
Ch. globularis	3	16	15	1	4	—	39

internodes, a feature which is readily observable in young actively growing material. Lucas & Smith (1973) showed that the precipitation occurred on the wall in regions where bicarbonate ions were being absorbed and hydroxyl ions released. Fig. 5.1*a* shows how this can be detected with glass electrodes and that a pH difference of three units can occur between adjacent areas. Fig. 5.1*b* shows that the pH changes occur only in the light. The regions of high pH can move along the walls prior to precipitation of calcium carbonate but become stabilised when the crystalline deposits form.

In lakes with high calcium bicarbonate content, phosphates can be precipitated leaving the water low in phosphorus and evidence from the work of Forsberg (1965*a*) and others has shown that charophytes do not tolerate high phosphate levels; indeed growth maxima have been reported at one part per million. The species which live in acidic and oligotrophic waters would likewise be in a relatively low phosphate environment. *Chara zeylandica* is reported (Taylor, 1959) in the sediment between roots of the mangrove *Rhizophora*, which occurs around brackish ponds in North America and a number of species are more abundant in coastal regions where ponds and channels receive salt by seepage or from wind-blown material. Some instances of the occurrence of charophytes in brackish communities in the Baltic Sea are mentioned in the previous chapter where they form associations alongside *Fucus*, *Furcellaria*, etc. Charophytes can also penetrate hypersaline waters, e.g., in the lagoonal system known as the Coorong in South Australia (Womersley, personal communication).

There are few modern data on the seasonal occurrence of charophytes but many can overwinter beneath ice and many propagate vegetatively from secondary protonemata and storage bulbils. Others are more closely linked with certain seasons and are re-established annually by germination of oospores which appear to require periods of rest, low temperature and low oxygen concentration, though each species needs much greater study and some of the accounts in the literature are conflicting. The beds of sediment beneath dense *Chara* and *Nitella* are often blackened and rich in hydrogen sulphide. The epipelic flora in such beds is very sparse (Round, unpublished observations).

Communities growing in various depths of water tend to be zoned and the charophytes are no exception but from the data scattered in the literature it is difficult to assemble any very precise features. Some species, e.g., *Chara aspera*, seem to grow only in shallow water though this itself is generally correlated also with sandy bottoms. In deeper water more silt is admixed, altering the rooting environ-

Fig. 5.1 *a, b*: typical pH values along the wall of *Chara corallina* after 2 h illumination (●) and dark (▲). The broken line indicates the pH of the surrounding solution. *c*: change in pH on transfer of *Chara* cells from dark → light → dark. (●) Dark pH response, (○) light pH response. (△) Steady state pH response after 2 h illumination. From Lucas & Smith, 1973.

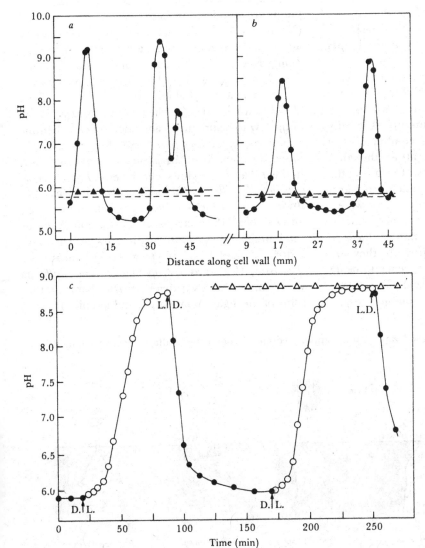

ment complementary to the decrease in light. Below 1.0–1.5 m *Nitella flexilis*, *Nitellopsis obtusa*, *Chara filiformis* and *C. denudata* are more prevalent and some are certainly 'shade species'. The deeper growing populations of the same species tend to be taller than those in shallow water (Corillion, 1957; Forsberg, 1965*b*), and actively growing charophytes are recorded down to at least 60 m in clear lakes, though in the majority of sites, whilst they tend to occur in deeper water than the rooted angiosperms, they are usually not found much below 10 m.

The occurrence of charophytes in alkaline waters suggests an ability to utilise bicarbonate ions and this has been proven experimentally by Smith (1968) though all the plants tested also took up carbon dioxide in much greater amount when grown at pH 6.5. The data suggest that growth of the plants may be somewhat slower in habitats where the pH is above 9.0.

The marine rhizobenthos has received equally little attention but it is obviously a highly productive community, especially in tropical regions. In cold and temperate waters this community is hardly represented, possibly owing to a combination of wave disturbance and low temperature, though I know of no experimental studies. Certainly the Bryopsidophyceae are major components in warm and tropical waters; few extend into colder waters and one assumes that temperature is a major factor in their distribution. The rhizomatous *Caulerpa* species extend the furthest and perhaps they are the most resistant to sediment movement, being anchored by tufts of rhizoids at intervals. These algae often overgrow rock and are not confined to soft sediment. In temperate estuaries algae, such as *Enteromorpha*, *Dasya*, etc., occur sporadically in the silt, but they are usually found to be anchored to pebbles or shells buried in the sediment and therefore not truly rhizobenthic, although, since they occur in sandy–silty regions, they will be considered as part of the rhizobenthos. There has been very little work on the stability of the rhizobenthos. In Bimini lagoon, the *Batophora* attached to buried shells is removed, together with the shells, when the current across the sediment reaches 50 cm s^{-1} whereas the genera anchored by extensive rhizoidal systems (*Penicillus*, *Halimeda*, *Udotea* and *Rhipocephalus*, Fig. 5.2) are not affected until the current reaches 80 cm/s (Scoffin, 1970). Nearby in the Bight of Abaco these plants are found at a density of about 20 plants m^{-2} (Neumann & Land, 1969). *Batophora* seems to be almost the only macroscopic algae in pools amongst mangroves, e.g. in Curaçao (van den Hoek, Colijn, Cortel-Breeman & Wanders, 1972); this genus is apparently a euryhaline species – growing from 20–23‰ to 120–160‰. The work in Curaçao is one of the few to distinguish associations of rhizobenthic organisms, e.g. of *Caulerpa verticillata* or *C. septularioides* under the mangrove *Rhizophora* and of various algae in the *Thalassia* meadows, characterised by *Halimeda opuntia* or *Penicillus capitatus* or *P. pyriformis* or *Goniolithon strictum* each of which can be very rich in species, with, for example, 113 species in one association. Another sand–mud binding community in Curaçao was dominated by small Rhodophyta growing beneath the *Thalassia*; *Centroceras–Polysiphonia–Spyridia–Griffithsia* were common here. van den Hoek *et al.* comment that some of the algae commonly found in epilithic sites,

Fig. 5.2. Macroscopic green algae living in and trapping sediment around the holdfasts. From Scoffin, 1970.

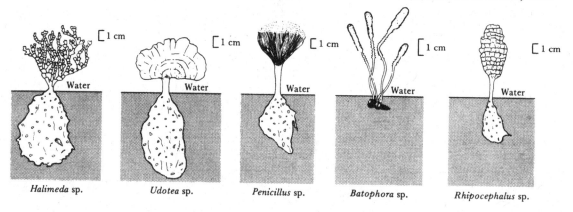

Halimeda sp. *Udotea* sp. *Penicillus* sp. *Batophora* sp. *Rhipocephalus* sp.

seem capable of growing attached to sand grains and connect these together with rhizoids and stolons to stabilise the habitat. The rhizobenthos also includes associations of filamentous algae occurring mainly in marine and estuarine situations, and in freshwater a similar niche is occupied by *Vaucheria*, some species of which are also marine. In Sweden the more alkaline lowland rivers have been designated *Vaucheria* streams owing to the predominance of this genus forming mats in the bethos; yet again these are neglected communities though very widespread and it is not clear whether or not they start growth on pebbles, etc. Scoffin records *Enteromorpha* commencing growth on shells in the sediment and spreading over an area of 1000 m² and to a depth of 1 cm in one month; it then remained healthy, but ceased spreading and after about three months it disappeared. These mats produce mucilage which binds sand grains and it is quite common to find this community raised slightly above the surrounding deposits (Fig. 5.3). The thalli are positively phototropic and can grow upwards through 2 cm of sand in a few days. In the Bahamas, *Polysiphonia havanensis* and *Laurencia intricata* form similar, but weaker mats, whilst blue-green algae (e.g. *Lyngbya* spp.) also grow in this manner, but are much less stable. Mats

composed of *Cladophoropsis* and of *Schizothrix* are also reported from Abaco. The *Cladophoropsis* is coated with *Thalassiothrix* (this genus is usually planktonic and its occurrence as an epiphyte needs checking) and together these algae secrete masses of mucilage to which sand attaches (Neumann, Gebelein & Scoffin, 1970).

The larger marine rhizobenthic algae secrete calcium carbonate as aragonite crystals and contribute large quantities to the sediment in certain tropical regions. Some aragonitic deposits previously thought by geologists to be inorganic precipitates may be formed in this way (Lowenstam, 1955). One method for detecting the origin of such sediments is to determine either the $^{18}O/^{16}O$ or $^{13}C/^{14}C$ ratios in the deposits. Values for inorganic oolites are restricted to a narrow range, but those produced by algae show considerable variation in the ratios. Lowenstam & Epstein (1957) showed with these isotope methods that considerable banks of aragonitic needle deposits in the Bahama–Florida–West Indies region were derived from algal sources.

Epipelon

The epipelon is an extremely widespread community occurring in all waters in regions where sedi-

Fig. 5.3. Mat-forming marine algae with *Enteromorpha* dominant, binding sand off Bimini. From Scoffin, 1970.

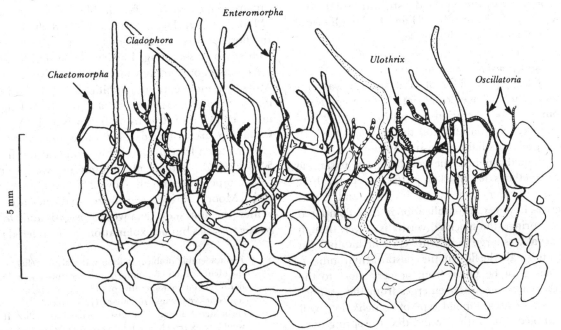

ments accumulate and on to which light penetrates. The species are almost all microscopic and the associations rich and widespread. They live on and in the surface millimetres of sediment and cannot withstand long periods of darkness and anaerobic conditions (Moss, 1977*b*).

Many hundreds of species are involved in the algal association of the epipelon; diatoms are often the most conspicuous in both marine and freshwater sediments, whilst Cyanophyceae, desmids and some flagellates are very common in freshwaters. In a recent study of small ponds, which were free of ice for only about four months, some 357 different kinds of algae were found in the epipelon (Moore, 1974*b*). There have been few attempts to distinguish plant associations after the pattern of the phytosociologically minded land ecologists, though there is no doubt that this can and should be done (*see* p. 34).

The species composition of the epipelon varies widely between habitats, though it tends to be fairly constant for any one type of water, e.g., Flensburg (1967) and Flensburg & Sparling (1973) recorded the algae from bog–mire water at sites in Canada and Sweden and in both instances found Desmids the dominant group (38 in Canada and 238 in the more extensively studied Swedish waters whereas only 5 and 50 diatoms were recorded). The figures obviously reflect intensity of collecting but the proportions indicate the expected distribution of the two groups in water of pH 3.0–4.21 for the Canadian and 5.5–7.0 for the Swedish samples.

Springs and flowing water

The epipelic flora of cold springs warrants further study since conditions of temperature (around 8–9 °C), aeration and nutrient supply tend to be relatively constant and the only factor which varies greatly is light intensity. Springs are widely distributed throughout the world and easy to sample by the methods described on p. 167, except for those which discharge underwater in lakes and rivers. There is, however, a complication in that, compared to hot springs, the species diversity is larger, though not usually as large as in other epipelic habitats. In order to obtain the flora in the constant environment, samples must be taken from or very close to the region where the water emerges from the ground, otherwise a stream–pond type of epipelic flora will be sampled. The author found that the spring flora

at two sites in England was composed mainly of diatoms and the dominants tended to be non-motile species, perhaps not a surprising finding in a habitat where there is constant flow. Both were in calcareous regions and the dominants were *Achnanthes lanceolata*, *Gomphonema intricatum* var. *pumila*, *Achnanthes exigua* var. *heterovalvata*, *Denticula tenuis* at Leuknor and *Fragilaria construens* var. *venter*, *F. pinnata*, *Achnanthes minutissima* var. *cryptocephala*, *A. lanceolata*, *Fragilaria leptostauron* var. *dubia* at Malham (Round, 1957*c*, 1960*c*). Since fine particles were washed out of such sites it is possible that some of the species are living attached to sand grains (i.e., are actually epipsammic) but others, e.g., *Fragilaria*, form filamentous colonies. Nevertheless close to the upwelling point motile, typical epipelic genera such as *Neidium dubium*, *Stauroneis smithii*, *Navicula pupula*, *N. hungarica* var. *capitata*, *N. radiosa*, *Cymatopleura soles*, *Surirella spiralis*, *Campylodiscus hibernicus* are found. The last three species tend to occur also on the deeper sediments of lakes, confined possibly by low temperature as in the springs.

The epipelon of rivers and streams also consists mainly of diatoms; a study of a small stream in Ontario where the water had a pH of 8.2 to 8.4 revealed that, contrary to the situation in standing waters, there were 321 diatoms and only 32 Chlorophyta which showed that the epipelon of streams is apparently not a very favourable site for desmids (Moore, 1974*a*). Diatoms in fact formed 93–99 % of the flora in these streams; this is a situation which, from my own sampling, I believe to be common for the epipelon of running water in lowland situations. Moving to more extreme conditions, Moore (1974*a*) found 200 diatom species out of a total of 240 algae in the epipelon of streams on Baffin Island. On this Canadian arctic island the flora is only actively growing from June to October at temperatures from 5 °C to 8 °C, though only the middle two months have temperatures much above 2 °C. At some sites, which Moore considers were frozen to the bed during winter, the dominant taxa were *Mougeotia* and *Oedogonium** species but no explanation can be offered for

---

* A considerable problem in fresh water algal ecology is the identification of unbranched filamentous algae. The floras rely on characteristics of the zygotes and/or reproductive organs but only rarely are these encountered during the course of ecological studies. It would be extremely helpful if vegetative keys could be

survival under such conditions, though freezing in other habitats does not necessarily kill vegetative growths of algae. The author (Round, 1959a) also found that filamentous species of Spirogyra, Zygnema, Mougeotia, and Oedogonium were common in rivers of the Scandinavian arctic. As in the springs mentioned above, the dominant diatoms were often non-motile species of the genera Achnanthes and Tabellaria and since these rivers were draining Precambrian rocks the species were those of acidic habitats. Dominants were Achnanthes kriegeri, A. marginulata, Eunotia tenella, Nitzschia palea, N. filiformis, N. ignorata, N. amphibia v. genuina and Tabellaria flocculosa. This flora is also characterised by a large number of Eunotia and Pinnularia species. Moore found that the growth rates were much higher than in comparable temperate situations, up to 3.2×10^5 cells $m^{-2}.day^{-1}$ were produced and standing crops commonly exceeded 8×10^6 cells cm^{-2}. He attributed the large crops to the absence of scouring and relative unimportance of grazing. In addition, epipelic populations in streams rarely seem to be nutrient limited. The nutrient contribution from both water and sediment has never to my knowledge been studied adequately for this community but presumably depletion in the water column would ultimately extend to the interstitial water of the sediments. Records of algae from arctic and antarctic sites all seem to point to massive crops and one cannot help but feel there is some relationship between this and the severity of conditions. Thus, if the crops are merely in a state of suspended animation during the eight months of winter, and also only slightly grazed but otherwise undepleted, each new season starts with quite the opposite situation from that in temperate zones where the winter months are periods when losses exceed gains. It is therefore a positive disadvantage to be in a situation of low turnover, whilst all the factors leading to loss of cells continue to operate. Compare this with the situation in another extreme habitat where standing crops are high, the tropical lake on or near the equator, where there is likewise no period of the year when losses grossly exceed production but instead there is a balance with large crops turning over rapidly (see p. 278, Lake George). Uptake of organic molecules during the dark, frozen

period may also be a factor in maintaining high populations, since such uptake systems have been shown to exist, especially in Chlorophyceae but also in some diatoms, e.g., the freshwater centric diatom, Cyclotella meneghiniana, and also some species of Pinnularia exhibit dark uptake of glucose (Lylis & Trainor, 1973). Clearly it is too early to generalise since very few diatoms have been tested and experiments need to be undertaken on natural systems but, as Lylis & Trainor comment, many of the species found to be facultative heterotrophs grow close to the sediment and therefore in regions of high organic content.

In contrast to the above figures for an arctic site the data from the temperate rivers Ivel and Gade showed only 2.6×10^6 cells m^{-2} and 1.6×10^6 cells cm^{-2} respectively and chlorophyll a concentrations of 0.68 g m^{-2} and 0.69 g m^{-2} respectively in spring (Edwards & Owens, 1965).

There are reports that species from the stream communities* appear at certain times of the day in the drift (material being carried downstream). The algae about to be considered should not be confused with the many species which appear in lists of river plankton, derived from the benthos, but simply torn off their substrata and washed away and therefore related to periods of high scour and not exhibiting rhythmic patterns. In 1954, Blum reported that the diatom Nitzschia palea, which normally grows on sediments, appeared in the water. This seems to be the first record of a regular occurrence of a non-planktonic alga in any quantity in the drift. Further detailed studies by Müller-Haeckel (1966, 1967, 1970, 1971a, b, 1973a, b) confirmed these results (Fig. 5.4) with peaks of cells in the drift around mid-day. At first sight it is surprising that such a daily periodicity exists, and even more surprising when one realises that some of the species are normally attached to substrata by mucilage pads. In an experimental system Müller-Haeckel (1971b) found that Synedra ulna was released from stones, drifted in the water then re-attached lower down the stream; this even occurred when the experiments were conducted in an artificial laboratory stream system in constant darkness (Fig. 5.5). Whilst diatoms were found to rise into the drift during the day-time,

devised and I see no reason why this cannot be achieved considering how much progress has been made on the equally difficult genus Chlorella.

* In order to avoid repetition the rhythmic phenomena shown by species from epipelic, epiphytic and epilithic communities will be discussed at this point.

Fig. 5.4. The appearance of three diatom species in the drift in control streams (continuous lines) and experimental ones (dashed lines), sampled at 2-h intervals over 72 h. *a*, *Ceratoneis arcus*; *b*, *Synedra minuscula*; *c*, *Achnanthes minutissima*. From Müller-Haeckel, 1973.

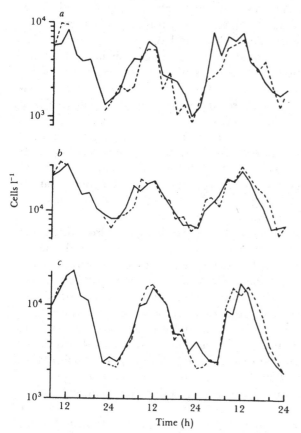

a species of the green chlorococcalean algae, *Monoraphidium*, appeared in the drift during the night (Fig. 5.6). Müller-Haeckel considers that cell division is a possible explanation for the release of the normally sessile algae into the drift; presumably the parent cell remains attached to the substratum and the daughter cell is released. Experiments in artificial channels suggested that a relatively short time (a few seconds) is spent in the drift and the downstream movement is quite small, probably less than 18 cm. Experiments in which wood or plastic plates were immersed in a stream for two hour intervals throughout 24 h revealed re-attachment with a peak around mid-day (Fig. 5.7) and a quite exceptional preference for attachment to wood (Müller-Haeckel 1970). This is the only study I know which involves such short-term exposures and is the only one to throw any light on the almost unconsidered problem of recolonisation by epiphytes. There have been very few estimates of cell densities of attached algae in streams and Müller-Haeckel gives figures of 1.2×10^9 cells of diatoms and the same number of unicellular green algae released

Fig. 5.6. The appearance of *Monoraphidium dybowskii* in the drift in control streams (continuous lines) and experimental ones (dashed lines) of the Kaltisjokk. From Müller-Haeckel, 1973a.

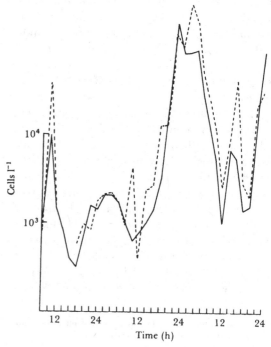

Fig. 5.5. The daily periodicity of colonisation of *Synedra ulna* cells in natural light/dark cycle (nLD) and constant darkness (DD) in an experimental system in the laboratory. From Müller-Haeckel, 1971b.

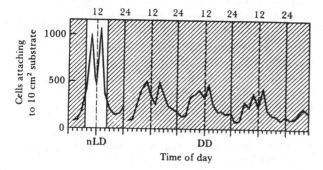

Fig. 5.7. Comparison of the diurnal colonisation
pattern of diatoms on wood (open columns) and
plexiglass (black columns) during a 24-h period.
a, total diatoms; *b*, *Synedra minuscula*; *c*, *Ceratoneis
arcus*; *d*, *Synedra ulna*. From Müller-Haeckel, 1970.

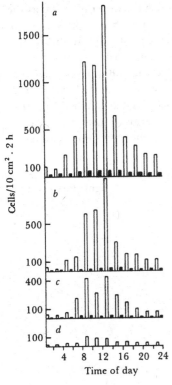

Fig. 5.8. *a*, the daily periodicity of algae in the drift of the Kaltisjokk, *a*, during June 1969 when water
temperature and solar radiation fluctuated regularly in stable weather and *b*, during June 1972 when the
weather was overcast with slight daily amplitude of temperature and solar radiation. *i*, *Monoraphidium dybowskii*,
ii, *Ceratoneis arcus*. From Müller-Haeckel, 1973*b*.

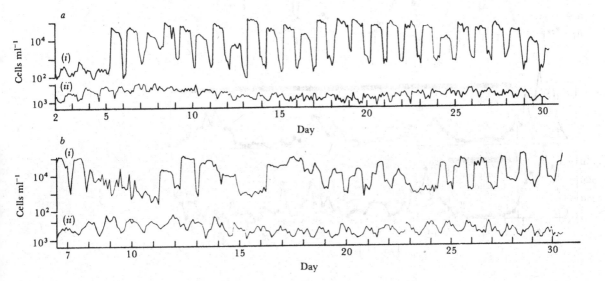

from 7.2 m² every 24 h, in her experimental channels. In earlier work (Müller-Haeckel, 1966), epipelic species of *Navicula*, *Nitzschia* and *Gyrosigma* were found in the drift, together with the centric diatom *Thalassiosira fluviatilis*, all with peaks in the day-time. The latter species is normally considered to be planktonic (however Müller-Haeckel found *Thalassiosira* as flocs on the sediment) and its increased representation during the day suggests that cell division is responsible for the pulsing. Blum (1954) thought that oxygen production by *Nitzschia palea* caused the cells to rise into the drift; the reasons for such rises are complex and whilst oxygen production may be one factor, others are cell division, absence

of reserve substances and lower silica content in newly divided cells, plus increased motility of some species during the light period. Further observations by Müller-Haeckel (1973b) showed that in the northern summer, when there was a pronounced daily amplitude of insolation (of the order of 1:100 at the site north of the Arctic Circle where there was still 500–600 lux around midnight), the daily drift of *Monoraphidium* was very marked but that of the diatom *Ceratoneis* was not (Fig. 5.8a). In another year when the weather was overcast and cloudy (1972) the amplitude of insolation was less and the rhythm of *Monoraphidium* was aperiodic until the weather changed (25 June), but the rhythmic activity of

Fig. 5.9. The rhythmic migration cycles of epipelic diatom populations from four streams near Bristol. Dark lines indicate times of total darkness. *a*, light:dark cycles; *b*, in constant darkness after one light:dark cycle; *c*, in constant light after one light:dark cycle. From Round & Happey, 1965.

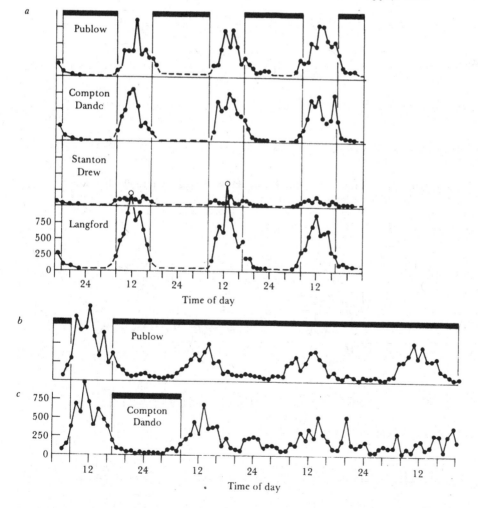

Ceratoneis arcus was more precise under these unstable conditions (Fig. 5.8*b*). Normally *Monoraphidium* appears in the drift during the night but in the continuous illumination of the arctic summer it shifts its phase and becomes a day-drift species. It is interesting that the normally day-active *Ceratoneis* and night-active *Monoraphidium* have such opposed behaviour in fine and in clouded weather. Such variations are imposed by differing responses to the entraining factor (Zeitgeber) which in these examples is almost certainly the normal alternate light–dark cycles. This rhythmic behaviour is one which would repay further investigation in temperate and tropical waters.

Similar rhythmic phenomena, but involving motility of stream diatoms, have been reported by Round & Happey (1965); the epipelic diatoms migrate upwards through the sediment during the light period (Fig. 5.9) and down into the sediment during the early part of the dark period and this rhythm is maintained in the laboratory under constant conditions and in darkness (*see* p. 179 for more details of this phenomenon). The fact that the upward movement commences during the hours of darkness suggests that a rhythm of movement independent of light is involved, but that light attracts diatoms to the surface, hence the peak around mid-day. Obviously when the diatoms are moving at the surface they are more prone to suspension in the drift.

The primary production of Red Cedar river, Michigan is mainly from benthic diatoms and large numbers of these are found in the drift. During one year Ball & Bahr (1975) found that some 633.9 metric tonnes drifted past the sampling station and in energy values this amounted to 634.4×10^6 cals. It is not clear whether or not the diatoms in the drift were actively growing.

Ponds and lakes

The sediments of ponds and lakes appear to the casual observer as a sterile environment, yet a very rich microscopic epipelic flora lives creeping over and between the silt or sand particles* and in the upper millimetre or so of interstitial water. The sediments are in motion, continuously moved by water movement and the activities of the numerous animals living on, and in, the sediment. This movement results in the steady sinking of heavy particles, including algae, into the sediment and only algae capable of positive phototactic movement are likely to survive. The algae are not only disturbed by the animals but many are actively grazed, though the precise effect of this on the population dynamics is unknown. The composition of the epipelic flora is determined by a complex of factors, of which the chemical composition of the overlying water is the dominant one, interacting with the chemical and physical nature of the sediment and with the degree of water movement. Superimposed on these factors are the effects of differing growth rates coupled with preferential grazing of some algal species. The epipelic flora can often be seen as a brown-green-blackish film on the sediments, the variation in colour being caused by the preponderance of diatoms, Chlorophyta/Euglenophyta or Cyanophyta, and Lund in 1942 established it as a community totally distinct from the epiphyton and plankton. Its active metabolism is revealed by the copious photosynthetically produced gas bubbles frequently seen on the surface of the film. Sometimes the surface layer of algae and mud is dislodged by trapped gas bubbles and patches float up into the overlying water; this is especially common when the weft of algae is predominantly filamentous Cyanophyta. Occasionally one finds species from these floating flocs recorded as planktonic species, which clearly they are not.

It is quite common for the epipelon to become so entangled that a 'carpet-like' growth results; this has been termed 'tapetic' by Wood (1965) but it is really only a vigorous development of the epipelic community. The work of Moore (1974*b*) revealed that in the epipelon of ponds in the Canadian arctic diatoms contributed 63 % of the total species and at

---

* There is a semantic problem here in that this essentially *motile* community can exist in sandy sediments, but it ought still to retain the descriptive term epipelon since epipsammon has been used for the community *attached* to sand grains (*see* p. 122) and the free-living flora living on sandy sediments is made up of species having identical life forms to those on muds and silts and indeed the same species can occur moving over the surface of sand, silt or organic mud. The only difference is a physico-chemical one involving size of particles and amount of admixed organic matter. This problem has also been discussed by Colocoloff & Colocoloff (1973) who separate the flora in sand into 'diatomes psamnique libres' and 'diatomes psammique attachés' but as stated above the 'libres' are simply epipelic.

Table 5.2. *The distribution of some common epipelic species on sediments of the English Lake District. From Round, 1958*

Increasing nutrient status →

Species	Low										High												
	Crummock Water	Wastwater	Ennerdale Water	Derwentwater	Brother's Water (S)	Brother's Water (N)	Bassenthwaite Lake	Loweswater	Ullswater (S)	Buttermere	Windermere III	Rydal Water	Coniston Water	Windermere II	Elterwater	Windermere I	Ullswater (N)	Blelham Tarn I	Loughrigg Tarn	Blelham Tarn III	Esthwaite Water	Blelham Tarn II	Grasmere
Caloneis silicula	13[a]	7	27	58	61	21	61	80	59	12	60	28	78	55	23	55	43	79	82	69	75	66	5
Neidium iridis	7	7	—	37	31	57	6	13	12	29	37	28	39	7	29	35	—	38	35	13	50	41	22
Frustulia rhomboides	33	40	53	5	—	14	—	—	—	53	—	5	—	—	—	2	—	—	—	—	—	—	—
Eucocconeis flexella	7	—	13	—	—	7	—	—	—	18	—	—	—	—	—	—	—	—	—	—	—	—	—
Neidium hitchcockii	—	—	—	10	—	—	—	—	—	—	20	—	—	2	—	35	—	15	94	10	36	5	—
Stauroneis smithii	—	—	—	—	—	—	11	—	—	—	2	—	—	—	—	22	—	59	—	5	86	23	—
Diploneis ovalis	—	—	—	—	—	—	—	13	6	—	2	—	—	—	—	5	14	20	76	20	—	—	—
Gymnodinium aeruginosum	—	—	—	10	—	—	—	7	—	—	10	24	39	17	24	10	—	26	6	41	29	23	23
Cryptomonas ovata	—	—	—	—	—	—	—	20	12	—	38	12	17	5	18	40	2	20	29	13	71	33	23
Synura uvella	15	—	—	10	—	—	—	—	6	—	13	18	6	13	—	5	6	10	18	7	21	13	18
Holopedium geminata	—	—	—	—	—	—	—	13	—	—	25	6	—	10	—	32	—	15	70	59	36	5	—
Aphanothece stagnina	—	—	—	—	—	—	—	—	—	11	28	—	—	15	—	2	—	14	24	—	7	38	—

[a] Numbers refer to the number of times a species was encountered in counts over a period of 18 months expressed as % of the possible total.

times could form 60–100% by cell numbers or 40–100% by volume in the flora. Very high standing crops were recorded in these ponds with cell densities frequently exceeding 60×10^6 cells cm^{-2}. This is 20–30 times the crops likely in temperate waters and Moore concluded that nutrient depletion was unlikely, light intensity was high and grazing very sparse, thus allowing such enormous growth even in water of low temperature. The epipelon assumes major importance in overall carbon fixation where planktonic growth is low, e.g., Miller & Reed (1975) report that the epipelon of ponds in Alaska contribute from 6 to 10 times more carbon fixation than the phytoplankton and Clesceri (1979) found that autotrophic activity in the sediments of Lake George at 9 m was 10^3 times greater than the heterotrophic activity. The epipelic algae of the Alaskan ponds were not inhibited at high light intensity as were the phytoplankton (Stanley & Daly, 1976) and the epipelic algae were also shown to have a higher temperature optimum than the plankton. The latter feature correlates with the higher temperatures of the surface sediments in these Arctic ponds. A direct correlation between temperature and saturating light intensity of photosynthesis was also shown for this community. The daily peak in photosynthesis occurs in late afternoon (*see* similar situations in the phytoplankton, p. 324).

Early work by the author revealed a distinct correlation between the epipelic flora and a classification of the English Lake District lakes based on chemistry of the sediments (Table 5.2) and although cells of many species can be found in lakes of both high and low nutrient status certain species and groups are clearly more abundant in one group or another (Round, 1957a, 1958b). Indicators of even more acidic and humic lakes can be recognised which do not occur even in the lowest nutrient status waters of the English Lake District but do occur in, for example Finnish lakes (Round, 1960a); such species are *Actinella punctata*, *Amphicampa hemicyclus*, *Eunotia bactriana*, *E. polyglyphis*, *E. septentrionalis*, *Fragilaria constricta*, *F. hungarica*, *F. polygonata*, *Tabellaria binalis* and *Teteracyclus emerginatus*. None of these are motile species; they occur in the flocculent material lying on the sediment and should perhaps be separated as a distinct community. In these papers the studies were made on acid cleaned sediment and hence cells from epiphytic and epilithic populations tend to be mixed amongst the true epipelic species. Motile examples would be *Anomoeoneis exilis*, *A. serians*, *Frustulia rhomboides*, *Cymbella gracilis*, *Stenopterobia intermedia* and *Pinnularia undulata*. Extending the survey in the opposite direction to calcareous lakes, e.g., in Ireland (Round, 1959b) no species of *Eunotia* are common and *Gyrosigma attenuation*, *Mastogloia smithii*, *Navicula oblonga*, *N. dicephala*, *N. placentula*, *N. reinhardtii*, *Nitzschia sigmoidea*, *N. denticula*, *Cymatopleura solea* and *C. elliptica* can be selected as very much more common in, if not confined to, alkaline lakes.

The epipelon of sandy fresh water sediments and beaches has been very neglected, especially the latter. Davies (1971) found that the algae were more abundant in a sandy beach of a lake in Michigan than on the submerged sands even during periods of drought. He found a maximum of numbers during December–March when the beach was frozen, and a smaller peak during May–July whilst the inverse held for the lake. In this habitat filamentous Cyanophyta were somewhat more abundant than diatoms.

Sampling

Sampling involves removal of the surface deposit. This can be done by drawing a length of glass tube across the sediment and allowing it to fill with a mixture of surface sediment and water, or by the use of more precise suction devices which draw off sediment from a defined area. Such area-based samples are essential for quantitative studies of cell numbers, chlorophyll *a* estimation or ^{14}C-uptake studies. The sampling of sediments beneath a metre or more of water requires more complex apparatus and often the help of divers and the paucity of data on this most important flora is partially related to sampling problems. The epipelon is mixed with the organic and inorganic sediment during collection and direct observation of the mixture under a compound microscope usually reveals only a few cells and therefore a technique to separate the algae from the sediment is essential. If the sample is allowed to settle and then the supernatant liquid removed by gentle suction the sediment can be placed in petri or similar dishes and the motile and positively phototactic algae will move to the surface from which they can be removed (Lund, 1942). But the simplest way to observe these algae is to place cover glasses on the sediment surface and remove these after 24 h. The motile algae will be found creeping over the under-surface of the glass which has been in contact with the sediment (Fig. 5.10). For more accurate measurements of cell numbers, for pigment analysis, or for further experimentation it is better to lay squares of cellulose tissue on the sediment surface. The algae move into the meshwork of cellulose fibres which can be removed and the algae washed out. Table 5.3 gives results from tests of the cover glass and tissue method showing the number of cells left behind after removal of the traps, compared with the numbers in sediment estimated directly by very lengthy, laborious counting. In these techniques it is important to remove just sufficient water from the sediment so that the cover glasses or tissues are in intimate contact with the sediment particles but not so close that they pick up large amounts of sediment; trials have to be made with

each sediment. Only a percentage of the population is recovered in this way but it can be as much as 87% (Hickman, 1969); it must be remembered therefore that all counts and chlorophyll *a* estimates of this community are too low and as yet I know of no absolute measurements of biomass or production. It is important also to take into account the rhythmic behaviour of these algae (*see* below), and to remove the cover glasses or tissues at the optimum time of day. An alternative method of cell counting is to remove aliquots of sediment which are mixed in a large volume of water and subsequently sedimented so that the algae can be counted in a chamber using an inverted microscope (Greundling, 1971: *see* p. 249). This technique produced similar overall results to the above method and is to be preferred if the sediments support large crops of non-motile algae, such as filamentous Chlorophyceae. It is usually difficult to observe even planktonic algae in counting

Fig. 5.10. *a*, *b*, the epipelon from marine (*a*) and freshwater (*b*) sites removed by the 'cover glass' technique: *c*, examples of freshwater benthic algae living on sediments. A, A', *Chara*; B, *Vaucheria*; C, *Closterium*; D, *Spirogyra*; E, *Micrasterias*; F, *Mougeotia*; G, *Zygnema*; H, *Euastrum*; I, *Oscillatoria*; J, *Merismopecdia*; K, *Holopedium*; L, *Chroococcus*; M, *Aphanacapsa*; N, *Cosmarium*; O, *Netrium*; P, *Spirotaenia*; Q, *Neidium*; R, *Stauroneis*; S, *Navicula*; T, *Pinnularia*; U, *Cymatopleura*; V, *Surirella*; W, *Campylodiscus*; X, *Nitzschia*; Y, *Frustulia*; Z, *Anomoeoneis*; AA, *Mastogloia*. All the algae illustrated, except A, are microscopic, though the filamentous genera may aggregate to form visible masses.

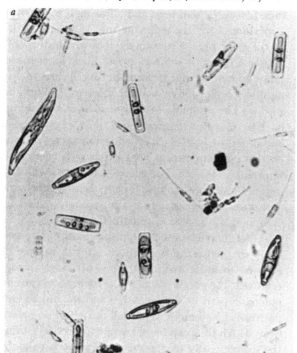

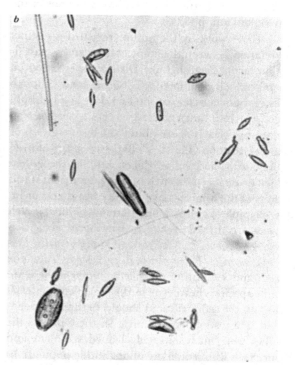

Fig. 5.10 (*cont.*)

c

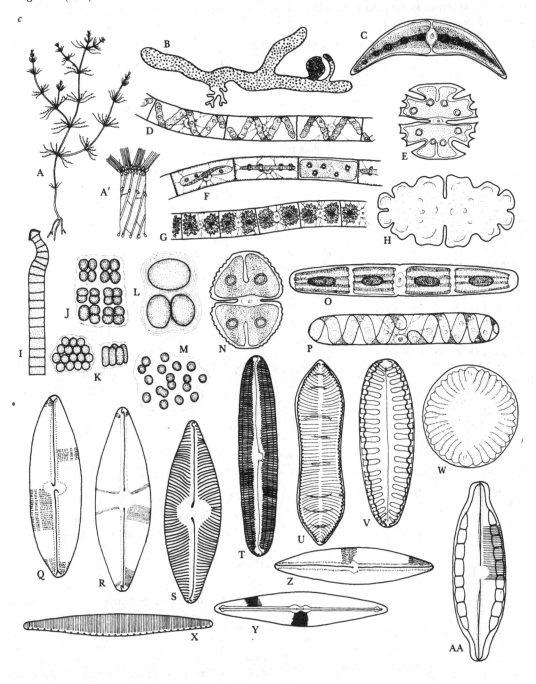

Table 5.3. *The efficiency of harvesting of diatoms from field samples, and proportional representation of species by the cover glass and tissue methods. Samples from Abbot's Pond, Somerset, 27 Nov 1963. From Eaton & Moss, 1966*

	Number present on exposed mud surface (e)		Number remaining on mud after removal of cover glass (c)		Number remaining on mud after removal of tissue trap (t)	
	Counts per 1000 dead valves	%	Counts per 1000 dead valves	%	Counts per 1000 dead valves	%
Gyrosigma acuminatum (Kz.) Rabh.	51	4.07	43	0.85	3	4.31
Navicula pupula Kz.	26	2.07	0	2.75	0	2.33
N. cryptocephala Kz.	393	31.45	44	37.05	32	32.35
N. salinarum v. *intermedia* (Grun.)	96	7.66	18	8.25	0	8.61
N. rhynchocephala Kz.	70	5.58	9	6.40	0	6.28
N. hungarica v. *capitata* (Ehr.) Cleve	52	4.15	18	3.59	0	4.66
N. menisculus Schum.	201	16.05	44	16.60	11	17.05
Nitzschia palea (Kz.) W. Sm.	192	15.33	63	13.65	11	16.25
N. sigmoidea (Ehr.) W. Sm.	22	1.75	3	2.01	2	1.79
N. flexa Schum.	47	3.75	20	2.86	0	4.21
Other spp.[a]	102	8.14	44	5.92	99	2.16
Total	1252		306		158	
% removed	—		75.5		87.5	
Actual no. live cells counted	1679		201		171	
Proportional representation of spp.	—		$\sum_1^{11} \min (e, c) = 91.43\%$		$\sum_1^{11} \min (e, t) = 94.02\%$	

[a] Mainly non-motile sedimented plankton.

Table 5.4. *Vertical distribution of the epipelic algae in the surface sediments of Marion Lake, B.C. Data are expressed as the mean percentage of the total algal cell volume of six collections. From Gruendling, 1971*

Station depth (m)	Mean percentage distribution Surface (cm)					Mean no. of cells counted per sample
	0.5 cm	1.0 cm	2.0 cm	3.0 cm	4.0 cm	
0.5	52.0	22.8	19.1	5.4	0.7	892
1.0	54.7	25.5	16.6	2.3	0.9	995
2.0	65.0	20.3	12.8	1.1	0.8	1279
3.0	74.9	20.8	3.3	1.0	—	302
4.0	83.3	14.3	2.3	0.1	—	274

chambers when any sediment is present and this technique must be difficult to apply to the epipelon.

Movement of the sediment undoubtedly produces some downward vertical mixing of the epipelic population but measurements by Gruendling on Marion Lake, British Columbia showed that 90% of the biomass lived in the uppermost 2 cm of sediment. This is probably a reflection of wave induced vertical mixing, since samples of the surface sediment taken in increasing water depth had progressively higher percentage of algae (Table 5.4). In estuarine sediment, where there is an anaerobic zone starting at a few millimetres depth, the epipelic algae are even more concentrated towards the surface (Palmer & Round, 1965; Fig. 5.11). There is evidence that some algae can maintain a chosen position in the sediments living either at, or near, the surface or at varying depths; this problem however needs very careful reinvestigation.

Marine epipelon

Intertidal epipelon, salt marshes. The sediment-living flora is richly developed in both the intertidal and subtidal zones where suitable substrata exist. It is very prominent on salt marshes in the coastal intertidal and also in estuarine situations. Few estimates are available of the algal contribution to the biomass of salt marshes. Jefferies (1972) considered 25% to be reasonable for a marsh on the Norfolk coast. The two major microscopic algal groups colonising sediments of salt marshes are blue-green algae and diatoms (*see* Fig. 5.13). Admiraal (1977*b*) comments that diatoms are the most important primary producers on tidal flats. At least three

Fig. 5.11. Vertical migration rhythm in *Euglena. a.* Typical vertical distribution of *Euglena* in the mud during a tidal cycle. Shaded blocks signify the percentage of cells at each depth. The wavy lines indicate the times of high tide. Note that the organisms begin to re-burrow back into the mud well in advance of the returning tide. *b.* The number of *Euglena*/mm² on the surface of mud samples maintained in the laboratory, in alternating light–dark cycles and away from any tidal influence. Stippling indicates dark periods. N, noon; M, midnight. *c.* Number of *Euglena* on the surface of mud samples maintained in continuous illumination (98 ft-c.), constant temperature (15 °C), and away from any tidal influence. Note that the period of the rhythm here, as in light–dark cycles, is diurnal, rather than tidal, as it is on the river banks. From Palmer & Round, 1965.

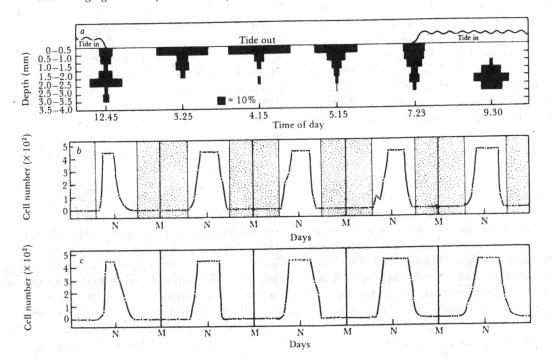

separate habitats exist on salt marshes, the flat marsh surface, the closed pools (pans) and the drainage channels. A preliminary study on the Dee estuary marshes (Round, 1960b) showed *Pleurosigma angulatum* var. *quadratum*, *Cylindrotheca gracilis*, *C. closterium*, *Diploneis didyma*, *Navicula gregaria* and *N. pygmaea* common in the pans, whilst *Surirella gemma* and *Gyrosigma wansbeckii* were more common in the channels. On the flat marsh itself there was a definite distinction between the flora of the upper-marsh, stiff, silty, sediments where *Navicula cincta* var. *heufleri*, *Nitzschia navicularis* and *N. obtusa* var. *scalpeliformis* were dominant and the sandy sediment of the lower marsh where *Pleurosigma aestuarii*, *P. elongatum*, *Caloneis formosa*, *Navicula viridula* and *Nitzschia panduriformis* achieve greater importance and the species of the upper marsh tend to be absent. Sullivan (1975) studied a similar flora on a Delaware salt marsh and found highly characteristic groups of species at various levels on the marsh. The three commonest species in each community were: amongst tall *Spartina alterniflora*, *Navicula phyllepta*, *N. spartinetensis*, *Stauroneis salina*; amongst dwarf *Spartina alterniflora*, *Denticula subtilis*, *Navicula binodulosa*, *N. salinicola*; amongst *Distichlis spicata*, *N. diserta*, *N. taraxa*, *Nitzschia filiformis*; on bare bank, *Navicula ammophila*, *N. cincta*, *N. salinicola*; and in a pan, *Cymbella pusilla*, *Navicula salinarum*, *Nitzschia obtusa* var. *scalpelliformis*. Thus on both sides of the Atlantic Ocean very distinct diatom floras are found close together on salt marshes yet the surfaces of the marshes are washed by the tide every day. Sullivan recorded 104 taxa and Drum & Webber (1966), working on a Massachusetts marsh, found 151 taxa many of which are euryhaline and were recorded also from the Des Moines river.

In this habitat the flora will be subjected to great fluctuations in salinity which the diatoms obviously tolerate. Recently, Admiraal (1977a) has shown that salinities between 4 and 60‰ resulted in no more than a 30‰ decrease in photosynthesis below the initial values at ambient salinities. As in so many situations, a salinity less than normal seems to be advantageous: thus Williams (1964) found that maximal rates of diatom cell divison (0.6–3.2 divisions day^{-1}) occurred at 20‰. Cooksey & Chansang (1976) also reported that isolates of the euryhaline diatom *Amphora coffeaeformis* grew well at salinities between 11 and 50‰ achieving 80% of their maximum growth rate. A further interesting aspect of this work was the finding that growth in low calcium media caused a sessation of motility. This raises the interesting problem of whether or not the diurnal rhythm of motility might be related in some way to calcium concentration in the cells.

Webber (1967) recorded 29 species of blue-green algae on a Massachusetts salt marsh and many of the common forms were heterocystous and known to fix atmospheric nitrogen. Stewart & Pugh (1963) found 33% of the species of a British salt marsh were heterocystous and capable of contributing considerable amounts of nitrogen to the habitat. In southern Delaware three cyanophytes dominate the marsh at all levels (*Microcoleus lyngbyaceous*, *Schizothrix calcicola* and *S. arenaria*) and Ralph (1977) considers that ecophenes of these three probably dominate the blue-green flora of many northern marshes. A very similar habitat in the tropics is the bare silt beneath mangroves and along channels through the swamps; van den Hoek *et al.* (1972) found a felt of *Microcoleus* (*M. chthonoplastes*, *M. tenerrimum*) and *Lyngbya aestuarii* in such habitats in Curaçao. *Vaucheria* often grows and binds the sediments on muddy marshes and has been recorded as far north as Ungava Bay (Wilce, 1959), but also binds mud in tropical mangrove swamps (Doty, 1957). These *Vaucheria* patches are often raised slightly above the surrounding marsh, presumably due to the trapping of sediment between the filaments. *Vaucheria* is equally common, binding sediments in shallow freshwater pools and ditches; it might in fact be thought of as the aquatic equivalent of the sward-forming grasses on the land. Simons (1975a, b) has made one of the most detailed studies of *Vaucheria* in the Dutch coastal estuarine habits and clearly shown the presence of discrete species groupings related to salinity and height above mean sea level. Fig. 5.12 summarises his data (Simons, 1975a) which, as far as they can be compared with sparse records from elsewhere, are valid for other regions. The non-diatom flora associated with some of these Dutch sites is given in Table 5.5. Simons (1975b) found a distinct seasonal distribution of the *Vaucheria* species, the majority being developed during the winter months under high moisture and low salinity conditions except at the lowest levels where the genus is found in summer, presumably because this is the only site he studied which is wet enough at that time of the year. However, only

V. velutina grows at the low level in summer and at the uppermost level only *V. intermedia*; these are the only two conspicuous summer species.

In intertidal estuarine situations the epipelon is rich, e.g., extensive green patches of *Euglena* spp. are common together with diatoms such as *Cylindrotheca*, *Navicula*, *Amphiprora* etc. In the sand at Barnstaple Harbour, Mass. *Hantzschia virgata* var. *intermedia* is abundant (Palmer & Round, 1967) and the author found the same species in a very similar estuarine habitat at Porto Novo on the east coast of India. This, I believe, illustrates how some species have very precise ecological requirements; in this instance the alga is associated with clean fine estuarine sand, whereas it has never been found on the highly organic silts of the Avon Estuary, England. Amspoker & McIntire (1978) also report that the sediment properties greatly affect the community structure in estuaries. Algal patches seem to be favourable sites also for the trapping and germination of salt-marsh angiosperms, e.g. *Salicornia europaea*, *Puccinellia maritima*, etc. Many salt marshes are extremely productive; thus Estrada, Valiela & Teal (1974) found algal chlorophyll levels of 200–500 mg m^{-2} in unshaded parts of a New England salt marsh. They found that fertilisation with sewage and urea did not increase the amount of algal chlorophyll but stimulated grass growth, which shaded the algae and actually reduced the total algal chlorophyll. Gallagher & Daiber (1973) report a tenfold diel amplitude in the photosynthetic rhythm of a diatom-dominated flora on a salt marsh but only a twofold amplitude in one in which filamentous green algae were abundant. Nitrogen fixation by blue-green algae plays an important role in some salt marshes, e.g., Carpenter, van Raalte & Valiela (1978) record 10–20 mg m^{-2} day^{-1} nitrogen fixed on a Massachusetts marsh. Fixation is most intensive in summer (Jones, 1974), optimal fixation occurs at low salinities and that in pools is approximately double that of the drier marsh surface. Jones found that maximal fixation occurs when the algae are uncovered but still moist and it is stimulated in the rhizosphere of the marsh angiosperms.

The diatoms on the surface of highly silted salt marshes, where the silt forms a sticky surface layer, do not seem to undergo the vertical migration movements common in looser sediments (Round, unpublished).

Fig. 5.12. The horizontal and vertical distribution of six groups of *Vaucheria* species along the salinity gradient of Dutch estuaries. Group 1 (vertically hatched): *V. velutina*, *V. subsimplex*; group 2 (obliquely hatched): *V. compacta*; group 3 (+): *V. arcassonensis*, *V. coronata*, *V. intermedia*, *V. minuta*; group 4 (●): *V. synandra*, *V. canalicularis*, *V. cruciata*, *V. erythrospora*, often only *V. synandra* present (○); group 5 (■): *V. frigida*, *V. bursata*, *V. canalicularis*, *V. cruciata*, often only *V. bursata* or *V. canalicularis*, or both, present (□); group 6 (△): *V. terrestris*, *V. dillwynii*.

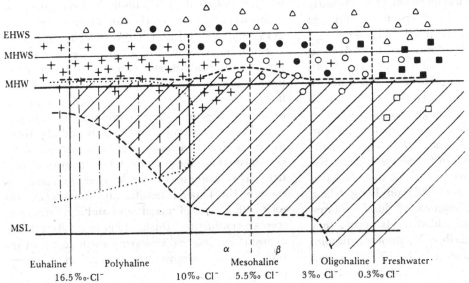

Table 5.5. *The green and blue-green algae accompanying* Vaucheria *on some Dutch sites*. From Simons, 1975*b*

Chlorophyceae	Cyanophyceae (filamentous forms)
Enteromorpha prolifera (O. F. Müller) J. Ag.	*Anabaena variabilis* Kütz
Enteromorpha torta (Mertens) Reinb.	*Anabaena torulosa* (Carm.) Lagerh.
Percursaria percursa (C. Ag.) Bory	*Calothrix aeruginea* (Kütz.) Thuret
Rhizoclonium riparium (Roth) Harvey	*Microcoleus chthonoplastes* (Mert.) Zanard
Ulothrix pseudoflacca Wille	*Microcoleus lynbyaceus* (Kütz.) Crouan
Ulothrix subflaccida Wille	*Microcoleus vaginatus* (Vauch.) Gom.
Xanthophyceae	*Nodularia harveyana* (Thwaites) Thuret
Vaucheria canalicularis (L.) Christensen	*Nostoc* sp.
Vaucheria synandra Wor.	*Oscillatoria brevis* (Kütz.) Crouan
Vaucheria intermedia Nordtst.	*Oscillatoria laetevirens* Crouan
Vaucheria sescuplicaria Christensen	*Oscillatoria margaritifera* Kütz.
Vaucheria erythrospora Christensen	*Oscillatoria nigro-viridis* (Thwaites) Gom.
Vaucheria arcassonensis Dangeard	*Oscillatoria tenuis* Ag.
Vaucheria coronata Nordtst.	*Phormidium corium* Gom.
Vaucheria velutina C. Ag.	*Schizothrix calcicola* (Ag.) Gom.
Vaucheria subsimplex Crouan frat.	*Spirulina subsalsa* Oerst.
Vaucheria littorea Hofm. ex C. Ag.	*Symploca funicularis* Setch. & Gardner
Vaucheria compacta (Collins) Coll. ex Taylor	Cyanophyceae (coccoid forms)
	Agmenellum quadruplicatum (Menegh.) Bréb.
	Anacystis dimidiata (Kütz.) Dr. & Daily
	Anacystis montana (Lightf.) Dr. & Daily
	Cocoochloris stagnina Spreng.
	Entophysalis deusta (Menegh.) Dr. & Daily

Intertidal epipelon, sandy shores. Of all the micro-habitats in which the epipelon develops, that of the sandy sea shore–beach is probably the most extreme and sometimes few species can be detected. On fairly exposed sandy beaches the author has found that there is a flora of highly motile flagelletes e.g., species of *Chroomonas*, *Cryptomonas* and *Amphidinium* together with a few naviculoid diatoms (Fig. 5.13).

Sands in more sheltered bays, and amongst rock outcrops in inlets, support a rich flora in which *Cryptomonas* spp. are exceedingly common together with *Amphidinium*, *Glenodinium*, and the diatoms *Pleurosigma*, *Amphora*, *Navicula*, *Diploneis*; in some areas the epipelon is rich in *Cylindrotheca* (*Nitzschia*) *closterium*. One other investigation of this habitat is that of Burkholder *et al.* (1965) on Long Island Sound where again a mixture of dinoflagellates, euglenoids, Cyanophyta and diatoms occur; they report 1×10^6 *Euglena* cells g^{-1} sediment; diatoms reach 1.3×10^7 and bacteria 5.64×10^8. They considered that the community was 'sun adapted' but nevertheless the daily curve of carbon fixation had a mid-day depression. Patmatmat (1968) also found that photosynthesis of this surface flora on a sand flat was not inhibited in full sunlight. Even when the sediment particles reach the size of fine gravel a rich flora may be found provided that water movement is gentle enough. Excessive grinding of particles prevents the establishment of a flora. It is difficult to predict exactly which sediments will have a rich flora, however it is relatively easy to determine the presence or absence by the technique given on p. 167.

In the intertidal of some sandy and also rocky shores the red alga *Audouinella* (*Rhodochorton*) *purpureum* forms mat-like growths which trap sand particles. It is sometimes common over the surface of rocks in the upper intertidal and can obscure the rock with a layer of mixed sand and algal filaments several centimetres thick. The lower layers are almost colourless and upward growth occurs at the surface in an almost 'stromatolite'-like growth pattern.

Subtidal epipelon. Of all the important algal habitats, I can think of none which is so understudied as that of the subtidal sediments. Yet in favourable habitats the bottom silts are coloured a rich brown with teeming masses of diatoms. In the cold waters off Barrow, Alaska a rich flora was recorded by Matheke & Horner (1974) in which *Berkeleya (Amphipleura) rutilans, Diploneis smithii, D. subcincta, Gyrosigma balticum, G. fasciola, G. tenuissimum, G. spenceri, Pleurosigma longum, P. stuxbergii, P. angulatum, Navicula directa, N.*

transistans, Nitzschia bilobata, N. (Cylindrotheca) closterium, N. longissima, Pinnularia quadratarea and several unidentified species of *Amphora, Navicula* and *Nitzschia* were present. The floristics of many sites need to be studied and compared to provide a good background for this flora. The ratio of sand to silt in the underwater sediments affects the flora just as.it does in the intertidal and Boucher (1975) found that the biomass measured as chlorophyll was very low on fine sand ($0.5 \ \mu g \ g^{-1}$), increased where silt was

Fig. 5.13. Examples of microscopic epipelic algae, especially those which live on beach and estuarine sediments, though some also live on other sediments. A, *Scytonema*; B, *Plectonema*; C, *Entophysalis*; D, *Mastigocoleus*; E, *Lyngbya*; F, *Schizothrix*; G, *Chroomonas*; H, *Euglena*; I, *Pleurosigma*; J, *Diploneis*; K, *Caloneis*; L, *Amphora*; M, *Hantzschia*; N, *Amphidinum*; O, *Glenodinium*.

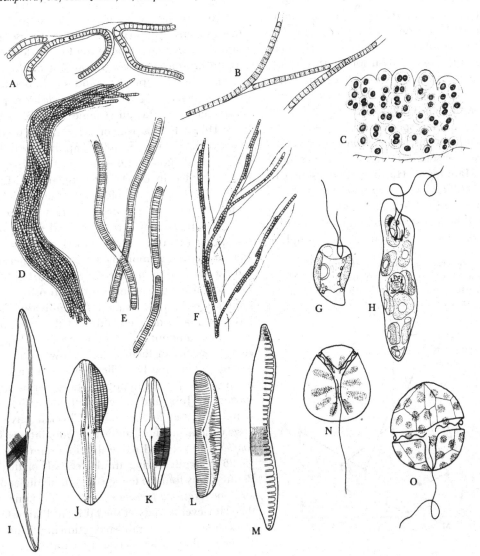

mixed into the sand and the grain size increased $(2.3–6.3\ \mu g\ g^{-1})$ and decreased again in the fine silt $(3.6\ \mu g\ g^{-1})$. In samples taken off La Jolla the author found that when the sediments were brought back to the laboratory and placed in petri dishes the surface would steadily brown with a rich coating of diatoms many cells deep; cover glasses placed on the surface at the beginning of the examination were completely brown on the underside when removed. The deeper the sediments the deeper the brown coloration, undoubtedly because of the increased pigment formed as an adaptation to depth. At this depth only blue-green light illuminates the flora and Vesk & Jeffrey (1974) have found that growth of the planktonic diatom *Stephanopyxis turris* in blue light dramatically increased the number of plastids so that they almost filled the region normally occupied by the vacuole; the number of thylakoid stacks per chloroplast also increased several fold with the reduction of both stroma and pyrenoid regions. The photosynthetic capacity of the cells was greatly increased and, although this work was concentrated mainly on a planktonic species, other species including pennate diatoms, were found to behave similarly (Jeffrey, personal communication). Earlier work (Strain, Manning & Harding, 1944) showed that *Cylindrotheca* (*Nitzschia*) *closterium* produced more diadinoxanthin in white light than in red. Obviously these effects need further study and correlation with ecological conditions. Contrarywise, the samples

taken by the author from the surf zone off La Jolla were often almost colourless though the diatoms were perfectly healthy. Diatoms form almost the total biomass in this marine epipelon with merely a few dinoflagellates and some blue-green filaments or colonies of *Holopedium*. These blue-greens were often tinted red, presumably by an excess of phycoerythrin induced by the low light intensity and changed spectral composition of the light under which they were living. A perceptive series of studies of this habitat are those of Bodeanu (1964, 1968, 1970, 1971) who recorded 77 pennate and nine centric diatoms on the sediments in the Black Sea; he found an average of 72×10^6 cells m^{-2} and appreciated the fact that this was a discrete assemblage unrelated to the plankton though it could contaminate the latter during storms. He also pointed out that these epipelic diatoms were a major food source for benthic molluscs and the crustacean *Idotea*. In shallow muddy water, especially in the tropics, the mud is stabilised by mats of Cyanophyceae e.g. *Lyngbya majuscula* forms patches on sediments in Hawaii (Round, unpublished).

Off La Jolla a peak of cell numbers was found at a depth of 25 m (Round, unpublished). This is somewhat shallower, but quite comparable, to the peak recorded in the Mediterranean off Marseille where Plante-Cuny (1969) found the maximum at 30 m, and showed, in a detailed floristic and seasonal survey, that the main growth period occurs during the spring months. Peak numbers occur in April–May and although more data are needed there seems to be a low period in June–July (Fig. 5.14); similar low numbers have been recorded in other epipelic populations in mid-summer. The individual species have optima at different times of the year, but an enormous amount of work is needed before the ecology of any of them is well known. One genus, *Amphora*, notoriously difficult from a taxonomic point of view, seems to oppose the trend; its growth started in July and it was still increasing when sampling ceased in September. Checking my earlier observations on the Californian populations I find *Amphora* also tended to be the dominant species at 15–25 m depth during the latter half of the year. Plante-Cuny found a few species still living at 300 m and one, *Navicula pennata*, at 360 m but none below this. However a study of the tables in Plante-Cuny's work shows a considerable reduction in cell numbers, as one would predict, between 50 and 100 m depth.

Fig. 5.14. The seasonal variation during 1964 in epipelic diatom counts at four stations (MI$_1$, MI$_2$, DC, DE$_1$) each at a different depth in the Gulf of Marseille. From Plante-Cuny, 1969.

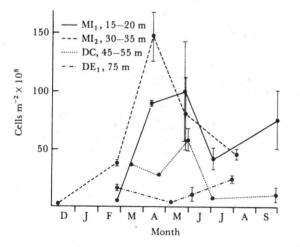

Table 5.6. *The distribution of pigments and carbon fixation of the epipelon at various depths in the Mediterranean Sea. From Plante-Cuny, 1969*

| Depth (m) | Chl a(mg m^{-2}) | Phaeopigment (mg m^{-2}) | Carbon fixed | |
			(mg mg Chl a^{-1} h^{-1})	(mg m^{-2} day^{-1})
5	31.52±6.45	18.9 ±2.18	1.09±0.10	410
15	34.03±10.17	32.24±6.80	0.58±0.18	232
25	21.49±15.74	47.87±22.16	0.22±0.21	40
38	2.74 (one measurement only)	32.24	0.29	6

Some notable features are the apparent preference for depths between 30 and 75 m by the large-celled genera *Campylodiscus* and *Surirella*. *Navicula pennata* seems to come into the flora around 50 m and is one of the few still found at depths around 100 m. Assuming that the effect of increasing depth will be the elimination of species, the data from Marseille show that, in general terms, at 20 m there is an average of 58 species, at 30–35 m 89 species, at 40–60 m 42 species, at 70–76 m 53 species, at 100 m 23 species, and between 170 and 300 m only 5 species. A total of 293 different diatom species were found off Marseille and this is about the size of flora to be expected in any well-studied region, for example, I found over 200 spp. off the coast of California. These reports of epipelic diatoms off Marseille certainly show the limited depth to which diatoms live on oceanic sediments, and the reports of such algae at enormous depths are totally suspect. The deep ocean must be a most inhospitable place for phototrophic organisms and, simply on theoretical grounds, one would not expect to find a flora in such a situation. Nevertheless there have been several reports, an interesting one, in 1973 (Malone, Garside, Anderson & Roels) of a collection from 6150 m in the north Atlantic. The cells were all diatoms and when suspended in antarctic bottom water they did not grow for three days, though during this period nitrate was removed from the water, and after day three growth commenced. These diatoms were coastal pennate species and as suggested by Malone *et al.* were transported to such depths. Other uniden-

tified, pigmented cells were also present and presumably the uptake of nitrogen was due to the fact that the cells were nitrogen-starved during their passage to and sojourn at, such depths. Alternatively other organisms were taking up the nitrate. There is nothing unusual about growth of organisms after long periods in the dark (cf. p. 551) but it is highly unlikely that any of the reports of live algae at great depths are of anything other than sedimented organisms which will be incapable of growth at the depths sampled. Some reports in fact record *freshwater* organisms which have been sedimented; this is a field where great care is required, coupled with skilled taxonomic and ecological knowledge of the reported taxa! It is however quite conceivable that colourless heterotrophic strains of algae could exist well below the photic zone but such have not been reported yet.

Recently, resting cells of an epipelic diatom *Amphora coffeaeformis* have been found in the deep ocean. Such stages could be induced by placing cells in the dark at 7 °C and growth would resume when they were returned to the light and 25 °C (Anderson, 1976).

Waves and currents will sort the sediments (and possibly also the algae) and so a complex set of factors in addition to those associated with light and temperature will operate on this epipelic flora. In the well-illuminated region the diatom flora is probably richest on the finer grained sediment (Plante-Cuny, 1969). Off Madagascar, the diatom flora showed a distinct increase in large celled species

Fig. 5.15. *a*. Field observations of the vertical migration rhythm of *Euglena*. Wavy lines represent the times of high tide. The times of sunrise and sunset are represented by the boundaries between the stippling and the undotted spaces. For ease in comparison the highest cell count for each cycle was designated as 100 and all the other values as percentages of this; in no case is 100% less than 10^4 cells cm^2. The shaded, horizontal bars in *v* signify 1 h periods during which the exposed mud was covered by opaque canisters. From Palmer & Round, 1965. *b*. the vertical migration rhythm of *Hantzschia virgata* in constant light (LL) and in alternating light and dark conditions (LD). Consecutive days run from top to bottom of each graph. State of tide on day of collection symbolised at the top of each graph, stippled areas show natural dark period, wavy lines indicate period of tidal cover. *i, ii*, show drift of migration through the day under constant laboratory conditions; *iii, iv*, show phase change from afternoon to morning appearance. x indicates time of collection. From Palmer & Round, 1967.

a

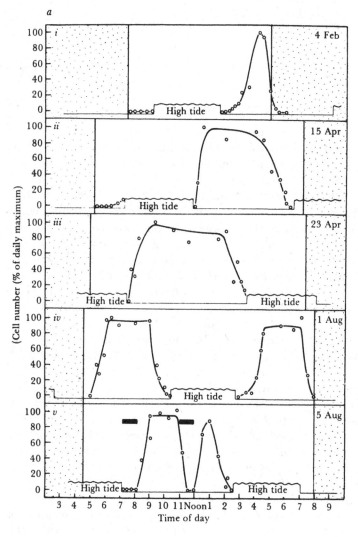

with increasing depth (Plante-Cuny, 1973). Many factors will be involved in the distribution and growth at various depths, e.g., the smaller species are least likely to suffer damage in the wave disturbed shallow waters, whilst in the poorly lighted deep waters the cells need to spread their chloroplasts over the maximum surface area. At 38 m in the Mediterranean there was little measurable light, very little chlorophyll was present and photosynthesis was low (Table 5.6). This suggests that the data quoted above based on occurrence of species at various depths involve only very small populations below 40 m.

Rhythmic motility

Allied to position in the sediment is the problem of diurnal movements in a vertical plane. Such movements have been known for over half a century, being first observed on intertidal sands where, during low tide, species of diatoms (Fauvel & Bohn, 1907) and dinoflagellates (Herdman, 1924) were observed to move to the surface and back into a subsurface position before the tide returned. This has been confirmed at many intertidal sites e.g., by Callame & Debyser (1954) and by Pomeroy (1959). Herdman found so many diatoms on the surface of the sand at Port Erin 'that on two occasions (in August) the photosynthetic activity was distinctly audible as a gentle sizzling...while the sand was frothy with bubbles of gas, presumably oxygen given

Fig. 5.15 b

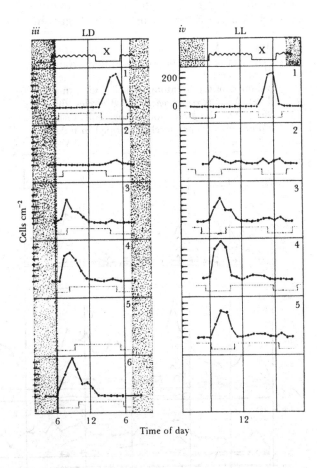

off by them'. The earlier workers did not quantify their observations, but the techniques of trapping the algae on the underside of cover glasses or in tissues (see above) has enabled more precise studies of these rhythms. Fig. 5.15 shows, diagrammatically, the movement of *Euglena* (*a*) and *Hantzschia* (*b*), from the estuarine river Avon (Bristol) and from the intertidal mud flats at Barnstaple, Massachusetts, respectively. Movement to the surface is fairly rapid when the sediments are exposed by the retreating tide and when light intensity is increasing after dawn. Once assembled at the surface the algae remain there until an hour or so *before* the tide returns over the sampling site, when they migrate downward. That the reaction is a positive phototactic response is shown by covering the sediment at different times; movement of the cells away from the surface results. This could be due to intermixing in the animal-moved sediment coupled with random movement of the algae since

in the dark there is no directional stimulus. Later in the cycle there is a loss of the positive phototactic reaction though it is still not clear what controls the actual downward movement. It could be that the cells change their response to one of negative phototaxis, or motility could cease which would inevitably result in burial. It is not related to decreasing illumination, except when the downward period happens to coincide with the approach of dusk. The rhythms are maintained in the laboratory if the sediment with its algal population is placed in constant light or alternating light and dark cycles and constant temperature (Fig. 5.9); the upward and downward movement is therefore presumed to be under the control of a biological clock system. The rhythm expressed in the laboratory by *Euglena* is a daily rhythm and not a tidal rhythm. In nature the basic daily rhythm is converted into a tidal rhythm and it has been suggested that this is caused by the

Fig. 5.16. Movement of diatoms into and out of sediment under laboratory conditions. Natural light intensity in foot-candles is given. The numbers of total diatoms (filled circles, thick line), *Navicula cryptocephala* (filled circles, thin line), *Nitzschia palea* (dotted line) and *Navicula rhynchocephala* (open circles) trapped in tissues at

the surface of the sediment over a period of 83 h in natural light–dark (L:D), constant light (L:L) and constant dark (D:D). Below each set of graphs of cell numbers is a graph of the percentage of paired cells (i.e. dividing) in the population. From Round & Eaton, 1966.

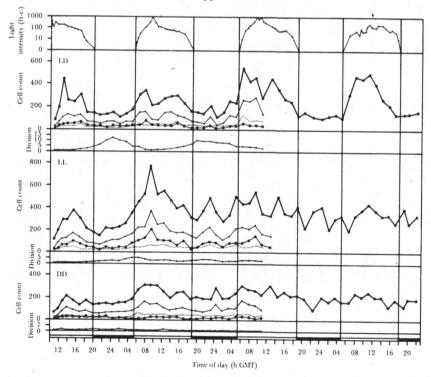

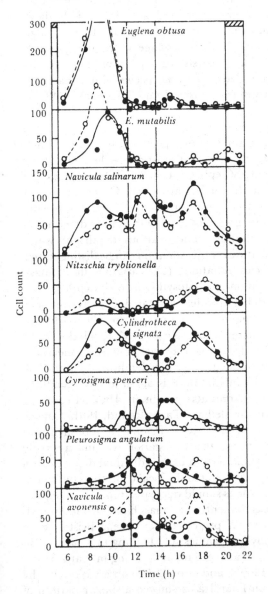

Fig. 5.17. Laboratory observations of the vertical migration pattern of *Euglena* and diatom species after a normal dark, night period (filled circles) and during continuous light (open circles). Cross-hatched areas indicate natural dark period and dotted wavy line the period of tidal cover experienced by the cells during the previous day in the natural habitat. From Round & Palmer, 1966.

periodic darkening of the sediments by the very *turbid* flood tides in the River Avon (Palmer & Round, 1965). This hypothesis gains strength from the observations of Perkins (1960) who found that the diatoms on similar sediments remained on the surface during the whole daylight period but the flood tides over the River Eden site were extremely *transparent*. Epipelic algae living on permanently submerged sediments were not suspected of under-going such diurnal movements and indeed it is less easy to detect these in nature. However, work by Round & Eaton (1966) on a pond (Fig. 5.16) and Round & Happey (1965) on a stream showed that the epipelic algae, when transferred to the laboratory, undergo movements similar to those of species living under tidal regimes. The rhythms differ, however, from those of the intertidal species in that upward movement commences during the dark phase (Fig. 5.9) and like many other rhythms, tends to become irregular in constant light. There has still to be a study of these non-tidal rhythms on the actual sediment though by analogy with the intertidal studies which can be followed in nature and as persistent rhythms in the laboratory, one can assume that a similar migration occurs in underwater sites. Most species seem to undergo these rhythms but only the intertidal *Euglena* and *Hantzschia* have been shown to remain for a short time in the supra-surface position; other species have varying responses when brought into the laboratory (Fig. 5.17). It is strange that whereas *Euglena*, in the laboratory, maintains the cycles with the timing which existed on the day of collection, *Hantzschia* alone has been shown to move through the day in approximate phase with the tides in nature (Fig. 5.15b), as was discovered but not quantified by Fauré-Fremiet (1951). To explain this phenomenon and especially the re-phasing effect, the model of Fig. 5.18 was constructed. It is possible that Herdman (1924) also observed similar phenom-ena since she reported 'synchronous movements of dinoflagellates in the laboratory with those on the beach'. One assumes that these movements have adaptive significance permitting maximum photo-synthesis during the surface phase but Taylor & Palmer (1963) found that *Hantzschia* was a 'shade species' and that sufficient light penetrated the sand to obviate the need for a trip to the surface. There are still many aspects of these rhythms which require classification: Sournia (1976b), for example, reported that high light intensity results in a downward movement of Cyanophyta into the sediments of coral lagoons.

A rhythmic change in the rate of movement in light–dark cycles (photokinesis) has recently been shown to occur in laboratory experiments on the epipelic desmid *Micrasterias* which although moving very slowly (2–3 mm h^{-1} at their fastest) showed clear peaks at mid-day (Neuscheler, 1967). A similar rhythm occurred in the orientation of the cells to the light which is obviously a necessary factor in movement towards the light. The threshold for initiation of photokinesis and phototaxis was about 100 times higher at night than during the day.

Light is obviously a most important factor in the expression of the vertical migration rhythm since

Fig. 5.18. Diagrammatic representation of the interaction of a 24.8-h bimodal vertical migration rhythm (here represented as a disk with opposing bulges; each bulge signifying the surface phase of the rhythm) and a 24-h suppression–expression rhythm (represented as an incomplete disk superimposed over the lunar-day rhythm). The shaded area of the disk is that part of the 24-h rhythm that suppresses the night-time phase of the migration rhythm, and the open segment the part that allows the expression of the day-time supra-surface phase. Because the supra-surface phase of the lunar rhythm occurs 50 min later each day it eventually falls under the influence of the suppressive portion of the solar-day rhythm. As this phase is inhibited the unexpressed early morning phase is expressed. The net result is an *apparent* rephase of the migration rhythm. From Palmer & Round, 1967.

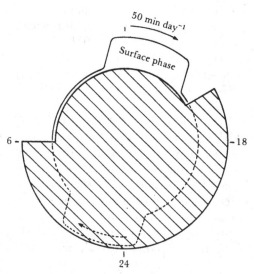

in the intertidal habitats cells never appear on the surface during night-time; also cells are prevented from rising to the surface by artificial darkening and cells already on the surface can be induced to move below the surface by artificial darkening. Light given during the normal night-time does not induce upward migration, though in the experiments on freshwater algae slight upward movement commences before dawn and accelerates after dawn. One may deduce from these results that the algae undergo a rhythmic change in responsiveness to light. In fact a persistent rhythm of phototaxis has been recorded in *Euglena* species with a maximum of responsiveness around mid-day (Pohl, 1948).

Depth distribution

The amount and spectral composition of light reaching the sediment must affect the depth distribution of the epipelic community though experimental studies on this are scarce. Other factors such as wave disturbance, grading of sediment, grazing etc., undoubtedly interact with light in determining the distribution under various depths of water. Three features of this zonation can be distinguished: (1) Species zonation; (2) Variation in biomass; (3) Variation in photosynthesis with depth.

Studies in widely separated sites have revealed that the shallowest depths are not necessarily the richest source of biomass e.g., in Marion Lake the 0.5 and 1.0 m depths have a lower biomass, as measured by cell volume, than the 2 m depth, whilst at the 3 m and 4 m points biomass is again low (Gruendling, 1971). Similar results were obtained by a cell counting technique in Blelham Tarn (Round, 1961*b*) and in Lake Itasca, Minnesota (Round, unpublished) but in Windermere the 1 m station was unexpectedly high. The biomass at each individual depth station will vary according to the local conditions but under ideal situations should decrease at increased depths, e.g., measurements of biomass based on chlorophyll *a*, in a small protected pond near Bristol, gave chlorophyll *a* values of 11.7 mg m^{-2} at 0–1 m 7.3 mg m^{-2} at 1–2.5 m, and 4.0 mg m^{-2} at 2.5–4 m (Moss, 1969*b*), and unpublished observations by the author on Lake Itasca showed a similar pattern of chlorophyll down to 10 m; the amount at 0.2 m was approximately two thirds of that at 2 m, where the maximal amount was recorded, at 5 m it was less than half and at 10 m it was about one fifth. The epipelic community receives light which has been spectrally altered by its passage through the water and through the overlying phytoplankton; dense growths of the latter will reduce considerably the depth penetration of the epipelic algae, thus Moss (1969*b*) showed that in Abbot's Pool the phytoplankton could reduce the compensation point to around 2 m depth. To do this needs 200 mg chlorophyll *a* m^{-2} which lowers the penetration of light such that at 2.2 m there is only 1 % of surface illumination, whereas with no chlorophyll this figure would be reached at 4.1 m. In the Neusiedler See turbidity in the very shallow water of this large lake reduces the epipelic population in summer but under ice in winter light penetration improves and a good flora develops (Dokulil, personal communication). However some algae seem capable of survival on the sediment at low light intensity, e.g. the blue-green alga *Arthrospira jenneri* and the diatom *Nitzschia flexa* formed large populations during the summer at depths below 2.5 m; here oxygen concentration was low and chemical features associated with this are probably also important (Moss, 1969*b*). Low oxygen tension itself may not be the operative factor since these algae have also been recorded, though less abundantly, in the well oxygenated shallow sediment. Survival without oxygen over a period of 10 days has been reported for Cyanophyta whilst green algae could not survive for more than 48 h without oxygen (Dokulil, 1971). In some ponds hydrogen sulphide-rich layers occur at the sediment surface and even here populations of algae, especially blue-greens and euglenoid flagellates, grow. Cohen, Padan & Shilo (1975) showed that *Oscillatoria limnetica* from such a layer could perform either oxygenic or anoxygenic photosynthesis with sodium sulphate as an electron donor. Above the *Oscillatoria* layer other accumulations of phototrophic sulphur bacteria proliferate. Clearly there is much to be discovered about the algae in these sulphide-rich layers, especially about their biochemical functioning.

There is no doubt that in some lakes a number of species occupy certain defined depths, e.g., in Lake Itasca *Navicula tuscula*, *Epithemia* and *Rhopalodia* were only common at the shallow 0.2 m station and *Oscillatoria*, *Anabaena* and *Spirulina* were more abundant between the 4 and 6 m depths. The effect of low oxygen concentration on populations in the hypolimnion of lakes during summer stagnation should be investigated, since microaerophyllic conditions induce nitrogen fixation in some Cyanophyta

grown under experimental conditions and large populations of Cyanophyta and some flagellates, e.g., *Trachelomonas*, often occur in these deoxygenated sites. Patmatmat (1968) reported algae in the anoxic zone of an intertidal sand flat and it would be interesting to know how metabolically active these algae are. His report of live algae down to 12 cm is again not surprising since they could be carried down by animal activity. Many Cyanophyta are reported living in the sulphide-rich layer of marine sediments and this community certainly needs reinvestigation (Fenchel & Riedl, 1970).

A series of photosynthesis determinations on samples of epipelon collected at 2 m depth in Lake Itasca, and incubated at 0.2, 1, 2, 3, 4 and 5 m gave slightly lower values at 0.2 than at 1 m and a steady decrease from 1 m to 5 m, where oxygen production was approximately one seventh of that at 1 m (Round, unpublished). Surprisingly, although the 2 m station was the richest, based on determination of cell numbers and chlorophyll *a*, it was always less favourable for photosynthesis than the shallower stations and the peak of biomass at 2 m is probably related to water movements which scour cells from shallower stations. Gruendling (1971), working on another lake, found a maximum of carbon fixation at 2.0 m (54.9 g m^{-2} yr^{-1}) and this fell to 23.6 at 4 m. As illustrated in Fig. 5.19 a short growing season with a peak of carbon fixation in August was recorded from this lake. An even shorter growing period occurs in Alaskan tundra ponds where carbon values of 4 to 10 g m^{-2} yr^{-1} have been determined (Stanley, 1976). In some shallow lakes the epipelon can contribute more carbon fixation than the phytoplankton, e.g., in a Lake Michigan dune pond 26 % of the total fixation was due to the epipelon and only 13 % to the plankton, the remainder being contributed by the macrophytes (Barko, Murphy & Wetzel, 1977).

There have been very few experimental studies involving the marine epipelon but Gargas (1971) showed that a population at 8 m in the Oresund was shade-adapted with an Ik value* of 8 klux whilst one living at 0.5 m had an Ik value of 21 klux (Fig. 5.20).

---

* The Ik value is a very convenient way of expressing the photosynthetic behaviour of a population relative to light intensity. It is the irradiance at which the saturation level of photosynthesis would be reached if the linear section of the photosynthesis–irradiance plot is extrapolated.

Gargas refers to the communities he studied as psammophilic (associated with sand grains) and pseudobenthic (free living). The latter is simply epipelon and the photosynthetic data are presumably obtained from the mixture of the two communities. Methods involving relatively undisturbed populations in cores placed in chambers in the laboratory have been devised (Gallagher & Daiber, 1973; Darley, Dunn, Holmes & Larew, 1976). On salt marshes, van Raalte, Stewart, Valiela & Carpenter (1974) recorded carbon fixation rates varying from 5 to 74 mg m^{-2} h^{-1} and Gallagher & Daiber (1973)

found a tenfold amplitude of fixation in diatom dominated cores and only a twofold amplitude when the cores supported a flora of filamentous green algae.

Communities associated with the sediment

It is clear that there is an additional association of species which live as a layer, loosely associated with the sediments but not creeping on and between the particles; indeed the majority of the species of this association are non-motile centric diatoms or filamentous algae. This assemblage bears a similar

Fig. 5.19. *a*, the weekly mean incident light, day length and temperature at two depths. *b*, the average weekly values for epibenthic algal production (open circles), community respiration (triangles) and bacterial respiration (filled circles) at various depths in Marion Lake, British Columbia. From Hargrave, 1969.

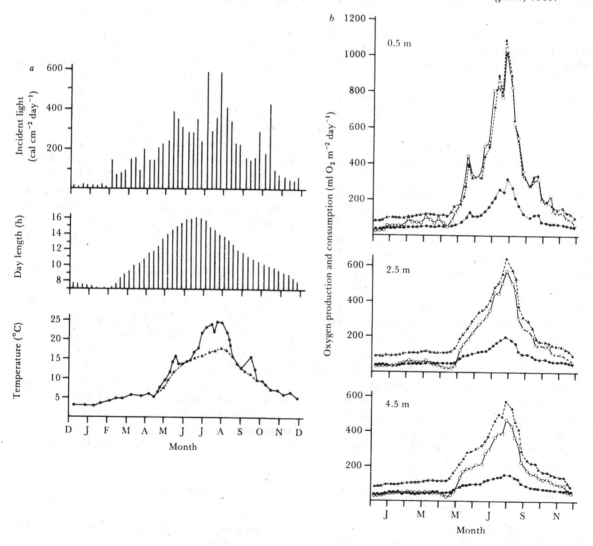

relationship to the epipelon as the metaphyton does to the epiphyton and the term plocon used by some workers is, I think, applicable. It has been described from the sediments off Heligoland by von Stosch (1956) where the dominants were *Paralia sulcata*, *Biddulphia rhombus* f. *trigona*, *Aulacodiscus argus*, *Actinocyclus ehrenbergii*, *Podosira stelliga*, *Biddulphia granulata*, *Actinophychus undulatus*, *Cerataulus smithii*, *Actinoptychus splendens*, *Cerataulus turgidus*, *Auliscus sculptus*, *Coscino-*

Fig. 5.20. The rate of photosynthesis of the sediment living community at Nivå Bay, as a function of light intensity on 13 May (*a*). The Ik values (filled circles) temperature (filled triangles) and light intensity (filled squares) plotted from May onwards at a deep station, Helsingør, 8 m depth (*b*) and at a shallow station, Nivå Bay, 0.3 m depth (*c*). From Gargas, 1971.

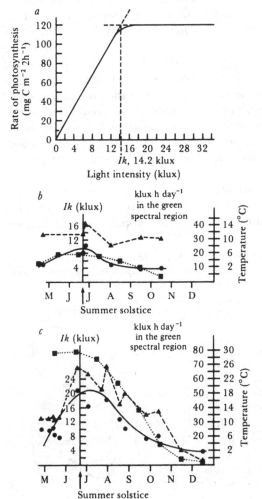

discus oculis-iridis, and *Triceratium favus*. A similar assemblage has been described from the Black Sea (Manea & Skolka, 1961) where *Coscinodiscus stellaris*, *Rhabdonema adriaticum*, *Melosira moniliformis*, *Biddulphia vesiculosa* and *Cerataulus turgidus* were the dominant species. Von Stosch comments that in late summer this association can be stirred up into the overlying water and greatly exceeds the true plankton in numbers. In many coastal regions species such as *Paralia sulcata*, *Bacillaria paradoxa* and *Cylindrotheca* (*Nitzschia*) *closterium* are common constituents of this community (*see* Fig. 5.21) which, whenever the sediments are disturbed, find their way into the inshore plankton: the literature gives the impression that the plankton is their true niche but this is not so. The number of large centric diatoms in the association is striking and diagnostic. There has always been a lack of ecological data on the non-planktonic centric diatoms which have been recorded only rarely in eological studies but at least now one habitat has been revealed. The extent of this community is totally unknown and certainly it does not occur on all sediments, e.g., it is absent at the stations sampled by the author off California and is not mentioned by Plante-Cuny (1969, 1973) in Madagascar or the Mediterranean. Presumably the community can only develop in waters where there is a back-and-forth bottom surge and not where flow is unidirectional.

Free living, macroscopic algae are not uncommon in a position similar to that occupied by the large centric diatoms just discussed. Sometimes these populations float up into the surface waters and maintain a more planktonic existence (e.g. *Sargassum* and *Phyllophora*, *see* p. 295). But in other habitats, aggregations of thalli can form 'algal balls' in which vegetative growth of a normally attached species continues in a drifting state adjacent to the sediments. Perhaps the best known are the balls of *Cladophora* found in some freshwater lakes where they can reach the size of tennis balls. They are so striking in a lake in Japan that they have achieved the fame of appearing on a Japanese postal stamp! They can also form in marine situations, e.g. the genus *Bryopsis* can grow in this manner. On the sandy sediments on the leeward side of a coral reef, species of non-calcified Bryopsidophyceae (*Caulerpa racemosa*, *Codium isthmoscladium*, *Codium taylori*, *Bryothamion seaforthii* and the brown alga *Dictyota divaricata*) have been reported growing on *Halimeda opuntia*, and, as overgrowth proceeded, the *Halimeda* decayed and the clump was freed from the sediment (Almodóvar & Rehm, 1971). The gentle movement of the water then rolled these algae into balls in which the algae continued to grow but in an unattached state. Loose spherical aggregations of *Ascophyllum nodosum* ecad *scorpioides* and *Furcellaria fastigiata* f. *aegagropila* occur on salt marshes and in protected tidal waters in Scotland (*see* Gibb, 1950, for detailed survey of *Ascophyllum*). The loose form of *Furcellaria* is apparently quite common in the Baltic Sea, where it is dredged up and used as fertiliser (Austin, 1960). Apparently a large number of species form such growths in the Adriatic Sea, and Austin quotes *Cystoseira barbata*, *Halopitys pinastroides*, *Rytiphloea tinctoria*, *Chondria tennuissima*, *Cladophora* spp., *Vidalia volubilis* and *Valonia utricularis* as being capable of such a life style in that region. Similar growths of *Ascophyllum* occur in the 'marine lake' Bras d'Or on Cape Breton Island, Nova Scotia together with free living ecads of the red algae *Ahnfeltia plicata* and *Gracilaria foliifera* var. *angustissima* (McLachlan & Edelstein, 1970–1). Such communities are obviously common in suitable freshwaters and marine habitats and as they do not fall into any of the named communities they should be considered as a distinct entity. They are of course also very common in lakes and ponds in the form of floating masses of *Oedogonium*, *Spirogyra*, *Mougeotia*, etc., which can, in some instances, form exceedingly dense masses rising up into the water and growing as a blanket over the surface. Few workers have recognised this community as a discrete entity, although Ilmavirta *et al.* (1977) recognised its importance in a Finnish lake and termed it 'littoral plankton' but commented that it contains many benthic and epiphytic algae. Perhaps the term 'littoral plankton' should be replaced by 'littoral detriplankton' since it does seem to be a collection of displaced species. *Hydrodictyon* also occurs in such a manner in slow flowing streams and backwaters.

Endopelon

Reports of subsurface growths of algae are rare, but Hoffmann (1942) and Gerlach (1954) both found an interesting population of blue-green algae beneath sediments along the North Sea intertidal. Here in this so-called 'Farbstreifensandwatt', *Microcystis*, *Gloeocapsa*, *Chroococcus*, *Merismopedia*, *Calothrix*, *Hydrocoryne*, *Nodularia*, *Anabaena*, *Spirulina*, *Oscillatoria*, *Phormidum*, *Lyngbya*, *Microcoleus* and *Hydrocoleus* occur. In the sand below these Cyanophyceae a zone of purple bacteria was found above the deeper anaerobic sand. Light penetration through wet sand is such that attenuation to between 1 and 2% of surface intensity occurs at around 5 mm depending on the colour, etc., and, interestingly, it is the longer red wavelengths which penetrate to this depth.

Stromatolites

Algal stromatolites are laminated, rock-like structures formed by the trapping, precipitation and

Fig. 5.21. Examples of diatoms living in loose associations with sediments. A, *Bacillaria paradoxa*; B, *Actinocyclus roperi*; C, *Aulacodiscus kittoni*, valve and girdle view; D, *Actinoptychus undulatus*; E, *Biddulphia rhombus*, girdle and valve view; F, *Cerataulus turgidus*, girdle view; G, *Paralia* (*Melosira*) *sulcata*, valve view.

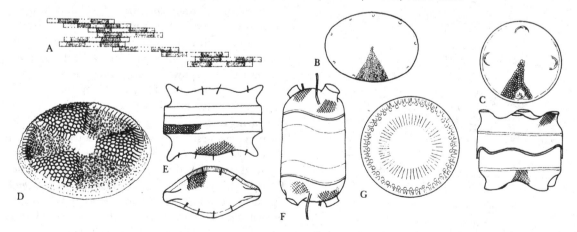

Fig. 5.22. Hamelin Pool, Shark Bay, Western Australia. *a*, lithified subfossil mound-shaped stromatolites in the foreground, with living Cyanophyta mats in the background now binding carbonate sediment to form stromatolites. Beach dunes in background: *b*, section through living Cyanophyta mats binding carbonate sediments to form stromatolites in the intertidal zone. Pencil in left hand corner indicates the scale. Photographs kindly supplied by Dr M. R. Walter.

binding of sand, silt and clay and algae. Filamentous and coccoid Cyanophyta are the commonest organisms involved and stromatolites form in many intertidal and subaerial tropical regions where wave disturbance is at a minimum, e.g., Florida, Bermuda, Bahamas, Persian Gulf and Western Australia. Few inland water sites are favourable for this type of growth though somewhat similar algal structures built up by a *Microcoleus* sp. are reported on the surface of an intermittently wet saline basin in Australia (Clarke & Teichert, 1946). The first report of modern stromatolites was made by Black (1933) at Andros Island, where he found a rubbery mat of *Schizothrix* in the intertidal region and cushion-like structures produced by *Scytonema* in brackish and fresh water. These deposits are laminated owing to cyclical deposition of particles and growth of the algae. Monty (1967) reported that a single algal lamina up to 600 μ thick is formed during a single day. Fossil stromatolites have been found in geological strata back to the Precambrian and most studies have been made by geologists and reported in geological journals; a particularly stimulating and detailed account is given by Walter (1972). The major stimuli to study have arisen from the need to relate the fossil forms to the present-day types and to study the environment under which they are produced, in order to be able to define more clearly the conditions under which the fossil structures formed. Ancient stromatolites do not contain any

actual fossilised algal remains but are merely the residual deposits (Logan, Rezak & Ginsburg, 1964). There has been very little awareness, or study, of these structures from a phycological standpoint. They bear a strong resemblance to the algal mats discussed above (p. 159) and are equally part of the rhizobenthos and a separate discussion is simply warranted by their geological importance.

One of the classic sites of formation is at Shark Bay in Western Australia and Fig. 5.22 shows this area. Logan (1961) described three different types of stromatolites; (1) flat lying, algal, laminated sediments underlying a terrain of continuous algal mat; (2) low sinuous domes and discs of laminated detrital sediment underlying relatively continuous algal mats; (3) discrete club-shaped and columnar structures of laminated detrital sediment with algal mats limited to the tops and sides; this is the type in Fig. 5.22 and the relationship of some of the types can be seen in Fig. 5.23. The various forms are due to interaction between the algal mat and the physical features of the environment. The Western Australian stromatolites form in the quiet, hypersaline waters (56–65‰ salinity) at the heads of bays and *Schizothrix fuscescens*, *Plectonema terebrans* and *Microcoleus chthonoplastes* have been identified in the algal layer.

In the Persian Gulf at Abu Dhabi the upper mud flats are covered by a rough rubbery mat of blue-green algae, which has also been termed a stromatolite (Bathurst, 1971). It is generally broken

Fig. 5.23. Diagrammatic representation of the stromatolite zone, Hamelin Pool, Shark Bay, Western Australia. From Logan, 1961.

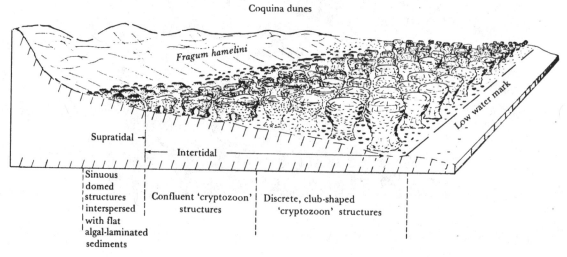

Coquina dunes

Fragum hamelini

Low water mark

Supratidal →

← Intertidal →

Sinuous domed structures interspersed with flat algal-laminated sediments

Confluent 'cryptozoon' structures

Discrete, club-shaped 'cryptozoon' structures

into polygons by shrinkage during desiccation but is also covered by some tides during each month, and variation in supply of sediment and of growth of the blue-green algal filaments gives rise to the laminated sediment. Kendall & Skipwith (1968) have investigated these stromatolites and recognised four distinct types occurring parallel to the waters edge. Nearest the edge is the 'cinder zone' resembling a warty black layer of volcanic cinder, then the 'polygonal zone', higher still the 'crinkle zone' (a leathery algal skin loosely attached to the underlying gypsum crystals) and uppermost the 'flat zone' (Fig. 5.24).

The parallel zoning of different types of stromatolite is yet another example of the linear zonation parallel to the sea edge within the intertidal. The algal components of these zones require detailed examination: in one area *Microcoleus chthonoplastes*, *Schizothrix cresswellii*, *Lyngbya aestuarii* and *Entophysalis deusta* have been identified in the mat but it is obvious from the comments in the geological literature that there are many other algae present. These structures form in an extreme environment where the temperature can fluctuate by 30 °C and the salinity vary between 47 and 196‰ (Kendall &

Fig. 5.24. Zonation of algal-stabilised sediments in a Persian Gulf Lagoon, and diagrammatic sections through these regions. 1, Lagoonal carbonate sands and (or) muds; 2, poorly laminated carbonate rich algal peat; 3, algal mat formed into polygons; 4, lagoonal sediments with gypsum crystals; 5, cinder zone algal peat with gypsum crystals; 6, polygonal zone algal peat with gypsum crystals; 7, mush of gypsum crystals and algal peat; 8, anhydrite nodules and layers in a matrix of wind blown carbonate and quartz; 9, halite crust formed into compressional polygons. From Kendall & Skipwith, 1968.

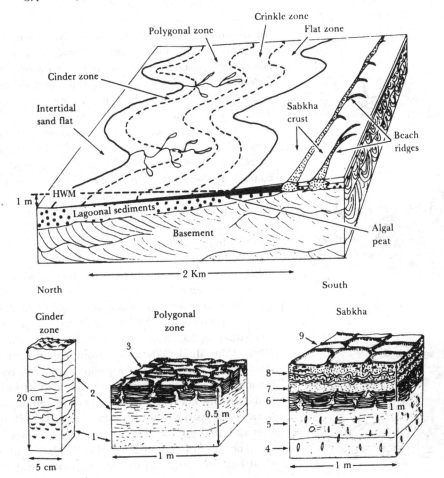

Skipwith, 1968); they are also salt encrusted and, in the Sabkha calcium carbonate encrusted. Similar mats are also reported from hypersaline coastal lagoons in Baja, California, and here diatoms are also common in the surface region (Horodyski, Bloes & Von der Haar, 1977).

The commonest cyanophyte recorded in stromatolites is *Schizothrix calcicola* which is said to occur in both marine and freshwater habitats. It is a species which does not calcify its sheaths but does bind particles (Sharp, 1969). Monty (1965, 1967) found this alga associated with subtidal, paired laminae and by scattering boiled carborundum as a marker, concluded that one hyaline and one sediment-rich lamina is laid down each day. The hyaline lamina is produced during daylight when the *Schizothrix* moves upwards into the overlying

water and forms a layer (lamina) of mucilage. Later in the day and during the night the *Schizothrix* filaments become packed in a horizontal plane and it is then that the carbonate particles are deposited. These movements appear to be related to light-stimulated movement and orientation. However, Gebelein (1969), using similar techniques but marking with ferric oxide, concluded that the algal rich laminae accumulated at night and the sediment rich laminae during the day, but he too noted a horizontal growth pattern of the filaments during the night (Fig. 5.25). In yet another publication it is merely said that the high tides deposit the sediment, and the algal capping grows out during low tide (Gebelein & Hoffman, 1968). Obviously further studies are required on this fascinating aspect of the ecology of these Cyanophyta. It seems that in subtidal situations, where little sediment is washed over the mats, only a relatively pure algal mat forms, adding evidence to the suggestion that this alga does not precipitate carbonates. The sediment trapped around the algae would however hardly build up the rock-hard stromatolites unless there was some cementing process at work. Both *Scytonema* and *Entophysalis* do appear to precipitate calcium carbonate in, and around, their sheaths, probably only another manifestation of the widespread precipitation of carbonate associated with photosynthetic removal of carbon dioxide (*see* p. 156). The calcite formed in stromatolites by this mechanism tends to contain a proportion of magnesium carbonate and this can be distinguished mineralogically from aragonitic magnesium calcites from other sources. A report from a study of the blue-green algal mats growing in Baffin Bay, Texas also suggests that there is precipitation from supersaturated seawater in the lower parts of the mat, possibly associated with bacterial activity. The precipitate is very fine microcrystalline aragonite (it was termed algal micrite) and the particles varied from 'clay sized', hence the term micrite, to 200 μm in diameter (Dalrymple, 1966). The algal material is laminated in typical stromatolite structure and a photograph of a core through the material resembles the banded deposits found in some lake cores. Gebelein (1969) found that the sand trapped by the algae was of a fairly uniform size and that in the domed structures, accretion could occur at rates up to 3 mm day^{-1}. The so-called algal biscuits could form in three months and the

Fig. 5.25. Diagrammatic representation of the light–dark cycle in stromatolite formation. *a*, *b*, daylight upward growth and sediment trapping (*Schizothrix calcicola*); *c*, dark horizontal growth and sediment binding (*Oscillatoria submembranacea*). From Gebelein, 1969.

a

b

c
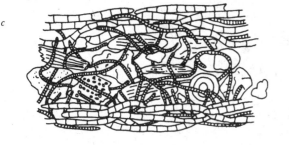

algal domes within a year at this rate of sediment trapping.

Clusters of carbonate sand grains cemented by micritic aragonite occur, e.g. around the Bahamas, and are termed grapestone. Little is known concerning their origin but algae acting both as precipitating and boring agents have been found associated with the granules and it is certain that algae are involved in their genesis, though their actual contribution is not known.

On shallow stable sediments off the Bahamas, and certainly common in other tropical areas, is an organic 'film' binding the sand grains and permeated by species of filamentous Cyanophyta, Chlorophyta, Rhodophyta with an intermixture of diatoms (Bathurst, 1971). Bathurst termed this the 'algal mat' and he reports that the pale green or brown of this sediment is observable both by diving and from the air. The top 0.25 cm of the mat tends to cohere and, under the microscope, is seen to be 'a colourless, transparent, elastic gel-like substance with the sand grains embedded in it'. This description indicates a mat very similar to that found by the author in freshwater (*see* p. 126). Scoffin (1970) reports growths of the blue-green alga *Schizothrix* in the mat and Bathurst reports that in places the sand grains have diatoms on them, 'about ten to each grain'. These 'algal mat' sediments cover many thousands of square kilometres and are stable, not only because of the feeble currents, but because of the binding capacity of the mucilage secreted by the algae. Bathurst comments that it is difficult to see how the animals living on these sandy seabeds could survive without feeding on this mat. Rooted in this 'algal mat', macroscopic siphonaceous algae also occur and their rhizoidal system aids the binding. There are frequent comments in the literature about *Lyngbya–Oscillatoria* mats occurring on the silt amongst mangrove roots; there is often a considerable amount of hydrogen sulphide production in the muds under these mats and the community structure may be controlled to some extent by this feature; certainly some Cyanophyta are tolerant of high sulphide levels. It is clear that the ecology of all these epipelic communities requires much deeper study.

Bunt *et al.* (1970) investigated similar cyanophycean communities between 20 and 60 m off Miami and recorded a reddish-brown coloured assemblage in which *Schizothrix calcicola, S. mexicana,*

S. tenerrima, Porphyrosiphon notarisii, P. kurzii, Microcoleus lyngbyaceous and *Spirulina subsalsa* occurred during the winter. The phycobilin pigments leached out of these algae during transport to the laboratory and so experiments on nitrogen fixation were made *in situ*; it was found that whilst cultures of *Nostoc* fixed nitrogen at these depths, the mainly non-heterocystous natural assemblage did not.

Siliceous stromatolites have been found in the hot spring and geyser effluents of Yellowstone National Park, where blue-green algae are present; in addition, silica is deposited around the algae and bacteria (Walter, Bauld & Brock, 1972). The silica is not biogenically accumulated but presumably the algal mat is essential as a framework.

Soil algae

In the words of Shields & Durrell (1964), 'Soil algae have suffered an obscurity'; this is certainly true if one compares the soil algal literature with that on aquatic algae or with the treatment in most texts purporting to discuss soil microbiology. Nevertheless much of the floristic information had been obtained prior to 1964, indeed Ehrenberg (1854) recorded many soil algae in the early part of the last century (*see* Fig. 5.26 from his *Mikrogeologie*).

One of the first and most important publications on soil algae is that of Petersen (1928)* on the flora of Iceland, in which he recognised the richness of the assemblages. This study is however a little misleading for workers in drier areas of the world since the very large number of diatoms recorded by Petersen are undoubtedly due to the maintenance of high soil moisture content under such wet climatic conditions.

There have been several unfortunate misconceptions in the literature about soil algae, one of which is to consider them as a fragment of the aquatic flora living in a less favourable environment. The underwater epipelic algal flora which is the limnic equivalent of the soil flora only extends slightly out of the water into the splash zone and comparison of the two floras shows relatively little

* This publication and the preceding one on 'The freshwater Cyanophyceae of Iceland' Petersen (1923) are almost entirely devoted to a discussion of species, taxonomic notes, etc., but amongst the comment is a wealth of important observations for anyone studying soil algae. The account was followed in 1935 by a further more extensive survey.

Fig. 5.26. Drawings from Ehrenberg's *Mikrogeologie* of algae from soil. *a*, *Hantzschia amphioxys* is clearly visible (2B, 4A): *b*, diatoms from dust collected off the West Coast of Africa in which *Stephanodiscus* (IVa) and *Melosira* (IVc, Va, b, c) are identifiable: *c*, diatoms, silicoflagellate, radiolarian and siliceous spicules from sea ice. *Actinocyclus* (5, 7, 13) and *Asteralampra* (3). This sample is of marine plankton frozen into sea ice and not of the bottom ice assemblage: *d*, diatoms, silicoflagellates, spicules and siliceous fragments from marine sediment. *Melosira*, *Actinocyclus*, *Actinoptychus* present.

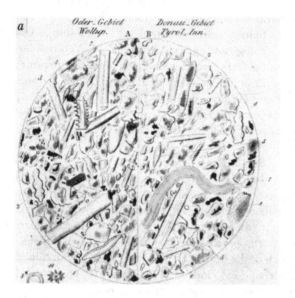

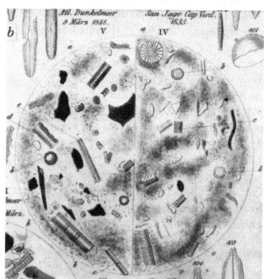

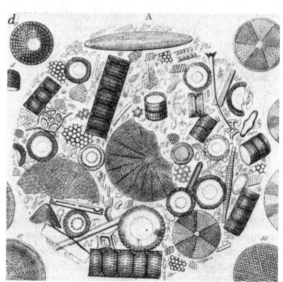

interchange. The soil algal flora contains less than 100 common species* whereas the freshwater epipelon probably supports 5–10 times this number of species. The difference also extends to the average number of species in individual samples, e.g., only four or five algae are to be expected in any one desert soil (Cameron, 1960), a figure which agrees with the cultural studies of Schwabe (1960) on Chilean desert soils, and only a few more would be found under less extreme conditions. Another example, but with the emphasis on diatoms alone, is given by Brendemuhl (1949) who found that, of 162 samples, only three had more than 20 diatom species, only 23 more than ten, and the average number per soil was from five to nine. Assuming approximately the same number of green and blue-green algae the total is likely to be around 20. The same kind of 'surface' habitat is involved on both soil and underwater sediments and so the life forms are similar in both habitats. Some soil species do live in aquatic sites but the majority of soil algae only develop massive populations on the surface of the soil and there are undoubtedly many species confined to the soil habitat. This confusion between species which merely exist and those which luxuriate in a habitat is common to many studies. A simple example is provided by the commonest soil diatom *Hantzschia amphioxys* which occurs as occasional cells on freshwater sediments but only forms rich brown coatings on soils and can almost invariably be cultured from any soil sample (e.g., Brendemuhl recorded it on 83% of her soil samples). Where species do occur in both habitats there is certainly a tendency, noted by Lund (1945–6), for the populations on soil to consist of small individuals, at least as far as diatoms are concerned. A certain amount of intermingling occurs with the algal flora of Bryophyta but here again the aerial parts of the large mosses do support a characteristic flora (*see* p. 214).

Another misconception is that the soil algae are present only in the form of resting stages. At certain times of the year even casual observations reveal active growth, visible as expanding macroscopic patches on some soils. Transference of samples of the flora to moist conditions usually results in immediate commencement of movement by diatoms and flagellates. There may nevertheless be periods when, in dry soils, release of zoospores is blocked since they are rapidly released on moistening, e.g., Trainor & McLean (1964) added cultured *Spongiochloris typica* to dry soil and then wetted samples at monthly intervals; every time zoospores were discharged within 24 h. I suspect that cell division has already occurred in such instances and wetting merely triggers the release mechanism, but experimental verification is needed.

The soil flora is selected from a limited range of algae within the Chlorophyta, Cyanophyta, Bacillariophyta and Xanthophyta, with occasional species from the Euglenophyta and Rhodophyta plus the controversial algae *Cyanidium* (which also occurs in thermal waters). Most of the species seem to be widespread and occur in arctic, temperate and antarctic soils, e.g., species of *Chlamydomonas*, *Carteria*, *Bracteococcus*, *Chlorosarcina*, *Hormidium*, *Stichococcus*, *Zygogonium*, *Vaucheria*, *Aphanocapsa*, *Anabaena*, *Cylindrocystis*, *Nostoc*, *Phormidium*, *Caloneis*, *Navicula*, *Pinnularia*, *Stauroneis* and *Hantzschia* (*see* Fig. 5.27). The studies of Lund (1945), Brendemuhl (1949) and Bock (1963) are particularly useful for identification of soil diatoms and Fig. 5.28 gives an overall picture of the occurrence of soil diatoms relative to pH. The soil diatoms are almost exclusively biraphid forms capable of movement and most other groups are slowly motile or form flagellate stages. In the tropics, the records suggest a greater abundance of Cyanophyta (54 species of Cyanophyta in south Arizona (Cameron, 1964)) with genera such as *Porphyrosiphon*, *Camptylonema*, *Schizothrix*, *Scytonema* and *Hapalosiphon* (Duvigneaud & Symoens, 1949) prominent. The regions of the world with the highest rainfall tend to occur in well defined regions between latitudes 20° N and 20° S and the wetness, coupled with high temperature, probably forms one of the most favourable habitats for soil (and other terrestrial) algae but I know of no studies in such sites.

So many soil algae exist as unicells or, in the case of Cyanophyta, as short lengths of filament that I am convinced that cultural studies are *essential* for the correct identification of these species. This is not the place to discuss problems of isolation, culture and

* This is simply an estimate from studying the literature and I have the impression that an even smaller number of species comprise the bulk of the biomass. For example Durrell (1964) recorded 62 species (16 Chlorophyta and 46 Cyanophyta) from a range of tropical soils and Mitra (1951) 80 species from Indian soils. However, Shtina (1960) reports about 900 species from Russian soils.

Fig. 5.27. Some microscopic terrestrial algae found on mineral soils or brown earths below the grass *Deschampsia antarctica* on Signy Island. A, *Nodularia*; B, *Oscillatoria*; C, *Nostoc*; D, *Phormidium*; E, *Phormidium*; F, *Microcoleus*; G, *Schizothrix*; H, *Monodus*; I, *Heterothrix*; J, *Achnanthes*; K, *Navicula*; L, *Pinnularia*; M, *Navicula*; N, *Hantzchia*; O, *Chlamydomonas*; P, *Chloromonas*; Q, *Rhopalocystis*; R, *Planktosphaerella*; S, *Myrmecia*; T, *Cylindrocystis*; U, *Bracteococcus*; V, *Chlorhormidium*; W, *Stichococcus*; X, *Dictyosphaerium*; Y, *Microthamnium*; Z, *Gongrosira*; AA, *Ourococcus*; BB, *Planophila*.

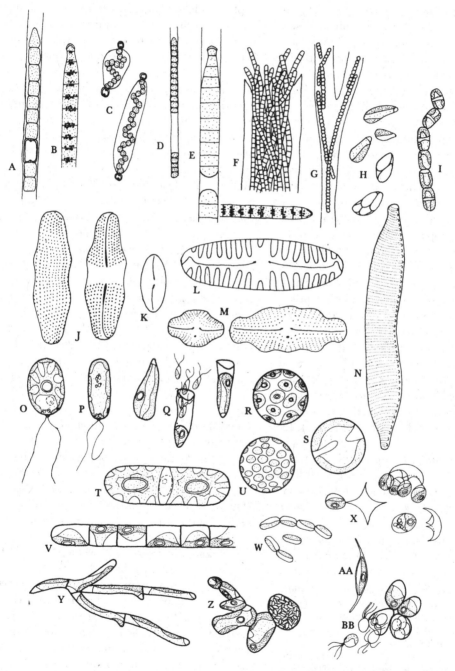

taxonomy of such algae but the excellent study of unicellular coccoid algae by Starr (1955) and the numerous publications from the Texas group under Bold should be consulted; a good review is that of Bold (1970).

A regrettable aspect of the literature is that algae of rice fields have been written about as though they were soil algae; a few may be, but during the flooded season limnic species develop and the problems fall into the realm of those of temporary waters (*see* p. 45).

Methods. Soil algae are not easy to investigate qualitatively or quantitatively. Direct observation of surface scrapings of soil mixed with water is tedious, but reveals many, often unidentifiable, algae in a non-quantitative manner. Culture of soil in water tends to favour the algae which can live submerged and is therefore likely to bias the results and even to encourage growth of contaminant species of algae which never actually grow on soils. Brendemuhl (1949), in a very detailed study of soil diatoms found that, of 116 species, 14 were only found in liquid culture and 20 only on agar. The liquid technique

Fig. 5.28. The percentage occurrence of species of soil diatoms in four pH ranges. White columns, pH 8.1–7.0; thick black lines, pH 6.9–6.0; thin black lines, pH 5.9–5.0, and dotted lines, pH 4.9–3.7. From Brendemuhl, 1949.

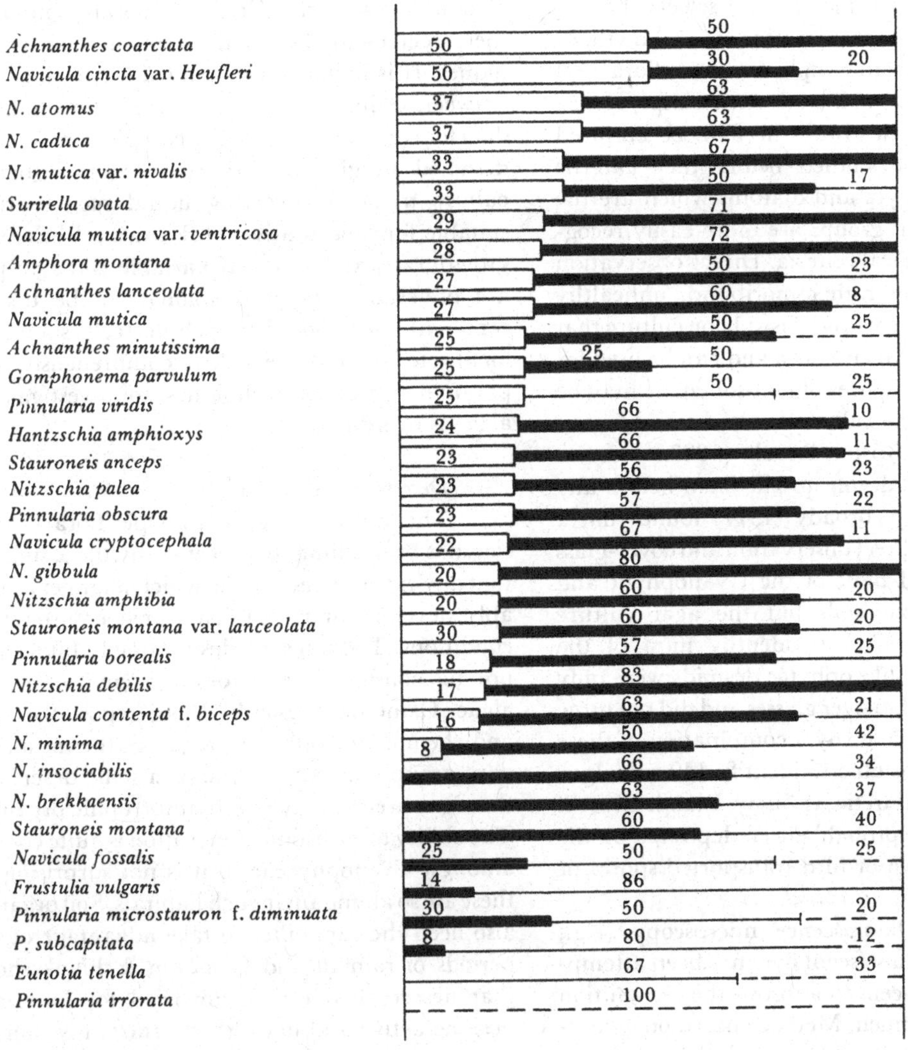

Species	White (pH 8.1–7.0)	Thick black (pH 6.9–6.0)	Thin black (pH 5.9–5.0)	Dotted (pH 4.9–3.7)
Achnanthes coarctata	50	50		
Navicula cincta var. *Heufleri*	50	30	20	
N. atomus	37	63		
N. caduca	37	63		
N. mutica var. *nivalis*	33	67		
Surirella ovata	33	50	17	
Navicula mutica var. *ventricosa*	29	71		
Amphora montana	28	72		
Achnanthes lanceolata	27	50	23	
Navicula mutica	27	60	8	
Achnanthes minutissima	25	50	25	
Gomphonema parvulum	25	25	50	
Pinnularia viridis	25	50	25	
Hantzschia amphioxys	24	66	10	
Stauroneis anceps	23	66	11	
Nitzschia palea	23	56	23	
Pinnularia obscura	23	57	22	
Navicula cryptocephala	22	67	11	
N. gibbula	20	80		
Nitzschia amphibia	20	60	20	
Stauroneis montana var. *lanceolata*	30	60	20	
Pinnularia borealis	18	57	25	
Nitzschia debilis	17	83		
Navicula contenta f. *biceps*	16	63	21	
N. minima	8	50	42	
N. insociabilis		66	34	
N. brekkaensis		63	37	
Stauroneis montana		60	40	
Navicula fossalis	25	50	25	
Frustulia vulgaris	14	86		
Pinnularia microstauron f. *diminuata*	30	50	20	
P. subcapitata	8	80	12	
Eunotia tenella		67	33	
Pinnularia irrorata		100		

is best reserved for the culture of individual species obtained by other means. A more natural culture technique is to spread dilute suspensions of soil on to agar plates which have been dried out. The water of the suspension is absorbed into the agar and leaves the soil algae on the surface in a humid habitat similar to that in nature. Addition of various nutrients to the agar will of course bias the growth of the species, but can be useful. A further method, used by Esmarch as long ago as 1910, is to place the soil in a petri dish, moisten it with distilled water and then place cover glasses or tissues (e.g., lens-cleaning tissues) on the surface; for one of the most detailed studies by this method *see* Lund (1945–6). The algal growths attach to the glasses or tissues and can be removed at intervals. This technique is the best one to enumerate diatoms since it only selects the live cells. As in other habitats a combination of methods is generally necessary to sample the entire flora. Soil algae are generally small and often simply green spheres. These almost always need to be cultured and their life cycle studied before they can be identified. Cyanophyta and diatoms which are the other two important groups are more easily recognised on morphological criteria. Direct observation of soils often reveals rather moribund, unhealthy looking cells or filaments and it is only on culture that the 'normal' appearance is seen and can be used for identification. Since spores of non-soil algae may be present on soils some culture methods may favour their growth and Bristol Roach (1927*a*) recommended six months drying to kill them before any cultures were set up. Broady (1977) found that a combination of the direct observation and cover-glass techniques revealed most of the Cyanophyta and diatoms on antarctic soils but the agar culture technique was necessary to identify most of the Xanthophyta and Chlorophyta; desmids were only regularly noted on the cover glasses and did not grow at all on the agar plates. By a combination of these methods Broady was able to identify 149 soil algae from soils of the antarctic at Signy Island. Most if not all of these have probably been deposited on the island as wind blown or bird transported spores or fragments.

The use of fluorescence microscopy (e.g., Tchan, 1952) and fluorescent dyes has been attempted but it does not seem to improve the recognition and counting very much. Methods based on chlorophyll extraction and estimation have been used but the amount of phaeopigments in soil must interfere seriously with this technique.

Undoubtedly the most extensive studies of soil algae have been those undertaken by Soviet soil scientists and Shtina (1960) considers that a direct method of observation yields more accurate quantitative data and much higher counts than techniques using aqueous nutrient media, though these should be used to obtain additional data. The direct method described by Shtina involves shaking the soil in water, allowing it to settle and then sedimenting the algae out from the supernatant liquid according to a set system of settling times. This presupposes that all the algal cells go into suspension and do not adhere to soil fragments. Shtina claims that counts of an order of magnitude or greater are obtained by the direct method over those of the culture techniques. It is important in soil studies to distinguish between techniques designed to obtain qualitative data for comparison of species complement and those designed to give quantitative results. Obviously culture methods involving liquid media are not suitable for quantitative work since selectivity and growth rates are extremely variable and agar plates are preferable, for each colony can be taken to represent one original propagule (cell, spore, fragment, etc.) within the soil. A comprehensive comparative survey of techniques to investigate soil algae is overdue.

Deserts

The study of the microscopic flora of desert soils has been stimulated recently by the assumption that any extra-terrestrial life which might exist probably develops under similar harsh environmental conditions. Though such desiccating habitats might not be considered a favourable environment for algae, Cameron & Blank (1966) report that algae and chemoautotrophic bacteria are the most abundant groups encountered, and in some desert areas the algae are the only photo-autotrophic organisms. The ability to withstand desiccation is quite common amongst Cyanophyceae so it is not surprising that these are so abundant in such habitats. Soil organisms also need the capability to take advantage of short periods of rainfall and Cameron & Blank showed that desert algal crusts, air-dried for four years, became active and new growth started within 24 h of wetting. Although the number of algae is reported to decrease after two to four years air drying, it is still possible to revive some species, e.g., *Protosiphon cinnamomeus*, *Schizothrix calcicola*, *Nostoc ellipsosporum* and *Microcoleus* spp. have been revived even after periods of drying for up to 70 years on herbarium sheets (Bristol, 1919). This ability of Cyanophyta to survive adverse conditions, respond rapidly to drying, wetting, heating, thawing etc. is not confined to soil Cyanophyta and it is generally assumed that their success in unusual situations is related to the properties of prokaryotic cells. Reports of diatoms living almost as long in dry soil seem to me to be rather suspicious especially when it is the common 'weed' *Nitzschia palea* which is recorded. The algae of desert soils tend to form surface aggregations, 'algal crusts' or 'rain crusts' (Fletcher & Martin, 1948) in which the filamentous Cyanophyta are the most conspicuous element and some green algae the next; diatoms do not seem to be common. Similar crusts of Cyanophyta are reported from cold desert soils in antarctica. These have been investigated in the scanning electron microscope, under which it is possible to recognise algal filaments and to see bacteria in association with the mucilage of the filaments (Cameron & Devaney, 1970). There have been almost no attempts to relate the algal flora of soils to the vegetation shading the soil and deserts are obviously favourable sites for such investigations. Durrell (1962) simply added a comment that he could find no particular relationship. This is surprising considering the well-known effects on germination and establishment of seedlings under desert shrubs. The species most frequently recorded in deserts are *Nostoc muscorum*, *Scytonema hofmanii*, *Schizothrix calcicola*, *Microcoleus vaginatus*, *M. chthonoplastes*, *Protococcus grevillei*, and *Protisiphon cinnamomeus*. The first two species fix atmospheric nitrogen (Cameron & Blank, 1966), and the high nitrogen content in desert crusts is certainly related to a continuous supply of nitrogen from the algae (Shields, Mitchell & Drouet, 1957). Cameron & Blank also report that, in desert soil, algae tend to become parasitised or lichenised. These algae have to withstand a very high temperature and Trainor (1966) reported that some algae in dry soil can withstand 100 °C for short periods. In a somewhat surprising series of experiments, Trainor & McLean (1964) found that dry soil containing *Spongiochloris* could be heated to 100 °C

and all this did was to slow the growth when the soil was remoistened. The culture which had originally been added to the soil, dried and heated, remained viable for the twelve months of the experimental study. The unheated soil always produced evidence of growth before the heated soils, which suggests that heat kills off the vegetative cells but not the resting spores, the latter probably requiring a little longer to germinate and produce visible populations. Less surprising was the finding of Booth (1946) that some soil Cyanophyta could withstand dry-air temperatures up to close to the thermal death point of 110–113 °C. Surface temperatures of desert soils are often 5–25 °C above air temperatures and in Death Valley, California a soil temperature of 95 °C has been recorded (reported in Cameron & Blank). It is highly probable that some of the tropical soil algae have elevated temperature optima for growth and indeed Fay & Fogg (1962) found that *Nostoc* (*Chlorogloea*) *fritschii* isolated from Indian soils can grow experimentally up to 50 °C, and produces its maximum yields at 40 °C, whereas below 20 °C growth is imperceptible. Algae living on some Russian desert soils (the 'takyrs') must be able to withstand greater overall extremes of temperature and moisture content than almost any group of algae. Lund (1967) quotes figures from Russian workers of summer temperatures about 87 °C and winter lows at around − 11.5 °C. These temperatures are of course important for survival and are not the range over which growth could occur. It might be expected that desert algae would metabolise at low water potential but in fact Brock (1975) found that *Microcoleus* was not particularly adapted to the drought-like conditions it is likely to live under. Photosynthesis was rapidly reduced as water potential was lowered and it appears that this alga could only grow rapidly during periods of rain. This study amplifies the earlier reports of Tchan & Whitehouse (1953) that growth is stopped at soil water holding capacity below 12%. The desert species are therefore likely to be selected for the ability to *survive* desiccation rather than ability to metabolise in dry conditions. *Cyanidium*, which is sensitive to water stress and normally lives in a layer 3–5 mm below the surface, seems able to adapt to varying moisture conditions since the most desiccation resistant populations were isolated from soils with the lowest water potential (Smith & Brock, 1973).

Table 5.7. *The percentage occurrence of algae in 120 samples of soils from tropical localities. Of the 62 species found, 16 were Chlorophyta and 46 Cyanophyta. Adapted from Durrell, 1964*

Percentage of soils which contained alga named	Alga
90	*Chlorococcum humicola*
56	*Phormidium tenue*
23	*Phytoconis botryoides*
22	*Hormidium subtilis*
21	*Nostoc muscorum*
21	*Chlorella vulgaris*
13	*Nostoc paludosum*
10	*Trochiscia reticulatum*
9	*Phormidium ambiguum*
7	*Phormidium uncinatum*
7	*Phormidium Retzii*
5	*Anabaena torulosa*

On deposits characterised by a particular mineral one might expect that an exceptional flora would develop; however few examples have been reported, though Shields *et al.* (1957) found an assemblage dominated by *Palmogloea protruberans* with occasional *Plectonema nostocorum* and *Nostoc* sp. on a gypsum sand deposit whilst on the surrounding desert soil a normal collection of Cyanophyta occurred.

Tropical soils exhibit a great range of conditions and certainly do not all qualify as desert soils and, as can be seen from Table 5.7 which is an overall compilation of algal occurrences on tropical soils, green algae are almost as frequently recorded as Cyanophyta.

The Antarctic

An extensive study over two and a half years has been made on the soil algal flora of Signy Island, in the South Orkney Islands between latitudes 60 and 61° S and south of the antarctic convergence (Broady, 1977). An island such as this affords interesting sites, since exposed glaciated material and older more mature sites are available close to one another and can be compared. Young raw mineral soils can be found alongside glaciers, etc., and these tended to be the sites with the smallest numbers of

soil algae (less than 20 species) whereas the richer mineral soils, especially those containing more calcareous material, all had between 20 and 55 species, and the total number of algal species on mineral soils on this antarctic island was 102. Other soils with less than 20 species were those on which mosses (*Polytrichum–Chorisodontium* and *Andreaea*) were growing or on which birds or seals were producing an organic mud. The mosses tended to colonise acid, base-poor soils and, coupled with the addition of moss humus to the soils, probably restricted the flora. The author has found that in British sites an accumulation of partially decayed plant material on top of the soil greatly reduces the flora, presumably since the algae are hardly in contact with the necessary soil mineral matter. On the other hand, the soils on Signy Island between the wet moss and liverwort clumps and under herbaceous vegetation had higher numbers of species. Cyanophyta, e.g., *Aphanocapsa*, *Nostoc*, *Pseudanabaena*, *Oscillatoria* and *Phormidium*, tended to be quite abundant on the soil under the herbs but the other soil algal groups were all well represented, e.g. *Monodus* and *Gloeobotrys* (Xanthophyta), *Navicula mutica*, *Pinnularia borealis* and *P. mesolepta* (Bacillariophyta), *Chlamydomonas*, *Chloromonas*, *Chlorella*, *Trochiscia*, *Cylindrocystis*, *Chlorhormidium* and *Stichococcus* (Chlorophyta).

Depth distribution

Several workers have studied the depth distribution of algae in soils and some have implied that surface and sub-surface associations occur. Friedmann *et al.* (1967) suggested that the former be termed epedaphic algae and the latter endedaphic. Obviously surface growths will be washed to lower levels and spores of surface forms may exist for many years below the surface but I am not sure that it has ever been proved that an *actively* growing population exists in the deeper layers. In fact, Tchan & Whitehouse (1953) reported a 99% decrease in numbers in the first 5 mm of soil, though obviously depth distribution will depend to some extent upon agricultural practices. Thus Willson & Forest (1957) found a similar flora down to the depth which the plough penetrated (18 cm) in cultivated fields, though even in such well mixed soil more algae were recorded at the surface.

The only experimentalist who has really tried to investigate movement through soil is Petersen

(1935) who dripped water through soil columns with and without earthworms, and found that a considerable number of cells of *Pleurochloris* could be washed through tubes 70 cm long. Interestingly, when the experiments were done with a small *Nitzschia* species, no cells were washed through the column and only when earthworms were present could a few *Nitzschia* cells be found at the bottom of the tube. *Nitzschia* is a motile diatom and presumably its upward phototactic movement was adequate to overcome the downward flow in these experiments.

It is difficult to see how photo-autotrophic organisms could actively metabolise at the low light intensities below a few millimetres of soil. Heterotrophy may rear its head but very careful studies are required to prove that it occurs in nature to a sufficient degree to permit the existence of truly endedaphic algae. There is no doubt that many soil algae can take up organic molecules in culture and this was shown in the excellent early work of Bristol Roach (1926, 1927b, 1928) who also showed that a strain of *Scenedesmus costulatus* var. *chlorelloides* could liquefy agar and, could presumably therefore break down substances external to the alga in soils. Later studies, e.g., Parker (1961) confirmed these results but demonstrated that the tendency was only widespread amongst Chlorococcalean soil algae. Parker, Bold & Deason (1960) found that the ability to grow in the dark on glucose was shown by all known species of *Bracteococcus*, *Spongiochloris* and *Dictyochloris*, some species of *Neochloris* and *Spongiococcum*, whilst in contrast all *Chlorococcum* spp. are obligate phototrophs.

Soil stabilisation and condition

Algae, especially the filamentous Cyanophyta which form confluent mucilaginous sheets, are frequently recorded as the primary stabilisers of bare eroded soils (Booth, 1941) and act as the pioneer community. The mucilage binds the soil particles, and reduces water loss and removal by wind. The binding property is retained when the blue-green algae move out of their sheaths, which remain behind and constitute a non-living framework (Durrell & Shields, 1961). If, however, the surface dries out, it cracks and the whole surface layer of soil may be lost. Although less conspicuous, some of the green algae also produce masses of mucilage, e.g., on burnt heathland, where *Gloeocystis vesiculosa*, *Scenedesmus*

obliqua, *Trochiscia aspera* and *Dactylococcus infusiorum* form a thin layer (Fritsch & Salisbury, 1915). The time taken to colonise an algal-free soil has rarely been investigated. Forest, Willson & England (1959) buried pots of sterilised soil back into their natural habitat and found that after five and a half months the flora on the surface of the pot soils approached that of the natural soil.

Algae contribute humus to all soils on which they grow and this could be important especially in deserts and on coral atolls. The very widespread excretion of organic compounds by algae must also occur amongst soil algae, indeed it may be even more extensive under conditions in which cells need to form a mucilage sheath in order to retain moisture. The effect of this on the soil is of some importance in providing organic substrates for other organisms but it also must assist in aggregation of soil particles. Recently Bailey, Mazurak & Rosowski (1973) have shown conclusively that growth of *Oscillatoria*, *Chlorella* and *Nostoc* on sterilised and homogenised soil resulted in an increase in the percentage of soil aggregates which were water-stable. The soils inoculated with *Oscillatoria* and *Nostoc* formed somewhat larger aggregates than those with the *Chlorella*; this is not surprising if the algae themselves contribute to the binding, since unicells are presumably less effective than filaments. Parker & Bold (1961) found that growth of soil algae was stimulated by the presence of bacteria and fungi and that the stimulation occurred in the light. The bacteria increased the growth of *Bracteococcus* some twentyfold and it was found that this was due to breakdown of substances in the soil by the bacteria and the release of nitrogen. In another example a species of *Streptomyces* not only increased the growth of a *Chlamydomonas* but enhanced its motility and induced akinete formation and the actinomycete was itself stimulated in growth and conidial production. A third example, given by Parker & Bold (1961), was of the interaction between a fungus and *Phormidium*, with the result that the latter was annihilated. Forest *et al.* (1959) reported that when *Mastigocladus paludosus* formed a crust diatoms were absent and the author (unpublished observations) has found a similar situation in patches of *Cylindrospermum*, which suggests either competition for space or some antagonistic effect. The whole problem of interactions between the soil flora (algae, fungi, bacteria) and the soil protozoa needs careful study. The investigation of the region around roots (rhizosphere) has been gathering momentum recently and there are reports that washings from roots stimulate algal growth, e.g. Cullimore & Woodbine (1963) noted increased soil algal growth alongside the roots of seedlings germinated on agar to which soil algal cultures were added, but little is known of such effects within natural soils although Hadfield (1960) reported that there was an increase in soil algae in the rhizosphere of tea plant roots. It is all too easy to show that a multitude of substances, both organic and inorganic, stimulate algae under cultural conditions but it is a much more difficult matter to prove the effect in nature though there is now ample evidence of secretion of organic compounds from roots. Much of this may not be immediately available to the algae which are only active in the surface millimetres where roots are not so common.

Similarly the interaction between fertilisation and soil flora is little studied. Stokes (1940) showed that adding calcium carbonate increased the soil algal flora and adding organic matter reduced it. The fertility of soils is often tested by soil scientists using pot cultures of angiosperms but Tchan, Balaam, Hawkes & Draette (1961) showed that exactly parallel results could be obtained by culturing algae in a soil–algal nutrient medium and determining the amount of algal pigment. This technique gave a very good indication of soil quality more rapidly than the pot plant technique.

The effects of herbicides on soil algae is a neglected field but Cullimore & McCann (1977) have shown that *Chlamydomonas, Chlorococcum, Hormidium, Palmella* and *Ulothrix* are all sensitive to 2,4-dichlorophenol indophenol, trifluralin, MCPA (2,methyl-4-chlorophenoxyacetic acid) and trichloracetic acid whilst *Chlorella, Lyngbya, Nostoc* and *Hantzschia* were most resistent.

Algae and soil types

The distribution of algae on different soil types still requires careful study along phytosociological lines but there is a widespread and almost certainly correct impression, based on many sets of isolated data obtained by non-comparable methods, that acid soils support good growths of chlorophyta (Lund, 1947) with the desmid genera *Cylindocystis* and *Mesotaenium* probably confined here together with *Euglena mutabilis*. On very acid soils only two

diatoms, *Caloneis fasciata* and *Pinnularia silvatica*, were found (Lund, 1945–6). These papers should be consulted for further details of British soil algae, which since that date have been grossly neglected. Dominance of Chlorophyta and Xanthophyta is also reported from arctic tundra soils which are acidic (Novichkova-Ivanova, 1972). As the alkalinity increases the Cyanophyta gradually achieve dominance, and nowhere is this better seen than on sand-dune soils where shell fragments give a base-rich status and where *Nostoc* colonies up to the size of golf balls can be found. Such soils investigated by the author (1957) also had a rich diatom flora in which *Epithemia zebra, Rhopalodia parallela, R. ventricosa* and *Nitzschia sinuata* var. *tabellaria* were striking indicators of base-rich conditions. The two diatoms normally considered ubiquitous on soils (*Hantzschia amphioxys*, and *Pinnularia borealis*) were virtually absent from these dune soils but were present amongst the moss leaves a few millimetres above the soil surface! As Shields & Durrell (1964) point out, most arid sites are alkaline and many moist sites are acid so it is not always easy to distinguish moisture factors from base status and pH. Nevertheless in liquid cultures Cyanophyta tend to dominate at high pH and Chlorophyta at low and the dune soils mentioned above are both moist and alkaline. I am not sure that Shields & Durrell's view, that physical properties influencing moisture level are more important to the algae than the chemical nature of the substratum, is correct. A further complication is that base-rich soils also tend to be rich in nitrates and phosphate and Lund (1945–6, 1967) found that soils rich in these anions were also richest in algae.

The data abstracted from Russian work by Lund (1967) show that on a cell count or biomass basis the soil algal flora of cultivated Russian soils is considerably higher than on virgin soils (Table 5.8). It is not clear whether this is due to fertilisation and working of the cultivated soils or whether they were originally selected because they were richer soils. One suspects a combination of causes and in my experience cultivation does greatly increase the biomass of algae, creating a more favourable environment whilst at the same time promoting growth of the crop plants. However, contrary to this, King & Ward (1977) have recently shown that undisturbed soils had a more diverse flora than disturbed soils, though when biomass is measured the disturbed soils

Table 5.8. *Abundance of algae in some virgin soils and in four cultivated soils of the USSR. From Lund, 1967*

Soils	Cells (g $\times 10^{-3}$)	Biomass (kg ha^{-1})
Virgin soils[a]		
Sod podsol	10–208	40–300
Podsol	5–30	7–20
Boggy podsol	6–50	10–45
Peaty-boggy	5–80	20–80
Alluvial meadow	52–300	80–450
West-Siberian chernozem	48	16
Dark chestnut	2150	187
Solonetz-meadow, deep columnar	988	515
Solonetz-steppe, deep columnar	392	74
Solonetz-crusty, nutty	5146	429
Cultivated chernozem		
Under red oats	755–2161	194–494
gooseberries	588–1702	134–436
strawberries	181–1578	55–525
Vegetable garden	390–2043	98–546

[a] In the virgin soils average abundance from the surface to 10 cm depth, except for the chestnut and solonetz soils where the depth range is 0–2 cm. In the cultivated chernozems under different crops the Cyanophyta and Chlorophyta were counted together and all the figures shown were obtained from samples 5 cm below the surface.

were more productive (compare the comment above on Russian soils). Soil disturbance favoured blue-green algae due to an increase in pH. An interesting study involving the growth of *Lolium perenne*, *Festuca rubra* and *Trifolium repens* singly and in combination in pots of soil with various fertilisers, coupled with a study of the soil algae which develop, was made by Knapp & Lieth (1952). They found that when calcium phosphate or potassium chloride was added there was good growth of Cyanophyta. With all three angiosperms in a single pot the algal flora was much reduced and they considered that the higher plants out-competed the algae for nutrients. Pots without fertiliser or with ammonium nitrate–calcium phosphate–potassium chloride produced little algal growth. Some earlier data of Stokes (1940) showed that liming of unmanured plots increased the number of algal cells to as many as 212 000 cells g^{-1} whilst manuring with organic matter reduced the numbers. This result is at variance with earlier work on the Broadbalk field at Rothamsted which had

been manured continuously since 1843 and which had more algae than the unmanured field (Bristol Roach, 1927a). The problem certainly needs further investigation and I suspect that the contribution of the Cyanophyta to soil fertility has been under-estimated. Rarely have algae been utilised by agri-culturalists in experiments yet they are a vital component of the soils and might be expected to yield many clues about soil fertility, nutrient deficiency, etc. A paper by Salăgeanu (1968) does describe a technique for testing the fertiliser requirements of soils by adding various combinations to soils to be tested, then applying a thin layer of sand, and finally covering with a filter paper on which a culture of *Chlamydomonas reinhardi* has been seeded. The growth of this alga was a good and rapid indicator of nutrient status of the soils and of the effects of various additions.

On soils manured by animal excreta, the two commonest algae are *Phormidium autumnale* which forms glistening black films and the green alga

Prasiola crispa. Amongst the thalli of the latter the diatom *Navicula nitrophila* is commonly found.

Since silicon is such an abundant element in soil, one might expect that diatoms would be very prolific in soils, yet they rarely exceed the counts of Cyanophyta and Chlorophyta. This may be related to the availability of soluble silica which, whilst in abundance in the moisture below the soil surface, may be flushed downwards from the superficial layer. Measurements of silicon and diatom growth on the surface of soils are badly needed as also are exact estimates of biomass of the various algal groups.

Nitrogen fixation

Much of the work on this phenomenon in nature has been conducted in rice fields which, as already mentioned, are not a soil environment but would be better classified as a temporary aquatic habitat. There is no doubt that fixation by Cyanophyta does occur also on the surface of soils but there are few positive data to show the exact contribution. Stewart (1970) summarises the data which vary from 26 to 400 times the nitrogen content for soils with Cyanophyceae crusts compared to those without the algae. Pankratova & Vakhrushev (1971) reported that nitrogen was accumulated by *Nostoc commune* at the rate of $3.3 \text{ kg ha}^{-1} \text{ yr}^{-1}$ in a loamy sand in Russia. Indian workers have found fixation in fields under sugar cane and maize. Seeding of algal inocula and creation of favourable conditions for algal growth may contribute to natural fertilisation and reduce the dependence on expensive nitrogenous fertilisers in some regions, but the exact contribution needs to be studied in the same way that the contribution from nitrogen fixing bacteria is being determined. Reports of stimulation of *Azotobacter* by Cyanophyta are confused by other reports that there is no effect: this reflects the need for more carefully controlled experiments in the field. Cameron & Fuller (1960) found that fallow soils increase in nitrogen and carbon contents and they showed clearly that *Nostoc*, *Scytonema* and *Anabaena* isolated from soils, fixed nitrogen.

Temporary algal associations

Temporary algal associations occur wherever water collects for short periods followed by drying out, for example in temporary pools on footpaths,

etc. These sites are colonised by what Comère (1913) termed 'formations pasagères' and whilst in some sites the soil algae themselves may develop profusely there is often a very rapid proliferation of species of flagellates. Euglenoids or volvocalean genera are common and when the soil is wetted these algae either divide rapidly, or perhaps release numerous motile cells from resting stages. This happens very quickly and the water becomes a 'green soup' showing that limnic species are involved. Sampling of rain water pools on paths frequently yields rich collections of the soil diatom *Hantzschia amphioxys* which seems to multiply rapidly under these conditions. Some diatoms form internal 'shells' under adverse conditions, e.g., in saline pans in desert areas, *Navicula cuspidata* forms these and can remain alive in this dry state for many years. Few algae have been especially reported from such temporary habitats although the branching chaetophoralean alga *Fritschiella tuberosa* was first recorded from soil in drying pools in India. It also grows profusely on the drying banks of the Blue Nile as the river recedes after summer floods (Brook, 1952*b*). This 'temporary' habitat occurs also alongside non-tropical rivers where *Protosiphon botryoides* often occurs. The dying stems and leaves of reeds deposited along water courses also provide a temporary habitat for algae.

Hot springs

The flora of hot springs has been studied fairly intensively because of interest in such an extreme habitat, but many studies are obsessed with the absolute temperature maxima for growth. Apart from this, springs and especially hot springs are fascinating steady-state habitats where the complication of high species diversity and fluctuating chemical conditions are minimised, thus allowing the study of interaction between a few species and factors. At high temperatures (and equally at high salinities) grazing animals are absent and loss of cells is mainly a physical phenomenon. These habitats are not merely heated (geothermal) waters but have the additional characteristic of great enrichment with some inorganic ions; sodium, potassium, carbonate, silicate, sulphide, sulphate and chloride make a rich medium, though still only with about 1‰ salinity.

At the highest temperatures the Cyanophyta alone are dominant e.g., Brock & Brock (1971) report *Mastigocladus laminosus* at temperatures up to

60–65 °C in New Zealand but the next most tolerant alga *Cyanidium caldarium* grows only up to 57 °C (Doemel & Brock, 1970) in areas as widely separated as Italy, Iceland, North America, Japan and New Zealand. *Mastigocladus* has been recorded occasionally at even higher temperatures e.g., 73–75 °C in the USA, which suggests that this taxa comprises more than one temperature-resistant form (cf., *Synechococcus*, Fig. 5.29). However, Castenholz & Wickstrom (1975) imply that *Mastigocladus* does not occur much above 60 °C and the records of Cyanophyta at temperatures around 85 °C (e.g. Mann & Schlichting, 1967) are almost certainly too high. The blue-green alga, *Synechococcus lividus* is also reported to tolerate a temperature of 74 °C (Peary & Castenholz, 1964; Brock, 1967) and it is the blue-green algae alone which survive at temperatures above 57 °C, some diatoms existing up to 50 °C and Chlorophyta only up to 48 °C. Peary & Castenholz found *S. lividus* along a stream at temperatures between 53 and 75 °C and isolated four separate strains along this gradient, each of which had characteristic tempera-

ture optima (Fig. 5.29). The growth rate of the lowest temperature strain was truly prodigous with 10 doublings per day. This rate decreased rapidly in the higher temperature strains which photosynthesised and grew much slower. The biology of hot spring algae may be still more complex since Sheridan (1976) has discovered that there are also genetically fixed sun and shade ecotypes of *Plectonema notatum*. These ecotypes had different chlorophyll contents (highest in the shade forms) and the high light clones were light saturated at higher light intensities and also had enhanced carboxylating enzyme activity compared with the low light clones.

According to Brock (1967), *S. lividus* is the commonest alga in western North American hot springs at temperatures between 50 and 73 °C, whilst diatoms are often dominant at temperatures between 30 and 40 °C. Fairchild & Sheridan (1974) report optimum growth (two divisions each day) of *Achnanthes exigua* at 40 °C and believe this is the highest optimum known for diatoms; 40 °C was also the temperature of the collection site. The concern with the upper temperature limits in this habitat has perhaps obscured the fact that the algae actually have their developmental and temperature optima somewhat lower, e.g. Brock & Brock (1966) found a peak of chlorophyll content in cores from the hot streams at 54 °C in Yellowstone and at 48 °C in Iceland and very much lower chlorophyll (i.e., biomass) at higher temperatures. They considered that temperature was the factor limiting growth and not available carbon dioxide or light.

Stewart (1970) showed that *Calothrix* from the Yellowstone thermal area could fix nitrogen over the temperature range 28–46 °C but not at higher temperatures. *Mastigocladus laminosus*, on the other hand, could fix nitrogen up to 54 °C, with an optimum near 42 °C. Since nitrogen is in relatively low supply in these hot spring waters such fixation may be important both for the blue-green algae and indirectly for others.

A further complicating factor in some hot springs is their extreme acidity and in such environments the enigmatic *Cyanidium caldarium* occurs; enigmatic since it cannot be satisfactorily classified in spite of considerable work. Its optimum pH for growth is 2.0, no growth occurs above pH 5.0 (Ascione, Southwick & Fresco, 1966) and it has a temperature optimum for ^{14}C incorporation of 45 °C

Fig. 5.29 Growth rates of eight clones of thermophilic *Synechococcus* sp. at 5 °C intervals. Clones are classified into strains each characterised by different temperature optima, 45, 48 and 53, Strain I; 55 and 60, Strain II; 66 and 71, Strain III; 75, Strain IV. No clones have absolutely identical temperature optima. From Peary & Castenholz, 1964.

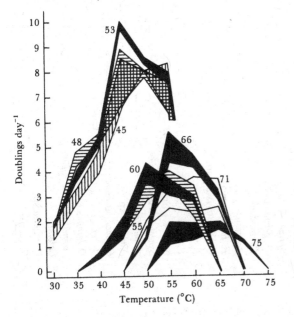

Table 5.9. *The algal composition along a temperature gradient at Jackson Hot Springs, Montana. From Kullberg, 1971*

Taxa	Temperature (°C)																						
	58.0	57.0	56.7	54.0	52.8	52.0	51.0	50.5	49.5	48.0	46.0	46.0	44.5	43.5	42.0	42.0	41.0	40.0	39.0	38.5	37.5	37.0	36.0
Synechococcus arcuatus	×	×	×	×																			
S. elongatus	×	×	×	×	×	×	×	×	×	×	×	×	×	×	×	×							
S. lividus	×	×	×		×	×	×	×	×	×	×	×	×	×		×							
S. viridissimus	×	×		×	×	×	×	×	×	×	×	×		×	×	×	×	×		×	×		×
S. eximus		×					×		×	×	×		×										
S. lividus var. *curvatus*		×																					
S. vescus		×		×		×			×	×													
S. vulcanus		×														×							
S. lividus var. *nanum*																		×					×
Oscillatoria boryana			×	×	×	×	×	×	×	×	×	×	×	×	×	×	×						
Mastigocladus laminosus				×		×											×						
S. elongatus var. *vestitus*																×							
Isocystis pallida					×	×	×	×	×	×	×	×	×	×	×	×	×	×	×				
Spirulina corakiana																×							
O. geminata					×	×	×	×	×	×	×	×	×	×	×	×	×			×			
Phormidium													×										
angustissimum					×	×	×	×	×	×	×	×	×	×	×	×	×	×	×	×	×		
P. laminosum					×	×	×		×	×	×	×	×	×	×	×	×	×	×	×			
Cylindrospermum sp.					×	×						×			×	×		×			×		
P. lignicola									×	×	×	×	×	×	×	×	×	×	×				
P. tenue										×			×		×								
Synechocystis minuscula														×	×	×	×						
Synechococcus cedrorum										×	×			×	×	×	×						
Achnanthes sp.												×	×	×	×	×	×	×	×	×	×	×	
Navicula sp.													×	×	×	×	×	×	×	×	×		
Chroococcus minutus														×	×	×	×		×	×			×
Aphanothece castagnei															×	×				×			
O. geminata var.													×		×		×		×				
tennella forma *minor*																			×	×	×		
Calothrix thermalis																				×	×	×	
Pinnularia sp.																		×					
O. brevis																						×	×

	Ulothrix sp.	Chroococcus minor	C. turgidus	Chamaesiphon prescottii	Spirogyra sp. (54 μ)	Oedogonium sp.	S. sp. (27 μ)	Gomphonema sp.	Denticula sp.	Aphanothece saxicola	Oscillatoria geminata var. fragilis forma breve	Pleurosigma sp.	Microcystis densa	C. minimus
	×		×		×	×							×	×
	×	×			×		×							
	×	×			×	×					×			
		×	×		×				×	×	×			
						×								
	×					×	×	×	×					
		×	×	×	×	×								
	×	×	×	×	×									

(Doemel & Brock, 1970). In acidic hot springs, blue-green algae are not usually found and *Cyanidium* is the dominant alga. The bacterium *Bacillus coagulans* and the fungus *Dactylaria gallopava* are common components of the mat-like growth associated with *Cyanidium*. Bacteria can tolerate temperatures well above those at which Cyanophyta grow, in fact some can survive brief exposure at boiling point. The biomass of the fungus and of *Cyanidium* seem to run parallel in the habitats sampled but the bacterial count did not appear to be correlated with the growth of these components (Belly, Tansey & Brock, 1973). The cells of *Cyanidium* excrete 2–6% of the total carbon fixed and, although not investigated, it seems likely that movement of fixed carbon into the bacteria and fungi occurred, since they grew well in mixed culture with the alga or with extracts of the dead alga. Some springs with high sulphide levels do not support the usual species and have bacteria only above 50 °C and the blue-green *Spirulina labrynthiformis* below this temperature (Castenholz, 1977). This blue-green is acclimatised to the high sulphide levels which poison other species.

Extracellular excretion of organic compounds has been reported also for *Synechococcus lividus* in short-term tracer experiments in the field (Bauld & Brock, 1974) and these workers showed that the excreted compounds were taken up by bacteria living in close association with the blue-green alga. Earlier, Moser & Brock (1971) had reported a layer of flexibacteria beneath *Synechococcus*, and these too may be in a mutualistic relationship with the blue-green algae.

An interesting problem is how these hot spring species become so widely dispersed when they have growth requirements which can hardly be met during aerial or water-borne transport. They might simply survive at the lower temperatures during transport since they certainly could not metabolise. A less likely alternative for their disjoint distribution is that they have evolved independently from non-thermophilic strains. It is interesting that Castenholz (1969) found only 6–8 species of Cyanophyta in hot springs in Iceland whereas there are many times this number in *continental* thermal waters and he suggests that the explanation may lie in inadequate dispersal, though he adds that low winter light intensity may present a further stress at the latitude of Iceland. A further possibility is that the thermophilic species

could be simply relict populations from the early stages of colonisation of the planet when similar environmental conditions would have been more widespread (Brock, 1967). There is evidence that the Icelandic thermal waters were unaffected at least by the last glaciation but for continuity to be maintained they need to have been unaffected by all previous glaciations, which is unlikely. Whilst thermophilic algae do not develop in regions downstream from hot springs, some 'propagules' have been recovered up to 100 km from the thermal region (Jackson & Castenholz, 1975) showing the viability of some cells for long periods under inclement conditions. These workers also found some thermophilic species in non-geothermal warm waters in Florida but experimentally the true hot spring species, could be selected out only by raising the culture temperature to 55 °C. A detailed study of the occurrence of species along the temperature gradient in the effluent from hot springs was made by Kullberg (1971) and Table 5.9 gives a representative example of the sequence of colonisation.

Snow and ice algae

Two rather disparate assemblages are dealt with in this section but they have in common the ability to grow at very low temperatures. The snow algal flora is composed of freshwater species which live on the surface of permanent snow. The ice flora on the other hand is marine and grows on the undersurface of pack ice at high latitudes. It may seem strange to include these in the chapter on sediments but the species encountered have the same general growth habit as those on organic and inorganic sediments. Matheke & Horner (1974) hold a similar view and comment: 'The ice is an inverted benthic habitat and many of the organisms living in the ice, especially the diatoms, behave like benthic organisms'. Some of the ice flora does fall to the bottom and continue growth on the sediments after break up of inshore ice but also some of the most abundant species do not and Matheke & Horner quite rightly comment that detailed studies are needed to determine the exact microhabitats involved.

Populations of algae are a constant and widespread feature of the surface of permanent snow fields in mountainous areas and of ice in the arctic and antarctic regions. Several community terms have been suggested, e.g. c(k)ryoplankton, but this does not seem appropriate, nor does the term c(k)ryoseston. Cryovegetation is probably the least disputable term whilst cryobiosis can be used when referring to the totality of organisms. Kol (1968) has given a most comprehensive account of cryovegetation and in a table recording the earlier reports of 'red snow' gives the date 1585 as the first record, in the Davis Straits. Not surprisingly perhaps, another record in 1875 was made by that remarkable naturalist, Charles Darwin, who first saw the phenomenon in 1835 in the mountains of South America and recognised that it was caused by a unicellular organism he identified as *Protococcus nivalis*. This alga, now known to be *Chlamydomonas nivalis*, is still the most frequently recorded alga from permanent snow fields. It is the aggregations of *Chlamydomonas nivalis* resting cells which store the pigment astaxanthin (3,3' dihydroxy-4',4' diketo-β-carotene) and give rise to red tinted snow which according to Kol (1968) was mentioned by Aristotle in *Meteorologies*. Although not many genera have been recorded or described from this habitat (*see* Fig. 5.30 for some examples), there are probably many confusions and many taxa, e.g. *Scotiella*, *Cryocystis*, *Carteria*, etc. need to be re-examined since they are likely to be stages in the life cycles of other volvocalean algae (Hoham, 1974). Hoham & Mullet (1978) have now shown that most *Scotiella* species are simply zygotes of *Chloromonas*. Some in fact may prove to be even further removed from their original taxonomic position, e.g. the taxon *Trachelomonas kolii* has been shown by Hoham (1974) to be a new genus of the Volvocales, *Chlainomonas*, and the common snow organism often classed as an alga '*Chionaster*' is in fact a fungus (Stein & Amundsen, 1967). A recent report of red snow from British Columbia involves yet another new species, *Sphaerellopsis rubra* (Stein & Brock, 1964).

Chlamydomonas nivalis tends to be concentrated on the snow surface, but Thomas (1972) also found cells down to 30 cm below the surface and since there was enough light for photosynthesis down to 50 cm, these cells may be active, although he tentatively concluded that carbon dioxide may limit *in situ* photosynthesis. Whereas algae at such depths in soil are probably in a resting stage, evidence that the deep populations in snow may be active is reported by Fogg (1967) who found actively motile cells of *Chlamydomonas nivalis* and green cells of *Chlorosphaera*

antarctica down to 25 cm in snow on Signy Island. As Fogg suggests, increase in cell numbers in snow is sometimes a result of concentration of cells at the surface as snow is removed by ablation (sublimation) leaving the algae behind. Experiments on ^{14}C fixation gave very low uptake rates, about 10 mg cm snow surface^{-2} day^{-1} which is below the lower end of the range quoted for extremely oligotrophic lakes in Swedish Lappland, 20–100 mg cm^{-2} day^{-1} (Rodhe, 1958). The important feature is that samples in which red spores were abundant did not differ from other samples as far as photosynthesis was concerned and so the spores themselves are not metabolically 'resting'. Thomas also reported that high light intensities at the surface inhibited ^{14}C uptake, so these organisms apparently have no protective mechanism in spite of living in such an exposed habitat and are affected in the same way as is the phytoplankton. Interestingly, uptake was also restricted by melted snow water. This could be due to high temperature or to reduction of carbon dioxide supply in the aqueous medium since at the pH of snow (4–5) the carbon source will be mainly carbon dioxide rather than bicarbonate. The surface of snow must be one of the more extreme of algal habitats with its low temperature (though summer temperature above the snow can be quite high), great fluctuations of insolation and low nutrient status. Surprisingly, it is one of the few extreme habitats which has escaped colonisation by cyanophyta. The source of nutrients is wind-blown dust and rain-trapped ions carried down by precipitation, a characteristic of which is that nitrogen is likely to be in short supply. This may explain the presence of astaxanthin in the snow flora since in other habitats (e.g., rock pools) carotenoid formation is enhanced at low nitrogen concentrations (Droop, 1955). Alternatively some workers have suggested that the red pigment is developed as a protection against short-wave radiation but the inhibition of photosynthesis at high light intensity belies this.

Fjerdingstad, Kemp, Fjeringstad & Vanggaard (1974) analysed *Chlamydomonas nivalis* from snow in Greenland and found that it was capable of concentrating many elements in its cells from the low concentration in the snow water; this was particularly noticeable with silica and iron.

The snow alga *Raphidium nivale* has been cultured and exhibits pleomorphism as do so many

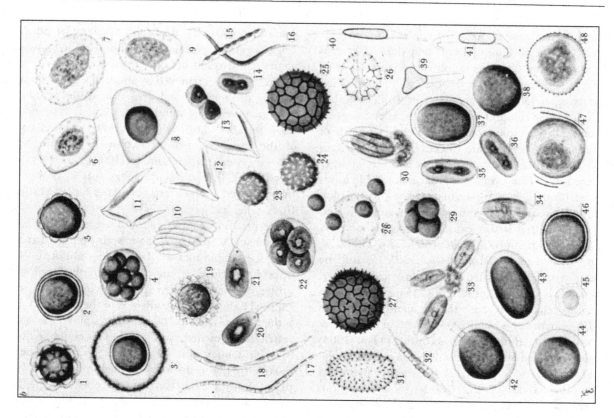

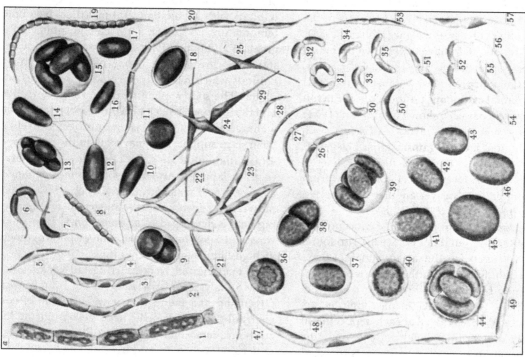

Table 5.10. *The species which occur in surface snows of different colours from sites in the South Orkney Islands. Red spores include those from* Chlamydomonas nivalis *and* Chlorosphaera antarctica *and numbers for these species are for green cells only. From Fogg 1967*

Species	Cell no./mm³ surface snows					
	Green snow, 30. i. 66, Meier Point, Coronation Island	Green snow, 13. ii. 66, Moraine Valley, Signy Island	Yellow snow, 4. ii. 66, Factory Cove, Signy Island	Yellow snow, 12. ii. 66, Moraine Valley, Signy Island	Red snow, 14. ii. 66, Moraine Valley, Signy Island	Red snow, 4. ii. 66, Factory Cove, Signy Island
Chlamydomonas nivalis	955	0	0	0	0	0
Chlorosphaera antarctica	15	0	0	1	3	0
Red spores	0	0	0	0	40	45
Raphidonema nivale	115	4770	364	289	2	320
Hormidium subtile[a]	768	0	0	0	0	0
Trochiscia antarctica	2	0	0	0	0	0
Chodatella brevispina	0	12	0	0	1	10
Scotiella antarctica	0	0	33	3	0	55
Ochromonas (?) sp.	32	12	2620	3	1	4170

[a] *Ulothrix subtilis* Kütz. of Fritsch.

other chlorococcalean algae. As expected it seems to be adapted both to low temperature (growth was greater at 5 °C than at 10–20 °C) and to high light intensity (Hoham, 1973). Earlier work by Chodat (quoted in Kol, 1968) reported that *Chlamydomonas nivalis* moved actively and copulated at temperatures around 0 °C but that the activity ceased around 4 °C. Apparently the optimum conditions for many snow algae are found in melt water at temperatures just above freezing point. 'Red snow' (*Chlamydomonas, Trochiscia, Scotiella*) is found in snow at a pH around 5–5.5 whereas 'green snow' caused by *Car-*

Fig. 5.30. Illustrations of some common snow algae from the comprehensive review by Kol (1968). *a*, 1, *Ancylonema nordenskoldii*; 2–3, *Raphidonema sabandum*. Filament division; 4–5, *Koliella nivalis*; 6–7, *Koliella tatrae*; 8, *Raphidonema brevirostre*; 9–18, *Chlamydomonas nivalis*; 9, cell division, 10, 12, 14, vegetative cells with two flagella; 11, round cells, 13, 15, cell division, 16–17, oval cells with one flagella, 18, giant cell with thick cell wall; 19, *Raphidonema sabandum*; 20, *Raphidonema nivale*; 21, *Koliella hernina*; 22–23, *Koliella chodatii*; 24–25, *Koliella viretii*; 26–29, *Koliella alpina*; 30–35, *Chodatia tetrallantoidea* various developmental stages; 36–46, *Chlamydomonas sanguinea*; 36, commencement of cell division; 37, oval cell with thick cell wall; 38–39, cell division; 40–42, cell with two flagella; 43, 46, oval cell; 44, cell division with mucilaginous cell wall; 45, giant cell; 47–49, *Koliella bernina*; 50–52, *Koliella tatrae*; 53–57, *Koliella helvetica*. *b*. 1–5, *Chlamydomonas nivalis*, 1, old zygote; 2, resting cell with thick laminate wall; 3, spherical cell with thick mucilage envelope; 4, single sporangium; 5, young zygote; 6–9, *Smithsonimonas abbotii*; 6, development of warts on the vegetative cell; 7, 9, resting stage; 8, vegetative cell; 10, *Scotiella polyptera* var. *megallanica*; 11, 12, *Koliella chodatii*; 13, 14, *Mesotaenium berggrenii*; 15, *Raphidonema brevisrostre*; 16, *Koliella tatrae*; 17, *Raphidonema nivale*; 18, *Raphidonema nivale*; 19–22, *Chlamydomonas yellowstonensis*; 19, zygote, 20, 21, vegetative cell, 22, cell division; 23–29, *Trochiscia americana*; 23, 24, young cell, the development of warts; 25, 27, resting zygote? 25, cell with ribs and bristles; 27, cell with ribs; 26, empty wall of an autosporangium; 28, release of autospores out of an autosporangium; 29, young autosporangium; 30, 33, 34, *Scotiella nivalis*; 31, *Cryocystis brevispina*; 32, *Raphidonema sabandum*; 35, 36, *Ancylonema nordenskioldii*; 37, 38, *Chlamydomonas sanguinea*; 39, *Chionaster nivalis*; 40, 41, *Chionaster bicornis*; 42–44, 46, *Chlamydomonas bolyaiana*; 42, vegetative cell; 44, resting stage; 45, *Tetraedron valdezi*; 47, 48, *Groenlandiella nivalis*.

Table 5.11. *Species recorded from the Antarctic sea ice (Bunt & Wood, 1963) and from the Arctic sea ice (Meguro, Ito & Fukushima, 1966)*

Antarctic species	Arctic species
Amphiprora kjellmanii	*Amphiprora kryophila*
A. oestrupii	*Gomphonema exigium* v. *arctica*
Biddulphia weisflogii	*Navicula algida*
Coscinodiscus subtilis	*N. crucigeroides*
Eucampia balaustium	*N. directa*
Fragilaria linearis	*N. gracilis* v. *inaequalis*
Nitzschia martiana	*N. kjellmanii*
N. seriata	*N. obtusa*
Pleurosigma antarcticum	*N. transitans* and v. *derasa, erosa*
Rhizosolenia alata	*N. trigocephala*
R. rostrata	*N. valida*
	Nitzschia lavuensis
	Pinnularia quadratarea and
	v. *biconstricta, capitata, stuxbergii*
	P. semiinflata and v. *decipiens*
	Pleurosigma stuxbergii and
	v. *rhomboides*
	Stenoneis incospicua

teria, *Chlamydomonas, Koliella, Raphidonema* and *Ulothrix* is found on snow of higher pH (6–6.5). These pH values depend on whether dust derived from siliceous or calcareous rocks is being deposited on the snow surface and although the pH values are quite close they do apparently result in the growth of quite different floras on the snow surface (Kol, 1968). Fogg (1967) gives a table of species occurrence in surface snow of different coloration (Table 5.10) but could not confirm Kol's report that green snow is characteristic of calcareous and red of siliceous snow fields. On the ice of glaciers the desmid *Ancylonema nordenskioldii* grows and, like its chlorophycean counterpart on the surface of bogs (*Zygogonium*), it also has purple vacuolar sap. Apparently *Ancylonema* is widespread in the northern hemisphere.

Temporary ice does not appear to support a true cryovegetation, though it has been rarely investigated. Schiller (1954) in one such study did find a flora which was composed of small coccoid and flagellate species such as are found in cold lakes (*see* p. 272); when cultured they grew best between 0.5 and 5 °C. These are almost certainly species from the water of the lake and not an exclusively ice flora.

Algae found frozen into the pack ice are apparently common in polar regions but may not contribute more than 2% to the total production (Bunt, 1967a). Nevertheless since the ice occupies a great area, about 2.6×10^6 km² according to Burkholder & Mandelli (1965) and fixes carbon at the rate of more than 900 mg m⁻³ h⁻¹, the total fixation around Antarctica alone would be 500 000 tons of carbon per day! The under-ice flora is rather loosely attached and can be easily washed away if the ice is disrupted (Bunt, 1963). This assemblage is quite separate from that of the plankton (Bunt & Wood, 1963) and its composition is listed in Table 5.11 together with a comparable flora from an arctic site (Meguro, Ito & Fukushima, 1966). Although said to be non-planktonic some of the antarctic species in the list grow planktonically in other regions. These algae grow in the brine cells between the ice crystals at a temperature of about − 1.75 °C and if the ice layer is melted and the chlorophyll estimated it is found to be about 100 times that of the underlying water (Meguro *et al.*, 1966). Figures of a similar order of magnitude are given by Clasby, Horner & Alexander (1973) for arctic sea ice. When the ice melts and the algae are released into the water they apparently cease photosynthesis. Possibly they are adapted to the brine cell environment or are senescent by this time, or are restricted by the higher water temperature. At least as far as salinity tolerance is concerned these algae can grow over a wide range of salinities from 5‰ (15‰ in one instance) up to 50‰ (Grant & Horner, 1976). Such high salinities are likely in brine cells in the ice. In the arctic, algae are also reported (Horner & Alexander, 1972) to form a layer, 2–3 cm thick, on the underside of one-year-old ice. This again is not a community which becomes planktonic when the ice melts. Diatoms are common (*Nitzschia frigida, Amphiprora hyperborea, Fragilariopsis oceanica, Cylindrotheca closterium*) but several small species of (*Proto?*)* *Peridinium, Eutreptiella, Platymonas* and *Cryptomonas* are also common. There is no doubt that taxonomic studies are needed on this community, since Bunt (1967b) isolated a number of algae into culture and recorded among them up to six unidentified *Fragilaria* species and four *Synedra* spp. The term 'bottom type plankton ice' has been used (Meguro, Ito & Fukushira,

---

* The marine species of *Peridinium* have recently been transferred to *Protoperidinium*.

1967) for the community in the lowermost layer of ice but I am sure that the word plankton should not be employed in this context.

This is yet another habitat where one might expect to find evidence of heterotrophy since the populations will be in the dark for many months. In fact, work on a *Synedra* sp. did show slow growth on glucose at low light intensities (8–10 lux) but then in later work (Bunt & Lee, 1972), no growth of other species could be detected on succinate, acetate, pyruvate, glycollate, citrate, lactate or glycine, again in low light. Horner & Alexander (1972) also determined that algae from the bottom of arctic sea ice did not take up labelled organic molecules but the bacteria did. On the other hand, several species from the ice flora are adapted to remarkably low light intensity (7 ft-c. compensation point) and are obligately psychrophilic* (low temperature requiring), e.g., about 10 °C is generally lethal (Bunt, 1967*b*, 1968*a*). An example is *Fragilaria sublinearis* which has its optimum for growth at 6 °C but can grow at −2 °C and in very low light (50–100 lux). The optimum temperature for respiration is higher than that for photosynthesis (Bunt, Owens & Hoch, 1966) and this is an obvious advantage in such low light–temperature habitats where conservation of energy is important.

Brown bands of material in the sea ice, such as were recorded by Fukushima (1961) and Meguro (1962) are thought to be derived from phytoplankton which becomes incorporated in the surface layers when the weight of snow depresses the blocks of ice. The phytoplankton is almost certainly the source of the large populations of *Phaeocystis* in some ice samples studied by Bunt (1968). It is not clear to what extent this community is metabolically active in the frozen state. It is of course protected from grazing and sinking losses and so, as Bunt & Lee (1970) comment, it may be an ideal situation in which to check the relationship between photosynthesis and yield; in fact at two sites they found organic carbon levels of about 4760 and 1100 mg m^{-2} in well-illuminated and shaded sites respectively and estimated photosynthetic efficiencies of 9 % and 0.3 %. The 9 % is a high figure but even higher ones are reported in some systems (*see* p. 464).

Summary

This chapter commences with a discussion of the *rhizobenthos* which in freshwaters consists almost entirely of Charophyta and in marine situations of the macroscopic Bryopsidophyceae with some small Rhodophyta growing by means of stolons and binding sand. Species of *Enteromorpha*, *Oscillatoria*, etc. also penetrate through sand and form a close turf over the surface. Many of the species of this community are calcified and contribute considerable amounts of calcareous debris to the sediments.

The *epipelon* is discussed in relation to its occurrence in springs, flowing water, lakes and marine coastal situations. The majority of the algae in these sites are motile since they must be capable of returning to the surface after burial. Species of flowing water epipelon have been reported in the *drift* of streams and the *diurnal* occurrence of these is discussed. Sampling techniques are given and some of the problems of quantitative measurements discussed. The marine epipelon is divided into that of *salt marsh* and of *sandy* shores and in the salt marsh section the extensive flora accompanying *Vaucheria* is noted. The lack of work on the very extensive subtidal epipelon which can extend down below 100 m is mentioned and general problems of depth distribution are discussed. Rhythmic upward and downward migration of epipelic species is dealt with in this chapter. These rhythms have been studied in both marine (estuarine) intertidal and freshwater situations and the control of migration shown to be under a biological clock system.

A short account of the communities of free living microscopic and macroscopic species growing as flocculent masses rolling over some sediments is briefly commented upon.

In some situations the algae growing on and in sediments, trap and bind mats of material, probably by deposition of calcium carbonate, to form *stromatolites*. These have been known for many years to geologists and are common in some strata but it is only recently that their natural growth has been studied. Cyanophyta are the predominant algae involved and stromatolites mainly form in warm, hypersaline waters.

---

* This term is used, in my opinion wrongly, by some workers, e.g., Reyssac & Roux (1972) to refer simply to low temperature species even though these may be tropical forms.

The algae growing on the surface of *soils* are in some ways similar to those of the freshwater epipelon (motile, unicellular, diatoms abundant, blue-green filaments present, etc.) but there is also a greater contribution by non-motile coccoid green algae and mucilaginous crust-forming blue-green algae. The flora and methods of study are discussed and then briefly the characteristics of desert and antarctic soils given as examples. Depth distribution, soil stabilisation, relationship to angiosperm growth and nitrogen fixation are noted.

The flora of *hot springs* is also mainly epipelic in nature and as in so many extreme habitats it is the blue-green algae which are abundant. The upper temperature limit and growth rates at different temperatures are discussed together with problems of dispersal.

Snow and *ice* algae though having little in common with the epipelon are treated here since they are associated with rather unstable surfaces. A *Chlamydomonas* species is the most abundant snow alga and colours the surface red (due to carotenoid pigment in the resting cells) whilst the under-ice flora is dominated by diatoms giving the ice a dirty brown coloration.

6 Phytobenthos: epiphyton, metaphyton, endophyton, epizoon, endozoon

In chapter 2 the phytobenthos was split into those communities growing on inorganic substrata (*see* chapters 3–5) and those on living substrata (i.e. on plants and animals) and these are the concern of this chapter. The tenacity of attachment of epibiota to the substratum varies and obviously the more actively moving hosts can only be colonised by algae having very strong holdfasts. The converse is however not true, since many cylindrical stemmed aquatic macrophytes which hardly move and live where water currents are slight also have very firmly attached epiphytes. This whole section of the benthos tends to be neglected in most ecological surveys mainly, one suspects, due to the difficulties of sampling.

The general terms epibiota (adj. epibiotic) or epiflora are sometimes used for the total community living on a surface but whenever possible it is desirable to use more specific epithets. Endobiotic applies to those algae living internally in other organisms. The usual word for epibiota on plants is *epiphyton*,[*] but if special plant organs act as host there is every reason to be even more specific, e.g. if leaves are involved the flora may be described as *epiphyllous*, if wood or bark as *epiphloeophyllous* (or *corticolous*) whilst for mangrove roots the term *rhizophyllous* has been used. These terms should be used as purely descriptive with no particular specificity being implied, unless or until this has been proven. The other descriptive terms listed in the subtitle have been even less commonly used but it is equally essential to specify position and host type.

---

[*] The term phyco-periphyton, which has been coined and used for epiphyton e.g. by Schlichting & Gearheart (1966), is as superfluous as the term periphyton itself. Equally the German term Aufwuchs whilst equally descriptive is better replaced by the widely accepted Latin designation which is also common in German literature.

Epiphyton
Aerial
The algal flora on terrestrial angiosperms has been very rarely recorded or studied. On leaves it is composed largely of genera of the Trentepohliales (*Trentepohlia, Cephaleuros, Phycopeltis*) which spread their branching filaments as felt-like wefts over leaf surfaces. They are often recognisable as light green or orange coloured patches on leaves of semi-tropical and tropical plants (*see* p. 27). The orange colour is due to carotenoid pigments dissolved in oil in the cells. Watson (1970) found a gradual increase in the number of *Trentepohlia* and *Phycopeltis* colonies over a period of three years on the fronds of the palm *Euterpe globosa*. Some of these green algae are associated with fungal filaments and constitute examples of loose lichen associations. They tend to occur in situations where the leaf surface is sufficiently moist for algal growth but not so wet as to encourage the growth of bryophytes over the leaf surface. Growth of a corticolous flora on tree bark is very common, yet has received hardly any attention; it is certainly not a simple flora of '*Pleurococcus*', e.g. Cox & Hightower (1972) found, in order of number of isolates, *Chlamydomonas, Chlorella, Chlorhormidium, Neochloris, Spongiochloris, Stichococcus, Tetracystis, Oscillatoria* and *Botrydiopsis* in cultures made from five different trees. Wylie & Schlichting (1973) report that gymnosperms support only Chlorophyta whilst other trees have Cyanophyta and a few diatoms. Petersen (1928), in his now classic work on aerial algae of Iceland, comments on the lack of a flora on trees and wooden buildings etc., in that country and attributes this to the extreme desiccation from wind which overrides the generally moist conditions which from the meteorological data suggests a suitable climate. Petersen was one of the first to recognise a whole series of unicellular species; he records

Table 6.1. *Association of algae on trees in a forest in Puerto Rico. From Foerster, 1971*

Host tree or tree fern	Algal association
Tabebruia rigida	*Stigonema/Chroococcus*
Tabebruia rigida	*Aulosira/Zygnemopsis*
Tabebruia rigida	*Phormidium/Frustulia*
Tabebruia rigida	*Stigonema/Frustulia*
Ocotea spathulata	*Hapalosiphon/Phormidium*
Ocotea spathulata	*Stigonema/Frustulia*
Ocotea spathulata	*Aulosira/Frustulia*
Miconia pachyphylla	*Stigonema/Chroococcus*
Cyathea pubescens	*Navicula/Frustulia*

*Apatococcus lobatus,** Botrydiopsis arhiza, Chlorella ellipsoidea, C. rugosa, Coccomyxa dispar, Desmococcus vulgaris,** Pleurococcus vulgaris, Stichococcus bacillaris* and *Troschiscia hirta* on the few wooden sites where algae grew at all. Here, as on most other subaerial objects, diatoms hardly figure in the flora probably owing to the lack of silica in the water moistening the habitat. Wylie & Schlichting list in order of frequency *Stichococcus bacillaris* (61%), *Protococcus viridis* (51%), *Chlorella* spp. (39%), *Chlorococcum* spp. (38%), *Trentepohlia aurea* (18%), *Coccomyxa dispar* (8%), *Trebouxia* (4%), *Hormidum* (2%), *Chlorosarcinopsis* (2%); *Nostoc sphaericum* (5%), *Anacystis* (5%), *Lyngbya* (4%), *Oscillatoria* (2%), *Tolypothrix* (2%), *Gloeocapsa* (1%), *Aphanothecae* (1%), *Gloeothecae* (1%), *Navicula* sp. (4%), *Hantzschia* (3%).

If the forest is wet enough it seems that algae normally associated with underwater habitats can colonise tree stems and Foerster (1971) found a zonation with diatoms near the base, Cyanophyta in the mid-zone and only *Chlorophyceae* in the upper zone of trees in an elfin cloud forest in Puerto Rico. He even found the estuarine diatom *Isthmia enervis* on one plant. Table 6.1 gives some of the associations described from trees and a tree fern in this forest. Foerster (1971) also found a mixture of *Diatoma–Stigonema–Pithophora* in the water in the leaves of the bromeliad *Vriesia sintensii* and this habitat is one which would repay further study though in the few bromeliads I have sampled no algae were present.

* See Brand (1925) for the first description of these genera.

A rather specialised aerial habitat is formed by moss and liverwort thalli, few of which yield no epiphytic algae when examined carefully. Green coccoid algae are quite abundant and have received little attention though I suspect some are found equally abundant on the soil surface. Blue-green algal colonies (*Aphanocapsa, Aphanothecae*) and filaments (*Lyngbya, Oscillatoria, Nostoc*) are also common. The third group of importance is the diatoms and samples from all parts of the world tend to reveal the same species complement. Beger (1928) was one of the first to study this habitat and even then pointed out the extent of studies on the moss fauna and the neglect of the habitat by botanists; he ranked the diatoms, in order of importance, *Pinnularia borealis, Hantzschia amphioxys, Melosira roeseana, Navicula contenta, Achnanthes minutissima, Navicula lanceolata, N. mutica, Achnanthes coarctata, Navicula cincta* and *Achnanthes exigua* and pointed out that in the wetter mosses these species tend to decrease. They seem therefore to have a definite requirement for low moisture content. Observations on the algal flora of dune mosses confirm this aspect in that 'hypnoid' forms growing in the damp 'slacks' supported some soil diatoms (contaminants?) and *Epithemia–Rhapalodia* (cf. other aquatics in dune slacks, p. 216) whereas the mosses of the dunes themselves were dominated by *Achnanthes coarctata, Pinnularia borealis* and *Hantzschia amphioxys* at both Harlech and Braunton (Round, 1958c). In samples collected in Ghana (Round, unpublished) the same genera were very common (Fig. 6.1) and these samples were taken mainly from high trees in the forest and could not have been contaminated by soil species.

Hantzschia amphioxys is also a common soil diatom but it is interesting that on these calcareous dunes it is scarce on the soil but abundant in the more acidic (?) moss clumps. These moss inhabiting diatoms are not like other epiphytic algae in that they are *motile* and the attached epiphytic habit exhibited by *Cocconeis, Gomphonema* etc. is hardly represented. Equally the non-motile araphid genera are also scarce and the assemblage seems to be characterised by very slight mobility. In a sense it is an extension of an epipelic type of flora on to plant surfaces but one which develops on the land and is almost unknown in the aquatic environment where in fact many more species are available to move up and colonise macrophytes; that they do not is

Fig. 6.1. Epiphytic diatoms from mosses collected from high branches of trees in Ghana. *A, B, Melosira roseana,* girdle and valve views; *C, Navicula mutica*; *D, Navicula* sp.; *E, Hantzschia amphioxys*; *F, Pinnularia borealis.* (Author, original.)

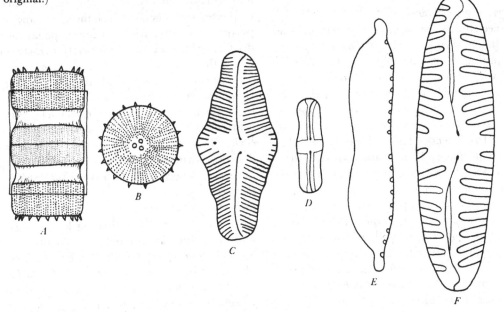

Fig. 6.2. Examples of epiphytic growth forms. *A,* Surface and side view of *Cocconeis* attached by a pad of mucilage; *B, Coleochaete,* growing as a disc of cells on the substratum; *C, Gomphonema,* attached by a mucilage pad and stalked; *D, Oedogonium,* attached by a pad produced from the basal cell; *E, Stigeoclonium,* upright filament attached by branching basal rhizoids. *C, D* and *E* are pedunculate genera and *A* and *B* are adnate.

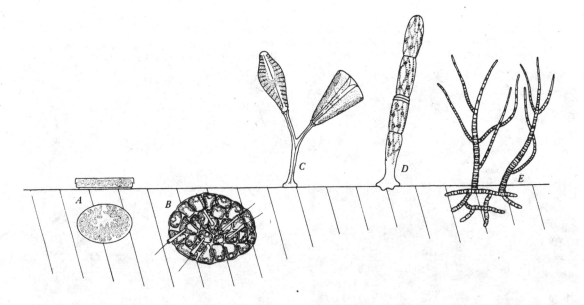

presumably related to their inability to withstand water movement past the hosts.

Streams, rivers and lakes

Detailed studies are required to determine the exact distribution of epiphytes in freshwaters – it will not be surprising if some are found to be adapted either to flowing or to 'still' waters. Two distinct life forms are exhibited by aquatic epiphytes: the adpressed habit (Fig. 6.2) with the cell, colony or filamentous system living in contact with the host epidermis and an extended habit with only the basal cell or a basal mucilage stalk in contact with the host. The former were termed sessile epibionts and the latter pedunculate epibionts by Margalef (1960) but I prefer the word adpressed rather than sessile since both kinds are sessile, although the adjective pedunculate seems highly appropriate. Examples of adpressed epiphytes are cells of *Cocconeis, Amphora, Epithemia, Coleochaete scutata, Aphanochaete* and of pedunculate epiphytes are *Gomphonema, Cymbella, Achnanthes, Oedogonium, Bulbochaete, Stigeocloneum,* etc. The adpressed type is probably living in a relatively non-turbulent environment and very much subject to secretion, etc., from the host whereas the pedunculate type is subjected to water flow past the thallus and may receive little in the way of nutrients,

inhibitors, etc. from the host. The pedunculate habit is perhaps also more subject to loss from mechanical and grazing effects. On many leaf surfaces the epiphytes are associated with inorganic and organic particulate matter which at times appears as abundant as the epiphytes themselves. *Elodea* leaves under the scanning electron microscope show very evident diatom and flocculent material (Fig. 6.3). Large macroscopic algae are hardly represented in the freshwater epiphyton but instead a large number of attached filamentous, colonial or unicellular algae occur. They are in dense wefts and so aggregated that they form visible masses around stems (e.g. on *Equisetum, Phragmites*) or on leaves (*Nuphar, Nymphaea, Ranunculus, Ceratophyllum*) or on leaves and roots of *Eichornia* and *Pistia*.

The undersides of floating leaves, e.g. *Lemna* and water lilies, are colonised but the upper surfaces are not. Some examples of the flora on *Lemna* are illustrated in Fig. 6.4. The diatom genera *Epithemia* and *Rhopalodia* are extremely common as epiphytes especially in base-rich water, e.g. in a permanent pond lying amongst sand dunes on the Lancashire coast the author found these dominant on *Equisetum, Mentha, Hydrocotyle, Littorella* and *Oenanthe* (Round, 1958a). In acidic habitats *Tabellaria, Eunotia, Actinella, Oedogonium* and *Bulbochaete* are common. Few attempts

Fig. 6.3. Scanning electron micrographs of the diatom community on the upper surface of *Elodea canadensis* leaves: *a*, young leaf (× 500) and *b* old leaf (× 1000). Photographs kindly supplied by D. Bell.

 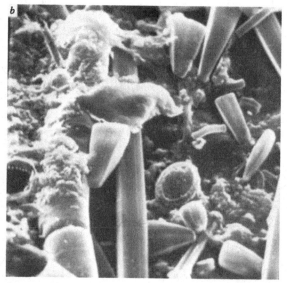

Fig. 6.4. Epiphytic diatoms from *Lemna* leaves and roots. *A, B, K, L, Achnanthes*; *C, Eunotia*; *D, I, Gomphonema*; *E, Fragilaria*; *F, Cocconeis*; *G, Navicula*; *H, Cymbella*; *J, Nitzschia*.

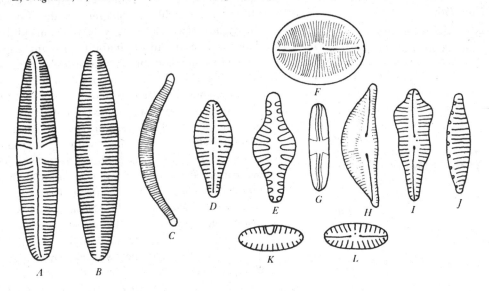

Fig. 6.5. Epiphytic algae growing on *Cladophora* filaments. The needle-like colonies are *Synedra*. The small stalked colonies on *a* are *Gomphonema*. The adnate cells on *b* and *c* are *Cocconeis* and the filamentous alga on *c* is the green alga *Aphanochaete* growing along the *Cladophora* cell.

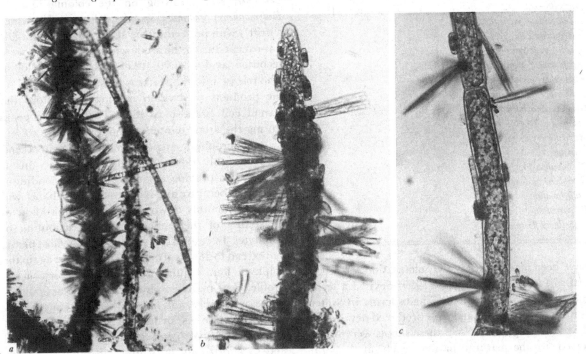

Table 6.2. *The order of occurrence and cover of the common epiphytes on* Cladophora glomerata. *From Chudbya, 1965, 1968*

Occurrence in order of dominance

Cell counts	Degree of cover
Diatoma elongatum v. *tenuis*	*Diatoma vulgare* v. *producta*
Synedra ulna	*Cocconeis placentula* v. *euglypta*
Fragilaria capucina	*Chamaesiphon incrustans*
Gomphonema olivaceum	*Diatoma elongatum* v. *tenuis*
Diatoma vulgare v. *producta*	*Gomphonema olivaceum*
D. vulgare v. *ehrenbergii*	*Cocconeis placentula*
Gomphonema olivaceum v.	*Cymbella ventricosa*
calcarea	*Achnanthes minutissima*
Diatoma vulgare	
Cocconeis placentula v.	
euglypta	
Nitzschia linearis	
Cymbella affinis	
Cocconeis placentula	
Diatoma vulgare v. *ovale*	
Nitzschia dissipata	
Amphora ovalis	
Cymbella ventricosa	
Chamaesiphon incrustans	
Navicula tuscula	
Synedra vaucheriae	
Gomphonema angustatum	
Achnanthes minutissima	
Rhoicosphenia curvata	
Gomphonema intricatim v.	
pumila	
Navicula cryptocepha v.	
veneta	
N. pupula v. *rostrata*	
Gomphonema longipes v.	
subclavatum	
G. parvulum	
Tabellaria flocculosa	

have been made to study the colonisation of 'clean' leaf surfaces though Prowse (1959) devised a technique for growing aquatic angiosperms in water containing silver ions and thus produced new shoots devoid of epiphytes. When these shoots were transferred to the natural habitat each angiosperm induced a distinctive flora: *Enhydrias* favoured *Oedogonium*, *Najas* favoured *Eunotia*, whilst *Utricularia* bore the heaviest growth of *Gomphonema*.

Algae, especially the filamentous genera, are themselves hosts to secondary epiphytic species, although the highly mucilaginous walls of many of the *Zygnemaphyceae* are relatively free from epiphytes. This may be due to the inability of many algae to attach to a mucilaginous surface or to the production of tannins which has been reported for these species. Tannins could not be detected in extracts of *Cladophora* though they could in extracts from *Spirogyra* and *Mougeotia* and these extracts were bacteriocidal though tests were not made with algal epiphytes (Pankow, 1961). Genera such as *Oedogonium*, *Bulbochaete* and *Cladophora* (*see* Fig. 6.5) are not conspicuously mucilaginous and are often richly coated with *Achnanthes*, *Cocconeis*, *Rhoicosphenia*, *Gomphonema*, *Characium*, *Aphanochaete*, *Microthamnion*, *Plectonema*, *Chamaesiphon* and *Epipyxis*. Chudyba (1965, 1968) found a total of 220 species of epiphytic algae on *Cladophora glomerata* in a river and of these 176 were diatoms, 27 *Chlorophyceae* and 19 *Cyanophyceae*. Table 6.2 gives the dominants based on cell counts and cover.

In streams in the Swedish arctic the epiphytic *Ceratoneis arcus* growing on the colonial diatom *Didymophenia geminata* is released into the water ('drift') with peak numbers appearing around midday except during the arctic winter when peaks were recorded at 4.00–6.00, 12.00, 18.00 and 22.00 h. The release into the water is presumably caused by the products of division moving away from the parent cell (*see* also p. 161). The multiple peaks during the arctic winter are not explained.

Macrophytic angiosperms in freshwaters tend to occur only down to depths of 10 m; this is sufficient to provide a range of habitat conditions (light, temperature and mechanical disturbance will vary considerably down the profile) but I know of no studies of their effect on the distribution of epiphytes. In contrast, in marine sites the host plants can extend to depths of over 100 m and here again the epiphytic flora requires detailed study down such a profile. How many epiphytes occur also as epilithophytes is simply not known but certainly a few species colonise one or the other, e.g. Kann (1973) found *Chaemaesiphon minutus* on the living substrata and *C. geitleri* only on stones and other inorganic substrata. Few workers have attempted to determine the contribution of epiphytic production to a lake system as a whole but there is an interesting study by Allen (1971) who found that an emergent angiosperm had an underwater flora which fixed carbon at 336 mg m^{-2} day^{-1} and a completely submerged angiosperm fixed 258 mg m^{-2} day^{-1}. The epiphytes as a whole contributed 31 % of the total littoral production and 21.4 % of the lake as a whole. Such high values are of course only possible in shallow lakes and although production may be intense in the shore zone of large lakes it is not likely to contribute more than 1–2 % to the total. Allen was also able to show that ^{14}C labelled products of *Najas* metabolism could be taken up by the epiphytes and without doubt there is a biochemical web of interaction in the dense epiphytic growth. Some very interesting work on the Norfolk Broads showed that the excessive development of epiphytes (and smothering of plants by filamentous algae such as *Spirogyra*) can lead to a dramatic decline in the macrophyte populations which have been all but eliminated in some waters. Previously it was thought that shading by phytoplankton was responsible for loss of the macrophytes but Phillips, Eminson & Moss (1978) believe that epiphyte loading is responsible and that when the macrophytes are removed the phytoplankton develops (compare p. 278 where it is suggested that macrophytic growth tends to suppress phytoplankton). The epiphytes stress the host plants by reducing the light level and this happens when the nutrient loading (mainly nitrogen is effective) increases. By studying the concentration of epiphytes in sediment cores taken from the various Broads Phillips *et al.* showed that the increase in epiphytic loading has occurred mainly in this century and is associated with changes in agricultural practice and the influx of tourists, sewage etc. Linking with this relationship between epiphytes and phytoplankton is Confer's (1972) observation that the filamentous epiphytes in an experimental tank system were in direct competition for phosphorus and this is probably one aspect of their deleterious effect on the phytoplankton. A further indication of the high activity of epiphytes is Brock's (1970) finding that photosynthesis of infested *Utricularia* in waters in the Florida Everglades was much greater than that of the host alone.

Marine habitats

Many of the macroscopic seaweeds* themselves support an attached flora of algae and nowhere is this more obvious than on the large laminarians,

the stipes of which are often covered with a mixture of red, brown and green macroscopic algae. Another very obvious site is the surface of sea grasses (*Posidonia, Zostera, Cymodocea, Thalassia, Amphibolis, Halodule,* etc.) where the brown and red epiphytes stand out from the green angiosperm leaves and crustose calcareous red algae form superficial discs on both leaf surfaces.

In general the degree of colonisation is positively correlated with the age of the host and protection from wave action. In rock pools, the older algae can be almost obscured by the growth of epiphytes, whilst plants growing on the exposed rocks are much less colonised and usually the epiphytes are confined to the lower, more protected parts of the hosts. An exception however is *Ascophyllum nodosum* which grows in the exposed intertidal and is host to *Polysiphonia (Vertebrata) lanosa*, though it is perhaps significant that this is the only common epiphyte on *Ascophyllum* and is confined to this host. Old *Fucus* stipes are frequently coated with 'ectocarpoid' algae and with *Enteromorpha*. The more spectacular epiphytic growths occurring on laminarian stipes include *Palmaria (Rhodymenia) palmata, Phycodrys rubens* and *Membranoptera alata* on shallow water plants, *Cryptopleura* and *Phycodrys* at moderate depth and, in deep water, only *Cryptopleura* (Smith, 1967). Smith also found that the epiphytes occurred on the narrow side of oval stipes but could offer no explanation for this. Off Heligoland the *Laminaria* forest supports a rich growth of epiphytic algae, e.g. *Membranoptera alata* and *Polysiphonia urceolata* on *L. hyperborea*, but at the lower depths this flora tends to be less well developed and there the *Laminaria* is covered with hydroids and Bryozoans. The stipes of *Laminaria digitata* and *L. saccharina* are less well colonised and then mainly on their lowermost parts (Lüning, 1970). Kain (1971) lists four species as being particularly important epiphytes on *L. hyperborea*; these are *Palmaria (Rhodymenia) palmata* (especially in shallow waters), *Phycodrys rubens* (more important in deep water), *Ptilota plumosa* and *Membranoptera alata*. In addition *Delesseria sanquinea, Odonthalia dentata, Cryptopleura ramosa, Plocamium, Pterosiphonia, Callophyllis, Ptilothamnium* and *Plumaria* also occur. Of 19 epiphytes recorded on *Saccorhiza* only two occurred on

---

* Some workers (e.g. Linskens, 1963*a, b*) use the term basiphyte for the host plant.

Table 6.3. *The seasonal occurrence of the common epiphytes on* Saccorhiza polyschides *at Port Erin, Isle of Man. From Norton & Burrows, 1969*

	June	July	Aug.	Sept.	Oct.	Nov.	Dec.	Jan.	Feb.	Mar.	Apr.	May	June	July
Ectocarpus siliculosus (Dillw.) Lyngb.	+	P	·	·	P	·	·	·	·	·	·	·	·	·
Rhodymenia palmata (L.) Grev.	+	+	+	+	+	·	·	·	·	·	·	+	+	+
Phycodrys rubens (L.) Batt.	·	+	+	+	+	·	·	·	·	·	·	·	·	·
Polysiphonia elongata (Huds.) Spreng.	·	+	+	+	+	·	·	·	·	·	·	·	·	·
Delesseria sanguinea (Huds.) Lamour.	·	+	·	·	·	·	·	·	·	·	·	+	·	·
Cryptopleura ramosa (Huds.) Kylin ex Newton	·	+	+	+	+	·	·	·	·	·	·	+	+	+
Ceramium rubrum (Huds.) C. Ag.	·	+	+	+	·	·	·	·	·	·	·	·	·	+
Saccorhiza polyschides (Lightf.) Batt.	·	+	+	+	+	·	·	·	·	·	·	·	+	+
Cutleria multifida (Sm.) Grev.	·	+	+	+	+	·	·	·	+	+	+	+	+	+
Ectocarpus fasciculatus Harv.	·	·	+	+	P	·	·	·	P	B	P	P	P	P
Chilionema ocellatum (Kütz.) Sauv.	·	·	P	·	·	·	·	·	·	·	·	+	+	P
Giffordia hincksiae (Harv.) Hamel	·	·	P	P	P	·	·	·	·	P	P	P	P	P
Litosiphon laminariae (Lyngb.) Harv.	·	·	U	B	+	·	·	·	·	·	·	+	P	U
Laminaria spp.	·	·	+	+	·	·	·	·	·	·	·	+	+	+
Epilithon membranaceum (Esper) Heydr.	·	·	·	·	·	·	+	·	·	T	+	·	·	·
Laminariocolax tomentosoides (Farl.) Kylin	·	·	·	·	·	·	·	·	P	P	·	+	P	·
Rhodophysema elegans (Crouan frat. ex J. Ag.) Dixon	·	·	·	·	·	·	·	·	·	·	·	+	+	+
Total number of species	2	9	11	11	11	0	1	0	3	5	4	11	12	11

P, *plurilocular sporangia*; U, *unilocular sporangia*; B, *plurilocular and unilocular sporangia*; T, *tetrasporangia*; +, *vegetative*.

Chorda filum and seven others grew on *Chorda*. These investigations did not take into account the microscopic unicellular algae and it is also possible that pollution reduces the number of epiphytes, e.g. Bellamy, John & Whittick (1968) report fewer epiphytes on laminarians (except *L. hyperborea*) than do other workers.

The depth distribution of an epiphyte below low water mark may not correspond to that of its host; South & Burrows (1967) comment on the decrease in epiphytes on *Chorda* in deep water. Norton & Burrows (1969) record that *Saccorhiza polyschides* has the greatest *abundance* and *diversity* of epiphytes when growing in calm water, though some species e.g. *Ectocarpus fasciculatus* were more abundant only on plants in turbulent water. Table 6.3 gives the epiphytic flora on this host. *Saccorhiza* is an annual plant and so the whole epiphytic flora is developed within a single year.

These large plants afford more than a single niche for colonisation and thus different epiphytic species may colonise the frond, stipe and holdfast. In addition the different parts of the host may be subjected to quite different current regimes e.g. the extremely large *Macrocystis pyrifera* has its frond extending up into surface waters whilst the holdfast may be 20–60 m below the surface and therefore in the surge zone (*see* p. 101).

The seasonal distribution of epiphytes is partly tied to the seasonal occurrence and annual or peren-

Table 6.4. *The annual sequence of colonisation of* Chorda filum *by common epiphytes at Rhosneigr, Anglesey, 1964. From South & Burrows, 1967*

Species	Apr.	May	June	July	Aug.	Sept.	Oct.	Nov.	Dec.	Jan.
Bulbocoleon piliferum	+	+	+	+	+	+	+	+	+	+
Acrochaete repens	+	+	+	+	+	+	+	+	+	+
Ectocarpus siliculosus	·	+	+	+	+	+	+	+	+	+
Litosiphon pusillus	·	+	+	+	+	+	+	·	·	·
Ceramium rubrum	·	+	+	+	+	+	+	+	+	+
Polysiphonia violacea	·	+	+	+	+	+	+	+	+	+
Sphacelaria pennata	·	·	+	+	+	+	+	+	+	+
Chylocladia verticillata	·	·	·	+	+	+	+	+	+	+
Dictyota dichotoma	·	·	·	+	+	+	+	+	+	+
Colpomenia peregrina	·	·	·	·	·	·	·	+	+	+
Melobesia minutula	·	·	·	·	·	·	·	·	+	+

nial nature of the host. However, superimposed on this is the fact that there is a succession of species colonising the host plant, an example of which is given in Table 6.4 for epiphytes of *Chorda filum*. This succession may be controlled by environmental factors such as water temperature or more subtle biological factors related to the surface and/or extracellular products of the host or to availability of sporulating stages of the epiphyte.

Since some of the epiphytes of plants such as *Chorda filum* are present on the previous year's growth, the transfer of the epibiota from old to new plants is possible. But this is apparently unlikely for species such as the brown alga, *Litosiphon pusillus* (known only on *Chorda*), which possibily overwinters as an inconspicuous juvenile thallus, originating from zoospores produced in unilocular sporangia. Yet other epiphytes, e.g., *Bulbocoleon piliferum*, can reproduce repeatedly during the year and re-infect the parent host or new hosts. Such studies emphasise the need for a thorough knowledge of life cycle, etc. when dealing with the ecology of these algae.

Relatively small hosts are also colonised but by small species: *Chaetomorpha area*, a uniaxial filament, can support a wide range of epiphytes apart from the ubiquitous diatoms, and Valet (1960), for example, records *Ectochaete, Entocladia, Ulvella, Pringsheimella, Erythrocladia, Erythrotrichia, Audouinella (Acrochaetium), Fosliella, Myrionema* and *Giraudya*. Many of these are small creeping filaments but some are branched systems of somewhat greater morphological complexity than the host. There is an almost universal epiphytism by diatoms on both large and small host species although the communities have never been adequately studied. Lee *et al.* (1975) point out the complexity on just one host, *Enteromorpha*, where they recorded, during the summer only, some 218 species on one salt marsh. The six dominants were *Fragilaria construens, Cocconeis scutellum, C. placentula, Achnanthes hauckii, A. pinnata* and *Amphora coffeaeformis* var. *acutiuscula*. These workers found that adding soil extract or acetone extracts of the thallus to agar cultures, produced greater numbers of species and colonies of diatoms from the *Enteromorpha* samples. They suggest a possible nutritional relationship and concluded that many organic substrates are recycled amongst the members of the community. The hardened mucilage of the tubes secreted by some diatoms also has an epiphytic flora, for example on *Berkeleya (Amphipleura) rutilans* growing on sediment after the breakup of the inshore ice, Matheke & Horner (1974) record *Licmophora ehrenbergii* and *L. oedipus* and Cox (1979) found *Synedra* in spring and autumn and *Licmophora abbreviata* and *Cocconeis scutellum* predominantly in the spring. No host specificity could be detected in a study of epiphytes in an intertidal estuary (Main & McIntire, 1974). Many algae extrude mucilage on to the surface and this undoubtedly affects colonisation by other algae, either providing substrates for the pioneer growth of bacteria and small coccoid algae or preventing the establishment of sporelings. Whatever inhibitory substances, etc. might be involved in natural situations, some epiphytes appear to be very resistant to poisons, e.g. Rautenberg (1961) found that species of *Ochlochaete*, *Streblonema*, *Microsporangium* and *Ascocyclus* could colonise test panels painted with extremely poisonous paints. The resistance may of course be apparent and no uptake be involved. Obviously all other 'paired' species e.g. *Litosiphon laminariae* on *Alaria esculenta* need to be checked and the exact factors which induce such obligate behaviour in nature determined.

This problem of host specificity is relatively unstudied and there is a general impression that many epiphytes colonise several hosts, but there is a great range from the extreme specificity of an alga to a single host through all stages to ubiquity. An interesting study of the diatom flora on *Halodule*, *Syringodium* and *Thalassia* in Mississippi Sound showed that at least the dominants (*Fragilaria hyalina*, *Mastogloia pusilla*, *Opephora pacifica*) were abundant on all three angiosperms (Sullivan, 1979). Some species grow both as epiphytes and epilithophytes, e.g. *Ectocarpus siliculosus* and *Giffordia mitchellae* off Port Aransas, Texas (Edwards & Kapraun, 1973); Edwards & Kapraun comment however that *most* epiphytes do not grow on rock and most show little host specificity. Recently the red algal epiphyte, *Smithora naidum*, which was thought to be an obligate epiphyte growing only on the sea grasses *Phyllospadix scouleri* and *Zostera marina*, has been grown under field conditions on artificial substrata (Harlin, 1973). This study showed the remarkable speed with which the alga could transfer from the natural host to the artificial substratum, small distromatic cushions being recorded on the latter after only one day in the field and development into upright blades after only two weeks. Seagrasses are not evenly colonised by *Smithora*: on *Phyllospadix* it grows over the entire leaf surface but only along the edges of *Zostera* leaves (Harlin, 1975). Harlin (1973) quotes similar work showing that *Myriotrichia subcorymbosum*, previously known only as an epiphyte on turtle grass (*Thalassia testudinum*), can also colonise plastic strips. Such results suggest that these algae do not require stimulatory substances from the algal hosts either for adhesion or subsequent growth. Their absence from other hosts may, however, be related to inhibitors formed by the basiphyte, but is it reasonable to conclude that all other possible hosts produce such substances? There is, however, a possibility that substances inhibitory to these epiphytes are given off

by algae and not by angiosperms, e.g., the halogenated compounds mentioned on p. 223.

The physical nature of the host surface may be a factor affecting selection, e.g., a comparison of *Laminaria digitata* and *L. hyperborea*, shows that the smooth stipe of the former has few epiphytes relative to the rough stipe of the latter. Linskens (1963*a*) measured the surface tension of some marine thalli and found that it varied from 25 dyne cm^{-1} (*Taonia*) to 355 dyne cm^{-1} (*Sebdenia*), but he could find no real relationship between this factor and the occurrence of epiphytes. The fact that the epiphytes tend to develop most abundantly in axils and constrictions (e.g. of *Corallina* and *Scytosiphon*) suggests trapping of spores and/or protection for germling stages.

In spite of the above experiments, showing that colonisation can occur on artificial substrata, the confinement of some species to one or a limited number of hosts is an ecological reality, requiring explanation. There must be a reason why, for example, four species (*Myrionema orbiculare*, *Giraudya sphacelaroides*, *Castagnea irregularis* and *C. cylindrica*) occur only on *Posidonia* in the Mediterranean whilst species such as *Melobesia lejolisii*, *M. farinosa* and *Dermatolithon* cf. *irregularis* occur on both *Posidonia* and other hosts (Van der Ben, 1971). Diatoms also exhibit some discrete preferences, e.g., the author found a species of *Amphora* attached only to the hairs protruding from the conceptacles of a *Fucus* spp. collected at Woods Hole, whilst Funk (1955) comments on the occurrence of *Navicula ostrearia* only amongst the hairs on the thallus of *Padina*. New growths are naturally free of epiphytes but colonisation has been rarely followed; it seems however to follow a similar pattern to that on artificial surfaces e.g. on *Nereocystis* the tube-dwelling diatom *Navicula grevillei* appeared first and was then followed by *Enteromorpha* and then *Antithamnion* (Markham, 1969). Presumably there is an even earlier coating of bacteria.

The host algae produce various extracellular products some of which have been shown to have antibiotic properties and could influence colonisation by epiphytes. Such an effect was shown to occur on *Sargassum*, where the branch tips are free of epiphytes; this is attributed to the production of tannins by the alga (Sieburth & Conover, 1965). The material secreted was found to have antibacterial properties, but this varied in samples taken on

Table 6.5. *Uptake of* ^{32}P *by epiphyte and basiphyte and transport to epiphyte and basiphyte. From Linskens, 1963*

Epiphyte labelled		Host plant not labelled		Ratio of specific activity in host and epiphyte
Species	Specific activity $(l\ min^{-1}\ mg^{-1})$	Species	Specific activity $(l\ min^{-1}\ mg^{-1})$	
Nitophyllum	93.1	*Codium*	15.1	5.4:1
Nitophyllum	119.0	*Codium*	17.3	7.9:1
Polysiphonia	11.6	*Codium*	6.3	1.8:1
Polysiphonia	42.3	*Codium*	17.3	2.4:1
Ectocarpus	30.0	*Codium*	21.9	1.4:1
Ulva	482.3	*Codium*	8.0	60.0:1
Basiphyte labelled				
Codium	11.48	*Nitophyllum*	50.01	1:4.3
Codium	10.02	*Nitophyllum*	45.75	1:4.5
Codium	17.24	*Polysiphonia*	87.29	1:5.7
Codium	19.63	*Polysiphonia*	96.75	1:4.9
Codium	13.34	*Ectocarpus*	40.28	1:3
Codium	—	*Ectocarpus*	38.0	—

different cruises, a factor which may be related to differences in temperature and/or the seasonal change in the number of physodes, which produce the tannin, in the cells. The yellow *gelbstoff*, phenolic substances, among which phloroglucinol and several unidentified flavenols of the catechin-type tannins have been found, are produced only by brown algae possessing physodes. These have been shown to be toxic to the unicellular algae *Monochrysis lutheri* and a *Porphyridium* sp. and also to have an effect on the growth of *Skeletonema*, *Amphidinium* and *Dunaliella* (Craigie & McLachlan, 1964). It was also suggested that growth of epiphytes may be adversely affected by such compounds (McLachlan & Craigie, 1964). Sieburth & Jensen (1969a) calculated that as much as 500–800 mg of phenolic material and 1.0–1.5 g of carbohydrate was exuded per kilogramme dry weight of algae in 24 h under laboratory conditions. Certainly the actively growing regions of fucoids are relatively free from epiphytes but it has still to be proven that these phenolic substances are effective at the *concentrations* found in nature. The epiphytic flora of the floating *Sargassum* is certainly species-poor compared with coastal *Sargassum*; Carpenter

(1970) found only thirteen species of diatoms and, of these, eight belonged to the genus *Mastogloia*, a diatom which lives inside mucilage bubbles attached to the substratum. In view of Linsken's (1966) and Harlin's (1973) work showing reciprocal transport between epiphyte and basiphyte (*see* p. 224) it would be valuable to investigate whether or not this transport aspect affects colonisation and presence or absence of epiphytes on certain hosts. It is interesting that Harlin (1973) found that there is an exchange of isotopes between host and epiphyte even though the epiphyte has no rhizoid, the epiphyte taking the substance in from the water before diffusion and water movement carries it away. Both workers also showed that there was transport not only between angiosperm host and alga but also between algae epiphytic on algae. All the evidence to date indicates very complex cycling of compounds within the dense 'forest' of epiphytes and attempts to relate epiphytic growth to environmental factors must take this into account. The study is of course complicated, since it may involve extracellular effects on spore attachment and/or growth of the germling stage, or at this early stage merely some surface property quite unrelated

Table 6.6. *The carbon and nitrogen uptake by roots of* Zostera marina *and its transport through the plant and into the leaf epiphytes. From McRoy & Goering, 1974*

	Carbon uptake (μg g^{-1} h^{-1})				Nitrogen uptake (10^2 % uptake h^{-1})			
	Roots and rhizomes	Stems	Leaves	Leaf epiphytes[a]	Roots and rhizomes	Stems	Leaves	Leaf epiphytes
Experiment I; ^{15}N added as NH$_4^+$								
Low [N]	2.69	1.76	0.07	(0.03)	3.35	0.39	0.32	7.8
High [N]	11.83	0.27	0.13	(0.01)	4.16	1.01	0.25	7.2
Experiment II; ^{15}N added as NO$_3^-$								
Low [N]	1.23	1.58	0.17	(0.01)	1.32	1.32	1.50	21.9
High [N]	5.91	0.24	0.08	(0.04)	1.82	1.33	1.88	6.5
Experiment III; ^{15}N added as (NH$_2$)$_2$CO								
Low [N]	—	—	—	—	0.025	0.033	0.051	8.7
High [N]	—	—	—	—	0.031	0.038	0.048	7.3

[a] Units for carbon uptake by epiphytes are μg per g leaf per hour.
Low [N], 33 μg atoms N l^{-1}. High [N], 67 μg atoms N l^{-1}.

to interchange of substances between basiphyte and epiphyte. One study of the sporeling stage (Beth & Merola, 1960) showed that swarmers of *Enteromorpha compressa* introduced into glass dishes containing pieces of *Dictyota dichotoma* or *Dictyopteris membranacea* grew on the glass dish but merely settled on the two brown algae and did not grow on them. However if the *Dictyopteris* was killed then the swarmers germinated and grew on dead tissue just as well as on glass. It is surprising how few experimental studies there have been on these interesting associations.

The problem of nutrient transference from basiphyte to epiphyte and vice versa was studied by Linskens (1963b) who showed that *Codium, Caulerpa* and *Cutleria* all take up radioactive phosphate and transport it from the base to the apex of the plant. Any epiphytes growing on a plant also received ^{32}P from the basiphyte and this transference was faster to *Polysiphonia* than to *Nitophyllum* and even slower to *Ectocarpus* (Table 6.5). The data also show that when the epiphyte was immersed in the solution containing ^{32}P, uptake was faster than into the basiphyte and there was also a slight transference of radioactive label to the basiphyte, but the main pathway seems to be in the opposite direction; similar results were obtained using *Himanthalia elongata* and *Pylaiella littoralis* (van den Ende & van Oorschot, 1963). Such experiments indicate that the relationships are more

complex than previously thought and it is not surprising therefore to find differences between attachment to live and inorganic substrata. Transport of photosynthate through laminarians (*see* p. 91) is further evidence in favour of metabolic sinks, which could influence colonisation, growth of epiphytes, etc. It would be interesting to compare growth rates of the same species on rock and on various plant surfaces, to identify and quantify the exchange of nutrients. In the case of marine epiphytes growing on *Zostera* it has been shown (McRoy & Goering, 1974) that the transfer of both radioactive carbon and nitrogen occurs from the sediment in which the *Zostera* is growing, through the macrophyte to the epiphyte (Table 6.6).

Where there is a clear specificity between epiphytes and host, e.g. of *Polysiphonia lanosa* and *Ascophyllum nodosum*, one might expect a flow of 'nutrients' from one to the other. However, Harlin & Craigie (1975) have shown that this is not the case for this pair, at least as far as transport of photosynthate is concerned. The *Polysiphonia* is quite capable of fixing its own carbon; obviously generalisations are dangerous in this field and every situation requires investigation. It is highly likely that the Rhodophyta which are such an exceptional group in their metabolism of the halogens chlorine, bromine, and iodine, produce substances incorporating these

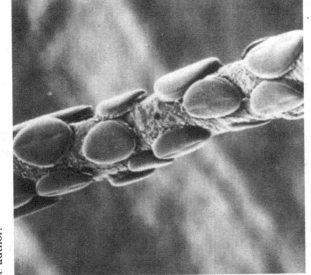

Fig. 6.6. Examples of diatoms epiphytic on blue-green algal filaments. Scanning electron micrographs taken by the author.

halogens which are released and affect epiphytes. The bromophenols formed by some species of the Rhodomelaceae are certainly toxic to some bacteria and other algae (*see* Fenical, 1975 for review). According to Fenical, some of the halogenated ketones formed by *Asparagopsis* and *Bonnemaisonia* are toxic to humans!

Large and small algae act as hosts to the smallest microscopic secondary epiphytes and whilst these may not be immediately obvious on a large *Laminaria* they can almost obscure filamentous species. Thus it is common to find species of *Cladophora*, *Plumaria*, *Polysiphonia*, *Ectocarpus*, etc. so densely covered by epiphytic diatoms that the green or red colour is overlain by brown. These diatoms attach by means of mucilage pads or stalks and their effect on the smaller host species must be every bit as drastic as that of the large epiphytes on the larger hosts. Epiphytes can occur also on the sheaths of Cyanophyta and Fig. 6.6. is a photograph of a *Lyngbya* filament colonised by diatoms. Again little is known about the distribution of the microscopic epiphytes; they appear at first sight to be relatively unspecific, e.g. Rautiainen & Ravanko (1972) found most of the common epiphytic species in all the belts of macroscopic algae in the Baltic, but comment that in pure *Cladophora* stands, *Cocconeis* is clearly dominant, and my own observations support the view that there is a greater degree of host specificity than generally recognised. Even on single *Cladophora* plants there is an apical–basal distribution, e.g. Jansson (1970) records *Synedra* and *Licmophora* on the upper branches, *Rhoicosphenia* and *Epithemia* near the base and *Cocconeis* equally all over. The herbivores grazing these epiphytes were also stratified, *Hydrobia* and *Chironomid* larvae near the base, and *Idotea* and *Proales* in the upper part; these samples were from brackish water, hence the strange mixture of genera. Other algal phyla involved in the marine epiphyton are the Cyanophyta, growing either as layers of coccoid cells or as filaments which tend to lodge in crevices, e.g. in the joints of *Corallina* species or they are scattered over the host. The Cyanophyta are particularly abundant epiphytes on tropical algae. Some Chlorophyta also occur e.g. coccoid *Chlorella*-like cells and creeping filaments belonging to the Chaetophorales.

The surfaces of sea grass leaves are coated by a layer of diatoms in some situations, whilst in others there may be a solid encrustation of calcareous Rhodophyta, e.g. *Fosliella farina*, *F. lejolisii* and *Melobesia membranacea* were recorded on *Thalassia* off Florida (Humm, 1964) and can reach such a density that the underlying leaf is totally obscured. Encrusting brown algae, e.g. *Ascocylus* and *Hecatonema* also colonise the leaf surfaces. The encrusting red algae are calcified and contribute a considerable amount of calcium carbonate to the sediment. The majority of epiphytic species on *Thalassia* are present throughout the year, e.g., the encrusting red algae, but a few such as *Cladosiphon occidentalis*, *Ulva lactuca*, *Ectocarpus confervoides*, *E. mitchellae* and several *Enteromorpha* spp. occur only during winter. Humm records a small number of species as exclusive or nearly so to sea grasses, e.g. *Eudesme zosterae*, *Cladosiphon occidentalis*, *Myriotrichia subcorymbosum* and *Stictosiphon subsimplex*. The total flora on these grasses is very large, 113 species on *Thalassia*, not including the more microscopic species such as flagellates and diatoms. The latter are often attached to the encrusting red and brown algae as secondary epiphytes. The blades of these grasses grow from a basal meristem and therefore the tips are the oldest parts and there tends to be a zonation along the leaf related to age, with the denser populations at the tips. The additional weight near the apices must cause a considerable strain on the plant and may be one of the factors contributing to the shedding of leaves. The shed leaves are frequently rolled up by tidal currents and end up as balls of fibrous material washed up on the shore. As long suspected, the epiphytic load affects the rate of photosynthesis of the host because of the decreased bicarbonate diffusion and light attenuation; the rate varies according to the bicarbonate concentration, at 0.2 meq/l^{-1} the reduction was 45% but it was only 15% at higher bicarbonate concentrations (1.17 meq) because of the increased diffusion gradient. At ambient sea conditions the reduction is about 31% (Sand-Jensen, 1977). These sea grasses form extensive meadows in many temperate–tropical environments and the algal growth on them contributes an appreciable amount to the total algal carbon fixation in shallow coastal seas. They often occur in silted regions where other macrophytic algae are lacking, and, together with the epipelic algae growing at such sites, replace the larger red and brown algal populations as the primary producers in the shallow water. The portioning of production between the sea grasses and the algal components was studied by Penhale (1977) and as might be expected the epiphytes fixed less carbon (0.65 mg g^{-1} h^{-1}) than eelgrass (0.88 mg g^{-1} h^{-1}) and on an annual areal basis, the figure for carbon fixed by eelgrass was 0.9 g m^{-2} day^{-1} and for carbon fixed by epiphytes was 0.2 g m^{-2} day^{-1}. These values are in line with earlier figures of Thayer & Adams (1975) who found carbon production was 300 g m^{-2} yr^{-1} for epiphytes and 350 g m^{-2} yr^{-1} for eelgrass at a site on the Atlantic coast of America. These together are very high rates of production and exceed the averages of corn, rice and hay crops in America. It has been suggested that excretions from the epiphytes are also recycled but Penhale & Smith (1977) showed that in the case of *Zostera* epiphytes only about 2% of their photosynthate is excreted.

The rhizomes of *Posidonia* growing in the Mediterranean also support a rich flora of Ceramiales and Boudouresque (1968) found a variation in the epiphytic flora with depth. At low light intensities a shade community of *Peysonnelia* and the bryopsidophycean alga *Udotea* develop whilst on the better illuminated rhizomes a *Plocamium* community is found.

Nitrogen fixing Cyanophyta (e.g. *Calothrix*) are often found on sea grasses and Goering & Parker (1972), using the acetylene reduction technique found that 2.4–16.5 μg nitrogen was fixed in the light each hour, per milligramme plant nitrogen. Capone & Taylor (1977) report low night-time fixation and a rapid increase at dawn. Such fixation could be important in these communities, which are highly productive and liable to nitrogen limitation. As might perhaps be expected the small epiphytes seem to metabolise faster than the host, e.g., Carpenter & Cox (1974) found that the cyanophyte *Dichothrix*, growing on *Sargassum* in the Sargasso Sea turns over carbon at a rate four times faster than does the *Sargassum*. Interestingly they found that both host and epiphyte had low carbon fixation rates (0.42 mg g^{-1} h^{-1} for *Sargassum* and 2.3 mg g^{-1} h^{-1} for *Dichothrix*) at southern stations, in water where vertical mixing of the water column was minimal, compared to more northerly stations (1.2 and 2.8 mg respectively) where there was greater mixing. However this could not be correlated with inorganic phosphorus or nitrogen levels. Conover & Sieburth (1964) found that the *Sargassum* population in the

centre of the Sargasso Sea gyre was devoid of epiphytes on the tips of their branches but north west of the centre, epiphytes became abundant, a finding which agrees with the fixation data.

Mangrove swamps

Mangroves produce two types of roots, those growing at an angle downward from the main-stem and the upright rhizophores (or pneumatophores) emerging from the sediment, and they both support algae. A rather special association of epiphytic algae occurs attached to the rhizophores of the mangroves *Avicennia*, *Rhizophora*, and *Laguncularis*. Børgesen (1911) suggested that this attached community be termed 'rhizophyllous'. In the shade near the trunk, where the water level fluctuates, there is a common association of red algae (*Caloglossa*, *Catenella*, *Bostrychia*, *Murrayella*) whilst below water level *Caulerpa verticillata* tends to coat the rhizophores except near the growing tips, which remain free of macroscopic algae (Almodóvar & Pagán, 1971). The rhizophores of the mangroves of some regions, e.g., the Hawaiian Islands, Mindora (Philippines), and the Thousand Islands (Indonesia) are devoid of algal growth, for as yet unexplained reasons. These areas are relatively free of intertidal mud, which is so common in the areas where the epiphytic algae grow, and which tends to coat the algae but not the rhizophores (Doty, 1957). The coating of mud is probably associated with mucilage produced by red algae; elsewhere it is common to find clay particles incorporated into the outer mucilage layers of many intertidal algae. In Puerto Rico, Almodóvar & Biebl (1962) recorded a distinct zonation of epiphytes on *Rhizophora* roots growing downwards from the branches; up to 1 ft above the high tide line *Catenella opuntia*, *Caloglossa leprieurii*, and *Bostrychia tenella* form a thick felt attached by short holdfasts. In the region of fluctuating water level *Murrayella periclados* and *Centroceras clavulatum* form a distinct community with some *Polysiphonia ferrulacea*, *Enteromorpha flexuosa* and *Rhizoclonium hookeri* intermixed. Roots growing downwards from the outermost branches and therefore in brighter sunlight than the roots nearer the trunk support a subtidal flora down to 4 ft below tide level, e.g. of *Acanthophora spicifera*, *Spyridia filamentosa*, *Hypnea musciformis*, *Laurentia obtusa*, *Wrangelia bicuspidata* and several species of *Valonia* and *Caulerpa*.

In the better illumination further out into the

channels other algae become established on the rhizophores (species of *Amphiroa*, *Valonia*, *Wrangelia*, *Acanthophora*, *Laurencia*, etc.). Directly beneath the rhizophores no macroscopic algae grow in the sediments but further out the flora develops with species of *Avrainvillia*, *Penicillus* and *Udotea* (a rhizobenthic association, *see* p. 158). The algae growing high on the rhizophores are subjected at times to freshwater and Almodóvar & Biebl (1962) found in experimental studies that these algae had the greatest osmotic resistance and could withstand immersion in distilled water whilst those on the submerged rhizophores had a much narrower osmotic tolerance.

On the roots of *Rhizophora mangle*, *Bostrychia montagnei* occurs in summer and *Monostroma oxyspermum* in winter (Taylor, 1959). On *Avicennia nitida*, *Rhizoclonium hookeri* occurs in the driest parts and *Bostrychia montagnei* on intertidal roots but in shade the algal flora is absent. A detailed account of the associations of *Bostrychia* and other species in various parts of the tropics is given by Post (1968).

Epiphytic bacteria

It is certainly no exaggeration to say that all natural populations of algae have epiphytic bacteria associated with their surfaces either in a casual manner or as specialised epiphytes. As Provasoli (1971) comments 'This epiphytism may conceal a relationship akin to symbiosis'. Filamentous bacteria of the *Leucothrix* type are common on marine algae forming hair-like outgrowths somewhat narrower than the outgrowths of the algae themselves (e.g. the trichoblasts of red algae). It is highly likely that the free-living bacterial flora is a major source of vitamin B compounds in nature and the importance of the bacteria so closely associated with algae needs careful study. Provasoli (1971) found that normal growth rate and morphology was not possible in some cultured marine algae unless these vitamins were added to the axenic cultures. The bacterial population on the surface of marine algae is more concentrated than in the surrounding water, e.g. 10^4–10^9 bacteria cm^{-2} on *Macrocystis* compared to 10^3–10^5 ml^{-1} in the water (Zobell, 1972). From the associated *Ascophyllum nodosum*–*Polysiphonia* (*Vertebrata*) *lanosa* some 25 bacteria and one yeast were isolated by Chan & McManus (1969). The populations of bacteria are different on the stipe, holdfast and lamina of *Ascophyllum* (Cundell, Sleeter & Mitchell,

1977) and the tips are relatively free. Scanning electron micrographs show a complete lawn of species covering the 3–4 yr old plants. Experimentalists should obviously be aware of this and its possible metabolic activity. Statements that bacteria occur on dead, but not on living diatoms, dinoflagellates and blue-green algae (Oppenheimer & Vance, 1960) need careful rechecking since electron microscope studies rarely fail to reveal attached bacteria which are not likely to attach during preparation for microscopy (but *see* also p. 396). These workers postulated that the algae produced bacteria-repelling electrokinetic potentials or antibiotic substances.

Epiplankton

A further interesting case of epiphytism which might be mentioned at this point is that of *Chlamydomonas*, *Stylosphaeridium*, *Phormidium*, *Nitzschia* etc., which frequently grow on and within the mucilage of freshwater planktonic algae such as *Microcystis*. The term epiplankton is sometimes applied to the algae growing in this habitat. However some workers (e.g. Petipa, Pavlova & Mironov, 1970) use the term epiplankton for the free-living forms of phytoplankton which live in the uppermost part of the euphotic zone, and the term bathyplankton for that in the lower part of the euphotic zone. This seems to me an unfortunate choice of words for 'epi' is usually used in the sense of 'on' 'upon' not 'upper' though by extension it might be so applied. A better term for an upper surface water community is epipelagic and it is certainly preferable though again does not quite convey the sense of a layer which is often 100–150 m thick. Some epiplankton species appear quite specific to their host, e.g. there is a *Chlamydomonas* species which lives in the thecae of *Dinobryon* colonies.

In addition to the obvious epiplankton there is a less conspicuous bacterial and fungal epiflora and also a rich epifauna of protozoa, etc. (*see* also p. 395), and these interact with the planktonic hosts to form a complex micro-ecosystem. Bursa (1968) recorded 30 species on a single filamentous cyanophyte (*Nodularia spumigena*) in the Baltic. The sheer volume of the total epiflora and fauna may 'sink' the host apart from any other effect.

Table 6.7. *Numbers and types of motile animals per dm² on triplicate samples of clean and variously encrusted surface kelp blades, collected 14 July 1958, La Jolla kelp bed. From Wing & Clendenning, 1971*

Animals	Clean mature blades			Lightly encrusted			Moderately encrusted			Heavily encrusted		
	a	b	c	a	b	c	a	b	c	a	b	c
Ostracods	31	45	35	38	5	59	12	192	268	209	589	739
Copepods	20	30	22	54	96	99	37	506	102	431	433	678
Gammarids	0	31	37	48	0	24	38	364	11	36	17	44
Caprellids	0	55	14	16	0	20	0	13	7	23	21	17
Polychaetes	19	10	10	12	0	20	69	16	66	85	165	117
Nematodes	0	0	0	0	10	0	0	0	0	27	56	85
Turbellaria	25	21	25	66	95	57	639	1005	291	1242	871	461
Molluscs	17	12	7	5	5	28	23	20	9	16	21	18
Miscellaneous	10	21	13	18	43	17	38	113	19	14	52	42
Totals	122	225	163	257	254	324	856	2229	683	2083	2225	2201

Epiphytic fauna

Associations between algae and animals without direct grazing are common yet rarely have the mutual effects been studied. Only brief mention can be made of these associations which would require a book written by a zoologist to cover the field. Sessile animals such as Bryozoa are often abundant on macrophytic algae, and must have some effect on the metabolism of the underlying alga. On *Macrocystis*, encrusting colonies of *Membranipora* hinder growth of the blade, producing distortion and stunting, whilst the weight of Bryozoans can lead to sinking of the fronds. Woolacott & North (1971) found 59 species of Bryozoans on Californian *Macrocystis*. Clendenning (1964) recorded 5 g dm^{-1} of *Membranipora* on completely encrusted *Macrocystis* and calculations of the total biomass on an average front (wt 1500 g; area 10000 cm²) vary from 400 g to 1000 g but in spite of this Woolacott & North found that photosynthesis is not reduced by the animal cover. Wing & Clendenning (1962) calculated an average surface area of 15 m² of *Macrocystis* frond per square metre of bottom (range 4–40 m). Green algae, growing within the *Membranipora* itself, actually raise the photosynthetic capacity of the combined system 50% above that of the kelp itself. Table 6.7 shows the numbers and types of animal associated with *Macrocystis* blades, but excludes microscopic species of protozoa, sessile bryozoa and hydroids, which are all common.

Studies of the animals associated with the *Macrocystis* holdfast reveal a considerable number of species, e.g. Ghelardi (1971) recorded twelve different amphipods, thirteen polychaetes and six isopods as occurring frequently in a small sample whilst the total number of species in live holdfasts was 128 and in the dead centres of holdfasts 156. Ghelardi also showed that a consistent and recurrent animal community was associated with young live holdfasts and with the dead centres.

The floating *Sargassum* is an obvious host for study since it is presumably uncontaminated by attached species from the epilithon, etc. though some confusion with the zooplankton may occur. Fine (1970) found 67 species of animals on 35 *Sargassum* samples and recorded a seasonal and geographical variation in the fauna. Wieser (1959) reports very varied numbers of animals on Mediterranean algae e.g. 580–2360 per g of supratidal Cyanophyta, 520–900 per g of supratidal *Cladophora–Enteromorpha*, 758–1464 on intertidal *Gelidum crinale*, 444–612 on *Laurencia papillosa*, and 90–488 on *Corallina mediterranea*. On subtidal populations down to 1 m depth he found, 408–447 on *Jania rubens*, 930 on *Gigartina acicularis*, 21–128 on *Cystoseira*, 134 on *Pterocladia clavatum*, at 3 m depth, 1965 on *Halopteris scoparia* and at 6 m depth on *Dictyota dichotoma* 2500–11670. The small numbers on *Cystoseira* and *Pterocladia* are rather surprising but some marine algae apparently produce poisons and Wieser quotes early workers who found

Table 6.8. *The fauna and flora of* Fucus serratus *plants at Ardkeen. From Boaden et al., 1975*

Protozoa
  Gromia oviformis Dujardin
Coelenterata
  Dynamena pumila (L.)*
  Gonothyraea loveni (Allman)*
  Obelia geniculata (L.)
  Actinia equina L.
  Sagartia spp.
  Tealia felina (L.)
Turbellaria
  Cycloporus papillosus Lang
  Leptoplana tremellaris (O. F. Müller)
  Oligocladus spp.
  Rhabdocoela: unidentified
Porifera
  Grantia compressa (Fabricius)*
  Halichondria panicea (Pallas)
  Halisarca dujardini Johnston
  Hymeniacidon perlevis (Montagu)
  Leucosolenia spp.
  Myxilla incrustans (Johnston)
  Spongelia fragilis
  Sycon ciliatum (Fabricius)
  Sycon coronatum (Ellis and Solander)*
Nemertini
  Unidentified spp.
Annelida
  Leptonereis glauca Claparède
  Phyllodoce maculata (L.)
  Syllidae: unidentified
  Spirorbis borealis Daudin*
  Spirorbis pagenstecheri Quatrefages*
  Spirorbis spirillum (L.)*

Crustacea
  Amphithoë rubricata (Montagu)
  Balanus spp.
  Calliopidae: unidentified
  Carcinus maenas (L.)
  Elminius modestus Darwin
  Gammaridae: unidentified
  Idotea granulosa Rathke
  Jaera nordmanni Rathke
  Jassa spp.
  Stenothoë monoculoides (Montagu)
  Tritaeta gibbosa (Bate)
  Leucothoë spp.
Pycnogonida
  Achelia simplex (Giltay)
  Nymphon gracile Leach
  Pycnogonum litorale (Strøm)
Mollusca
  Acanthodoris pilosa (O. F. Müller)
  Acteonia semestra Quatrefages
  Calliostoma zizyphinum (L.)
  Doto coronata (Gmelin)
  Doto pinnatifida (Montagu)
  Gibbula cineraria (L.)
  Nassarius incrassatus (Strøm)
  Littorina obtusata (L.)
  Onchidoris muricata (O. F. Müller)
  Pelta coronata Quatrefages
  Polycera spp.
  Lasaea rubra (Montagu)
  Musculus marmoratus (Forbes)
  Mytilus edulis L.
Polyzoa
  Alcyonidium gelatinosum (L.)
  Alcyonidium hirsutum (Fleming)*

  Bowerbankia imbricata (Adams)
  Bugula stolonifera Ryland
  Celleporella hyalina (L.)
  Celleporina hassalli Johnston
  Electra pilosa (L.)*
  Flustrellidra hispida (Fabricius)*
  Membranipora membranacea (L.)*
  Mucronella spp.
  Schizoporella unicornis (Johnston)
  Scruparia chelata (L.)
Echinodermata
  Asterina gibbosa (Pennant)
  Amphipholis squamata (Delle Chiaje)
Tunicata
  Ascidia mentula O. F. Müller
  Ascidiella spp.
  Botrylloides leachi (Savigny)
  Botryllus schlosseri (Pallas)
  Ciona intestinalis (L.)
  Dendrodoa grossularia (van Beneden)
  Didemnum maculosum (Milne-Edwards)*
  Polyclinum aurantium Milne-Edwards*
  Sidnyum turbinatum Savigny
Algae
  Ceramium spp.
  Chondrus crispus Stackh.
  Cladophora spp.
  Ectocarpus spp.
  Enteromorpha spp.
  Fucus vesiculosus L.
  Griffithsia spp.
  Laminaria digitata (Huds.) Lamour.
  Lithothamnion spp.
  Polysiphonia spp.
  Sphacelaria cirrhosa (Roth.) Ag.
  Ulva lactuca L.

* Indicates the commonest animal species

Table 6.9. *The distribution of the common animals on plants of* Saccorhiza *growing on the Port Erin breakwater, Isle of Man. From Norton, 1971*

Species	Nature[a] of count	Lamina	Stipe	On bulb	Under bulb	Inside bulb
Campanularia sp.	%	0	0	8.8	10.9	6
Obelia geniculata	%	6.7	2	0	0.2	0
Lepidonotus squamatus	No.	15	0	14	29	18
Platynereis dumerili	No.	8	0	1	25	29
Jassa falcata	%	0.2	0	5.4	8.2	6.2
Galathea dispersa	No.	18	0	5	26	23
Porcellana longicornis	No.	8	0	20	50	24
Patina pellucida	No.	167	19	16	21	85
Gibbula cineraria	No.	227	0	8	1	1
Ophiothrix fragilis	No.	0	0	7	13	6
Celleporella hyalina	%	0.2	0.5	4.2	10.5	2.2
Callopora lineata	%	0.3	1.1	0.9	13.1	5.9
Membranipora membranacea	%	18.6	17.7	9.6	2.6	0.6
Scrupocellaria reptans	%	0.2	0	7	8.4	1.6
Electra pilosa	%	1.6	7.6	7.2	5	2.8
Escharoides coccineus	%	0	0.2	0.2	9.2	2.8
Ascidiella aspersa	No.	0	0	1	48	8
Liparis montagui	No.	13	1	2	5	14

[a] No., the total number of individuals recorded on that region of the host plant. Sample size 700 plants. %, the percentage of host plants on which the animal species occupied that region.

that *Ceramium rubrum*, *Halidrys siliquosa* and *Dictyota dichotoma* yield substances poisonous to *Idotea*, yet *Dictyota* had the richest fauna at this station. The number of animals per square metre was calculated to be in the region of 10×10^6 at 1 m depth in the Mediterranean which compares well with other estimates e.g. 12×10^6 in the Bristol Channel, 2×10^6 in the Danish Waddensee and 10^6 at Banyuls.

Boaden, O'Connor & Seed (1975) studied the epibiota of *Fucus serratus* plants and found a total of 91 taxa associated with this common alga (Table 6.8). One of the most intensive studies of animals associated with algae is the work at Lough Ine by Kitching and associates (summarised in Kitching & Ebling, 1967).

Different parts of such macrophytes as *Saccorhiza* are colonised by different animals and on this host the lamina is dominated by *Patina* and *Gibbula* whilst the bulb has a much more varied fauna with some obvious preference for position either on, under or inside (Table 6.9). On *Sargassum* in Japan a distinct seasonal trend of associated non-attached animals has been recorded (Mukai, 1971) with numbers as high as $130-266 \times 10^3$ per plant. The majority were benthic copepods but nematodes were also common as detritus accumulated on the *Sargassum*. This fauna is not strictly epiphytic but is related to the cover afforded by the host alga and provides yet another aspect of algal ecology.

Initial colonisation and microdistribution

The primary colonisers of submerged surfaces are generally thought to be bacteria, but Sieburth & Thomas (1973) have recently shown that the diatom *Cocconeis scutellum* is a primary coloniser on eel grass (*Zostera marina*) and coats the surface with a uni-algal mat. Only when the mat is complete can other algae get a foothold. As Sieburth & Thomas noted, the mat gradually forms a crust presumably due to incorporation of particles into the mucilage. I have found that *Cocconeis* cells often become embedded in a crustal material which can occur for instance on filaments of blue-green algae (Fig. 6.6). Sieburth & Thomas speculate that the *Zostera* may provide a unique surface for the *Cocconeis*, although this can hardly be so since it is very common as an epiphyte on many different hosts. I think it is unlikely that *Cocconeis* secretes a special substance for attachment; it is probably the same mucopolysaccharide as that secreted by other diatoms, though little work has yet been undertaken on these compounds. Sieburth & Thomas's suggestion that some substance is released from the *Zostera* which prevents all but *Cocconeis* from attaching is perhaps more likely, but, if this is so, other organisms also secrete this or similar substances, since *Cocconeis* is by no means confined to *Zostera* in marine habitats and occurs in pure stands on a variety of hosts. Other species of *Cocconeis* are equally abundant as primary mat-forming epiphytes in freshwaters and I suspect the success of the species is due to a particular adaptation of its morphology to certain host surfaces; for example, it may be that the mucilage pad can only attach to surfaces free of certain other mucilages or to compounds with particular polar groups.

Very few studies have considered the small scale spatial microdistribution of epiphytic algae on host surfaces but those that have suggest significant distributions, e.g., Allen (1973) found that colonies of *Phycopeltus expansa* were associated with the margins of the epidermal cells of the orchid *Calyptrochilum* and he thought that supply of water and light may be important factors in determining the distribution. On underwater plants the grooves in the leaf surfaces are usually colonised first (Godward, 1934; Düringer, 1958) and also points close to the midrib. Godward (1934) made the interesting observation that epiphytes not only settle in far greater numbers on well illuminated substrata but also on slides provided with a green rather than a black background. Düringer noted that green algae favoured the upper leaf surfaces and diatoms the lower, and also that diatoms tend to start their growth at the apices and green algae at the bases of the leaves.

Metaphyton

The metaphyton encompasses the algal flora which is present between the epiphytes as a loose collection of non-motile or slightly motile organisms, without any obvious mode of attachment. These algae are readily observable even in casual observa-

tions of the epiflora on plants such as *Equisetum*; they occur between the firmly attached epiphytes and in many cases are not in the water surrounding the host plant but in the mucilage secreted by the epiphytes. It is likely that many species generally considered epiphytic are in reality metaphytic, e.g. the common *Tabellaria flocculosa* var. *flocculosa*.

Knudsen (1957) in her careful study of this genus came to the conclusion that the colonies could not be attached to the host by the apical mucilage pads which hold the individual cells together and she supposed that it occupied its position merely by entanglement amongst other algae. However, other genera of similar morphology e.g. *Diatoma*, can be seen to attach by means of the mucilage pad produced by the basal cell of the filament: each genus requires careful study! In some habitats, especially in acidic waters, most of the epiphytes themselves are also not projecting freely into the water but are immersed in the loose mucilage which may extend several millimetres from the host surface. Mucilage tends to diffuse into the water at its outer surface and there is not usually a discrete bounding layer. The unattached community was clearly recognised and the term metaphyton* first proposed by Behre in 1956, but since then little has been added to his study. Behre's concept was simply of the complete mass of unicellular algae free living amongst the microscopic algae and higher plants but in no way rigidly attached to a substratum. Flocculent masses with little relationship to the plant surfaces are thus involved but, from my observations, this assemblage is purely an outward extension from the plant surfaces and its origin always seems to be from cells

---

* The term pseudoperiphyton has been used for this same assemblage; it has all the disadvantages of the term periphyton (*see* p. 41) and should not be used. Ilmavirta *et al.* (1977) use the term 'pseudoplankton' which they define as 'derived in variable proportions from true plankton and detached epiphyton – found between the stems of emergent macrophytes' for a somewhat similar assemblage, though I believe that a true metaphyton is hardly contaminated by plankton and these workers may be dealing with yet another detrital community. That this assemblage is nevertheless very important in the economy of small lakes is shown by the fact that in Pääjärvi its production is tenfold that of the *Equisetum* epiphyton and Ilmavirta *et al.* reckon that if this is typical of the production in all the macrophyte stands of the lake then its production would amount to $\frac{2}{3}$ of that of the plankton.

which commence growth adjacent to the host plant whether this be a higher plant, a filamentous or thalloid alga. The metaphyton can also occur amongst masses of filamentous Zygnemaphyceae floating above the sediments. Behre was of the opinion that since it was an association present all the year round it could be used to characterise the flora of the body of water; within limits this is so, but the metaphyton does vary somewhat according to the plant with which it is associated. Some unpublished data of Dr J. Brown, working in Bristol, showed how extensive this flora is around the stems of *Equisetum*, and both she and Behre found that there were many more species in this association than in the epiphyton or plankton. By washing the stems with jets of water the metaphyton could be separated from the epiphyton which was often found to be so rigidly attached that only scraping with a sharp scalpel would remove it. Other methods of removal use acid hydrolysis (Gough & Woelkerling, 1976), but these are not ideal, since all the algae are killed and no estimate of live cells can be achieved. Some of the diatoms in metaphytonic associations are apparently motile, i.e. they have raphe systems, but they appear to move only slightly, perhaps due to the difficulty of adhesion* to the surrounding mucilage. Brown (1970) found that the primary epiphyton growing on *Equisetum* consisted mainly of *Oedogonium*, *Bulbochaete*, *Eunotia*, *Anomoeoneis* and *Ulvella*. The *Anomoeoneis* is a surprising find since it has no apparent attachment mechanism. On the two filamentous genera Brown found a series of secondary epiphytes of which *Tabellaria*, *Achnanthes*, *Eunotia*, *Characium*, *Epipyxis*, *Lagynion* and *Chaetosphaeridium* were the commonest. The metaphyton living between these primary and secondary epiphytes was dominated by *Scenedesmus*, *Oocystis*, many small unidentified green algae, *Mallomonas*, *Dinobryon*, *Pinnularia*, *Frustulia*, *Stenopterobia*, *Cymbella*, *Navicula*, *Gymnodinium*, *Ulothrix*, *Mougeotia*, *Oscillatoria*, *Staurodesmus*, *Staurastrum*, *Cosmarium*, *Euastrum*, *Closterium*, *Gonatozygon* and *Spondylosium*. This is very similar to the flora described by Behre (1956) from north German lakes but these have an even larger flora probably determined by the large number of niches provided by the greater depth, size, availability of macrophytic hosts, etc., compared to the single very

---

* Motility of diatoms is dependent on a degree of adhesion to a surface (Harper & Harper, 1967).

specialised habitat in Priddy Pool investigated by Brown. Although not all species recorded by Behre are epiphytic–metaphytic a very large proportion of species in his lists do belong to these communities and he records a total of 953 species, varieties and forms in the nine lakes he investigated, with the individual lake flora varying from 105 to 397 species. Certainly this is no exaggeration of the total flora to be expected in small lakes and the epipelic and planktonic floras were less intensively studied. From my own observations I am inclined to agree with the view expressed by Behre, that the metaphyton when developed in a lake is the richest of all the communities. It also tends to be perennial, though with definite changes in total biomass as that of the host plant changes during the year.

The metaphyton is only common on certain types of host and under certain conditions of water chemistry and at a low rate of water flow. It does however form a very conspicuous element of the flora in oligotrophic lakes and in bog pools. In a study of the algae in some Finnish rivers a very large number of desmids, especially *Cosmarium* and *Staurastrum* spp. were found on angiosperms and certainly must be considered as forming a river metaphyton (Round, 1959a). It is also present in marine habitats though less easy to recognise; the major components are diatoms, most of which are motile e.g., species of *Navicula*, *Mastogloia*, *Diploneis*, *Caloneis* and *Nitzschia*.

Whenever motile or non-motile unattached algae are found in association with distinctly attached epiphytic species further investigation should be made to ascertain the exact microhabitat of the species. Metaphyton probably occurs also amongst the epilithic community especially where this grows in relatively still waters, although I know of no investigation of this. A loosely attached community which I assume is similar to the metaphyton is reported by Cattaneo & Kalff (1978) to develop on plant surfaces in Lake Memphremagog. It is less well developed on artificial surfaces placed in the lake, but firmly attached species were best developed on plastic 'plants'.

Movement of the water in a pond or lake tends to remove the looser attached metaphyton and deposit it on the sediments; it should not then be included in the epipelon since in general the metaphytonic species are non-motile and cannot survive burial. In some sites it tends to be washed out into the plankton and in shallow lakes it is certainly difficult to distinguish the metaphytonic species, especially some of the desmids, from the planktonic unless both communities are being studied simultaneously.

Over vast regions of the cooler parts of the world, but extending also into temperate mountains, there are large areas of damp moorland and bogs with pools, small streams and lakes of varying size. The peaty bottom sediments support rich algal growths often of flocculent or jelly-like masses. The individual habitats are difficult to typify but in general they seem to support a flora which is partially epipelic but has many similarities with the metaphyton. When the shallow pools dry out the sediment is often obscured by a purplish sheet of the zygnemaphycean alga, *Zygogonium ericetorium*, which has a violet coloured vacuolar sap. In regions where the supply of water is mainly rainfall and not minerally enriched inflow, only a small algal flora occurs, characterised by the cyanophyte *Chroococcus turgidus*, the diatoms *Frustulia saxonica*, *Navicula subtillisima*, *Eunotia exigua*, the chlorophyceans *Gloeocystis vesiculosa*, *Oocystis solitaria*, xanthophycean *Chlorobotrys polychloris*, euglenoid *Euglena mutabilis* and a range of small desmids (*Cylindrocystis*, *Netrium oblongum*, *Cosmarium curcurbita*, *C. sphagnicolum*, *Euastrium binale* f. *Gutwinskii*, *Closterium acutum* var. *variabilis*, *C. pronum*, *Pleurotaenium minutum*, *Xanthidium antilopaeum*). These also occur in more mineral bog waters and as Fetzmann (1961) points out this extreme acid, low mineral water flora is characterised merely by the absence of a great range of species which occur in the more eutrophic bogs (compare the numbers of species found by Behre in waters of varying pH, p. 233). There are many minor habitats in such regions and they require detailed sociological study, e.g. Fetzmann found some sixteen *Closterium* species in a small stream rich in iron deposits and here again masses of *Frustulia saxonica* and *Navicula subtilissima* also occurred. The latter two diatoms occur repeatedly in the literature on these habitats, linked with various desmids (species of *Cosmarium*, *Euastrum*, *Staurastrum*, *Micrasterias*, *Gymnozygon*) and often *Chroococcus* to form associations all of which possibly have affinities to the metaphyton around the angiosperm stems.

Sampling techniques

Quantitative sampling of the epiphyton and metaphyton is difficult and is undoubtedly one of the major reasons for the relative lack of studies. The first essential is a study of the host and methods must be devised to estimate its biomass per unit area; the methods used in higher plant ecology or in rocky shore studies can be used. The longevity and growth rates of host species must be determined, and changes in availability of host surfaces determined as new plants succeed the old. In freshwaters the easiest plants to investigate are those with cylindrical stems, e.g., *Equisetum*, in which the new stems can be detected in April/May near Bristol (Brown, 1970) and can be followed through until their decay in the July of the next year, i.e. overlapping the next years new growth. The area of stem per square metre of bottom available for algal colonisation increases from May–September, then remains relatively constant until February, after which it declines. Changes in diameter and/or surface area also need to be taken into account and in the case of broadleaved plants the area must be estimated by measurement. Removal of the host from the water sometimes results in washing away the metaphyton and any loosely held epiphytes; this occurs also when large marine algae are sampled under certain conditions. Whenever possible, it is better to enclose the host sample in a glass or plastic tube or plastic bag whilst under the water and then cut off the sample at the base and cork the tube or tie the bag. In the case of sampling *Equisetum*, this always resulted in a considerable 'sediment' of metaphyton in the tube. The remaining metaphyton and the loosely attached epiphyton can be removed by shaking the sample vigorously, by agitating or sonicating it or by using fine jets of water. Hollow stems such as those of *Equisetum* can be slid over glass rods and rotated in jets of water to wash out the metaphyton. Table 6.10 shows the steady removal of algae in this way during three washings followed by a final scraping with a scalpel, and demonstrates that although a large amount of *Oedogonium* and *Bulbochaete* is removed by washing an even greater amount is firmly attached; that removed by washing is probably broken off. The fairly large population of diatoms left after washing includes most of the adpressed epiphytic flora whilst the washed off species live in the metaphyton. Only the desmids as a group are totally unattached and easily

Table 6.10. *Algae removed from* Equisetum *stems by fine jets of water and finally by scraping with a scalpel. Cell numbers* $\times 10^6$ *per* m^2 *of host. From Brown, 1970*

	Number of algae removed			
	1st washing	2nd washing	3rd washing	Final scraping
Oedogonium, Bulbochaete	1240	560	160	2760
Diatoms	7080	4440	1200	3600
Desmids	4840	760	80	80
Other algae	200	1480	80	240

removed by washing. The most rigidly attached epiphyton can rarely be moved without scraping the host with a scalpel or by brushing. The algae thus removed can be thoroughly mixed and sub-samples taken for cell counting or for chlorophyll *a* extraction, ^{14}C incubation, etc.

Endophyton

The endophytonic algal community grows between the cells of other plants or in cavities within plants. Well known associations occur in the liverworts *Blasia, Anthoceros, Cavicularia*, where blue-green algae of the genus *Nostoc* live in cavities within the thalli. The particular species *N. sphaericum* fixes atmospheric nitrogen which is transferred to the bryophytes (Bond & Scott, 1955). The water fern *Azolla* also supports nitrogen fixing *Anabaena* species in cavities on the lower side of the leaves; the *Azolla* is able to absorb nutrients more easily from the *Anabaena* than from the medium and its growth is stimulated by the alga (Ashton & Walmsley, 1976). Amongst the gymposperms the cycads (*Cycas, Macrozamia, Stangeria* and *Encephalartos*) have coralloid root nodules containing blue-green algae which almost certainly fix nitrogen though they have been little investigated. Algal inclusions in angiosperms are rare but *Gunnera* does have a *Nostoc* species in swellings of the leaf bases. All these associations are in cavities and, as far as I know, there are no instances of algae growing in between the cells of spore or seed plants; the cellulose cell walls seem to form an impenetrable barrier and a fungal mode of growth and nutrition is required to insinuate these plants. Some fungi of course have affinities to algae and may have evolved from photosynthetic epiphytic forms to non-photosynthetic haustorial organisms in these sites on leaves.

Many macroscopic marine algae, e.g. *Laurencia, Furcellaria, Polyidies, Gelidium* and *Gastroclonium* often have filamentous growths of the brown alga *Streblonema* proliferating between the cells of the peripheral tissue. That such endophytes are widespread is shown by White & Boney's (1969) study of the red alga *Acrochaetium* (now *Audouinella*); species of this alga are found also in *Heterosiphonia plumosa* plants collected from sites all around the British Isles. The endophyte (*A. endophyticum*) tends to occur in the basal part of the plant and not in the apices, and this basal position is in fact common to many associated algae both endophytic and epiphytic. White & Boney found that both the host and the endophyte could be cultured separately and kept living for up to three years during which time re-infection could be achieved by mixing the cultures. One peculiar feature and one which can also be seen in natural material is that though the host may decay the endophyte often remains healthy. Endophytes such as *Audouinella* can grow out into the medium in culture and Boney (personal communication) considers that there is considerable plasticity between endophytic and epiphytic modes of life. Plasticity is also obvious in that *A. endophyticum* re-infecting rhizoids of *Heterosiphonia* can take on an '*Erythrocladia*'-like growth form; yet another instance of the taxonomic problems posed by the plasticity of growth form of red algae. There is a tendency to describe epibiotic species as separate entities according to the host infected yet White & Boney found that *A. infestans* and *A. endophyticum* could grow on hosts other than the *Obelia* and *Heterosiphonia* from which they were isolated. On alternative algal hosts however they tended to grow as epiphytes rather than endophytes. Later, White & Boney (1970) found that *A. endophyticum* was also common in

Polysiphonia elongata and very similar plants were also found in *Ceramium rubrum*, *Dasya* sp. and *Porphyra umbilicalis*. The endophytic *A. bonnemaisoniae* grows both in the thallus and cystocarp of *Bonnemaisonia asparagoides* (Fig. 6.7). Jacob (1961) records an interesting endophytic community of blue-green algae inside the globular thalli of *Codium bursa* growing in the Mediterranean. The alga *Phormidium codicola* is dominant but *P. agile*, *Microcoleus wuiteri*, *Plectonema adriaticum*, *Lyngbya gracilis* and *L. agardhii* also occur. The Cyanophyta are red pigmented presumably due to chromatic adaptation to light climate inside the *Codium*. The outside of the plant has an epiphytic community in which *Sphacelaria tribulata* and *S. plumula* are most abundant.

Planktonic algae producing mucilage often have other algae growing within the mucilage (Fig. 6.8): a form of endophytism.

Recently diatoms have been found in the mucilage of conceptacles of *Ascophyllum nodosum* (Hasle, 1968) and in the mucilage between the filaments forming the thallus of the red alga *Dumontia incrassata* (Baardseth & Taasen, 1973; Taasen, 1974) and two new species *Navicula endophytica* and

Fig. 6.7. Filaments of the endophyte *Achrochaetium* (= *Audouinella*) *bonnemaisoniae* growing within the cystocarp of *Bonnemaisonia asparagoides*. a, b, c, low power magnifications of filaments in the cystocarps; d, e, higher power magnification of filaments. From Boney, 1972.

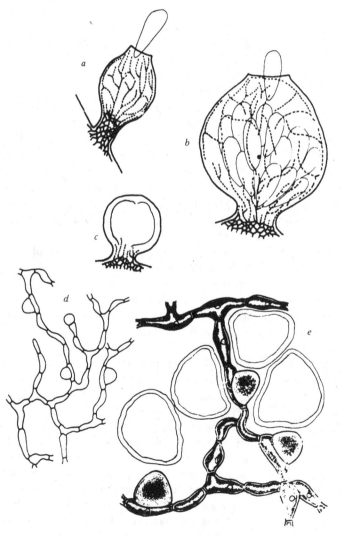

Fig. 6.8. A blue-green alga forming short filaments within a colony of the blue-green alga *Gomphosphaeria*. Photograph by Dr H. M. Canter.

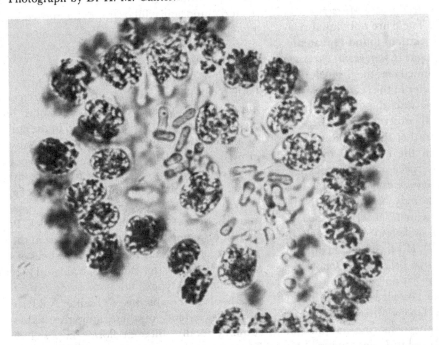

Fig. 6.9. The planktonic alga *Staurastrum* sp. showing rod-like bacteria embedded in mucilage. Centrally a colony of *Pandorina*. Photograph kindly supplied by Dr H. M. Canter.

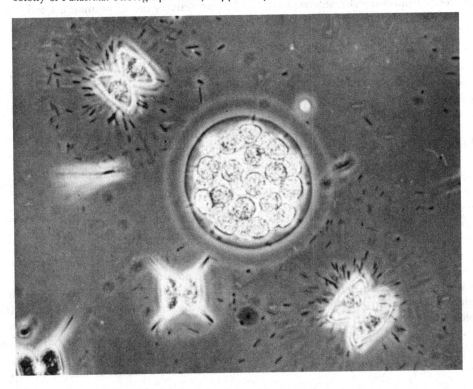

N. dumontiae have been described. It is of course quite common to find *Nitzschia* species within the mucilage of planktonic algae, e.g. in freshwater *Microcystis* and in the marine *Phaeocystis*. These are not casual relationships but regular associations found repeatedly. The freshwater green algae *Chaetophora incrassata* also has a flora within its mucilage; the first coloniser is a *Nitzschia* species but later in the season *Epithemia* and *Rhopalodia* also penetrate the mucilage and multiply there (personal observation). *Nitzschia* species also occur in the rather special endotubular situation, growing within the hardened mucilage tubes secreted by some species of diatoms (e.g. *Cymbella*, *Berkeleya*, *Amphipleura* and *Navicula*). This association with mucilage may be linked with the reported heterotrophic tendencies of *Nitzschia*, a genus which is reputed to favour waters rich in nitrogen (Cholnoky, 1968). However whether or not this is a case of symbiosis as believed by Cholnoky (1929) and earlier workers has not, as far as I know, been proved. Certainly the host alga can live without the endophyte. A green '*Chlorochytrium*'-like alga is also often present in the tubes of marine diatoms and a similar alga is also reported in *Lemna* species. This *Chlorochytrium* is sometimes extremely common in diatom material collected around the British coasts in the spring, but further work is required on its status, possibly it is a phase of another green alga.

There is of course the complex situation of algae whose spores germinate on the surface of other algae and then commence to penetrate the host forming an endophytic holdfast (e.g. *Polysiphonia lanosa* on *Ascophyllum nodosum*) whilst at the same time developing an essentially epiphytic vegetative thallus subjected to the stresses of the epiphytic habitat. These plants may benefit from both endophytic and epiphytic conditions (but *see* p. 223).

Endophytic bacteria are certainly present in the mucilage of many algae, e.g., they are often conspicuous as short rods between the blocks of mucilage produced by planktonic desmids (Fig. 6.9) and also between the cells of many macroscopic marine algae. In some of the latter they may produce galls; a bacterium which is similar to *Agrobacterium tumefasciens* and which causes 'brown gall' in higher plants has been isolated from algae such as *Chondrus crispus*. The marine planktonic algae *Phaeocystis pouchettii*, however, produces acrylic acid which is a bacteriocidal agent (Sieburth, 1964, 1968); it would be of interest to study the effect of this in nature since the mucilage of *Phaeocystis* harbours both bacteria and unicellular epiphytic and endophytic algae.

Epizoon

Many animals, especially sedentary species with hard shells, are host to numerous common algae. This is obvious on any rocky shore but the algae are often only those also found in the epilithon. A noticeable feature however is the abundant growth on the shells whilst the rock or encrusting calcareous algae are relatively free from attached algae. This is almost certainly due to lack of grazing activity over the live animals whilst these very animals are grazing the algae from the rock. Possibly there is also some growth stimulation from the waste products of the host. The flora on the shells of species such as *Patella* (*see*, for example, Table 6.11) may, in fact, reflect the type of flora one could expect in the rock pools in the absence of grazing. Analysis of the stomach contents of these animals (Table 6.12) shows a totally different flora composed of small filamentous algae and diatoms taken off the rock surfaces.

More actively motile animals are less likely to have algae attached, although some do. For example filamentous algae are quite common on the beaks of parrot fishes in tropical regions. There seems to be no host specificity and the common algae in one Pacific site were *Ectocarpus indicans*, *Acrochaetium* spp., *Polysiphonia scopulorum* and *Microcoleus lyngbyaceous* (Tsuda, Larson & Lujan, 1972). *Cladophora* and *Ectocarpus* have also been recorded from the corners of the mouths of parrot fish (*Scarus guacamaia* and *S. vetula*) off Bermuda (Winn & Bardach, 1960). The diatom *Synedra affinis* has been found on free-living marine nematodes (Tietjen, 1971).

Both freshwater and marine Copepoda are frequently found with blue-green algae or diatoms attached to the appendages. A genus of diatoms (*Pseudohimantidium*) is a characteristic component of the epizootic flora on the marine copepod *Corycaeus* and is found in the Adriatic Sea, through the Atlantic Ocean and into the Pacific Ocean; it has been mis-identified several times as a *Cymbella* (which is freshwater apart from one exceptional species, *C. pusilla*), as *Licmophora* or *Amphora* and recently as a new genus *Sameioneis* which is almost certainly a

Table 6.11. *The algal flora colonising the shells of* Patella *near Endoume. From Koumans-Goedbloed, 1965*

	Rough coast	Calm coast
Cyanophyceae		
Entophysalis deusta (Men.) Drouet & Daily	2	2
Mastigocoleus testarum Lag. ex Born. et Flah.	2	2
Plectonema batteres Gom.	+	+
Plectonema terebrans Born. et Flah.	1	1
Lyngbya lutea Gom. ex Gom.	+	.
Lyngbya aestuarii Liebm. ex Gom.	.	1
Schizothrix calcicola (Ag.) Gom.	1	1
Calothrix crustacea Thuret ex Born. et Flah.	.	+
Hydrocoleum lyngbyaceum Kütz.	1	1
Chlorophyceae		
Disc-like Chlorophyceae	1	1
Blidingia minima (Näg.) Kylin	2	2
Enteromorpha compressa (L.) Grev.	1	2
Ulva sp.	1	2
Cladophora albida (Huds.) Kütz.	1	1
Pilinia rimosa Kütz.	+	+
Eugomontia sacculata Kornmann	.	+
Ulothrix pseudoflacca Wille	1	1
Chaetomorpha capillaris (Kütz.) Boergs var. *crispa* (Schoesboe) Feldm.	+	+
Codium vermilara (Olivi) Delle Chiaje	.	+
Phaeophyceae		
Prostrate ectocarpoids	.	+
Ectocarpus confervoides (Roth.) Le Jol.	1	1
Scytosiphon lomentania (Lyngbye) Endl.	+	+
Petalonia fascia (Mull.) Kuntze	+	.
Rhodophyceae		
'Conchocelis' stage of *Porphyra* and *Bangia*	+	.
Bangia fusco-purpurea (Dillw.) Lyngb.	2	2
Porphyra sp.	1	1
Ceramium tenerrimum (Martins) Okamura	1	.
Ceramium ciliatum (Elvis) Ducluz.	.	+
Gelidium latifolium (Grev.) Thur. et Born.	.	+
Pterocladia capillacea (Gmel.) Born. et Thur.	.	1
Erythrotrichia carnea (Dillw.) J. Ag.	.	1
Corallina officinalis L.	2	.
Lithophyllum incrustans Phil.	2	.

synonym of *Pseudohimantidium* (Russell, personal communication).

Cribb (1969*a*) found crusts of calcareous algae on the plastron of the marine turtle *Eretmochelys* and a fine fur of filamentous algae, 38 species in all, over the plastron and carapace. A crustose calcareous alga *Fosliella farinosa* has been found growing on the spines of the sea urchin *Heterocentrotus trigonarius* (Lawrence & Dawes, 1969). The freshwater turtle *Chelodina* has *Basicladia ramulosa* growing on it in Australia (Ducker, 1958) and in North America several species of turtle are colonised by *Basicladia*

Table 6.12. *The algae in the stomach contents of* Patella. *From Koumans-Goedbloed, 1965*

From calm (A) and rough (B) coasts	28 Nov.		24 Jan.		12 Feb.		22 Mar.		26 Apr.		17May	
	A	B	A	B	A	B	A	B	A	B	A	B
Entophysalis deusta (Men.) Drouet & Daily	+	·	·	+	+	+	+	+	+	·	+	+
Lyngbya lutea Gom. ex Gom.	·	·	+	·	·	·	+	·	·	·	·	·
Lyngbya aestuarii Liebm. ex Gom.	?	·	·	·	·	·	·	·	·	+	·	·
Enteromorpha sp.	·	·	?	·	·	·	+	·	+	·	·	·
Blidingia minima (Näg.) Kylin	+	?	·	·	+	·	+	·	+	?	+	+
Ulothrix pseudoflacca Wille	·	·	·	·	+	·	·	·	·	·	·	·
Ulva sp.	·	·	·	·	·	·	+	·	·	+	·	·
Bangia fusco-purpurea (Dillw.) Lyngb.	+	·	·	·	+	+	+	+	+	·	·	·
Porphyra sp.	·	·	·	·	·	·	+	·	·	·	·	·
Licmophora communis (Hab.) Grun.	·	·	+	·	·	·	+	·	·	·	·	·
Licmophora ehrenbergii (Kg.) Grun.	+	·	+	+	2	2	·	·	·	·	·	·
Licmophora dalmatica (Kg.) Grun.	·	?	·	·	·	2	+	·	·	·	·	·
Licmophora flabellata (Carm.) Ag.	·	·	·	·	·	·	2	·	2	·	·	·
Licmophora gracilis (Ehr.) Grun.	+	·	+	+	·	2	+	·	+	·	·	+
Licmophora juergensii Ag.	·	·	·	·	·	·	+	·	·	·	·	+
Achnanthes brevipes Ag.	+	+	+	·	+	+	+	+	+	+	+	+
Achnanthes longipes Ag.	+	+	·	·	+	+	1	+	+	·	·	·
Achnanthes brevipes var. parvula (Kütz.) Cleve	+	1	+	·	+	+	2	+	2	+	+	+
Dimerogramma fulvum (Greg.) Ralfs	+	+	+	·	2	2	+	+	+	+	+	+
Cocconeis scutellum Ehr.	+	+	+	·	+	+	+	+	+	+	+	·
Cocconeis molesta Kg.	·	·	·	·	·	·	·	·	·	1	·	·
Grammatophora gibberula Kg.	+	2	+	·	+	+	+	+	+	·	2	·
Opephora marina (Greg.) Petit	+	+	·	+	+	+	·	+	+	·	+	+
Opephora pacifica (Grun.) Petit	·	·	·	+	+	+	+	·	·	·	·	+
Rhoicosphenia curvata (Kg.) Grun.	+	2	+	+	+	+	+	+	+	·	·	·
Thalassionema nitzschioides Grun.	+	·	+	+	+	+	·	·	+	·	·	·
Synedra crystallina (Ag.) Kütz.	·	·	·	·	·	·	·	+	1	·	+	·
Diploneis suborbicularis var. constricta Hustedt	·	·	·	·	·	·	+	·	+	·	·	·
Bacillaria paradoxa Gmelin	+	·	+	·	·	·	+	·	+	·	·	·
Trachyneis aspera Ehr.	·	·	·	·	·	·	·	·	+	·	·	·
Amphora exigua Greg.	·	·	+	·	+	·	+	+	+	·	·	·
Amphora inflexa Greg.	·	·	·	1	·	·	·	·	·	·	·	·
Campylodiscus thuretii Bréb. Diat.	+	·	+	·	·	·	·	·	·	·	·	·
Navicula gracilis var. schizonemoides van Heurck	+	2	+	·	+	2	+	+	2	+	+	+
Actinocyclus ehrenbergii Ralfs	·	·	·	·	·	·	·	+	+	+	·	·
Hyalodiscus scoticus (Kütz.) Grun.	·	·	+	·	·	·	·	·	·	+	·	·

and *Dermatophyton* (Edgreen, Edgreen & Tiffany, 1953) but some species of turtle have no algal epiphytes; this seems to some extent to be governed by the length of time the turtles spend basking in the sun.

A characteristic epizooic flora occurs on the skins of whales and attempts were made to use the abundance of this flora to determine the time the whales had spent in antarctic waters; the flora was richest on whales which had lived some time in the cold waters and least on animals caught in warm waters (Bennett, 1920). The algae form yellowish streaks and patches on the whale skins and these consist of mucilaginous masses of the diatom *Cocconeis ceticola*. This is the common and apparently exclusively 'Ceticolous' species but others (*Licmophora lyngbya, Cocconeis imperatrix, C. gautei, C. wheeleri, Navicula* sp. and *Gyrosigma arcticum*) are also reported (Hart, 1935).

Endozoon

Endozooic relationships of algae have been recorded infrequently except for those which are symbiotic in nature and dealt with in Chapter 9. The freshwater colonial ciliate *Ophrydium* has a *Nitzschia* species living in its mucilage (Round, unpublished). How many unknown associations there are is anyone's guess, recently, for example, there has been a report that a *Euglena* species inhabits the hind gut of the nymphs of the damsel fly during winter only (Willey, Bowen & Durban, 1970). Earlier a green euglenoid genus *Euglenamorpha* was discovered in the rectum of tadpoles (Wenrich, 1924) and a very recent publication (Rosowski & Willey, 1975) has shown that a species of *Colacium* lives in the rectum of damsel-fly larvae.

Red algae such as *Audouinella* (= *Acrochaetium*) *infestans* occur within the perisarc of *Obelia*. White & Boney (1970) found that cultured material of *A. endophytium* normally found in algal hosts could infect *Obelia* in culture. In nature the filaments project out into the water just beyond the perisarc.

There is a widespread community of algae living on and boring into the shells of both freshwater and marine mollusca and a few other animal groups (Frémy, 1936). I am not quite certain where to include these algae since they occur in both living and non-living calcareous shells. In the latter they are perhaps more closely allied to the endolithic flora of calcareous rocks and even in the living molluscs they could be regarded as only colonising the 'non-living' part though since this increases as the animal develops I include them here. Van den Hoek (1958) made a clear distinction between algae perforating the shells and species simply growing on the surface and I am sure this is correct since the epizooic are simply using the shell as an attachment site and they occur equally on other substrata. Parke & Moore (1935) found that colonisation of *Balanus* shells begins when the animal is 4–6 months old, and the first alga to infect was the blue-green *Plectonema*

tenebrans followed by *Hyella caespitosa* and *Gomontia polyrhiza*; *Ostreobium quekettii, Mastigocoleus testarium* and *Microchaete grisea* were also recorded. The depth of penetration gradually increases as the season progresses and the activity of these algae can be seen by the appearance of oxygen bubbles on the shells. Wilkinson & Burrows (1972a) recorded a similar sequence in *Spirorbis* growing on *Himanthalia* with *Plectonema/Hyella* first, followed by *Conchocelis* and later 'Codiolum' stages. They commented that *Plectonema* seems to be the most commonly found boring alga. All the shells investigated on transects in the Isle of Man were infected by all the boring algae, except for *Littorina littoralis* which was usually attacked only by *Tellamia intricata*; interestingly this algae lives only in the periostracum and does not penetrate the calcareous part of the shell (Wilkinson & Burrows, 1972a). When the periostracum of *L. littoralis* is eroded away other algae penetrate the remains. There was, however, a vertical zonation of boring algae on the shore, with *Entocladia perforans* only in the upper third, *Ostreobium quekettii* in the lower third and *Codiolum* stages throughout the intertidal. The species *Eugomontia sacculata* was found mainly in the subtidal and mostly in estuaries. Nielsen (1972) discovered that whilst most of the algae grew well in culture, *Mastigocoleus testarum* only grew well when fragments of shell were added to the media.

There are great taxonomic problems involved with these shell boring algae and Wilkinson & Burrows (1972b) consider that at least six entities could be confused under the name *Gomontia polyrhiza*; four could be identified, *Eugomontia sacculata, Entocladia perforans*, 'Codiolum' stage of *Codiolum polyrhizum* and the 'Codiolum' stage of *Monostroma grevillei*.

Wilkinson (1975) points out that the boring sporophyte of *Monostroma grevillei* shows as great a range of morphological variation when cultured in different host shell species, as is found between the unicellular phases belonging to the life histories of *M. grevillei, Gomontia polyrhiza* and *Eugomontia sacculata*. Differences in the morphology and physiology can also occur in intertidal and subtidal populations of *Eugomontia sacculata* (Wilkinson, 1974), for example, shallow estuarine plants survived in culture under higher light intensity than those from the subtidal and plants from the latter habitat could not survive reduced estuarine salinity.

Summary

This chapter deals with the algae attached to living substrata, i.e. other plants and animals. The former is the *epiphyton* and latter *epizoon*. Algae growing within plants are the *endophyton* and within animals *endozoon*. A loosely associated algal community generally found between the epiphyton is known as the *metaphyton*.

Epiphyton can grow on aerial substrata, e.g. leaves and bark of trees and the number of species in these associations are relatively small. Within the moist tropics the community is well developed. Epiphyton of flowing waters, lakes and the macrophytic marine angiosperms and algae is dealt with briefly. In all these sites, host specificity, the relationship between the colonising algae and the host surfaces still requires much study. Also the relationships between the secretions and nutrient exchanges between epiphytes and between epiphyte and host is important. The term *basiphyte* has been used for the hosts. Nitrogen fixation occurs where certain blue-green algae grow epiphytically.

Epiphytes are common in the specialised habitat of the plants of *mangrove* swamps.

Epiphytic *bacteria* are more common than generally recognised. There is also a very distinctive epiphytic fauna, especially on the large algae though also on colonial planktonic genera.

The problem of initial colonisation has been studied and bacteria found generally to precede diatoms and filamentous forms.

The metaphyton has only been studied around a few freshwater plants but it is undoubtedly an important contributor to carbon fixation.

Endophyton occurs in cavities within various macroscopic plants whilst marine macrophytes often have filamentous genera creeping between the cells.

Marine animals in particular and especially those with hard shells often have well developed algal epizoon; such a community is found even on the skin of whales. Some algae actually penetrate the shells of aquatic animals.

A few algae have been reported to live in the guts of aquatic animals and the interior of the perisarc of coelenterates can support algae. Most recently algae of the new group Protochlorophyta have been found within Ascidians (*see* p. 563).

7 Phytoplankton

The open water or pelagic environment always contains particulate matter kept in suspension by water movement, known collectively as seston and divisible into living (bioseston) and non-living (abioseston or tripton). The components of the bioseston are the plankton and nekton and prefixes denote algal (phyto-) or animal (zoo-) types. Nekton encompasses the larger organisms, such as fish, which move independently of turbulent water movement. Most techniques for sampling phytoplankton really collect seston though in the oceans biologically derived particles dominate this detrital phase (Lal, 1977). In the sea and in most large inland waters the bulk of living matter found in the open water is phytoplankton and hence the biological importance is immense; yet live plankton is still only a very small part of the seston and the amounts vary. Gillbricht (1952) estimated 4.2% but Holm-Hansen (1969) only 1%. The majority of phytoplankton species are non-motile and therefore at the mercy of turbulence within the upper water mass and of their own sinking potential in still water; some, however, are motile by means of flagella e.g. *Chlamydomonas*, *Volvox*, dinoflagellates, coccolithophorids, etc. These can move only to a limited extent but in relatively still conditions are capable of upward and downward movement though, as far as it is known, they do not move with the distinct lateral mobility shown by the nekton. The phytoplankton are not suspended in a clear solution, as seems to be the case when water from a lake or the open ocean is observed in a collecting bottle or down a microscope, but they drift in a complex medium of inorganic and organic particulate and colloidal matter. Part of this is derived from flow off the land, and part from secretion, excretion and death of the totality of organisms in the water. The proportions derived from these sources vary in each water mass and with changing season but the important point is that phytoplankton drifts in a medium which is either a dilute, or concentrated solution of faeces and dissolved organic matter. 'Let us face the fact that the aquatic medium is really vegetable soup' (Darnell, 1967). Attempts to investigate the spatial relationship of the particles are rare and indeed very difficult to undertake since they involve studying a series of cubes of water from the surface downwards.

Separation of the living and non-living material is difficult and therefore great care has to be taken if measurements of chemical elements within natural populations are attempted. Equally most studies of water are made on samples from which the plankton has been removed by filtration. This also removes much of the colloidal and flocculent chemical matter and so at least in rich waters the chemical data obtained do *not* adequately express the natural chemical environment. Obviously if the absolute number of organisms per unit volume is to be determined an unconcentrated sample must first be obtained and nets are of little use for this. Water from an unproductive station will reveal few, if any, phytoplankton cells when unconcentrated small samples are taken and observed directly under the microscope, whilst at the other extreme, water from a small productive pond may reveal twenty or more organisms in a low-power field.

Terminology of the phytoplankton

Many types of plankton have been distinguished since Hensen (1887) first defined it as the drifting organic material (live and dead) in the water.

Phytoplankton encompasses a surprising range of cell size and cell volume from the largest forms visible to the naked eye, e.g. *Volvox* (500–1500 μm) in freshwater and large *Coscinodiscus* species in the ocean (*C. asteromphalus*, for example, can be 200 μm in diameter after sexual reproduction) to algae as small as 1 μm in diameter and to even smaller bacteria, which are a relatively neglected element of this association. Fig. 7.1 shows some of these species drawn to the same scale. The larger forms are often referred to as net plankton since they are retained by the mesh of commonly used phytoplankton nets (150–200 meshes to the inch; other workers quote aperture size from 20 to 90 μm). A better term for this net plankton is microplankton; it varies from 0.06 to 0.5 mm (60–500 μm) but forms

down to 0.01 mm (10 μm) are retained in nets by clogging of the pores (Hutchinson, 1967). Nannoplankton, 0.005 mm–0.06 mm (5–60 μm), is that which would normally pass through nets but is collected in considerable quantity by normal phytoplankton nets, again due to the clogging of the pores. The smallest forms are sometimes referred to as the μ-plankton (μ-algae, μ-flagellates) or ultraplankton and range from 0.5 to 5 μm.* Obviously, divisions

* The term ultranannoplankton has been coined for the smallest forms (Harrison, Azam, Renger & Eppley, 1977).

Fig. 7.1. Some microscopic marine and freshwater phytoplankton drawn to about the same scale.

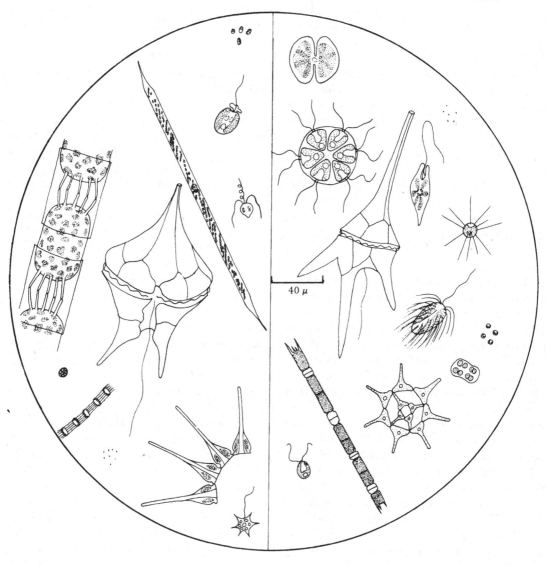

40 μ

are arbitrary, and workers such as Drebes (1974) give ranges up to 5 μm (ultra-), 5–20 μm (nanno-) and above 10 μm (micro-) but all too often only the micro-plankton is considered since this is obtained by the common techniques of sampling. The majority of the *conspicuous* diatom plankton is in the size range 20–100 μm but there are genera, e.g., *Ethmodiscus*, reaching 2 mm diameter and cells of *Rhizosolenia* can reach 1–1.5 mm in length. The latter forms colonies which can therefore be quite large, and the small cells of *Chaetoceros* and *Bacteriastrum* which are 10 to 50 μm in diameter but have spines at least twice this length can form chains of cells perhaps up to 1 mm in length. There may be some significance in the fact that freshwater planktonic diatoms tend to be at the lower end of the size range and colony formation does not produce such large aggregates as amongst the marine species. Dinoflagellates tend to vary in size between 10 and 80 μm with a small number of genera, e.g., *Peridinium* and *Ceratium*, reaching overall sizes up to 350 μm but never attaining the size of the largest diatoms. *Ceratium* can form colonies and these are naturally somewhat larger. In both groups there are many very small-celled forms which have been very inadequately described and studied but the mean of the conspicuous forms probably falls around 40–50 μm. The Prymnesiophyta on the other hand are small flagellates or coccoid cells which rarely reach 20 μm and many are less than 10 μm along the largest axis. A few, bearing coccoliths, are elongated and with their coccoliths may form larger structures, but they are relatively rare. There is a general impression that marine planktonic species as a whole are larger than freshwater but a recent analysis (Stull, 1975) has shown that there is no great difference when complete populations are compared though there are a few large marine species; these are, however, generally scarce and hardly affect the overall distribution of size classes.

The term *euplankton* (= *holoplankton*) is usually used to refer to the permanent planktonic assemblage of organisms which completes its life cycle suspended in the water. It is often extremely difficult to prove that the organisms never settle out or form spores. Lund (1949), however, studied *Asterionella formosa* and found by sampling the water column extensively throughout a year, that a few cells were always present and there is therefore no need to postulate a resting phase. I know of no similar extensive collecting of oceanic species to illustrate this point; it is difficult to see how the oceanic plankton can maintain itself without being euplanktonic, which in no way denies the formation of spores (e.g. in *Chaetoceros*, Fig. 7.23), resting stages, etc. but simply implies that these are always in circulation at least in the open ocean. It must be assumed that there is little chance of resuspension of algae from the deep water or deep sediments. The interesting results of Wall & Dale (1968*b*), who have germinated the spores of some dinoflagellates collected from ocean sediments, do however show *survival* in deep water. We do not know whether these are simply the 'lost' spores or whether germination in the ocean depths and return to the surface water is a possibility, for instance by vertical transport on nekton. Wall (1971) considers that the spores in the deep sediments may have been transported there by turbidity currents.

Species which spend part of each season resting on the sediments comprise the *meroplankton* (e.g. *Melosira* and *Gloeotrichia* discussed on p. 435). Species which undergo sexual reproduction, e.g. desmids and many Chrysophyceae, are presumably meroplanktonic since the resulting zygospores are too heavy to be maintained in the water column. The statospores of *Dinobryon* have recently been shown to be capable of germinating immediately after they have been formed though those which reach the sediments do not germinate until after the ice melt of the next spring (Sheath, Hellebust & Sawa Takasi, 1975). Some coastal (neritic) plankton almost certainly live on the sediments for part of the year. Hutchinson (1967) points out that the number of planktonic species does not increase as lake area increases, indeed a decrease may occur, which is the reverse of the situation for small islands where the diversity of bird, mammal and ant faunas increases with increasing area.* This may be an indication that more freshwater species are meroplanktonic than generally recognised since increase in size presumably is of no consequence to euplanktonic but a disadvantage to meroplanktonic species which have a more intimate association with the sediments. Conversely large shallow lakes ought to have large numbers of species unless shallowness imposes a constraint on the euplanktonic organisms. Only

* Lakes are comparable to islands from a dispersal–geographical aspect; they are islands of water.

detailed long-term investigations will solve these queries.

In the open ocean it is unlikely that casual species derived from other habitats would be collected; the only casual species might be those derived from the flora attached to detritus or transported by the nekton. Such casual species in any water mass are referred to as *pseudoplankton* (= *tychoplankton*), a word often incorrectly applied to meroplankton. Pseudoplankton is quite common in lakes, especially after storms, and very common in rivers owing to inwash from ponds, etc. and to suspension of scoured benthic species. Coastal (neritic) plankton is also often contaminated by pseudoplankton derived from the epipelon (e.g. species of *Pleurosigma*, *Navicula* and *Amphora*) or from the epiphyton (e.g. species of *Licmophora*). Inclusion of such species produces impressive lists but is totally misleading, unless qualified, and such data should never be used for characterising water types or for diversity studies. A small number of species belonging to genera much more common in the epipelon are truly planktonic in some waters (mainly tropical) so considerable problems can arise. In rivers and coastal regions, such pseudoplankton may undergo some cell divisions whilst suspended and may have a temporary effect on the habitat (e.g. decreasing light penetration, removing nutrients etc.). A few species, e.g. *Cylindrotheca* (= *Nitzschia*) *closterium* and *Bacillaria paradoxa* are regularly found in neritic plankton but are certainly more abundant on the shallow sediments and I believe they may be representatives of a small group of organisms which can live in either habitat.

The prefixes *limno-*, *heleo-*, *potamo-* are used to refer to the plankton of lakes, ponds and rivers respectively and marine plankton is often subdivided into *neritic* which occurs in coastal waters and *oceanic* in the open ocean. All these types are characterised by certain species assemblages.

There is a flora both in marine and freshwaters associated with flocs of organic detritus up to a millimetre or more in diameter, sometimes termed 'marine snow' by oceanographers. The algae are not planktonic species but are forms associated with surfaces (epiphytes, epilithophytes) living in a rather unusual site. This flora has rarely received any attention and no special term exists for it, though the term *detritiplankton* could be used.

A few planktonic species are known to commence their growth attached to other species, e.g. *Phaeocystis pouchetii* is often found in the seas around the British Isles as single or small groups of cells adhering to planktonic diatoms. In freshwaters a few planktonic species may start growth or even form small populations amongst the littoral angiosperms, but there is little evidence for the statement in some texts that large growths occur along the shores and then invade the open water, for few organisms are adapted to life in more than one habitat!

Sampling

There are essentially two stages in studying plankton; first, representative samples must be obtained from the natural habitat, and second, laboratory techniques must be employed to determine species composition, biomass, chlorophyll concentration, amounts of carbon fixed and other properties of the plankton.* Obviously such techniques involve subsampling and then extrapolation to the situation in nature with all the attendant uncertainities. *In situ* devices using electrical probes can however be used for certain measurements of environmental factors, such as dissolved oxygen, pH, turbidity, and it is desirable to develop techniques to measure others. Environmental factors can be measured down a depth profile and sometimes semi-permanent probes can be placed in the profile and continuous records obtained. Another technique involves pumping samples from a range of depths and continuous measurement of nutrients, fluorimetric estimations of chlorophyll, etc. in shipboard analysers.

Obtaining samples of the plankton alone is complicated by spatial distribution of the components, which may be scattered down a profile a few metres in depth or down to 150 m or more. Usually however the bulk of the phytoplankton occurs in the upper 50 m (however *see* p. 351). If the water is completely mixed then a single sample from any depth may yield a representative assemblage; however a completely homogeneous dispersion is probably impossible to find. When a single component, *Ceratium*, was considered in detail (Heaney, 1976), it was found to be non-uniform in both vertical and horizontal planes even in a small lake and frequently vertically stratified so that a sample from the surface may contain a different association of species from

* For details of methodology *see* Vollenweider (1974).

one at one metre depth and this may differ from one at 5 m and so on even down to 60 m in Lake Victoria (Fig. 7.2). This situation can be overcome by sampling a core of water: Lund (1949) devised a very simple method in lakes merely by lowering a weighted hose pipe vertically down the water column and allowing it to fill, closing the top, hauling it up and pouring the contents into a bottle (Fig. 7.3). Obviously this technique is more difficult in deep or rough water and it would be difficult to sample the deep populations of Lake Victoria by this method. Under such conditions a water bottle sampler is used; this is lowered to the appropriate depth and then closed, and the samples in effect take short cores of water from known depths and these can then be studied separately or bulked. This technique is valuable when stratification is being studied whilst the tube technique is valuable when the gross changes in the water column are being investigated. If large volumes of water are required, and this is sometimes necessary at sea in regions of low production, water can be pumped up from various depths or the depth profile can be sampled by using continuous pumping as the intake is gradually lowered. Pumping from known depths can continue as the ship steams since fins can be added to the lower end of the tube to maintain a depth. Most techniques yield the amount or *concentration* of material in a volume of water but it is often more valuable to have a measure of *cover*

especially when comparative studies between different areas and with other vegetation types, including terrestrial vegetation, are attempted. Cover in relation to phytoplankton involves a measure of the biomass (cell numbers, carbon content, chlorophyll content, etc.) under unit area and this information must be calculated from the data obtained from the subsamples.

It is often desirable to obtain a concentrated sample of phytoplankton in order to study the species in quantity and also to obtain the rare species; samples can be achieved by towing a net through the water. There are many very delicate species in the phytoplankton and compaction at the base of nets damages these. To obtain such species it is often preferable to take water bottle samples and carefully concentrate the sample by settling. Even filtering can be damaging especially to nannoplankton and this damage is increased if vacuum pumps are used. If the water column is mixed then a horizontal tow will suffice but if it is stratified a weighted net can be lowered to a known depth, closed at a lesser depth and then hauled to the surface thereby sampling parts of the water column. Nets towed at set depths are rarely used for phytoplankton studies and the elaborations used for zooplankton work are not usually necessary. When collecting plankton, care should be taken to prevent over-concentration which merely leads to rapid

Fig. 7.2. The vertical distribution of some phytoplankton species down the water column in Lake Victoria. 'Individuals' are cells except for *Botryococcus* and *Pediastrum* which are colonies. The photosynthetic rate is shown as a fraction (P/P_m) of the maximum rate (P_m). From Talling, 1957.

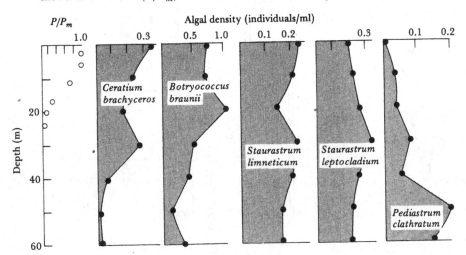

decay before the sample can be investigated. It is often instructive to leave the sample to settle in a cylinder overnight, to allow the non-motile forms to sediment to the bottom, while the motile ones tend to swim in the centre, to one side or to the surface and the buoyant forms float to the surface (Fig. 7.4). Such simple techniques can be used sometimes to separate species or groups of organisms for subsequent analysis of cell constituents, etc., and often yield valuable data for the ecologist since they give a measure of the constituents in nature as opposed to the much more commonly quoted data from cultivated populations.

The Hardy (continuous) plankton recorder is

Fig. 7.3. Diagrammatic procedures for sampling and subsequent treatment of phytoplankton.

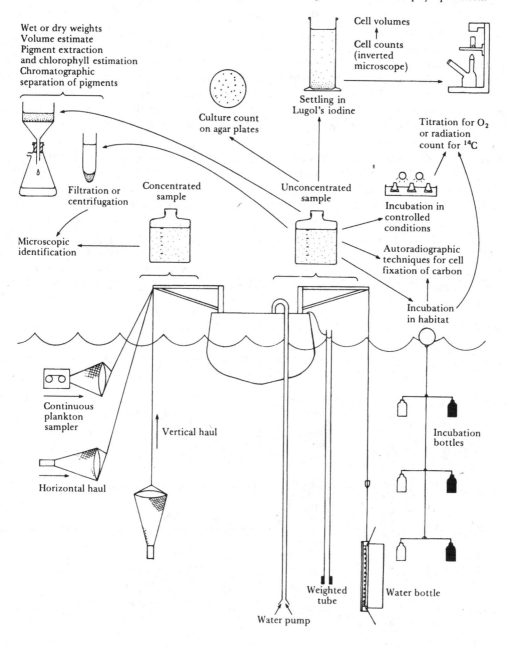

a device for obtaining samples to study the horizontal distribution of plankton and is essentially a system to filter plankton from the sea and deposit it on a roll of collecting material. The recorder is towed behind a ship at a known or fluctuating depth and in effect samples a line transect which can then be analysed

Fig. 7.4. The movement of some freshwater phytoplankton left overnight in a cylinder. At the top float *Microcystis*, *Anabaena* and *Oscillatoria*, in the centre *Chlamydomonas* and *Pandorina* are swimming, and deposited at the bottom are *Melosira*, *Asterionella* and *Fragilaria*.

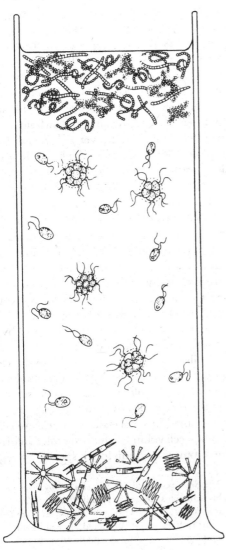

at a later date. Such a technique has only been used by a small number of investigators and examples of the results are given in Figs. 7.17 and 7.18. This is, of course, a modified net sampling technique and can therefore only be used for crude quantitative studies, very valuable for plotting geographical distribution of large species but not for obtaining information on the phytoplankton as a whole. Nets are not suitable even for fine qualitative studies (R. R. Anderson, 1969) since the variability of samples is too great. The only way to obtain detailed qualitative information is to use sedimentation techniques on known volumes of water. Features such as species diversity cannot be computed from net data.

Another synoptic technique was adopted by Anderson & Munson (1972), who obtained repeated ocean-wide data by sampling the seawater drawn from the main injection system in the engine room of merchant ships sailing between Seattle and Yokohama. The engine room intake was about 9 m below the surface but varied somewhat according to the ballast of the ship. The water was filtered through HA Millipore filters and refrigerated prior to chlorophyll *a* estimations. Millipore filtration also has its complications, for fine species are often destroyed and the preservation of the filters can damage some species.

From the above comments it should be clear that there are great problems in sampling and if all the sizes, classes and groups of organisms need to be sampled, several techniques may have to be employed for each sample.

Comparison of phytoplankton data in the literature is often difficult since so many unstandardised techniques are employed, even after the initial collection, which itself can introduce considerable bias into the data. Most approaches can be classified either as studies of biomass or of production(ivity) and ideally the latter should be expressed relative to the former.

Biomass is the living matter present in a sector of the environment and is measured either in terms of cell numbers, mass (volume) or in biochemical terms. Cells are enumerated using microscopic techniques, the most accurate and satisfactory of which are those based on the inverted microscope (Ütermohl technique). This involves sedimenting known volumes of water using Lugol's solution (iodine in potassium iodide) and identifying and counting the cells present. The technique gives a large amount of information in the form of cells per volume but requires expert taxonomic knowledge, yet its skilled use yields some of the most valuable data, e.g. the autecological studies on *Asterionella* (Lund, 1949), *Melosira* (Lund, 1954, 1955) and desmids (Lund, 1971). Methods using counting chambers (e.g. Sedgwick Rafter cells, haemocytometers), are sometimes used but are more difficult to handle and not so accurate.

Counting by direct microscopic methods is of limited use for the nannoplankton and virtually impossible for the μ-plankton. The latter are mainly unrecognisable by light microscopy and it is necessary to employ culture techniques which involve plating aliquots of water onto agar plates and then counting the colonies. This also yields large numbers of cells for microscopy and subsequent identification but introduces all the problems involved with selectivity of media, variable growth rates, etc. It is a satisfactory

technique for flagellate and coccoid freshwater green algae (Happey, 1970; Happey-Wood, 1976) but has not yet been developed for more delicate flagellates e.g. Chrysophyceae, Prymnesiophyceae. The fact that the nannoplankton is small should not lead to its neglect since in many waters it is responsible for more carbon fixation than the more immediately obvious microplankton,* e.g. in a north temperate lake Kalff (1972) found that on no occasion did net plankton yield more than half of the photosynthate and on an annual basis 70–80% (total carbon 82–78 m m^{-2}) was attributable to the nannoplankton. Similar results have been recorded in Lake Biwa (Tanaka, Nakanishi & Kadota, 1974) and Fig. 7.5 shows that this situation was a feature of the whole water column, both on a chlorophyll *a* basis and from measurements of carbon fixation. Even higher fixation values (95–98% of total) for the nannoplankton were reported by Rodhe, Vollenweider & Nauwerk (1956) during the spring outburst after the ice melt in Lake Erken. Data from estuarine–marine stations provide further evidence of the importance of small species, e.g. McCarthy, Rowland Taylor & Loftus (1974) found that over a two-year study in Chesapeake Bay the nannoplankton (in this case species passing through a 35 μm mesh net) was responsible for 89.6% of the carbon fixation. At times they found that this activity was concentrated in fractions passing even a 10 μm mesh net which means that most of the fixation is by organisms not *normally* seen or difficult to see in routine phytoplankton studies. In the open ocean, especially in oligotrophic regions, the nannoplankton are often the most abundant organisms (Hulbert, Ryther & Guillard, 1960). Pomeroy (1974) gives a table which shows that over 90% of total fixation is by forms smaller than 60 μm in diameter.

Whilst cell counts give valuable data on the proportions of species present in the plankton, and from general phycological knowledge some rough idea of the relative bulks of the various components can be gleaned, it is necesssary to measure cells and to calculate cell volumes if more detailed information of the biomass of individual species is required.

Fig. 7.5. The vertical distribution of primary productivity (O) and chlorophyll *a* ($\triangle$) in the water column in Lake Biwa. Each curve shows the larger contribution by the nannoplankton relative to the net plankton down the whole profile. From Tanaka *et al.*, 1974.

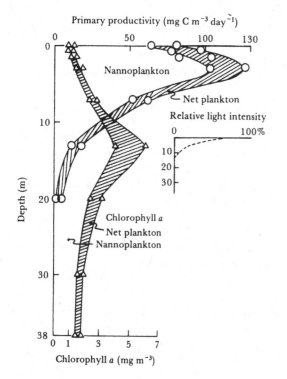

* This was recognised as early as 1912 by Gran; many of his comments were so perceptive that they are equally true today and his writing in *The Depths of the Ocean* by Murray & Hjort (1912) should be read by all contemplating a study of aquatic ecology.

Tables 7.7–7.9 show how cell counts of small organisms over-emphasise their importance in the plankton. Zeitzschel (1970) found that, on average, phytoplankton in the Gulf of California consisted of 72 % of naked flagellates and about 10 % of diatoms whilst coccolithophorids made up less than 5 % in estimates based on cell numbers; but the 72 % of naked flagellates contained only 10 % of the phytoplankton carbon whereas the diatoms contributed 51 %. Nevertheless the true significance of such figures is obscured until rates of cell division, loss and nutrient uptake are determined for each species. Generally speaking the smaller forms exist in high numbers and low volume but have high turnover rates. However there are also seasonal variations making rules a little difficult to apply, e.g. in Narragansett Bay, Durbin, Krawiec & Smayda (1975) found that the net fraction was most important in carbon fixation during winter and spring whilst in summer the nannoplankton accounted for 46.6 % of the annual biomass of chlorophyll *a* and 50.8 % of the total production. It is clearly desirable to express such data on an annual basis although in Narragansett Bay 88 % of the phytoplankton production took place in summer.

More rarely, surface area of cells has been estimated, usually for comparison with nutrient uptake data. Under some circumstances, e.g. when the flora consists of relatively few species fairly simple in form, species can be counted using an electronic particle counter (e.g. a Coulter counter) which can be set to record the number of particles within certain size ranges. This has been used successfully for river plankton where the 'pill-box like' genera *Cyclotella* and *Stephanodiscus* are common (Evans & McGill, 1970; Evans, 1971). Filaments tend to block the apertures and if individual species need to be estimated then the technique has to be combined with microscopic examination, estimation of cell volume and comparison of these with the peaks of particle size counted. Modern developments involve automatic counting systems and matching species against predetermined computerised outlines.

Biomass estimates involving dry weight, wet

Fig. 7.6. Cellular ATP contents of 30 algal cultures as a function of the organic carbon per cell. From Holm-Hansen, 1970.

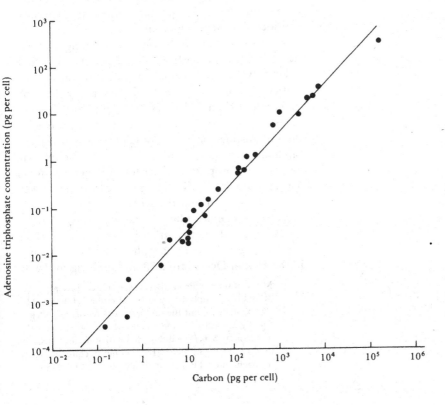

weight (e.g. in Fig. 10.27) or packed volume are more rarely made in ecological studies owing to the difficulty of separating phytoplankton, zooplankton and detritus, but are more common in experimental studies. Similarly, measurements utilising the turbidity caused by the cells in the sample are more common in culture techniques but occasionally such measurements are made at depths in lakes or at sea to detect layering of communities (*see* p. 349).

Biochemical measurements of biomass involve estimation of carbon, nitrogen, particulate organic matter, pigments (usually chlorophyll *a* but others can be very valuable) and most recently by estimating ATP.* A valuable review of methods is that of Strickland (1960). Of these constituents only chlorophyll is exclusive to plants and hence has been most frequently used; it does however vary somewhat from population to population, between 'sun' and 'shade' plants, etc. The ratio of chlorophyll to carbon can vary by a factor of nine or more. The

Fig. 7.7. Depth distribution of total particulate organic carbon (solid line) and organic carbon in live cells (interrupted line) estimated as ATP down an ocean profile (32° 41′ N, 117° 35′ W). From Holm-Hansen, 1970.

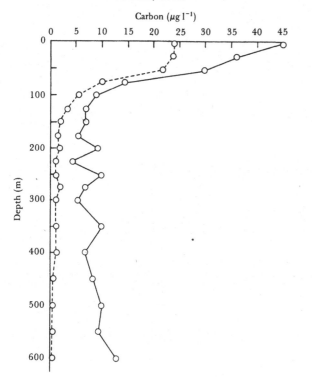

ATP content does not vary to the same extent (Fig. 7.6) and in micro-organisms averages 0.4% of the organic carbon content. According to Holm-Hansen & Paerl (1972) multiplying the ATP values by a factor of 250 gives a figure for carbon content. ATP estimates cannot distinguish between the various sources, and both plant and animal will be estimated simultaneously: this is a source of error when biomass is estimated down a depth profile (Figs. 7.7, 7.8). Unlike cell counting, chlorophyll *a* estimations provide a very rapid and simple method for comparing the amounts of photosynthetic material in the plankton, though it must be appreciated that the actual amounts of pigment per cell vary with the algal groups, e.g. the diatoms contain about twice as much chlorophyll on a dry weight basis as do dinoflagellates according to Gillbricht (1952). There does seem to be a rather general phylogenetic similarity between genera as far as physiological responses are concerned (Lewis, 1977). There are reports that the chlorophyll *a* content of cells varies seasonally, but a recent study (Jones & Ilmavirta, 1978) on a Finnish lake could not detect any variation. Another important source of error in these estimates is the inclusion of chlorophyll degradation products (phaeophytin, phaeophorbide, chlorophyllide); these are generally considered to be negligible in open oceanic water but this view may have to be revised in the light of the data from Kawarada & Sano (1972) who found that chlorophyll exceeded phaeopigments at only three stations out of 29 in the north Pacific along the 155° E meridian from 10° to 48° N. The pigments were measured by fluorescence techniques and since these do not distinguish chlorophyllide *a* from chlorophyll *a*, the degradation products are probably an even greater proportion of the total pigment content. Yet another study (Tominaga, 1971) is a transect across the Indian and the Antarctic Oceans which revealed that phaeophytin almost always followed the chlorophyll *a* distribution but at a higher value, e.g. average values of phaeophytin and chlorophyll in the Indian Ocean are 0.39 and 0.17 mg m^{-3} and

* Many of the problems discussed in this chapter are placed here simply because the experimental material was planktonic, but the problems and results are often equally applicable to the ecology of the algae discussed in Chapters 3–6. The difficulties of determining carbon in natural systems and its relationship to chlorophyll, nitrogen and ATP are admirably discussed by Banse (1977).

Fig. 7.8. The vertical distribution of ATP, particulate carbon, chlorophyll *a* and carbon assimilation, *a*, at an offshore station and, *b*, at an inshore station in Lake Ontario. From Stadelmann & Munawar, 1974.

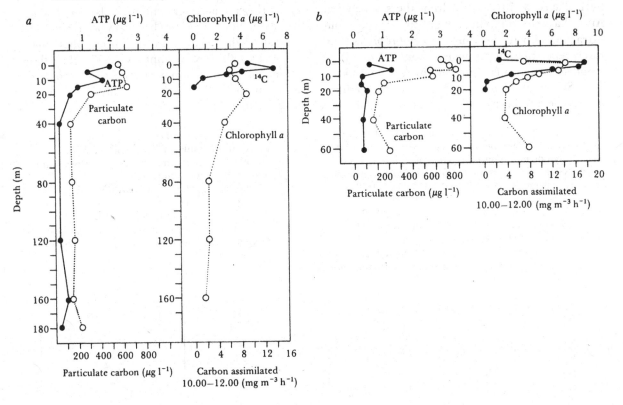

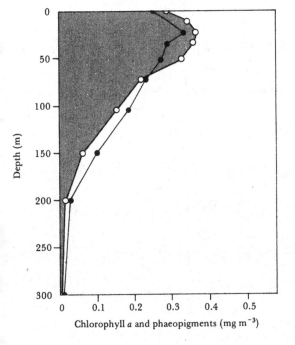

Fig. 7.9. The vertical distribution of chlorophyll *a* (stippled area) and phaeopigments (filled circles) based on average values from all stations during a cruise from Australia to Antarctica. From El-Sayed & Jitts, 1973.

in the Antarctic 0.51 and 0.24 mg m^{-3} respectively. Values of a similar order of magnitude are reported by El-Sayed & Jitts (1973) from the sea between Australia and Antarctica (Fig. 7.9). Errors from extraction of pigments similar to chlorophyll *a* are rarely reported from freshwaters though Rott (1978) found such pigments a problem in Austrian lakes.

Spectrophotometric and fluorescence methods have generally been used for pigment estimation though errors arise not only from interference by the degradation products but from uncertainty in measuring the accessory chlorophylls (b, c_1 and c_2); new extinction coefficients have now been calculated, increasing the accuracy of the spectrophotometric method (Jeffrey & Humphrey, 1975). The fluorescence method does not distinguish chlorophyll *a* from chlorophyllide *a* and phaeophytin *a* is the sum of phaeophytin *a* and phaeophorbide *a*. Jeffrey

Fig. 7.10. *a*. Chromatographic 'map' of major photosynthetic pigments separated from coastal phytoplankton on thin layers of cellulose. Solvent system: first dimension, 25% chloroform in light petroleum (60–80 °C); second dimension, 2% *n*-propanol in light petroleum (60–80 °C). Pigment fractions: 1: carotene (orange); 2: chlorophyll *a* (blue-green); 3: diadinoxanthin (yellow); 4: fucoxanthin (orange); 5: neofucoxanthin (orange); 6: chlorophyll *c* (light green); 7: pheophytin *a* (grey); 8: chlorophyllide *a* (blue-green); 9: pheophorbide *a* (grey); 10: astaxanthin (pink); 11: chlorophyll *b* (olive green); 12: neoxanthin (yellow); 13: peridinin (red); 14: neoperidinin (red).
b. Chromatograms of photosynthetic pigments through the water column, 8 miles off Sydney, Australia, Station 1, 1 October 1970. Cross-hatched areas, chlorophyll *a*; stippled areas, breakdown products of chlorophyll *a*. From Jeffrey, 1974.

a

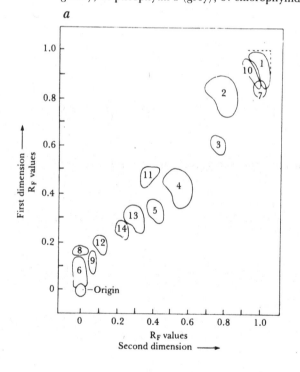

b

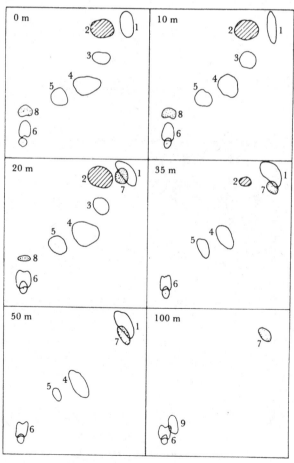

(1974) has developed a thin layer chromatographic separation of the pigments occurring in the water column which distinguishes between the various active and degradation products. Fig. 7.10a gives an example of the pigment separation and 'chromatographic map' and Fig. 7.10b an actual set of results from a depth profile. Fig. 7.11 shows their distribution in samples from the surface to 100 m depth at an oceanic station. Since most estimations in the literature are of chlorophyll a only, they give no information at all on the composition of the phytoplankton, since this pigment is present in all groups. The chromatographic method gives many more data since chlorophyll b is present in Chlorophyta, chlorophyll $c_1 + c_2$ in diatoms, Chrysophyta and Xanthophyta, c_2 in Dinophyta and Cryptophyta, fucoxanthin and neofucoxanthin in diatoms and Chrysophyta, peridinin and neoperidin in Dinophyta,* neoxanthin in Chlorophyta and astaxanthin in copepods. As can be seen from Fig. 7.11 the pigments of the diatoms were present throughout the water column, but the green algal pigments were present only between 8 and 28 m. Senescent diatoms were present from 15 to 75 m (presence of chloro-

phyllide a) and patches of copepods occurred in the water column (presence of astaxanthin). Phaeophorbide occurred below the level of the animals and probably came from faecal pellets of animals feeding on diatoms. The occurrence of a zone of green algal pigments is interesting in that green algae are not normally thought of as components of the marine phytoplankton. Data from photosynthetically active chlorophyll a were compared with carbon fixation–depth profiles and in all cases maximum fixation occurred higher in the water column than did the chlorophyll maximum (Fig. 7.12). These results again show that there is rarely an exact correlation between chlorophyll a distribution and carbon fixation but the chromatographic method greatly extends the ecological information which can be obtained from pigment analyses. Chlorophyll c is stable in the ocean under conditions which cause the breakdown of chlorophyll a and Jeffrey (1974) suggests that the chlorophyll $c:a$ ratio might be used more often as an index of the amount of detritus present rather than the more vague chlorophyll a: phaeopigment ratio. Certainly this chromatographic technique is an extremely valuable one and should be more widely used.

Production–biomass quotients have been devised to indicate assimilation activity; the most common of these are assimilation number (photosynthetic index), measured as milligrams carbon assimilated per milligram chlorophyll a in unit time

* Jeffrey *et al.* (1975) have shown that there are a number of dinoflagellates which contain fucoxanthin and these have been shown also to contain chlorophyll c_1 and c_2. Those containing only c_2 have peridinin as the major carotenoid. See also Jeffrey (1976) for a review of these pigments in algae as a whole.

Fig. 7.11. The distribution of algal pigments at different depth at a station off Sydney. Cross hatched columns, pigments from living cells. Black bars, degraded pigments. From Jeffrey, 1974.

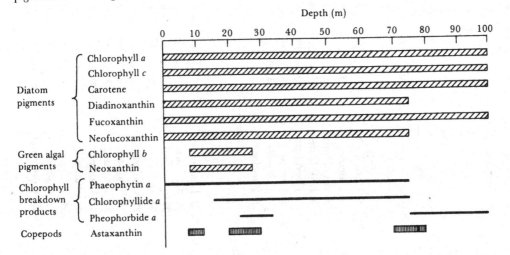

Table 7.1. *The activity coefficients of Cryptomonas erosa, Ceratium hirundinella and Cyclotella comensis in some Austrian lakes, together with some data on the situation at the time of sampling. From Findenegg, 1971*

Lake	Date	Depth interval (m)	Av. temp. (°C)	Surface irradiance (m cal.cm⁻² min⁻¹)	For the optimal depth interval					Average values	
					Depth (m)	Fresh weight (mg m⁻³)	Species contribution (%)	Activity coefficient	Irradiance	Activity coefficient	Irradiance
Cryptomonas erosa											
1 Bodensee	13. ix. 63	0–5	16	730	1	500	60	2.67	350	1.59	272
2 Bodensee	14. x. 65	0–5	12	880	1	200	90	11.37	484	4.20	364
3 Bod. Unter.	26. vi. 63	0–5	17	580	1	210	83	4.23	400	1.66	313
4 Bod. Unter.	1. vi. 64	0–3	18	810	1	250	75	6.06	420	5.12	390
5 Überlingsee	30. vi. 65	0–8	20–11	1210	1	32	94	18.30	814	10.42	525
6 Mondsee	28. iv. 69	0–5	6–5	280	1	180	70	3.90	170	2.41	123
7 Wallersee	4. viii. 70	0–2	21	1150	1	720	65	3.00	365	1.13	215
8 Zürichsee	30. v. 64	0–5	17	1120	2	110	68	3.00	550	2.50	405
								6.57	444	3.63	326
Ceratium hirundinella											
25 Längsee	8.viii.69	0–3	23	900	0	150	75	4.40	900	3.00	515
26 Mondsee	25. viii. 66	0–5	17	230	0	140	79	3.18	230	1.00	104
27 Mondsee	22. vii. 68	0–5	16	280	0	1600	93	1.47	280	0.50	120
28 Mondsee	24. ix. 70	0–5	16	960	2	500	81	0.52	430	0.49	470
29 Obertrumer	1. x. 65	0–5	15	920	3	220	75	2.30	95	5.30	334
30 Traunsee	9. vii. 68	1–3	20	1320	3	450	97	1.09	580	1.05	915
31 Traunsee	13. vii. 70	0–2	16	1100	2	200	91	1.73	430	0.55	750
32 Wallersee	29. vii. 69	0–3	23	1250	2	1500	78	0.58	300	0.40	577
33 Wallersee	7. x. 69	1–3	15	910	1	2100	81	0.28	480	0.02	370
34 Wallersee	4. viii. 70	3–8	18	1150	3	7700	82	0.14	40	0.17	10
35 Wallersee	9. xi. 70	0–1	9	210	0	7140	98	0.09	210	0.06	142
36 Wörthersee	28. vii. 67	0–8	23	880	5	260	70	2.04	160	1.48	446
								1.48	346	1.18	396

					Cyclotella comensis						
41 Millstätter See	26. viii. 70	0–5	20	1050	2	2030	92	1.54	408	1.15	499
42 Feldsee	19. vi. 67	0–5	15	320	0	550	85	0.61	156	0.21	173
43 Feldsee	13. vii. 67	0–5	19	1280	2	1000	90	0.80	570	0.20	560
44 Feldsee	18. vii. 69	0–8	19	1105	2	1800	86	0.64	500	0.50	483
45 Keutschacher See	7. vi. 68	0–5	19	310	0	650	95	0.55	310	0.53	160
46 Ossiacher See	20. vii. 65	0–3	22	1020	2	510	88	3.71	290	2.80	500
47 Ossiacher See	23. vi. 66	0–5	22	1260	2	1200	84	1.02	500	0.58	600
48 Ossiacher See	11. vi. 68	0–5	18	980	3	2000	82	0.93	212	0.59	441
49 Ossiacher See	12. vi. 69	0–5	18	900	3	2100	95	0.92	140	0.83	420
50 Ossiacher See	4. viii. 69	0–8	22	920	2	1150	89	1.28	450	0.55	417
51 Ossiacher See	14. vii. 70	0–5	24	1120	3	1000	90	2.34	320	1.10	562
52 Zeller See	12. x. 65	0–8	13	820	3	460	90	1.00	225	0.87	342
53 Zeller See	28. v. 68	0–8	13	1050	2	390	98	0.87	550	0.61	507
								1.22	352	0.78	430

and the activity coefficient, measured as milligrams carbon assimilated per milligram carbon biomass in unit time. Both these quotients vary considerably, e.g. during a single year the assimilation number varied between 0.5 and 8.5 for *in situ* incubation in the surface water of Abbot's Pool with higher values in summer and lowest in the autumn (Hickman, 1973). This compares well with figures from other waters, e.g. Berman & Pollingher (1974) record a range from 0.4 to 7.8 in Lake Kinneret. Findenegg (1971) found that the activity coefficient in some alpine and pre-alpine lakes, measured when single algae were more or less dominant, varied tenfold between the least active species (*Oscillatoria*) and the most active (*Cryptomonas*), whilst even greater variations have been recorded in lakes over periods of time (Table 7.1). This is not surprising when one considers the range in carbon contents of single algae at different times, e.g. Serruya *et al.* (1974) found carbon:nitrogen:phosphorus ratios of *Peridinium cinctum* f. *westii* to be 235:18:1 in the early stages of the bloom and 645:25:1 in the later stages. The general assumption that chlorophyll is equally efficient at all stages is not valid and the efficiency varies with the nutritional state of the population (Senft, 1978).

Production as total carbon fixed in unit time and productivity or rate of carbon fixation in unit

time is measured generally either by the uptake of ^{14}C from labelled bicarbonate or by measuring the oxygen output. In both instances dark controls are run to determine the amount of respiration. Many safeguards are necessary in both methods and techniques are continually being improved: Bienfang & Gundersen (1977) developed a technique for filling bottles and injecting ^{14}C into them at defined depths without bringing samples to the surface and exposing the algae to unnatural light conditions. In the ^{14}C method, errors can arise on filtration when small cells break or leak and then release some of the fixed carbon into the filtrate; Schindler & Holmgren (1971) found that this effect alone could add an error of 1.2–6.1 % to the uncorrected value with the larger error in lakes in which nannoplankton was dominant. It is often assumed that the dark bottle controls will show a similar amount of respiration down a depth profile but in some lakes there can be considerable variations (Gerletti, 1968) and therefore dark bottles should be used at all depths. Many observations have been made in which the ^{14}C incubations have been done at fixed depths but obviously no cell in nature remains at a fixed depth and experiments in which bottles were circulated up and down the profile yielded fixation rates 19–87 % higher than estimates based on fixed bottles (Marra, 1978).

Another factor which needs to be considered

Fig. 7.12. Depth profiles of photosynthetically active chlorophyll *a*, and photosynthetic carbon fixation at four different stations. Solid lines, chlorophyll *a*; interrupted lines, ^{14}C uptake. From Jeffrey, 1974.

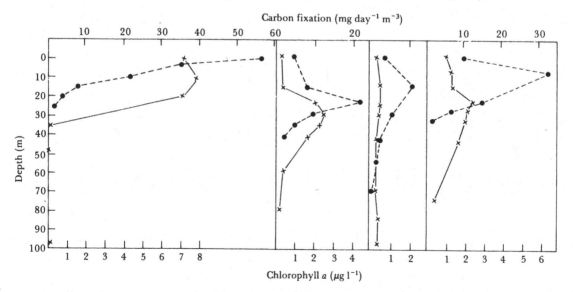

is photorespiration; this can occur under conditions of low carbon dioxide and high oxygen concentrations such as have been recorded in eutrophic sites on warm summer days or in bottles if dense blooms and long incubation periods are used. Lex, Silvester & Stewart (1972) showed experimentally that the blue-green alga *Anabaena cylindrica*, performs photorespiration and since this involves oxygen uptake in the light it would decrease the rate of oxygen evolution measured by the Winkler technique; Burris (1977) has shown that eight other species of algae, including macroalgae, can photorespire. If

Fig. 7.13. *a*, the photosynthetic response of the spring diatom assemblage of Lake Ontario to short-term rising and falling light regimes. The photosynthesis follows the upper curve when light intensity is increasing and the lower curve when light is decreasing. Samples taken from lake under sunny conditions (diamonds) and cloudy condition (circles). *b*, daily photosynthesis on a cloudless day for the top 10 m showing the mid-day depression. From Harris, 1973.

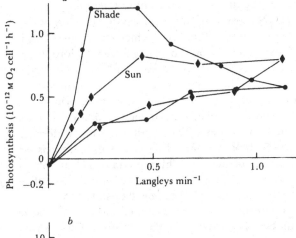

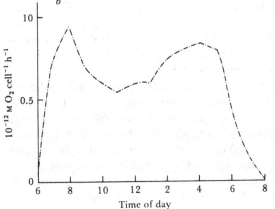

carbon dioxide is also evolved during photorespiration (as it is in higher plants) measurements of ^{14}C fixation will be misleadingly lowered. Lex *et al.* also showed that oxygen uptake in the light may exceed oxygen uptake in the dark and would lead to errors when net photosynthesis is estimated by subtracting the values obtained in dark bottles from those in the light. Evidence for photorespiration has also been obtained in experiments by Harris (1973) in which he subjected phytoplankton to increasing and decreasing light regimes and found a hysteresis effect (Fig. 7.13) which was attributed in part to photorespiration. The hysteresis was confirmed by Falkowski & Owens (1978) who thought it may be due to changes in light quality as the intensity was changed. It must be emphasised however that some of these effects are based on experiments with unialgal cultures and it is necessary to check whether such effects occur in nature.

More rarely, changes in pH have been used as a measure of carbon dioxide uptake (e.g. Jackson & McFadden, 1954; Verduin, 1959). These techniques almost always involved subsampling and short time exposures, and methods of continuous assessment of photosynthesis are badly needed: the technology is available though the techniques may be expensive.

Thus plankton which appears at first sight to be a simple community can only be studied satisfactorily by using a variety of sampling techniques, both in nature and in the laboratory. These together yield the essential data on species composition and numerical data on cell concentrations, chlorophyll, etc. In spite of all the problems, this community has been investigated more intensively than any other, since it is still considered one of the easiest to sample, yet I doubt whether the total phytoplankton of any lake or ocean site has ever been completely enumerated: contrast this with the situation in higher plant ecology where the species composition is often known down to the last detail. One other striking feature appearing from various sources in the literature is that it seems that limitations in the experimental techniques lead to a considerable underestimation of the amount of carbon fixation by phytoplankton (and other algal communities). In fact these underestimates have been put at between 25 and 80% (Harris & Piccinin, 1977) figures which, if true, have important implications for models of energy flow.

Phytoplankton populations

Very large numbers of species are involved both in freshwaters and in the sea, e.g. Skuja (1948) records 440 species in the phytoplankton of Lake Erken and close on 1000 species are recorded for the Pacific Ocean by Semina & Tarkhova (1972). This latter figure was made up mainly of diatoms and dinoflagellates. The only other conspicuous marine microplanktonic forms are the spherical green cells belonging to the Prasinophyta (*Halosphaera*, *Pterosperma*) and the bundles of filaments of the Cyanophyte genus *Trichodesmium* (*Oscillatoria*): both of these groups tend to float to the surface, the former buoyed up by oil globules and the latter by gas vacuoles in the cells. The nannoplankton is almost entirely composed of small flagellate cells belonging to the Prymnesiophyta, i.e. forms which in addition to possessing two flagella also have a haptonema. This group now contains the genera of the *Prymnesiales* (= Coccolithophoridaceae) since many of these have been shown to possess a haptonema. Some are caught in nets but many are too small and pass through the meshes or, being delicate, are damaged beyond recognition or are destroyed by preservatives (e.g. formalin, which is not an ideal preservative for phytoplankton) and their numerical abundance is rarely determined. The section of the Prymnesiophyta bearing calcareous plates (coccoliths) are more obvious than the delicate forms bearing organic scales, e.g. *Chrysochromulina*, but the latter can make up a considerable amount of the biomass in some seas. A few small diatoms, dinoflagellates and other groups (e.g. *Dictyocha*) occur in the marine nannoplankton but detailed studies are still needed. A surprising number of major algal groups are virtually absent from the marine plankton; thus there are few records and these need careful checking of the Chlorophyta, Chrysophyta, Xanthophyta, Euglenophyta or Cyanophyta* (excluding *Trichodesmium*). A few species of the Cryptophyta and an unknown number of flagellate species of the Prasinophyta occur in small numbers but no genera of the Phaeophyta or Rhodophyta are small enough or have a suitable morphology to form plankton. Floating beds of species of these two latter groups occur and in a strict sense they should be regarded

---

* Since this chapter was completed, very abundant populations of bacterial sized Cyanophyta have been discovered in the ocean and shown to be important in carbon fixation.

as macro-plankton (*see* p. 295).

Species tend to occur in groups throughout natural communities and it ought to be possible to distinguish associations of species in the plankton. Some workers assume that there is only one continuous spectrum of species: this may be so in some small, well mixed lakes but it is not easy to check. At sea so few stations are occupied more than once and rarely long enough to yield seasonal data that it is difficult to build up the necessary information. Observations from some detailed surveys and from the continuous plankton recorder certainly suggest that there are discrete associations. These associations appear to be linked with geographical zones (currents, water masses) rather than with subtle differences in water chemistry, etc., though these may also prove to have some influence when they are fully analysed. Only in a few places where fresh and saltwater mix is there any intermingling of freshwater and marine phytoplankton; this has been recorded in the southern Baltic (Ringer, 1973), and presumably the marine and freshwater elements are at the limits of their salinity tolerances.

No group of algae dominates the freshwater phytoplankton (Fig. 7.14) which, taken over all, is a mixture of diatoms, dinoflagellates, Chlorophyta, Cyanophyta, Chrysophyta and Xanthophyta. Species of the Prymnesiophyta and Prasinophyta occur, but seem to be relatively rare, whilst the Euglenophyta tend only to be abundant in small bodies of water. The most striking difference between marine and freshwater phytoplankton is the multitude of species of Chlorophyta and Cyanophyta in the latter coupled with the almost complete absence of coccolith bearing flagellates. A further characteristic of freshwater plankton is the large number of colonial forms encountered (Fig. 7.15), a feature making estimation of cell numbers extemely difficult. Since the range in chemical composition of freshwaters is so much greater than that of the sea, more easily recognisable associations of species occur. The work of Teiling (1916) and Naumann (1917) resulted in the recognition of two major assemblages, one of oligotrophic waters with Chrysophyta and desmids dominant (Caledonian phytoplankton) and another of eutrophic waters, with diatoms and Cyanophyta dominant (Baltic phytoplankton). However this is an over-simplification and a series of associations can be recognised

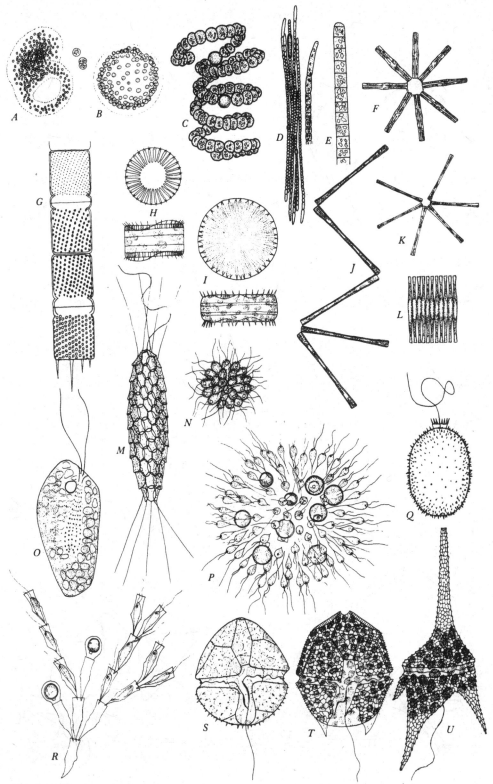

Fig. 7.14. For legend see page 263.

Fig. 7.14 (*cont.*)

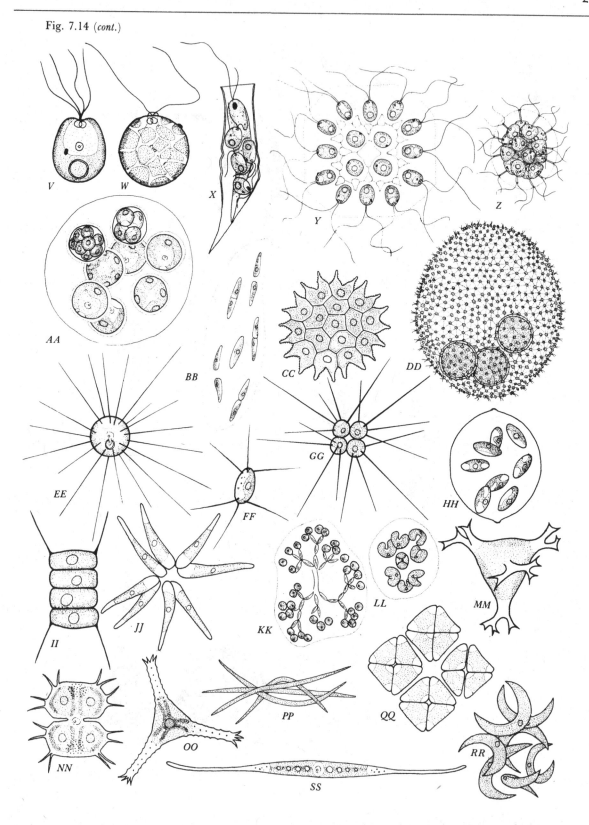

V   W   X   Y   Z

AA   BB   CC   DD

EE   FF   GG   HH

II   JJ   KK   LL   MM

NN   OO   PP   QQ   RR

SS

from both water types and yet others from mesotrophic, dystrophic, inland saline waters, etc. There is a considerable amount of work still to be done analysing these associations along plant sociological lines.

Although the numbers quoted at the beginning of this section seem high, most estimates of diversity are in fact low since they are based on lists prepared from cell counts derived from identifications made from simple morphological criteria determined by light microscopy. Extensive studies with the electron microscope are already yielding more variants especially amongst the small species and these can be shown to be discrete entities, occurring for example, in separate geographical regions without inter-

gradation between populations (Round, unpublished). In addition species, which have been given the same name in collections from different regions often exhibit differences in physiology and are in fact physiological races. An example of the distribution of different forms of a species in freshwaters is given in Fig. 7.16 showing the distribution of *Tabellaria* in the English Lake District (Knudsen, 1955). Some of these were recorded in Irish lakes in 1906 (West & West) and were found to be still present in the same waters in 1953 (Round & Brook, 1959), so these apparently minor variants are fairly stable features of the plankton. Most recently, Koppen (1975) has continued the study of these varieties and distinguished further North American forms, which

Fig. 7.15. Phytoplankton (× 160) from Windermere South Basin, December 1965. Photograph by Dr H. M. Canter.

Fig. 7.14. Some freshwater planktonic algae, all of which, except *Volvox*, are microscopic. *A, Microcystis*; *B, Coelosphaerium*; *C, Anabaena*; *D, Aphanizomenon*; *E, Oscillatoria*; *F, Tabellaria*; *G, Melosira*; *H, Cyclotella*; *I, Stephanodiscus*; *J, Diatoma*; *K, Asterionella*; *L, Fragilaria*; *M, Mallomonas*; *N, Synura*; *O, Cryptomonas*; *P, Uroglena*; *Q, Trachelomonas*; *R, Dinobryon*; *S, Gymnodinium*; *T, Peridinium*; *U, Ceratium*; *V, Carteria*; *W, Chlamydomonas*; *X, Chlamydomonas* in the theca of *Dinobryon*; *Y, Gonium*; *Z, Pandorina*; *AA, Sphaerocystis*; *BB, Elakatothrix*; *CC, Pediastrum*; *DD, Volvox*; *EE, Golenkinia*; *FF, Lagerheimia*; *GG, Micractinium*; *HH, Oocystis*; *II, Scenedesmus*; *JJ, Actinastrum*; *KK, Dictyosphaerium*; *LL, Kirchneriella*; *MM, Tetraedron*; *NN, Xanthidium*; *OO, Staurastrum*; *PP, Ankistrodesmus*; *QQ, Crucigenia*; *RR, Selenastrum*; *SS, Closterium*.

suggests that even this common planktonic genus still requires detailed study. An equally interesting example from the marine plankton is afforded by the diatom, *Rhizosolenia styliformis*, which occurs in at least three varieties which Robinson & Waller (1966) showed to have a distinct distribution in the North Sea and North Atlantic Ocean (Fig. 7.17). In addition to the three varieties they showed that there were geographically separated populations, which could be distinguished on size, so that in this area alone there were some seven populations whose distribution is given in Fig. 7.18. This work shows the value of long term plankton recorder data and analysis of fine detail, and is a nice example of the information content referred to by Margalef & Estrada (*see* p. 2).

More complex systems for the classification of lakes, based on the composition of the phytoplankton, have been proposed by Thunmark (1945) and Nygaard (1949): these are merely numerical attempts to express degrees of oligotrophy and eutrophy (*see* p. 526 for a discussion of these concepts) from a consideration of species complement rather than from nutrient levels. Thunmark devised an index based on the ratio of the number of species of

$$\frac{\text{Chlorococcales}}{\text{desmids}}$$

and Nygaard added to this a Cyanophycean index

$$\frac{\text{Cyanophyceae}}{\text{desmids}},$$

a diatom index

$$\frac{\text{centric diatoms}}{\text{pennate diatoms}},$$

a Euglenophycean index

$$\frac{\text{Euglenophyceae}}{\text{Cyanophyceae, Chlorophyceae}},$$

and a compound index

$$\frac{\text{Cyanophyceae} + \text{Chlorococales} + \text{centric diatoms} + \text{Euglenophyceae}}{\text{desmids}}.$$

These indices will vary somewhat according to the

Fig. 7.16. The different morphological strains of *Tabellaria flocculosa* found in the plankton of lakes in the English Lake District. From Knudsen, 1955.

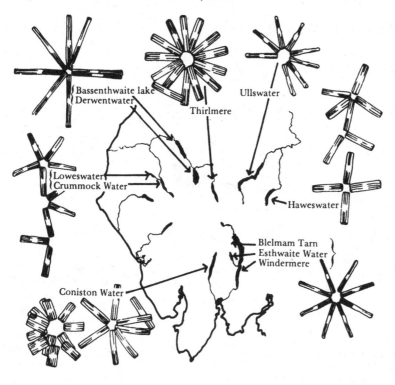

Fig. 7.17. The distribution of the three figured varieties of *Rhizosolenia styliformis* in 1959. Data from continuous plankton recorder counts. *a*, *R. styliformis* var. *styliformis*; *b*, *R. styliformis* var. *semispina*; *c*, *R. styliformis* var. *oceanica*. From Robinson & Waller, 1966. Crosses, no *Rhizosolenia* found; dots, less than 1100; open circles, 1100–3299; half-filled circles, 3300–7699; filled circles, more than 7700 per sample.

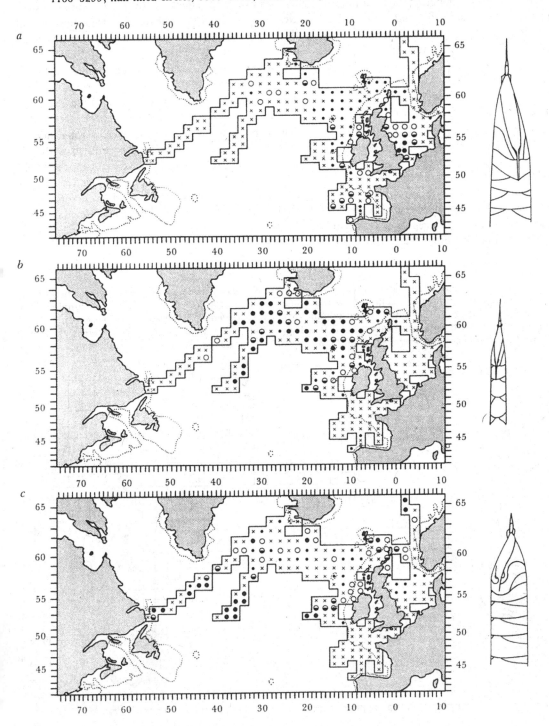

type and time of collection and Coesel (1975) warns of the dangers involved in using desmids which are more widely distributed in waters of all types than is usually recognised. In general the compound index seems to be the most useful and if this is less than 0.2 it indicates a dystrophic state, up to 1.0 an oligotrophic water, 1.0–3.0 mesotrophic and above 3.0 eutrophic. Using the compound index on a series of Irish loughs a reasonable correlation between the index and the general drainage basin types was established, separating the lakes into those lying in peat and those on limestone (Round & Brook, 1959); combining this information with an analysis of the dominant species gave the associations in Table 7.2 showing the trend from Cyanophyceae in the eutrophic to desmids in the oligotrophic. It was concluded however that the mesotrophic category was not a useful grouping, in this region at least, and that on species dominance Lough Gara and Lough

Levally rest with the oligotrophic and Lough Corrib with the eutrophic. This study was made on only single samples from the lakes and obviously it would be desirable to check it over an annual cycle. Nevertheless it does indicate that an assessment of the trophic status can be made from a single sample, a feature which is of some value when field surveys are undertaken under difficult conditions.

A system based on indicator species also has merit though as one can see from Table 7.2 some species have a wide range (e.g. *Asterionella*) and a system based on species with narrow tolerances is needed. This should possibly be coupled with cell counts since in all ecology the extremes are relatively easy to typify but the mass of intermediate types requires much greater study and occurrence plus abundance adds considerably more information.

Fig. 7.18. The distributions of the populations of *Rhizosolenia styliformis* in the areas sampled by the continuous plankton recorder. Only statistical rectangles with average numbers greater than 22 000 cells per sample in any month were used except for the 'small'-celled populations of var. *styliformis* (*styliformis* B and C) when statistical rectangles with average numbers of 8000 cells per sample were used. A, B, C, *R. styliformis* var. *styliformis*; D, *R. styliformis* var. *semispina*; E, F, G, *R. styliformis* var. *oceanica*. From Robinson & Waller, 1966.

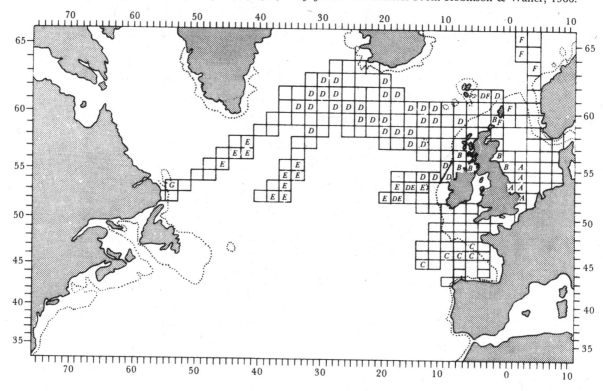

Table 7.2.a. *The dominant species of phytoplankton and the Compound Index in Irish loughs of various water type. From Round & Brook, 1959*

Lake type and lake	Dominant species	Compound index
Eutrophic		
Key	*Oscillatoria agardhii, Coelosphaerium naegelianum*	21.0
Arrow	*Melosira granulata, Volvox aureus*	9.5
Gill	*Fragilaria crotonensis*	9.0
Colgagh	*Peridinium voltzii, Ceratium hirundinella*	8.0
Erne	*Melosira granulata, Fragilaria crotonensis, Coelosphaerium naegelianum*	7.5
Rea	*Asterionella formosa*	6.7
Bunny	*Asterionella formosa, Dinobryon divergens*	6.5
Carrick	*Melosira italica* subsp. *subarctica, Tabellaria flocculosa* var. *asterionelloides, Hyalotheca mucosa*	5.0
Cullin	*Botryococcus braunii, Fragilaria crotoensis, Asterionella formosa*	5.0
Caherglassan	*Eudorina elegans, Asterionella formosa*	4.5
Glencar	*Synedra acus* var. *angustissima, Melosira italica* subsp. *subarctica*	4.0
Conn	*Oscillatoria agardhii*	3.5
Melvin	*Coelosphaerium naegelianum, Microcystis aeruginosa*	3.0
Talt	*Asterionella formosa*	3.0
Mesotrophic		
Levally	*Asterionella formosa, Dinobryon divergens*	1.1
Gara	*Tabellaria flocculosa* var. *asterionelloides*	1.0
Corrib	(North and South): *Synedra acus* var. *angustissima*	1.0, 1.1
Oligotrophic and dystrophic		
Beltra	*Tabellaria flocculosa* var. *asterionelloides, Dinobryon divergens* var. *schauinslandii*	0.7
Aunierin	*Spondylosum planum, Staurodesmus triangularis* var. *inflatus, Staurastrum longipes*	0.6
Derryclare	*Staurastrum anatinum*	0.55
Kylemore	*Staurastrum anatinum, Staurodesmus cuspidatus* var. *canadense*	0.55
Bofin	*Staurastrum longipes, Cosmarium contractum*	0.5
Ardderry	*Asterionella formosa, Dinobryon divergens* var. *schauinslandii*	0.4
Oorid	*Staurastrum longipes, Staurodesmus triangularis* var. *inflatus*	0.3
Ballynahinch	*Staurastrum anatinum, Asterionella formosa*	0.2

Rivers

Flowing waters develop phytoplankton crops only in the slower moving parts; in the River Nile the quantity was shown to be inversely proportional to the current speed (Brook & Rzòska, 1954). Diatoms and coccoid Chlorophyta tend to be the common organisms, and Szemes (1967) reported that diatoms were the dominant forms in the River Danube throughout its length, both in number of species and of cells. Of the diatoms the small centric species of the genera *Stephanodiscus* and *Cyclotella* are undoubtedly the most common (e.g. in the River Thames (Evans, 1971; Lack, 1971), in the Rivers Lee, Stour and Severn (Swale, 1964, 1969) and in the River Avon (Aykulu, 1978)). Lack (1971) recorded 67 000 cells ml^{-1} of centric diatoms at the

Table 7.2.b. *The average number of species of the various algal groups in the different categories. From Round and Brook, 1959*

Taxonomic group	Average number of species per lough		
	Eutrophic loughs	Meso-trophic loughs	Oligo-trophic loughs
Chlorococcales	3.5	1.8	2.5
Cyanophyceae	7.2	5.2	3.1
Centric diatoms	2.7	1.8	0.9
Desmideae	2.4	8.6	15.5
Pennate diatoms	5.6	5.4	3.7
Chrysophyceae	1.2	2.8	3.3
Dinophyceae	1.3	1.4	1.4

peak of growth in the River Thames when they formed 94% of the population.* Roeder (1977) showed that there was an inverse relationship between rate of flow and the number of centric diatoms in the plankton; diatoms were common only when the flow rate was less than 60 ft^{-3} s^{-1}. The only other group of algae regularly recorded in any quantity in river plankton are the flagellate Volvocales and coccoid Chlorococcales of the Chlorophyta (e.g., *Chlamydomonas, Gonium, Pandorina, Pediastrum, Scenedesmus, Ankistrodesmus, Crucigenia, Lagerheimia, Golenkinia, Micractinium, Actinastrum, Dictyosphaerium*). Species of Cyanophyta (*Microcystis, Anabaena*), Euglenophyta (*Euglena, Phacus, Trachelomonas*), Dinophyta, Cryptophyta and Chrysophyta are relatively rare and only very occasionally form large populations in rivers. In a comparison of three river systems in Hungary, Uherkovich (1969) found *Cyclotella–Nitzschia acicularis–Synedra acus–Actinastrum hantzschii* dominant in the River Danube, *Cyclotella–N. acicularis–S. ulna* and *Scenedesmus* spp. in the River Theiss and *Ceratoneis arcus–Cyclotella–Diatoma vulgare–Synedra ulna* in the River Drau. The latter river, which is the fastest flowing, is clearly regularly contaminated from the epilithon which yields the *Ceratoneis–Diatoma* populations, whilst the occur-

rence of *Synedra* and *Nitzschia* probably indicates contamination from the epiphyton and epipelon.

There have been few synoptic accounts of river plankton but one excellent review is the United States National Water Quality Network survey of 65 river stations throughout the United States (Williams, 1962). Sampling was direct, i.e., from water samples and not, as is often common in river studies, from artificial substrata, so the results reflect the actual populations drifting in the rivers and show the penetration of brackish species (e.g. *Cyclotella striata* and *Coscinodiscus denarius*) up the lower reaches. The northern rivers tend to have populations dominated by *Stephanodiscus, Melosira, Synedra* and *Fragilaria* whilst the southern rivers flowing through arid regions and with waters rich in calcium carbonate and other dissolved salts, are rich in pennate diatoms, e.g. *Amphiprora alata, A. paludosa, Caloneis amphisbaena, Pleurosigma delicatula, Surrirella brightwellii, S. striatula* and *Diploneis smithii*, whilst in the southeast *Cyclotella pseudostelligera* and *Melosira distans* var. *alpigena* are characteristic of certain stations. The latter two species are undoubtedly planktonic and perhaps associated with a supply of certain organic (peaty) components which are absent from the waters of other regions. The waters flowing through the arid states are notable for the fact that the species quoted above are all motile diatoms equally common in the epipelic community and one suspects that they are recruited from the sediments, though at some of the stations there is also a fairly rich centric diatom population co-existing in the water with the pennate diatoms. This is a very significant study since it is the only one of which I am aware, which shows that river plankton has very distinct geographical elements. In a later paper, Williams (1972) reports the penetration of Lake Michigan diatoms many miles up the polluted Illinois River.

Early workers (e.g., Krieger, 1927; Butcher, 1932, 1940) believed that the river phytoplankton originated from headwaters, lakes and pools along the river or was recruited from the sediments. There seems to be little doubt that in fact all these and others (e.g. epiphytic populations) contribute cells to the (pseudo)plankton and that in certain rivers, or sections of rivers, plankton is mainly a miscellaneous collection from such sites (Swanson & Bachmann, 1976; Roeder, 1977), carried along in the stream and

* Numbers of diatom species in the range 140–190 recorded for rivers by Patrick (1963) and for the River Avon near Bristol (Moore, 1976) are not of purely planktonic forms.

perhaps dividing to some extent in the open water. It is equally certain that other genera, e.g. *Stephanodiscus, Chlamydomonas, Ankistrodesmus, Crucigenia*, etc. remain in the open water for considerable lengths of time and divide many times, often to form rich blooms, and must be regarded as euplanktonic. However, these algae are also trapped amongst debris and epiphytes and doubtless these masses form an inoculum which is maintained in benthic habitats (Aykulu, 1978). In fact, on theoretical grounds it is difficult to see how in a flowing system any species can remain in the water unless a few cells are trapped in backwaters or between vegetation. One should not overlook the fact, that, as in lakes (*see* p. 245), there may always be the occasional cell in the water mass ready to re-commence growth when conditions become favourable.

The bulk of work on rivers has been done using glass slides held in various devices (diatometers) but these do *not* measure the phytoplankton and works such as that of Williams (1962) quoted above are much more valuable. Slides measure the populations mainly of epiphytic and epilithic species which happen to attach to the glass; van Landingham (1964) summed it up. 'The diatometer method may be good for studying diatom populations under certain conditions; nevertheless at best it is artificial. Diatometers are unsuitable for plankton diatom studies; in fact they are unsatisfactory for the study of sessile diatom populations. For plankton diatom population studies, perhaps the most satisfactory method is the volumetric sample method.' Few would disagree, especially with the last statement and it is unfortunate for phycology that so much effort has been concentrated on such an unsuitable method when so much more useful information could have been obtained about the actual planktonic populations.

Flooding and changes in turbidity are obviously of greater significance in rivers than in lakes but effects on the flora are obscure; often the result is the report of a new species in the riverine plankton, for example the inclusion of *Nitzschia sigmoidea* in early lists of plankton from the River Thames is purely the result of removal of this species from the sediments during floods. The comprehensive data on the River Danube (Schallgruber, 1944) listed many of the benthic forms though they were recognised by the author. Care should be taken to distinguish the algae which move temporarily and diurnally into the river water (*see* p. 161 for a discussion of these forms).

Perusal of the literature suggests that a few species of algae are common to phytoplankton of both rivers and lakes; it would be interesting to check the exact identity of these. Comparisons of floras are notoriously difficult owing to the lack of detailed taxonomic studies on many species, even those that are common. One species which is reported from both lakes and rivers is *Asterionella formosa* but other reports indicate that lake populations of this species are rapidly lost in fairly fast flowing outflows from lakes, and rate of flow may be the key to survival. Ruttner (1956) found that 90% of lake plankton had disappeared at a station one kilometre down the outflowing watercourse. There is little doubt that this is a general feature and wherever the water filters through beds of macrophytic plants the lake plankton is reduced even faster; it can be decimated by passage through only 20 m of such a bed (Chandler, 1937).

If laminar flow occurs in a watercourse then theoretically no planktonic population could maintain itself and a state close to this certainly exists in the upper reaches of many streams where the depth is relatively shallow and vertical water movements are slight. In wider, deeper stretches the contours lead to turbulent flow and back flow in many places, which slow the flow very considerably and allow time for the growth of populations. There is also a concept known as 'ageing' of the water and Shelford & Eddy (1929) comment that river plankton develops in water 20 days old. However this is more likely to be a factor of flow, supply of inoculum, etc., and little to do with regeneration of nutrients which is what 'ageing' usually implies. Growth experiments with typical river algae and water from the upper reaches are needed.

The seasonal cycles of river plankton are not so well pronounced as in lakes or the oceans but there are clear winter lows and spring blooms (e.g., of diatom in the River Danube (Schallgruber, 1944) and autumn blooms (*see* also p. 448). Whilst rate of flow is clearly an important factor in rivers (Fig. 7.19), as Lack (1971) points out the effect of increase in flow is often seen only as a decrease in density of the population (i.e. a dilution of the population) whilst the total number of cells at any point in the river is similar. Rivers tend to be rich in nutrients and though instances of limitation have been recorded it is likely to have a subsidiary effect on annual cycles. However, as Fig. 7.19 shows, the growth of diatoms is accompanied by falls in silica concentration though there is still apparently sufficient to sustain considerable further growth and therefore it is unlikely to limit diatom growth in rivers as it does at times in lakes (Fig. 10.11) and in the ocean (Fig. 10.13).

Measurements of chlorophyll concentration in the River Thames show greater fluctuations than do cell counts but the peaks and troughs coincide. Clearly the occurrence of large amounts of phaeopigments is something which must be taken into account in any study of algal pigments in rivers since the load of plant detritus obviously yields great quantities of these inactive pigments, especially in winter, e.g., in the River Thames on one occasion they reached 95% of all the pigments and over the year averaged 32% of total chlorophyll pigments (Kowalczewski & Lack, 1971).

Photosynthesis in the River Thames was almost entirely confined to the upper two metres and for much of the year production was concentrated in the upper metre. Oxygen production by the phytoplankton was enhanced in laboratory determinations when the bottles were rotated to prevent settling and it may be that this is more important in experiments on river algae than in those from standing waters. Again, unlike the general situation in lakes and oceans, there is often a net loss of oxygen from the water especially during winter, e.g., in the River Thames, whilst in the River Kennet, with a very low phytoplankton content, the respiratory removal of oxygen exceeded that of algal production over most of the year (Lack, 1971).

Plankton of tropical rivers has received little attention though the River Ganges in north India is reported to be dominanted by *Synedra* and *Fragilaria* and not the usual centric species of the temperate zone (Pahwa & Mehrotra, 1966), and the filamentous Zygnemaphycean genera *Mougeotia* and *Spiro-*

Fig. 7.19. The variation *a* in total cell numbers, *b*, in centric diatoms (filled circles) and silica (open circles), *c*, pennate diatoms, *d*, chlorophyceae, *e*, all other groups, *f*, temperature (filled circles) and discharge (vertical bars) in the River Thames. From Lack, 1971.

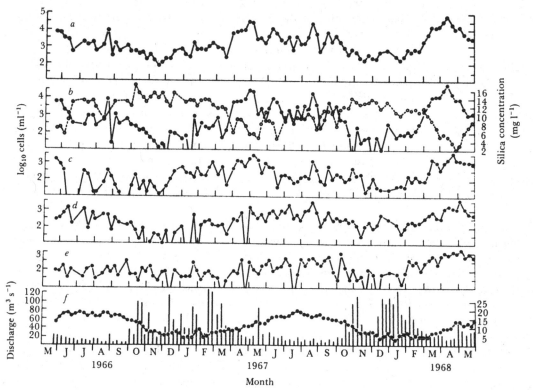

Table 7.3. *The levels of oxygen, biological oxygen demand, turbidity, suspended solids, chlorophyll and particulate carbon in four London reservoirs in mid-summer and the concomitant levels in the River Thames which supplies these reservoirs. From the Metropolitan Water Board, 45th Annual Report*

Reservoir	Dissolved oxygen (mg l⁻¹) July 5	12	19	26	August 2	9	16	Biological oxygen demand (mg l⁻¹) July 5	12	19	26	August 2	9	16
Wraysbury	7.5	7.4	7.4	6.5	7.2	7.6	7.3	0.8	1.4	1.9	1.4	1.1	1.6	1.1
Knight-Bessborough	6.3	7.5	7.0	5.7	5.6	7.6	6.4	1.3	1.9	2.3	1.9	1.4	1.7	0.6
Queen Elizabeth II	7.3	7.6	8.0	7.8	8.1	8.2	7.5	1.1	0.7	3.4	1.5	3.6	1.2	3.0
Queen Mary	7.7	9.2	9.4	7.9	8.2	8.2	7.6	1.6	1.7	1.8	2.1	1.7	0.8	0.9
R. Thames	8.8	9.1	8.3	8.3	8.9	8.4	9.1	1.9	2.1	4.1	1.4	2.1	2.4	2.9

Reservoir	Turbidity units July 5	12	19	26	August 2	9	16	Suspended solids (mg l⁻¹) July 5	12	19	26	August 2	9	16
Wraysbury	0.4	0.3	0.8	0.5	1.3	0.8	0.7	1.1	1.2	1.8	1.2	1.6	1.0	1.1
Knight-Bessborough	0.6	0.9	1.3	0.6	1.2	1.4	1.3	1.4	1.6	1.9	1.4	1.8	1.4	0.6
Queen Elizabeth II	0.6	0.6	1.5	0.6	0.9	1.7	1.3	1.5	1.4	2.2	2.8	—	1.2	3.0
Queen Mary	—	—	1.5	1.1	1.1	1.7	1.2	4.6	2.0	2.5	2.5	3.0	1.0	0.9
R. Thames	9.0	8.2	14.0	13.2	12.4	11.2	11.5	9.0	7.0	7.0	8.0	8.0	7.0	12.0

Reservoir	Chlorophyll a (μg l⁻¹) July 5	12	19	26	August 2	9	16	Particulate carbon (μg l⁻¹) July 5	12	19	26	August 2	9	16
Wraysbury	3.2	3.5	3.4	1.1	11.9	2.7	1.3	459	358	465	333	823	335	310
Knight-Bessborough	2.5	5.8	4.1	1.2	4.8	4.6	2.4	432	436	481	321	643	494	347
Queen Elizabeth II	—	2.2	2.2	2.8	11.8	5.9	1.0	—	300	301	393	830	619	589
Queen Mary	—	2.6	3.2	8.9	1.4	2.1	2.2	—	806	477	760	481	296	328
R. Thames	23.1	27.8	41.6	17.4	24.4	20.4	39.2	1327	1600	1973	1569	1993	1380	1351

gyra are also common. Except for *Melosira granulata* and its var. *angustissima*, centric species are also absent from the River Nile near Khartoum (Rzóska, Brook & Prowse, 1955) and when the rate of flow is reduced blue-green algae (*Anabaena flos-aquae* var. *intermedia* f. *spiroides*, *Raphidiopsis curvata* and *Lyngbya limnetica*) take over the dominant position in the phytoplankton. Even in a flowing system such as the River Nile, there can be a variation in cell numbers from the top to the bottom of the water profile, with, for example, a concentration of *Melosira* at the surface (Abdel Karim & Saeed, 1978). There is evidence (Rzóska *et al.*, 1955) that in the Nile, after a period of high rate of flow when plankton is scarce,

the diatoms appear first and as flow rate decreases blue-green algae become abundant. Similar abundance of *Melosira granulata* is reported in the River Volta, Ghana (Biswas, 1968) together with chlorococcalean and cyanophycean algae, and here also growth of phytoplankton is restricted to the dry season, during reduced flow.

The phytoplankton of estuaries is now receiving greater attention (*see* review by Round, 1979) as the problems of pollution of estuaries increase. In many estuaries the combined effects of turbulence and turbidity produce most unfavourable conditions for phytoplankton growth e.g., in the Hudson and Newport estuaries and probably in most large British estuaries. In less turbid estuaries diverse floras of small diatom species and dinoflagellates are common and as in the open ocean it is the nannoplankton which contributes the greatest amount of chlorophyll and production e.g. 81% and 94% respectively for the Chesapeake Bay estuary (McCarthy *et al.*, 1978). The extent of the seaward penetration of river plankton and the possibility of a truly 'estuarine' phytoplankton are problems which remain relatively unworked. The literature tends to indicate a greater penetration and contribution of neritic marine species in estuaries, but physiological characterisation of the populations needs to be undertaken. Welch, Emery & Matsuda (1972) studied the growth of phytoplankton and the production of algae on glass slides in an estuary and found that the greatest growth was at the point where the salt wedge mixed with the down-flowing freshwater. It is generally found that there is a sharp discontinuity in the flora at the 5‰ salinity level (Amspoker & McIntire, 1978).

In many parts of the world river water is abstracted and stored in the reservoirs of waterworks prior to filtration. Such storage might be expected to lead to increases in the planktonic algae. However experience of the Metropolitan Water Board in the London reservoirs fed with Thames water, which often contains large planktonic populations, has shown that over the past 27 years the river diatoms are eliminated in a 3–4 day passage of the water across the reservoirs and they find that for their reservoirs a retention time of 7–10 days allows the river flora to die out without the subsequent appearance of a lake plankton. Table 7.3 shows that the effect is very pronounced even in large reservoirs with retention times between 20 and 120 days and this incidentally also shows the value of such storage in reducing the turbidity, suspended solids and particulate carbon in the water. Since there are now many schemes for storing water and moving large volumes from one catchment area to another it is interesting that the Metropolitan Water Board data show that dissolved oxygen and B.O.D.* do not necessarily deteriorate with storage and that river diatoms do not colonise reservoirs. Therefore, con-

* B.O.D., biological oxygen demand, is a measure of the organic material in waters and is used largely by bacteriologists of river authorities and water and sewage works to measure the purity of the waters.

cern over deleterious algal and chemical effects from mass movements of water may be premature although studies comparable to those of the Metropolitan Water Board should be undertaken at many more sites. If the through flow time is too short and/or the reservoir is too shallow, the river phytoplankton can in fact proliferate in the reservoir and cause blockage problems in the subsequent filtration process; but the usually prolific diatoms can be eliminated by giving the inflowing water doses of between 0.20 and 1.5 mg l^{-1} of chlorine and the cost of this is negligible compared to costs of filtration (Metropolitan Water Board Report, 1971–3).

Lakes

No two lakes are alike, and so no two floras are alike, although there are certain broad categories of lakes in which the basic conditions are similar. A convenient grouping is into polar/alpine, temperate and tropical lakes since these represent environments controlled by differing climatic conditions. Consideration of the phytoplankton in such lakes emphasises the importance of physico-chemical variables on the floral composition. The recognition of oligotrophic or eutrophic states, for example, must be tempered with the climatic factor since it is unlikely that, for example, oligotrophic waters will have similar assemblages in arctic, temperate and tropical climates. Attempts to define oligotrophic and other assemblages on a worldwide basis only lead to confusion.

Arctic, antarctic and alpine lakes. Lakes in semi-arctic or high mountain regions are only ice-free for a short period in summer and during this time plankton growth can be rapid, although there are also many records of populations growing earlier in the year beneath the ice, and in fact Maeda & Ichimura (1973) found that plankton can be more abundant under winter ice than in comparable lakes without ice. This under-ice plankton tends to be dominated by small flagellates. In the northern Swedish Lake Torne Träsk, for example, Skuja (1964) recorded *Chromulina, Mallomonas, Dinobryon, Rhodomonas, Cryptomonas, Gymnodinium* and *Peridinium* starting to grow in April to May under the ice whilst in the deeper water non-flagellate coccoid species of *Chlorella* and *Stichococcus* occur; it is only during the ice-free period that typical diatom and desmid floras develop in

Torne Träsk. In an Alaskan lake Alexander & Barsdate (1971), found *Chlorella/Chlamydomonas*, and *Glenodinium* before the ice melted, followed by *Anabaena flos-aquae*, then as the water warmed *Aphanizomenon flos-aquae* took over the dominance. Neither of these studies, nor the extensive alpine investigations of Pechlaner (1971), record diatoms or desmids in any quantity in such waters. Pechlaner found only a small number of species, varying from 12 to 36, in his high alpine lakes and this seems to be a general feature of this habitat: in summer diatoms and desmids tended to occur only in the deeper waters (Pechlaner, 1967). The occurrence of *Anabaena* followed by *Aphanizomenon* has also been noted in a lake in Saskatchewan by Hammer (1964), who found *Anabaena* developing at temperatures of 14 to 16 °C and *Aphanizomenon* only when the lakes reached 20 °C; at intermediate temperatures *Microcystis* occurred. In summer in an alpine lake, Kosswig (1967) found only species of Chlorococcales (*Sphaerocystis schroeteri, Crucigenia rectangularis, Oocystis submarina* var. *variabilis, Elakatothrix gelatinosa, Planetococcus* sp., *Characium limneticum* and *Ankistrodesmus acicularis*) but towards the end of summer these were replaced by Cryptomonads. Kalff (1967a) reported that in two small ponds in Alaska an average of 84.5% of the nannoplankton was less than 10 μm in diameter and there were very few organisms larger than 20 μm. Tilzer (1973) found a dinoflagellate (*Gymnodinium uberrimum*) dominant throughout the year in the Vorderer Finstertaler See. Although not arctic lakes, lakes of the Experimental Lakes Area in Northwestern Ontario have ice cover from November to April and support a flora very similar to that of alpine lakes; Schindler *et al.* (1973) noted that they were dominated by Chrysophyceae all year (*Chrysochromulina parva, Chromulina* spp., *Chrysococcus* spp., *Kephyrion* spp., *Pseudokephyrion* spp. and *Botryococcus braunii* in winter and in spring and summer *Dinobryon* spp., *Chrysochromulina parva, Mallomonas pumila, Chrysoikos skujae, Pseudokephyrion* spp. and *Kephyrion* spp., with a mixture of these two groups in autumn).

One result of ice formation on a lake can be a steady reduction of oxygen concentration (due to low photosynthetic activity, lack of exchange with the atmosphere and continuing respiratory removal); Willen (1961) found that hydrogen sulphide accumulated in the bottom water of Ösbysjön, Sweden, and at this stage Euglenophyta and bacteria

became more abundant there and Chlorophyta and Dinophyta moved up to a position just under the ice. This illustrates a chemotactic attraction of some species to the bottom waters and a phototactic response of others higher up the water column. Snow accumulation on the ice results in reduced light levels and is followed by upward migration of some species, e.g., *Synura carolinianum* (Wright, 1964).

Kalff (1967a) found that an ice core collected before the thaw contained in each litre some 1.4×10^6 cysts and cells, and these may form an appreciable inoculum for rapid growth when the ice thaws. In two arctic ponds Kalff (1967b) found annual carbon production varying from 380 mg m^{-2} to 850 mg m^{-2}. This is an extremely small amount of fixation compared to productive lakes where figures of 200–600 g m^{-2} yr^{-1} have been recorded (*see* Table 11.3).

Fogg & Horne (1970) comment on the paucity of phytoplankton in antarctic lakes compared to that in the nearby inshore seawater; chlorophyll *a* content of 3.6 mg m^{-3} was recorded in a lake on Signy Island and this value may be compared with 30–40 mg m^{-3} in Orwell Bight. Since these sites have comparable climatic conditions, Fogg & Horne thought the difference may be possibly due to differences in penetration of ultraviolet radiation; they quote Rodhe, Hobbie & Wright's (1966) finding that 10% of the incident radiation between 300 and 400 mμ may be transmitted to a depth of 28 m in a clear mountain lake. Wright (1964) reported that 20% of photosynthetically available radiation penetrated the ice but only 2% penetrated heavy snow cover. In spite of this low light level, he found small populations immediately below the ice and he attributed the small size of the population to excess light rather than low temperature. When samples were incubated in this sub-ice layer, Wright found that fixation was reduced by 83% compared to that at one metre depth, so that there is both a lower population and reduced synthesis in this under-ice water. Under snow, the reverse effect was noted and there was a 300% increase in fixation immediately below the ice compared to the fixation at one metre depth, showing the absence of a high light effect when snow cover reduces the radiation. Wright concluded that the species living in ice-covered lakes were extreme 'shade' forms which could photosynthesise at 0.07% incident radiation, so it is not surprising that they are inhibited in the high light layer beneath the ice. Similar inhibition due to high light immediately beneath the ice was recorded in Lake Haruna by Maeda & Ichimura (1973) who found that the species living immediately beneath the ice were light saturated at 10 klux and inhibited at 20 klux whilst deeper samples were saturated at 7 klux and inhibited at 10 klux. As in other studies the optimum temperature for photosynthesis was higher (10–16 °C) than the *in situ* temperature. Maeda & Ichimura report, however, that flagellates such as *Cryptomonas* have a lower temperature optimum than diatoms. In spite of ice cover and low light intensity, much of the carbon fixation during the year can occur beneath the ice; Hobbie (1964) recorded 49% and 83% of the annual fixation occurring beneath the ice in an Alaskan lake during two years. The effect of the continuous light of the polar summer has rarely been studied but Goldman, Mason & Wood (1963) measured carbon fixation during the 24 h of daylight and found a peak at midnight (Fig. 7.20) when light intensity was at its lowest (less than 0.2 langley min^{-1})* which indicates an adaptation of the phytoplankton to the rhythm of light intensity.

Fig. 7.20. Photosynthesis, measured as carbon fixed, in two antarctic lakes (Lake Skua and Lake Alga) during a day of continuous light. From Goldman *et al.*, 1963.

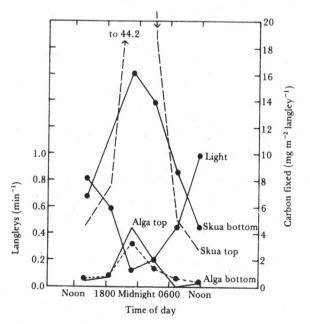

Immediately after the ice melts on lakes which have a fairly long period of ice cover there tends to be a burst of phytoplankton production, e.g. in Lake Erken (Rodhe *et al.*, 1956), and Lake Pääjärvi (Figs. 7.21, 7.22).

There are lakes in polar regions which have thick *permanent* ice cover and even such apparently inhospitable habitats can support an algal flora, e.g. from Lake Miers in South Victoria Land, Antarctica, some 34 taxa of algae were described by Baker (1967). The algae presumably entered the water via the small streams which flow in summer into the narrow melt zone around the edge of the lake, since the remainder of the lake is permanently covered with about six metres of ice. The algae occur in a deep (15–18 m) region of convective water which is heated to about 5.25 °C by storage of solar energy in a density stratified layer. Some antarctic lakes have an even more definite saline layer at the bottom (Goldman, 1964). The flora consisted of Cyanophyta, diatoms and a few green algae, all of which must have been living at extremely low light intensities. Goldman calculated that the optimum light intensity was 0.3 langley min^{-1} and he also found that adding nitrate to the water almost doubled the rate of photosynthesis. The possibility of some form of heterotrophy under these conditions must be borne in mind and has been raised many times in the literature; certainly some algae have been shown to be capable of induction of the necessary aids to heterotrophy, such as glucose transport systems developed in algae kept in the dark but not when the same algae are grown in light (Hellebust, 1972). The Cyanophyta in the antarctic lakes may be growing very slowly by heterotrophy since Khoja & Whitton (1971) have shown that many blue-green algae can grow in the dark, albeit very slowly. In laboratory experiments, Wright (1964) did show heterotrophic growth on acetate of two *Cryptomonas* species from an ice-covered lake, a feature confirmed by Maeda & Ichimura (1972) on species from a similar situation in Japan. Nevertheless neither Rodhe *et al.* (1966) nor several other workers have been able to show any real degree of heterotrophy by algae in such lakes.

Heat can be stored in the bottom sediments of lakes and forms an important source of energy for convective currents during the period of ice cover. These currents may be important in preventing

* One langley: 1 cal cm^{-2}.

Fig. 7.21. The fixation of carbon in Lake Erken in 1955 (outer scale, closed circles) and under laboratory conditions at 700 lux (inner scale, open circles). From Rodhe *et al.*, 1956.

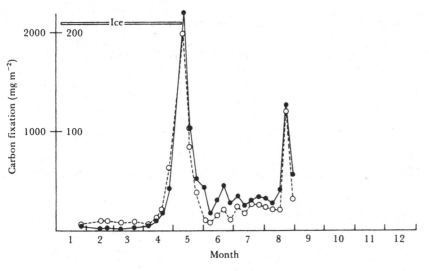

Fig. 7.22. *a*, the seasonal primary production and, *b*, the total biomass of phytoplankton at two stations, in Lake Pääjärvi in 1970–1. The solid lines give the results from Station 1 and the stippled areas those from Station 4. From Ilmavirta & Kotimaa, 1974.

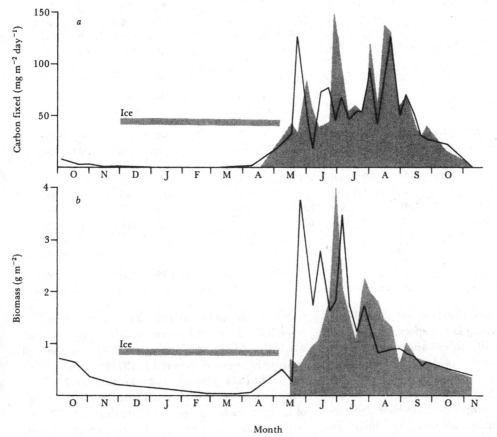

Table 7.4. *The average standing crops of algae in the lakes of the English Lake District separated into small μ-algae, large algae, and Cyanophyta. Data are means of differing numbers in the various years. Gorham et al., 1974*

	(A) μ-algae cells (ml⁻¹)			(B) 'Large algae' cells (ml⁻¹)			(C) Cyanophyta indiv. (ml⁻¹)				Approx. dry wt (rounded to two figures)		
	1955–6	1961–3	Mean	1949–51	1961–3	Mean	1949–51	1961–3	Mean	(A)	(B) (μg l⁻¹)	(C)	Total
Group 1													
Wastwater	713	586	650	4	2	3	0	0	0	14	9	0	23
Thirlmere	919	454	687	19	6	13	+	0	0	15	40	0	55
Buttermere	569	442	506	20	17	19	0	0	0	11	59	0	70
Brother's Water	·	·	·	35	11	23	0	0	0	·	71	0	85
Ennerdale Water	782	401	592	35	17	26	0	0	0	13	80	0	93
Crummock Water	793	842	818	24	28	26	1	+	1	18	80	9	110
Mean										14	57	2	73
Group 2													
Hawes Water	1056	831	944	162	56	109	+	1	1	21	340	9	370
Elterwater	·	284	·	233	55	144	10	+	5	6	440	47	490
Coniston Water	3341	1958	2650	45	431	238	0	0	0	58	730	0	790
Grasmere	·	·	·	(325)	234	280	1	+	1	·	860	9	910
Rydal Water	·	·	·	75	595	335	1	0	1	·	1000	9	1000
Windermere N Basin	2424	2633	2529	360	500	430	1	3	2	56	1300	19	1400
Derwentwater	1947	1608	1778	103	1079	591	+	1	1	39	1800	9	1800
Mean (rounded)										36	920	15	970
Group 3													
Loweswater	2102	2271	2187	376	820	598	17	37	27	48	1800	260	2100
Windermere S Basin	6054	3438	4746	369	1228	799	28	20	24	100	2500	230	2800
Blelham Tarn	5270	4261	4766	749	1470	1110	3	12	8	110	3400	76	3600
Loughrigg Tarn	·	·	·	1161	·	·	7	·	·	·	3600	66	3700
Ullswater	1292	1800	1546	1387	1016	1202	+	14	7	34	3700	66	3800
Bassenthwaite Lake	5966	3238	4602	1010	1658	1334	+	4	2	100	4100	19	4200
Esthwaite Water	3317	914	2116	1438	2260	1849	191	37	114	47	5700	1100	6800
Mean (rounded)										73	3500	260	3800

sedimentation of the nanoplankton but are unlikely to be important in keeping larger species in circulation and in fact inability to remain in suspension may explain the absence of large cells in this environment.

Temperate lakes. Many temperate lakes have a microplankton in which diatoms are easily the dominant species – this applies to some of the largest lakes, e.g. Lake Baikal and the Laurentian lakes where in Lake Michigan, diatoms comprise 65–99% of the cell counts (Schelske & Stoermer, 1971).

The bulk of the work on lake plankton has been done in temperate climates. There are many classes of lakes in such regions (*see* Hutchinson, 1957) but basically they are characterised by isothermal conditions during winter, often with short periods of ice formation followed by summer thermal stratification, prior to 'overturn' in the autumn and the return to isothermal conditions (*see* Fig. 1.7 for a typical annual cycle of heating and cooling). Such a series of events, coupled with annual climatic variables, imposes an immense stress on the phytoplankton, resulting in numerous changes of composition during the year. This complicates the floristics and contrasts with the fairly simple situation in the high latitude lakes and with the infinitely more 'monotonous' situation of the tropics. Compounding the situation is the great range of chemical conditions encountered in temperate regions leading to dystrophic, oligotrophic, mesotrophic and eutrophic situations (*see* p. 527) each with characteristic phytoplankton assemblages. In the series oligotrophic through to eutrophic in the English Lake District there seems to be a trend from phytoplankton dominated by 'μ-algae' to ones dominated by 'large algae' (*see* Table 7.4) plus an increase in the abundance of Cyanophyta in the nutrient rich waters. Elterwater is clearly out of place in this series, as it is also from levels of calcium, and it should be placed in the Group 3 series. An almost identical grouping of lakes was proposed by Round (1957*a*) and there Elterwater comes in the more productive series (*see* also p. 166). Large algae may appear to dominate eutrophic waters, e.g. Abbot's Pool, but when appropriate culture techniques were applied to this small lake a series of peaks of small coccoid and flagellate Chlorophyta were found (Happey-Wood, 1976). The lakes of the Lake District also represent three classes of average standing crop, lowest in the most oligotrophic, highest in the eutrophic (Table 7.4); these are also correlated with calcium, total anion, total cation concentrations, etc. If this type of survey is extended into more nutrient-poor waters or into richer ones the extremes will no doubt be extended but the *overall* pattern will be maintained. There have been arguments about the value of classification systems for lakes but clearly when sufficient data are available valuable generalisations are possible.

Superimposed on the biomass–phytoplankton size pattern, there is of course a pattern of algal associations, e.g. in the dystrophic waters* the Chrysophyta, especially species of *Dinobryon, Syrura, Mallomonas,* etc. are common, together with the diatom *Melosira distans*; in oligotrophic waters a desmid plankton (*Staurastrum, Staurodesmus, Cosmarium, Hyalotheca, Desmidium, Sphaerozosma,* etc.) is common and was first described by Teiling (1916) and termed the Caledonian type since it occurred in waters on acidic rocks. Lakes on the eastern Canadian shield are not exactly temperate but here again Ostrofsky & Duthie (1975) found that *Dinobryon,* with some *Mallomonas,* could make up 50% of the biomass with the diatoms *Asterionella formosa* and *Tabellaria fenestrata* next most common and *Rhizosolenia eriensis, Melosira islandica, M. italica* subsp. *subarctica, Synedra ulna* and *Cyclotella* spp. also present. Numerous other diatoms were also identified but they never contributed greatly to the biomass. No one association is characteristic of the oligotrophic waters but rather a spectrum, though with certain constraints, e.g. Cyanophyta are not abundant. Chlorophyta such as *Sphaerocystis* and *Gloeocystis* dominate some waters, whilst the dinoflagellate *Peridinium willei* is abundant in others. Apparently *Botryococcus* can also be dominant in some oligotrophic lakes and according to Hutchinson (1967) the chlorococcalean alga *Oocystis* is also characteristic of certain waters.

Mesotrophic waters are more difficult to define though *Peridinium* spp. (other than *P. willei*) seem to be common in this type; these and other dinoflagellates tend to be common only in autumn and winter.

Eutrophic waters are characterised by blooms of Cyanophyta, especially in late summer and species of *Microcystis, Anabaena, Aphanizomenon, Lyngbya, Oscillatoria* and *Gloeotrichia* are common. The diatoms *Asterionella formosa, Fragilaria crotonensis, Melosira granulata, Stephanodiscus astraea, S. niagarae* tend to be common and in midsummer, green flagellate algae such as *Pandorina morum* and *Chlamydomonas* spp. also occur. Eutrophic lakes also often have large summer growths of Chlorococcales (e.g. *Pediastrum, Scenedesmus, Dictyosphaerium, Ankistrodesmus, Crucigenia, Tetraedron, Chlorella, Kirchneriella,* etc.) and these become especially abundant in small lakes and ponds. A few desmids are characteristic of eutrophic lakes and these are important indicator organisms especially when the absence of the bulk of desmid species is taken into account.

* Dystrophic waters are those with nutrient-poor, brown, peaty water. Some authorities dislike the term dystrophic and use chthoniotrophic. These waters are very common in the subarctic bog and peat lands of the world and locally elsewhere in areas draining highly organic and peaty soils.

There is some evidence (*see* Phillips *et al.*, 1978) that phytoplankton populations may be low in lakes rich in macrophyte growth. The relationship is not yet understood but it has been suggested that the macrophytes both remove nutrients from the water, which they certainly do as well as absorbing nutrients via the roots, and also secrete substances which might act as chelators of essential nutrients or as anti-metabolites.

The above brief comments on floristics of different waters can be supplemented greatly by consulting Hutchinson (1967).

Tropical lakes. The phytoplankton of tropical lakes has certain unexpected features when compared with temperate lakes, in particular the occurrence of diatom genera which are benthic in distribution in cooler water lakes, e.g., species of *Cymbella*, *Denticula*, *Gomphonema*, *Nitzschia*, *Surirella* and *Cymatopleura*. There are so many records of these raphe-bearing diatoms in tropical lake plankton that there can be no doubt that they are truly planktonic and not drawn up from the sediments. The only record of such large species in non-tropical lakes is that of the occurrence of *Cymatopleura solea* var. *apiculata* and *Amphiprora ornata* in Gull Lake in Michigan (Moss, 1972*b*) and the buoyancy characteristics of these species urgently needs study. Richardson (1968) suggests that the virtual restriction of the *Nitzschia* dominated plankton to the tropics is due to its requirement for a rich organic environment. Sufficient decomposition to supply the organic substances within the photic zone is perhaps only possible in the tropics, though why its occurrence is restricted to deep lakes is not clear, nor why the genus is generally absent from tropical lakes producing blooms of Cyanophyta. *Nitzschia* species can be collected from the plankton of temperate lakes but when detailed studies are made it is usually found that they are epiphytic forms growing in the mucilage of colonial Cyanophyta, etc. Needle-like diatoms do occur in temperate lakes in the form of *Synedra* species (which also occur in the tropics) so it is unlikely that a selection of a particular morphological form is involved. Lakes with *Nitzschia* type phytoplankton tend to have waters of high alkalinity whereas the low alkalinity waters tend to favour a *Melosira*-dominated phytoplankton which unlike the *Nitzschia* type extends into waters dominated by Cyanophyta

and also into temperate waters. Some of the *Melosira* species are found over a wide geographical range whilst others are confined to the tropics or to the temperate zones. It would be of great interest to determine whether the diatom species growing in the benthos of temperate latitudes are the same as those planktonic in the tropics; some are certainly different but few lakes have been examined in sufficient detail.

Richardson (1968) points out that an admirable way to collect plankton samples for comparative study is to sample the sediments, since this yields the flora from all the seasons and indeed integrates that of several years, so that anomalous collections are avoided. It can only be satisfactorily applied to the diatoms, however, since most other groups tend to become too fragmented for recognition. It does also add the complication of deciding which species are benthic and which planktonic. Still it could avoid the misleading comparisons sometimes made between lakes, based on samples obtained at different times of the year.

Tropical lakes tend to have lower species diversity than temperate lakes (Lewis, 1978) with Chlorophyta tending to be dominant. In Lake Lanao, Lewis found that the species composition through the year was fairly stable (compare Lake George) and this stability is probably a feature of tropical assemblages. In contrast to the species diversity pattern, Lewis estimates that the biomass of tropical lakes is almost twice that of temperate lakes with an average value of 13.8 g m^{-2} as opposed to 7.5 g m^{-2} for temperate lakes.

One of the most intensively studied tropical lakes is Lake George, lying on the equator in Uganda. It is shallow and has a permanent and dense crop of phytoplankton with Cyanophyta (29 spp.) comprising 70–80% of the biomass, the commonest genera being *Anabaena*, *Anabaenopsis*, *Aphanocapsa*, *Aphanizomenon*, *Lyngbya* and *Microcystis*. There is a very depauperate flora with 11–14 species of diatoms, of which *Nitzschia*, *Synedra* and *Melosira* are most conspicuous and 18 species of green algae, with *Pediastrum* and *Scenedesmus* the most abundant. Unlike the sequence in temperate lakes there is here only about a twofold annual variation in chlorophyll *a* concentration with slight peaks during the seasons of maximum rainfall (April–May and September–November).

Fig. 7.23. Marine phytoplankton – some common microscopic genera. *A, B, Ceratium; C, Gymnodinium; D, Peridinium; E, Phalacroma; F, Gonyaulax; G, Dinophysis; H, Pyrocystis; I, Coccolithus; J, Rhizosolenia; K, Ditylum; L, Biddulphia; M, Asterionella; N, Chaetoceros; O, Thalassiosira; P, Bacteriastrum; Q, Thalassiothrix; R, Nitzschia; S, Phaeocystis; T, Eucampia; U, Skeletonema; V, Coscinosira; W, Coscinodiscus; X, Trichodesmium; Y, Halosphaera; Z, Chaetoceros* (filament with spore in one cell and spore isolated).

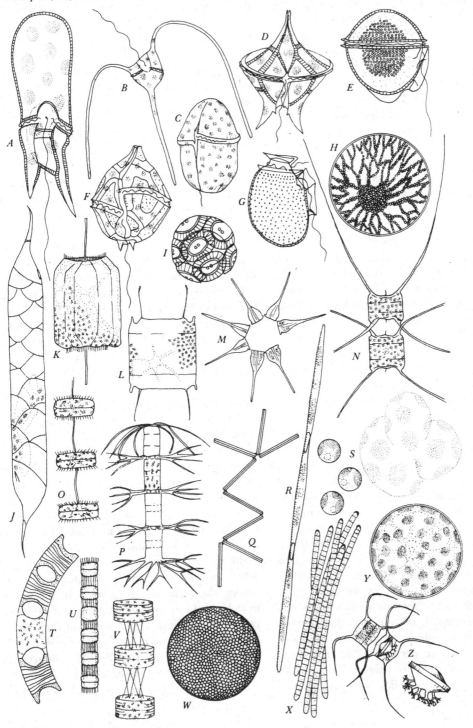

The exceptional abundance of Cyanophyta in some shallow tropical lakes (e.g. Lake Chad, Lake George, Lake Nakuru) seems to be connected with the possession of buoyancy mechanisms by these algae. Daytime temperature stratification produces well-defined epilimnia and hypolimnia and sinking of non-buoyant forms will be rapid in such warm hypolimnia. The daily repetition of such conditions appears to prevent the build-up of large populations of heavy species though needle-like diatoms can persist. In lakes such as Nakuru and Elmenteita there is a great excess of sodium but no calcium in the waters, and this may account for the unbalanced flora in which the Cyanophyta *Spirulina platensis*, *S. laxissima*, *Chroococcus minutus* and *Anabaenopsis arnoldii* and the diatoms *Nitzschia frustulum*, *N. sigma* and *Navicula elkab* are dominant (Melack & Kilham, 1974). Other diatoms with distinct halophilous tendencies which occur in these east African lakes are *Cyclotella meneghiniana*, *Coscinodiscus rudolfi*, *Anomoeoneis sphaerophora* and *Rhopalodia gibberula*. In these lakes the lowest pH level is 8.3 and it can rise during active photosynthesis to 10.6 (Hecky & Kilham, 1973). The lakes in which sodium chloride is the dominant salt tend to be in the arid areas and it is interesting that some of the diatom species in them are identical to the species in rivers flowing through arid areas in the United States (*see* above). In the more humid regions of the tropics sodium carbonate and bicarbonate are the dominant salts and a more balanced flora exists in these lakes.

Oceans

Considering the vast extent of the oceans it is quite certain that on a global basis it is the phytoplankton (Fig. 7.23) in the surface veneer which is responsible for the bulk of algal carbon fixation. Indeed some workers consider that the amount of carbon fixed in the ocean about equals the total on the land (Steemann-Nielsen, 1952); earlier Rabinowitch (1945) put the oceanic production even higher at 80% of the total. Slightly before this, Riley (1944) gave estimates of terrestrial fixation as $20 \pm 5 \times 10^9$ tons per year and oceanic fixation as $126 \pm 82 \times 10^9$ tons per year thus putting oceanic production even higher. A more recent estimate of carbon fixation for the ocean is 31×10^9 tons per year (Platt & Subba Rao, 1975). Really definitive figures are badly needed to put oceanic production into perspective.

As Riley pointed out, carbon fixation in the ocean is a continuous, if at times slow, process whereas on the land it is seasonal. The phytoplankton in the oceans cannot be considered as a single unit and, as in the section above, water masses can be divided according to climatic zones, each with characteristic phytoplankton. There is, however, an additional complication in that inshore (neritic) and open ocean (oceanic) waters also differ in their phytoplankton assemblages, chemistry etc., and these will be considered briefly, prior to the discussion of the climatic zones.

Inshore and offshore phytoplankton. There has never been a completely satisfactory explanation for the difference in composition of the neritic and oceanic phytoplankton assemblages. The differences may in part be due to a selection of meroplankton, in the neritic zone, i.e. species having bottom living resting stages, though this is difficult to prove. Another contributory factor is the supply of organic matter and silt particles derived from the land and from resuspension of decaying matter off the coastal sediments. Certainly some species which are common in the neritic plankton can assimilate organic molecules, such as urea, and this has been shown for the diatoms *Skeletonema costatum*, *Amphiprora alata*, *Chaetoceros simplex* and the haptophyte *Chrysochromulina* (Carpenter, Remsen & Watson, 1972). Some common *Gonyaulax* species have also been shown to be stimulated by humic acid fractions of varying molecular weights (Fig. 7.24), and these acids are likely to be present in greater quantity in inshore waters (Prakash, 1971). Such pieces of evidence are not in themselves proof that organic substances are the discriminating factor between neritic and oceanic plankton but they cannot be ignored, especially since Cooksey *et al.* (1975) have shown that soluble material extracted from mangrove sediments may sometimes stimulate and sometimes inhibit heterotrophic growth of benthic diatoms; tests on autotrophic growth are clearly required. Certainly production is greater in the coastal zone, e.g. in a study in the Baltic Sea near Helsinki the inshore waters fixed carbon at the rate of 80–192 g m^{-2} yr^{-1} whilst in the open Baltic Sea carbon fixation was only 15–36 g m^{-2} yr^{-1} (Bagge & Lehmusluoto, 1971). Similarly, in the Caribbean Sea the greatest amount of phytoplankton is found in inshore waters off Nicaragua and

Honduras (Hulburt, 1968). He quotes figures seven times as high for the shallow water as for the open sea, whilst Malone (1971a) found a comparable situation with the net plankton though the nanoplankton was the most important component in both regions. Similar increases are noted around atolls, reefs etc., where only slight upwelling and little run-off is

likely: this has been termed the 'island mass effect' by Doty & Oguri, 1956 and has been confirmed by Gilmartin & Revelante (1974). Much more work is needed to explain the higher productivities of inshore waters but Doty & Morrison's (1954) suggestion, that the benthic algae accumulate inorganic nutrients from the nutrient-poor water and on their

Fig. 7.24. *a*, the response of two *Gonyaulax* species (open circles, *G. tamarensis*; filled circles, *G. catenella*) to various quantities of humic acid and, *b*, the relative yields of three species to various molecular weight fractions of humic acid added at 4 μg ml⁻¹ concentration. From Prakash, 1971.

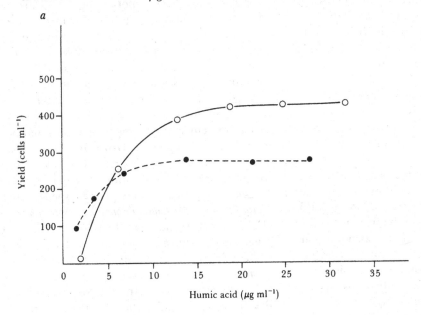

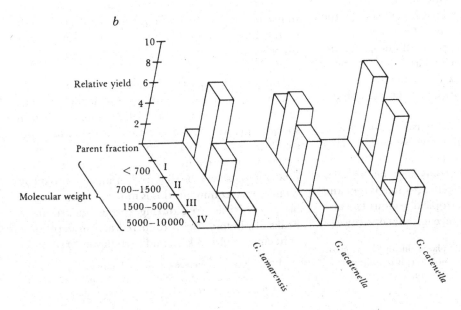

decay the nutrients are released and become available to the plankton, merits further study. The increase in nitrogen levels due to fixation by blue-green algae could also be important especially in view of the recent studies showing considerable fixation on reefs. Fixation in the benthos around non-reef girdled islands needs checking: epiphytic and epilithic Cyanophyta are commoner than generally assumed!

Planktonic diatoms occurring in more turbid coastal and estuarine water are often found with small clay-like particles attached to the thin layer of mucilage which occurs on the outside of all diatoms. This material may be of significance in their nutrition though I know of no studies on this aspect. Such binding may increase sedimentation rates and strip the water of fine particles.

One of the world's commonest neritic diatoms is *Skeletonema costatum** which is very common in the brackish conditions near river mouths. Ferguson-Wood (1971) reported its sedimentation and resuspension and subsequent growth to form blooms. Such ability to live 'sedimented' and in suspension is obviously a feature of many neritic species; another common example is *Bacillaria paradoxa*. However this is a motile species and there is probably a distinction to be made between benthic species spending part of their time suspended and planktonic species (e.g. *Skeletonema*) which have no motility mechanisms and manage to exist for periods in the benthos.

A small number of inshore species have life cycles in which a benthic filamentous phase alternates with a flagellate planktonic stage, e.g., the coccolithophorid *Hymenomonas* (*Cricosphaera*) has been shown to be the motile phase of an *Apistonema*-type growth. It is important that many more of the filamentous growths earlier attributed to the Chrysophyta, but more likely to be Prymnesiophyta, be investigated in detail.

Studies of buoyancy of neritic and oceanic species are needed: I have the impression that many of the heavier diatoms are especially common in neritic habitats, and at sea not only does the thickness of the valve decrease but the cells are often larger and 'bubble'-like, especially in tropical regions (*see* also p. 286). The relative absence of the open-ocean

species in inshore waters may be related to chemical conditions and also the the simple physical inability to withstand the damaging circulation from water to sediment and back. Little thought has been given to this problem but certainly there are many delicate naked flagellates in the open ocean which presumably cannot withstand the turbulence of the coastal region (*see* also p. 10 for the effect of wave motion on species).

Differences between inshore and offshore stations have also been reported for large lakes e.g. Munawar & Nauwerck (1971) found on average 25% higher biomass in the inshore regions of Lake Ontario than in the central waters. In addition the offshore waters tended to have a single peak of phytoplankton production whilst the inshore had multiple peaks (Fig. 7.25). Holland (1969) showed that the inshore 'sidearm' Green Bay phytoplankton had a different species composition from that of the main Lake Michigan. In Green Bay, *Stephanodiscus niagarae* dominated the biomass at all times, whilst *Melosira granulata*, *Stephanodiscus* (*Melosira*) *binderana*,[†] *Fragilaria capucina* and some pennate species, *Amphipleura pellucida*, *Amphiprora ornata* and *Surirella* species also occurred. In the main body of the lake on the other hand, *Asterionella formosa*, *Cyclotella michigiana*, *C. glomerata*, *C. stelligera*, *Fragilaria crotonensis*, *Melosira islandica*, *Stephanodiscus tenuis*, *Cyclotella bodanica* and *C. comta* were common. In addition Holland found that the generation times for the species were longer in the offshore waters, but times could also lengthen in inshore waters, for example on the Wisconsin side of the lake: apparently this is simply due to the differing trophic levels.

Cold oceans. In regions where sea-ice forms, plankton production can continue under the ice, e.g. Sorokin & Konovalova (1973) found thirteen species, of which *Thalassiosira nordenskioldii* and *Chaetoceros pseudocrinitus* formed 80% of the biomass, under 60 cm of ice in a bay in the Japan Sea. The total biomass was 5–10 g m^{-3}, and, as Sorokin & Konovalova comment, this is much greater than the biomass in this area during spring and summer. Peaks of photosynthesis occurred at 2 m depth in a light climate of 1–3 klux and carbon was fixed at the

---

* Some workers have recently placed this in *Stephanopyxis* but have been shown to be incorrect; it is a distinct genus as demonstrated by electron microscope techniques (Round, 1973*b*).

† This very common taxon in the Laurentian lakes is certainly not a *Melosira* species (*see* Round, 1972*b* for a discussion based on electron microscope data).

rate of 100 mg m^{-2} day^{-1} at a shallow station and 200–300 mg m^{-2} day^{-1} at deeper stations. However although the biomass was greater in winter than in summer, the summer rates of fixation were 2–3 times higher. Experiments showed that the optimum temperature for fixation by these under-ice populations was 9 °C at 2 klux illumination, so although light was adequate, in nature the temperature (-1.7 to 1.8 °C) was far below optimum. At a comparable latitude in the Atlantic Ocean off Maine, *Th. nordenskioldii* was also recorded in late winter, followed by a spring population of *Chaetoceros socialis–Ch. debilis* and in summer *Skeletonema costatum* became dominant (Hulburt & Corwin, 1970). Two interesting points arise from this work. The occurrence of *Th. nordenskioldii* in two oceans separated by a land

barrier prompts thoughts about the maintenance of geographically separated races of the same taxon, and whether they really are identical, morphologically and physiologically. The second point is the summer occurrence of *Skeletonema costatum* which in more temperate waters tends to be a spring species: timing of growth is clearly a feature related to general climatic regimes. Hulburt & Corwin (1970) comment that, at the site with continually mixed conditions in the water column, diatoms persist throughout the summer, whereas in the Gulf of Maine, coccolithophorids replace diatoms in the summer community.

Experimental studies have shown that *Th. nordenskioldii* grows well at low temperatures and low light intensities, but if the light intensity is raised

Fig. 7.25. The surface phytoplankton volumes at inshore (full line) and offshore (interrupted line) stations in Lake Ontario during 1970. From Munawar & Nauwerck, 1971.

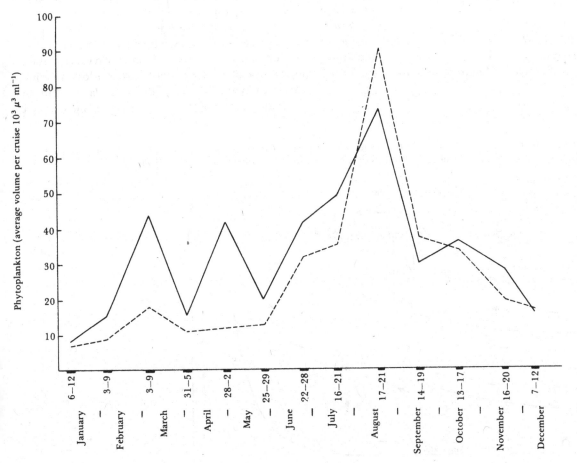

temperature becomes a critical factor (Jitts *et al.*, 1964). On the other hand *Skeletonema* has a broad tolerance of both light and temperature; such experimental results fit neatly with the observations from nature.

Steyaert (1973*a*, *b*) makes the point that in a transect from Africa to Antarctica the populations consist of ⅔ diatoms, with *Nitzschia* dominant in the subtropical zone, whilst south of this *Chaetoceros* becomes more important and in the subantarctic *Fragilariopsis* is more abundant, though some *Nitzschia* spp. can occur. A very interesting study along a more restricted transect was made by Walsh (1971) who sampled along the ecotone* across the antarctic convergence and found distinct communities to the north and south (Fig. 7.26). Within the convergence the southern community (with indicator species *Chaetoceros bulbosus*, *Ch. dichaeta*, *Dactyliosolen antarcticus*, and *Eucampia zoodicus*) sinks and can be traced northwards underneath the northern population (indicated by *Rhizosolenia delicatula*, *Amphidinium amphidinioides*, *Oxytoxum variabile* and *Phalacroma pulchellum*). A most detailed study of the phytoplankton in the Pacific sector of Antarctica was made by Hasle

* An ecotone is any region in the environment where two separate communities merge.

(1969), who recognised eight groupings of species, although basically there are four major distributions working from the north towards the continent. The most northerly grouping contains widespread warm-water species e.g. *Roperia tessellata*, *Ceratium fusus*, *Umbellosphaera tenuis* and *Dictyocha fibula* which have their *southern* limit far north in the subantarctic zone; a second group (*Thalassionema nitzschioides*, *Ceratium pentagonum* and *Coccolithus huxleyi* extends into the antarctic convergence; a third group ranges from the southern subantarctic zone and into the northern antarctic zone (e.g. *Rhizosolenia simplex* and *Thalassiothrix antarctica*), a fourth group is restricted to the southernmost part of the antarctic zone (e.g. *Charcotia actinocyclus*, *Fragilariopsis sublinearis*, *Melosira sphaerica* and *Nitzschia lecointei*). The preponderence of *Corethron–Chaetoceros* and *Fragilariopsis–Nitzschia* associations in the Southern Ocean was noted by Hendey (1937) in working through material from RRS *Discovery*.

Another study of under-ice production coupled with a study of the flora attached to the ice is that of McRoy *et al.* (1972) who found an average carbon fixation of 20.9 mg m⁻² day⁻¹ from a low standing stock of phytoplankton in the north Pacific. But on the under surface of the ice they found a much higher

Fig. 7.26. The movement of water masses in the southern ocean showing the sinking of northward flowing antarctic surface water beneath subantarctic surface water. Modified from Walsh, 1971.

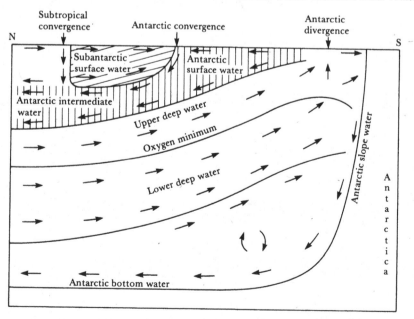

stock of algae which fixed carbon at 44.4 mg m^{-3} day^{-1} as opposed to 1.20 mg m^{-3} day^{-1} in the free water beneath the ice. The cryophytic population was found on the southernmost ice and continued northward for an undetermined distance: McRoy *et al.* comment on the high probability that this population plays an important role in the food webs of this zone, which is a refuge for many species of marine birds and mammals. Several authors consider that algae frozen into ice (*see* p. 206 and Fig. 5.26) are a source of inoculum for growth during the short summer months. Walsh (1969) reported that, after break-up of the ice, these forms can comprise 99%

of the species in the water column. There is however some confusion in the literature since others state that the species frozen into the ice are *not* planktonic (*see* chapter 5).

The number of species in the phytoplankton of cold regions is small compared to that of the tropics e.g. Semina & Tarkhova (1972) record 94 diatoms and 43 peridinians (dinoflagellates) in the subarctic Pacific water compared with 192 and 614 species in the tropical Pacific. North of 40° N diatoms were dominant and south of this dinoflagellates. Semina (1972) also made one of the few studies of cell size relative to water mass and found that cells were small

Fig. 7.27. Cell size areas for the Pacific phytoplankton. I and VII are areas with the smallest cells, II and VI areas with intermediate sized cells, and III and V ones with the largest cells. Area II has a mosaic of cell sizes. From Semina, 1972.

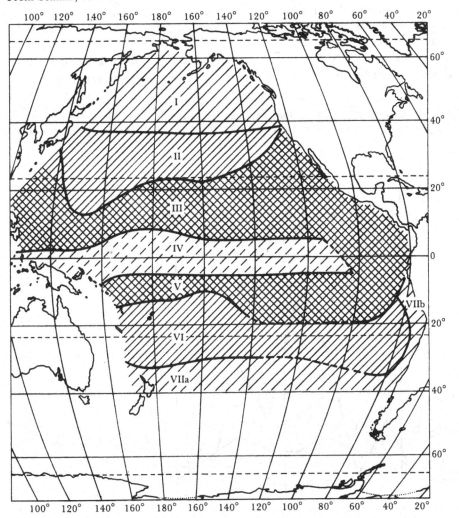

in the subarctic, intermediate in the temperate zone and large in the tropics (Fig. 7.27) though in spite of the larger size in the warm waters the greatest productivity is still in the colder waters. Hendey (1937) had earlier noted the size difference between the warm and cold waters and also the greater silicification of the cold water forms.

The euphotic zone extends only to about 30 m in antarctic waters but in this region, when the data are integrated over the whole water column, the standing crop of algae is considered by Walsh to be of a similar order of magnitude to that in the Strait of Florida (Fig. 7.28). As far as productivity is concerned the antarctic is probably, at most, equal to this tropical region but over an annual cycle probably less productive. Other estimates for antarctic waters suggest they are about six times as productive as the world average (carbon production taken as 0.15 g m^{-2} day^{-1}) but the waters south of the convergence only cover about 5% of the world ocean area and on a per unit area basis they are about 400 times as productive (El Sayed, 1968).

Phytoplankton of cold oceans must be capable of surviving long periods in the dark and Antia (1976) found that about $\frac{1}{3}$ of the algal species he tested could survive for six months in the dark at 2 °C and another $\frac{1}{3}$ could survive for one year. But if the temperature was raised to 20 °C the dark tolerance times were greatly reduced. Antia was not working with cold-water forms, which are possibly more resistant.

Tropical oceans. There are more pronounced differences between neritic and oceanic populations in tropical regions than in temperate or cold oceans. These differences are related to the frequent upwelling zones along tropical coasts contrasting with the more or less permanently stratified and thus nutrient-deficient offshore zones. The upwelling water tends to be cooler and in this a more temperate flora can develop, e.g. off the west coast of Africa (Ivory Coast) a cool-water flora dominated by diatoms (*Chaetoceros, Rhizosolenia, Bacteriastrum,* and *Coscinodiscus*) develops between July and October,

Fig. 7.28. The seasonal variation of phytoplankton standing crop in the Straits of Florida during 1965 and 1966 and in the Southern Ocean (dashed line, observed data; dotted line, postulated). From Walsh, 1969.

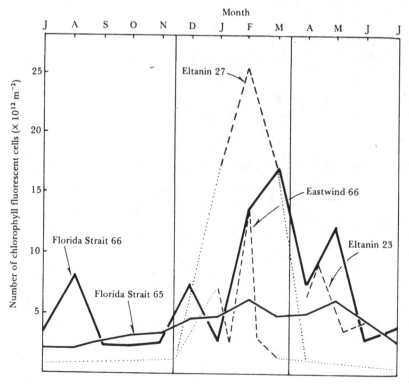

whilst during the remainder of the year a warm-water flora dominated by dinoflagellates is present (*Ceratium, Pyrocystis, Peridinium* and *Dinophysis* (Reyssac & Roux, 1972). It is not easy to distinguish a truly tropical element except for the large inflated *Ceratium* species referred to on p. 351. This is undoubtedly due to the transport of so many species into temperate zones, though I believe there is a truly tropical element, some species of which are *Stephanopyxis palmeriana, Planktoniella sol, Gossleriella tropica, Hemidiscus cuneiformis, Astrolampra marylandica*: the two last named were noted by Hendey (1937) to be abundant in warm waters and they are certainly conspicuous in the Gulf of California (Round, 1967) together with *Pseudoeunotia doliolus* and *Roperia tessellata*. Taylor (1973) discusses the problem of a distinct tropical dinoflagellate flora and in his paper *Heteraulacus* (= *Goniodoma*) *polyedricum, Ceratocorys horrida, Ceratium carriense, C. massiliense* var. *macroceroides, C. trichoceros* and *Pyrocystis pseudonoctiluca* are reported to form a tropical complex at least in the Indo-West Pacific region.

Direct measurements show low levels of nutrients in the central regions of tropical oceans though this does not always reflect low productivity since rapid recycling may be occurring. However if nutrient enrichment experiments are performed these should indicate whether or not there is nutrient limitation. In one such study, the addition of nitrogen, phosphorus and especially iron, to samples from the nutrient-poor region north of Hawaii resulted in large increases in productivity whereas similar experiments using upwelling water or cold water from near the Aleutian Islands produced no effect (Glooschenko & Curl, 1971). The tropical communities tend to be dominated by diatoms with coccolithophorids next in abundance and dinoflagellates relatively scarce.* However, in bays it is possible to find high cell numbers of dinoflagellates, e.g., in Bahia Phosforente off Puerto Rico, but in moving out to sea diatoms replace flagellates as dominants. Another feature is that near tropical estuaries large centric diatoms tend to be abundant: perhaps these are associated with sediments similar to the populations recorded off Heligoland by von Stosch (1956).

---

* This refers to cell counts. On a species diversity basis there are many more dinoflagellates in tropical oceans than in cold oceans.

The Mediterranean is perhaps intermediate between temperate and tropical oceans in that it certainly has cooler water, and a winter flora in which the silicoflagellate *Dictyocha fibula* is abundant at temperatures below 15 °C (Nival, 1965). However, Travers & Travers (1968) record this species throughout a longer season in the Gulf of Marseille though still with peaks in the cooler months (March–April) after which they comment that it aestivates in the deeper water. This is one of the few mentions of aestivation or hibernation in the algal literature though such phenomena must be common. In the Sargasso Sea *Coccolithus huxleyi* is the predominant species throughout six months of the year and coccolithophorids in general are abundant. Only during spring do diatoms achieve any degree of dominance (Hulburt, Ryther & Guillard, 1960). The spring does however seem to be the peak period for primary production in this area and according to Menzel & Ryther (1961a) tends to be associated with the period of greatest mixing of the water column.

In tropical seas the Cyanophyte, *Trichodesmium thiebaudii* is fairly common (Hulburt, 1968; Steven, Brooks & Moore, 1970); Steven *et al.* suggest that there is a 2–4 month cycle of this species in the water off Barbados and this cyclical development is reflected in the fluctuating chlorophyll values. There are areas many miles across containing rich blooms of *Trichodesmium* forming a layer 50 cm deep in areas such as the Coral Sea, Arafura Sea and Banda Sea. The Red Sea is perhaps the one most associated, at least in name, with this alga since its name is derived from the appearance of the surface waters when *Trichodesmium* is blooming.

The most impressive large-scale study of any oceanic area has been the recent international study of the Indian Ocean which involved 20 nations and 40 research vessels during the 1960s. Regretably, however, many of the data seem never to have been published and large areas of the ocean were visited too infrequently whilst others received excessive attention. The phytoplankton productivity varies considerably between the periods of the southwest monsoon and northeast monsoon and Fig. 7.29a, b give the average values for these periods and clearly shows the much larger carbon fixation during the southwest monsoon (Fig. 7.29a) with the surprisingly high average value of 0.5 g m^{-2} day^{-1}. During the northeast monsoon (Fig. 7.29*b*) this drops to 0.15 g m^{-2} day^{-1}. Maximum production, as expected, occurs in the upwelling regions and along the coasts. The seasonal variation is also shown in work by Jitts (1969) along the 110° E meridian and Fig. 7.73 shows that the higher values occur in the tropics but during the late summer period, when a thermocline is established, values are low. The photosynthetic layer averaged 85 m in this region (range from 60 m in late summer to 130 m in spring).

There is some evidence that phytoplankton blooms in waters around India are related to periods of monsoon rains which lower the salinity of the surface waters (Subrahmanyan, 1960; Qasim, Bhattathiri & Devassy, 1972). Since experimentally the uptake of ^{14}C by many species has been found to be

Fig. 7.29. The fixation of carbon in the Indian Ocean during *a*, southwest monsoon and *b*, northeast monsoon. Carbon fixation measured as g m^{-2} day^{-1}. 1, no observations; 2, < 0.15; 3, 0.15–0.38; 4, 0.38–0.75; 5, 0.75–1.45; 6, > 1.45. From Kabanova (1968) reproduced in Cushing, 1971.

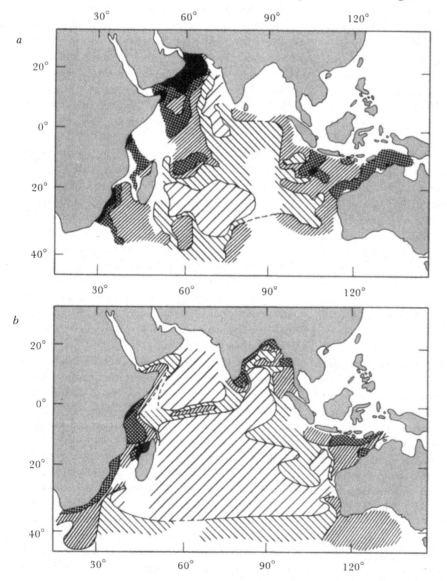

at its maximum at salinities below 33‰ the dilution factor alone may be responsible. In inshore waters the lowering of salinity by runoff from the land is complicated by the addition of nutrients so that even greater increases in algal growth are to be expected, though they may be reduced by high turbidity or deepening of the mixed layer by wind action (Huntsman & Barber, 1977). In fact, Flemer (1970) found that the highest chlorophyll values occurred in the low salinity areas, in a somewhat more temperate region, Chesapeake Bay, where there is a range of salinity conditions.

Temperate oceans. In spite of the accessibility of these oceans the flora is in many ways less well documented than those of the tropical and polar waters. I have tried to determine which species are characteristic of such waters but the literature is very confusing, perhaps not surprisingly since warm and cold water forms tend to intermingle in this region. Species such as *Corothron hystrix, Thalassiosira subtilis, Coscinodiscus excentricus, C. radiatus, C. concinnus, Rhizosolenia styliformis, R. hebetata, Chaetoceros densus, C. borealis, C. decipiens, Dinophysis acuta, Gymnodinium abbreviatum, Pyrophacus horologicum, Peridinium conicum, P. curtipes, P. pallidum, P. punctulatum, P. pyriforme, P. steinii, P. thorianum, Gonyaulax digitale, Ceratium macroceros, C. tripos,* are mentioned in Drebes (1974) as specifically oceanic and also occurring in the region around Heligoland but the bulk of species for this

Fig. 7.30. The distribution of chlorophyll (mg m⁻³) and temperature along transects at sea during different cruises, *a* and *c* showing fairly constant values for both. *b* shows an inverse relationship between temperature and chlorophyll and *d* shows a peak of chlorophyll over a zone of temperature change. From Lorenzen, 1971.

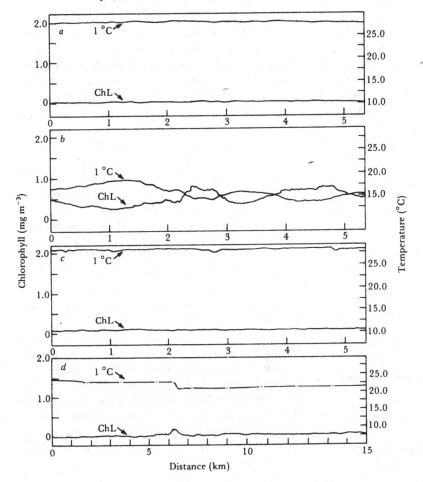

region are neritic. Other aspects involving temperate plankton are dealt with in the chapters or sections concerning seasonal cycles, nutrient relationships, etc.

Horizontal distribution. It is quite clear that phytoplankton is not evenly dispersed across stretches of ocean, even within relatively homogeneous climatic zones. Very few studies have been made of this aspect, and those that have suggest that, rather than a gradual clinal variation, the phytoplankton is distributed in patches. Along an eight mile transect sampled at 0.1 mile intervals there was, for example, considerable variation in chlorophyll levels and patches varying in extent from 0.5 to 1.5 miles in diameter were found off the coast (Platt, Dickie & Trites, 1970). Even smaller scale patchiness was recorded by Bernard & Rampi (1965) who found, for example, considerable variations of the whole population over a mere 230 m. Variations can be very diverse as Fig. 7.30 shows: sometimes almost constant chlorophyll *a* values are recorded, at other times considerable fluctuations coupled also with temperature changes, and sometimes a slight jump in the chlorophyll level occurs as a temperature discontinuity is traversed. Similar results were also obtained by Denman (1976) and Fasham & Pugh (1976). The chlorophyll levels in this work were recorded automatically using the chlorophyll fluorescence technique (Lorenzen, 1971). Recently,

Fig. 7.31. The horizontal distribution of population density expressed as cells ml⁻¹, along 64 m transect in surface water, for four species of algae.

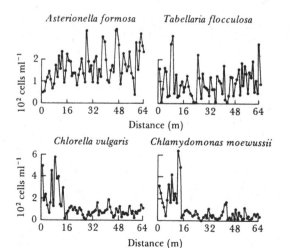

Richards (personal communication) found that *Asterionella* and *Tabellaria* counts varied considerably in samples taken from the surface waters across a large lake in north Wales and pattern analysis revealed peaks at defined intervals (Fig. 7.31); interestingly the peaks for *Asterionella* occurred at the same interval in two different lakes. Such small-scale patchiness is difficult if not impossible to detect by methods for locating discrete water masses, e.g. the routine temperature and salinity determinations made by most oceanographic ships. Patches of a similar order of magnitude were reported in estuarine phytoplankton by McAlice (1970) who found statistically significant differences in samples collected at 10 cm intervals. Bainbridge (1957) found that most patches of marine phytoplankton were oval shaped and had an average size of 10 by 40 miles but could range from a few feet across to 30–40 miles by 120–140 miles. Bainbridge gives an interesting account of historical records starting with those of Captain James Cook and, not surprisingly, of Darwin who noted bands of water coloured by *Trichodesmium*. Platt (1972) thought that patches ten metres to one kilometre in extent might be controlled by physical transport processes. There is considerable evidence of patches of discoloured water in the North Sea (28–100 miles by 15–45 miles) in which the diatoms *Rhizosolenia* and *Biddulphia* are abundant; these were studied earlier this century in relation to the herring fishery. Lucas & MacNae (1940) reported 'very marked patches (of diatoms, *Phaeocystis* and dinoflagellates) with sharp boundaries, yet within these general patches are equally defined sub-patches of the individual species'. Fig. 7.32 shows plots of the lateral distributions across the North Sea showing some of these patches. It is now possible with satellite infra-red scanning to pick up very precisely the patterns of phytoplankton growth and to see how often the rich blooms are associated with oceanic fronts (Holligan, personal communication).

Great changes are often recorded in passing from the inshore sea into estuaries and ultimately into freshwater: Overbeck (1962*a*) studied the nannoplankton (in fact on the classification given on p. 244 he studied the μ-algae) in the region of Rügen and found a great increase in these forms as sampling moved into the freshwater zone though even in the marine areas they were more abundant than the net plankton.

The horizontal distribution need not be in the surface water but can be at lower depths: for example, Venrick (1972) records a horizontal pattern of *Nitzschia turgiduloides* along the top of the thermocline where internal waves produce a vertical displacement which may occur on a scale as small as one mile. It has been shown clearly only rarely that phytoplankton distribution follows such movements but these could obviously be important under relatively stable thermal conditions both at sea and in large lakes such as the Laurentian lakes (*see* Glooschenko & Blanton, 1977). In these lakes there is certainly a horizontal component to the overall phytoplankton growth, clearly shown in the seasonal development of the phytoplankton biomass (measured as cell volume, Fig. 7.33) in Lake Ontario (Munawar & Nauwerck, 1971) and the variation in the species composition in different regions of the arm (Green Bay) of Lake Michigan (Holland & Claflin, 1975). Wind drift of surface layers is quite common and there seems to be some degree of slipping of surface layers over lower water (Baker & Baker, 1976).

Collection of phytoplankton can occur in 'wind-rows' which result from Langmuir convection cells produced by wind-induced motion of the water. This produces circulating cells of water (Fig. 7.34) and aggregation of the phytoplankton (or detritus, oil, etc.) along the lines of convergence which may extend from one centimetre to several metres. Howard & Menzies (1969) report that *Sargassum* is also aligned in rows by such a mechanism. When floating or swimming species are present, wind-induced water movements can produce aggregations of cells even in lakes, e.g. in Esthwaite Water (George & Heaney, 1978) where the situation is complicated by the diel migration of forms such as *Ceratium* which moves from the metalimnion to the surface during the day in this lake (*see* also p. 330). This type of movement is small-scale and is important in keeping the plankton in motion in the surface waters and thus overcoming the natural tendency of most organisms to sink.

Species diversity changes along transects at sea and Hulburt (1963) found a few cells of many species in poor shelf waters during summer on the western Atlantic seaboard, but in winter growth conditions improved and a few successful species grew and dominated the flora.

Rock pools. The rock pool flora in the supratidal region is obviously subjected to some of the most drastic changes that can be conceived for a natural phytoplankton population, in that extremes of temperature, salinity, nutrients and gases are all involved. Algae such as *Brachiomonas submarina, Oxyrrhis marina, Cryptomonas salina, Dunaliella* spp., *Platymonas tetrahele, Chlamydomonas ovalis, C. globosa, C. subcaudata, Euglena robertilami, Cryptomonas (Rhomomonas) baltica, Hemiselmis virescens, Nannochloris* oculata* (Droop, 1953; Bourelly, 1958) must withstand these conditions and additionally have a well developed osmoregulation system which is a vital

* This name may have to be changed if the species proves to be a member of the Eustigmatophyceae, as some species of the genus undoubtedly are.

Fig. 7.32. The occurrence of patches of dinoflagellates (black areas) and *Phaeocystis* (broken line) on transects across the North Sea in 1935. From Lucas & Macnae, 1940.

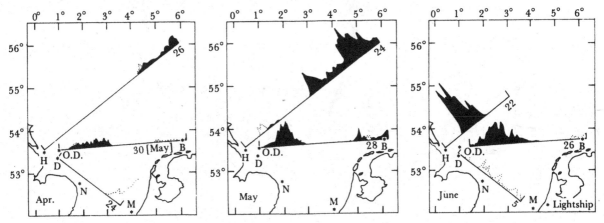

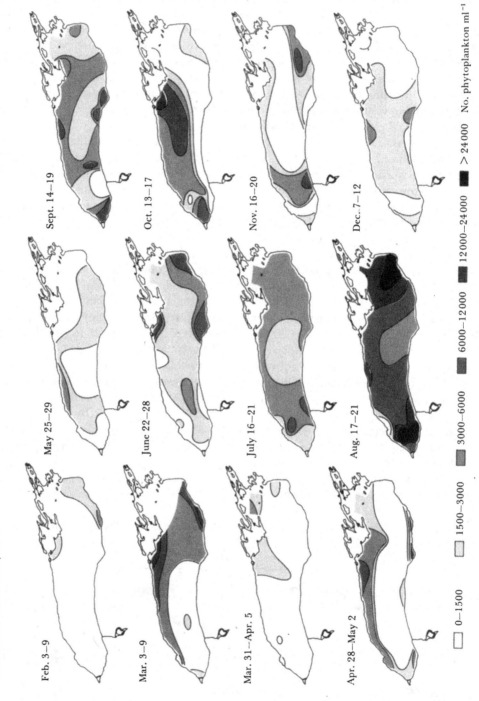

Fig. 7.33. The seasonal development of the surface phytoplankton in Lake Ontario during 1970. From Munawar & Nauwerck, 1971.

factor in survival. In *Dunaliella tertiolecta*, increasing the salt concentration under experimental conditions from 0.17 M to 2.7 M produced an increase in the photosynthetically formed glycerol from 13% to 65% (Wegmann, 1971) and Ben-Amotz (1975) showed that the amount of glycerol was positively correlated with the external salt concentration. In a similar type of experiment using *Monochrysis lutheri*, addition of 0.5 mole l^{-1} of sodium chloride caused an 80–90% increase in cyclohexanetetrol (Craigie, 1969). Both glycerol and cyclohexanetetrol may act as osmoregulators and, according to Borowitzka & Brown (1974), the glycerol in *Dunaliella* species is even more important in maintaining enzyme activity under conditions of high extracellular salt concentration and hence low (thermodynamic) water activity.

Changes in the tide pool environment are likely to be rapid, e.g. when a high wave suddenly floods a pool which has not been flooded for some days or when rain dilutes the pool. Hence it is interesting that Craigie found the increase in cyclohexanetetrol occurred within four hours of addition of salt, whilst dilution resulted in a new steady state within ten minutes. One of the few other experimental approaches to the ecology of the rock-pool plankton is that of Droop (1958) who cultured species in concentrations of magnesium from 0.4 to 4000 mg l^{-1}, of calcium from 0.2 to 2000 mg l^{-1}, and of sodium from 300 to 24000 mg l^{-1} and

measured growth by means of optical density (Fig. 7.35). For comparison he also cultured the neritic diatom *Skeletonema costatum* which although usually considered to be euryhaline is distinctly stenohaline compared to the rock-pool species. The results show remarkable variations in response for species colonising the same habitat; they all tolerate a wide range of sodium ion concentration but in all cases (even in *Skeletonema*) the optimum for growth is below that in seawater (10 720 mg l^{-1}) but much above that of freshwater. *Nannochloris* is interesting in having a high magnesium ion requirement which is higher than the concentration in seawater. There is generally little growth in the experiments where sodium values are considerably above those of seawater or where calcium and magnesium ion concentrations are low.

Dunaliella grows under a wide range of sodium concentrations and the most recent studies (Borowitzka & Brown, 1974) show that sodium ion is excluded from *D. tertiolacta* (marine) and *D. viridis* (halophilic) but there is moderate uptake of potassium ion. Exclusion of sodium ion is known for other brackish and marine algae, e.g., *Ulva lactuca* contains very high levels of potassium relative to sodium ion (West & Pitman, 1967) virtually reversing the ratio in seawater. Ben-Amotz (1975) showed that the intracellular glycerol was in osmotic balance with the medium over the range 0.6 M to 2.1 M NaCl.

The extreme environment of the rock pool obviously imposes considerable heat stress on the organisms and it is interesting that Ukeles (1961) found that the upper limit for *Dunaliella* probably exceeds 38 °C.

The rock-pool flora develops in a medium which is often very rich in organic matter (derived from birds, decaying seaweed, seepage off the land etc.) and so perhaps it is not surprising that Lewin (1972) found that of 55 clones of diatoms isolated from such habitats, some 60% were auxotrophic (i.e. required an exogenous supply of vitamins). Antia, Berland *et al.* (1975) showed that *Hemiselmis virescens* was an exceptional alga in that it did not utilise nitrite or nitrate but could take up the animal excreted ammonia and urea, suggesting a genetic adaptation to this niche where excreted nitrogen dominates the supply.

Rock pools can have a neustonic flora, as yet little investigated, and Bourelly (1958) even found a

Fig. 7.34. The surface water circulation producing wind-rows and Langmuir spirals. U, Upwelling water; D, Downwelling water.

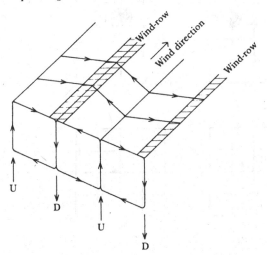

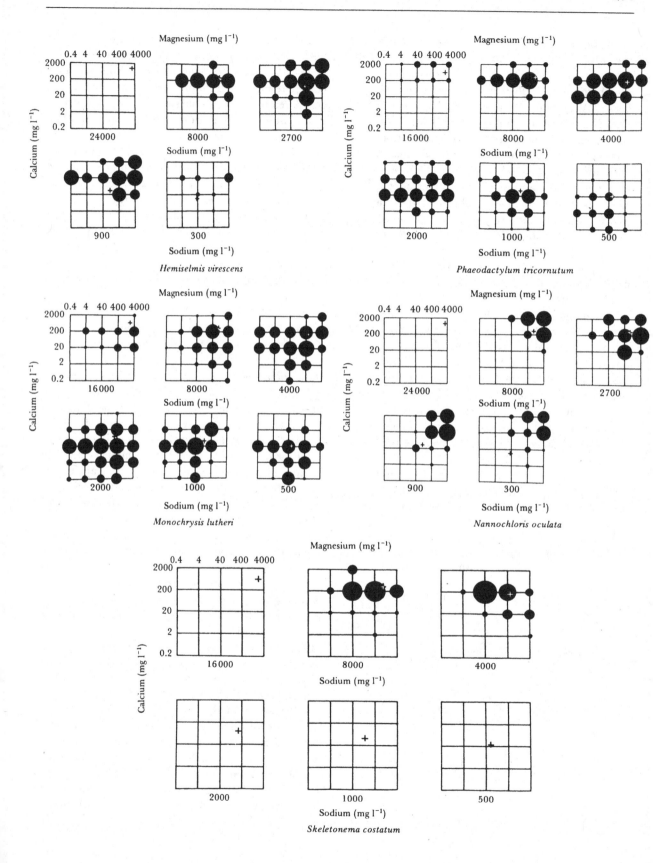

Hemiselmis virescens

Phaeodactylum tricornutum

Monochrysis lutheri

Nannochloris oculata

Skeletonema costatum

new species of the common genus *Chlorella* living in the neuston of one supratidal pool. In freshwater rock pools at inland sites the Volvocalean flagellate *Haematococcus pluvialis* is common and occasionally the colonial flagellate *Stephanosphaera* occurs. These are especially common in solution hollows on limestone but can also occur on acidic rocks and then the base and organic matter content is high often from bird droppings (Swale, 1966).

Upward movement in rock pools of mucilaginous masses of *Peridinium gregarium*, buoyed up by oxygen bubbles is reported by Lombard & Capon (1971) who found that during the afternoon and evening the cells sink leaving behind the mucilage. Such activity may keep the cells away from the grazing copepods which exhibit the opposite sequence of diurnal movement!

Surf plankton. Along some coasts (e.g. those of Oregon and Washington and part of New Zealand and

Fig. 7.36. The cell numbers of *Chaetoceros armatum* in the surf at Copalis Beach during a 24-h cycle (26–27 October 1971). Two values at each sampling time are of numbers at two stations 100 yd apart. From Lewin & Hruby, 1973.

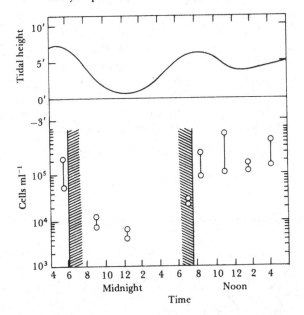

Australia) a rather exceptional planktonic community has been found associated with the surf zone. This is not a region where one would expect to find the rather delicate forms of the open water plankton and indeed they seem to be replaced by a small number of, presumably, specially adapted species. From about 1927 until the 1950s a large robust centric diatom *Aulacodiscus kittoni* was present in the waters along Copalis Beach (Washington State) in such quantities that large masses floated upon the surface and were left on the shore by the receding tide. More recently this species has been superceded by two much more delicate forms, *Chaetoceros armatum* and *Asterionella socialis* (Lewin & Norris, 1970), though the latter was present earlier this century (J. C. Lewin, 1975).

Lewin & Hruby (1973) found that *C. armatum* occurred as floating masses from early morning until about 4.0 p.m. when they sank and became dispersed throughout the water column (Fig. 7.36). The flotation appears to be due to a change in buoyancy of the cells during the 24-h period and it is noteworthy that the increase in buoyancy commences before sunrise (cf. data on diurnal movements, p. 330). Lewin & Hruby postulate that this change in buoyancy is an adaptation to maintain the diatoms in a surface position where they can utilise the low light intensity. However the associated *Asterionella socialis* does not undergo this rhythmic change in buoyancy. As Lewin & Mackas (1972) point out this is an unusual community in that one or two species predominate throughout the year without any seasonal succession. Whilst this book was in proof, I found *C. armatum* in the surf along the Severn Estuary.

'Sargasso Sea'. The classical 'Sargasso Sea' is a region in the Atlantic Ocean where a floating population of *Sargassum natans* (80%) and *S. fluitans* (10%) is maintained in a gyre extending between 20° N to 40° N and 35° W to 77° W. It was first noticed by Columbus in 1492 on his westward voyage and he even noted the small berry-like bladders on the side branches of this alga (Parr, 1939). It is a

Fig. 7.35. Growth responses of species of marine algae to varying concentrations of sodium, calcium and magnesium. The yield is proportional to the diameter of the circle in each case, and the crosses indicate the positions at which the ions are present in the same proportion as in seawater. The single sodium value beneath each frame is the concentration used in that set of experiments. From Droop, 1958.

self-supporting population reproducing by vegetative means and forming a planktonic community together with its associated epiphytic species. The algae form an essentially floating community with very little vertical dispersion (Parr, 1939). Parr's estimate of 2–5½ tons of weed per square mile is very low in relation to other planktonic populations in the sea. There has been relatively little work on these floating populations, and only recently has a measure of net production been obtained (Hanson, 1977). He reports carbon fixation at a rate of 328 ± 114 ng (g dry wt)$^{-1}$ h^{-1} at various stations and says that up to 50% of the gross production can be lost as dissolved organic carbon.

Other areas of floating *Sargassum* occur in the South China Sea, around Japan, the Philippines and in the Red Sea. In quiet water bays along the Texas coast, free floating masses of *Gracilaria verrucosa*, *Digenia simplex* and *Chondria corphilia* are reported (Edwards & Kapraun, 1973) and *Phyllophora* grows in a similar manner in the Black Sea. Apart from the publication by Parr (1939) these planktonic masses of macrophytic species have been virtually unstudied. Most of the literature is in rather obscure journals but a semi-popular account is to be found in Deacon (1942).

Neuston

The interface between a water surface and the atmosphere forms a very special habitat occupied by a small number of aquatic angiosperms (e.g. *Lemna*), animals (e.g. *Gerris*), algae, bacteria and protozoa. These organisms are collectively termed the pleuston and the microscopic forms have been separated off as neuston.* The species which occupy this habitat can either live on the upper surface of the film

(epineuston) or on the lower surface (hyponeuston). They have been relatively neglected and the best known is the epineustonic *Chromophyton* (*Chromulina*) *rosanoffii*, which secretes a small mucilage stalk raising the alga above the water surface (Fig. 7.37). The chromatophore can move in the cells (as in many algae) and takes up a position on the side away from the incident light so that the front of the cell apparently acts as a minute lens: further investigation is needed to confirm this statement, however, since it may be that the chromatophore lies in such a position to protect it from direct insolation. The most extensive review of these organisms in freshwaters is that of Valkanov (1968) and he lists the following small number of genera as occurring in the habit: *Nautococcus, Nautococcopsis, Botrydiopsis, Eremosphaera, Emergococcus, Kremastochloris, Kremastochrysis, Kremastochrysopsis, Chromophyton, Arnaudovia* and *Rhexinema*. Most of these are relatively little studied genera belonging to the Chlorophyta and Chrysophyta and no other group seems to have colonised this habitat. Some of the algae have special 'umbrella'-like extensions of the cell wall which no doubt assist in flotation and stabilisation at the air–water interface (Fig. 7.35). The interest in this community is not only in the unusual and modified algae colonising the site but also in the accumulation of cells and chemicals in the microlayer. Thus Parker & Wodehouse (1972) report increased algal cell counts from the surface film compared to subsurface samples and, also, very much increased concentrations of ammonium and nitrate nitrogen, phosphate phosphorus and surfactant substance in the layer. In

* There seems to be little merit in distinguishing cyanoneuston (dominated by Cyanophyta), etc. as suggested by Bursa (1968).

Fig. 7.37. Examples of some microscopic species. *A, Coleochlamys apoda; B, Nautococcus emersus; C, Krematochlamys conus; D, Nautocapsa neustophila; E, Chromulina rosanoffii.*

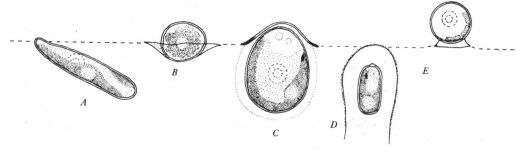

another paper Parker & Hatcher (1974) record mainly chlorococcales and euglenoids, and more rarely Bacillariophyta in the surface microlayer; none of these algae are really in any way adapted to life in the neuston. They are possibly part of the flora studied by Happey (1970) and should not be included in the neuston. The neuston has been studied mainly on small protected lakes where the surface film can take on a brown, red or green coloration but it is quite clear that even under oceanic conditions there is a concentration of organisms at the surface and special collecting sledges have been designed to skim off this very thin surface veneer. In one such study Harvey (1966) found a great concentration of the dinoflagellate *Prorocentrum micans* in the surface layer, 31 270 cells as opposed to 3900 at 10 cm and 1000 at 13 m. Diatoms on the other hand were low at the surface (930) and high at 10 cm (37 770) and at 13 m (16 000). Parker & Barsom (1970) give a general review of techniques to investigate the surface film. Yet another report (Manzi, Stofan & Dupuy, 1977) indicated surface concentrations of diatoms up to 10–15 fold those at 1.5 m in an estuarine situation. The neuston should not be confused with the algae which form 'water blooms' simply by aggregating at the surface and which do *not* possess any adaptations for life on the surface film: this stricture may apply to the *Prorocentrum* just quoted and certainly does apply to the commoner cyanophyceaen blooms. The metabolic rates of these microlayer collections require study: how do they, for example, overcome the usual surface inhibition of photosynthesis?

Sinking and floating

As illustrated in Fig. 7.4 a sample of plankton left to stand in the laboratory rapidly stratifies with the forms which are heavier than the water at the bottom, those lighter at the top and the flagellates between. Without taking very elaborate precautions there are always slight convective movements of water in laboratory vessels and as a result very small species may remain in suspension for a long time. But such simple observations do show that once the wind stress is removed from the surface of the water the majority of algae sink, and only a small number float, suspend or swim at some level. This indicates that wind induced turbulent mixing is *all* that maintains the bulk of the phytoplankton in its environment.

The effect of the wind stress is also shown when ice forms on the surface, resulting again in the heavy forms sinking (e.g. *Melosira*, Fig. 10.17) and buoyant or motile forms moving up to form a layer just below the ice. Measurements of rates of sinking are not easy to obtain and the methodology is complex (*see* Bienfang, Laws & Johnson (1977) for a review of some aspects).

Very few measurements have been made of the density of cytoplasm but those that have suggest a figure just above 1.0 (1.03–1.10). If the density of the wall material is added to that of the cytoplasm, the total density is increased markedly, since the silica walls of diatoms have densities around 2.6, coccolithophorid scales 2.70–2.95, whilst cellulose, such as is found in dinoflagellates and desmids, has a density of about 1.5 (Smayda, 1970). In the case of coccolithophorids, a direct comparison can be made between sinking rates of naked cells and those coated with coccoliths; Eppley, Holmes & Strickland (1967) found that the latter sank five times as fast as the naked cells. It is obvious that unless special features occur to reduce the density below 1.0 all algal cells will sink: such sinking prevents the surface waters from becoming packed with cells. The increased sinking rate of senescent and dead cells is also fortunate in 'winnowing' out these from the photosynthetic zone. The variation in the rate of sinking at different stages during growth is quite considerable, e.g. Boleyn (1972) found that *Ditylum brightwellii* sank at 0.26 m day^{-1} during the log phase of growth and at 2.75 m day^{-1} at senescence. Light or dark does not seem to affect the sinking rate but increase in nutrient concentration reduces the rate. Flotation devices occur in the form of the gas-filled vacuoles of the Cyanophyta and in the oil-packed globules in *Halosphaera* in the ocean and in *Botryococcus* in freshwaters.

Archimedes principle applies to planktonic species and hence the resultant force of gravity on the cells will be

$$F = gkd^3(p^1 - p) \qquad (7.1)$$

where kd^3 is cell (or filament) volume, d is a linear measure, g the acceleration due to gravity and p^1 and p the densities of the organism and liquid respectively. The expression $(p^1 - p)$ is termed the excess density and a cell will remain in suspension when this is zero but usually $p^1 > p$ and some other force must be

Fig. 7.38. Blue-green algae and desmids mounted in Indian Ink to show mucilage sheaths. *A*, *Microcystis* (×115); *B*, *Anabaena* (×560) with heterocyst and two akinetes; *C*, *Anabaena*, coiled and straight filaments and *Gomphosphaeria* (×145); *D*, *Cosmarium* (×560); *E*, *Xanthidium*, dividing cell (×525); *F*, *Arthrodesmus* (×850); *G*, *H*, *I*, *Staurastrum* (×525). Photographs kindly supplied by Dr H. M. Canter.

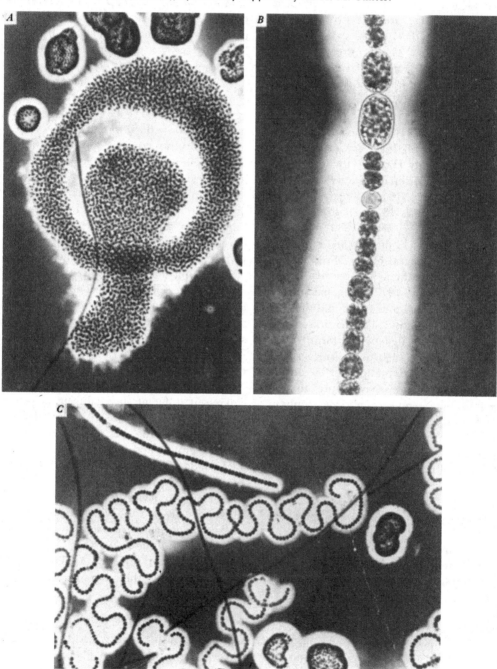

applied to prevent the cell sinking. $p^1 < p$ only occurs in the forms with gas vacuoles or oil.

The rate of sinking may be a somewhat theoretical concept as far as many algae are concerned since under most natural conditions a considerable amount of turbulence will be experienced. However, in extremely sheltered situations, in very shallow lagoons and in the deepest part of the photic zone, it is of importance. Also, the heavier the alga, the greater the sinking potential. There is undoubtedly

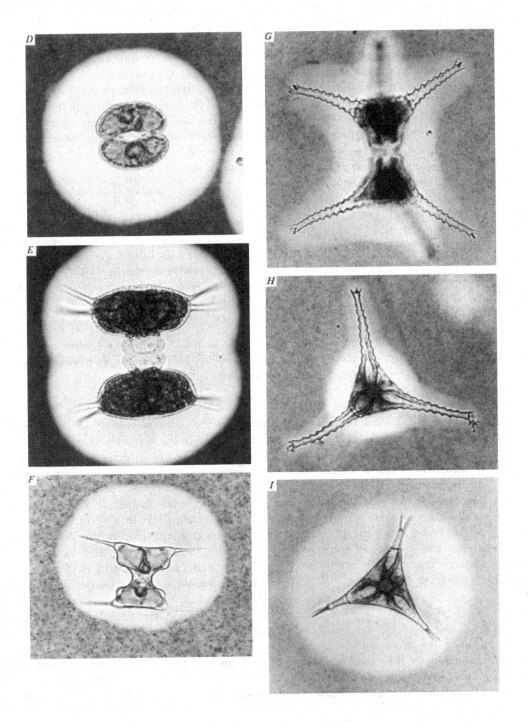

a vertical zonation of species in many waters (*see* p. 348) and it is likely that differential sinking rates or buoyancy mechanisms coupled with density stratification of the water, play a part in maintaining these zonations. Hulburt (1963) comments that small species and especially small flagellates are often common in sheltered bays and this could be due to their decreased rate of sinking in still waters: the influence of buoyancy on the composition of populations has not yet been adequately explored. Seasonal succession of some species in sheltered ponds may be related to interactions between turbulence and sinking rates of phytoplankton, e.g. Moss (1969*b*) considered that the presence of non-motile phytoplankton (*Asterionella*, *Stephanodiscus*) was related to turbulent conditions whilst during stratification and less turbulent conditions the plankton was dominated by flagellates (*Pandorina*, *Cryptomonas*). As pointed out by Knoechel & Kalff (1975) the sinking rate of *Tabellaria fenestrata* was 31 cm day⁻¹ in early summer and this had a negligible effect on the sedimentation of the population but the rate increased later in summer. As the thermocline develops the sinking speed will increase as turbulence in the epilimnion decreases and then sinking becomes an important consideration (turbulence shows an inverse relationship to depth). Flagellate algae which can swim will also become stratified whenever conditions are calm enough for their speed of swimming to overcome vertical mixing by current movement.

Planktonic algae are diversely shaped and, with the exception of the few spherical forms, can present various planes to the vertical direction of sinking (though it is more likely that a sideways and downward drift is involved). Thus a *Cosmarium* species with little mucilage around it could sink with three differently shaped faces towards the water below it. Many species however secrete mucilage sheaths (*see* Fig. 7.38) and tend to present spherical profiles to the medium. The velocity of spheroid particles sinking through liquid medium can be expressed by Stokes' equation:

$$v = \frac{2}{9} gr^2 \frac{p^1 - p}{\eta} \tag{7.2}$$

where r is the radius, g the acceleration due to gravity, and η the viscosity of the medium. The viscous drag on a plankter is thus altered by changes in viscosity and this in turn decreases with increasing

temperature, a 2% decrease in viscosity occurring for every degree Centigrade rise. Hence, theoretically, sinking increases during spring and summer and is at its lowest rate in midwinter. Smayda (1970) calculated that the sinking rate of a cell 20 μm in diameter increases approximately 4% for every degree rise in temperature, simply due to the temperature effect on viscosity. There are reports however that the electrical properties of the cell wall also influence the viscous drag.

Many phytoplankters are morphologically complex and the drag of these through the water will not be accounted for in Stokes' equation. The morphological complexities, wings, spines, etc., add up to a factor of 'form resistance' (R) which can be incorporated thus:

$$v = \frac{p^1 - p}{\eta \cdot R} \tag{7.3}$$

R is extremely variable whereas $p^1 - p$ is not. Attempts to quantify R have been made using a shape correction factor in Stokes' equation (Hutchinson, 1967) so that a non-spherical body sinks according to

$$v_a = \frac{2gr^2}{9} \frac{(p^1 - p)}{\eta \cdot \phi_r} \tag{7.4}$$

where ϕ_r is the 'coefficient of form resistance'. Various methods of calculating ϕ_r have been devised (Hutchinson, 1967):

$$\phi_r = v_s / v_a \tag{7.5}$$

where v_s is the terminal velocity of an equivalent sphere and v_a is the terminal velocity of a non-spheroidal body of similar density and volume.

Form resistance depends also on the alignment of cells and, as Lund (1959, 1965) has shown, some algae which are needle-shaped (*Oscillatoria*, *Melosira*) tend to align themselves in a vertical plane when observed in a constant temperature chamber at 0°, thus exposing the least surface area to the force of gravity. But under natural conditions such a position will not be maintained and at times they may fall when lying horizontally and therefore the viscous drag of the water will be maximal. Many planktonic algae have complex shapes (e.g. *Ceratium*) or form complex colonies, e.g., *Asterionella* and it is extremely difficult to calculate their form resistance under the varying conditions in nature or evaluate the advantages of the various shapes. For unicells the general trend is for sinking rates to increase as average cell diameter increases; in other words, large organisms sink faster than small. However large organisms seem to sink proportionately less rapidly (Eppley, Holmes & Strickland, 1967).

Clearly then it is probably of very little value to quote the sinking rate of a planktonic species without giving cell dimensions; thus, some preliminary data provided by Dr C. Reynolds indicate that *Stephanodiscus astraea* cells of diameter 31.4 ($\pm$3.7) μm have a dry weight of 3303 μg for 10^6 cells and sink at 5.4 ($\pm$1.7) μm s^{-1} and cells of 43.3 ($\pm$4.2) μm diameter with a dry weight of 6279 μg for 10^6 cells sink at 8.65 ($\pm$1.02) μm s^{-1}. This is interesting in that the larger cells are presumably post-auxospore cells. The upper size limit for this species, according to the floras, is 70 μm and post-auxospore cells of 55 μm diameter have been recorded (Round, unpublished). One might expect, therefore, a greater loss by sedimentation immediately after auxospore formation and massive blooms only of moderate or small-size populations. The difference in size is shown in Fig. 7.39.

During growth of a community containing diatoms, cells of very different size will be present, e.g., the common large marine *Coscinodiscus asterom-*

phalus can reduce from a valve diameter of 200 μm down to 55 to 60 μm at which stage it is no longer capable of mitosis (Werner, 1971). It might be expected that the different sizes would divide at different rates but this is not so for *C. asteromphalus*, at least in culture, where at 24 °C there is one division per day and at 18 °C there are 0.6 divisions per day. Werner calculates that cells take 90 days to reduce from 200 μm to 60 μm at 24 °C and 165 days at 18 °C. In nature, during this time, conditions governing buoyancy may change drastically but in fact there are few data on the effects of such changes on natural populations of single species. Smayda (1971b) found that a large *Coscinodiscus* species would sink about 35 m day^{-1}: this is a high rate and would presumably result in fairly rapid loss of the population from surface waters during periods of calm. Eppley, Holmes & Strickland (1967) thought it difficult to account for the occurrence of such large species as *Coscinodiscus* or *Ethmodiscus* unless some of the cells were neutrally buoyant. On the other hand a small celled, although colonial, diatom (*Thalassiothrix nitzschioides*) sank as slowly as 0.8 m day^{-1}.

Species which exist as colonies tend to sink faster than unicells (Smayda, 1970). This is shown in Table 7.5 for *Bacteriastrum hyalinum* and applies

Fig. 7.39. *A*, small vegetative cells of *Stephanodiscus* and an immediate post-auxospore cell with the two large primary hemispherical valves and (*B*) after the first division of the post-auxospore cell revealing a valve (arrowed) of the new size range. Another small vegetative cell is lying on the new valve. Photograph, author.

Table 7.5. *The influence of colony formation on diatom sinking behaviour. From Smayda, 1970*

Bacteriastrum hyalinum		Skeletonema costatum	
Colony size (cells)	Mean sinking rate (m day^{-1})	Colony size (cells)	Mean sinking rate (m day^{-1})
1–5	0.79	2–5	0.73
6–10	1.31	6–10	0.55
11–15	3.21	11–20	0.32
		20	0.13

also to *Chaetoceros*, *Thalassiosira gravida*, *Nitzschia* (colonial forms), *Leptocylindricus* and *Asterionella* but, as the table shows, not to *Skeletonema costatum*. Smayda suggested that this anomaly may be due to micro-turbulent conditions between the siliceous spines connecting the cells: however my observations on *Skeletonema* with the scanning electron microscope show that girdle bands often cover the spines and

Fig. 7.40. Two cells of *Skeletonema* linked by processes from which girdle bands have been lost (centre) and retained, but torn in preparation (above and below). Scanning electron micrograph, author.

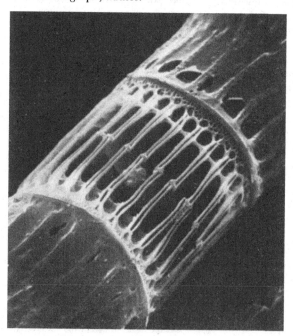

thus form a more closed cylinder than light microscopic observations would suggest (Fig. 7.40).

Structures such as spines (e.g. in *Chaetoceros*), are likely to increase sinking rates rather than retard them as suggested by some authors. In addition freshwater genera such as *Staurastrum* have spines but these are buried in a sphere of mucilage (Fig. 7.38) and hence the spines only add excess weight to the sphere. The dinoflagellate *Ceratium hirundinella* has been shown to vary the length of its horns seasonally but the relationship of this to sinking is not clear. The rather delicate spines on the end cells of the coenobium of *Scenedesmus* reduce the sinking rate of this freshwater genus and the longer bristles growing from the cells of some species definitely aid flotation (Conway & Trainor, 1972). The chitin threads produced by *Thalassiosira* can be removed by treatment with chitinase and such cells sink 1.7 times as fast as control cells (Walsby & Xypolyta, 1977). Since the formation of bristles of *Scenedesmus* depends upon the concentration of the nutrient media, these algae may possess differing capacities for flotation as the season progresses: studies under natural conditions are now needed to show the efficacy of this aid to flotation.

The problem of buoyancy really involves only the small number of algae which float, since the density of the others is varied but always with $p^1 > p$ and they therefore sink. It has been generally assumed that fat is an agent of buoyancy, presumably since oil floats on water, and certainly algae such as *Halosphaera* and *Botryococcus* do float in still water. However, Smayda (1970) points out that this role of fat has been frequently questioned and, indeed, in many algae fat accummulation is a feature of senescence and such algae tend *not* to float but sink. Certainly where studies have been made, ageing is always accompanied by sinking. A few unusual instances of oily films on the sea surface are accompanied by, and possibly related to, blooms of diatoms (e.g. *Coscinodiscus concinnus*) but Smayda calculated that even taking into consideration the extreme values for fat content the excess density of the diatoms would be reduced only by 3.5% and they would still sink. It is necessary therefore to seek some other explanation of buoyancy, even for fat filled cells, and one of the most discussed is that of ionic adjustment of the liquid in the vacuole. By eliminating the heavy ions the excess density can be reduced, but few algae have been shown to operate such a system. One is the colourless heterotrophic *Noctiluca*, which is able to exclude all the heavy ions, especially sulphate, has a high intracellular concentration of sodium relative to potassium ions, which are heavier by about 40%, and has a relatively high concentration of light ammonium ions. Similar mechanisms have been postulated for the diatom *Ditylum* (Gross & Zeuthen, 1948) and for *Ethmodiscus*, the most massive of all planktonic diatoms which is up to 2000 μm in diameter and which excludes magnesium ions completely (Beklemishev, Petrikova & Semina, 1961); however, there is much confusion in the literature and because of the low salt concentration in freshwaters it is an unlikely mechanism for freshwater plankton (Lund, 1959), yet these are equally buoyant. Diurnal variations in sinking rate are reported (Eppley, Holmes & Paasche, 1967), e.g., maximum settling of *Ditylum* occurs after eight hours in the dark: since *Ditylum* is one of the few diatoms studied in which p^1 approximately equals, p, small changes in ionic composition could facilitate changes in buoyancy, especially since uptake of ions also exhibits a diurnal variation (*see* p. 329). Anderson & Sweeney (1977, 1978) showed that *Ditylum* does have its cell density increased by 3.4 mg ml^{-1} by the end of the light period and this can account for an increase in settling rate of about 0.3 m day^{-1}. This should increase the settling rate during the course of the light period and is not quite in line with the results of Eppley, Holmes & Paasche. The dark period is also the time during which cell elongation and chromatophore division is occurring. Swift, Stuart & Meunier (1976) showed that buoyancy states in *Pyrocystis* vary between reproductive cells (negatively buoyant) and newly formed cells (buoyant) and the buoyancy is due to less dense vacuolar sap (Kahn & Swift, 1978). Working with the same genus the Russian workers Sukhanova & Rudyakof (1973) report that the vegetative cells were found at a mean depth of 70 m but the reproductive were around 120 m. A decreased rate of sinking in rich media has been shown in culture conditions (Smayda, 1970) and this factor has also been proposed to explain the aggregation of cells at precise depths in the ocean (p. 348).

Little is known of the residence time of senescent or dead cells in the photic zone but ultimately they all sink, even in water which is permanently turbulent. Eppley, Holmes & Strickland (1967) found that on cessation of logarithmic growth there was an abrupt change and sinking rates increased about fourfold (confirmed now for freshwater species by Titman & Kilham, 1976). In colonial genera, e.g., *Asterionella*, *Chaetoceros*, etc., any increase in the number of dead cells (e.g., as a result of parasitism, *see* p. 460) should increase the sinking rate of the whole colony. These dead cells are probably lost as they circulate downwards into the regions of decreased turbulence and/or are incorporated into faecal pellets. Schrader (1971) showed that phytoplankton is readily incorporated into pellets but that below 500 m the pellets break up. So whilst sinking rate is increased in the upper water it presumably reduces lower down. The fairly rapid sedimentation in regions such as the Gulf of California (deduced from a study of the surface populations and remains in the sediments (Round, 1967)) indirectly indicates some acceleration mechanism even in the deeper water. The increased sinking rate of dead cells is of course indirect evidence for the importance of the living protoplast over that of the form of the cells in control of buoyancy. We have insufficient evidence about the fate of dead cells and do not know what percentage is lost by sinking and what percentage by grazing; indirect evidence, such as the rapid disappearance of dying populations in freshwaters followed immediately by their appearance on the sediments suggests sinking rather than grazing, at least in shallow waters.

A stationary alga actively metabolising will deplete nutrients from a shell of surrounding water, change the pH of this water and build up a shell of extra-cellular products; all of these changes will tend to depress cell metabolism. A steady fall or rise through the water will greatly reduce the shells and hence will be beneficial to metabolism. Aggregation of plankters in close proximity can be harmful since nutrients are more easily depleted, extracellular products accumulate, and self-shading reduces the rate of photosynthesis. Such aggregations do occur close to the surface when gas vacuolate Cyanophyta or oil bearing *Botryococcus* float upwards and form a water bloom in summer (in winter turbulence is such that water blooms rarely form). Under water bloom conditions it is not uncommon to find many moribund cells.

There is often an inhibition of carbon fixation

in the surface waters (*see* p. 343) and this is a further hazard to these aggregations at the surface. Oxygen supersaturation also occurs and since nitrogen fixation by Cyanophyta decreases as oxygen content increases such surface aggregation may also be detrimental to the metabolism of nitrogen. Thus the simple view that floating is beneficial since light intensity is greater at the surface may be offset by other, adverse, factors.

It must not be assumed that all vacuolate Cyanophyta float to the surface. As Lund (1959) and Baker & Brook (1971) demonstrated, some filamentous species collect in deeper water just above the thermocline (Fig. 7.41).

Unlike the situation in the fat-forming algae, the presence of gas vacuoles in the cells of Cyanophyta have been shown to be adequate to account fully for density and buoyancy changes (Walsby, 1969). The upward flotation of the cells is as dependent on Stokes' law as the sinking of others and therefore the rate varies as the square of the radius, i.e. the larger the body the faster the upward movement. In fact, the large colonial species, which tend to be so prominent in water blooms, may rise so rapidly and be so unwieldy that they overcompensate and collect in a relatively unfavourable surface position. Colonies of *Gloeotrichia echinulata*

rise at rates up to 0.23 mm s^{-1} whereas the smaller colonies of *Anabaena circinalis* achieve only 0.03 mm s^{-1} (Fogg *et al.*, 1973). The mechanism by which the cells adjust their position in the water column is assumed to be by adjusting the gas vacuole:cell volume ratio. *Oscillatoria rubescens* and *O. agardhii* var. *isothrix* are the species most commonly found in definite sub-surface layers in lakes; they often form very stable layers stretching across the entire lake (Fig. 7.41). Others form more temporary layers, e.g. *Microcystis* and *Anabaenopsis* species concentrate in the surface waters of Lake George in the morning and sink to the bottom of this shallow tropical lake every evening (Ganf, 1974*c*). Such rapid changes in buoyancy could be accounted for by collapse of the gas vacuoles which occurs during intensive photosynthesis and their regeneration in the dim light at the bottom of the lake. The aggregation of vacuolate Cyanophyta in blooms at the surface results in massive rates of photosynthesis, e.g. 8 g O$_2$ m^{-2} day^{-1} was recorded at the peak of an *Aphanizomenon* bloom (Barica, 1975). Such rates can deplete the water of carbon dioxide and therefore reduced photosynthesis; the excess photosynthesis which is required to bring about the collapse of the vacuoles cannot occur and so the population is maintained as a bloom under detrimental surface

Fig. 7.41. Optical density profiles across Deming Lake, Minnesota. From Baker & Brook, 1971.

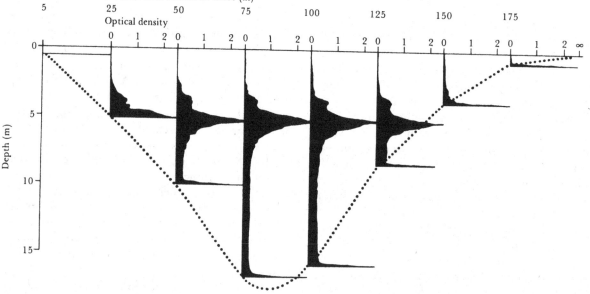

conditions and the death of the cells results (Walsby, personal communication). Collapse of gas vacuoles brought about by a rise of turgor pressure at high light intensity has been recorded by Dinsdale & Walsby (1972) and can occur within an hour, so it could be effective in such diel movements (Fig. 7.42 summarises a possible sequence; *see also* Kanopta *et al.*, 1978). The pressure required to collapse *Anabaena* gas vacuoles is about 6 bars, but only 2–3 bars when the cells are turgid since the cell turgor presure is 3–4 bars (Walsby, 1971). On the other other hand, in the blue-green marine genus *Trichodesmium*, which often forms layers at 15–25 m depth and can occur in appreciable amounts down to 200 m, the ability to float back to the surface depends on the presence of gas vacuoles able to resist the high pressures at these depths, about 20 bars. In fact Walsby (1978) found that *Trichodesmium* gas vacuoles had mean critical collapse pressures of 12 bars (*T. erythraeum*), 34 bars (*T. contortum*) and 37 bars (*T. thiebautii*): these are far too high to be collapsed by increasing cell turgor pressure and the theory put forward for freshwater blue-greens cannot apply to this common marine form although its colonies are positively or neutrally buoyant. Maintenance of a midwater position is presumably due to the subtle combination of gas vacuole formation and discharge which achieves a balance in such a low light intensity zone where photosynthesis proceeds at a reduced rate (*see* p.

343). These midwater species are often filamentous, a morphology which reduces the rate of floating and sinking compared to that of colonies. As with other groups of algae a downward movement of populations of Cyanophyta has been recorded as the season progresses, e.g. in Lake Lucerne the *Oscillatoria* population moves down from 7.5 m in June to 12.5 m in July accompanied by decreases in phosphate concentration (Zimmermann, 1969).

Braarud (1976) thought that the swimming ability of dinoflagellates gives them a competitive advantage in the still waters of inner parts of the Hardangerfjord where they predominate over the diatoms. The latter are more abundant in the outer, more turbulent, waters. Moss (1977a) has also suggested that the ability of *Pandorina marum* to swim upwards in the stratified waters in summer is one of the factors in its restriction to that period. There is little doubt that cell movements at different times of the year and the associated sinking and floating states have been neglected in assessing the factors involved in phytoplankton succession.

'*Water blooms*' *and* '*red tides*'

The aggregation of masses of cells and filaments of algae in surface waters is involved in these two phenomena. Water blooms are more frequently observed in freshwater lakes, usually when masses of blue-green algae float at the surface on calm days,

Fig. 7.42. A possible sequence of events in buoyancy regulation by planktonic Cyanophyta. From Reynolds & Walsby, 1975.

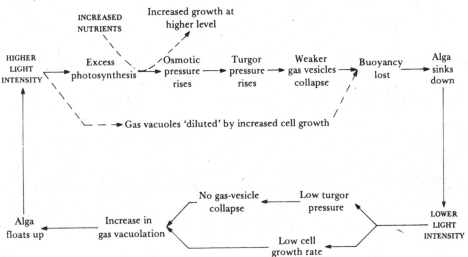

Fig. 7.43. The seasonal changes in the vertical distribution of *Microcystis* colonies in Rostherne Mere, 1972–3. From Reynolds & Rogers, 1976.

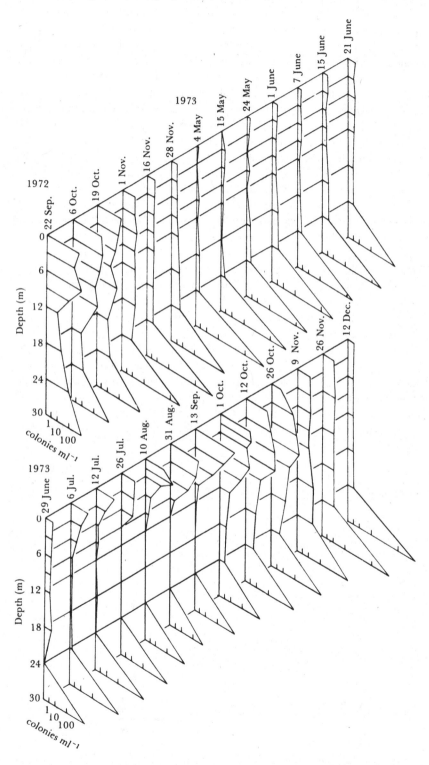

but have also been recorded in the ocean and again a blue-green algae (*Trichodesmium*) is usually involved (Ferguson-Wood, 1965; Cribb, 1969c). In the Baltic Sea the blue-green alga *Nodularia spumigena* forms blooms (Bursa, 1968) but this is in water of very low salinity. The 'Red Sea' and the 'Vermilion Sea' (El Mar Rojo) (the old name for the Gulf of California) are names originating from the coloration of the water by organisms such as *Trichodesmium* or dino-flagellates. The term 'bloom' perhaps implies vigor-ous growth but it is probable that many of the algae in these blooms are senescent, though as shown by Gerloff & Skoog (1954) the cells are not nutrient-limited and indeed often contain luxury amounts of nitrogen and phosphorus (in *Microcystis aeruginosa*). Reynolds (1972) found concentrations of *Anabaena circinalis* filaments of 16 000 ml^{-1} common in blooms in Shropshire. This is equivalent to about 500 000 cells ml^{-1} or a dry weight of about 20 g m^{-3}. Growth of such populations occurs in the water column followed by accumulation at the surface and although this may give the appearance of dominance by blue-green algae, counts often show that other algae are even more abundant in the water column as a whole (Reynolds, 1973a, b). In temperate waters, conspicuous bloom formation usually occurs only in late summer but in tropical lakes it can be a permanent feature. The 'bloom' of *Spirulina* in Lake Chad is collected and eaten by the population around the lake. It is interesting that in Rostherne Mere there always seems to be a deep-lying popula-tion of *Microcystis aeruginosa*, even when that in the uppermost water is forming a bloom (Fig. 7.43; Reynolds & Rogers, 1976). The 'breaking' of the meres (small lakes) in Cheshire is the local term for the sudden appearance of water blooms.

'Red tides' are a coastal marine phenomenon caused not so much by the aggregation due to buoyancy but by rapid cell division stimulated by particular environmental conditions. Bainbridge (1957) gives several fascinating accounts from writ-ings of the early naturalists. The algae involved are species of Dinophyta usually of the genera *Gymno-dinium* or *Gonyaulax* which, having red pigments dissolved in their oil globules, impart a reddish tinge to the sea. I wonder how often other genera prolifer-ate to the same extent but are not detected since their coloration is less striking? These algae occur in many oceans but their growth to dangerous proportions

tends to occur in tropical and sub-tropical zones; they are dangerous since they produce toxins lethal to fish and man but not to shellfish in which they tend to accumulate. The toxin (saxitoxin) from *Gonyaulax* blocks nerve conduction by interference with the initial increase in sodium permeability of the mem-brane (Bhakuni & Silva, 1974). Recently they have also been reported around Britain though it is rare for conditions to be favourable for such rapid growth of the species at such a latitude. Off the coast of the USA regular examination of phytoplankton samples and bioassays of the blooms is maintained and shellfish sales are banned, when the blooms become severe. The 'blooming' of the algae is reported to occur during the rainy season when influx of organic matter stimulates growth in waters off Florida (Col-lier, 1958). However a source of inoculum is necessary and this could be from cysts deposited on the sediments. Anderson & Wall (1978) have found such accumulations of cysts in areas where toxic blooms had occurred in the previous spring; the cysts could be stored in the dark for six months and an increase in temperature might then be sufficient to initiate germination. A fascinating account of the progress of a bloom in the waters of Chesapeake Bay is given by Tyler & Seliger (1978). They detail its beginning at the mouth of the Bay, followed by the deep transport of the cells in a layer below the pycnocline some 240 km up the Bay until the lower water is mixed into the surface and the bloom occurs. A vitamin-B$_{12}$-producing bacterium was found associa-ted with the *Gymnodinium breve* cultured from this site and its growth promoting effect may be an added factor in development of the bloom. Some 'red tides' are caused by the ciliate *Mesodinium* (= *Cyclotrichium*) which is reddish from a symbiotic cryptomonad in the cells.

Light adaptation

'Sun' and 'shade' plankton

Adaptation to high or low light regimes has been known for many years: algae live over a much wider range of light intensities than terrestrial plants and so are likely to show greater variation. The adaptation could be brought about either by varying, singly or in combination, the concentration of pig-ment and the concentration of enzymes (Steemann-Nielsen, 1974). According to Steemann-Nielsen, adaptation to light involves changes in pigment concentration although, in the case of *Chlorella*, the alga decreases its size as illumination increases, and this will have the effect of increasing the enzyme concentration per unit weight. When, however, there is an adaptation to a shift in temperature only the concentration of enzymes is altered, increasing as temperature is lowered. In nature, 'shade'-adapted algae often also grow at lower temperatures than 'sun'-adapted, so that two effects are operative at the same time. At very low levels of illumination it would be impossible for an alga to function if the temperature was not also low.

There have been few attempts to investigate this phenomenon in nature, an interesting exception being the study by Ryther & Menzel (1959) on the plankton of the Sargasso Sea. In winter the water was isothermal to depths below the photic zone and the algae behaved as 'sun-forms' being light

Fig. 7.44. Photosynthesis in Sargasso Sea phytoplankton from depths to which 100% (open circles), 10% (half filled circles) and 1% (filled circles) of the surface light penetrated in, *a*, November and, *b*, October. From Ryther & Menzel, 1959.

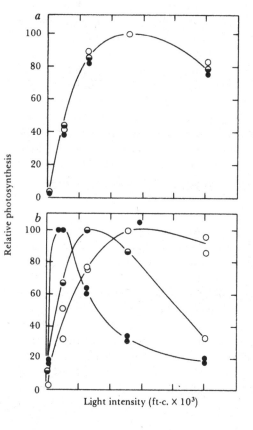

saturated at 5000 ft-c. In summer, when the water was stratified, the surface plankton was composed of 'sun forms' but those living at a depth around the 1% light level acted as 'shade-forms', reaching light saturation below 1000 ft-c. Those living at the 10% light level had an intermediate reaction (Fig. 7.44). Further confirmation of the presence of 'shade forms' comes from the recent work on the deep chlorophyll layer in the central gyre of the North Pacific where Eppley *et al.* (1973) found that the phytoplankton in the mixed layer (down to 60 m) was saturated at 23% of surface light (roughly the light level at 60 m) but at 120 m was saturated at the ambient level which was 1% of the surface value.

Temperature adaptation

Braarud (1961) showed that the relative growth rates of dinoflagellates cultured from warm water had temperature optima differing from those of the same species from cooler waters. Therefore it appears that there is an adaptation to local ecological conditions. Similar results are obtained from different species of the same genus, e.g. of the *Chlorella* spp. illustrated in Fig. 7.45*a*. Hulburt & Guillard (1968) found that the cosmopolitan *Skeletonema costatum* exists as cold and warm-water races but the tropical *S. tropicum* has no cold-water race. Different algal groups have different optima and maxima for growth, e.g., Ukeles (1961) showed that the maximum for Chrysophyta tends to be low (24°–27°): if this applies generally to this group it could in part explain their greater abundance in Arctic and Alpine situations. Records of Chrysophyta from warm lakes are rare but Pollingher (personal communication) tells me that *Mallomonas* occurs in the subtropical Lake Kinneret only in winter and even then it is rare. However, low temperature certainly does not seem to affect the spring outburst adversely, e.g., in Lake Erken more than half the growth takes place below 3 °C during which time there is a noticeable decrease in phosphate and nitrate (Pechlaner, 1970). Similar large growths, e.g., of *Skeletonema costatum*, occur at low winter temperatures in December in Narragansett Bay (Pratt, 1965). Semina (1972) reported a relationship between cell size and temperature: the smaller cells occur in cold water and the larger in warm (*see* also p. 285). The rate of the light reaction of photosynthesis is measured by the slope of the curve when fixation is plotted against light intensity (Yentsch, 1974) and is altered by temperature (Fig. 7.45*b*) although it has usually been assumed that temperature has little effect on this photochemical reaction (e.g. by Steemann-Nielsen & Jørgensen, 1968).

A form of temperature adaptation is the development of resting spores in *Thalassiosira* and

Detonula when the temperature is lowered (though the effect is also seen at low nitrogen levels). The spores were viable for very long periods at low temperatures (compare the survival of cells at low temperatures) and they may aid survival in polar regions (Durbin, 1978).

Nitrogen fixation

Fig. 7.45. *a*. Relative growth rates ($\log_{10}$ day^{-1} units) at approximately 15000 lux of three strains of *Chlorella* as a function of temperature. *Chlorella pyrenoidosa* (○) cold water strain, *C. pyrenoidosa* (◑) Emerson strain and *C. sorokiniana* (●) high temperature strain. From Fogg, 1969; *b*. Fixation of ^{14}C as a function of light and temperature in *Nannochloris atomus* (i) and *Cyclotella nana* (ii). The latter organism is now known as *Thalassiosira pseudonana*. Arrows denote critical value. From Yentsch, 1974.

Nitrogen fixation in natural environments has been recognised for many years but only recently have reliable techniques become available to measure it; first ^{15}N$_2$ studies were developed and later the simpler technique of determining nitrogenase activity by the acetylene reduction method was perfected (*see* review by Stewart, 1971). One of the most important sites for nitrogen fixation in the hydrosphere is the open water, especially of lakes, where heterocystous Cyanophyta, especially *Gloeotrichia*, *Anabaena* and *Aphanizomenon*, can be very abundant (Stewart *et al.*, 1971). In the sea it has been assumed that only *Trichodesmium* is likely to be effective but recently the endophytic blue-green alga *Richelia* growing in *Rhizosolenia* has been shown to fix nitrogen at the rate of 800 μg m^{-2} and the authors (Mague, Weare & Holm-Hansen, 1974) suggest that this could add appreciable amounts of nitrogen to areas such as the north Pacific gyre. Later work (Mague, Mague & Holm-Hansen, 1977) showed that the amount fixed could only supply about 3% of the daily total requirement in the central North Pacific, though *Trichodesmium* itself could fix its total

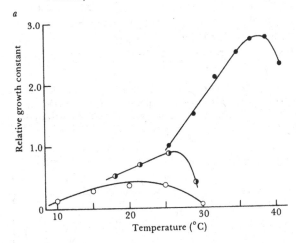

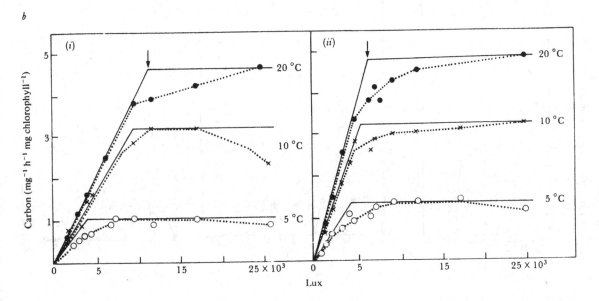

requirement. Although *Trichodesmium* is abundant in the Sargasso and Caribbean Seas, here also its total fixation (average value 1.3 mg day^{-1}) does not contribute greatly to the total nitrogen required. The non-heterocystous *Trichodesmium* fixes nitrogen in the light-coloured cells in the centres of the bundles of filaments (Carpenter & Price, IV, 1976) and when turbulence breaks the bundles, oxygen penetrates to the colourless cells, and reduces the rate of nitrogen fixation. This may explain why rapid growth and bloom conditions occur only in calm seas.

In lakes which have bloom-forming quantities of Cyanophyta this source of fixed nitrogen can be considerable, e.g. Horne & Goldman (1972) estimated that 500 Mg of nitrogen was added to Clear Lake in 1970 from this source; in fact this was 43% of the total, while 48% came from the inflow and 9% from rainfall. They also found that increased nitrogen fixation accompanied increase in heterocysts in the water mass (Fig. 7.46). In laboratory

experiments, heterocysts have been shown to be the major sites of nitrogen fixation in the filaments and Ogawa & Carr (1969) showed that in experimental systems the number of heterocysts was highest in the absence of nitrate-nitrogen and also that phosphate was not essential for heterocyst development. It is interesting therefore that both *Anabaena* and *Aphanizomenon* are found in the summer in Lake Erie when nitrate-nitrogen is near zero. There is however experimental evidence that *Anabaena flos-aquae* grown in a chemostat under conditions which prevented heterocyst formation were still able to fix nitrogen (Kurz & La Rue, 1971) confirming that vegetative cells are a minor site of fixation. Large numbers of heterocysts in the water mass are not the only necessity for nitrogen fixation since, according to Horne, Dillard, Fujita & Goldman (1972), there must be a relatively low concentration of nitrate and ammonia, a high concentration of phosphate and a moderately high concentration of dissolved organic nitrogen and

Fig. 7.46. The seasonal variation in nitrogen fixation and heterocyst numbers in the three basins of Clear Lake. Nitrogen fixation (filled triangles). Heterocyst numbers (for *Aphanizomenon*, open circles and *Anabaena*, crosses) and heterocysts as percentage of total cell number (closed circles for *Aphanizomenon* and open triangles for *Anabaena*). From Horne & Goldman, 1972.

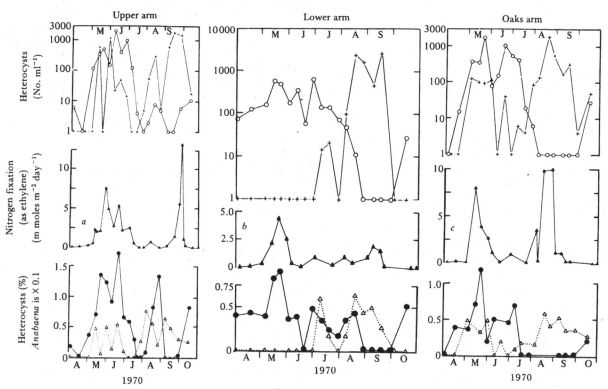

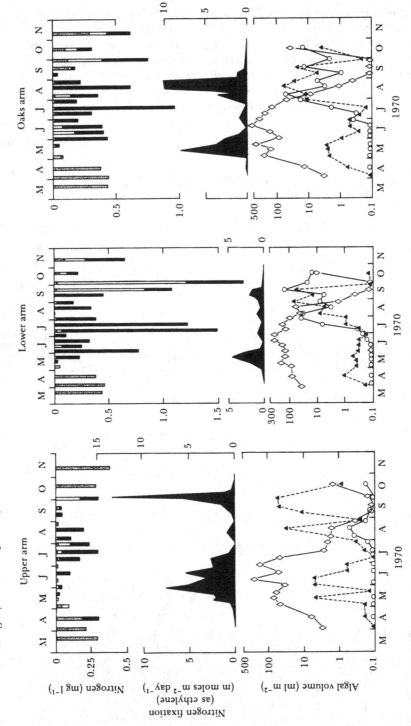

Fig. 7.47. Seasonal variations in nitrogen fixation, inorganic nitrogen and blue-green algae in the three basins of Clear Lake. Histogram shows surface nitrate (stippled), bottom-water ammonia (black), and surface ammonia (open). Cell volumes of *Aphanizomenon* (diamonds), *Anabaena solitaria* and *Anabaena* sp. (filled triangles) and *Microcystis* (open circles) are also shown. From Horne & Goldman, 1972.

only when all these conditions are satisfied will much nitrogen fixation be detected. These conditions tend to occur in late summer–autumn in eutrophic lakes but the high levels of organic nitrogen necessary may never occur in oligotrophic lakes. Stratification can provide the necessary conditions in the surface waters of either lake type and result in periods of nitrogen fixation, e.g., after the spring bloom or at irregular intervals during summer. Low oxygen tension is also advantageous even to the heterocystous species (Stewart & Pearson, 1970). Horne & Fogg (1970) found that almost all the nitrogen fixation (0.037–0.287 g m^{-2}) in the English Lake District lakes occurred in summer–autumn but contributed less than 1 % of the nitrogen input. In Clear Lake there can be both early and late summer periods of fixation (Figs. 7.46, 7.47, 7.48; Horne & Goldman, 1972). In tropical Lake George, on the other hand, Horne & Viner (1971) found that 33 % of the lake's annual nitrogen budget, amounting to some 44 kg ha^{-1} yr^{-1} came from this source. Even when estimates of annual input are low the localised

Fig. 7.48. Seasonal variation in nitrogen fixation per unit volume for the three basins of Clear Lake. Nitrogen isopleths at every 30 μmoles ethylene l^{-1} h^{-1}. From Horne & Goldman, 1972.

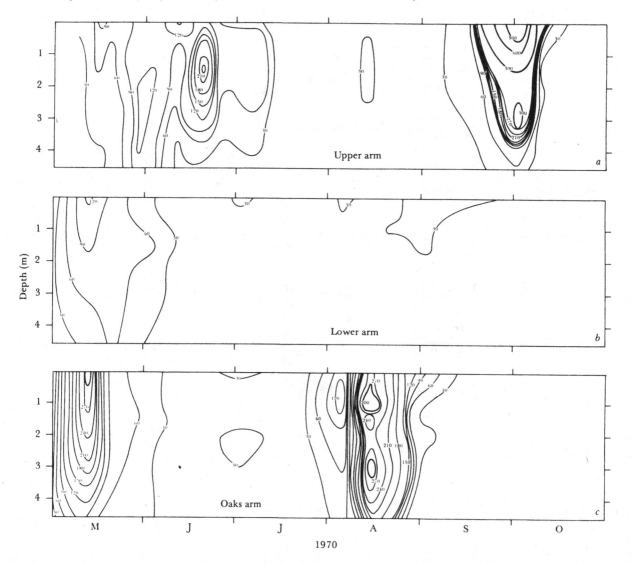

effect both in time and space of such metabolic activity may have great significance for the species performing fixation and for any closely associated species. Bacteria may be a minor source of nitrogen fixation but all the evidence, at least in the ocean, seems to favour cyanophyceae as fixers (Carpenter & McCarthy, 1975). Clearly the amounts fixed in lakes are very variable. The total amount of nitrogen fixed in the upper 5 m of Lake Mendota was calculated as 2.4 kg ha^{-1} yr^{-1} and the total for the lake approximately 9456 kg yr^{-1} (Stewart *et al.*, 1971): this is not a large figure when compared to inputs from other sources but it may be important at a time, i.e., midsummer, when those other sources are least effective. A year later other workers (Torrey & Lee, 1976) found approximately three times the amount of nitrogen fixation in this lake.

Fig. 7.49. Acetylene reduction and total nitrogen variation during one day (6 October 1969) in Green Bay. *a*, acetylene reduction (l^{-1} h^{-1}, open circles) and C$_2$H$_2$ reducing efficiency of the algae (filled circles) in surface waters; *b*, total nitrogen in the surface waters; *c*, acetylene reduction (l^{-1} h^{-1}, open squares) and C$_2$H$_2$ reducing efficiency (filled squares) at 2 m; *d*, total nitrogen at 2 m. From Stewart *et al.*, 1971.

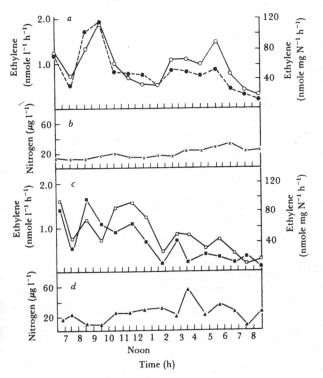

The nitrogenase activity (which is light-dependent) shows diurnal variation with peaks in the morning and afternoon in surface waters (Fig. 7.49; Stewart *et al.*, 1971; Peterson, Friberg & Burris, 1977), while Saino & Hattori (1978) found a 200-fold variation in fixation by *Trichodesmium* between night and day. The mid-day decrease might be due to photoinhibition or to high oxygen levels arising from photosynthesis. In samples from the central North Pacific gyre increase in oxygen tension did result in lower fixation (Mague *et al.*, 1977).

Euphotic zone, mixed depth, critical depth

The depth of the euphotic zone is often considered to be three to four times the Secchi disc depth (Strickland, Solorzano & Eppley, 1970) or the depth at which nutrients start to increase (Steele & Yentsch, 1960). The other common delineation is the depth at which only 1% of the surface illumination remains. Such physical or chemical measurements can often be made more rapidly than biological measurements, especially when ships occupy a station for a short period of time, though it must be remembered that such rapid observations do not necessarily reflect even the average depth of the euphotic zone over a period of time. On physical characteristics of light penetration the zone cannot extend much below 140 m in freshwaters or in the sea and every particle of colouring or particulate matter reduces this depth; in fact in one of the most transparent parts of the oceans, the Sargasso Sea, the euphotic zone extends to 120 m. Since it is essentially the zone in which algal photosynthesis is possible its extent is best determined by *in situ* determinations of photosynthesis. A reasonable average value for lakes and shallow seas is given by the depth to which benthic plants, especially epipelic, epilithic or epipsammic associations occur: in Lake Windermere, for example, this is around 6–8 m (Round, 1961*b*), a figure which agrees well with data from measurements of phytoplankton photosynthesis profiles (Talling, 1971 and earlier publications). In contrast, in the Pacific Ocean off La Jolla there was still a rich growth of epipelic species at 35 m (Round, unpublished) and Plante-Cuny (1969) found species down to 100 m (*see* p. 176).

Determinations of the rate of algal production with depth can be achieved by suspending either cultures of known species or collections of mixed plankton at selected depths and measuring the oxygen produced, the ^{14}C fixed or counting the cells. Examples are given in Fig. 7.50 for a freshwater and for a marine alga. However it can be argued that these cultures are not adapted to the light conditions at the various depths and there is no doubt that adaptation does occur under experimental and natural conditions; Fig. 7.51, for example, shows that the surface population of arctic summer plankton was saturated at 13000 lux, that from 25 m at 9000 lux and that from 50 m at 3500 lux. In these experiments the composition of the plankton was not investigated so it may be that different species were involved although the authors consider it most likely that in this region the same species occur down the profile and are adapted to 'sun' and 'shade' (*see* p. 307).

As Talling (1970) points out, much valuable work could be done exploring the effect of the underwater light field by using biological sensors, e.g. algal cultures of known photosynthetic characteristics used as 'phytometers' to measure the response under the natural conditions.

There are two other important 'depth' parameters: the depth of the 'mixed layer' and the 'critical depth'. The mixed layer is the depth through which the phytoplankton is circulating and as with the euphotic layer it can be measured by physical means (temperature, salinity, etc). Determination of the extent of the mixed layer can be valuable under some conditions, e.g. in very turbulent waters, but is it a less valuable measure in tropical areas where stratification of populations occurs. If the mixed depth is great compared to the euphotic depth algae will spend a large amount of

Fig. 7.50. *a.* The growth of *Asterionella formosa* in bottles suspended at depths between 0.5 and 7.0 m in Lake Windermere and in the laboratory. Cell numbers are plotted on a logarithmic scale, over 10 days. *b.* Oxygen production by *Chaetoceros affinis* suspended at various depths (solid line) and at different temperatures (interrupted line). From Lund, 1949 and Talling, 1960.

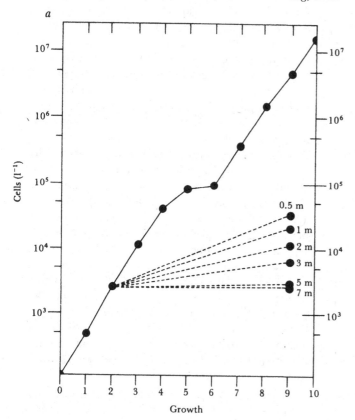

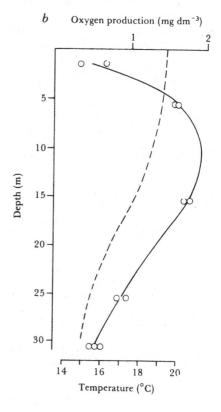

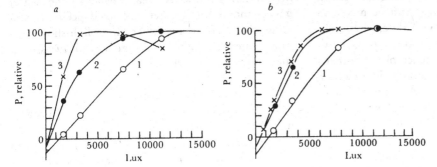

Fig. 7.51. *a*. Light intensity and relative rates of net photosynthesis in arctic summer plankton from three different depths; 1, surface; 2, 25 m; 3, 50 m. Dana Station 10561. *b*. Light intensity and relative rates of net photosynthesis in arctic summer plankton from three different depths: 1, surface; 2, 27 m; 3, 52 m. Dana Station 10540. From Steemann-Nielsen & Hansen, 1959.

Fig. 7.52. A comparison of the critical depth and the depth of mixing at two stations in the north Pacific, *a*, Strait of Georgia, west of 124° W and *b*, Station P. Solid bars, maximum and minimum critical depths, dashed bars, depth of mixed layer. The asterisk indicates the time of formation of the seasonal thermocline. From Parsons & Le Brasseur, 1966.

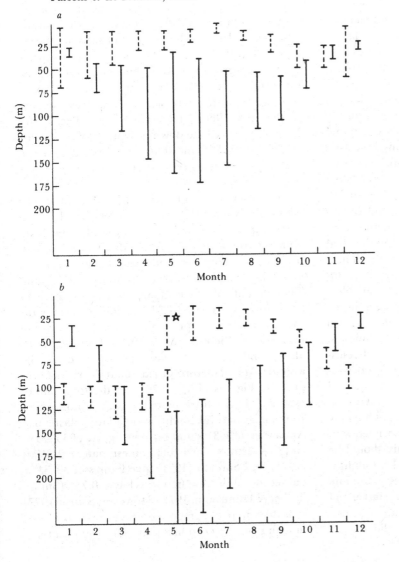

time in semi-darkness, and losses will exceed gains. Even in lakes the phytoplankton can be mixed to great depths, e.g., 450 m in Lake Tahoe, where cells were always found to be viable whatever their depth of origin, and they could commence photosynthesis immediately when placed in a suitable light climate (Tilzer, Paerl & Goldman, 1977). As the mixed layer decreases in summer, a point will be reached when the algae remain long enough in the well-lit upper waters for photosynthetic gains to overcome losses from all other sources and the population will start to increase. This depth has been termed the 'critical depth', and as can be seen from Fig. 7.52 it lies at deeper and deeper points (especially at the open ocean station) during spring and early summer and then rises again later in the year. It is not synonymous with the 'compensation depth' since this is merely a point at which light intensity is sufficient for photosynthesis just to balance respiration and varies from species to species. Since any one cell normally spends part of each 24 h in the dark the population is not harmed by extension of the dark period so long as this time does not become so excessive that loss by grazing, etc. occurs. In fact it is often reckoned that in cold seas the 'critical depth' can be anything from five to ten times the compensation depth (Anderson & Banse, 1963). The mixed depth in lakes is often two to five times the euphotic depth (*see* Fig. 10.35).

To follow the time course of natural phytoplankton growth in the euphotic zone is complicated by the fact that the water masses carrying the algal associations move laterally and vertically. These variations can be smoothed out by extensive sampling and vertical variations in populations are often eliminated by sampling a 'core' of water e.g. in the work on *Asterionella* (p. 540) or by enclosing water in a plastic bag (*see* p. 339). Neither of these techniques, nor that of enclosing samples in bottles and resuspending them, exactly reproduce the natural conditions of circulation. In small lakes the lateral movement of 'parcels' of water may be slight and in any case the seasonal succession in the plankton may introduce greater changes than those associated with movement of water masses. Every study should be designed to take into account the natural variability; it can either be smoothed out by extensive sampling and bulking of data or the minutiae of 'parcels' of water must be studied. Since these 'parcels' exist, many measurements are needed for valid overall

production studies. At sea, however, the water masses move much faster and fixed position sampling, such as from a weathership or lightship, gives instantaneous figures for different water masses as they move past the ship.

Extracellular secretion

The phenomenon of secretion of organic matter from algae has mainly been studied in phytoplankton and many compounds have been shown to be secreted. (Valuable reviews are those of Hellebust, 1967 and Fogg, 1971.) Secretion mostly takes place through the cell wall but in the case of the flagellate *Ochromonas danica* vesicles are released into the medium and this involves the release of both acid and alkaline phosphatase, as well as amino acids, proteins, vitamins, DNA, RNA and carbohydrates (Aaronson, 1971). The amount of secretion varies greatly and it seems that the release is low in sparse populations growing actively, but at high cell density, inhibiting light intensity or limiting carbon dioxide concentration the secretion can increase greatly (Nalewajko, 1966; Nalewajko & Schindler, 1976). In a study of release down a depth profile, Wallen & Geen (1971*b*) found that there was a decrease in percentage of released carbon and that this was directly proportional to the size of the ethanol soluble fraction in the cells, i.e., to the pool into which newly fixed ^{14}C passes. Wetzel *et al.* (1972) in a key paper on detrital carbon in hard water lakes found that 7.3 mg carbon m^{-2} day^{-1} was released by the phytoplankton and over a year the mean was 5.7% of phytoplankton primary production. Though difficult to compare, a recent figure of 0.10–0.13 mg carbon m^{-3} h^{-1} obtained in laboratory experiments is probably of the same order of magnitude (Wiebe & Smith, 1977). Wiebe & Smith also found that the released carbon could be assimilated by heterotrophs at about the rate of production. Figures of up to 70% of the total photoassimilated carbon are quoted by various authorities. In the near natural conditions of a large plastic bag Antia *et al.* (1963) found extreme figures of 35–40% and this seems to be a common estimate, e.g. Watt (1971) and Sorokin (1971) give figures of 30–31% but recent studies give figures as low as 8.7% (Smith, Barber & Huntsman, 1977). However, Sharp (1977) considers that the reported levels may be high, though Fogg (1977) thinks that the earlier results are substantially correct. As might be expected from the studies of secretion by symbiotic algae (*see* p. 387) the various algal groups tend to secrete different compounds, e.g. glycerol and mannitol are secreted by Chlorophyta, Chrysophyta and Prymnesiophyta, glutamic acid and arabinose by Dinophyta and glycerol and a little mannitol by Bacillariophyta (Hellebust, 1965). One of the first substances identified in extracellular material was glycolic acid (Fogg & Nalewajko, 1964) but Hellebust found that it was a significant component secreted by only four of 22 algae tested. It comprised 50% of secretion from *Skeletonema costatum*; this alga does not take up glycollate though some others, such as *Thalassiosira nana* and some Cyanophyta, do. The presence of such organic molecules in nature may enable rapid growth of certain algae, since it is reported that when they are added to cultures, the lag phase is reduced (Fogg, 1963). There are reports of the excretion of acrylic acid by various algal groups and its antibiotic activity (*see* p. 222) and it is secreted by at least one planktonic alga *Phaeocystis pouchetii* (Guillard & Hellebust, 1971).

Obviously this secretion can give rise to misleading results when ^{14}C studies are made on populations which release large amounts of labelled compounds. Stephens & North (1971) also found that *Platymonas* and *Nitzschia* absorbed labelled amino acids when these were fed but returned the amino-nitrogen moiety to the medium whilst retaining the carbon skeleton.

Uptake and activity of extracellular products and other organic molecules

As just mentioned, glycolate is not only secreted but it is also absorbed by some plankton. Uptake of the secreted organic compound is probably slight in the light (McKinley (1977) found little uptake of ^{3}H-glucose), but in the dark facultative heterotrophic mechanisms can develop in the cells; White (1974), for example, showed that glucose and galactose uptake ability in *Cyclotella cryptica* and glucose uptake ability in a *Coscinodiscus* sp. are rapidly acquired in the dark even in the *absence* of the sugars in the medium. Yet other compounds which do not support heterotrophic growth can be taken up by these two diatoms and used in respiration. Under normal circumstances the concentration of these substrates is very low in natural waters, e.g. Saunders (1972a)

quoted figures of 10^{-7} M for amino acids and acetate and 10^{-8} M for glucose in freshwaters and these are much lower than the levels in experimental studies. The energy required to concentrate substances which are present in such small quantities is excessive and unlike the bacteria, which have uptake mechanisms adapated to low concentrations, algae do not have the permease and general extracellular enzyme systems developed to the same degree. Some active concentrating mechanism is therefore necessary and Saunders (1972b) showed a slight active uptake of amino acids in *Oscillatoria agardhii* var. *isothrix* in the light but none in the dark. As so frequently reported, he could detect no dark uptake in lakes but some glucose uptake by this and other Cyanophyta living near the bottom of the photic zone did occur. Wright & Hobbie (1966) found in fact that although *Scenedesmus* can take up glucose and acetate it is not capable of growth on these substrates at *natural* concentrations, whereas bacteria take up the major fraction of adsorbed organic material in nature. According to Ukeleles & Rose (1976) microscopic marine forms have extremely specific and *limited* abilities for the uptake and metabolism of organic compounds. To test the ability of phytoplankton species to take up organic compounds Pollingher & Berman (1976) used an autoradiographic technique akin to that described on p. 320; they found that only a small proportion of the species in the plankton of Lake Kinneret had the ability to take organic compounds into the cells. But since the tests were done on natural samples not all the species are necessarily in an active metabolic state and checking at many times in the year may be necessary. It is easier to work on a cultured species, and Berman, Hadas & Kaplan (1977) found that *Pediastrum duplex* will take up a number of organic substances both in the light and in the dark, but that they only enhanced growth at low light intensities. A similar effect of glycolate on *Skeletonema* growing at low light is reported by Pant & Fogg (1976).

There have been intermittent reports of stimulatory or inhibitory effects of algae on one another since the turn of the century but very little concrete experimental or natural evidence. Jörgensen (1956) showed that *Asterionella* did not grow well in a filtrate from *Nitzschia palea* and the latter was inhibited by filtrates from *Scenedesmus* and *Chlorella* but his theory that epiphytic diatom populations grow only when the planktonic diatoms have ceased growth seems unlikely to be connected with inhibitors. Talling (1957) was unable to show any distinct effects of filtrates of *Asterionella* or *Fragilaria* on one another (Fig. 7.53). Yet, in another study, Proctor (1957) grew five algae in ten two-membered combinations and found that no two species grew as well in mixed as in single culture; *Chlamydomonas* invariably became the dominant and fat-like substances released on death of the *Chlamydomonas* cells was thought to be responsible for suppression of the other algae. Kroes (1971) performed similar experiments but transferred the filtrates and kept the algae apart; under these conditions *Chlorococcum* inhibits *Chlamydomonas* but not vice versa. This illustrates the variability one finds in the experiments using closely related algae. Gauthier, Bernhard & Aubert (1978) found that *Asterionella japonica* and *Chaetoceros lauderi* produce a lipid antibiotic and that the amount of this in the cells is increased when the diatoms are grown with *Prorocentrum micans*.

Cessation of growth of phytoplankton populations has sometimes been attributed to inhibitors produced by the phytoplankton species themselves, although there is little proof of this occurring in nature (but see below). However, in culture, Harris (1970) showed that the freshwater planktonic green

alga *Platydorina caudata* produces an inhibiting substance which can be detected in the culture after 15 days and after 25 days it completely inhibits further growth of *Platydorina* (auto-inhibition). It did not inhibit the growth of other colonial Volvocales (*Volvox*, *Pleodorina*, *Eudorina* or *Pandorina*). Further work, illustrated in Table 7.6, showed that filtrates of some species reduced the growth of others (Harris, 1971). Amongst diatoms, Denffer (1948) reports a mitotic inhibitor produced by cultures of *Nitzschia palea* which is auto-inhibitory. Such effects may be operative during dense phytoplankton blooms or in small bodies of water, e.g. ponds and rock pools. Interaction between species in still water can be reversed by turbulence caused by aeration of the medium (Bakus, 1973) suggesting that in nature these effects may be eliminated in fast flowing or turbulent situations.

The observations that between May and October in Narragansett Bay the diatom *Skeletonema costatum* and the flagellate *Olisthodiscus luteus* alternate but almost never coincide let Pratt (1966b) to test the species against one another in culture and he found that each species inhibits its own growth when re-inoculated into media which had supported the species (cf. *Platydorina*). But the situation is not simple since the *Skeletonema* medium does not inhibit the

Fig. 7.53. Growth of *Asterionella formosa* (a) and *Fragilaria crotoneneis* (b) in uni-algal cultures containing varying additions of filtrate from dense cultures of the other diatom species. ●, No filtrate; ◑, 10% filtrate; ○, 20% (a), 25% (b) filtrate. From Talling, 1957.

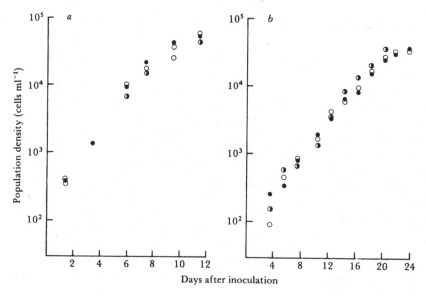

Table 7.6. *Inhibition of growth in several volvocacean genera by culture filtrates from members of the same family. From Harris, 1971*

Culture filtrate from	Inoculum										
	E. cylindrica	*P. charkowiensis*	*G. pectorale*	*P. morum*	*V. globator*	*E. illinoisensis*	*V. pringsheimii*	*V. tertius*	*E. elegans*	*E. californica*	*P. caudata*
Eudorina cylindrica	O	I	—	I	I	I	I	I	—	—	—
Pandorina charkowiensis	—	O	—	I	—	—	I	—	I	—	—
Gonium pectorale	—	I	O	—	I	—	I	I	—	—	I
Pandorina morum	I	I	I	O	I	I	I	I	I	—	I
Volvox globator	—	I	—	I	O	I	I	—	I	—	—
Eudorina illinoisensis	I	I	—	I	I	O	—	I	I	—	—
Volvulina pringsheimii	—	I	I	I	I	—	A	I	I	—	I
Volvox tertius	—	—	—	—	—	I	I	O	—	—	—
Eudorina elegans	—	I	I	I	I	—	I	—	O	—	—
Eudorina california	I	I	—	I	I	—	I	I	I	O	—
Platydorina caudata	—	I	—	—	I	—	I	I	—	—	A

I, at least 20% inhibition of growth; A, autoinhibition; O, no autoinhibition.

Fig. 7.54. Growth of *Olisthodiscus* in dilutions of *Skeletonema*-conditioned medium with nutrients restored. 0%, control (i.e. fresh medium); 100%, undiluted conditioned medium. From Pratt, 1966*b*.

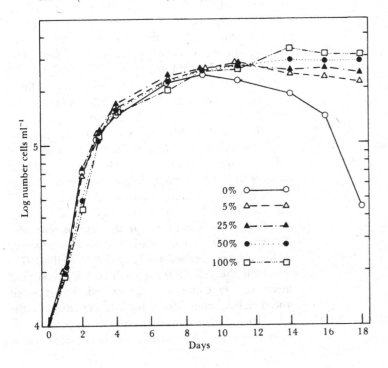

growth of *Olisthodiscus* (Fig. 7.54) whereas *Skeletonema* is inhibited by media in which dense growths of *Olisthodiscus* have been maintained, but is stimulated by medium from small growths of *Olisthodiscus* (Figs. 7.55). So Pratt suggested that the *Olisthodiscus* may inhibit the *Skeletonema* in nature but the diatom achieves its dominance by its superior growth rate. Substances released from cells and producing effects on others are sometimes termed ectocrines. The technique of separating the cells in two media by a membrane allowing the passage of excreted materials has been used by Huntsman & Barber (1975) in an ingenious experiment with an exponentially growing population on one side and a small inoculum of the same species on the other; the substances diffusing across the membrane depressed the growth rate of

the new culture but at the same time *reduced* the lag phase. This suggests that both self-inhibitory and self-stimulatory substances are released by the actively dividing population.

A nice example of inhibition by natural lake water is the finding that Linsley Pond water inhibited growth of diatom cultures and also that the filtrates from cultures of Linsley Pond Cyanophyta had the same effect (Keating, 1978). However filtrates from Cyanophyta from other lakes had less effect on the diatoms of Linsley Pond and Cyanophyta from Linsley Pond also had less effect on diatoms from other lakes. This study reveals very subtle 'symbiotic' effects which are worthy of deeper study.

Carbon fixation by species

Almost all studies of primary productivity have been based on mixed samples and, as Stull *et al.* (1973), comment 'Our understanding of the relationships between primary productivity and phytoplankton composition is incomplete without an ability to express community carbon fixation as a function of the carbon fixation rates of each species within the community'.

The ^{14}C technique applied to any type of algal association measures the uptake into the whole population but gives no information on the variation between individual species unless a portion of the material is filtered and the filter covered in autoradiographic stripping film which can then be developed and the silver grains counted as described by Stull *et al.* (1973) and Watt (1971). Table 7.7 shows the results Watt obtained, with two dinoflagellates and two diatoms contributing the main carbon fixation, but with these figures bearing little relationship to the cell counts, and again showing that the small species contribute much more fixation than the cell count data would suggest. He also showed that some of the dinoflagellates had a considerable range in their rates of fixation whilst the diatoms did not; this was assumed to indicate the retention of senescent dinoflagellates in the water mass whereas senescent diatoms are lost. An alternative method to the above technique was devised by Maguire & Neill (1971) in which the cells, after growth in ^{14}C-containing medium, are concentrated, washed, resuspended and then subjected to mild sonication to disperse the cells. In a dark room, drops of the suspension are then transferred to microscope slides on to which

Fig. 7.55. Growth of *Skeletonema* in dilutions of *Olisthodiscus*-conditioned medium with nutrients added. 0% is the control (i.e. fresh medium), 100% is the undiluted *Olisthodiscus* medium. From Pratt, 1966*b*.

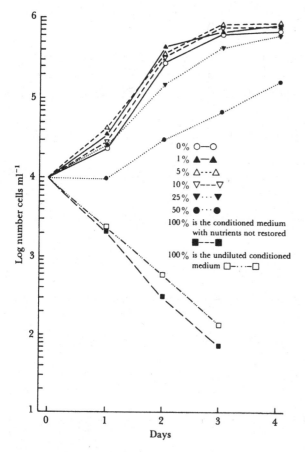

Table 7.7. *The cell counts and autoradiographic estimates of primary productivity of each species at one station in the northwest Atlantic Ocean* (a) *and* (b) *the range of primary productivity of five species. From Watt, 1971*

(a)

Latitude 42° 44′ N 1 May 1969 Species	Longitude 63° 25′ W Surface temp. 5.8 °C Cells (l^{-1})	Slope water Rate of photosynthesis (μg C m^{-3} h^{-1})
Ceratium longipes	44	240
Ceratium fusus	8	11.0
Ceratium furca	6	3.5
Peridinium ovatum	78	145
Peridinium pellucidum	12	23.0
Peridinium monospinum	10	0.4
Peridinium oceanicum	6	0.8
Gymnodinium arcticum	4	0.4
Gymnodinium sp.	8	0.3
Amphidinium sp.	66	2.6
Dinophysis lenticula	6	3.1
Thalassiosira nordenskioldii	33	121
Coscinodiscus sp.	10	39.1
Chaetoceros sp.	24	87.1
Phaeocystis sp.	6	4.5
Trochiscia clevei	8	0.0
Trochiscia sp.	38	0.1
Unid. coccolithophore sp.	142	18.9
Unid. sp. (10 μm sphere)	33	1.8
Unid. sp. (12 μm sphere)	21	7.0

(b)

| | | Carbon (μg h^{-1}) | | | |
| | | Per cell | | Per 10^3 μm^3 | |
	Number of observations	Highest	Lowest	Highest	Lowest
Ceratium longipes	7	5.44	0.02	0.030	0.000
Ceratium fusus	6	1.38	0.03	0.014	0.000
Peridinium brevipes	4	0.11	0.01	0.064	0.004
Nitzschia seriata	6	0.02	0.01	0.018	0.008
Coccolithus huxleyii	7	0.15	0.12	0.820	0.670

Kodak AR-10 stripping film is placed. Subsequent treatment follows standard autoradiographic techniques and the silver grains on the film are counted under oil immersion microscopy (Table 7.8). A modification of this technique is track autoradiography in which the number of tracks left by the emission of the β particles passing through the silver containing emulsion are counted (Knoechel & Kalff, 1975). Using this technique, Knoechel & Kalff found that the fixation rates of individual cells of

Table 7.8. *Standing crop and productivity of surface phytoplankton in Lake Livingston (total productivity was measured by liquid scintillation counts and partitioned by autoradiographic counts). From Maguire & Neill, 1971*

Type of cell	No. of cells ml ± 10%	Total biovolume (μ^3 ml^{-1})	Mean grain no./exposure day μ^{-3}	g C fixed m^{-3} h^{-1}	% of total C fixed	% of total biovolume	% C fixed / % biovolume
Scenedesmus sp. (non-reproductive)	1.9×10^3	1.3×10^5	0.060	2.7×10^{-3}	12	0.9	13.3
Scenedesmus sp. (reproductive)	9.2×10^1	1.1×10^6	0.002	8.3×10^{-4}	4	7.2	0.56
Chlorococcum sp.	4.7×10^2	2.6×10^5	0.007	6.1×10^{-4}	3	1.8	1.67
Hyalotheca sp. (?)	1.4×10^3	1.0×10^5	0.018	6.1×10^{-4}	3	0.7	4.29
Chlorella sp.	3.2×10^4	1.2×10^7	0.004	1.7×10^{-2}	77	84.0	0.92
Unicellular blue-green	9.1×10^4	7.9×10^5	0.001	2.4×10^{-4}	1	5.4	0.19
Total	1.3×10^5	1.5×10^7		2.2×10^{-2}	100	100	

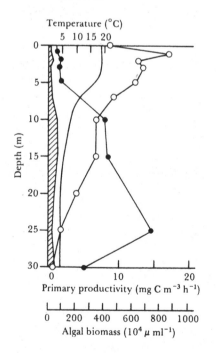

Fig. 7.56. Vertical profiles of temperature, total algal biomass, primary productivity and dark carbon fixation for Castle Lake on 28 June 1970. Solid line, temperature; filled circles, algal biomass; open circles, light primary production; cross-hatched area, dark fixation. From Stull *et al.*, 1973.

Tabellaria fenestrata actually increased relative to those of *Anabaena planktonica* during summer and that the *Tabellaria* was not declining in vitality but that loss of cells from sinking in late summer was the factor responsible for the decrease in carbon fixation. The same technique has also been used to follow ^{14}C uptake over short (2 h) incubations at various depths (Knoechel & Kalff, 1978).

As long suspected only a proportion of the cells recorded in routine counts of phytoplankton is metabolically active and in the most recent study the percentages varied from 28 to 42 in an estuarine site (Faust & Correll, 1977).

Stull *et al.* (1973) found that in one day in Castle Lake the curves of primary productivity and algal biomass showed an inverse relationship (Fig. 7.56); this is in fact a common occurrence and simply indicates that biomass measurements are not necessarily an index of rate of production. The data from the autoradiographic approach are given in Table 7.9. The renewal rate is the ratio of carbon fixed per hour, to cell carbon content, and represents the percentage of carbon being added or replaced by each hour's photosynthesis. The inverse of the renewal rate is the time (renewal time) it would require a cell photosynthesising at its observed rate to replace completely its cell carbon. Thus those algae with long renewal times are probably not metabolically active and it is only those with short times that are contributing appreciably to carbon fixation. It is interesting that of the diatoms only *C. meneghinania*

Table 7.9. *Carbon content, rate of carbon fixation, renewal rate and renewal time for* *algal species at 0 m on 28 June 1970 in Castle Lake. See text for further details.* *From Stull* et al., *1973*

Species	Carbon content (pg cell^{-1})	Rate of carbon fixation (pg h^{-1} cell^{-1})	Renewal rate (% per hour)	Renewal time (h)
Unknown 1	2.5	1.5	60.0	1.7
Unknown 2	31.0	1.7	5.5	18.0
Chlorophyceae				
Botryococcus braunii	5.0	2.6	54.0	2.0
Cosmarium bioculatum	270.0	43.0	16.0	6.4
Oocystis lacustris	22.0	19.0	86.0	1.2
O. naegelii	13.0	4.7	37.0	2.7
Staurastrum brevispinum	440.0	0.0	0.0	—
Chrysophyceae				
Dinobryon sertularia	92.0	12.0	13.0	7.7
Mallomonas sp.	160.0	0.14	0.09	1100.0
Bacillariophyceae				
Achnanthes linearis	22.0	0.19	0.86	120.0
Cyclotella meneghiniana	9.1	1.1	13.0	7.9
Navicula cryptocephala	36.0	0.0075	0.02	5000.0
N. lanceolata	110.0	0.93	0.87	110.0
N. minima	20.0	0.12	0.06	170.0
N. radiosa	170.0	2.5	1.4	71.0
Synedra radians	36.0	0.68	1.8	56.0
Tabellaria fenestrata	230.0	2.2	0.7	130.0
T. flocculosa	170.0	2.5	1.5	67.0
Cyanophyceae				
Chroococcus limneticus	35.0	2.2	6.2	16.0
Gloeocapsa granosa	29.0	4.4	15.0	6.7
Merismopedia glauca	5.4	0.98	18.0	5.5
Microcystis aeruginosa	60.0	3.3	5.6	18.0

has a short renewal time and of the species listed only *C. meneghiniana, S. radians* and possibly the *Tabellaria* spp., are actually planktonic. Even the *Tabellaria* spp. may be derived from epiphyton where they are at times some of the most abundant species; without identification of the form or variety of *Tabellaria* it is not possible to pinpoint its origin. *Cyclotella* was the most numerous organism in the phytoplankton, especially in the epilimnion on that day, and together with *Microcystis* was the most important carbon fixing organism in the shallow water. However at 1 m although *Cyclotella* and *Microcystis* still comprised the largest amount of biomass, the most productive alga was a *Peridinium* sp. which contributed 32% of the community primary productivity but only 1%

of the biomass. Thus, as has been long suspected, a species comprising only a small proportion of the biomass may make an inordinately large contribution to the carbon fixation. Conversely organisms present as large components of the biomass can contribute very little to the carbon fixation. An interesting observation arising from this work is that *Cyclotella meneghiniana* has a renewal time of about eight hours in the lake, whereas a time of 70 h has been reported for cell division in culture; yet another example of the danger of extrapolation from laboratory to natural conditions. The renewal times for *Botryococcus* and *Oocystis* seem to be remarkably low in this experiment and it would be interesting to compare them with cell division rates. Calculations for the renewal time of phytoplankton carbon in Lake Michigan range from 1 to 6 days (av. 2.6) (Parker, Conway & Yaguchi, 1977*b*), which is probably longer than the average in Castle Lake but not surprising considering the difference in lake type. Another example of variation between laboratory study and lake is the division rate of *Peridinium* which, in culture is approximately 11 days, but in Lake Kinneret is 17–40 days (Pollingher, personal communication).

In view of the often reported higher primary production per unit biomass of communities dominated by small cells over those dominated by large cells a plot of mean renewal time of the Castle Lake species against estimated carbon content (Table 7.9) shows that in general the small cells (i.e. those with low carbon content) are more active than the larger cells. Whilst the high photosynthetic rate of small cells was confirmed by Paerl & Mackenzie (1977) they also found that the net phytoplankton contribution increased during the afternoon and evening. In addition they noted a higher rate of loss of fixed carbon by the nannoplankton during darkness. Such results again show the difficulty of selecting times for uptake–release experiments. It is interesting to note that in laboratory microecosystems the diurnal course of carbon dioxide can show morning peaks of uptake and evening peaks of output (Beyer, 1965). This feature has been confirmed by Faust & Correll (1977) who showed that cells smaller than 10 μm from the Rhode River were not active.

Diurnal effects on metabolism, movement and luminescence

There is an obvious variation in photosynthetic activity between day and night, but less obvious is a disparity in the amount of carbon fixed during succeeding intervals in the light period. There have been many confusing accounts of this rhythm of activity but more recent work has substantiated the early discovery that photosynthesis of natural populations is not a steady process during the natural light period. Experimental studies have confirmed the rhythmic activity of photosynthesis, respiration and many other metabolic functions of cells. Mac-Caull & Platt (1977) suggest that rhythms result from intrinsic oscillations within cells coupled with environmental factors but the subject is complex and texts on biological rhythms should be consulted.

Sournia (1974) has reviewed the circadian periodicity of marine phytoplankton.

Eppley & Strickland (1968) comment that the diurnal variation in photosynthetic capacity is 'a great nuisance to field work as it introduces difficulties in the correct measurement of *in situ* productivity'. On the other hand, as Eppley *et al.* (1967) point out, algae in the field grow in an alternating light–dark climate and therefore it is sensible to use the same conditions during laboratory experiments. It is neither desirable nor practical to run *in situ* oxygen or ^{14}C determinations on plankton populations during the whole light period and generally a morning 2–4 h experiment is performed and the data extrapolated to the whole light period. Such extrapolation was always suspect (Verduin, 1957), and it is now clear that misleading values can be derived in this way and daily rates should be computed only from a sequential series of short exposures. However, errors derivable from diurnal variations in computing integral production are of the order of 4 to 20 % and Fee (1975) considers that it is not necessary to incorporate a factor for this into equations for integral daily production since errors

Fig. 7.57. Daily productivity data as carbon fixation (mg C m^{-3} h^{-1}) in Marion Lake. Bottles exposed for 4 h at 1 m depth. Median times of the 4-h exposures are plotted. From Efford, 1967.

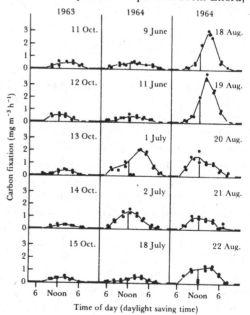

Fig. 7.58. Diurnal phytoplankton production curves (solid lines) for a number of lakes based on series of short-term consecutive experiments (histograms). Ordinate units are mg Cm⁻³ h⁻¹. Temperature and surface radiation (g cal cm⁻²) are shown above each series. Asterisked numbers represent phytoplankton biomass (mg m⁻³). The activity coefficient (mg C day⁻¹ mg⁻¹ freshweight) is shown in box by each graph. From Anderson, 1974.

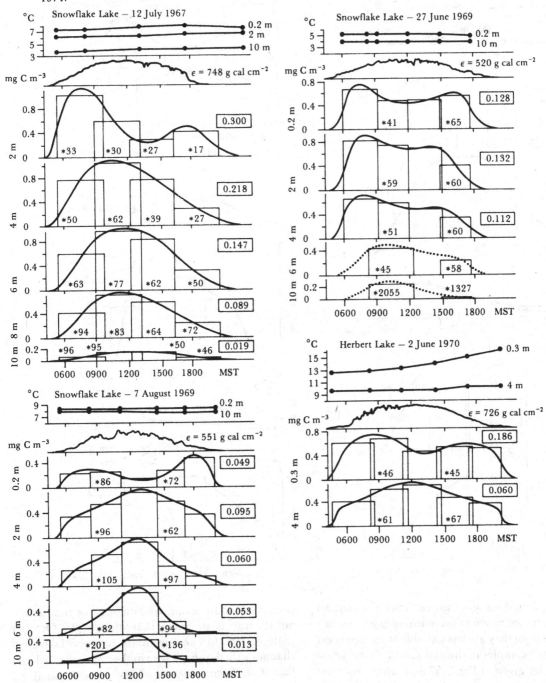

(*continued on page 326*)

Fig. 7.58 (*cont.*)

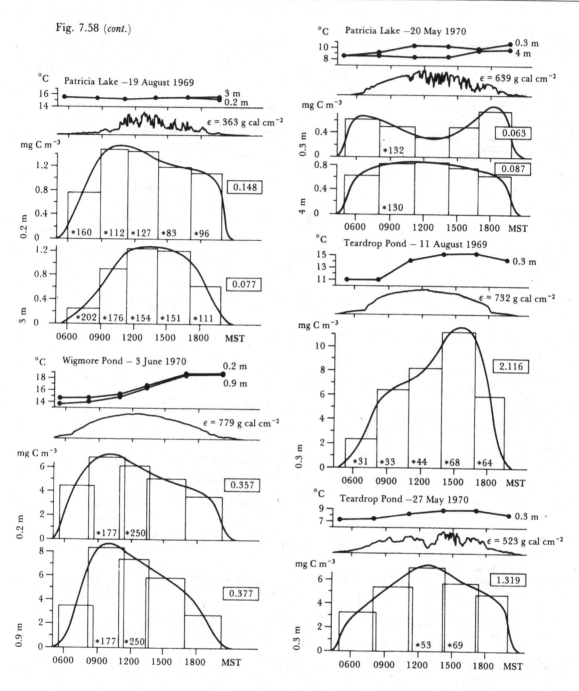

from sampling, data analysis, etc. exceed them: he did however recommend that morning exposures are preferable since they are less variable than afternoon ones. Some examples of diurnal productivity measurements are given in Fig. 7.57 and, apart from the considerable changes from dawn to dusk, the daily variation is high and illustrates the dangers of

extrapolation from limited data, e.g. the production on 18 July is no more than that of 15 October (Efford, 1967). The variability between lakes and changes with depth are seen in the data collected into Fig. 7.58 from a series of lakes investigated by Anderson (1974). The data of these two workers suggest that on certain days the computation of daily rates from a single short-term experiment would be misleading.

There is also a diurnal variation in the total amount of photosynthetic pigment present in the cells of the phytoplankton with a tendency to a peak around 08.00 h and a lowest point in the evening (Yentsch & Ryther, 1957; Yentsch & Scagel, 1958). The variation is not large but will obviously affect attempts at extremely precise measurements of biomass or calculations of fixation related to pigment content; in one of the early studies, for example, Doty & Oguri (1957) found a 5–7 fold variation in net productivity during the day. Early scepticism over rhythmic changes which arose from the enormous difficulties of measuring small metabolic changes in mixed natural populations have now been dispelled by experimental studies. Eppley & Renger (1974), in fact, showed very considerable circadian variations in chlorophyll *a* content and photosynthetic rate (maximum at midday and minimum at midnight) during the growth of a small marine diatom. Chlorophyll synthesis was much greater between 05.40 h and 12.00 h than between 12.00 h and 17.50 h.

The amplitude of the diurnal variation varies latitudinally as first suggested by Doty (1959), though later workers, e.g., Jitts (1965), did not find such a large change (4.5–1.5) as Doty. The relationship is inverse with latitude, i.e. a decrease from the equator northwards or southwards. This is confirmed by work in freshwater lakes on the equator where again the diurnal variation is certainly quite marked, e.g. in Lake George (Fig. 7.59; Ganf & Horne, 1975).

In general, peaks of carbon fixation seem to occur between mid-morning and early afternoon; the spread is probably due to slightly different patterns of response by different organisms in the mixed natural populations. Lehman, Botkin & Likens (1975) consider that the late afternoon depression is due to internal end-product inhibition. All the diurnal rhythmic phenomena have a species component which determines the time of the peak reaction. In cells which divide once a day the peak times tend to be in the latter part of the day, e.g. afternoon–early evening (Fig. 7.60) was recorded for diatoms by Eppley *et al.* (1971), and night-time division is common in dinoflagellates (Sweeney & Hastings, 1958; Weiler & Chisholm, 1976). Weiler & Chisholm also showed that division took 3–3.2 h (*see* Fig. 7.61). In the Gulf Stream, Doyle & Poore (1974) found at least nine dinoflagellate species dividing between 03.00 and 06.00 h, a feature which shows a higher degree of population synchrony than that of the three species investigated by Weiler & Chisholm. They thought that the diurnal nutrient

Fig. 7.59. The diurnal oxygen production (dashed line), carbon (continuous line), and nitrogen fixation (dotted line) in Lake George. Horizontal lines indicate the time span of each experiment. Integrated values obtained by planimetric integration from depth profiles. From Ganf & Horne, 1975.

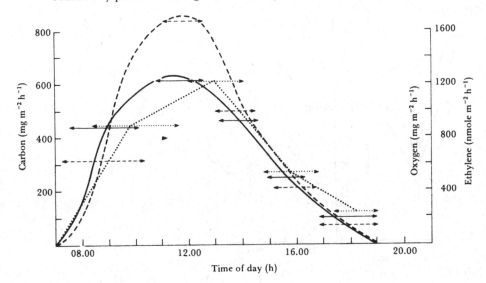

Fig. 7.60. *a*, the increase in diatom cell numbers of California coastal surface seawater phytoplankton enriched with different nitrogen source; *b*, rate of nitrogen assimilation as the decline in the nitrogen concentration in the medium and the rate of photosynthesis as carbon assimilated (μg l^{-1} h); *c*, rate of phosphate disappearance in the culture media supplied with different nitrogen sources. Phosphate assimilation represents the decline in phosphate in the medium; *d*, ratio of cell carbon to cell chlorophyll during the above experiments. From Eppley *et al.*, 1971. *a, b, d*: ○, NO$_3^-$; △, NH$_4^+$; □, used; ●, ▲, ■, PS rate. *c*: ○, nitrate culture; △, ammonia culture; □, urea culture.

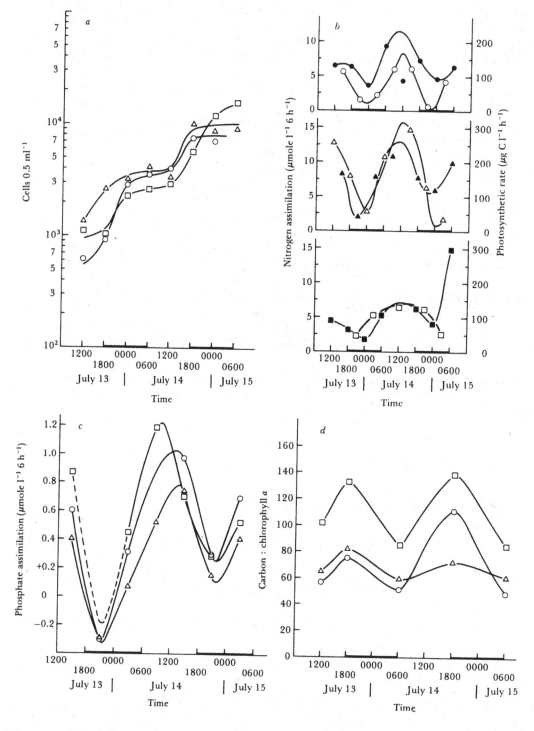

uptake rhythm acting through interspecific competition is the coupling and synchronising agent in nature. Evidence on the time of day of division of diatoms is a little conflicting but Azam & Chisholm (1976) concluded from silicon uptake studies that it was a night-time function and observations by Round (1978) on *Amphora* species in Lake Kinneret certainly confirm this by microscopic observations over a 24 h period.

The uptake of nutrients by phytoplankton also exhibits diurnal periodicity; Fig. 7.62 from work by Goering, Nelson & Carter (1973) shows that silicate (*see* also Azam & Chisholm, 1976), ammonia and nitrate uptake and nitrate reductase activity (*see* also

Fig. 7.61. The percentage of dinoflagellate cells (*a*, *Ceratium dens*; *b*, *C.* furca; *c*, *Dinophysis fortii*) undergoing division in water from Santa Monica Bay, California. ●, paired nuclei; ○, recently divided. From Weiler & Chisholm, 1976.

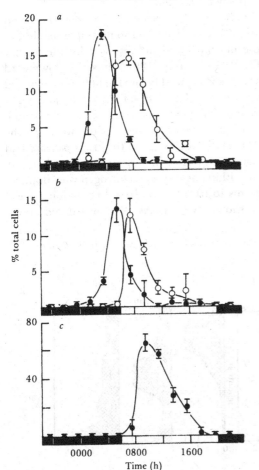

Fig. 7.60) all peak in the latter half of the light period.* Similar results but with a slightly earlier peak (midday) were found by Eppley *et al.* (1971) in nutrient enrichment experiments on nutrient-depleted Pacific water. Harrison *et al.* (1977) found a mid-day peak in phosphate uptake by coastal marine plankton. The uptake of silicate is light dependent and can be described by a Michaelis–Menten type equation. In work by Eppley *et al.* (Fig. 7.60) the decline in rate occurred before night-time and increased before sunrise which is interesting in relation to other periodicities involving biological rhythms (*see* p. 179). In spite of fluctuations in the ratio of cell carbon to cell chlorophyll there was no diurnal variation in chlorophyll synthesis in these mixed cultures composed mainly of diatoms. Goering *et al.* (1973), thought that in an upwelling region with lush nutrient content (e.g. off Peru) as the silicate concentration declines, uptake in the photic zone becomes silicate-limited if enough other nutrients are present, but at the bottom of the photic zone, uptake of silicate is limited by light.

Malone (1971*a*) reports that the marine net plankton of oligotrophic tropical waters always had a peak of carbon fixation in the afternoon but the nannoplankton peaked in the morning in nitrate-

* It is now known that at least each diatom species studied has its own characteristic time course for silicon uptake, coupled to the time course for cell division (Chisholm, Azam & Eppley, 1978).

Fig. 7.62. The uptake of silicate, nitrate and ammonium (as μg atom) and the nitrate reductase activity with time, by natural populations of marine phytoplankton. From Goering *et al.*, 1973.

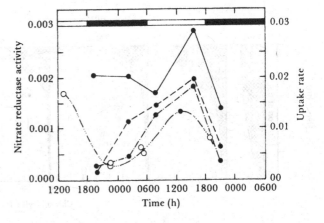

depleted water but in the afternoon in nitrate-rich water; this finding is in agreement with those of Eppley *et al.* (1971) experimenting with nutrient enrichments and with results from the nutrient-poor Sargasso Sea. Thus the phasing of the uptake cycle is partly a function of uptake kinetics; it is, however, somewhat surprising that Malone (1971*a*) did not find diurnal rhythms in the more eutrophic California current water. Paerl & Mackenzie (1977) found net plankton more variable.

Diurnal vertical migrations of phytoplankton probably occur whenever the water column is sufficiently stable for flotation, sinking or swimming motion to overcome turbulent water movement. Ideally they could be studied under ice, which eliminates wind-induced water movement, and in fact Wright (1964) did record upward movement of algal populations in the late afternoon. There is evidence that some flagellates seek out a zone of optimum illumination; in tertiary oxidation ponds, for example, where there is a planktonic flora in spite of the artificiality of the system, *Euglena rostifera* moves into the water during the day but only into regions which do not exceed 175 c ft^{-2} and never into the surface 6 in of water (Hartley & Weiss, 1970). Similarly, in experiments in a large deep tank Eppley *et al.* (1968) found that some dinoflagellates did not migrate up into the surface half metre. Movement of *Cachonina niei* is shown in Fig. 7.63; this alga was moving through the thermocline, which was situated at 2.5–3.0 m depth at this time and the cells avoided the deoxygenated 'hypolimnion' which started at 12 m. In nature, similar movements

have been recorded for a range of dinoflagellate species in Oslo fjord where the cells are concentrated in the uppermost 5 m but migrate towards the surface during the daytime, often collecting around 2 m depth (Hasle, 1954). As Talling (1971) points out, this movement might be expected to enhance photosynthetic productivity but the effects of self-shading in the dense surface population may in fact reduce the apparent advantages. Vertical migrations of dinoflagellates have possibly led to serious underestimates of primary productivity (Larson, 1978).

Another example of a species which clearly undergoes vertical migrations but which forms its peaks below the surface is *Peridinium cinctum* f. *westii* in Lake Kinneret (Berman & Rodhe, 1971); this species also shows a descending peak as the day progresses (Fig. 7.64) resulting in a significant clearing of the surface water.

Eppley *et al.* (1968) report rates of movement of 1–2 m h^{-1} in their deep tank experiments and their dinoflagellates showed some responses similar to those of epipelic algae observed by the author and Dr Palmer (*see* p. 180); thus the downward movement commenced before the light was switched off and the dinoflagellate *Cachonina niei* also commenced movement before the start of the light period. One, perhaps unexpected, result of this work was the finding that nitrogen starved *Gonyaulax polyedra* had no diurnal migration but recovered the ability to migrate within one day when nitrogen was supplied. This seems to be a species specific reaction since it was not shown by *Cachonina*. An upward and down-

Fig. 7.63. Chlorophyll profiles representing the movement of the dinoflagellate *Cachonina niei* in a 10 m deep tank. From Eppley, Holm-Hansen & Strickland, 1968.

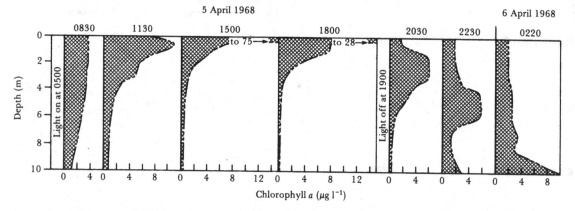

ward migration also occurred in the dark; all the data again indicate the operation of a biological clock system and in addition the dark results suggest a responsiveness to a gravitational field. The latter aspect of rhythms needs detailed study since there are many indications of gravitational effects on algae but little concrete experimental evidence. The ability to migrate may have some nutritional significance since it enables an alga to reach nutrient-rich pools; Harrison (1976) thought it may be important in 'red tide' organisms (e.g. *Gonyaulax polyedra*) which are liable to collect in surface layers and deplete nutrients. Vertical migration would give such algae a competitive advantage especially when they can move into nutrient-rich waters below the thermocline, as can *Cachonina* (Kamykowski & Zentara, 1977).

Experiments with the dinoflagellate *Gyrodinium dorsum* revealed the usual daytime movement towards light but also a feature which does not seem to have been recorded before, the cessation of motility when the light stimulus is switched on and a subsequent phototactic response (Forward & Davenport, 1970).

Movement of flagellates against gravity has to overcome the excess density of the cells (*see* p. 297); this is not likely to be great for naked species but will be more pronounced when a wall (e.g. in dinoflagellates), theca (e.g. *Trachelomonas*, *Chrysococcus*) or scales (e.g. coccolithophorids, *Mallomonas*) are present. In addition the cells have to overcome any downward currents and so it is perhaps not surprising that such movements are only detectable when stable conditions occur for example in Abbot's Pool (Fig. 7.65). A vast range of swimming speeds of dinoflagellates has been reported. The time taken to traverse one millimetre varies from two seconds for *Dinophysis acuta*, through 3–10 s for *Peridinium* spp. and 4–16 s for *Ceratium* spp., to 30–120 s for large,

Fig. 7.64. Vertical turbidometric profiles showing the aggregation of *Peridinium* cells at various depths, and a count of *Peridinium* cells at 11.50 h. Bottom left graph, temperature–depth profiles at different times; centre, light penetration; bottom right, carbon assimilation in the water column. The experiments were carried out at Lake Kinneret, Station A, May 1969. From Berman & Rodhe, 1971.

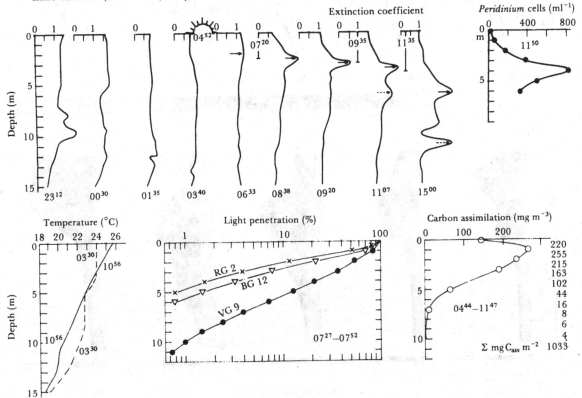

long-armed, *Ceratium* spp. (Peters, 1929), although doubts have been cast on the flagella-motivated swimming to account for the upward movements, which could be related merely to changes in buoyancy. However, Blasco (1978) has now shown that dinoflagellates can move against water movement and he also reports a negative reaction to light around noon which is a feature recorded also for some *Euglena* spp. and often noted by the author working with benthic populations.

Indeed movement of non-motile phytoplankton has also been recorded, e.g. of *Scenedesmus quadricauda*, and is caused by physical factors (convection, etc.) resulting in sinking during the day and upward movement at night (Overbeck, 1962a). Similar movement also occurs in tropical lakes, e.g. in Lake George where forms with gas vacuoles are also involved (Ganf, 1974a, c).

The emission of light (bioluminescence) when stimulated is a feature of some planktonic dino-

Fig. 7.65. The vertical distribution of temperature, oxygen and cells of *Chrysococcus diaphanus* down the depth profile of Abbot's Pool during a 24-h period. ■, live cells ml^{-1}; ◩, dead cells ml^{-1}. From Happey & Moss, 1967.

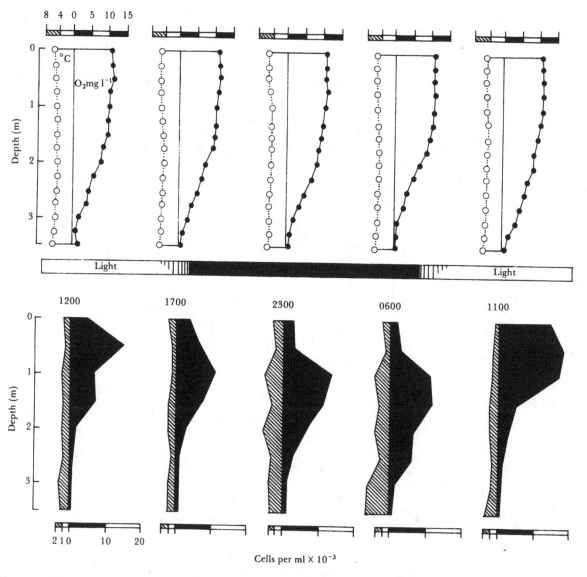

Cells per ml × 10^{-3}

flagellates, a feature noted a century ago by John Murray, the naturalist on the *Challenger* Expedition. A general characteristic of this phenomenon is the circadian rhythm of emission (Yentsch, Backus & Wing, 1964); it is hundreds to thousands times greater during the dark period than during the light. As far as I can determine this is a feature of dinoflagellates only amongst algae and is confined to marine planktonic species. A recent investigation of *Dissodinium lunula, Pyrocystis acuta, P. fusiformis* and *P. noctiluca* showed that they all produced the same colour bioluminescence as other dinoflagellates with an intensity peak at 474 nm (Swift, Biggley & Seliger, 1973). These workers measured the total emission by stirring a cell suspension in front of a photomultiplier tube and integrating the light emitted; the largest species, *P. noctiluca* and *P. fusiformis*, emitted $37–87 \times 10^9$ and $23–69 \times 10^9$ photons per

Fig. 7.66. *a*, Flashing rate of organisms in surface water from Eel Pond transferred to the laboratory, during August, kept in the dark and stimulated for 30 to 60 s at intervals. *b, c, d*, the effect of light inhibition on flashing at different times of the day. Arrows indicate times of start of 15-min periods of exposure to light. Upper dashed line connects rates after complete recovery; lower dashed line connects rates after light exposure. From Kelly & Katona, 1966.

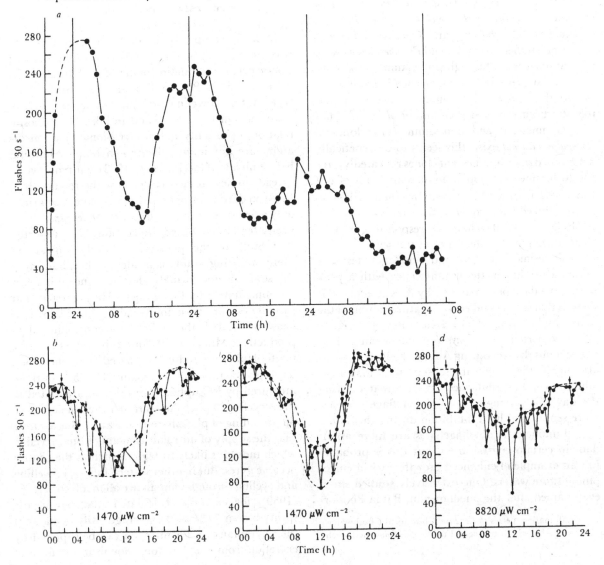

cell, respectively, which is about 1000 times that emitted by the much studied *Gonyaulax polyedra*. The flash is a very brief one, lasting in the case of *Gonyaulax* for about 90 ms (Hastings & Sweeney, 1958). Only a small number of dinoflagellates are luminescent but there are many which are still untested. The luminescent species appear at first sight to be mainly tropical in distribution though this may not prove to be entirely true since the large species are more intensely luminiscent and it is these which are confined to the tropics. Certainly there are luminescent species in the sea at Woods Hole (Kelly & Katona, 1966) and nine of twelve species identified from the Eel Pond (an inlet of the sea) were luminescent (*Gonyaulax digitale*, *G. spinifera*, *Protoperidinium* (*Peridinium*) *claudicans*,* *P. conicum*, *P. granii*, *P. leonis*, *P. oceanicum*, *P. subinerme*, and *Ceratium fusus*) and some unidentified *Gonyaulax* and *Protoperidinium* were also luminescent. This high proportion at one sampling site suggests that luminescence is more widespread than previously thought. However, it has recently been reported (Schmidt *et al.*, 1978) that both luminescent and non-luminescent clones of *Gonyaulax excavata* exist; this seems to be genetically determined since luminescent clones repeatedly gave rise to luminescent populations. Both kinds of cell were toxic, so at least in this species these attributes are not linked. According to the survey by Swift *et al.* (1973) the dinoflagellates are responsible for most of the diffuse bioluminescence of the sea.

Experiments on cultured dinoflagellates have shown a rhythm in the luminescence with a peak during the dark period and Kelly & Katona (1966) showed that samples taken from nature exhibited the same rhythm (Fig. 7.66). Later, however, Kelly (1968) reported that *Gonyaulax* only exhibited its endogenous rhythm during August. Kelly & Katona also made the interesting observation that light inhibition of the flashing activity was greater during the day than at night, i.e. the dinoflagellates are more sensitive to light inhibition during the day.

Luminescent dinoflagellates are more abundant in certain coastal areas, but this is probably just an example of enhanced growth rates in eutrophic inshore waters. One extensively studied site is even named after the phenomenon, Bahia Phosfor-

---

* All marine species of *Peridinium*, but not the freshwater ones, have been transferred to *Protoperidinium* by Balech (1974).

ente in Puerto Rico. Some of the flagellates in Bahia Phosforente have been shown to migrate vertically in the water column and unlike others which have a daytime migration to the surface these forms migrate to the surface at night (Taylor *et al.*, 1966). At other times, however, when freshwater flowed across the surface of the bay the peak of bioluminescence would be detected in deeper water, in the thermocline region. Seliger *et al.* (1970) found that high concentrations of *Pyrodinium* in the surface waters of the bay were only possible in situations where physical surges of water carried the cells to regions where flushing was negligible.

Nutrient relationships of the phytoplankton

Most studies show that nitrogen tends to become limiting in the oceans before phosphorus; this seems to be a function of the ratio of the elements in seawater ($N:P = 15:1$) and of the more rapid regeneration (cycling) of phosphate relative to nitrogen. On the other hand in lakes the opposite relationship usually holds. Occasionally, when diatoms are dominant, silicate can become limiting before nitrate (Goering *et al.*, 1973) but more widespread studies are needed to show the relationships between diatom growth and silicate limitation since only a few species (e.g. *Asterionella*, *Skeletonema*) have been adequately studied. Golterman (1972) reported that 80% of the phosphate rapidly appears in solution during autolysis of algae; phosphate from this source is not normally detected since it is taken up immediately by living algae. He suggested that the rate of mineralisation (oxidative breakdown of dead material) is the main factor controlling algal production. Minas (1970) found that estimations of phytoplankton production based on phosphate levels give values only approximately 25% of those measured by ^{14}C uptake and this was attributed to the recycling of phosphorus rather than dependence on upwelling of phosphate-rich water. This assumes that the supply of nutrients is mainly from recycling which indeed is likely in regions such as the central oceanic gyres. But in other waters, inflow, upwelling and cycling through animals are all involved; Harris (1959), for example, calculated that during the spring bloom 77% of the average daily demand of the phytoplankton for nitrogen can be supplied by excretion from zooplankton, though in the form of ammonia, and in the rest of the year the figure is 40%. In stratified tropical oceans the phytoplankton in the upper waters can be utilising ammonia and those lower down the water column in the discontinuity layer can be living on nitrate (Goering, Wallen & Nauman, 1970).

Nutrient depletion certainly occurs in many waters but one cannot draw general conclusions from a small area; thus Anderson & Munson (1972) showed that the coastal waters of the North Pacific Ocean exhibit nutrient depletion in the spring but the subarctic waters do not, owing to entrainment of nutrient (nitrate) rich, deep water into the upper zone, coupled with the slow rate of removal by primary producers. South of the subarctic water, nutrient depletion does occur in ocean sites, presumably due to the establishment of a thermocline cutting off supplies from below. Evidence is accumulating that in spite of the obvious growth-promoting value of upwelling water, e.g. Roels, Gerard & Bé (1971) found that deep water off the Virgin Islands is much more productive than surface water (Fig. 7.67), though it is not always immediately stimulatory. Similarly, Barber & Ryther (1969) found that newly upwelled Cromwell Current water, rich in nutrients, only supported low plankton growth. The growth lag when using this water could be reduced by adding chelators (EDTA) or trace metals. This is akin to the 'ageing' of water which has long been

known to be beneficial to algal growth and which is thought to be due probably to release of organic material from organisms; these substances could act as chelators. As Thomas (1970) showed, there is no detectable nitrate or nitrite in some of the nutrient-poor Pacific waters, though there is ammonia and amino-nitrogen present which allows a reduced rate of production. There is ample evidence that some planktonic algae can utilise organic nitrogen molecules, e.g. McCarthy (1974) found that half of 35 isolates could utilise urea and Antia, Berland *et al.* (1975) showed the utilisation not only of urea but also of glycine and hypoxanthine. Experiments using the small diatom *Thalassiosira nana* in nitrogen-limited continuous culture have shown that nitrate reductase and the ability to utilise nitrate and ammonia increases with increasing nitrogen limitation whilst glutamic dehydrogenase activity, photosynthetic rate, N:C ratio and chlorophyll:carbon ratio all decrease with increasing nitrogen limitation, showing the adaptability of some of these planktonic organisms (Eppley & Renger, 1974). An indication that growth of phytoplankton could continue when nitrate was depleted had been obtained in the earlier experiments using seawater in a large plastic bag (McAllister *et al.*, 1961).

In order to study the concentrations of nutrients present under natural conditions it is necessary to separate phytoplankton species from the remaining

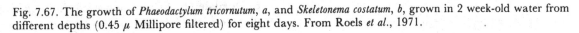

Fig. 7.67. The growth of *Phaeodactylum tricornutum*, *a*, and *Skeletonema costatum*, *b*, grown in 2 week-old water from different depths (0.45 μ Millipore filtered) for eight days. From Roels *et al.*, 1971.

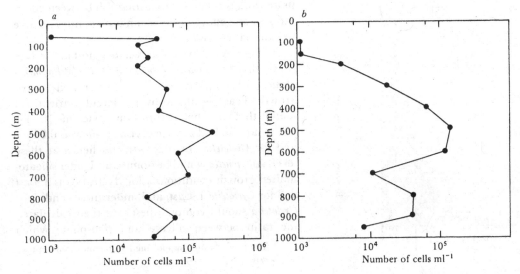

seston. This is difficult and the few estimations have been made at times when populations were virtually unialgal. Mixed populations would give very varied results since different species not only contain different pools of nutrients but would be in different physiological states. There have been very few determinations of the incorporation of nutrients into algal populations (most measure only the concentration of nutrients in filtered water) but one by Pechlaner (1970) clearly shows the transference from

Fig. 7.68. Measurements of phosphorus (*a*) and nitrogen (*b*) in lake water (horizontal lines) and phytoplankton (stippled area). From Pechlaner, 1970.

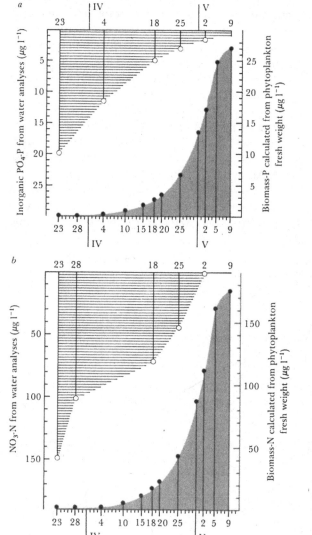

water to algal population (Fig. 7.68). A promising method for separation of algal cells is by continuous particle electrophoresis where all the cells move to the anode and mixtures can to some extent be separated (Bayne & Lawrence, 1972); this method is worthy of further study since dead cells, clay particles and bacteria can be separated from the live algae. Information is needed about the concentrations of nutrients, both internally and in the external medium since, as the recent study of Droop (1974) has shown, both are important in studies of uptake kinetics. Cell division rate is dependent solely on the internal concentration (Lehman *et al.*, 1975). In addition we now know that there are intraspecific differences in the uptake constants for some diatoms (Carpenter & Guillard, 1971) indicating adaptation to high or low nitrogen concentrations. The half-saturation constant* for nitrate in estuarine strains of three diatom species was 1.5 μm whereas for low nutrient areas it was 0.75 μm, a figure which is of a similar order of magnitude to the nitrate concentration in the Sargasso Sea. These differences were not short-term but were found to be stable over many years in culture, showing that they are almost certainly genetic adaptations to nutrient levels. They may be related to the presence of a nitrate/chloride activated ATP-ase on the cell membrane of some but probably not all marine species which, when present, allows the uptake of nitrate against a concentration gradient (Falkowski, 1975). The data of MacIsaac & Dugdale (1969) also suggested that oligotrophic plankton was adapted to low nutrient levels but more work is necessary to distinguish between adaptations and genotypic determination of rates relative to external concentrations.

Titman (1976) has shown in a most interesting study that *Asterionella formosa* and *Cyclotella meneghiniana* can only co-exist in a water body when growth of each is limited by a different nutrient. He found that the half saturation constant (k) for phosphate-limited growth of *Asterionella* was 0.04 μM and for *Cyclotella* was 0.25 μM and hence at these levels *Asterionella* would be dominant. Under silicate-limited growth conditions k for *Asterionella* is 3.9 μM and for *Cyclotella* 1.4 μM, and under these conditions *Cyclotella* should compete best. He then calculated the ratio between silicate and phosphate which

* The concentration supporting half the maximum uptake rate.

should produce equal growth rates of both algae and found that these lie between 5.6 and 97; experiments with concentrations between these ratios did in fact allow co-existence. Above 97 both species are limited by phosphate but *Asterionella* is the better competitor for the resource and becomes dominant whilst below 5.6 *Cyclotella* becomes dominant since both species are silicate-limited and *Cyclotella* can compete better for silica.

The cell contents of nitrogen and phosphorus in *Microcystis aeroginosa* increase with increase in the external supply and such luxury uptake has been

Fig. 7.69. Cross-section scale plan of plastic bag used for *in situ* phytoplankton. From Strickland & Terhune, 1961.

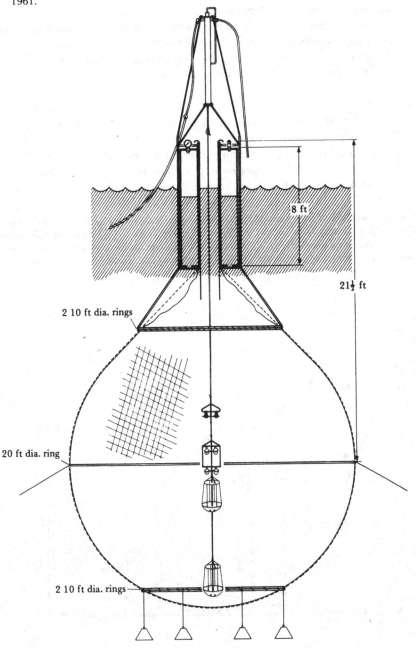

demonstrated for many algae, making a simple analysis of the cellular content a rather meaningless fact.

Many phytoplankton species undergo sexual reproduction or form resting spores, yet little is known concerning the ecological factors including nutritional ones, which exert control over these processes. Diatoms reproduce sexually forming auxospores, from which new large cells are released. Steele (1965) showed that for auxospore formation in the common marine diatom *Ditylum*, the medium must be *deficient* in manganese. Inhibition or stimulation effects by trace elements have received little attention though Goldman *et al.* (1968) found that traces of cobalt inhibited photosynthesis of lake phytoplankton and molybdenum stimulated it; they considered that inhibition effects could be shown at concentrations as low as 1 μg l^{-1}. This aspect would repay detailed study. In *Stephanopyxis* sexuality is not nutritionally controlled but the sexual stages are formed when the day length is over 16 h, whereas at

8 h only resting spores form. There is also a complicating light factor, for at 30.3 ft-c. only auxospore mother cells form and at 200–300 ft-c. only male cells are produced. Earlier, von Stosch (1954) showed that *Lithodesmium undulatum* produces only oogonia under continuous illumination, but low light and alternating light and dark is required for the formation of male cells. The incidence and significance of sexual reproduction in phytoplankton under natural conditions is hardly known; some diatoms, e.g. *Asterionella*, never seem to be reported in anything other than the vegetative state and one can be almost certain that sexuality is not a complicating factor in their growth cycle.

Evidence is accumulating that the concentration of the nutrient medium may affect the morphology of planktonic species; this work is mainly experimental but it may be that, in waters of differing nutrient status, certain forms of a single species may be induced. This is of course only likely in pleomorphic organisms, i.e. species which can

Fig. 7.70. *a*, changes in the algal weights of the main phytoplankton species developing in the plastic bag. —●—, *Skeletonema costatum*; —○—, *Thalassiosira rotula*; --■--, *T. nordenskioldii*; —□—, *Gyrodinium fulvum*; ——, *Glenodinium danicum*; ——, *Nitzschia delicatissima*. *b*, concomitant changes in the concentration of nitrate (+), nitrite, (--), ammonia (-●-), reactive phosphate ((--|--)) and reactive silicate (-○-). From McAllister *et al.*, 1961.

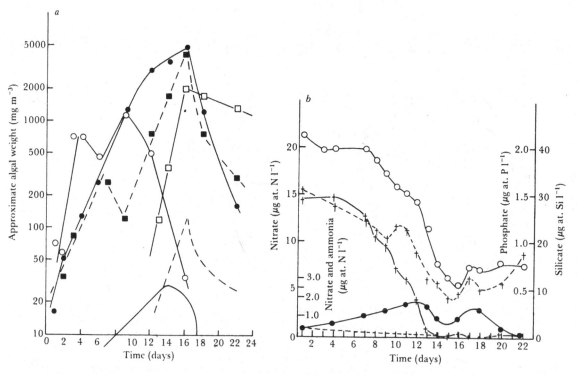

vary their morphology. A marine example is *Phaeo-dactylum tricornutum*, which forms oval cells in media with less than 15 mg l⁻¹ calcium and fusiform cells when the level is raised (Cooksey & Cooksey, 1974). Schultz (1971) found that the species specific pattern of cell wall structure of *Cyclotella cryptica* was formed in waters of salinity from 4.3‰ to near full seawater but at salinity around 1.4‰ it assumed the patterning of the common freshwater *C. meneghiniana*. But clones of *C. meneghiniana* could not be induced to

assume the '*cryptica*' form. Such polymorphism in some species raises many problems in taxonomy even if the ecotypes can be readily recognised in nature.

Trainor & Shubert (1974) working with the freshwater alga *Scenedemus* found that spineless colonies were formed in media with approx. 30 mg l⁻¹ inorganic salts. Some *Scenedesmus* strains maintain their colony form in dilute media but when soil extract is added they produce unicells which would be allocated to the genera *Chodatella* or *Franceia*.

A relationship between nutrient status and geographical distribution can be detected for some species, e.g., there is a negative correlation between phosphate concentration in the sea and the distribution of *Ceratium* sp. (Graham, 1941) which results in much greater species diversity in tropical seas.

The uptake of nutrients by phytoplankton under semi-natural conditions has been studied using a variety of techniques, e.g., growth in large plastic bags (Fig. 7.69) either in the sea (Strickland & Terhune, 1961) or in lakes (Schelske *et al.*, 1972). The experiments using the 'plastic' bag yielded extremely valuable data some of which are summarised in Figs. 7.70 to 7.72. The technique involved filling the bag with filtered seawater (125 m³) then adding an inoculum of seawater (4 m³) from which the larger zooplankton species had been removed. The diatom–dinoflagellate bloom which developed in the bag was then followed (Fig. 7.70). The rapid removal of nitrate is shown in Figs. 7.70*b* and 7.72 but growth continued and indeed peaked after day 13, showing that a few divisions occurred as the cells used up their intracellular store of nitrogen; in fact nearly half of the bloom biomass was produced *after* nitrate depletion. Fig. 7.71*a* shows net production of carbon measured by five independent methods and is of great interest for it is one of the rare examples where different methods are compared: the graph of pigment content (Fig. 7.71*b*) shows peaking before maximal production. The drop in the cell count, even when carbon fixation is obviously continuing is due to sedimentation of the cells in a population which is becoming senescent (cf. p. 303). Space does not allow me to do justice to these experiments and I strongly suggest the original works be consulted. Another technique developed by Strickland, Holm-Hansen, Eppley & Linn (1969) used a deep 3 m × 10 m tank illuminated from a searchlight and with various measuring and sampling devices

Fig. 7.71. *a*. The net production of carbon in the plastic bag measured by five independent methods, *b*. The changes in the concentration of plant pigments during this experiment. From McAllister *et al.*, 1961.

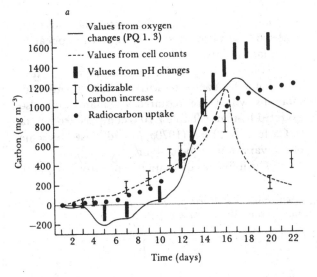

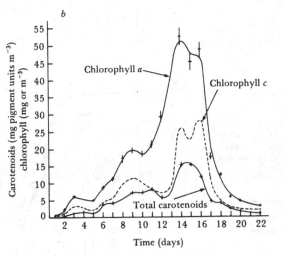

Table 7.10. *Vitamins required and secreted by various algae. From Carlucci, 1970a, b*

Alga	Vitamin requirements	Vitamins excreted
Dunaliella tertiolecta	None	B_{12}, thiamine, biotin
Phaeodactylum tricornatum	None	thiamine, biotin
Stephanopyxis turris	B_{12}	thiamine, biotin
Skeletonema costatum	B_{12}	thiamine, biotin
Gonyaulax polyedra	B_{12}	thiamine, biotin
Coccolithus huxleyi	thiamine	biotin (with low and high concentrations of thiamine in medium) B_{12} and biotin (with high concentration of thiamine in medium)

incorporated; an expensive but valuable device and one which unlike the bag experiments was used with unialgal populations. It enabled studies in a simulated 10 m column of seawater which had also first to be filtered and then large amounts of inoculum added (e.g. 130 l of a culture of *Ditylum brightwellii*). Fig. 7.72 shows the course of uptake of nitrogen compounds with time, and that ammonia-nitrogen is assimilated first by all the organisms cultured. Strickland and his collaborators also showed that it was not until almost all the ammonia had been depleted that nitrate reductase began to appear in any quantity in the cells.

Nutrient relationships of the phytoplankton would not be complete without brief mention of vitamins which are required by many and also excreted by many. Table 7.10 is compiled from data in Carlucci & Bowes (1970*a*, *b*) and illustrates some of the variations in behaviour.

The influence of vitamin concentration on the

Fig. 7.72. Changes of particulate and dissolved nitrogen with time in cultures of *Ditylum brightwellii*, (*a*), *Cachonina niei*, (*b*), and *Gonyaulax polyedra*, (*c*), in a deep tank. From Strickland *et al.*, 1969.

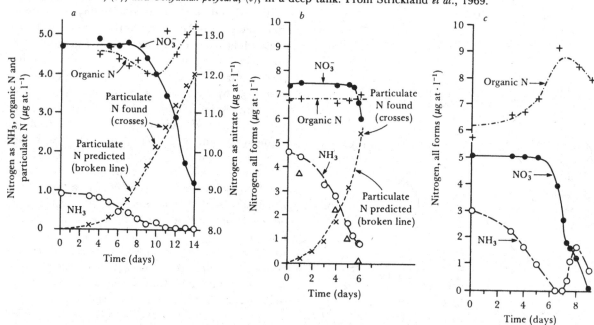

natural populations has received little attention, though Carlucci (1974) considers that the population as a whole is not affected but individual species may be; for example, *Gonyaulax polyedra* growth could be correlated with a decrease in vitamin B_{12} content of the water. It was found that the vitamin content of the water was high when phytoplankton concentration was high, suggesting that cells were actively secreting. This may be related to rapid growth or merely be a product of senescence. In the open ocean, where nutrient and climatic factors are more stable, changes in vitamins may be of greater importance and recently Jeffrey & Carpenter (1974) have postulated that vitamins may be important in annual cycles of phytoplankton in waters off Australia. In freshwaters Parker (1968) recorded a bloom of a *Chlamydomonas* species in a small pond after a rain storm which increased the vitamin B_{12} content of the water. Menzel & Spaeth (1962) could not show increased carbon fixation by enrichment experiments with vitamin B_{12} but thought that its concentration may affect the species composition of the plankton. The utilisation by algae of vitamins secreted by other algae has been shown (Carlucci & Bowes, 1970a, b). Vitamin production by bacteria has usually been thought to be the major source in the sea but obviously where phytoplankton populations are high they can contribute as much or more. Vitamin B_{12} is also required by some Cyanophyta, e.g. Van Baalen (1967) found that 50% of the species he studied required this vitamin.

Daisley (1969) analysed the vitamin B_{12} concentrations in various lakes of the English Lake district and found that the values were highest in the most productive, high nutrient lakes. Decrease in the vitamin B_{12} concentration in lake waters during diatom blooms has been recorded (Ohwada & Taga, 1973); this is accompanied by increased thiamine and biotin concentrations in the particulate and dissolved components, in general agreement with the results in the experimental studies of Carlucci. The Japanese workers suggested that algae might supply the vitamin to fish in their fishponds. Such data are very difficult to interpret when algae are both producers and users of vitamins. The levels may be important in determining the presence or absence of certain algae and in determining the turnover times of the populations. Swift & Guillard (1978), for example, found that vitamin B_{12} eliminated the lag phase of growth in axenic culture, but many more detailed studies will be required to elucidate relationships under natural conditions.

Guillard & Cassie (1963) found that the requirements of diatoms for vitamin B_{12} varied from 5 to 18.4 molecules of B_{12} h^{-1} u^3 of diatoms. They also showed one of the rare instances of abnormality caused by vitamin starvation, in *Skeletonema costatum*. Haines & Guillard (1974) demonstrated that diatoms which required vitamin B_{12} could be cultured in the presence of vitamin B_{12}-secreting heterotrophic bacteria but the diatom yields were not as great as in axenic cultures with added vitamin B_{12}. The bacteria utilised the diatom products and so, as long suspected, a web of stimulation is confirmed.

Attempts to study nutrient relationships often involve growing some populations in nutrient-poor water and parallel series in water to which nutrients have been added. Using Pacific Ocean water, Thomas (1969) found that nitrate and ammonia would stimulate growth in nutrient-poor water but there was no detectable limitation in nutrient-rich water. Smayda (1971a) used *Thalassiosira (Cyclotella) (pseudo)nana* in similar experiments on Atlantic Ocean water and found that enrichment of surface water invariably inhibited growth but enrichment of deep water increased growth. Here silicate promoted growth more frequently than other elements added singly. When he tried to culture *Thalassiosira* in unenriched water from 100 m depth the diatom lysed but no explanation could be given for this effect. In later experiments with Narragansett Bay water, he found that growth in unenriched water was poor from June to February and no growth accompanied by lysis occurred after February; this is strange since it is the time of the plankton diatom bloom in the Bay. Reduction of growth to 10% occurred when nitrogen was omitted from the enrichment and only during the November to December nitrite–nitrate maximum were additions of nitrogen ineffective (Smayda, 1974). Smayda also recorded rather unusual effects when silica was omitted, e.g., during August omission limited growth but in spring when the silica concentration was lower in the water the omission did not limit growth but addition actually inhibited it. He suggested that it is nutrient turnover and not the ambient concentration which is of greater importance. Menzel & Ryther (1961a) also found that adding nutrients to Sargasso Sea water had little effect on ^{14}C assimilation but a trace metal mix increased the assimilation several fold and iron was shown to be the effective element.

Nutrient regeneration from phytoplankton

Ultimately all the material incorporated into the phytoplankton will be released in some form, though a proportion of it is lost to subsequent algal populations owing to sedimentation (*see* chapter 12). However a vast amount of material must be re-cycled, yet few workers have studied the breakdown part of the cycle of algal production. Grill & Richards (1964) allowed diatoms to decompose in the dark and measured some of the breakdown products, finding that both particulate phosphorus and nitrogen increased at first and then later fell. Orthophosphate however decreased during the initial period and then increased and since this is likely to be the form taken up by other phytoplankton one cannot assume that nutrients are available immediately after the death of the cells. The initial release will depend upon the history of the cells; if they have an excess of nutrients in the cytoplasm, and this is quite common in algae, the nutrients will be released rapidly and therefore be available immediately for the growth of other algae, but if the cells have been growing under nutrient-poor conditions there will be no excess 'pool' and regeneration will only occur when the organic matter is enzymatically degraded (Foree, Jewell & McCarty, 1971). Foree *et al.* distinguished three stages of regeneration from algae placed in the dark. During the first 24 h there may be either a release or absorption of nitrogen and phosphorus, followed by a stationary phase of several days and then a period of active nutrient release, maybe lasting for up to several 100 days. They found that a large fraction ($\frac{1}{3}$) of the initial nitrogen and phosphorus was not regenerated but remained un-decomposed. The ranges of regeneration are, as one might expect, extremely wide, whether under aerobic or anaerobic conditions, but all the likely regeneration was completed in 200 days. Golterman (1960) reported that under sterile conditions 70–80% of the phosphorus compounds leave the cells within a few days but that 70–80% of nitrogen compounds remain in the cell. In another series of experiments in which UV-radiation was used to kill *Scenedesmus*

obliquus, Golterman (1964) found that 50% of the phosphorus was released by autolysis in the first few hours; about 25% is not released at all and about 90% of this was incorporated in nucleic acids. The nitrogen was released more slowly.

Other data suggest that the organic nitrogen is reduced to about 70% after the first 30 days of aerobic microbial decomposition though the cell wall is resistant to bacterial decay (Otsuki & Hanya, 1972). There are obviously many problems and contradictory statements in this field, some of them probably simply due to variation in techniques.

Algal cell walls often contain cellulose or other polysaccharides and it is unlikely that much of this is decomposed until the walls reach the sediments. It would be interesting to study the distribution of cellulolytic bacteria in the water and sediments but even casual observations show little breakdown of walls of algae such as dinoflagellates or desmids until they reach the sediments. In fact there can be a considerable amount of cellulosic material in the non-living organic detritus in the water column.

Depth–photosynthesis profiles

MacIsaac & Dugdale (1969) considered that there was a crossover point in the ocean at a depth where the incoming radiation was reduced to 10% of the surface value; below this point light is limiting growth and above it nutrients are limiting. Little consideration has been given to the operation of different factors down a depth profile but Semina (1967) comments that most of the fixation in the open ocean occurs in the upper 50 m and this is admirably demonstrated over a range of latitudes in Fig. 7.73 taken from Jitts' 1969 study along the 110° E meridian.

Since many variables affecting carbon fixation by phytoplankton change with depth, and even in a completely mixed situation a key factor, light, will always attenuate with depth, no cell of the phytoplankton experiences anything like a relatively constant environment. To determine the effect of depth on photosynthesis it is usual to enclose samples in bottles which are then incubated at fixed depths or incubated on the ship or in the laboratory under conditions simulating fixed depths. Literally hundreds of such studies have been made and Figs. 7.74 and 7.81 show some representative results. One study shows the comparison with ATP and total particulate carbon but this adds very little to the picture gained from chlorophyll values (Vollenweider, Munawar & Stadelmann, 1974). Integration of the values from each depth interval yields a reasonable measure of carbon fixation within the photic zone assuming that in nature there is a randomised circulation of cells in the zone. If the movement is rapid, one can perhaps assume that the cells do not have time to acclimatise to low or high light intensity as they move through the column and therefore 'sun' and 'shade' species are not present.

As can be seen from Fig. 7.74, the shape of the curve for fixation versus depth in lakes varies and Findenegg (1964) was of the opinion that three types of curve could be distinguished. The first occurs in rich lakes where there is a maximum in the upper epilimnion, the second has no distinct maximum (although over short periods an upper, epilimnetic maximum may occur) and the third has an upper and lower maximum and is common in lakes where blue-green algae collect in the metalimnion. I would be inclined to add a fourth type, which is that found in high mountain lakes with only a deep maximum. The common, mid-summer feature, of lower fixation in the subsurface layer has been attributed to the deleterious effects of high light intensity in this zone. Jitts, Morel & Saijo (1976) showed that a 2 mm thick glass cover over their experimental tanks would increase photosynthesis by 50%, which is strong evidence for the damaging effect of UV-radiation. It may not be such a drastic feature for individual cells in nature, since they will be circulated in and out of the 'damaging' region. However during 'bloom' conditions it can have a serious effect on the population and cause not merely a reduction in the rate of fixation but also lead to death of the populations; indeed samples taken from the surface blooms are rarely viable (Reynolds, 1967; Fitzgerald, 1967).

Fig. 7.73. Vertical sections along the 110° E meridian of the daily rate of production, in mg C day^{-1} m^{-3} × 100 for each of six cruises. The depth of the deepest sample (––) is also contoured as an approximation to the depth of the euphotic layer. From Jitts, 1969.

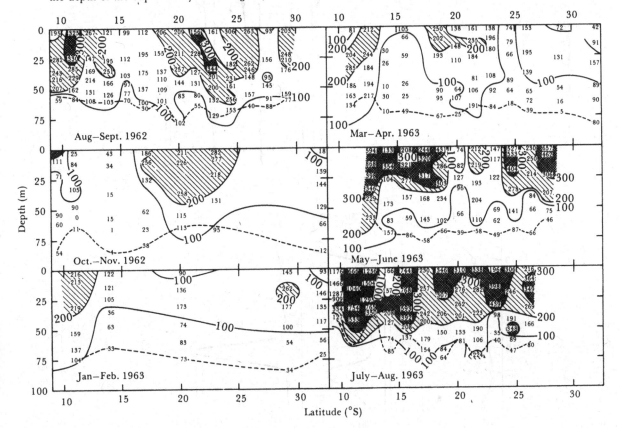

Surface inhibition has been shown to occur only around mid-day, in a Japanese lake (Ichimura, Nagasawa & Tanaka, 1968), in Lake Ontario (Fig. 7.13; Harris, 1973), and in some mountain lakes (Anderson, 1974) suggesting that it is a function of light intensity, as the dip in carbon fixation was not recorded at 15 m. This phenomenon was of course shown in the early, classic work of Marshall & Orr (1928) in the sea off Scotland using the diatom *Coscinosira polychorda* and again by Jenkin (1937), in the English Channel using *Coscinodiscus excentricus*; their data are shown in Fig. 7.75. It was also demonstrated in an interesting early paper by Ålvik (1934) in Norwegian fjords almost cut off from the sea where he used now-abandoned photographic techniques to measure light penetration. In spite of the fact that such surface 'inhibition' has been known for at least 50 yr there is still much confusion surrounding it and definitive studies are badly needed. Talling (1971) listed the following seven

Fig. 7.74. The three types of vertical carbon fixation profiles recognised by Findenegg. *a*, Ossiacher See; *b*, Attersee; *c*, Längsee. From Findenegg, 1964.

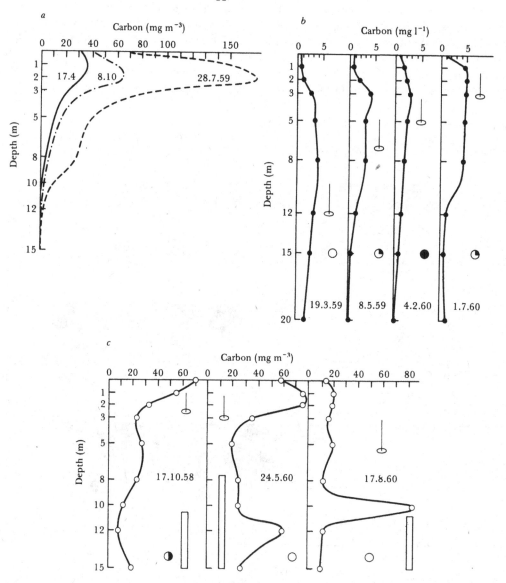

points which need to be considered: (i) lack of vertical circulation in the experimental systems – that is, unnatural confinement in bottles near the surface, (ii) probability of pronounced and little known spectral sensitivity, (iii) probability that at least some examples reflect a decreased population in the surface waters, (iv) varying degrees of reversibility of the inhibition, (v) the possible participation of low energy photoprocesses, (vi) the usual lack of pronounced surface inhibition in long-term experiments and (vii) possible 'adaptation'.

Another technique for studying the effects of depth is the collection of natural populations from a range of depths and incubation of each at each sample depth (Hickman, 1976). Fig. 7.76 shows some results obtained with the use of this technique. Features immediately revealed are that samples

Fig. 7.75. Diurnal variation of photosynthetic rates on sunny summer days measured at various depths for *Coscinosira polychorda* (*a*) in the sea of south-west Scotland. From Marshall & Orr, 1928. *b*, cultured *Coscinodiscus excentricus* in the sea of southwest England. From Jenkin, 1937.

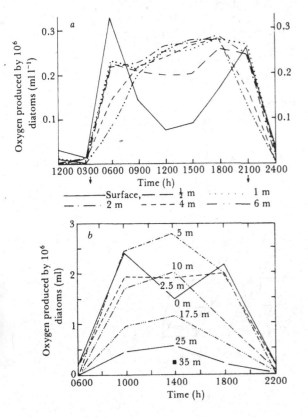

containing abundant chlorophyll, e.g., those from 3 m on 2 April, do not necessarily increase their fixation when brought nearer the surface whereas, invariably, samples incubated at lower depths show decreased fixation. Experiments of this type are necessary to distinguish between populations *existing* at different depths but being either active, dividing and metabolising actively, populations simply maintaining themselves but showing little active growth and those which cannot recommence growth but are still floating in the water. Only the experiments of 22 December show what one might deduce as a normal situation, yet such a situation is rarely found.

Stratification of phytoplanktonic populations has been known for a long time but is only now being studied in detail. Whenever this is a common feature, attempts to compute the carbon fixation by suspending bottles at various depths will merely reflect light attenuation rather than actual carbon fixation. It is sometimes thought that peaks only occur in deep layers (*see* below) where density stratification stabilises the layers but several peaks can occur down a profile (e.g. in Fig. 7.77) and Ryther & Hulburt (1960) showed that north Atlantic water completely mixed down to 150 m can still maintain stratified populations (Fig. 7.78). The mechanics of this require further study. In these instances, to study the precise carbon fixation down the profile, the phytoplankton must be collected from specific depths and then incubated at these depths, as in the experiments illustrated in Fig. 7.76. The accumulation of cells often occurs just above, within or below the thermocline and this may be simply a function of sinking of the population, e.g. Kiefer *et al.* (1972) record such an accumulation below the thermocline in Lake Tahoe. The association was of a typical group of common genera, *Asterionella*, *Fragilaria*, *Melosira*, *Stephanodiscus*, *Dinobryon*, *Elakatothrix*, and it was assumed that they had collected at the 100–200 m depth due to reduction of the sinking rate as the cells moved out of the nitrogen limited euphotic zone into the lower, nutrient-rich, lower temperature water. Similarly in Lake Kinneret, Cyanophyceae collect in the metalimnion (thermocline) where they can be ten times as abundant as in the epilimnion; Serruya (1972) records that in this region there is also an accumulation of hydrogen sulphide and increased bacterial action, which could produce carbon dioxide to enhance the growth of the algae. In spite of the deep chlorophyll peak in Lake Tahoe the maximum carbon fixation occurred at 10–15 m (0.45 μg carbon l^{-1} h^{-1}) and only 0.02 μg carbon l^{-1} h^{-1} was recorded at 100 m (Holm-Hansen & Paerl, 1972). Populations lying at a depth in the water are often dominated by

flagellates, e.g. species of *Cryptomonas* in alpine lakes and in a Japanese lake where Ichimura *et al.* (1968) attributed the high oxygen level in the metalimnion to the photosynthesis of this alga.

Because of these layering problems Baker & Brook (1971) reverted to measurements using a

Fig. 7.76. The content of chlorophyll (mg chlorophyll a m^{-3}) at the surface, 1, 2 and 3 m depth in phytoplankton samples from Abbot's Pool on five different dates. The values are given for carbon fixed (mg h^{-1} m^{-3}) by samples from these depths suspended at each depth. Figures below are the fixation values per mg chlorophyll a h^{-1}. From Hickman, 1976.

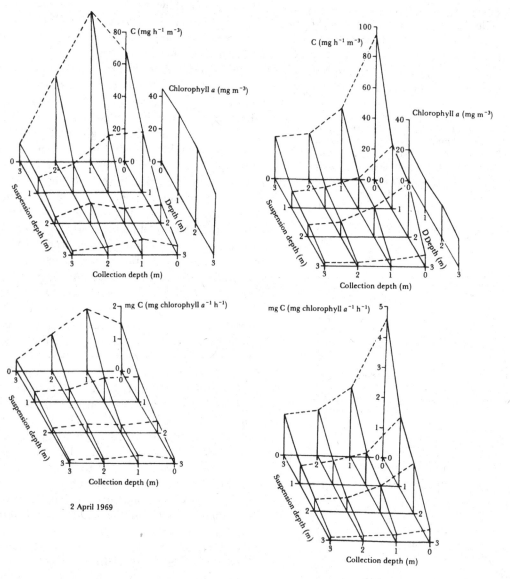

2 April 1969

6 May 1969

turbidometer to detect the fine strata (Fig. 7.79); the turbid layers must of course also be sampled and checked for phytoplankton since, under some conditions, suspended detritus, etc. give a similar reading on the meter. By coupling such observations with photosynthesis studies Baker, Brook & Klemer (1969) have shown that the fine strata of the blue-green alga *Oscillatoria agardhii* var. *isothrix* are actively photosynthesising (Fig. 7.80) and in fact in summer can provide 50% of primary production and in winter, when stratification occurs under ice, 100% of the planktonic primary production. However in spite of collecting in a layer in the metalimnion the algae can photosynthesise well at higher light intensities and in fact when suspended in bottles and the photosynthesis determined by oxygen evolution a typical rate against depth curve is produced with a peak at 1 m (Fig. 7.81). Similar results were

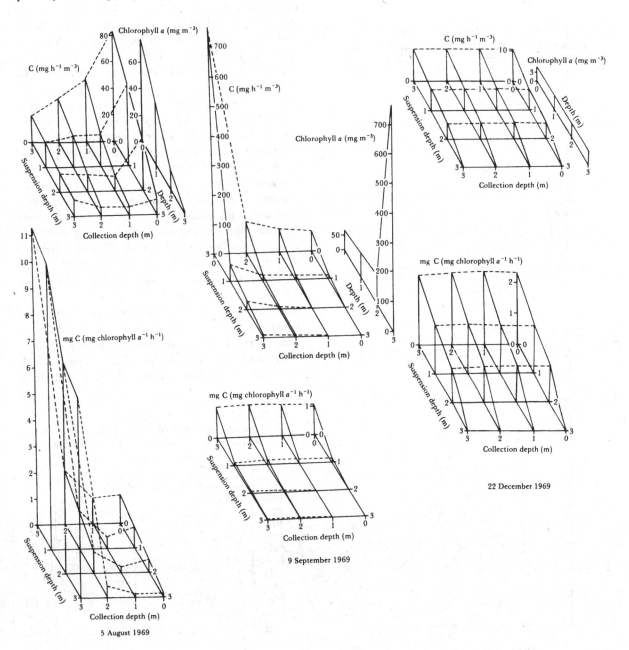

5 August 1969

9 September 1969

22 December 1969

Fig. 7.77. *a*, the vertical distribution of chlorophyll *a* concentration (open circles) and phytoplankton volume (horizontal bars) in Lake Tahoe on 30 September 1970. *b*, temperature (continuous line) and nitrate-nitrogen concentration (open circles). From Kiefer *et al.*, 1972.

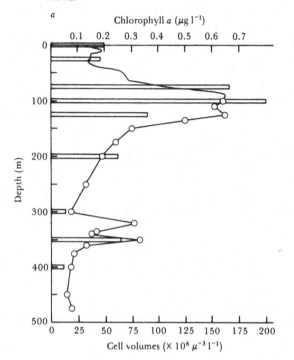

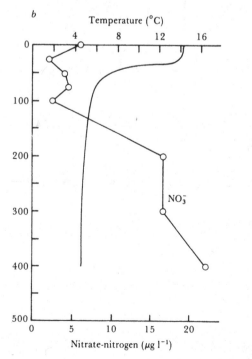

reported for a *Mallomonas* population which was stratified at 9 m in virtual darkness.

In the oceans, deep living populations have been recorded in the Pacific (G. C. Anderson, 1969) and Indian Oceans (Bienfang & Gundersen, 1977) and in the Sargasso Sea (Menzel & Ryther, 1960; Vaccaro & Ryther, 1960) Gulf of Mexico, etc. but the one for which most data exist is the Pacific. Fig. 7.82 shows how this deep layer appears when continuous vertical profiles of chlorophyll fluorescence are plotted and Fig. 7.83 presents phaeophytin and chlorophyll data for one station (Venrick, McGowan & Mantyla, 1973). These workers summarise the explanations which have been given by various authorities to explain this deep layer, i.e., concentration of detrital chlorophyll in the pycnocline (region of rapid change in density), differential zooplankton grazing (there is certainly grazing in the upper waters which may sharpen the profile), increase in the chlorophyll–carbon ratio in the cells, horizontal advection and layering of different water masses and populations, sinking of active or senescent cells and *in situ* production. Until recently this deep layer of chlorophyll was thought to be a tropical and semi-tropical feature but recently Pingree, Maddock & Butler (1977) and Holligan & Harbour (1977) have found these layers around the British Isles; Fig. 7.84 illustrates such layers at shallow depths associated with the thermocline. Some reports of concentrations of cells at great depths, e.g. of Coccolithophorids in the Mediterranean, must be treated with caution since these may not represent viable populations (Fournier, 1968, 1970). Revelante &

Fig. 7.78. Vertical distribution of the most abundant species of phytoplankton in the north Atlantic when the water was physically and chemically almost homogeneous. From Ryther & Hulbert, 1960.

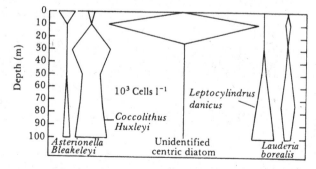

Gilmartin (1973) considered that there has to be a pycnocline above the chlorophyll rich layer and this certainly occurs at the station from which Fig. 7.83 is taken. This increase in chlorophyll content in such a deep layer agrees with expectations from work such as that of Jeffrey (p. 353) and with Anderson's (1969) finding that cell counts increased two-fold in the deep layer but chlorophyll increased three-fold. Talling (1966) found a similar increase in chlorophyll content in cells of *Asterionella* growing at a depth in a lake, so this effect is not confined to very deep layers. This would suggest 'shade-adaptation' which

Fig. 7.79. Optical density and temperature profiles in, *a*, Arco Lake and, *b*, Josephine Lake, northern Minnesota. In *a*, the optical density maxima are due, from top to bottom, to *Anabaena* sp., Cryptomonads and *Merismopedia trolleri*, and are shown on the profiles. The dotted lines give the total cell volume. In *b*, the peak at 5 m depth is due to *Oscillatoria redekei* and the deeper one to *O. agardhii* and Cryptomonads. From Baker & Brook, 1971.

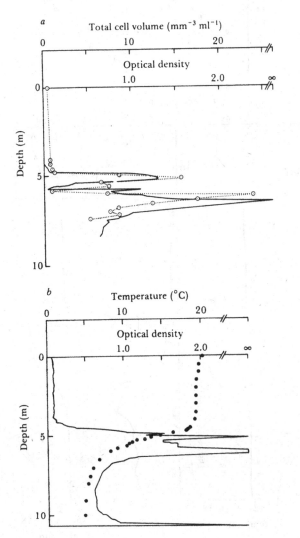

Fig. 7.80. The diurnal productivity curves of an *Oscillatoria agardhii* population at different light intensities from surface to 4% of incident illumination. From Baker *et al.*, 1969.

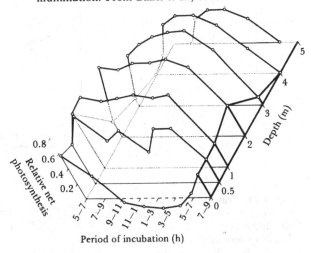

Fig. 7.81. Net production of the *Oscillatoria agardhii* population at each depth during daylight hours. Full line, production; dashed line, incident light. From Baker *et al.*, 1969.

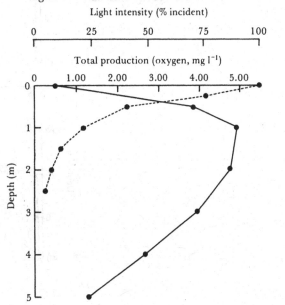

Fig. 7.82. Continuous vertical profiles of *in vivo* fluorescence of chlorophyll. *a*, a simple maximum layer in the North Central Pacific; surface chlorophyll concentration 0.02 mg m⁻³; *b*, a simple maximum layer in the South Central Pacific; surface chlorophyll concentration 0.01 mg m⁻³; *c*, a double maximum layer in the North Central Pacific; surface chlorophyll concentration 0.02 mg m⁻³. From Venrick *et al.*, 1973.

Fig. 7.83. Vertical profiles of chlorophyll and phaeophytin observed in October 1969 in the vicinity of latitude 25° S, longitude 155° W. Estimated depth of the 1 % light level indicated by the horizontal bar and cross-hatching. From Venrick *et al.*, 1973.

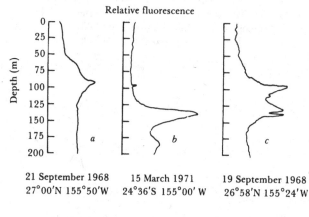

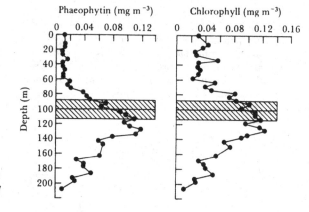

21 September 1968   15 March 1971   19 September 1968
27°00′N 155°50′W   24°36′S 155°00′W   26°58′N 155°24′W

Fig. 7.84. Temperature and chlorophyll *a* fluorescence records down the vertical profile at station E1 (*a*, *b*, *c*) in the English Channel at various stages of the tidal cycle on 20 August 1975. Continuous line, fluorescence as intake is lowered; dashed line, fluorescence as intake is raised. *d*. Surface measurement between E1 and the Eddystone Lighthouse. From Holligan & Harbour, 1977.

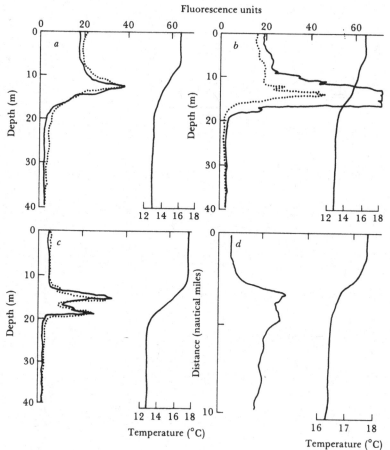

was presupposed by Saijo, Izuica & Asaoka (1969), who also considered that diatom cell division rates were greater in the low light intensity than in the inhibitory surface layers. A relationship to light intensity may also be discerned from Findenegg's (1947) graphs of *Oscillatoria rubescens* which peaks below the epilimnion in summer rising to the surface in autumn. The maximum chlorophyll concentration may occur in a layer less than 5 m thick and therefore routine water bottle sampling with the bottles separated by 15 m or more is not a suitable technique. Fig. 7.85 from Strickland (1968) shows the different profiles of chlorophyll *a* determined from water bottle samples and from continuous recordings down the depth profile with a fluorescence technique. Strickland estimated that water bottle sampling may underestimate the biomass by up to 25%. The profiles on 28 June and 5 July show very striking depth variations and not merely a single deep maximum as in Fig. 7.83. Microscopic examination has revealed significant increases in diatom cells in deep layers (Venrick, 1971): 80 species of

diatoms were identified in one series of observations. Of 64 species occurring at 125 m, 22 occurred only in samples from that depth. The occurrence of characteristic deep living dinoflagellate species has long been known* and Graham (1941) considered that at least 20 of the 58 *Ceratium* species found during the extensive cruise of the 'Carnegie' were deep living. These shade species have thinner walls, expanded horns and horns packed with chromatophores and they tend to be confined to tropical and sub-tropical waters. Surface living species never have such cell expansions. Deep phytoplankton-rich layers are not isolated phenomena but occur over large areas of the Pacific Ocean, especially beneath the central gyres where production is very low in the surface waters. As Table 7.11 shows, the maximum occurred below the depth penetrated by 1% of the surface radiation and since this is generally assumed to be the lower

* Much careful work is still needed on this problem and F. J. R. Taylor (1973) warns that 'a tropical "shade flora" of dinoflagellate species is still not entirely unequivocal'.

Fig. 7.85. Chlorophyll *a* profiles plotted as actual results from continuous measurements by an Autoanalyser (continuous line) and as smoothed curves (dashed line) based on samples taken at discrete depths by water bottles in 1967. *a*, 24 May; *b*, 28 June; *c*, 5 July; *d*, 26 July. From Strickland, 1968.

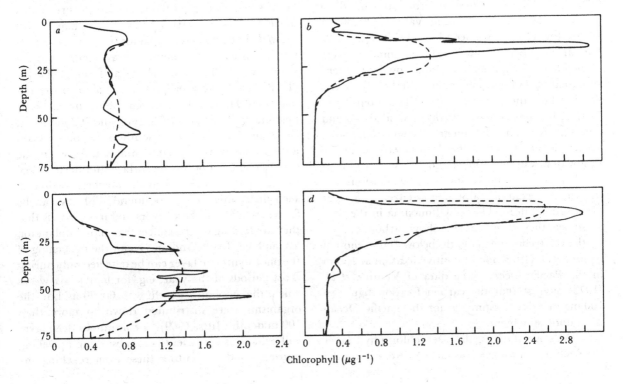

Table 7.11. *Mean value and 95% confidence limits of the mean for data relative to the vertical distribution of light, chlorophyll* a, *and productivity at two stations in the North and South Central Pacific Ocean. From Venrick et al., 1973*

Position	Date	Depth of 1% light (m) (1)	Chlorophyll *a*					Productivity	
			Depth of maximum (m) (2)	Surface concentration (mg m⁻³) (3)	Concentration at (2) (mg m⁻³) (4)	Water column total 0–200 m (mg m⁻²) (5)	% of (5) below (1) %	Total above (1) (mgC m⁻² h⁻¹)	Total below (1) (mgC m⁻³ h⁻¹)
27° N,	September 1968	79 ±5	104 ±8	0.03 ±0.01	0.16 ±0.03	11.92 ±2.62	73.5 ±8.3	16.26 ±2.25	
155° W	September 1969	73 ±5	111 ±9	0.09 ±0.03	0.11 ±0.02	11.67 ±1.32	53.7 ±4.6	31.74 ±7.35	> 2.28
25° S,	October 1969	100 ±13	122 ±11	0.03 ±0.01	0.13 ±0.02	12.35 ±3.22	58.5 ±6.4	12.87 ±9.08	> 3.28
155° W	March 1971	132 ±14	140 ±9	0.01 ±0.00	0.11 ±0.05	8.13 ±1.47	58.3 ±9.5	11.80	> 1.20

limit of the euphotic zone it is pertinent to question how much carbon fixation can occur below this level. Table 7.11 shows that although it is not great, it can vary between 7% and 20% of the total fixed in the column above; carbon fixation rates of 0.13 mg m⁻³ h⁻¹ have been recorded at some deep stations. Comparable fixation values (20–30%) were estimated for this deep layer by Revelante & Gilmartin (1973). Growth rates of some tropical plankters can be quite high even at low light intensities, e.g. Thomas (1966) reports that a *Chaetoceros* divided four times in 24 h at a light intensity of 10 ft-c., and that a small *Nannochloris* divided four and a half times and in addition grew heterotrophically, a possible advantage in the deep layers. As in the freshwater examples quoted above, it is desirable to check *in situ* fixation before any conclusions are drawn. For example, although Menzel & Ryther (1960) found deep chlorophyll maxima in the Sargasso Sea, they were unable to show carbon fixation by the ¹⁴C technique at depths below 100 m though Eppley *et al.* (1973) found positive fixation at 150 m in the Pacific Ocean. The data of Venrick *et al.* (1973) suggest that the carbon fixation figure of 100 mg m⁻² day⁻¹ estimated for the Pacific Ocean is an understatement.

At some stations, the deep chlorophyll layer has been shown to be a seasonal feature, e.g. on the western edge of the California Current where there is a mixed layer in February and at this time the chlorophyll is homogeneous throughout the surface 50 m. But once stratification commences, the deep water maximum forms only when a density gradient isolates this layer from the wind-driven surface turbulence. Reid, Stewart, Eppley & Goodman (1978) showed that these layers can form in coastal waters off California where they occur as bands of phytoplankton parallel to the shore and Schulenberger (1978) has shown a deep chlorophyll layer extending from Hawaii to the Kuroshio current with maxima up to 231 μg m⁻³ in gyres and 763 μg m⁻³ at the edge of a gyre. In the southern region off California there was a clear distinction between the species composition in the surface and deep layers, but further offshore and in the northern region no such distribution could be found. Fish fry can be found localised in these layers and it is obvious that they are feeding on the concentrated phytoplankton assemblage. In the Indian Ocean the deepening of the photosynthetic layer can be detected only during short periods of the year, e.g. for four to six weeks when the euphotic depth was 60–70 m but the organisms were distributed down to more than 100 m depth (Jitts, 1969). A continuous deep layer will behave as a nutrient trap, absorbing upwelling nutrients and preventing these even reaching the upper layers (Revelante & Gilmartin, 1973). This study supports Steele & Yentsch's (1960) conclusion that nutrient depletion in the surface water leads to a reduction in buoyancy and the impoverished cells sink and then accumulate at the top of the nutricline (the region where nutrient concentration is increasing) where sinking rate is reduced again by uptake of nutrients (cf. the similar features in Lake Tahoe). Vacarro & Ryther (1960) found an increase in nitrate and nitrite in the region of the deep chlorophyll maxima in the Sargasso Sea. Semina (in Bogorov, 1967) commented also on the growth of *Planktoniella* and *Gossleriella* at depths when phosphate phosphorus is limiting in the upper water and she believed that their distribution is related to nutrient availability rather than to light intensity.

Depth sampling of phytoplankton must take into consideration any internal wave movements (internal seiches). These can result in sampling of water in the region of the thermocline, from the bottom of the epilimnion or various points through the thermocline, or even in the upper hypolimnion as the waves displace populations. Such internal waves are common in the sea and also in large inland lakes (*see* Fig. 1.9) though they are probably of little significance in small bodies of water.

Koblentz-Mishke *et al.* (1972) found that the photosynthetic efficiency of marine phytoplankton increases with depth, and is accompanied by a shift in the composition of the pigments towards those absorbing blue light. Wallen & Geen (1971*a*) showed that in blue light increased amounts of pigments, ribonucleic acids and deoxyribonucleic acids were formed compared with the amounts in white light, and carotenoids were increased in green light. This has now been shown to be associated with increased number of plastids, thylakoids stacks per chloroplast and photosynthetic capacity of the cells in the case of *Stephanopyxis turris* grown in blue light and compared with similar cells in white light (Vesk & Jeffrey, 1974; Jeffrey, personal communication). Chlorophyll $a+c$ concentration was 80–100% greater in the *Stephanopyxis* cells grown in blue light (400 μwatts cm^{-2}); the carotenoid values also showed a similar increase such that pigment ratios remained virtually constant. Chloroplast polar lipids also increased by about 40% in blue-light grown cells. Such cells fix approximately 40% more carbon on a chlorophyll basis under low light conditions than do white-grown cells, cultured in white light. In white light, the blue-grown cells are not so efficient at fixing carbon (Jeffrey, personal communication). This implies that although the pigment ratios remain similar, the overall efficiency of fixation is increased in blue-light grown cells. The rate of respiration was shown to be lower in blue light (Wallen & Geen, 1971*b*) whilst the relative growth constants of *Cyclotella nana* in blue, white and green light were 0.37, 0.29 and 0.25 and of *Dunaliella tertiolecta* were 0.41, 0.31 and 0.29.

Earlier data (Strain *et al.*, 1944) had indicated that pigments and photosynthetic functioning could be altered by varying the light regime, e.g., that *Cylindrotheca* (*Nitzschia*) *closterium* produced more diadinoxanthin in white than in red light, and an earlier report by Baatz (1941) suggested that growth of marine diatoms was greatest in green, moderate in blue and least in red light. Yet another study (McLeod & Rhee, 1970) suggests that blue light arrests cell division in some planktonic diatoms, *Thalassiosira fluvialilis*, *T. nana*, *Cyclotella cryptica* and *Skeletonema costatum*. Jørgensen & Steemann-Nielsen (1965) found that *Skeletonema* grown at low temperatures (2 °C) had increased pigment and enzyme complements when grown with a short photoperiod (9 h) and decreased ones in continuous light; such findings may help to explain the functioning of algae in deep, cold waters.

Workers in the last century were greatly concerned with the problem of chromatic adaptation, i.e. variation of the pigment content as the light climate changed and there was considerable controversy over this problem, especially with regard to the Cyanophyta and their depth distribution. It still elicits studies, e.g. Hattori & Fujita (1959) showed that *Tolypothrix tenuis* alters its phycocyanin-phycoerythrin ratio under intensity-dependent conditions. Certainly there are great variations in the colours exhibited by species of Cyanophyta with the most striking effect that deep populations tend to a bright pink colour. It has still to be shown conclusively that any algae change their pigment ratios in a complementary adaptation to the light quality and Brody & Emerson (1959) could only show that the unicellular *Porphyridium cruentum* changed its pigment in the direction of complementary adaptation.

Artificial breakdown of the thermocline

The effects of formation of a thermocline have already been mentioned (p. 19) and two of the most significant are the rapid utilisation of nutrients in the epilimnion and the accumulation of nutrients (accompanied by declining oxygen concentration) in the hypolimnion. At first sight the breakdown of the thermocline should lead to increased plankton production since the nutrients in the whole water column would be available though light would be limiting for part of the day in deep basins. In a sense, upwelling areas of the oceans are comparable sites and under such conditions increased algal growth certainly occurs. Paradoxically, experiments to maintain isothermal conditions in closed basins have been undertaken by water purification authorities in order to *decrease* algal growth. It is also done to maintain oxygenation of the water and prevent hydrogen sulphide accumulation in the bottom waters of reservoirs. Two techniques have been used: the first, and the only one possible in natural basins, is to move the water by such means as air lift pumps or by release of air bubbles through perforated hosepipes laid along the bottom. The second technique can only be incorporated when building reservoirs; the inlet jets can be designed to give differing velocities and be angled to differing degrees with the result that the water is kept in circulation. An intriguing example of such a system is given by the Metropolitan Water Board Report (Windle-Taylor, 1971–3) where, in 1971 and in subsequent years, in the Queen Elizabeth II reservoir, the water supply was from high-velocity jets angled at $22\frac{1}{2}°$. From February to the end of July mixed conditions were maintained and the growth of diatoms and other plankton was sparse though the incoming Thames water contained silicon dioxide (12 mg l^{-1}), nitrate nitrogen (6.9 mg l^{-1}) and phosphate phosphorus (2.2 mg l^{-1}), which would certainly be sufficient for large growths under the stratified conditions which would normally occur. It is not clear exactly how this repression of growth is induced. A further interesting example of advantageous manipulation of the ecosystem occurred in August of 1971 when, in spite of maintaining mixed conditions, a population of the large centric diatom *Coscinodiscus* occurred (this is a rarely recorded genus in freshwaters) and to remove this the high velocity jets were closed for two weeks and the decreasing circulation resulted in sedimentation of the large cells.

Phytoplankton–Bacteria–Zooplankton relationships

Several workers have investigated the relationships between algal populations at various depths and the abundance of other groups of organisms. Many studies show that peaks of algal cell numbers down the depth profile are coincident with peaks of bacteria, e.g. Tanaka, Nakanishi & Kadota (1974) found a peak of bacteria and phytoplankton which used glycolate at 3–7 m depth in Lake Biwa (Fig. 7.86): it is interesting that one of the substances excreted by some algae is glycollic acid (*see* p. 316) and that Pant & Fogg (1976) found that glycollate uptake by *Skeletonema* is enhanced by the presence of bacteria. Bacteria are found on the surface of planktonic algae although, according to Overbeck & Babenzien (1964) most *Cyclotella* and *Stephanodiscus* are free from bacteria, a feature I can confirm from

Fig. 7.86. Bacterial biomass, filled triangles; chlorophyll *a*, open circles; photosynthesis, open squares; excretion of organic substances, filled circles, in the water column of Lake Biwa. Relative light intensity, ------. From Tanaka, Nakanishi & Kadota, 1974.

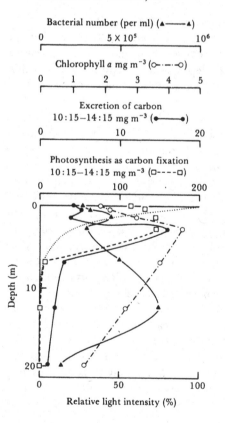

scanning electron microscope studies. Most of the bacteria occur free or on inanimate particles in the water but supported nutritionally by the excretion and autolysis of the phytoplankton (Overbeck & Babenzien, 1964). Bell, Lang & Mitchell (1974) grew *Skeletonema costatum* in the presence of a marine *Spirillium* and a marine *Pseudomonad* (Fig. 7.87) and found that the former declined as the alga moved from the log phase to the stationary phase of growth but then recovered as lysis of the algal cells occurred, whereas the *Pseudomonad* increased its growth in the presence of *Skeletonema*. Both bacteria could take up the extracellular material secreted by the diatom but the *Pseudomonas* showed a 40-fold higher rate over the *Spirillium*.

The formation of algal blooms by blue-green algae has been related to the presence of bacteria which supply carbon dioxide which might be utilised by the algae (Kuentzel, 1969). However Fogg & Walsby (1971) think that it is more likely that the relationship is brought about by oxygen consumption by the bacteria. Fogg (1969) points out that Cyanophyta blooms are more common in lakes with oxygen depleted hypolimnia and a correlation with dissolved organic matter is perhaps explicable 'on the basis that oxygen depletion is proportional to the latter'.

Summary

The classification of the various particles suspended, floating, moving and sinking through the water column is reviewed. Techniques of sampling the phytoplankton range from towing nets through the water to collecting discrete volumes, pumping water through filters, measuring cell numbers and volumes, chlorophyll and carotenoid contents, ATP contents, wet and dry weights and volumes, carbon, nitrogen, phosphorus contents, etc. and some of the problems involving such diverse measures are discussed. The composition of phytoplankton associations of various waters shows certain general characteristics and these, together with the use of

Fig. 7.87. *a*, the growth in batch culture of *Skeletonema* (filled squares) and a marine pseudomonad alone (open triangles) and in the presence of the alga (filled triangles). *b*, the growth in continuous culture of a marine *Spirillium* alone (open circles) and in the presence of steady state culture of *Skeletonema* (filled circles). Algal population fluctuation shown on left. Bacterial 'death rates' have been calculated and the difference between these and the culture dilution rate are indicated. From Bell *et al.*, 1974.

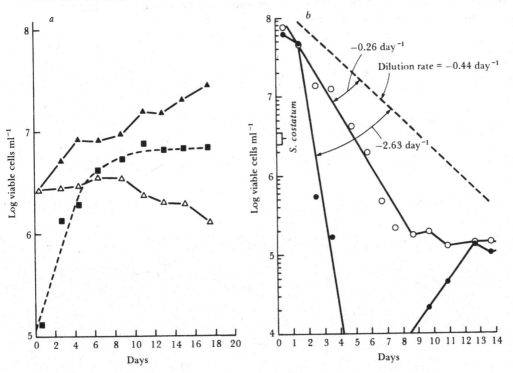

phytoplankton indices, are dealt with prior to a more detailed account of the phytoplankton of rivers, lakes (divided into cold water, temperate and tropical) and oceans (encompassing neritic–oceanic, cold, temperate and tropical, horizontal distribution). Phytoplankton also occurs in some surprising situations, e.g. in rock pools and in the surf, whilst other floating populations occur in the Sargasso Sea and associated with the surface films of all waters. Phytoplankton decays and the nutrients regenerated from this source are of great importance to the succeeding populations.

Sinking and floating of phytoplankton is an important topic and the theory, measurements and importance of these aspects are dealt with. Collections of rapidly growing populations in the surface waters form 'red tide' and 'water blooms'.

Phytoplankton populations are intimately linked with the stratification of waters and the effects of artificial breakdown of the thermocline, problems of 'sun' and 'shade' species, deep lying populations and carbon fixation down the depth profile are all discussed. The concepts of euphotic zone, mixed zone, critical depth, etc. are also involved in phytoplankton studies.

As in many other algal communities some of the species fix atmospheric nitrogen and the contribution of this to the water column is of some importance.

Secretion of organic molecules and uptake of these or others has received much attention from students of the phytoplankton and some of the effects are reviewed.

Whilst carbon fixation and oxygen output studies have been used extensively on phytoplankton associations, very little is known of the carbon fixation of individual species or of associations at depths other than those of collection and these too are discussed. Especially in the last decade, the effects of diurnal (diel, circadian, etc.) rhythms on metabolic uptake of nutrients, on vertical migration movement and on bioluminescence have become recognised and have made an important contribution to phytoplankton studies. These and depth profile studies have generally tended to show that all earlier estimates of phytoplankton production are too low.

Nutrient relationships of the phytoplankton have also been studied intensively over the last few decades and the results now enable some fairly accurate predictions to be made.

The relationship between phytoplankton, zooplankton and bacteria are intimately linked and no discussion of the plankton would be complete without mention of these. Grazing, parasitism and similar phenomena are dealt with in Chapter 9.

The twin problems of how a species spreads its progeny and the factors influencing its establishment and growth in geographically defined regions are major ecological topics. The continuity of species in habitats is a further aspect of ecological study. Evidence is often very difficult to find except for the macroscopic algae which are preserved on herbarium sheets and for some, mainly microscopic, which produce resistant walls or spores and, being thus preserved, can be obtained from sediments. Continuity over a long time scale is dealt with in the chapter on palaeoecology but of equal interest are the records of continuous occurrence in the more recent past. Many freshwater unicellular algae are widespread in both northern and southern temperate zones, often with a break in the tropics, whilst marine species have a tendency to distinct northern, tropical and southern elements. In both freshwater and marine environments, air-borne dispersal is a possibility. Marine species, however, have the opportunity of dispersion in a liquid medium across the warm tropical oceans, though high water temperatures may be a greater barrier to dispersal than warm air is to wind-borne freshwater species.

Distribution of the larger marine coastal species is more restricted, possibly owing to the relative lack of resistant spore stages and of any kind of aerial transport. Studies of dispersal require a knowledge of the life history of species and of the mechanisms of transport, whilst phytogeography is concerned with plotting the geographical limits of individual species, and with the factors which restrict species. There is no doubt that algal phytogeography is dependent upon a precise knowledge of species and, for some of the most common genera, such as *Fucus* and *Sargassum*, the taxonomic problems are considerable and all data must be examined very critically.* The geo-graphical limits of species are not difficult to determine, though the effort may be time-consuming and few species are adequately studied, whilst the determination of the limiting factors is certainly no simple task. It is all too easy to invoke temperature, salinity, and other factors, but experimental evidence is rarely forthcoming. There is no doubt also that ecotypes have evolved, so that the growth and dispersal of a species living in the North Atlantic Ocean, and another morphologically similar one living in the North Pacific Ocean, are not necessarily controlled by the same range of factors. Experimental verification is essential in studies of this kind and is equally necessary to test the relationship of morphological variants of some of the macroscopic forms to those of the species, e.g. the *cucullata* varieties of *Laminaria* and *Saccorhiza* (*see* p. 80).

It is also true that the information needed to compile distribution maps is limited by the fact that phycologists tend to work in restricted regions (usually near large research centres) and the amount of work on certain algal groups is sparse. Nevertheless there are many valid examples of confined distribution patterns especially of macrophytic marine, benthic species.

Dispersal

The occurrence of so many common freshwater species throughout the world is a reflection of ease

---

* A recent example is the demonstration that *Phyllogigas*, *Himantothallus* and *Phaeoglossum* are all growth forms of *H. grandifolius* and that this genus belongs in the Desmarestiales not Laminariales (Moe & Silva, 1977). The antarctic marine flora now has no members of the Laminariales. Modern techniques are also leading to valuable re-assessments of taxonomic entities e.g. Cheney & Babbel (1978) confirm by electrophoretic studies that four common species of *Eucheuma* can be reduced to two.

Table 8.1. *Genera of algae cultured from the atmosphere over Denton, Texas. Modified from Schlichting (1962)*

Ankistrodesmus	*Neochloris*-like alga
Aphanocapsa	*Oscillatoria*
Bracteacoccus	*Phormidium*
Chlorella	*Plectonema*
Chlorococcum	*Prasiola*-like alga
Coscinodiscus	*Protococcus*
Euglena	*Scenedesmus*
Nannochloris	*Spongiochloris*-like alga
Navicula	*Tetraedron*
Nitzschia	

of transport, yet for the majority there is no information on dispersal mechanisms.

Aerial transport

Air-borne algal spores and fragments have long been known to exist, e.g. Schlichting (1964, 1969) recorded some 187 taxa of algae and protozoa which had been discovered in air and he himself found about eight cells in each cubic foot. A total of 43 species of algae and protozoa have been cultured from 38 384 c. ft of air (Table 8.1).* Brown, Larson & Bold (1964) made counts of algal numbers in filtered air and found up to 3000 algae m^{-3}, the highest numbers in air laden with dust; in cleaner air, pollen and fungi were more abundant. It is interesting that this work and that of others with cultures from rain, e.g. Maguire (1963) reveal that most of the captured species are either nannoplanktonic forms or are soil algae. There seems to be no record from cultural studies of the large euplanktonic algae (e.g. *Asterionella* or the desmids) being air borne. However, Geissler & Gerloff (1965) have found live vegetative algae on air filters and Javorski & Lund (1970) found that *Asterionella formosa* could live on damp mud for two weeks, so presumably in damp weather such particles could be transported. I believe there is slight evidence for the aerial transport of planktonic diatoms in that, occasionally, live cells of species of *Cyclotella*, *Stephanodiscus* and *Melosira* have been recovered from moss and liverwort growth high

---

* The term 'aeroplankton' has appeared in the literature but is of doubtful value, since it may imply a degree of permanence and growth which has never been demonstrated.

---

up on trees, from leaf surfaces and from the water trapped in Bromeliads of tropical mist forests. In the moist conditions in these habitats the planktonic species may even divide. Live marine phytoplankton (*Pyramimonas*, *Coccolithus*) have been collected from the air 23 m above the ground and 60 km from the sea (Maynard, 1968), suggesting that at sea the plankton can be dispersed both above and below the water surface.

A rare chance to investigate dispersal on to virgin land occurred with the eruption of Surtsey in 1963. In 1968, samples were collected and investigated by Behre & Schwabe (1970) and although only a fairly small flora had reached the island and colonised the cooling volcanic material, they identified 106 taxa which must have reached the island by aerial transport since its eruption. All the species were microscopically small, and many were typical soil forms (49 of 69 diatoms); halophilic and filamentous species were lacking (except possibly some Cyanophyta), species were all occupying places where water collected and there was a high proportion of thermophilic species (Table 8.2). The large number of diatoms is striking, as is also the fact that almost all were also recorded at the same time from nearby Iceland.

Areas such as the Antarctic, which lack land bridges, can only be supplied with freshwater inoculum by airborne means, and the records, which are surprisingly numerous for such an isolated area (*see* Hirano, 1965) show very few endemic species and a very large number of cosmopolitan forms (*see* also p. 198). This is an area which deserves more detailed study, since it should indicate very clearly which species or groups are least capable of aerial transport; examination of the lists for Antarctica, for example, shows extreme paucity of Chrysophyta, Dinophyta and Charophyta, to name just a few groups. On the other hand, Cyanophyta, Bacillariophyta and coccoid Chlorophyta are common though the Cyanophyta increase in number of species from the offshore islands to the mainland, whilst the other two groups decrease dramatically.

Records of aerial transport of algae must be treated with great care since airborne dust often contains dead diatom frustules; in the Atlantic off West Africa, for example, the winds (Harmattan) carry dust from the Sahara bearing large amounts of *Melosira granulata* and other diatoms, which is even

Table 8.2. *The species cultured from soils or collected directly on Surtsey in 1968.*
From Behre & Schwabe, 1970

Cyanophyceae
  Aphanocapsa grevillei
  Anabaena variabilis
  Pseudanabaena sp.
  Oscillatoria amphibia
  Phormidium autumnale
  Phormidium mucicola
  Plectonema A
  Plectonema B (*gracillimum*)
  Schizothrix lardacea
Dinophyceae
  Gymnodinium sp.
Euglenophyceae
  Euglena mutabilis
Xanthophyceae
  cf. *Pleurochloris magna*
Chlorophyceae
  Chlamydomonas asymmetrica var. *gallica*
  Chl. augustae
  Chl. foraminata
  Chl. gloeopara
  Chl. intermedia
  Chl. pseudintermedia
  Chl. sp. ad *perpusilla*
  Chlorella minutissima
  Chl. vulgaris
  Chl. sp.
  Muriella terrestris
  Chlorococcalean sp. *a*
  Chlorococcalean sp. *b*
  Gloeotila protogenita
  Stichococcus bacillaris
  St. minor
Bacillariophyceae
  Gomphonema angustatum + var. *productum*
  G. constrictum
  G. parvulum + var. *micropus*
  Epithemia sorex
  Hantzschia amphioxys
  Nitzschia amphibia
  N. communis var. *hyalina*
  N. fonticola
  N. frustulum
  N. frustulum var. *perpusilla*
  N. hantzschiana
  N. microcephala
  N. palea
  N. perminuta
  N. recta
  N. subtilis
  Surirella ovata

Melosira italica
Cyclotella meneghiniana
C. striata
Stephanodiscus astraea
  + var. *minutulus*
Coscinodiscus lineatus
Diatoma elongatum var. *tenue*
Fragilaria construens
  + var. *venter*
Fr. pinnata
Synedra ulna
Eunotia exigua
Eu. lunaris + var. *subarcuata*
Eu. pectinalis var. *minor*
Cocconeis pediculus
C. placentula + var. *euglypta*
Achnanthes affinis
A. hungarica
A. lanceolata
A. minutissima
  + var. *cryptocephala*
Stauroneis anceps
St. Borrichii f. *subcapitata*
Navicula atomus
N. avenacea
N. bacillum
N. clementis
N. cocconeiformis
N. cryptocephala
N. cryptocephala var. *veneta*
N. decussis
N. cf. *dismutica*
N. gregaria
N. hungarica
N. integra
N. minima
N. mutica + f. *cohnii*
N. oppugnata
N. pelliculosa
N. pupula f. *rectangularis*
N. salinarum
N. seminulum
N. vitabunda
Pinnularia borealis
P. intermedia
P. interrupta f. *minor*
Caloneis fasciata
Amphora ovalis var. *pediculus*
Cymbella sinuata var. *antiqua*
C. ventricosa
Gomphonema acuminatum
  + var. *coronatum*

deposited in sufficient quantity to be recognisable in cores from the Atlantic (Kolbe, 1955). The first records of algae in these winds is in Ehrenberg's *Microgeologie* (1854) and Fig. 5.26 is taken from that classic publication. Dust was investigated by Hustedt (1921) and again by Folger, Burkle & Heezen (1967) but little new has been added since the early records. Geissler & Gerloff (1965) not only found diatom frustules in air whenever they sampled (100–7795 in 500 m³ over a ten-day period) but they also found them in various tissues of the human body; these enter via the air in the lungs and also through the wall of the gut.

The foams formed along the shores of lakes and along seashores contain viable algae and this is another way of 'escape' from the liquid phase.

Animals

Obviously any aquatic animals or other animals alighting on or swimming through water, are potential dispersal agents. If the surface of aquatic beetles is washed and the washings spread on an agar plate, then a whole series of algae can be cultured (Schlichting & Sides, 1969; Schlichting & Milliger, 1969). As many as 91 micro-organisms were cultured from the washings of the external surface of the hemipteran *Lethocerus uhleri*. Generally however the counts are not as high as this and the numbers cultured from the guts tend to be 2–7 times greater than those recoverable from the outer surfaces (Milliger & Schlichting, 1968). Equally the excreta of aquatic insects, fish, aquatic birds, etc. will usually yield a range of viable algae which have passed through the gut without digestion. Thus, Atkinson (1972) found *Scenedesmus*, other Chlorococcales and small naviculoid diatoms in cultures from the guts of wild fowl. She also fed live plankton to mallard but could only retrieve *Melosira* live from the faeces. Velasquez (1940) found viable algae in faeces of the gizzard shad and, since this is a migratory fish, transport will be quite extensive. In habitats such as rock pools the continual passage of the algae through the aquatic fauna almost certainly has a selective effect; the smaller the size of the habitat the greater such effects tend to be. However in all these instances the larger euplanktonic forms tend to be eliminated.

Direct flow

There are many examples of lakes flowing into other lakes with the inevitable transference of species, but rarely is even this continual source of inoculum of any importance, e.g. Belcher & Storey (1968) found that when peaks of cell numbers occurred in an upper body of water (Mere Mere), cells could always be found in the lower water (Rostherne Mere), but they never greatly influenced the composition of the population. Sampling at ends of the connecting stream showed a large decrease of the Mere Mere plankton during flow down the stream and this is a feature which has been shown many times (*see* p. 269) and emphasises the subtle link between plankton and turbulent motion in standing water. Lentic plankton does not seem to be capable of survival under the more laminar flow conditions of the stream environment.

Studies in restricted areas where the operation of all the above dispersal mechanisms is at a maximum might be expected to reveal a high degree of similarity in the flora of neighbouring lakes, yet Behre (1966) found that, of 953 algal taxa occurring in nine small North German lakes, only 21 species were common to all nine, and even some of these were regarded as 'collective' species and so the forms common to all lakes are even fewer. This is of course simply a reflection of the 'chemical selectivity' of the waters, superimposed on the dispersal pattern.

Artificial dispersal

Dispersal on ships' hulls and by man who transports foods and animals, has obviously been an important factor in the past and is likely to increase in the future. The recent arrival of *Ulva reticulata* and the earlier arrival of *Ulva fasciata* on Hawaii is presumed to have been on ships' hulls (Doty, 1973*b*). However, this genus is absent from the Central Pacific atolls and high islands which it may just not have reached, although the lack of the necessary high nutrient water may be an added factor preventing establishment. Certainly it has increased in quantity in Hawaii as the population has increased (*see* p. 532). Other recent introductions in Hawaii are *Acanthophora spicifera* and *Nematocystis decipiens*; the latter has been very carefully followed since its appearance in 1963 and its spread to different hosts and positions on the shore plotted (Doty, 1973*b*). The recent appearance of the very distinctive diatom, *Hydrosera triquetra*, in the Thames estuary is probably another example of ship transport since it is otherwise known only from tropical estuaries; it is surprising that conditions in the River Thames are suitable! Whilst certain genera, e.g. *Enteromorpha*, readily attach to ships and germlings of other genera live amongst them, it is doubtful whether any large alga could withstand the long periods of rapid water flow along the hull.

Current drift

Macroscopic algae even as large as *Macrocystis* are reported to drift in the open sea after detachment during storms and/or after deterioration of the holdfast, often accelerated by grazing (North, 1971). Identification of drift specimens of this genus must be treated with caution since the fronds are somewhat variable and the holdfast is the only reliable characteristic of the species (Womersley, 1954). Other long distance transports are the records of West Indian *Sargassum* spp. off the southwest coast of England and of *Gelidium cartilagineum*, a species with its northern limit in the Canary Islands, off Shetland, Heligoland and Schleswig Holstein (Dixon, 1963). Drift is not confined to northerly flow and John (1974) found clumps of the European *Ascophyllum nodosum* in the Atlantic Ocean south of the equator. Dispersal of populations of coastal marine algae is generally presumed to be mainly by release and dissemination of spores although, as Dixon (1965) comments, vegetative fragments may be of equal importance. Algae which form small balloon-like thalli are obvious contenders for dispersal in currents. One such, *Colpomenia peregrina* has spread from the English Channel, where it was recorded at the turn of the century, throughout the British Isles and in the last decades to Denmark and Norway. Obviously one must ask why such an alga is spreading only during this century. Presumably its spread is related to climatic changes and it is interesting that over the same period the flora of the arctic island of Spitsbergen has also increased both in species number and biomass. This may be a temperature effect but, as Dixon stresses, matters may not be as simple, for at the same time other species seem to be spreading southwards.

That there are many subtleties in dispersal is shown by the fact that *Ulva reticulata*, which sheds pieces of its thallus after formation of zoospores or gametes, can shed the zoospore patches before all the zoospores are released and presumably the patch can drift away and then release zoospores (Kemp, personal communication). There is surprisingly little information on the longevity of spores; they often appear to the microscopist as a rather delicate structure but the walls are probably as resistant as those of many floating algae.

Russell (1967a) showed that species of Ectocarpaceae, which are normally attached plants, can be kept in a free-living state in the laboratory for some two months, during which time they can reproduce vegetatively and also form sporangia; this indicates a considerable potential for dispersal without the host or substratum. Indeed, Burrows (1958) records many species (*see* p. 101) existing as loose lying populations in Port Erin Bay, so viability of such species is not dependent upon attachment.

However many coastal algae are infertile over large geographical areas or at the extremes of their range. Their dispersal is dependent upon vegetative propagation of fragments and it is worth noting that many Phaeophyta and most Rhodophyta exhibit remarkable capabilities for regeneration from fragments, and even from single cells. Careful examination, especially of the lower parts and rhizoidal regions of many of these algae, often reveals regenerating thalli. The limited spread of some species beyond the range in which they are fertile, indicates that conditions for vegetative growth are wider than for reproduction and such features merit careful study. However species such as *Gelidium* which is virtually sterile all around the British Isles might need rather special conditions for reproduction; alternatively the material around Britain may be a non-reproductive population in which the loss of reproductive capacity is perhaps genetic rather than environmentally controlled. There is an obvious need for experimentation here.

Many red algae have a more extensive spread of the tetrasporophyte generation than of the gametophyte, e.g. *Ceramium shuttleworthianum* and *C. flabelligerum* around the British Isles (Edwards, 1973). The absence of the gametophytes from North Scotland is attributed to inhibition of gamete production. At even more northerly stations no sporangia form at all. The occurrence of these plants must be due to vegetative propagation, or from spores drifting northwards. The tetrasporophytes were shown to be capable of producing viable gametophytes in culture, which suggests that environmental conditions are preventing the establishment of the gametophytes. There is also the possibility that meiosis is inhibited in the northern tetrasporic plants and the spores perpetuate the diploid generation; this has been reported for *Antithamnion boreale* (Sundene, 1962*b*).

There is still much to be learnt of the life cycles of marine algae, e.g. only recently has the tetrasporophyte stage of *Ahnfeltia* been discovered as an encrusting stage (Farnham & Fletcher, 1976), whereas that of the even commoner *Gigartina stellata* has not yet been recorded for British coasts although if it is similar to the Pacific material it must be sought for as an encrusting *Petrocelis*-like stage.*

A Rhodophycean alga which has been carefully studied is *Audouinella* (*Rhodochorton*) *purpureum* which Knaggs (1967) found was confined to localities with surface temperatures below 20 °C. Subsequent work (West, 1972) shows that although plants can be grown at temperatures outside their normal range, e.g. cold-water clones can be maintained at 20 °C and warm-water clones can be maintained at 5 °C, the former will not sporulate above 20 °C and the warm-water clone will not sporulate at 5 °C. However it is even more complicated in that two cold-water clones (from Alaska and Washington) would not sporulate above 10 °C but one cold-water strain (from Chile) would sporulate from 5 to 15 °C. West concluded that this temperature dependence is probably genetically selected in each clone according to the locality from which it was taken. This and other evidence (e.g., salinity tolerance of *Audouinella*, *see* p. 94 and photoperiodic effects, *see* p. 94) show that morphologically similar, but genetically different populations occur in geographically separated sites.

Appearance

The actual date of appearance of a species in a region for the first time is often impossible to determine but a few examples are documented. For example, the red alga *Asparagopsis armata* is generally regarded as originating from Australia. It was first recorded in the British Isles near Galway Harbour in 1939 in its tetrasporangial phase (*Falkenbergia rufolanosa*) and later as the gametangial plant at Muigh Ines, near Canna in 1941. It was not until 1949 that it was found in England, at Lundy Island (*Falkenbergia*) and in 1950 at the Lizard (*Asparagopsis*). The closely related *Bonnemaisonia hamifera* originates from Japan and became established in the North Atlantic at the end of the nineteenth century. Both *Asparagopsis* and *Bonnemaisonia* are only occasionally fertile in any stage and its seems likely that their spread is by vegetative means. There is a consensus of opinion that the climatic variation around an area such as the British Isles is so slight and so gradual that no real physical boundaries exist. However this is a rather negative approach and many examples can be found of restricted occurrence, especially of Lusitanian species extending northwards, e.g. *Phyllophora palmettoides* reaches the southwest and penetrates up the west coast to Anglesey whilst the Northern species *Phyllophora truncata* extends from its circumpolar distribution to Caernarvon and Northumberland on the eastern seaboard and to New Jersey on the western seaboard; obviously it is necessary to consider species which are more abundant elsewhere if one is to detect subtleties of geographical limits in such a small area as the British Isles. A further factor which may influence spread of species especially near their limits is the existence of regions of unsuitable substrata, e.g. the east coast of England between the Humber and Thames is very silted and has comparatively few sites for good epilithic growth and so a species which depends on small 'leaps' for its dispersal my be stopped by such a region. However, silted coasts are also frequently neglected by phycologists studying epilithic algae, yet when they are visited it is surprising how many species can be found attached to isolated substrata; good examples are recorded for the coast of Holland.

Another recent, well-documented case of rapid dispersal is that of *Codium fragile* subsp. *tomentosoides* on the Atlantic coast of North America. It appeared in the late fifties and is now expanding from its considerable base, encompassing Maine to New Jersey (Malinowski & Ramus, 1973). This is an alga which can grow on oyster shells and it has already caused considerable commercial problems by its rapid spread along the New England coastline. Being an introduced species it has no known predator (Malinowski & Ramus, 1973), though it does not appear to compete well with other benthic algae. It

---

* Such a stage has been produced in culture from British material (Irvine, personal communication).

is not known for certain how this species reached the North American coast, but transport on commercial shellfish from Europe is suggested (Churchill & Moeller, 1972). One of the most recently documented accounts of dispersal is that of the Japanese *Sargassum muticum* along the Pacific coast of Canada and the United States and its recent appearance in the British Isles off the Isle of Wight from where it is now spreading along the English Channel. Its presumed mode of transport is in the packing around commercial oysters and its spread is creating serious problems since it can outcompete some of the indigenous species and grow so prolifically that it clogs up shallow harbours, etc. A similar case of dominance via competition is the growth of *Caulerpa filifermis* on the outer edge of rock platforms north of Sydney since its first recording in 1923 (May, 1976). Examples of the first appearances of some microscopic species in lakes of the English Lake District are given on p. 532, and other records can be found from very careful study of post-glacial deposits and occasionally, when they are very recent, they can be linked with observations on the live flora, e.g. the sudden occurrence of a small *Stephanodiscus* sp. in Blelham Tarn (Haworth, personal communication). Another *Stephanodiscus* which is well documented is *S. binderanus* which began to grow in the Laurentian Lakes in the 1930s where it was presumably introduced by vessels from Europe. A further example from this region is the appearance of *Bangia atropurpurea* on rocks in Lake Michigan in 1968 and by 1976 it was colonising all suitable substrata around $\frac{2}{3}$ of the lake and had displaced the common green alga *Ulothrix zonata* (Lin & Blum, 1977).

Continuity

If a species produces resistant wall material, either in the vegetative or in the spore stage, there is a possibility of preservation in sediments and its continuity in the habitat can be checked. Otherwise the only means of establishing the continued presence of a species is from scientific (historical) records. Few such reliable records extend beyond the last hundred years. Whitton (1974) quotes the occurrence of *Chara cuspida* in a Durham pond where it has been the dominant organism since 1775, a feature which indicates the resistance of such a species to increasing enrichment of its environment. Mere lists of species are not completely reliable indications of continuity;

however, if accompanied by illustrations, they can be used, e.g. the author found that the forms of some planktonic algae in Irish lakes sampled in 1953 were identical to those photographed by the West brothers in 1906. Such records coupled with those from post-glacial deposits suggest that, in general, algal floras are stable over great lengths of time, except when drastic or rapid changes in the environment occur. Similar observations have been made in flowing water: a study of the River Tees in 1963–5 (Whitton & Dalpra, 1968) showed very few changes in the attached flora when compared with data from an earlier survey (Butcher, Longwell & Pentelow, 1937): only five obvious changes in abundance were noticed among the 148 species recorded and none involved dominant species. Firth & Hartley (1971) found that the diatom floras of three small Pennine streams were virtually the same over a period of 32 years.

Few attempts have been made to study the continuity of marine epilithic floras: one suspects that they are recognisable over long periods of time but data are sparse. Recently, Michanek (1967) re-sampled the sites in Gullmar Fjord where Gislen did his classic study in 1926 and found that similar communities were present, though he found that there were others which had not been described earlier: possibly the differences were due to the use of different sampling techniques in the two studies.

A most remarkable record of continuity is that described by Powell (1966) for a patch, 0.25 m² in area, of *Codium adhaerens*; this was recorded on the lower shore of a small island in Western Inverness-shire in 1927 and re-found by Powell in 1965 on the same two adjacent boulders! It can only have survived because the shore is very protected and the boulders relatively immobile. What is even more remarkable is that *C. adhaerens* is extremely rare in the British Isles and at this point is at its northern limit of distribution. If it can maintain itself for at least 38 years at its geographical limit, why then has it not spread southwards? Powell noted that no gametangia have been detected here or at its other Scottish sites. In all probability the plants are sterile and since this *Codium* sp. does not have any form of vegetative propagation it is difficult to understand how it is distributed.

Although so few records of continuity over more than 100 years are available, wherever it has

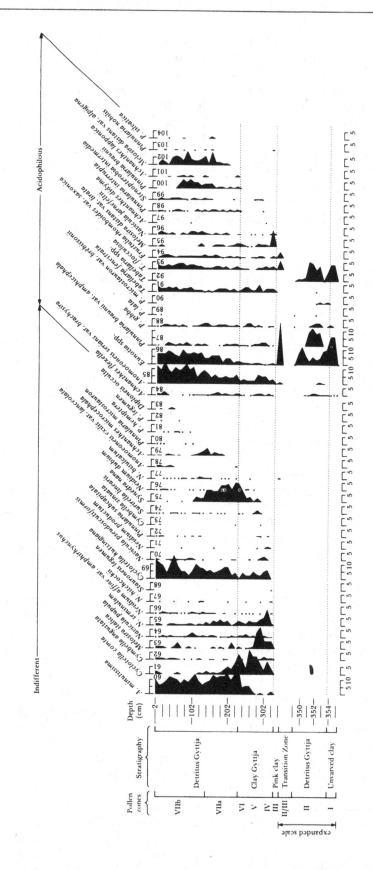

Fig. 8.1. A section of the diagram of diatom remains in a core from Blea Tarn, showing the constant occurrence of many species from the end of the late glacial (zone III) to the present day. From Haworth, 1969.

been possible to check the algal flora in a site it does seem to be relatively stable over much longer periods of time, a feature which contradicts the impression of most casual work. Stability is to be expected in systems which are continually open to inoculation and it is in fact probably difficult to introduce new algae (at least microscopic species) into sites where they do not already exist. The concept of a fairly rapidly changing flora as lakes and rivers mature during the post-glacial period is misleading (*see* also p. 526). An example of such continuity is revealed in the study of Blea Tarn (Lake District) which is situated in an area of minimum disturbance by man and Fig. 8.1 shows how constant the flora has been (Haworth, 1969). A similar constancy is recorded for Lough Neagh where Battarbee (1973*b*) noted that *Melosira italica* subsp. *subarctica* extends back well over 4000 yr. On the other hand, Lake Windermere and Esthwaite Water, being more prone to disturbance, have undergone slightly more change. During the last few decades, changes induced by man have accelerated, and one must not expect the stability and continuity of the last 10000 years to continue (cf. the rapidly changing acidity of rainfall in some areas and the possibly more rapid climatic fluctuations of recent times).

Three groups are particularly well suited to studies of continuity in freshwaters: diatoms because of their siliceous walls, desmids since they are well preserved and also often well illustrated, and charophytes which can be preserved as herbarium material and therefore readily checked. All three, however, have considerable taxonomic confusions and reliance on mere names in the literature is dangerous. Comparison of data collected at single times is obviously inadequate to indicate absence from a locality. Indeed, collections over periods of years are necessary, e.g. Fig. 10.21 shows that very few freshwater algae are present throughout the year and some occur in only one year in three. The reasons for such disjunct appearances are unknown; it is unlikely that lack of transport is the factor.

Phytogeography

There are two aspects of phytogeography which are interrelated but sometimes get confused. The first of these is the determination of the broad algal phytogeographic zones of the world and the second the geographic limits of individual taxa.

The phytogeographic zones are areas in which relatively homogeneous floras occur separated from one another by floristic discontinuities. Ideally they are based on a consideration of the whole flora. The floristic discontinuities *may* be coincident with topographical discontinuity, e.g. the separation of Antarctica from other land masses or they may be discontinuities caused by physical features of habitats, e.g. long stretches of sandy coasts preventing the spread of epilithic species, or temperature, salinity, boundaries, etc. Topographical discontinuities (Fig. 8.2) in the North Atlantic, e.g. between Norway, the Faroes and Iceland do not result in very large differences in the species complements and can be thought of as subzones within the general scheme. The floristic data determining the discontinuities must reveal a number (unspecified) of species confined to each zone. It is not possible to specify the number, which will vary according to the richness of the flora, but a tabulation such as that in Fig. 8.3 reveals 'steps' and 'holes' indicative of some fairly major and other lesser changes in environmental conditions. If a number of species react in a similar manner, some disappearing and others appearing, subzones within the broad phytogeographical framework can be determined and these are often termed provinces. Study of the ranges of individual species, especially in a limited area, gives the undoubtedly correct impression of a continuum of change with hardly any two species having exactly the same extension (e.g. Fig. 8.4 which shows the northeast Pacific *Laminaria* flora) and the limits of individual species cannot be expected to coincide with the broad phytogeographic zones which are abstractions from the total floristics of the regions.

Marine algae: macrophytes

In spite of the fact that dispersal of algae by water currents, ships, animals, air-borne particles etc. is fairly extensive, there are in fact few universally distributed algae. This can be attributed to the relatively narrow ecological requirements of most algae and to the barriers which are imposed between geographical zones; a simple example is the occurrence of distinct arctic and antarctic algae, in spite of the North–South continuity of the Atlantic and Pacific Oceans. On the simplest basis this can be explained by invoking a temperature barrier to passage through the waters of the tropics.

Phytogeographic studies of marine algae are relatively rare, a fact which is partly attributable to the difficulties of undertaking such studies. Determining distributions from the literature is fraught with difficulty and strictly any study should be based on checked material, either in herbaria or by actual collection; both approaches are time consuming. Dixon (1963) gives excellent examples of the dangers of relying on published records, e.g. the occurrence of *Pterocladia capillacea* at Scarborough on the York-

Fig. 8.2. *a.* The phytogeographic regions and provinces along the northern Atlantic coasts. *A*, warm temperate Carolina region; *B*, tropical Western Atlantic region; *C*, Canaries province; *D*, Mediterranean province; *E*, Lusitanian province of warm temperate Mediterranean–Atlantic region; *F*, Eastern province; *G*, Western province of cold temperate Atlantic–Boreal region; *H*, Arctic region. *b.* Surface temperatures in February (thin lines) and August (thick lines). From van den Hoek, 1975.

a

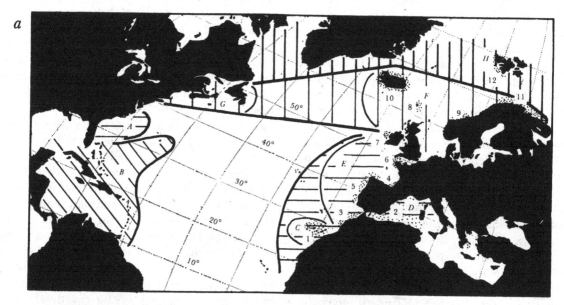

b

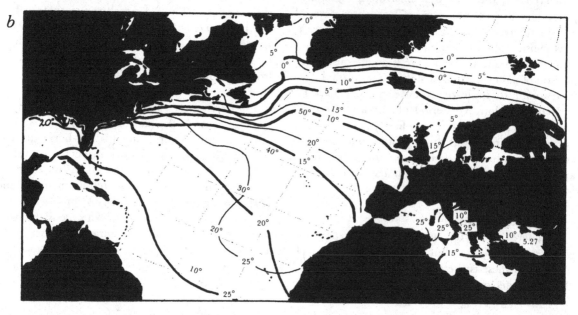

shire coast is based on the repetition of records of *Fucus pinnatus* which was assumed to be *P. capillacea* but has now been shown to be another taxon and this misidentification has put the distribution of *P. capillacea* some 300 miles north of its true limit. Another example is that of *Helminthocladia hudsonii*, which has been reported from various localities in the British Isles as far north as the Shetlands, but Dixon has shown that these reports are based on no less than six different taxa and in fact the northernmost limit of *H. hudsonii* is Tangier not Shetland!

Phytogeographic zones can be discerned, at least in the sea, by fairly detailed analysis of species (*see* p. 376 and Lawson (1978) for a discussion in which he concludes that reciprocal averaging was

the best technique whilst the best classificatory technique was indicator species analysis) but for a general discussion it is advantageous if zones can be designated by means of relatively simple physical features, which in themselves have physiological significance for the algae. Temperature is the common parameter which is often used, although it has the unfortunate feature, especially in temperate regions, that it is seasonally variable; however this factor itself may be utilised since species may be delimited by summer highs or winter lows. The highs and lows probably have a greater impact on the shore vegetation than on the plankton, and on the shore, air temperature may override sea temperature.

In the marine environment the chemical conditions are very similar over much of the surface and along many coasts, but temperature is a very pronounced variable: average water temperatures vary from around 0 °C to about 27 °C and such temperatures are bound to affect algal distribution. The magnitude of the annual fluctuation of temperature may also be an important factor and clearly the range which algae must tolerate is small in polar and equatorial waters and greatest in the temperate zone; it is of course small compared to the range

Fig. 8.3. The distribution of benthic macrophytic algal species along the Atlantic coast from Morocco to Spitzbergen. Each horizontal bar represents one species. Numbers refer to the zones on the Atlantic coast shown in Fig. 8.2. From van den Hoek, 1975.

Fig. 8.4. The distribution of *Laminaria* species in the northeast Pacific. Broken lines indicate regions where intervening populations are not known. 1, *L. groenlandica*; 2, *L. longipes*; 3, *L. dentigera*; 4, *L. yezoensis*; 5, *L saccharina*; 6, *L. setchelli*; 7, *L. ephemera*; 8, *L. complanata*; 9, *L. sinclairii*; 10, *L. farlowii*. From Druehl, 1968.

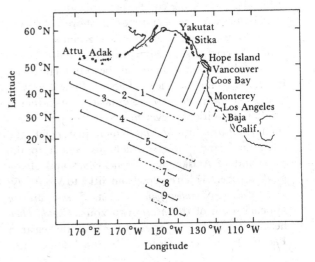

which land plants must tolerate. These ranges are for open water and they must be extended considerably for intertidal situations. Since it has been shown experimentally (*see* p. 111) that subtidal algae are more susceptible than intertidal algae to extremes of temperature, it has been argued that geographic distributions controlled by temperature are likely to be most distinctive in the subtidal flora; unfortunately this is often the least studied zone. Scagel (1963) found that the subtidal and lower intertidal species were the best indicators of the phytogeographic zones along the northeastern Pacific coast since they appeared to have the least flexibility. On the other hand the lower intertidal and subtidal zones are regions where annual fluctuations are minimal and therefore it could equally be argued that geographic extension in the subtidal zone would be potentially much greater than in the intertidal zone. Obviously there is an interplay between the resistance of the plants to extremes of heat and cold and the ability of algae to adapt to temperature ranges which are less subject to extremes. Adey (1971) found that the crustose coralline algae (Rhodophyta) of the North Atlantic Ocean could be separated into two species groups according to their winter temperature requirements, a subarctic group confined to winter water temperatures in the range 1–4 °C (*Clathromorphum circumscriptum*, *Leptophyton laeve*, *Lithothamnium glaciale*, *Phymatolithon laavigatum*) and a boreal group requiring higher winter temperatures (*Lithophyllum orbiculatum*, *Phymatolithon polymorphum*, *Lithothamnium sonderi*, *Heteroderma* spp.).

The phytogeography of the macroscopic epilithic algae along the coasts is complicated by their association with certain types of shore substrata, degree of wave action and vertical zonation on the shore. Nevertheless, over wide geographic ranges, comparable habitats can usually be found. However, in studying the northeast Pacific, Scagel (1963) found no sharp floristic boundaries but rather a gradual transition: nevertheless there are species such as *Arthrothamus bifidus* which just reach the Aleutian Islands, others which just reach to the mainland of Alaska, e.g. *Chorda filum* and *Alaria fistulosa*, others which extend from Sitka to Monterey, e.g. *Macrocystis integrifolia*, whilst *Eisenia arborea* appears south of the *Macrocystis* zone. *Chorda filum* here is a subarctic species though in the eastern Pacific Ocean it extends its distribution to the south

of Japan and in the Atlantic Ocean it extends in an arch from New Jersey to the coast of Portugal. Later Druehl (1970) was able to define limits for the distribution of the large laminarians on the eastern Pacific seaboard (Fig. 8.5) and these data can be compared with those discerned from plankton. Thus there is a southern segment with its great diversity of species belonging to the Lessoniaceae covering the planktonic transition zone whilst the equatorial zone south of Puerto San Eugenia has only one species of this family (*Eisenia arborea*) extending into it for any distance. A difficulty with all coastal phytogeography is that isolated records often occur outside the main zone of growth; thus *E. arborea* also occurs as isolated populations off British Columbia. In such instances it is desirable to check on the physiological tolerance of the two populations, since although morphologically similar they may be physiologically distinct. A number of laminarian species die out between Monterey and Point Conception and this is at about the northern limit of the central water mass defined by the plankton assemblages. Species such as *Pterygophora californica* and *Agarum fimbriatum* extend up to the region of the Queen Charlotte Islands and then die out, whilst northern species such as *Chorda filum*, *Laminaria longipes*, *Alaria fistulosa*, *A. praelonga* extend southwards to this point (Fig. 8.5). Similar boundaries detectable from water masses, plankton asemblages and coastal species also occur midway along the eastern Australian coast and between Honshu and Hokkaido in Japan.

In the Pacific area there is a distinct loss of species as one moves from west to east, e.g. *Sargassum* apparently reduces from 20 to 30 species around the large western islands to three in Hawaii and there are three *Bornetella* species in the west and only one in Hawaii (Doty, 1973b). The Philippines would appear to be the centre from which many species have spread outwards in the Pacific Ocean. The section *Eu-Sargassum* is tropical/sub-tropical in distribution, whilst other sections extend as far as the cooler waters of southwest Australia. Likewise there is a proliferation of *Caulerpa* species in this region, mixing with the endemic species of the area and the elements which have penetrated northwards from the colder Antarctic regions. This is thus one of the most interesting areas for a marine phycologist since there is a merging of tropical and antarctic species coupled with a high degree of endemism. Fortunately,

Fig. 8.5. *a*, distribution of laminarian species in the southern segment of the phytogeographical regions recognised in the northeast Pacific (*b*). From Druehl, 1970.

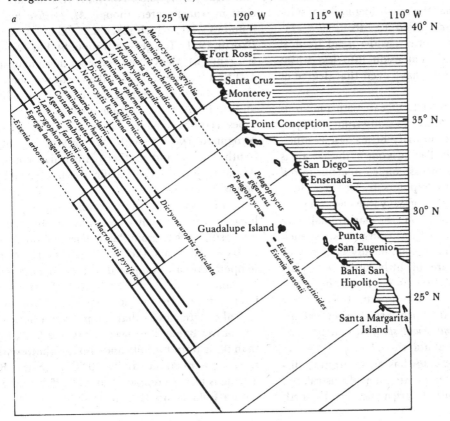

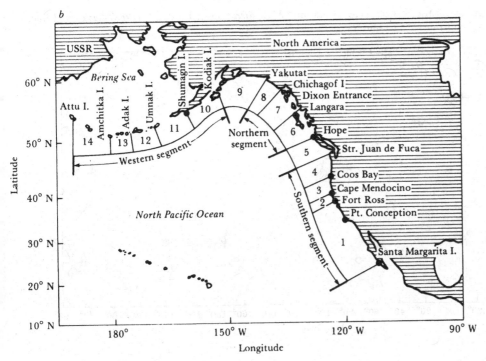

it is being studied extensively by Womersley and his students in Adelaide. Without going into any details it is perfectly clear that the macroscopic Bryopsidophyceae (e.g. *Halimeda, Penicillus, Avrainvillea, Tydemania, Caulerpa,* etc.) form a widespread tropical group with only slight extension into less warm water and not all species have even that degree of tolerance.

Areas such as the Pacific Ocean, with numerous isolated islands, are difficult to study from a biogeographical aspect; the linearity of a coast is much easier. Nevertheless the Pacific is one of the few areas where the data are being assembled and put into a form to be handled by computers, by Dr Doty. Such a massive undertaking is of great value and will yield the only detailed data on the latitudinal and longitudinal spread of algae in any area. In such work it is obvious that taxonomically confused or widespread genera are of little use: Doty lists *Sargassum, Ceramium, Udotea, Eucheuma* as being hopeless for such studies in the present state of their taxonomy but distinctive species such as *Tydemannia expeditioneis,* and *Acanthophora spicifera* can be used and have a clear distribution.

The genus *Macrocystis* has an interesting distribution; it occurs in temperate zones of Australasia, South Africa, and South America and extends north of the equator only along the Pacific coast of North America (*see* Fig. 8.6). Normally it seems unable to exist in warm water, though *M. pyrifera* has been found in water which reaches 26 °C in late summer (North, 1971). It is generally thought that *Macrocystis* is a genus which evolved in the southern hemisphere and that its extension into the north was via a cold water 'bridge' connecting the coastal zones of North and South America during quarternary or earlier times (Hubbs, 1952). There are three generally recognised species, with *M. angustifolia* only on South African and southern Australian coasts and *M. integrifolia* off the Pacific coasts of North and South America.

Macrocystis pyrifera exhibits slight morphological variation when populations from warm-water sites such as from Bahia Tortugas off the coast of Baja California (28° N, 20–26 °C summer sea temperature) are compared with the populations off La Jolla, 600 km north (33° N, 18–22 °C summer sea temperature). In addition there are clear physiological differences in that the photosynthetic capacity of the Bahia Tortugas plants was 50% higher than those from La Jolla and also the photosynthetic rates were maximal at 25–30 °C from the Bahia Tortugas plants compared to a 20–25 °C range for the La Jolla plants (North, 1972).

Fig. 8.6. The distribution of *Macrocystis* species. Solid line, *M. pyrifera*; dashed line, *M. integrifolia*; dotted line, *M. angustifolia*. Compiled from various sources.

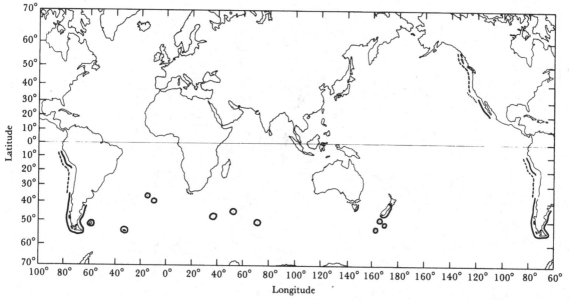

Fig. 8.7. The world distribution of three genera of coralline red algae, illustrating generalised Arctic, Pan tropical and Antarctic distributions. *a*, *Chlathromorphum* sp., *b*, *Lithoporella* sp.; *c*, *Pseudolithophyllum* sp. From Adey, 1970.

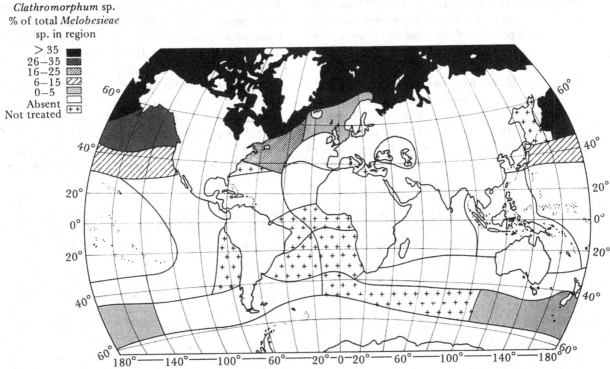

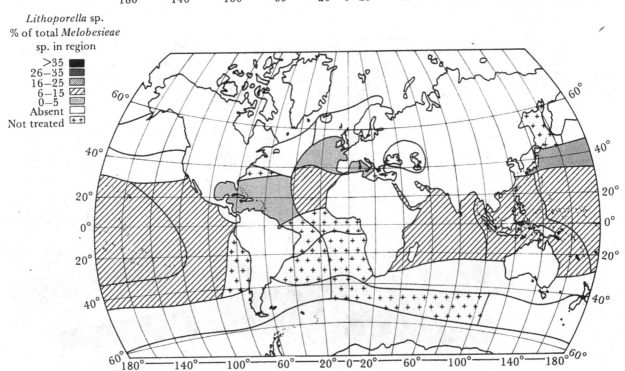

· The distribution of coralline red algae (Fig. 8.7) shows a very distinct latitudinal separation of the genera with arctic (*Clathromorphum*) pan-tropical (*Lithoporella*) and antarctic elements (*Pseudolithophyllum*). The spread of the predominantly antarctic genus and bipolarity of the *Clathromorphum* are the result of the method of analysis considering whole genera: the distribution of individual species is almost certainly blurring the picture.

To create a framework for further study many more compilations such as that shown in Table 8.3 from Powell (1963) are needed: such tables should preferably be based, as this example is, on detailed knowledge of the genera in order to discriminate amongst the data in the literature. This table shows the lack of penetration of many species into the Baltic Sea and the Mediterranean Sea. Information on abundance is the next requirement, e.g. of *Ascophyllum* at its most southern station in the Azores, where it is necessary to determine whether or not this is simply a record of a few isolated plants or of a reasonable, stable population.

Many elements of the temperate west European flora die out before the coast of Morocco is reached; Feldmann (1955) noted, for example, that *Pelvetia, Ascophyllum, Fucus serratus, Himanthalia, Laminaria digitata* and *L. hyperborea* are absent from that coast whilst *Fucus vesiculosus* and *F. spiralis* are rare. The restriction of most laminarian genera to cool and cold waters may be a simple temperature effect, since Sundene (1962c) found that *Alaria esculenta* is almost confined to waters where the mean August temperature is below 16 °C and when transplanted to regions of higher temperature the alga deteriorates, whereas transplanting to water of lower temperature does not impair growth. *Laminaria hyperborea* is at its southernmost limit in Portugal where summer temperatures reach 19–20 °C and Kain (1971) found that the optimum temperature for gametophytes and sporophytes is 10–17 °C; above 17 °C they survive, but when 20 °C is reached no sporophytes survive under experimental conditions.

There is a group of species which form a distinctly northern element in Europe and reach

Fig. 8.7. (*cont.*)

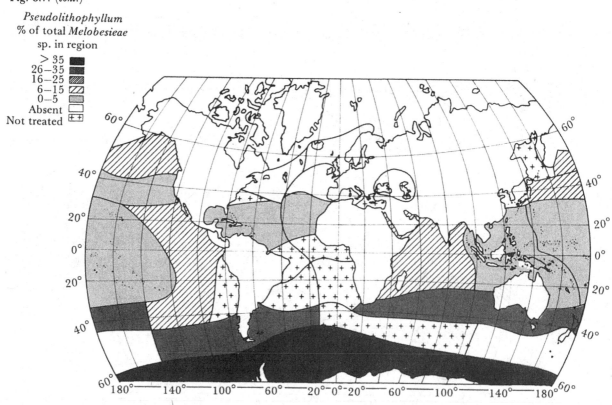

c   *Pseudolithophyllum*
% of total *Melobesieae*
sp. in region

> 35
26–35
16–25
6–15
0–5
Absent
Not treated

only the north of the British Isles, e.g. the two subspecies of *Fucus distichus*; one, subsp. *anceps*, occurs only on exposed coasts whilst the other, subsp. *edentatus*, occurs only in sheltered waters (Powell, 1957). Equally a few species appear only in the southernmost parts of the British Isles e.g. *Pterosiphonia complanata* (Norton & Parkes, 1972). As pointed out by van den Hoek (1967) various species drop out of the European flora as one moves either northward or southward, but he was not convinced of the reality of the phytogeographical regions designated by some workers, though he did recognise a discontinuity between northwest Brittany and the Faroes and a distinct northern Norway/Spitzbergen group. However, this work was based on literature surveys of several isolated regions and needs extending, though the complexities, especially of accurate identification, are enormous, e.g. in the latest survey of the British Phaeophyta 197 species are recognised (Russell & Fletcher, 1975); each species has a definite range and it is only when a number are found to have approximately similar limits that boundaries can be drawn with any confidence. Later, van den Hoek (1975) extended his study to the eastern seaboard and using the Jaccard similarity index and cluster analysis techniques to analyse the data, he determined the phytogeographic zones illustrated in Fig. 8.2; these correlate well with temperature limits and the zoogeographic zones of Briggs (1974). The number of algal species in this North Atlantic region diminishes from south to north as does that of animals. Whilst the general zonation is clearly correlated with temperature the subtle differences which can be found in lagoons, shallow bays etc. must be taken into account, e.g. the occurrence of warm-water species in summer in northern bays when temperatures rise.

The Mediterranean Sea itself is an interesting area where superimposed upon a cosmopolitan group of species there are pantropical, atlantic, indo-pacific (entering via the Red Sea) and mediterranean-lusitanian elements all contributing to the flora (Feldmann, 1958). In addition there are endemic species, some of which are trapped in the various 'diverticula', e.g. *Fucus versoides* in the Adriatic. In the northern section of the Adriatic there are also various ecotypes of *Cystoseira* and the monotypic red algal group Yadronellaceae (Lovrić, 1972).

Although still not studied in detail (*see* comments in Lipkin, 1972) the eastern Mediterranean–Suez Canal–northern Red Sea is an interesting region where but for the paucity of early records a study of migration and appearance could be made. The interesting account by Lipkin records 47 species common to the eastern Mediterranean–northern Red Sea, whilst 24 species occur only in the northern Red Sea and 14 species in the eastern Mediterranean only. The penetration of species through the canal seems to be mainly in a direction from the Red Sea into the Mediterranean and no species of truly Mediterranean or Atlanto–Mediterranean origin have succeeded in migrating southwards. Lipkin could find no differences in the species common to both the eastern Mediterranean and the northern Red Sea and concluded that various migrations had occurred (there were artificial canals prior to the construction of the present Suez Canal) but the floras had not been separated long enough to allow independent evolution. An earlier view that the indo-pacific species penetrated into the Mediterranean at the time when both basins were part of the Tethyan Sea now seems unlikely. More and more indo-pacific species are being found in the eastern Mediterranean indicating a continuing recent migration. Lipkin does not think there is too great a problem for migration of Red Sea species through the Canal in spite of some variations in salinity but once these species are in the eastern Mediterranean they face considerable difficulties in spreading to the remainder of that basin.

There is a rather large gap in our knowledge of the flora south of Morocco until the Guinea Coast is reached where Lawson and his coworkers have provided almost the only data. Lawson (1978) discusses the phytogeographical boundaries down the West African coast. Fig. 8.8 summarises the data into geographical zones for this coast and for the American coast. Cap Vert in Senegal is regarded as the southernmost limit of cold-water species (*see* also the map of van den Hoek, 1975) and the northern limit of the tropical flora. South of Cap Vert the muddy and mangrove-fringed coasts of Gambia, Guinea-Bissau and Guinea may present a physical barrier to the dispersal of the rocky shore algae and Cap Verga in Guinea may be the northernmost point of dispersal of some of the tropical forms. Species such as *Padina australis, P. mexicana, Acanthophora muscoides,*

Table 8.3. *The distribution of Fucaceae and Himanthaliaceae in the Northern hemisphere. Drift records excluded. From Powell, 1963*

	W. Norway	Nordland	Finnmark	S.W. & S. Iceland	N. & E. Iceland	W. Greenland	E. Greenland	Jan Mayen & Bear Island	Svalbard (Spitsbergen)	White Sea	W. Barents Sea	E. Barents Sea & W. Novaya Zemlya	Siberian Arctic Ocean	Atlantic USA	Atlantic Canada	American Arctic Ocean	Pacific Canada & USA	Aleutian Islands	Japan (N.)	Kurill Islands	Kamchatka	Bering Sea
Fucus distichus																						
subsp. *evanescens*	·	·	·	+	+	+	+	+	+	+	+	+	+	+	+	+	+	+	+	+	+	+
subsp. *edentatus*	+	+	+	+	+	+	+?	·	?	+	+	·	·	+	+	·	+	·	+	+	+	·
subsp. *distichus*	+	+	+	+	+	+	·	·	·	+	+	+	·	+	+	·	·	·	·	·	·	·
subsp. *anceps*	+	+	+	+	?	·	·	·	·	·	+	·	·	·	?	·	·	·	·	·	·	·
Ascophyllum nodosum	+	+	+	+	+	+	+	+	·	+	+	·	·	+	+	·	·	·	·	·	·	·
Fucus vesiculosus	+	+	+	+	+	+	+	+	+	+	+	+	·	+	+	·	·	·	·	·	·	·
Fucus serratus	+	+	+	+	+	·	·	·	·	+	+	·	·	·	+	·	·	·	·	·	·	·
Fucus spiralis	·	·	·	·	·	·	·	·	·	·	·	·	·	+	+	·	·	·	·	·	·	·
Fucus virsoides	+	+	+	+	+	·	·	·	·	+	+	+	·	·	·	·	·	·	·	·	·	·
Pelvetia canaliculata	+	+	+	·	·	·	·	·	·	·	·	·	·	·	·	·	·	·	·	·	·	·
Fucus ceranoides	+	+	+?	·	·	·	·	·	·	·	·	·	·	·	·	·	·	·	·	·	·	·
Himanthalia elongata	·	·	·	·	·	·	·	·	·	·	·	·	·	·	·	·	·	·	·	·	·	·
Pelvetia fastigiata	·	·	·	·	·	·	·	·	·	·	·	·	·	·	·	·	+	·	·	·	?+	·
Pelvetiopsis limitata	·	·	·	·	·	·	·	·	·	·	·	·	·	·	·	·	+	·	+	+	·	·
Hesperophycus harveyanus	·	·	·	·	·	·	·	·	·	·	·	·	·	·	·	·	+	·	·	·	·	·

Table 8.3 (*cont.*)

	S. Norway (Norway)	Oslofjord (Norway)	W. Sweden	Copenhagen & Malmö	S.E. Sweden	Inner Baltic Sea	Faeroe Islands	Shetland Islands (British Isles)	Orkney Islands (British Isles)	N. & W. Scotland (British Isles)	N. & W. Ireland (British Isles)	England & Wales (British Isles)	France	Iberian Peninsula (Atlantic)	Azores	North Africa (Morocco)	Canary Islands	Adriatic Sea	Belgium	Netherlands	German North Sea	W. & N. Denmark
Fucus distichus	·	·	·	·	·	·	·	·	·	·	·	·	·	·	·	·	·	·	·	·	·	·
subsp. *evanescens*	⊙	⊙	⊙	⊙	·	·	+	+	·	+	·	·	·	·	·	·	·	·	·	·	·	·
subsp. *edentatus*	·	·	·	·	·	·	·	·	·	·	·	·	·	·	·	·	·	·	·	·	·	·
subsp. *distichus*	·	·	·	·	·	·	+	+	·	+	·	·	·	·	·	·	·	·	·	·	·	·
subsp. *anceps*	·	·	·	·	·	·	+	+	+	+	+	+	·	·	·	·	·	·	·	·	·	·
Ascophyllum nodosum	+	+	+	·	·	·	+	+	+	+	+	+	+	+	+	·	⊙	·	?	+	+	+
Fucus vesiculosus	+	+	+	+	+	+	+	+	+	+	+	+	+	+	·	+	·	·	+	+	+	+
Fucus serratus	+	+	+	+	+	·	+	+	+	+	+	+	+	+	·	·	·	·	+	+	+	+
Fucus spiralis	+	+	+	·	·	·	+	+	+	+	+	+	+	+	+	+	+	·	+	+	+	+
Fucus virsoides	·	·	·	·	·	·	·	·	·	·	·	·	·	·	·	·	·	+	·	·	·	·
Pelvetia canaliculata	+	·	·	·	·	·	+	+	+	+	+	+	+	+	?	·	·	·	?	+	?	·
Fucus ceranoides	·	·	·	·	·	·	·	+	+	+	+	+	+	+	·	·	·	·	+	+	·	·
Himanthalia elongata	+	·	·	·	·	·	+	+	+	+	+	+	+	+	·	·	·	·	·	·	?	·
Pelvetia fastigiata	·	·	·	·	·	·	·	·	·	·	·	·	·	·	·	·	·	·	·	·	·	·
Pelvetiopsis limitata	·	·	·	·	·	·	·	·	·	·	·	·	·	·	·	·	·	·	·	·	·	·
Hesperophycus harveyanus	·	·	·	·	·	·	·	·	·	·	·	·	·	·	·	·	·	·	·	·	·	·

+ present; · probably absent; ? indicates that the evidence for presence (or absence) is either uncertain or unknown; ⊙ indicates recent extensions of range.

Corallina pilulifera, Bryothamnion seaforthii and *Laurencia intermedia* do not extend from the Gulf of Guinea into Senegal and plants of more northerly distribution such as *Cystoseira* sp. *Halimeda discoidea, Dilophus fasciola, Ecklonia muratii, Gelidiopsis intricata, Gelidium sesquipedale* and *Porphyra umbilicalis* do not extend into the Gulf of Guinea. The Canary Isles have communities dominated by *Cystoseira* (Johnston, 1969) and are therefore a southern extension of a Mediterranean-type flora. The Gulf of Guinea flora spreads southwards into Angola according to the very recent surveys of Lawson, John & Price (Lawson and John, 1977; Lawson, 1978) for here the flora is overwhelmingly similar to that of the Gulf. A small element of this Angolan flora occurs to the north of the Gulf (e.g. *Bryopsis balbisana, B. corymbosa, Halimeda tuna* and *Griffithsia opuntioides*) but not in the Gulf itself. In addition a few species with more southerly distribution penetrate into Angola (e.g. *Acrosorium maculatum, Plocamium becheri* and *P. suhrii*). South of Angola is the main transition region to the colder water flora of Southwest Africa, where species such as *Chondria capensis, Iridaea capensis, Pterosiphonia gloiophylla, Tayloriella virgata* and *Laminaria schinzii* occur.

The Gulf of Guinea has only 21% of its algal species common to the eastern Atlantic boreal-antiboreal province and only 28% common to the Mediterranean: of these many are cosmopolitan species. On the other hand the Gulf of Guinea flora shows closer affinities with the flora of the western tropical Atlantic, nearly two thirds of the species of the Guinea coast occurring on tropical American shores. A few species are confined to these two regions on opposite sides of the Atlantic Ocean (*Dohrniella antillarum, Halymenia duchassaingii, Waldoia antillarum, Spermothamnion investiens* and *Sargassum filipendula*). Slightly over half the Guinea Coast species are also common to the tropical Indian and Pacific Ocean coasts. As in other tropical areas the ratio of Rhodophyta to Phaeophyta is high (4.3) and similar to that of the western tropical Atlantic (4.8): for cold waters the ratio tends to be between 1 and 1.5.

Whilst there are many examples of extension of the range of marine species, reductions are less well recorded but undoubtedly there are concommitant shrinking of distributions e.g. Edelstein *et al.* (1971–3) report that *Fucus serratus* in the maritime provinces of Canada has disappeared from large areas of Prince Edward Island and the Gulf of St Lawrence shore of New Brunswick but has extended to other areas: the reasons for these movements are not at all clear. The degree of endemism in each flora is a valuable clue to the relative isolation of each region. As with so many aspects of phytogeography the data are lacking for so many areas and the percentage of endemics recorded depends on the activity of collectors thus it is very high in the south western region of Australia. This has been revealed by the extensive study of this area, and it is likely that further studies will merely add more endemic species to the already impressive total, although this does not disguise the fact that it is almost certainly a region of decided endemism. Few other areas are so favourably placed but there are still many areas with peculiarities of currents, barriers, etc., which would repay extensive study especially of the subtidal flora, e.g. the southern coasts of South America, South Africa and the relatively enclosed arms of the Red Sea and Persian Gulf.

Fig. 8.8. Geographical boundaries derived from a mathematical analysis of the distribution of coastal macroscopic algae on both sides of the Atlantic. WTAM, warm temperate American; WTAF, warm temperate African; TAF, tropical African; TAM, tropical American; TTA, tropical transitional African; BAF, boreal antiboreal African. From Lawson, 1978.

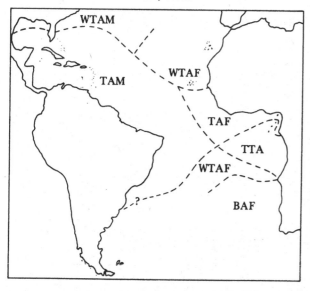

Marine algae: microphytes

Few world-wide studies of oceanic diatoms have been undertaken and such data are only likely to come from detailed studies of individual species, or groups of species, with a keen regard to their taxonomy. Disregarding the so-called cosmopolitan species, it is clear that many distinct oceanic provinces can be recognised from their phytoplankton populations and that these are closely related to latitudinal temperature variations. One of the major problems of oceanic distribution is the possibility of the occurrence of bipolar species but as Hasle (1969) comments there is still a shortage of critical studies on taxonomy and distribution at the specific level 'and until these are made reliable, conclusions regarding the existence of bipolarity in marine phytoplankton and the degree of endemism in the Southern Ocean can hardly be drawn'. In fact all species present in both polar regions may simply be cosmopolitan species (Smayda, 1958); nevertheless all such records

Fig. 8.9. The distribution of, *a*, *Ceratium lunula*, *b*, *C. cephalotum* and, *c*, *C. tripos atlanticum* in the equatorial water mass, the central water mass and the transition zone in the Pacific Ocean. The extent of distribution of the three dinoflagellates is defined by cross-hatching over distribution contours of various animal species. Modified from McGowan, 1971.

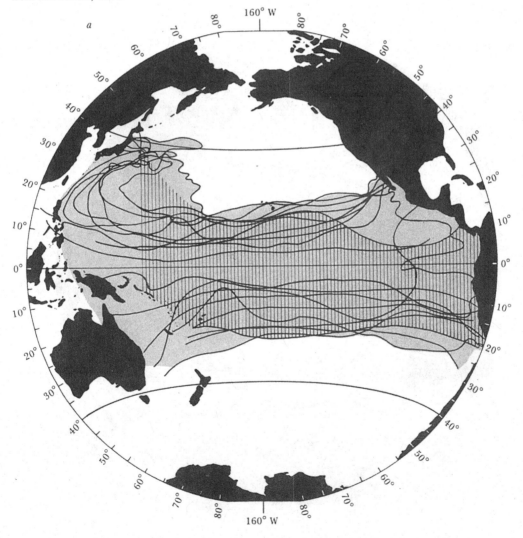

require not only morphological but physiological study since they may constitute physiological races. *Thalassiosira antarctica* is probably the only well studied example of a supposedly bipolar diatom. Previous studies of the antarctic region suggest that about 60% of the diatoms are southern hemisphere species (Hustedt, 1958) and Manguin (1960) recorded 30 endemic species. There can be no doubt that an antarctic element exists. It is clear that the problem of comparison of data from various authors is fraught with danger and perhaps the best hope is that studies of individual species and genera will be undertaken since comparative studies of whole floras from arctic–tropical–antarctic zones are almost impossible for any one person.

Wherever currents carry water from one geographical zone to another the indicator species are also carried, e.g. *Corethron criophilum*, an indicator species of antarctic water can be carried northward as far as 4° S in the Peru Current. But in spite of the continuity of the oceans and the extensive movement by currents there are records of disjunct distributions e.g. the colonial planktonic diatom *Planktoniella muriformis* occurs off the coast of Ghana (Round, 1972*a*) and a population with only slight morphological variations occurs in the Pacific from San Diego south to Punta Abbreojos off Baja California (Smayda, 1975); we have no idea how this distribution arose. An excellent review and compilation of regional diatom floras is that of Guillard & Kilham (1977).

Fig. 8.9 *b*

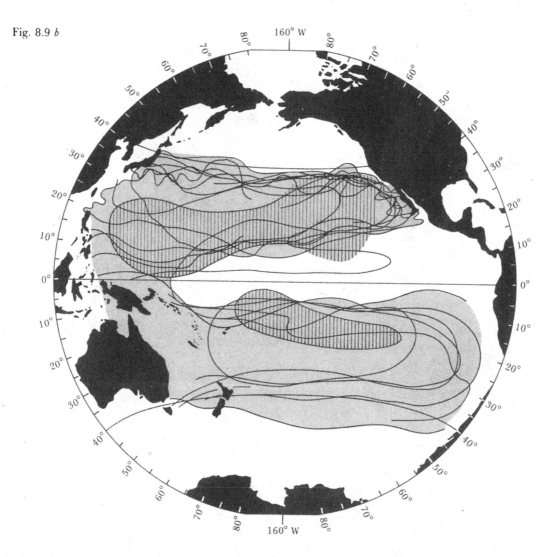

Amongst the coccolithophorids *Emiliania* (*Coccolithus*) *huxleyi* is common in all the oceans and extends into arctic and antarctic waters. On the other hand *Coccolithus pelagicus* is a North Atlantic cool-water species and has not been recorded in similar water masses in the southern hemisphere. *Rhabdosphaera claviger* on the other hand is an example of a species confined to warm waters and *Cyclococcolithus leptoporus* is bipolar (McIntyre & Bé, 1967).

The ranges of phytoplankton species in the ocean can be compared with the ranges of the fauna and the two often appear to be remarkably similar: the distributions of three *Ceratium* species in the Pacific Ocean are plotted together with those of a number of animals in Figs. 8.9a, b, c. Clearly *Ceratium*

tripos subsp. *atlanticum* is a transition zone species occurring in both hemispheres on either side of the 40° latitude line but extending northward in the Humbolt–Peru currents; *Ceratium cephalotum* is characteristic of both central water masses, but in the northern hemisphere the populations reach the coast of North America but not of South America; *Ceratium lunula* is characteristic of the equatorial water mass with a northward extension in the Kuroshio current. Down the coast of North America these species groupings suggest that four coastal zones might also be distinguished i.e. a northern zone above that of the transition species, a zone below this and 40° N, a third zone from 40° N to about 30° N and a final southern zone below this. The work of Druehl (1970)

Fig. 8.9 *c*

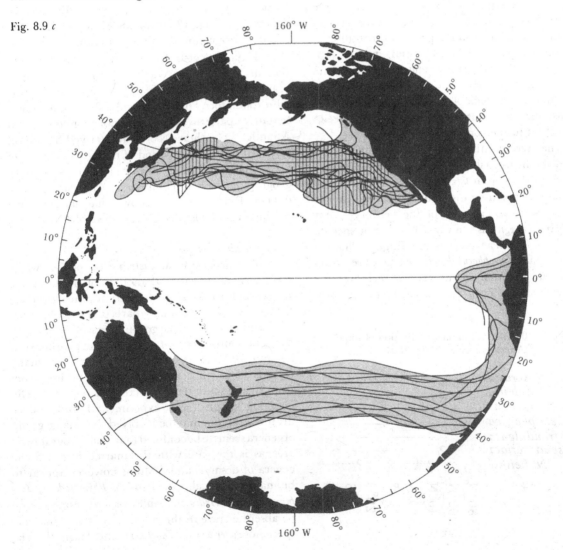

on the distribution of Laminariales down the eastern Pacific coast in fact shows a complex pattern (Fig. 8.5) with 14 zones distinguishable between Attu Island in the Aleutians to Santa Margarita in Baja California: such greater complexity must be expected in the more variable coastal environment. A study of the exceedingly difficult, colonial *Nitzschia* species (Hasle, 1972) reveals four zones of distribution (Fig. 8.10) but there are some problems of speciation still to be resolved. The arctic, antarctic, and a temperate–tropical distribution can obviously be correlated to temperature zones, whilst *N. pseudoseriata*, although a warm-water species, is somehow confined to the southern hemisphere. Attached epiphytic diatom species are perhaps easier to study from a phytogeographical standpoint but at the moment there is a dearth of data. One species however which has a fairly precise distribution is *Climacosphenia* which is only found in the subtropical–tropical zone. It is a reflection on the state of knowledge of these and the moving epipelic species that I can think of no other reasonably well circumscribed distribution yet hundreds of species are involved in the communities.

The recent distribution of dinoflagellates around the British Isles has been studied and compared with data from earlier work, and for many species a rather similar distribution occurs today to that recorded at the turn of the century. Thus *Dinophysis norwegica* is mainly a North Sea species, *Peridinium oblongum* a west coast and English Channel species, whilst *Pyrophacus horologium* is more widespread but absent from the central region of the English Channel. Fig. 8.11, from data kindly supplied by Professor J. D. Dodge, shows some of the complexities of distribution in a small area, including the growth of *Dinophysis norwegica* in this area (Dodge, 1977). Work on dinoflagellate cysts (Ried, 1975a) in the sediments around the British coasts has also revealed distinct patterns of distribution in eleven regions, some of which are only small bays. The patterns can be correlated with water masses, defined by other workers, using chaetognaths as indicators. In addition there was some evidence of movement of sedimented spores in relation to tidal radiation and convergence patterns. These tidal movements are features which have not been taken into account by workers studying distribution and it would be interesting to discover the extent of their influence. Reid found about 50 types of cyst in the bottom deposits (*see* Fig. 13.20 for an example) but only about 15 are common in the plankton.

Data collected by the continuous plankton recorder are extremely valuable for plotting the geographical extent of larger marine species and the self-explanatory Figs. 8.12a and b show some representative patterns of distribution for the north Atlantic. Since these records are derived from data collected over a long period of time the inclusion of occasional species carried out of their true phytogeographic zone is eliminated. Analysis of these data by Reid (1977) has shown that diatoms have declined in the last decade, north of 59° N.

Freshwater algae

The geographical distribution of freshwater species is much wider than that of most marine species; at least this is the impression gained from the listing of species from many parts of the world. Such lists tend to contain large numbers of apparently cosmopolitan species (some reliable lists give between 50 and 70% as cosmopolitan). However, many records of north temperate species in lists from tropical and south temperate regions require careful checking coupled with experimental studies since distinct ecotypes may be involved. No major group of genera seems to be confined to any one geographical area as is the case with the marine forms. Some genera of desmids and diatoms however appear to be mainly tropical, e.g. *Amscotia*, *Phymatodocis*, *Triplastrum*, and *Hydrosera*, whilst other genera tend to be absent or rare in the tropics, e.g. the planktonic species of *Asterionella*, *Tabellaria*, and *Fragilaria*. The

Fig. 8.10. The latitudinal distribution of some oceanic *Nitzschia* species. From Hasle, 1972.

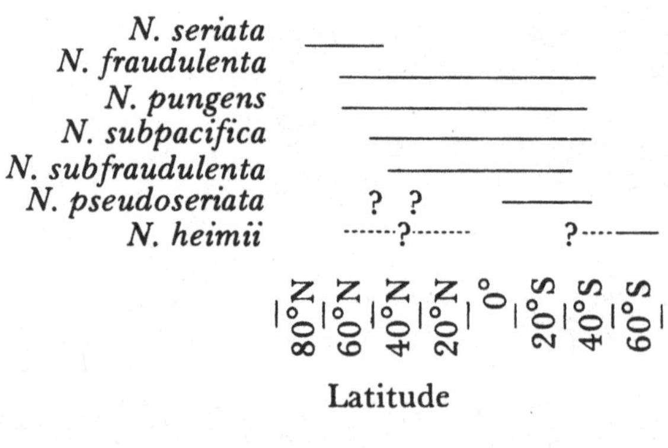

Latitude

geographical spread of some freshwater species into Australian waters seems to be restricted, possibly by the salinity of many of the waters; thus *Asterionella*, *Stephanodiscus* and *Aphanizomenon* appear to be rare or absent from these waters. The increased salinity factor may apply also to other arid regions e.g. in the rivers of the south western states of North America (*see* p. 268).

It is tempting to suggest that temperature (upper limit) is the likely controlling factor restricting certain species to particular geographical regions. There is some supporting evidence for this, e.g. Lund (personal communication) found that growth of *Asterionella formosa* was not possible at high temperatures and would be excluded from tropical lakes, just as it is from temperate lakes in summer. Other workers have noted the increase in temperate species in high mountain lakes in the tropics accompanied by a decrease in the tropical elements. Such records are further evidence of very widespread dispersal of species across regions which are normally considered unfavourable.

Many taxa are described in the literature as cold-water species merely because the original

Fig. 8.11. Charts of the distribution of the dinoflagellates in the North Sea, March–June 1971 and May 1974. *C. fur.*, *Ceratium furca*; *C. lon.*, *C. longipes*; *C. mac.*, *C. macroceros*; *C. tri.*, *C. tripos*; *D. nor.*, *Dinophysis norvegica*; *G. tam.*, *Gonyaulax tamarensis*; *P. dep.*, *Protoperidinium depressum*; *P. ov.*, *P. ovatum*; *P. sub.*, *P. subinerme*. From Dodge, 1977.

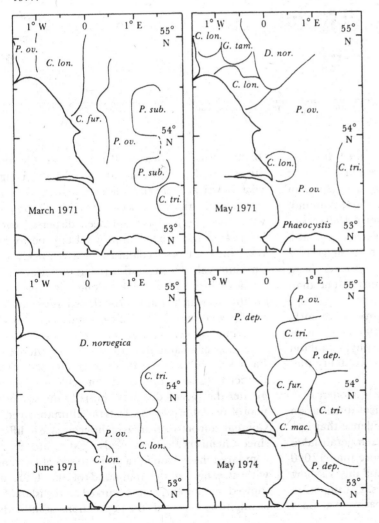

Fig. 8.12. *a*, the distribution of three species of *Ceratium* in the north Atlantic. *b*, the distribution of four planktonic diatoms in the north Atlantic. From Edinburgh Oceanographic Laboratory, 1973.

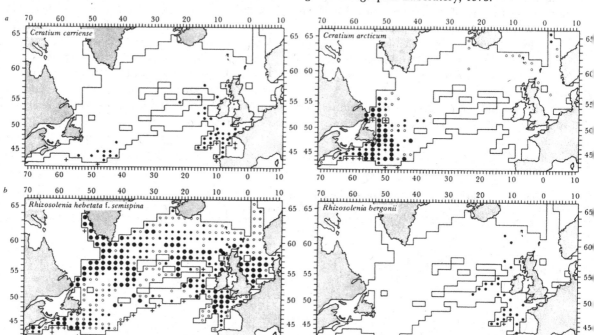

author found them in such a habitat; the brief notes on habitat in most floras should be treated with great suspicion. However recent work on Alaskan Chrysophyta (Hilliard & Asmund, 1963; Asmund, 1968) reveals a large number of species of *Mallomonas, Synura, Dinobryon, Epipyxis*, etc., in such waters and comparison with data from elsewhere suggests a fairly definite association. It may however be a relationship with chemical conditions rather than temperature, since the pools in which they are found are generally acidic and peaty but may have fairly high summer temperatures. Lund (1964) comments that there are no known stenothermal phytoplankton species though he does not deny the possibility that some may be found.

The time element is one which is often considered in relation to the geographical distribution of land plants and there is some evidence that it is important for freshwater lakes. Most temperate lakes have a very short history of little more than 12 000 years since the retreat of the last ice and this is a short time for the evolution of new species. On the other hand there are some ancient lake basins, e.g. Lake

Baikal, Caspian Sea in Russia, Lake Ochrid in Jugoslavia, Lake Biwa in Japan and Lake Tanganyika in Africa where endemic species, or species with restricted distributions, occur. It is difficult to explain why these species have not been dispersed over a larger area in such a long time, whilst the post-glacial lakes have all been colonised in a relatively short time interval. One suggestion is that they are endemic species which have evolved along with the special conditions which characterise these basins: unfortunately we do not have sufficient data coupled with experimental verification of correlations.

Size of an area has rarely been considered, but Patrick (1967) showed that the greater the area of artificial surface presented for colonisation the greater the species diversity. Equally the size of the pool of available species affects the ultimate diversity. Comparison between areas of differing size is difficult since chemistry, etc. may also vary, but Patrick compared the stream flora of the island of Dominica with streams on the mainland of the USA and showed that fewer species occurred on the island but, as might be expected, the populations of individual

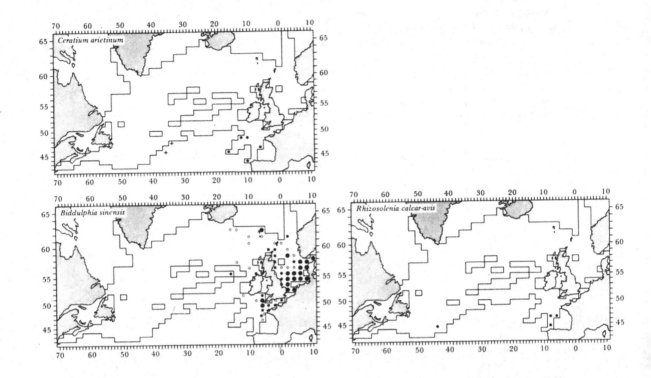

species were greater and fewer species maintained small populations. Patrick concluded that a high invasion rate maintains a number of species with small populations: the relationships between size of area, species pool, invasion rates, etc. has, as far as I know, rarely been considered in algal ecology and most examples of the interplay of these variables are found in animal studies.

Speciation and phytogeography

At first sight it might seem that at least in the open ocean there is a possibility of universal distribution, and therefore of widespread genetic interchange, but closer observation of the current systems shows that many areas are relatively isolated. Along the coasts this isolation is even greater since both open water and temperature barriers exist. Freshwater species have temperature and spatial barriers to dispersal but clearly many species overcome these by animal transport. Theoretically, animal and especially bird transport, should occur for marine species; there is however much less likelihood of such transport of large coastal species,

which have very short-lived unicellular stages, so that one might expect to find the greatest degree of separation and also endemism amongst them. This is certainly so for groups such as the Laminariales and Fucales and Table 8.4 gives the spread of some genera and species of Fucales.

It is generally belived that speciation in sexually reproducing organisms occurs allopatrically. This involves the splitting of a parent population into two spatially isolated segments, separated to such a degree that gene flow between the populations is eliminated. If the separation is maintained over a long enough period, sufficient genetic changes will be built into each population, such that if ultimately brought together they would be incompatible. Separation of algal populations will have occurred as the continents have drifted away from one another and also as the circulation of the water masses has changed. The data are very scanty, since many geographical areas are relatively unknown and in many cases there is doubt about the identity of species. But it is possible to identify some regions where populations are 'trapped', and where specia-

Table 3.1. *The world distribution of genera of the Fucales. From Nizamuddin, 1970*

	Atlantic Ocean														Indian Ocean (Arabian Sea)													Pacific Ocean														Antarctic / Pacific Coast of America						
Genera	South America	West Indies	Bahamas	Caribbean Sea	S.E. N. America	N.E. America	Iceland	Greenland	Sargasso Sea	W. Africa	W. Europe	Mediterranean Sea	Adriatic Sea	S. Africa	E. Africa	Madagascar	Red Sea	India	Pakistan	Aden	Persian Gulf	India (Bay of Bengal)	Ceylon	Malaya	Singapore	W. Sumatra	S.W. Australia	Japan	Korea	China	Formosa	Hong Kong	Philippines	Pacific Islands	Hawaii	N.E. Australia	Indonesia	New Zealand	W. coast of N. America	Guadeloupe Island	Galapagos Islands	S.E. Australia, N.S.W.	S. Australia	Victoria	Tasmania	S. Shetland Islands	Auckland Islands	Campbell Islands
Acrocarpia																											+									+						+	+	+	+			
Acystis																				+																												
Ascophyllum											+																														+							
Axillariella											+																																				+	
Bifurcaria														+																																		
Bifurcariopsis														+																																		
Carpoglossum																																												+	+			
Carpophyllum														+																						+		+							+		+	
Caulocystis																																				+						+	+	+	+			
Coccophora																												+	+									+									+	+
Cystophora	+																										+									+						+	+	+	+			
Cystophyllum																											+									+						+	+	+	+			
Cystoseira	+										+	+	+	+	+	+	+	+				+					+					+							+	+		+	+	+				
Cystosphaera																																														+		
Fucus					+	+	+	+			+	+	+															+											+									
Halidrys											+																	+										+										
Hesperophycus																																							+									
Himanthalia											+																											+										
Hizikia																													+			+																
Hormophysa														+	+	+	+	+																														
Hormosira																																						+										
Landsburgia																											+										+										+	
Marginariella																																						+									+	
Myriodesma																											+															+	+	+	+			
Pelvetia											+																	+	+			+							+									
Pelvetiopsis																																							+									
Phyllospora																											+															+	+	+	+			
Platythalia																											+															+	+	+	+			
Sargassum	+	+		+	+				+	+	+			+	+	+	+	+				+	+	+	+	+	+	+		+	+	+	+	+	+	+	+					+	+	+	+			
Scaberia																																										+	+	+	+			
Scytothalia																											+																+	+	+			
Serococcus														+				+									+															+	+	+	+			
Stokeyia																		+																														
Stolomophora															+	+	+	+				+	+	+			+						+			+												+
Turbinaria	+													+	+	+	+	+				+	+	+	+	+	+						+	+	+	+												
Xiphophora																																	+					+										

tion is presumably proceeding. For example, in the southern hemisphere the temperate southern coasts of Australia, parts of New Zealand, the southern tip of Africa and South America are separated one from another, and from Antarctica, by a large mass of cold water (the tip of South America is the least isolated) and from any northern populations by warm equatorial water. The North Pacific and North Atlantic populations are also separated in both coastal and oceanic regions, the South Atlantic central water mass is separated from the Indo-Pacific water mass and finally the Arctic Seas are quite separate from the Antarctic. There is a tendency for groups of species to be mainly northern or southern hemisphere in distribution, e.g. the genus *Laminaria* (45 taxa) has a group of species in the North Atlantic, another in the North Pacific, one only in the Mediterranean and two only in the South Atlantic (Kain, 1971). In all these and in many smaller areas, e.g. the inner Adriatic, Black Sea, Sea of Azov, Eastern Mediterranean and Red Sea, the populations are relatively isolated. On the land there are isolated and relict populations in some large lakes, e.g. Lake Baikal, the Aral Sea and the Caspian Sea. In the Caspian Sea, which was cut off from the ocean in the early Pliocene and is the largest inland water, 16 endemic species, plus a further 16 which are endemic to the Ponto-Caspian Basin occur (Karayeva & Makarova, 1973). The long evolution of species isolated from the sea has led to morphological peculiarities becoming fixed, presumably genetically, e.g. the striae of the diatoms are generally more numerous and more delicate than in the same taxa in the Black Sea and Mediterranean. In *Mastogloia smithii* var. *gracilis* one end is bifurcate in the Caspian population (Karayeva & Makarova, 1973). It is not, however, necessary to go to such ancient water bodies to find morphological variation fixed into the population e.g. in many waters the diatoms have deformity in the valves and these remain as features of the species over many years.

The Caspian Sea has an average salinity of 12.80–12.85‰ and the water is enriched in sulphates and carbonates. Such a salinity might be expected to favour the growth of many marine species which under cultural conditions (*see* p. 293) grow well at salinities much below that of seawater. Eight marine diatom genera occur in the Caspian Sea (*Actinoptychus, Biddulphia, Coscinodiscus* * *Dimerogamma,*

Grammatophora, Licmophora, Rhizosolenia, * and *Thalassionema*) but marine species of other genera are also represented, e.g. of *Thalassiosira, Chaetoceros, Navicula, Nitzschia, Amphora* and *Pleurosigma*; the first two are in fact almost totally marine in distribution. Some of the marine species have adapted to the almost fresh water of the Northern Caspian. The largest group of species is classified as brackish and these species together with the marine make up $\frac{2}{3}$ of the diatom flora. It is, however, a species-poor flora with only 286 species and varieties (77 planktonic and 209 benthic). The freshwater forms are assumed to have been transported into the Caspian Sea since its isolation. Attempts to trace the history of the Caspian flora have been made: originally it was thought that the flora was derived from northern seas, an opinion reinforced by zoological evidence, but more recently both faunal and floral studies have shown the presence of Mediterranean species and it is now considered more likely that it derives from the ancient flora of the Upper Tertiary, Ponto-Caspian basin with the addition of a second influx of Pliocene species entering through the Manych Strait during temporary connections with the Black Sea and Sea of Azov.

Algal studies very rarely include crossing experiments but one recent attempt (Luning *et al.*, 1978) has shown that *Laminaria longicruris* from Nova Scotia can be crossed with *L. saccharina* from Europe although the thalli were not grown to sporogenesis and so fertility of the hybrids could not be checked. Many more such studies are required.

Summary

Dispersal of freshwater species is relatively simple though there is little in the literature on the exact mechanisms for many genera. *Aerial, animal, artificial* means and *current* drift are discussed.

The first appearance of a species in a particular area is rarely ascertainable though there are recent examples e.g. of *Asparagopsis, Bonnemaisonia, Codium* spp., *Sargassum* sp. Equally difficult to study is the problem of continuity of species and again there are very few examples in the literature. However for both aspects the fossil record can provide interesting clues.

Phytogeography involves the determination of the extent of the spread of individual species and also

* A few freshwater species of these genera alone are known.

the classification of regions into distinct biological zones based on species complements. These are discussed for marine macrophytes with examples from Pacific and Atlantic sites. The distribution of the Fucaceae is given in some detail and the interesting Eastern Mediterranean–Red Sea flora very briefly mentioned.

Microscopic species of the oceanic phytoplankton also occur in discrete geographical zones, and even greater care is needed when studying them, since dead or senescent populations can be carried great distances in current systems. The Pacific Ocean again is one of the best documented areas. Detailed collections of dinoflagellates around the British coasts coupled with data from plankton recorder surveys are revealing the small scale patterns which occur even in a restricted area such as this.

Freshwater species are more widely distributed than marine, yet some distinctive patterns can be determined. Speciation and phytogeography is as yet a very incompletely studied aspect but one which would repay further study.

9 Symbiosis, parasitism and grazing

This chapter deals with the organisms and processes involved in intimate association and transfer of substances more or less directly between algae, animals, fungi, and bacteria. Grazing in its widest sense involves almost all algae and carries the photosynthetically fixed carbon to the secondary producers. The only other pathway for the dissemination of this algal material is via the detritus pathway (*see* p. 463).

Symbiosis

The concept of symbiosis, the living together of two different organisms in a close association, has been used in the sense of each organism benefiting from the presence of the other (mutualistic association) but also in the sense of one organism living on and destroying another. The latter sense is considered here as parasitism. Examples of saprophytic growth by algae are very rare but are briefly discussed. There is also the very common symbiotic state of commensalism, a looser association of organisms using each other for attachment purposes, which is dealt with in chapter 6. Until recently most of these commensal associations of algae were regarded simply as convenient attachment relationships but, wherever there is any degree of specificity between alga and host, a degree of symbiosis is indicated and indeed proved by the interchange of nutrients. In commensal associations there is no morphological modification of either partner whereas in symbiosis there are varying degrees of modification.

The actual occurrence of algal cells within or between the cells of a host is often termed endosymbiosis and has been known for approximately a century (Muscatine, 1974). In most of the associations the alga is to some degree modified by the host, e.g. *Platymonas* in *Convoluta* becomes a naked protoplast with the loss of flagella, flagellar pit, eyespot and theca (Oschman, 1966), although in at least one example (*Amphidinium klebsii* in the turbellarian *Amphiscolops langerhansi*) there is no apparent modification of the algal symbiont. Taylor (1971) speculates that this might be an indication of a recently formed symbiosis, a view which is perhaps strengthened by the fact that the turbellarian partner feeds voraciously throughout its life and can be reared to maturity in the absence of the alga. However such animals do not achieve sexual maturity since they are apparently dependent on an algal product for reproduction (Taylor, 1971, 1973). Implicit in the concept of symbiosis is the fact that the alga is not digested by the host and indeed it seems that most hosts, being carnivorous, are incapable of digesting their algae: they can, however, eject the algae under certain conditions. A few animals, e.g. the herbivorous tridachnids, do digest algae and the Saccoglossans have established a system of symbiosis based on conservation of chloroplasts extracted from algae. It is clear that the degree of interdependence of the hosts and algae is very variable, some hosts, e.g. *Phaenocora typhlops* seem to be only slightly benefited by their algal symbiont (Young & Eaton, 1975), whilst in others, e.g. *Convoluta roscoffensis*, the alga is essential (*see* p. 390). The cyanophyte genus *Aphanocapsa* occurs relatively unaltered in marine sponges and is liable to be digested by its host, which suggests a rather looser symbiotic relationship. However other Cyanophyta* e.g. *Cyanocyta korschikoffiana* (in *Cyanophora*) and *Skujapeltis nuda* (in *Glaucocystis*) are modified to the extent that their cell walls are reduced along with some of the internal vacuoles, granules, etc. The degree of modification of the algae can therefore vary according to the host type and presumably according to the degree of specialisation of the symbionts. The symbiosis can be looked upon

* See, however, the footnote on p. 389.

387

simply as an ecological phenomenon and the modification of the algae simply as the formation of
ecotypes within their chosen microhabitat.

The major changes when an alga enters into
a symbiotic relationship are a slow-down of its
growth rate whilst its photosynthetic rate is unaffected. There is, therefore, a production of excess
carbohydrates which are leaked to the host. When
the alga is cultured free from the host less carbohydrate is released than during symbiosis (*see* review
by Smith, Muscatine & Lewis, 1969).

There are probably few strictly parasitic algae;
although some red algae have lost their pigment and
live on other algae, usually other Rhodophyta. In
this chapter parasitism includes also the much more
common and very neglected topic of parasitic fungi
and protozoa living upon algae.

Algal–animal symbiosis

According to Smith *et al.* (1969) approximately
150 genera of invertebrates from eight phyla (but see
below) possess algal symbionts. The distinction in
the early literature into 'zoochlorellae' and 'zooxanthellae' according to their colour can now be
abandoned in favour of proper generic designation
since the green symbionts are forms of Chlorophyta,
e.g. *Chlorella* in *Paramecium* and *Chlorhydra viridissima*
(= *Hydra viridis*),* or of Prasinophyta, e.g. *Platymonas
convolutae* in *Convoluta roscoffensis*. A new 'green'
symbiont, *Prochloron*, has been described recently
(Lewin, 1975, 1976, 1977) which has green algal
pigments but is otherwise rather like a blue-green
alga. It is associated with Ascidians and has been
placed in a new phylum Prochlorophyta. Another
possible species of this genus has been found in the
cloacal cavity of *Diplosoma* and physiological experiments have shown this to have some 'blue-green'
features (Thinh & Griffiths, 1977). The yellow-brown
symbionts are most commonly dinoflagellates, *Gymnodinium* (= *Symbiodinium see* also Loeblich &
Sherley, 1979) *microadriaticum*, *Endodinium chattonii*
and *Amphidinium klebsii* in corals, the chondrophore
Velella velella, and the turbellarian *Amphiscolops* (=
Convoluta) *langerhansii*. The only other brown symbioses recorded are the diatom *Licmophora hyalina* in
Convoluta convoluta (Ax & Apelt, 1965), and an
unknown diatom protoplast in the foraminiferan

Amphistegina lessonii (Berthold, 1978). Species of
Fragilaria, *Nitzschia* and a new diatom, as yet
undescribed, have been found in *Heterostegina* and
Amphistegina (Lee, McEnery, Shilo & Reiss, 1979),
and these are the only known cases of participation
of diatom species with Protozoa. Re-infection has not
been achieved in studies on *Licmophora*, and the
problem needs further checking (Taylor, 1972*a*). It
is commonly stated in the literature that there is one
common species of *Gymnodinium*, symbiotic in a large
number of different animals. This needs very careful
checking since it seems highly unlikely that only one
of the numerous dinoflagellates should adopt this
habit and also that only one is acceptable to such a
very wide range of animals. A recent report (Febvre
& Febvre-Chevalier, 1979) of a dinoflagellate and a
prymnesiophyte in species of *Acantharia* indicates that
study will reveal further combinations of species.
Symbiosis with genera of the Ctenophora, Annelida,
Polyzoa, Echinodermata and Chordata requires
validation since preliminary investigations reported
by D. L. Taylor (1973) have failed to substantiate
symbiotic relationships. A comprehensive listing of
hosts and symbionts is given in Droop (1963) but all
lists should be treated with considerable caution
(D. L. Taylor, 1973) until re-inoculation studies
have been performed.

Much early work was observational and conclusions that the brown symbionts were members of
the Dinophyta were based on scanty evidence but
now there are ample data from ultrastructural, life
cycle and biochemical studies to substantiate this
fact. For example, Jeffrey & Haxo (1968) found
pigments in the mantle of five species of tridachnid
clams, eight zooxantharian and alcyonarian corals
and one hydrozoan coral from the Great Barrier
Reefs which were identical to those in the free-living
dinoflagellate *Amphidinium*. In this study no pigment
degradation products were found, an indication that
senescent algal cells are either released or broken
down elsewhere in the animals. The pigment complement was not identical in all the organisms: the
clam endosymbionts and the free-living dinoflagellates have a much higher proportion of chlorophyll
c than the coral endosymbionts. The actual biomass
of the symbiotic algae has rarely been ascertained,
though one figure of 15% of the total for *Zoanthus
flosmarinus* is quoted by von Holt & von Holt (1968).

Some brain corals (e.g. *Favia*, *Porites* and *Gone-*

---

* See Muscatine (1974) for a discussion of the taxonomy
and the reasons for favouring *Chlorhydra*.

astrea) have in addition to the outer layer of dino-flagellates, an inner layer of green filamentous algae belonging to the green siphonaceous genus *Ostreobium* (*O. reineckei*). This is reported to grow at light intensities of only 0.1% of surface illumination (Halldal, 1968) and it is presumably a much deeper growing association in those corals than the associations recognised by Odum & Odum (1955) which occur in a much more superficial position in their diagrams. There may be more than a single green layer and the layer(s) disappear when the polyp tissue above them dies, suggesting a fairly intimate association between the plants and animals. Jeffrey (1968) compared the pigments of this alga with those of *Codium* and *Halimeda* and showed that they were similar in all three. The green layer was considered by Odum & Odum (1955) to contribute even more algal biomass to corals than the zooxanthellae, a view which has been questioned by Goreau & Goreau (1960) who found such an algal layer much less developed in *Manicina areolata* growing in Jamaica. Some workers have concluded that these are 'boring' green algae, which weaken the coral skeleton and lead to its breakdown: this may ultimately happen but earlier the algae appear to be most important as photosynthetic units and in no way to impair the animal activity. In addition, in dead coral quite different algae bore into the skeleton. The ecology and physiology of these two endosymbiotic systems within corals needs detailed study. The very fascinating work of Odum & Odum (1955) indicates a much greater algal component in their Pacific atoll reef corals and further investigation is needed in other comparable sites, since the coral reef system of the Indo-Pacific region does seem to differ considerably from those of the Atlantic or from the Barrier Reef off Australia.

The symbiotic marine Dinophyta appear to be relatively easily cultured under autotrophic conditions in artificial media, supplemented by vitamins, and have no obligate dependence on their hosts. The freshwater symbionts are however more dependent on their hosts and some have never been cultured independently e.g. *Chlorella* from *Chlorhydra* (Pardy, 1974). Other symbionts which are in a rather loose association with their hosts are usually the easiest to culture separately, e.g. the *Chlorella* from sponges (Lewin, 1966). Within the host, however, it is clear that ammonia, phosphate and organic molecules

released by the host are taken up by the algae. In addition, the symbionts seem to require a further supply of nutrients since it can be shown that there is also nutrient uptake from the medium. In *Hydra* fed with ^{14}C-labelled *Artemia* some 23–34% of the total incorporated ^{14}C is transferred from the *Hydra* to the symbiont in 48 h (Cook, 1972). The reverse transference of ^{14}C from algae to coelenterate was shown by Muscatine & Hand (1958); clearly mutualistic effects. In some systems there appears to be an intricate relationship between feeding on zooplankton and maintenance of the algal symbiont since, although it seems illogical, starved corals expel their symbionts (Yonge & Nicholls, 1931*b*). It was unfortunate that Yonge (1940) concluded that 'the zooxanthellae play no part in nutrition of the corals', a view he was later to reverse.

The only other algal group regularly recorded in symbiotic relationships is the Cyanophyta, cells of which (often termed cyanelles) occur in some marine sponges (*Verongia* and *Ircinia*). The rather enigmatic unicells *Cyanophora paradoxa* (a flagellate) and *Glaucocystis nostochinearum** (an apochlorotic unicellular green alga) also contain cyanophycean endosymbionts. Blue-green 'phaeosomes' are recorded (F. J. R. Taylor, 1973) in some of the colourless marine Dinophyta (*Ornithocercus, Parahistioneis, Histoneis*) and since Cyanophyta have been cultured from some related genera, these also need detailed study.

It is striking that the Chlorophyta (e.g. *Chlorella*) are the common symbionts of freshwater animals (protozoa, sponges, *Hydra* and some rhabdoceolous turbellarian flat worms) whilst Dinophyta are only found in marine animals. A word of caution on the identity of the symbiotic *Chlorella* is given by D. L. Taylor (1973) since definite criteria are difficult to establish for this genus and a *Chlorella*-like cell has been recorded now in a marine mollusc (Cooke, 1975). If it could be isolated into culture then a biochemical study might yield valuable data for comparison with the free living species. It may be of some evolutionary and taxonomic significance

---

* There is confusion over *Glaucocystis*, for recent workers refer to a 'cyanophycean' inclusion, yet Robinson & Preston (1971) concluded that it could be a red alga, a primitive dinoflagellate or a representative of some completely different and perhaps now unrecognisable taxonomic group. They did not favour a blue-green alga within a green alga as suggested by some workers.

that the Prasinophyta which are separated from the Chlorophyta by most workers are symbiotic in the marine sphere only.

All these symbionts photosynthesise and release organic molecules to the animal tissues. Maltose is released from *Chlorella* (Cernichiari, Muscatine & Smith, 1969), although glucose is reported in the freshwater sponge *Spongilla lacustris* which is said to harbour *Chlorella*, glycerol from the dinoflagellates (Muscatine, 1967) and lactic acid and amino acids from *Platymonas* (Gooday, 1970; Taylor, 1971). Movement of photosynthate from the algal symbiont into the host is more important in *Convoluta roscoffensis* since in this association the alga supplies all the nutritional requirements of the animal (closed symbiosis). Such a closed symbiosis is found also in adult radiolaria whereas in most other symbiotic systems the animals feed normally (open or facultative symbiosis – host feeding) and the alga is merely an additional souce of nutrients. Taylor (1971) showed that the secretion of photosynthate from *Platymonas* was considerably enhanced at pH 5.5 compared with that at pH 7.8 and also enhanced at high cell densities. The form of the extracellular carbohydrate secreted by the alga is invariably different from that of the main intracellular product which, for example, is sucrose in the *Chlorella* species (Cernichiari *et al.*, 1969). As might be expected there is considerable variation in the amounts released from superficially similar algae; thus six *Chlorella* strains isolated from *Paramecium* secreted from 5 to 85% of the total carbon assimilated (Smith *et al.*, 1969) a fact which perhaps indicates the existence of different physiological races of the symbiont. In unialgal culture, the algal symbionts excrete little of their photosynthate. The host also exhibits some control over secretion, as is shown by the fact that homogenates of the host tissues increase excretion by cultured symbiotic dinoflagellates (up to 40% of the photosynthate), but homogenates from aposymbiotic species, i.e. hosts living without algal symbionts, have no effect (quoted in Smith *et al.*, 1969). However this stimulatory effect does not extend to the *Chlorella* from *Chlorhydra*.

Whilst the symbiotic relationship is, in general, of one alga with one animal it is possible to infect one animal with several closely related species, although usually not with great success as far as growth of the animal is concerned (Provasoli, Yamasu & Manton,

1968). Provasoli *et al.* also showed that when algae of two different species were presented to *Convoluta* the true symbiont invariably succeeded in establishing itself in the host. An attractive mechanism is thought to exist and early work (Keeble & Gamble, 1907) showed that the flagellates were strongly attracted to empty *Convoluta* egg cases. D. L. Taylor (1973) reports the interesting case of the anemone *Aiptasia* which can be infected with its symbiont when this is introduced by feeding with mixed plankton containing the symbiont but not with a pure culture of the symbiont.

The ecological significance of symbiosis is unknown for all but the dinoflagellate–anemone and coral system but owing to the widespread occurrence of these organisms it is likely to be the one of greatest ecological importance. Unfortunately, there are few studies on rates of carbon fixation and subsequent translocation and utilisation of photosynthate in these symbioses and equally few studies on the overall productivity of coral reef systems (*see* p. 150). Using oxygen and carbon dioxide measurements in flow studies across isolated stands of coral reefs, Odum & Odum (1955) calculated extremely high values for symbiotic production, some 18 g carbon m^{-2} day^{-1}. *In vitro* studies by Kanwisher & Wainwright (1967) yielded equally surprising values (2.7–10.2 g carbon m^{-2} day^{-1}) for hermatypic corals and gorgonians whilst Kohn & Helfrich's (1957) figure of 2900 g carbon m^{-2} yr^{-1} for Hawaiian reefs is of the same order: they comment that apart from some turtle grass communities, it is the greatest marine production system. D. L. Taylor (1973) carried the technique further by incubating small corals in polythene bags at various depths and monitoring oxygen production (Table 9.1) and using both oxygen and ^{14}C methods obtained figures comparable to those of earlier workers. However, other workers report lower figures e.g. Scott & Jitts (1977) found only 0.9 g carbon m^{-2} day^{-1} for the symbiont but this was about three times the rate of the adjacent phytoplankton. The hard corals give results similar to those quoted above but the two soft corals (*Mussa* and *Scolymia*) exhibit much lower rates of fixation; this agrees with Muscatine (1974*b*) who noted that, whilst the algae in reef corals may account for 50% of the protein nitrogen, in temperate sea anemones they contribute much less. Between 24% and 40% of the ^{14}C fixed by the symbionts in *Zoanthus* is

Table 9.1. *Productivity of six species of coral from four different depths based on* in situ *oxygen production. Data converted to g carbon m^{-2} day^{-1}. From D. L. Taylor, 1973*

Species	Depth (m)			
	5	10	15	20
Porites astreoides	13.3	13.9	13.8	12.4
Montastrea annularis	9.8	12.5	14.2	12.1
M. cavernosa	10.6	11.8	11.7	10.3
Siderastrea radians	12.1	12.6	13.7	13.1
Mussa angulosa	1.9	2.5	4.1	3.0
Scolymia lacera	2.7	3.7	5.9	3.7

transferred to the animal after 3 h of photosynthesis. In *Anthopleura*, 40–50% of the photosynthate moved into the host and in *Palythoa* only 20–25%. The photosynthate moves principally into the animal lipid and protein components (Trench, 1971*a*), although Muscatine (1967) found that up to 40% of the labelled photosynthate in corals and *Tridachna* was liberated as glycerol. Trench (1971*b*) also reported mainly glycerol but some alanine, glucose and fumaric, succinic, and glycolic acids as well as two further unidentified organic acids were also released. The amount of glycerol excreted could be increased if homogenates of the hosts were added to the seawater in which the symbionts were grown: but homogenates of *aposymbiotic* animals were ineffective (Trench, 1971*c*). The longer the symbiont was kept in culture the closer its secretory pattern became to the free-living species, which suggests that the symbiotic algae are modified by the host, but not irreversibly. Some workers, e.g. Franzisket (1969, 1970) claim that corals can achieve net growth in the absence of particulate food and certainly some anthozoans (*Xenia hicksoni*, *Clavularia hamra* and *Zoanthus sociatus*) seem to be virtually autotrophic and to have lost some of the structures and functions associated with animal feeding. The mantle of Tridachnid clams is also extremely well-developed for an autotrophic mode of nutrition and admirably exposes the photosynthetic tissue to light. In these animals the symbionts lie free in the blood space and often adpressed to blood cells (Kawaguti, 1966). On the other hand the reef corals, are 'superbly efficient and voracious carnivores' (Goreau, Goreau & Yonge,

1971) and do not seem to have become adapted to an autotrophic mode of feeding (except that algal cells are present). The general concensus of opinion is that corals feed on plankton, detritus, etc. but derive additional nutrients from the symbionts. The exact amounts, rates, etc., need to be studied in many species and under varied conditions: it would not be surprising to find all variations from purely autotrophic to heterotrophic. The different stages must also be studied since the planulae larvae of corals, e.g. *Favia fragum* require both light and a food source for growth (Lewis, 1974). A recent study on the freshwater sponge *Spongilla lacustris* suggests that this is a very efficient photosynthetic system with gross fixation measured at 11.5 mg C g^{-1} ash-free dry weight of sponge each hour (net 8.1) by the oxygen method and 5.4 mg by the ^{14}C technique (Gilbert & Allen, 1973).

In addition to the contribution made by the symbionts to organic production on the reef, there is the added feature of increased rate of coral calcification due to algal photosynthesis. Until recently the calcification process was regarded as a simple series of reactions (Goreau, 1963):

$$Ca^{2+} + 2HCO_3^- \rightleftharpoons Ca(HCO_3)_2$$
$$Ca(HCO_3)_2 \rightleftharpoons CaCO_3 + H_2CO_3$$

The velocity of these reactions is increased by photosynthetic fixation of carbon dioxide at or near to the site of calcification. This simple series of reactions has been questioned (Muscatine, 1974) but nevertheless it has been shown quite clearly that photosynthesis enhances the rate of calcification in *Acropora* (Pearce

& Muscatine, 1971) and that photosynthetic inhibitors reduce the rate to about that of the dark level in *Pocillopora* (Vandermeulen, Davis & Muscatine, 1972). Recent work by Crossland & Barnes (1976) suggests that the endosymbionts provide substrates for mitochondrial ATP production and this stimulates ammonium ion uptake from sites of calcification and aids cycling of urea which is involved in the calcification process. The urea is transported to the sites of calcification as calcium salts of allantoins formed by the conversion of glycolate secreted by the symbionts, via glyoxylate, which then combines with urea to form the allantoins. Work by Barnes & Taylor (1973), on the relationship between carbon fixation and light intensity, revealed the intriguing situation that corals growing at high light intensities, e.g., *Montastrea*, fix less carbon than those grown at moderate intensity, whether they are collected from 9, 15, or 35.5 m depth. This is a feature comparable to the light inhibition effect on phytoplankton in surface waters (p. 344). Barnes & Taylor also found that corals collected at 35.5 m depth had a higher rate of carbon fixation than those from shallower depths: this could be due to variation in pigment content, pigment ratio, and/or more efficient fixation such as has been noted in cultures of diatoms (p. 353). These features need investigation, especially since Drew (1973) reported a de-

crease in the fixation with increasing depth (Fig. 9.1). In addition, measurements of the uptake of ^{45}Ca at various light intensities showed an increase, as light intensity increased up to saturating values, but a decrease above these values. Vandermeulen & Muscatine (1974) confirm the effect of light in accelerating the Ca-deposition some 9–10 fold. Chalker & Taylor (1975) report a 2.8–4.0 times enhancement at 10500 lux. The light enhanced calcification is blocked by inhibitors and uncouplers of oxidative phosphorylation, but dark rates are unaffected. Barnes & Taylor (1973) concluded that there was a more intimate relationship between photosynthesis and calcification of corals than previously thought and that this was further evidence that the simple chemical equations for calcification given above are inadequate since an energy requiring process involving active transport of carbon and carbonate ions is involved. There is, in addition, a relationship between growth patterns (morphology) of hermatypic corals and their photosynthetic activity; this is an indirect effect of the transport of photosynthate to certain sites where calcification is then enhanced. Goreau (1963) found that in hermatypic corals the average light:dark ratio of calcification was 5.9 (earlier experiments had given a figure of 9.02, which is probably more accurate; this is high by comparison with the ratio for calcareous algae (1.6)). Although more studies are required, Goreau estimated that the primary productivity of the coral symbionts may be two or three times as high per unit of nitrogen as that of the algae *Padina* and *Liagora*. If this is true then it indicates a real benefit to both partners.

Goreau & Goreau (1959) showed that coral containing symbionts calcified about 19 times faster than that which had lost its algae. However they also showed (1960) that the transfer of labelled photosynthetic carbon from algae to host was slight and insufficient to supply the animals' nutritional requirement. Yonge & Nicholls (1931a) found that corals devoid of zooxanthellae excreted large quantities of phosphorus whereas those containing the algae do not; the complete organisms even take up phosphorus from the water. They could however be made to excrete phosphurus if they were kept in the dark. Simkiss (1964) thought that the uptake of orthophosphate by the algae removed an otherwise potential inhibitor (crystal poison) of calcification.

Fig. 9.1. The depth variation in gross photosynthetic rates measured as ^{14}C fixation (squares) and oxygen output (triangles) in three coral species, *a, Acropora*; *b, Millepora*; *c, Goniastrea*. From Drew, 1973.

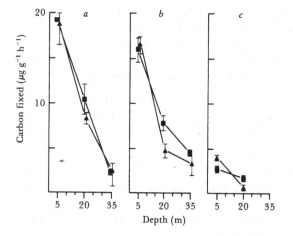

Gorgonian coelenterates (sea fans) also contain endosymbionts and are notably free from epiphytes and predators: this feature also has been attributed to the presence of the symbionts which are reported to form crassin acetate (Cierszko, 1962). The terpene certainly inhibits the development of sea urchin eggs and is toxic to parrot fish but not to one of the few fish (*Cyphoma gibbosa*) which feeds on the gorgonian.

Another symbiotic relationship involving host deposition of calcium carbonate is that between algae (both Chlorophyta and Dinophyta are reported) and Foraminifera. Lee & Zucker (1969) reported that both ^{14}C uptake and ^{45}Ca deposition were greater in light than dark and that here also there is a connection between products of photosynthesis and calcification.

Some foraminifera (e.g. *Amphisorus* (*Orbitolites*) *hemprichii*, *A. duplex*, *Marginopora vertebralis*) contain

Fig. 9.2. Effect of light intensity on the growth of green and bleached paramecia in medium initially filtered free of bacteria. Solid lines, stock 3–25 (green); dotted lines, stock 3*W* (bleached). Bright light, *LD* (460–580 ft-c.); dim light, *LD* (80–135 ft-c.); *DD*, continuous darkness. Vertical bars denote one standard error. From Karakashian, 1963.

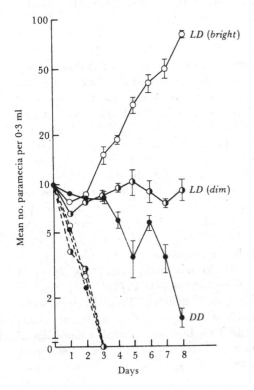

symbiotic dinoflagellates belonging to the genus *Gymnodinium* and Plante-Cuny (1973) reports 200–250 μg chlorophyll *a* g^{-1} foraminiferan and recently Sournia (1976*a*) found that the foraminiferan *Amphistegina lessonii* was responsible for the major carbon fixation on an atoll sandy lagoon. The foraminiferan *Archais* feeds on bacteria and diatoms (*Cylindrotheca closterium*, *Navicula diversistriata* and *Nitzschia* sp. were all taken up from culture) and so has an additional algal source of nutrients. Boltovsky (1963) reports many benthic genera of foraminifera containing *Chlorella*-like symbionts, whilst the symbiont in *A. angulatus* has recently been shown to be a *Chlamydomonas* sp. (Lee, Crockett, Hagen & Stone, 1974) and hence the Foraminifera, like the Turbellaria, are an unusual group of animals in possessing symbionts from several totally different algal phyla. In addition, Berthold (1978) has reported a cryptomonad (*Cryptomonas schaudinii*), two more dinoflagellates (*Pyrocystis* and *Dissodinium*) and a red alga. The effect of the symbiont on growth of the foraminiferan *Heterostegina depressa* shows clearly that light above 300 lux is depressive. This species grows in the benthos and large numbers are recorded in shallow waters. In similar habitats other species tend to be fewer per unit area as the size of the protozoan increases; this does not apply to *Heterostegina* presumably because it is independent of ingested food.

The colour of Foraminifera is not always due to symbiotic algae although Murray (1963) did find *Elphidium* fed with *Phaeodactylum* had brown cytoplasm and when fed with *Tetraselmis* it turned green.

The association of *Chlorella* and the protozoan *Paramecium* is well-known but numerous other protozoa also have symbiotic relationships, e.g. the freshwater ciliates, *Frontonia*, *Prorodon*, *Climacostomum* and *Euplotes*. The *Chlorella* symbiotic with these ciliates could be sub-divided into four species and these into subgroups showing that it is not simply a case of a single symbiont in many hosts (Sud, 1969). The algal partner in freshwater generally seems to be a *Chlorella* and when an animal group occurs in both freshwater and marine sites the freshwater generally have *Chlorella* (e.g. the turbellarian *Phaeonocora* (Eaton & Young, 1975)) whilst the marine turbellaria have Prasinophyta or Dinophyta and possibly other groups.

The symbiosis between *Chlorella* and *Paramecium bursaria* is perhaps the most intensively studied of the algal–protozoal type. Each *Paramecium* has several hundred *Chlorella* cells within its cytoplasm contained within special individual vacuoles and transmitted during division and conjugation (Karakashian, 1963). *Paramecium* also feeds by ingesting bacteria and Karakashian showed the effects of light and bacterial concentration on growth of the protozoa (Figs. 9.2, 9.3); the protozoa plus alga becomes partially independent of the bacterial food supply. The aposymbiotic *Paramecium* rapidly declines in culture when the bacteria are depleted. The *Chlorella* isolated from *Paramecium* releases large quantities of maltose into the medium when cultured free: this compares with very slight release by free-living *Chlorella* species which mainly secrete glycolic acid. Brown & Nielsen (1974) showed that this release of maltose also occurred within the *Paramecium*. The growth of *Paramecium* with symbionts in the absence of a bacterial source of food is governed by the rate

Fig. 9.3. Effect of bacterial concentration on growth of green and bleached paramecia, *a*, in *LD* (230–330 ft-c.) and *b*, *DD*. Solid lines, stock 3–25 (green); dotted lines, stock 3*W* (bleached). Bacterial concentrations were adjusted by dilution. Dilution factors appear under appropriate growth curves in graph. From Karakashian, 1963.

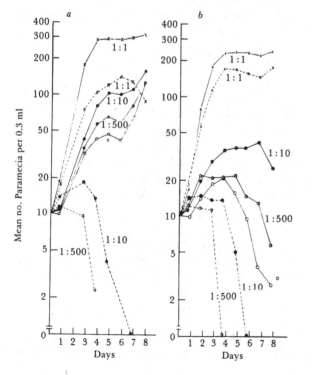

of photosynthesis and by the number of *Chlorella* per *Paramecium* which is a specific characteristic of the various strains of the protozoan. The division of the symbiotic *Chlorella* when the *Paramecium* are maintained in the dark (Karakashian, 1963), is presumably a consequence of their heterotrophic abilities. The ecological significance of the phototrophic or heterotrophic capacity and the contribution of these attributes to the ecology of this protozoan, or symbiotic protozoa in general, and their contribution to carbon fixation in freshwater habitats is not known.

Turning to a marine group of protozoa, the Radiolaria, one finds that the symbionts are again dinoflagellates (*Amphidinum* in *Colozoum* and *Sphaerozoum* (Taylor, 1974). There have been isolated reports that the primary production by these symbionts exceeds that of the free-living phytoplankton in some tropical oceans. The subject clearly requires re-investigation.

A problem which is still not completely understood is how the animal host recognises the right symbiont and then incorporates it into its own tissues. Work on *Chlorhydra* (*Hydra*) *viridis* by Pardy & Muscatine (1973) has shown that aposymbiotic *Hydra* can be re-infected by *Chlorella* from other *Hydra* and from one strain of *Paramecium* but other types of algae are rejected. Apparently the alga comes into contact with the digestive cells in the coelenteron of *Hydra*, and remains in contact with them during 'engulfment' (phagocytosis); this is followed by a brief 'recognition period' and if the alga is suitable it is incorporated whilst foreign algae and unwanted particles are rejected. The *Chlorella* is then transported to the base of the digestive cell where it undergoes a period of growth, including autospore formation within its individual vesicle (cf. *Paramecium*). Rejection of unsuitable algae occurs at the surface of the digestive cell in *Hydra* but in *Paramecium* the unsuitable algae are ingested into the vacuoles and then rejected. As in *Paramecium* there is a transference of the nutrients from *Hydra* to *Chlorella* when the animals are maintained in the dark, but the *Chlorella* cells do not divide until the *Hydra* is returned to the light (Cook, 1972).

Goetsch & Scheuring (1926) reported symbiotic *Chlorella* in the mantle of the freshwater clams *Anodonta* and *Unio* but only a small proportion of animals in a population were affected. They report earlier records of this symbiosis but the exact nature needs re-investigating since only a few animals exhibit it.

A rather unusual symbiosis is that between a green, un-named,* uniflagellate alga and the colourless Dinoflagellate *Noctiluca miliaris*; in this system the flagellates seem to be unaffected by their sojourn in the *Noctiluca* and swim when released. Sweeney (1971) reports the rather surprising number of 6–12 000 flagellates within a single *Noctiluca*. The *Noctiluca* can also ingest and feed on *Dunaliella*, yet another example of selectivity. In the absence of the *Dunaliella* the host can utilise products from the symbiotic flagellate but in the dark the symbiont is very rapidly lost. The evidence points to an even looser symbiosis than that between *Amphidium* and *Amphiscolops* (p. 387).

Loose associations between algae and animals must be fairly common but they are only slightly documented. Recently a widespread association has been reported between macroscopic growths of *Nostoc parmeloides* and midge larvae in Colorado streams (Todd, 1971). The community is at a peak in February to April when the midges emerge after eating the bulk of the algal colonies, which then tend to break down. Yet another recently found association is that between the gametophytes of *Desmarestia* and the Sea Pen (*Ptilosarcus gurneyi*): the filamentous gametophyte is embedded in the tissues of the Sea Pen. This is thought to be the first record of the *Desmarestia* gametophytes (Dube & Ball, 1971) and shows how intensively one may have to search for some stages of algal life-cycles.

A somewhat different example of animal–algal interaction is the finding that the Sea Hare (*Aplysia californica*) requires the thallus of *Laurencia pacifica* as a surface for its larval stage and without this the larvae fail to metamorphose and then die (Kriegstein, Castellucci & Kandel, 1974).

Many algal remnants are used by larvae, etc. to build cases or thecae (e.g. Tintinnids) and an interesting case is that of a trichopteran larva (*Nectopsyche*) which builds cone-shaped cases with the diatom *Terpsinoe* (Wallace, Sherberger & Sherberger, 1976).

Chloroplast symbiosis

A further type of symbiosis was discovered in 1965 by Kawaguti & Yamasu when they showed by electron microscopy that the green bodies in the digestive gland of the opisthobranch mollusc *Elysia atroviridis* were indistinguishable from the chloroplasts of *Codium fragile*. Later workers found that the chloroplasts were not being digested and also do not divide (at least in *Elysia viridis*,† Taylor, 1968, 1970) and indeed remain functional for at least three months when *E. viridis* is starved in the light (Hinde & Smith, 1972). Similar symbioses occur in *Tridachia crispata*, *Tridachiella diomedea* and *Placobranchus ianthobapsus* and photosynthetic studies (Trench, Green & Bystrom, 1969) showed fixation of ^{14}C accompanied by oxygen production and release of labelled hexose and hexosamine moieties in the mucus from the pedal gland, which indicates that the products of photosynthesis are utilised by the animals (Trench, Trench & Muscatine, 1970, 1972). Not all Sacoglossans maintain the chloroplasts as symbiotic organelles, e.g. in *Placidia dendritica* the activity is lost within 24 h and two species, *Berthelinia chloris* and *Hermaeina smithi* do not use the chloroplasts except as food (Greene, 1974). As far as I am aware, this is the only group of animals feeding suctorially on algae and presumably digesting or rejecting the non-chloroplast fraction of the cytoplasm but retaining the chloroplasts in an intact and active state in the digestive diverticula.

All the Sacoglossans so far studied feed on Bryopsidophyceae with the exception of *Hermaeina bifida* which obtains its plastids from the rhodophyta *Griffithsia* or *Delesseria* (Greene, 1970). The animals cannot transfer the chloroplast from one generation to the next via the egg cells and must obtain a new supply in each generation by feeding (Greene, 1974). Certain differences have been noted between the incorporation of ^{14}C into the chloroplasts symbiotic in these animals and into the chloroplast in the alga; one difference is that in the animal the ^{14}C enters the carotenes of the plastids but hardly into the chlorophylls (Trench & Smith, 1970). The rate of carbon fixation by the chloroplasts in *Elysia* is about 60% of that in *Codium* (Trench, Boyle & Smith, 1973) and the incorporation of the ^{14}C is into galactose and glycolic acid in *Elysia* but into sucrose and glucose in *Codium*: both the galactose and sucrose are, however, synthesised in the tissues

---

* Sweeney (1976) found that this was a new species of the prasinophyceaen genus *Pednomonas*.

† There is confusion in the classification, *E. viridis* and *E. atroviridis* may be identical.

outside the chloroplasts. The substances released by the chloroplasts within *Elysia* are glucose or galactose (Trench, Boyle & Smith, 1974). Greene (1970) reported that isolated chloroplasts of *Codium* leaked a compound tentatively identified as glycolic acid into the medium. Trench (1975) reviews work on this aspect and concludes that glucose is leaked from the chloroplasts to the mollusc which then converts it to galactose. Work on this symbiotic system has been undertaken only under laboratory conditions and productivity studies in nature are needed. Ecologically there has to be an intimate association of mollusc and alga and it would be interesting to study the geographical distribution of the 'couplets'.

Algal–fungal symbiosis

The other extensive symbiotic association involving algae is that between Chlorophyta or Cyanophyta and ascomycetous fungi to form lichens. There are some 17 000 different lichen species: each contains a single alga (often termed the phycobiont) of which the commonest genera are *Trebouxia* (Chlorococcales), *Trentepohlia* (variously classified either in the Chaetophorales or more likely in the Trentepohliales) and *Nostoc* (Nostocales). The lichens containing *Trentepohlia* tend to be more abundant in the tropics. The algal cells are located in a subsurface layer often adpressed to specialised hyphae and sometimes penetrated by haustoria. As with the algal–animal symbioses the algae actively photosynthesise and pass carbohydrate to the fungal mycelium. Smith & Drew (1965) showed that in *Peltigera polydactyla* 40 % of the photosynthetically fixed carbon was translocated, although work with some lichens has also shown very little transport. Again as with the algal–animal symbioses the movement tends to be mainly of a single carbohydrate, glucose from *Nostoc*, *Calothrix* or *Scytonema*, ribitol from *Trebouxia*, *Myrmecia* and *Coccomyxa*, sorbitol from *Hyalococcus* and erythritol from *Trentepohlia* (Drew & Smith, 1967; Richardson, Hill & Smith, 1968; Smith *et al.*, 1969). These are transformed to mannitol and to a lesser extent to arabitol in the fungal component. Whereas the symbiosis between animals, especially corals and algae, leads to highly productive systems, that of lichens is remarkable for its low productivity at least if growth of the lichen is any measure of productivity. Growth rates of lichens are low but the small amount of data on rates of photosynthesis of the phycobiont

suggest that this is not appreciably less than that for free-living algae (Smith, 1975). Could it be that this particular symbiosis slows down growth as an adaptation to an extreme environment where water and nutrient supply are often minimal, though not necessarily limiting? Lichens also grow in streams and along margins of lakes and its seems that in these more or less permanently wet sites growth is still slow: more data are needed. However since the symbionts of some lichens are species of Cyanophyceae which fix nitrogen one would expect an additional supply of nutrient from this source and indeed Bond & Scott (1955) found a distinct increase in labelled nitrogen in two lichens *Collema granosum* and *Leptogium lichenoides* which have *Nostoc* as the phycobiont. Recently, Horne (1972) has shown that the lichen *Collema pulposum* growing in the antarctic increased the nitrogen biomass in the sites where it grew by amounts varying between 0.8 and 2.3 % each year which could be important in this area where available nitrogen may be the only major nutrient likely to limit plant growth. Millbank (1975) has shown that the fixation takes place in the heterocysts in spite of the fact that these only occur at a frequency of about 3 %; the nitrogen fixed is transferred to the fungus.

Algal–bacterial symbiosis

Observations of both freshwater and marine algae rarely fail to produce microscopic evidence of bacteria. Sometimes quite obvious filamentous bacteria occur, but the thalli and mucilage of all algae have a less obvious coating of coccoid and bacilloid bacteria. The filamentous genus *Leucothrix mucor* seems to have a world-wide distribution on marine algae and other organisms (Johnson *et al.*, 1971). Uptake of organic acids into algae is certainly influenced by uptake into the associated bacterial flora. Freshwater algae, such as desmids, which produce coherent cones of mucilage frequently exhibit chains of bateria growing in the interstices between adjacent cones. Bacteria are found in the mucilage surrounding the chrysophyte *Chrysostephanosphaera* (Geitler, 1948) and also in the mucilage of *Volvox aureus* (Hamburger, 1958). In the case of *Volvox* the bacteria (strains of *Pseudomonas fluorescens*) seem to be essential to the flagellate since daughter colonies dissected out from the center of the parent colony were not viable unless supplied with the bacteria:

this could be the effect of an organic factor supplied by the bacteria. Two strains are involved, one of which gave good growth of the *Volvox* but small colonies which did not reproduce asexually and the other produced large but chlorotic colonies and no more than two generations. This suggests that several factors are produced by the bacteria. That this symbiosis is widespread was shown by the isolation of the same bacterial strains from several different *Volvox* sources and their specificity was shown by inability to infect the *Volvox* with other bacterial species. The fact that bacteria are so commonly found in close association with algae when sections are made for electron microscopy, even in some of the so-called bacteria-free cultures, suggests a very close relationship between the organisms. Another comparable observation is that of Machlis (1973) who found that male and female *Oedogonium* plants in axenic culture only rarely produced gametes whereas if they were grown in the presence of *Corynebacterium* or *Pseudomonas putrida* the development of reproductive cells was much increased. The marine planktonic alga *Asterionella japonica* could only be grown satisfactorily when bacteria were present (Kain & Fogg, 1958). Such effects must also occur in nature. Bacteria are even found endophytically, e.g. inside the nuclei of euglenoids (Leedale, 1969) and within the cytoplasm of dinoflagellates such as *Gymnodinium*, *Ceratium*, *Amphidinium* and *Prorocentrum* (Gold & Pollingher, 1971) At the moment however they do not seem to be quite as abundant within algae or as conspicuous on the outside as they are in some protozoa (Ball, 1969). If there is a true symbiosis between algae and bacteria we have not even scratched the surface of the problem and if the situation described for *Volvox* is at all common the ecological implications are extensive. Waite & Mitchell (1976) thought that a decline in the productivity of *Ulva lactuca* in heated waters was due to a reduction of growth-stimulating compounds from bacteria: they also isolated bacteria capable of degrading *Ulva* cell walls. On the other hand, some algae produce antibiotics which inhibit the growth of bacteria, though the tests have usually been on common non-marine bacteria, e.g. Hornsey & Hide (1974) tested 151 British marine algae on *Staphylococcus aureus*, *Escherichia coli*, *Bacillus subtilis*, *Streptococcus pyogenes* and *Proteus morganii* and found particularly good antibacterial properties in most of the Rhodomelaceae, and in *Chondrus crispus*, *Dilsea*

carnosa, *Gloiosiphonia capillaris*, *Sphondylothamnion multifidum*, *Desmarestia aculeata*, *D. lingulata*, *Laminaria digitata*, *L. saccharia*, *Dictyopteris membranacea*, *Cystoseira baccata*, *C. tamarisifolia* and *Halidrys siliquosus*. Later, Hornsey & Hide (1976) showed distinct seasonal variation in antibiotic production. Four patterns were recognised: (a), a *Polysiphonia*-type exhibiting uniform production throughout the year; (b) a *Laminaria*-type of maximal production in the winter months (cf. the growth patterns of this alga) which is also shown by *Chondrus crispus*, *Laurencia pinnatifida* and *Ulva lactuca*; (c), a *Dictyota*-type having maximal production in summer and also shown by *Ascophyllum nodosum* and *Dilsea carnosa*; and (d), a *Codium*-type which is of spring activity and also exhibited by *Halidrys siliquosa*. Of the green algae tested only *Codium* had much activity. Such interesting results should now stimulate studies on the effects, if any, of these presumably 'extracellular' products in nature and on the epiphytic bacterial and algal floras of the algae. Unicellular algae, e.g. *Skeletonema costatum* have been shown to have anticoliform activity in cultural studies (Sieburth & Pratt, 1962) and this type of activity is considered important in water purification systems, e.g., the growth of *Nitzschia palea* on filters is reported to reduce the coliform count. Antibacterial activity is present in the water where *S. costatum* have been blooming. Actually the situation is exceedingly complex since *S. costatum* also produces an extracellular compound stimulatory to bacteria (Bell *et al.*, 1974) as well as exhibiting antibacterial activity against *Escherichia coli* (Sieburth, 1965). Tests on freshwater algae show very few (a *Spirogyra* sp. and *Euglena viridis*) to have any antibacterial activity (Stangenberg, 1968*b*).

The association of bacteria with Cyanophyta may, under certain conditions result in a stimulation of photosynthesis especially in waters where carbon dioxide concentrations are extremely low. Lange (1971) thought that when carbon dioxide was limiting the addition of organic substrates which increased the bacterial metabolism may also have enhanced the Cyanophyta growth. There is however considerable doubt as to how often or effective this would be in natural populations: it may only be operative in dense blooms and when pH is raised considerably.

Saprophytism

Heterotrophic algae living on decaying organic matter are rare but one instance is that of colourless diatoms of the genus *Nitzschia* which have been isolated from rotting seaweed. The first record of this saprophytic diatom was made by Cohn as early as 1854 when it was identified as *Synedra putrida* and it has been recorded several times since. Recently Lewin & Lewin (1967) have described two other colourless species of *Nitzschia* and have shown from experimental studies that they can be maintained in artificial seawater supplemented with mineral nutrients, thiamin and cobalamin. Lactate or succinate could serve as the sole organic carbon source for all three species; two could also use glucose or glutamate.

Parasitism

Two different aspects of parasitism require consideration, parasitic algae and parasitism of algae by other organisms. The first is not at all common and involves a number of algae belonging mainly to the Rhodophyta which are colourless and parasitic on other Rhodophyta. Goff (1976) reports that there may be as many as 40 parasitic genera.

The ecological effects of parasitism and grazing are similar in that the growth rates of certain algal populations are changed, the balance of species is altered and a direct energy flow to the secondary producers is established, coupled with the formation of organic detritus from the damaged algal cells. This algal detritus can then be further degraded by bacteria or utilised directly by other secondary producers. Grazing has always been assumed to be of considerable importance in the energy flow from phytoplankton to zooplankton but there are surprisingly few experimental data on this topic. Equally there can be no doubt about the importance of grazing on rocky shores, a feature which has been recently indirectly highlighted by the adverse effects of pollutants and clearing agents on the animals which normally graze on algae. Epipelic populations are grazed by small invertebrates and also by fish, especially in tropical seas, lakes and reservoirs where some fish are purely algal feeders. Whilst grazing involves the ingestion of algal material, parasitism involves the penetration of the algal cell by the infecting species and often the complete 'grazing' of the cytoplasm leaving only the wall behind. A large number of species parasitic on algae are now known but the subject is less well studied than grazing.

Parasitic algae

Parasitic algae are relatively rare though it is a little difficult to distinguish between some instances of endophytism (e.g. those of *Chlorochytrium*) where some pathological reactions of the host have been reported, and the more definite cases of parasitism. Some organisms first described as chytridiaceous fungi are apparently more correctly classified as colourless parasitic algae (e.g. *Synchytrium borreriae*) and the fairly widespread but controversial *Rhodochytrium* which occurs on a range of tropical plants. Another alga lacking photosynthetic pigments is *Phyllosiphon*; this spreads its branching coenocytic filaments through the intercellular spaces of some tropical plants. Strangely it forms aplanospores which contain chloroplasts. All these, and perhaps the commonest of all, *Cephaleuros*, are tropical in distribution. The latter genus belongs to the Trentepohliaceae, a family which contains only subaerial algae. *Cephaleuros* is the only alga which is recognised by plant pathologists: it causes a disease known as 'red rust' on tea, coffee, cocoa, citrus fruits, nutmeg, rubber, guava, mango, avocado and oil palm. The alga forms disc-like growths generally between the cuticle and epidermal cells, with occasional penetration between the palisade and mesophyll cells. The thalli do possess plastids and photosynthetic pigments. *Cephaleuros* acts also as the phycobiont in some tropical lichens. In spite of the importance of *Cephaleuros* as a pathogen little is known of its ecology other than its wind-borne dispersal and confinement to the tropics and subtropics: much of the literature on these parasitic algae is contained in rather rare tropical agricultural journals (*see* Joubert & Rijkenberg, 1971).

The only other group containing parasitic genera is the Rhodophyta but their parasitism may be of a special kind in that the hosts are always other red algae. Such a relationship of closely related species is often termed alloparasitism. The species tend to form small colourless pustules on the surface, e.g. *Halosacciocolax* on *Rhodymenia*, the lower cells penetrating between the host cortical cells (Guiry, 1974a). In the case of *Harveyella* growing on *Odonthalia* and *Rhodomela* the parasite causes damage to the host tissue but also stimulates the host to increase its

Fig. 9.4. *a*. an immature sporangium of the chytrid *Rhizophidium nobile* on a *Ceratium* resting spore (Mag: ×1450). *b*, *c*, *Cosmarium* killed by a biflagellate fungus *Myzocytium*. Two infections present, note collapsed walls of fungal thalli inside the alga and external exit tubes. *c*, shows also the sphere of mucilage around the *Cosmarium* (Mag: *b* × 1600; *c* × 550). *d*, a cyst of *Vampyrella* attached to a planktonic species of *Ulothrix*. The feeding *Vampyrella* has extracted the contents of five algal cells and their undigested remains occur as a black mass in the cyst. When digestion is complete the contents of the cyst give rise to two new *Vampyrella* which seek out further algal cells. Photographs kindly supplied by Dr H. Canter.

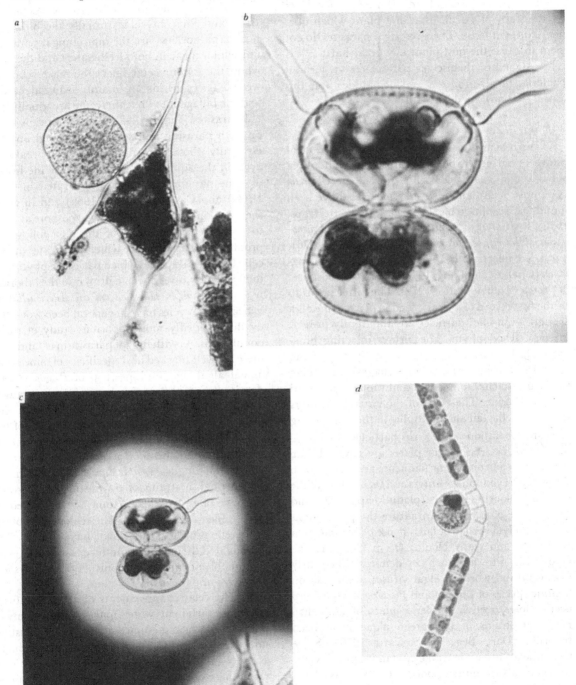

bicarbonate uptake (Goff, 1976). Some are even calcified, e.g. the recently described genus *Ezo* which occurs on the calcareous alga *Lithophyllum yessoense* (Adey, Masaki & Akioka, 1974). Here also the lower cells of the parasite penetrate the calcified host thallus. *Ezo* is one of the few genera of parasites which is also calcified and would form a good subject for the study of calcification in a plant which does not photosynthesise. These red algal parasites do not seem to cause the host plant any great harm.

Algae are themselves parasitised by viruses, bacteria, fungi, protozoa and nematodes and this aspect will now be dealt with.

Viruses

Viruses are the most recently discovered algal parasites and seem to be common only in Cyanophyta though one has also been reported in *Chlorella* (Gromov, 1968), and virus-like particles have been found in germlings of *Oedogonium* (Pickett-Heaps, 1972), in *Hydrurus* (Hoffman, 1978) and in *Hymenomonas*, *Micromonas* and *Cryptomonas* (Pienaar, 1976). Virus-like infections have been described in *Porphyridium* (Chapman & Lang, 1973) and in the zoospores of *Chorda* (Toth & Wilce, 1972). The latter seems to be the first record of a viral infection of a marine alga though 'viral-like' particles are apparently present in some Rhodophyta. The viruses infecting blue-green algae (cyanophages) are virtually identical to those infecting bacteria (bacteriophages). The tail of the virus attaches to the alga and the angular head remains outside. The DNA of the virus then enters the host via the tail and multiplies in the cell causing lysis and liberation of the virus particles. This viral action can be seen if agar plates are covered with cyanophycean colonies or filaments are seeded with virus. Each cyanophage enters and sets off lysis of the cells and colourless patches (plaques) appear around each centre of infection. In nature the populations of Cyanophyta need to be fairly dense before infection is obvious and water-blooms (*see* p. 305) are the favourable site for such epidemics. The early discovery and work on algal viruses was made on benthic species of *Lyngbya* and *Plectonema* which also occur close together as dense mats of filaments. Resistant strains of blue-green algae have been reported (Daft, Begg & Stewart, 1970). Since Cyanophyta are undesirable contaminants of water reservoirs, swimming pools, etc. it has been

suggested that some measure of control may be possible by adding virus cultures to the waters. The ecological effects of viral infections during natural growth of bloom-forming species need to be studied.

Fungi

Fungi of the groups Chytridiomycetae (chytrids) and Phycomycetae (usually the biflagellate, holocarpic species) are the main fungal parasites of unicellular algae in both freshwaters and the oceans whilst the Ascomycetae infect the macroscopic seaweeds. The chytrids are mainly external parasites and the biflagellate Phycomycetae are usually internal parasites (Lund, 1957). Further work is needed on both parasitic and saprophytic fungi on algae, especially those contributing to the break-down cycle in the aquatic environment. The life histories of some parasitic species are illustrated in Canter (1979) and Canter & Lund (1969) and in general the infection is brought about by a zoospore attaching to the mucilage or wall of the alga followed by production of a germ tube which penetrates the algal cell. The fungal species which parasitise phytoplankton have only rarely been cultured in the laboratory (e.g. *Rhizophidium planktonicum* on *Asterionella*) and hence the life cycles have in general been worked out by the decidedly more laborious study of natural populations. As with fungal parasitism of land plants there is a well-marked host specificity of some of these fungal–algal relationships (e.g. of *Rhizophidium* on *Asterionella*, which has now been elegantly shown in culture (Canter & Jaworski, 1978)), whilst other fungi attack numerous hosts. However few of these inter-relationships are simple and these workers also showed that some races of *Asterionella* are susceptible to the fungus *Zygorhizidium* and others are not, yet all English strains of *Asterionella* so far isolated and the one American strain so far tested are susceptible to *Rhizophidium*. The ecological problems of perennation of the fungi have hardly been investigated though most produce resting spores and presumably are meroplanktonic. Fig. 9.4 shows some examples of parasitism of algae.

The relative movements of algae and fungal spores in turbulent water makes contact between parasite and host a chance affair. This almost certainly explains why fungal infections amounting to 25% of an *Asterionella* population can only occur when there are eight or more diatom cells (approx.

one colony) in one millilitre and one cell in ten millilitres in the case of desmids (Canter & Lund, 1969). From this it can be seen why epidemics are more common in the richer lakes (Lund, 1959). In view of the chances of infection it is a little surprising to find that algae which occur only sporadically are more or less severely parasitised by their own specific parasites and this suggests that resting spores are involved. These, however, would have to germinate just when the host alga was blooming and therefore

Fig. 9.5. Changes in the abundance of six desmids in relation to fungal parasitism. The percentages of the cells of a given desmid bearing parasites are indicated by the numbers against the graphs. Windermere, South Basin, 1952. *a*, *Staurastrum pseudopelagicum*; *b*, continuous line, *Spondylosium planum*, dotted line, *Staurodesmus megacanthus* var. *jaculiferus*; *c*, continuous line, *Staurastrum cingulum*; dashed line, *S. lunatum*; dotted line, *Cosmarium contractum* var. *ellipsoideum*. From Canter & Lund, 1969.

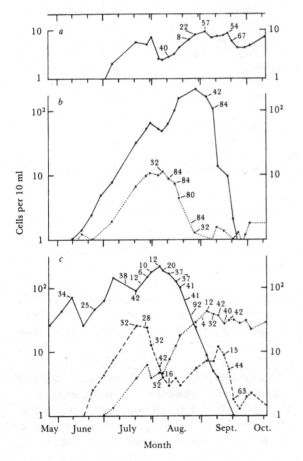

be under some correlated environmental control factor.

Most algal genera occurring in the freshwater phytoplankton have been found to be susceptible to fungal parasites but only rarely has the effect on natural populations been recorded. In fact the work of Canter in the English Lake District has highlighted this aspect of limnology and Fig. 9.5 illustrates the dramatic effects of such parasitism on species of desmids where, in most instances, fairly dramatic declines of algal populations occur. The frequency of chytrid epidemics during one year and the high rate of infection can be seen from Table 9.2. The epidemics do not however alter the normal seasonal periodicity (*see* Figs. 10.1, 10.15) and in all cases the algal populations have to increase significantly before the parasite gets a hold. However, as Canter & Lund (1969) point out, desmids would be much more abundant in Windermere in the absence of the chytrid and biflagellate parasites. The desmids occur in summer and early autumn, at a time when potentially a large number of species can grow rapidly in spite of the fact that this is often a time of low nutrient supply to the lakes and possibly therefore a time of competition for nutrients. The great species diversity at this time may, in part, be due to parasitism and grazing of separate species thus allowing others to proliferate before they in turn are attacked or grazed (Canter & Lund, 1968). That species recover from extensive parasitism seems surprising, until one remembers that division of a small number of cells rapidly re-establishes a population. At other times of the year, however, competition between species coupled with declining temperatures and/or day-length may prevent recovery by the population.

In the sea, fungal parasites have been less frequently studied though species of *Ectrogella* infect a number of diatoms and the biflagellate genus *Lagenisma* infects *Coscinodiscus* sp. and recently has been found to grow to epidemic proportions.

Parasitism in general does not seem to be related to 'unhealthy' populations; indeed, what is an 'unhealthy' algal population? Possibly one such 'unhealthy' condition might be when cell division is slowing down: it is, however, not known for certain at what stage of plankton growth most parasitic attacks occur. In fact, apparently, very healthy algal cells are readily parasitised but during the

Table 9.2. *The epidemics occurring in some of the English Lake District lakes and the lowland reservoir, Farmoor, during one year*

In Windermere, Northern and Southern basins. In Blelham Tarn the epidemics occurred in water held in a large plastic, experimental enclosure. *Asterionella* is known to be parasitised by three species of chytrid and each one of these produced an epidemic in Grasmere in 1973. From Canter, 1974

Date	Lake or reservoir	Alga	Parasite	Infection (%)
1973				
20 Feb.	Farmoor	*Asterionella*	Chytrid	66
26 Mar.	Grasmere	*Asterionella*	Chytrid	98
3 Apr.	Esthwaite	*Asterionella*	Chytrid	33
14 May	Grasmere	*Asterionella*	Chytrid	50
11 June	Grasmere	*Sphaerocystis*	Chytrid	53
25 June	Grasmere	*Asterionella*	Chytrid	85
1 July	Grasmere	*Radiococcus*	Protozoa	46
8 July	Esthwaite	*Tabellaria*	Chytrid	90
12 July	Blelham B.	*Cosmarium*	Biflagellate fungus	60
24 July	Blelham B.	*Staurastrum*	Chytrid	72
24 July	Windermere S.	*Staurastrum*	Chytrid	84
24 July	Esthwaite	*Asterionella*	Chytrid	60
30 July	Grasmere	*Staurastrum*	Biflagellate fungus	90
6 Aug.	Grasmere	*Asterionella*	Chytrid	81
6 Aug.	Blelham Tarn	*Oscillatoria*	Chytrid	59
7 Aug.	Windermere N.	*Staurastrum*	Chytrid and biflagellate fungus	57
14 Aug.	Windermere N.	*Chrysocapsa*	Chytrid	—
18 Sept.	Grasmere	*Eudorina*	Protozoa	—
20 Sept.	Grasmere	*Asterionella*	Chytrid	43
25 Sept.	Windermere S.	*Geminella*	Protozoa	—
1 Oct.	Grasmere	*Staurastrum*	Biflagellate fungus	69
1 Oct.	Grasmere	*Dictyosphaerium*	Protozoa	96
1 Oct.	Farmoor	*Staurastrum*	Biflagellate fungus	90
2 Oct.	Windermere N.	*Cosmarium*	Biflagellate fungus	52
2 Oct.	Windermere N.	*Staurastrum*	Chytrid	47
2 Oct.	Windermere N.	*Xanthidium*	Biflagellate fungus	76
6 Nov.	Blelham B.	*Oscillatoria*	Chytrid	65
1974				
2 Jan.	Farmoor	*Asterionella*	Chytrid	66

fast growing period few epidemics seem to take hold.

There are remarkably few studies of infections of intertidal algae but two ascomycete species, *Mycosphaerella ascophylli* and *M. pelvetiae*, appear to be confined to *Ascophyllum* and *Pelvetia* respectively. The latter is sometimes thought to be a 'lichenised' association, whilst both are host-specific (Kohlmeyer & Kohlmeyer, 1972). These workers comment that the *Ascophyllum*/*Mycosphaerella* association is rather

akin to an ectotrophic mycorrhizal system. In these and other fungal/algal associations the penetration of fungal hyphae is apparently always intercellular.

Protozoa

Protozoa of the genera *Pseudospora*, *Vampyrella*, *Aphelidium* and *Amoeboaphelidium* parasitise freshwater phytoplankton and spread from one alga to another, rather like the chytrid fungi, by motile stages in the life history. The parasites either suck out the cell contents via a filopodium or enter the algal cells and digest all the cytoplasm. Like some fungi they show a high degree of host specificity e.g. *Aphelidium* utilised only six strains, all coccoid, of Chlorophyta out of 248 strains tested (Gromov & Mamkaeva, 1969).

Bacteria

Bacteria of the Myxobacteria occur as pathogens of algae and like the viruses tend to be most abundant on Cyanophyta causing lysis of water-bloom genera, e.g. of *Anabaena*, *Aphanizomenon* and *Microcystis* (Daft & Stewart, 1971), although over 40 strains were susceptible. The bacteria involved are aerobic, gram-negative rods, non-flagellate but motile and pigmented. The bacteria themselves are necessary for lysis which cannot be achieved by filtrates from bacterial cutlures. Heterocysts within the filaments of the Cyanophyta were not attacked and it was shown that the bacteria have no ability to degrade cellulose, a component of the cyst wall. The blue-green algal cells have little or no cellulose and thus no protection from the bacteria: this is interesting ecologically since many other algae do have cellulose walls and also appear to be relatively immune to bacteria. The relationship of these bacteria to the often catastrophic decline of blue-green algal blooms needs study. Bacteria are often seen in algal cells but usually it is not possible to decide whether or not they have entered before or after damage to the cell though Turner & Friedman (1974) report endosymbiotic bacteria in *Penicillus* and Kochert & Olsen (1970) also found bacteria in *Volvox* vegetative cells as well as gonidia and sperm. Within the mucilage sheaths and on the surface of the mucilage of many planktonic and benthic algae numerous bacteria (usually rod-shaped) can be seen and these are probably not parasitic but rather symbiotic (*see* p. 396). Bacterial pathogens of the macroscopic seaweeds are reported, e.g. in *Macrocystis* a bacterial rot occurs.

Growth of bacteria on algae is certainly very common if not universal on macroscopic genera but very little is known of the ecology of these associations and some, as mentioned on p. 396, may be symbiotic whilst others are simply degrading dead and excreted material. Mann (1972*b*) reports unpublished work of McBride suggesting that there are between 6.8 and 20.7×10^6 epiphytic bacteria per gram dry weight of *Chondrus crispus* and Chan & McManus (1969) reported 10^4–10^7 bacteria per gram of *Ascophyllum nodosum*. Sieburth (1965) found that the common bacteria associated with *Polysiphonia lanosa* were *Vibrio* and these were common also on *Sargassum* though in the Sargasso Sea, where growth of epiphytic algae was less, the bacterial populations consisted of 50% *Vibrio* and 50% *Pseudomonas*.

Droop & Elson (1966) investigated some diatoms for traces of bacteria but could find none on *Coscinodiscus concinnus* and only 3–5 bacteria per 100 *Skeletonema* cells. A week later the population on the *Skeletonema* had increased to 32 and on the sixth day of the stationary phase of *Skeletonema* growth the bacteria suddenly increased: but by then the alga was unhealthy. Droop & Elsdon were concerned with the possibility that the spring growth of the diatoms was linked to bacterial metabolism but could find no evidence for this. There are however reports of anti-bacterial activity by *Skeletonema* (Sieburth, 1965) and it is obvious that the whole topic requires clarification.

Nematodes

Galls caused by parasitic nematodes are recorded on macroscopic marine algae, e.g. on *Ascophyllum* and *Fucus* (Coles, 1958) but little work seems to have been undertaken on the effect of these on the populations. Similar infestations are reported on *Chondrus crispus* and *Furcellaria fastigiata* (Dixon, 1973).

Grazing

Grazing on microscopic algae

Animals which graze on algae have been termed algivorous and there is no doubt about the vast passage of algae through animals both in the planktonic and benthic habitats. Cushing (1971) estimated that herbivores consume half or more of their body weight per day and that algae reproduce at 1–1.5 divisions per day: though only approximations, these figures express what must be the norm of algal growth outpacing that of animals. On the other hand relatively low rates of algal 'turnover' (6–7 times per year) are estimated to feed the zooplankton in Lake George (Ganf & Viner, 1973): presumably other organisms must crop the system. Cushing (1963) made the distinction between 'division rates' for individual algae and 'productive rates' for the rate at which the whole standing stock reproduces itself: certainly the two should not be confused. Grazing undoubtedly influences the specific composition of algal assemblages though there are few data on this aspect. It is often considered that grazing increases the species diversity and given not too great a grazing pressure and a fairly extensive environment this is probably the common situation. However, in restricted habitats, especially where recruitment is slow, grazing can have the opposite effect and decrease the species diversity.

Grazing is also one factor which influences the doubling times of phytoplankton species measured in nature compared to measurements made in the laboratory or in ship-board simulators, e.g. McAllister, Parsons & Strickland (1960) found a doubling time of 4 to 5 days in the northeast Pacific but only 1.75 days on the ship.

The diet of the Peruvian anchovy is reported to be 98% planktonic diatoms (Rojas de Mendiola, 1958), though more recent evidence suggests that copepods are also taken (Villanueva, Jordan & Burd, 1969; Rojas de Mendiola, 1971). Ryther *et al.* (1970) calculate that the annual fish production in the area off Peru must consume more than 10^{13} g of phytoplankton carbon annually and that an annual rate of 10^3 g carbon m^{-2} is needed to satisfy the anchovy production alone; they estimated a mean rate of net organic production of 10 g carbon m^{-2} day^{-1} which would only support $\frac{1}{3}$ of the anchovy needs. None would be left over for the other animals which are estimated to require at least as much again as the anchovy. The figure of 10 g carbon m^{-2} was obtained over a 5-day period only and therefore may not be representative for the year but it is an extremely *high* figure for organic production measured by the ^{14}C technique, though similar figures were obtained whenever the ship approached within 10 miles of the Peruvian coast.* This upwelling area is not a region of continuous upwelling but rather of a series of plumes of water in which production is high. Walsh & Dugdale (1972) suggest that the fish find these plumes and stay with them, feeding on the rich plankton growths. In this area of high algal production there is also a tongue of high carbon fixation extending out into the

* To maintain this rate would require continual upwelling of nutrient-rich water and good light conditions.

Pacific alongside the South Equatorial current (under which the Cromwell current runs) and here the algal production coincides with sperm whale catches (Fig. 9.6). According to calculations by McAllister *et al.* (1960), grazing fish appear to be the restraining factor on algal populations at a north Pacific station. Without this depletion the phytoplankton could theoretically increase up to 50 times its observed size, for nutrient starvation seems unlikely in these latitudes. Since phytoplankton can be markedly stratified in the water column one might expect that fish would seek out these layers and indeed, in experimental trials, Lasker (1975) found that populations of *Gymnodinium splendens* from the 15-m deep chlorophyll maximum off the California coast were actively grazed by the northern anchovy (*Engraulis mordax*) but *Chaetoceros* and *Thalassiosira* were not. Laboratory studies showed that the anchovy accepted *Gymnodinium*, *Gonyaulax*, *Prorocentrum micans* and *Peridinium trochoideum* without preference in laboratory experiments but did not take *Chlamydomonas*, *Dunaliella*, *Leptocylindrus*, *Thalassiosira* or *Ditylum* (Scura & Jerde, 1977). There is some evidence that the fish can take preferentially the darker coloured cells of *Peridinium*. The striking effect of plankton-rich water on the growth rate of an animal is shown in work by Paul, Paul & Nevé (1978), in which they pumped upwelling water into a pond in which they grew *Mytilis*. These molluscs grew from 8.8 mm to 30.2 mm in 90 days whereas in the natural habitat they take four years to reach this size.

The interesting experiments devised by Strickland (*see* p. 339) using phytoplankton enclosed in large plastic bags suspended in the ocean showed that, when the zooplankton were filtered off, the algal stock increased by about two orders of magnitude which again points to the restraining effect of zooplankton. In a somewhat similar experiment using lake water, Porter (1972) found that the algal groups varied in their response: the small (2–3 μm) algal cells increased in number except for the gelatinous green algae which decreased, whilst the larger algae were unaffected by the absence of grazers. This is yet a further example perhaps of selectivity and the difficulty of predicting results. Cushing & Vúcetìc (1963) followed the development of a patch of the copepod *Calanus finmarchius* for three months as it drifted in the North Sea and found that the volume of phytoplankton reached a peak and then declined

Fig. 9.6. Plot of the sperm whale catches in the Eastern Pacific, 1761–1920, *a*, April–September; *b*, October–March. Each point represents the capture of one or more sperm whales. From Forsbergh & Joseph, 1964.

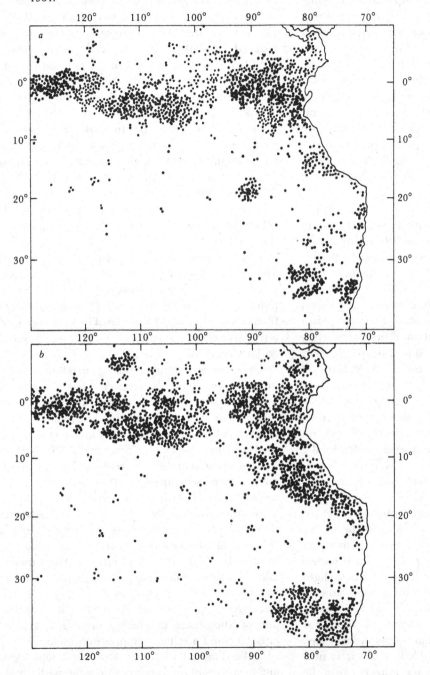

whilst phosphate, silicate and nitrate were still plentiful. The decline was brought about by heavy grazing, much of which was superfluous feeding, merely destroying the algae without extracting much nutriment. *Calanus* has been reported to select and grasp large algae in its feeding appendages! Cushing & Vúcetìc (1963) also reported that *Calanus* breaks open algal cells and spills the contents which then stimulate the growth of other algae, but in the light of experiments by Corner *et al.* (1972) this seems unlikely: when *Calanus* is fed on *Biddulphia* much of the algal material passes undigested through the gut. Richman & Rogers (1969) report that *Calanus* selectively grazes paired (i.e. recently divided) cells of *Ditylum* during the night but unpaired cells were grazed at the same rate throughout the experimental period. However nauplii of *Calanus helgolandicus* are reported not to grow on a diet of *Ditylum* (Mullin & Brooks, 1970), another example of ecological preference at different stages in the life cycle. An interesting experimental approach is that reported by Martin (1970) who presented two diatoms, *Skeletonema costatum* and *Rhizosolenia delicatula*, to grazers and found that the former was removed first: only when the cells of *S. costatum* were reduced below 300 cells ml^{-1} was the *Rhizosolenia* taken. A density of 300 cells ml^{-1} provides in fact a negligible number of cells to the predators which, through sheer lack of cells, would be forced to seek other sources. *Skeletonema* is smaller than *Rhizosolenia* so the general feature that larger cells are grazed preferentially (Parsons, LeBrasseur, Fulton & Kennedy, 1969) does not apply: this may be a cultural effect or some subtle chemical discrimination.

Grazing is likely to have a drastic effect on extremely dense populations which are becoming nutrient limited; this is frequently observed in sewage ponds where intense growth in rich waters results in depletion of a limiting nutrient or light limitation by self-shading; the subsequent slowing down of the algal division rate allows grazing (and/or parasitism) to deplete the population. For many years it was assumed that zooplankton grazed *around* dense phytoplankton blooms but it is now more generally thought that the grazers cause the patchiness of the phytoplankton rather than the phytoplankton excluding the grazers. A fascinating historical account of the arguments about exclusion versus grazing is given by Hardy & Gunther (1935);

they reported that herring which are ready to spawn and are not feeding avoid dense phytoplankton swarms. Certainly when phytoplankton is abundant, animals are often scarce and vice versa and Ryther (1954) found that the filtering rates of *Daphnia* were reduced by substances derived from algae both outside the animal and within the gut and that the effects were minimal with actively growing algae and maximal when senescent algae were present Senescent algae are of course very likely in dense blooms and this may be an important exclusion factor. The rate of grazing is sometimes enormous, e.g. Enright (1969) suggests that inshore phytoplankton could suffer a grazer loss of 40–50% in 24 h. Under such conditions he suggests that patchiness of the phytoplankton is essential to allow the cells to grow until the patch is 'discovered' again by the feeders.

The volumes of water grazed by some animals are certainly large and if the removal of organisms is effective it is not surprising to find 'empty' patches, e.g. the pelagic crab *Pleuroncodes planipes* filters some 540 l day^{-1} and one of its food organisms is the large celled *Coscinodiscus* (Longhurst, Lorenzen & Thomas, 1967).

When considering grazing in relation to the productivity of phytoplankton, values of carbon fixation expressed in grams per square metre of water surface can be misleading: in shallow waters the value may be comparable to that in deep waters but in the former the food is concentrated in perhaps 5–10 m whereas in the open water the same amount may be dispersed through 100 m. In fact there are several studies relating fish catches to depth and chlorophyll content of lakes and these generally show a decrease in catches as depth increases and chlorophyll dispersion increases, e.g. Saijo & Sakamoto (1970). In certain circumstances there is also a concentration in a particular layer in deep waters (*see* p. 345), and only by expressing the values for narrow depth intervals coupled with an estimation of abundance of grazing organisms at each interval could a reliable comparison be made.

Feeding on phytoplankton by zooplankton and herbivorous fish is not synonymous with uptake since a proportion of algal material often passes unharmed through the guts, especially when feeding occurs in rich phytoplankton blooms. Some early records failed to take into account the pH changes which take place in fish guts during feeding, for instance the common African freshwater fish *Tilapia nilotica* does not feed at night and when it commences to feed the gut is at pH 7; at this pH the algae are not affected and may pass into the intestine and out as viable algae. However as feeding proceeds the acid-producing cells of the stomach reduce the pH to below pH 2 and then breakdown and digestion of the algae occurs (Moriarty, 1973). Experiments in which ^{14}C-labelled phytoplankton were fed to *T. nilotica* and *Haplochromis nigripinnis* which, when over 5–6 cm in length, feed on mixed plankton and *Microcystis* respectively, showed that the *Tilapia* assimilated on average about 43 % of the carbon per day and the *Haplochromis* about 66 %. The figures vary according to the alga ingested, e.g., up to 70–80 % from *Microcystis*, *Anabaena* and *Nitzschia* and only 50 % from *Chlorella* (Moriarty & Moriarty, 1973). Similar amounts, 23 % of the carbon from *Spirogyra* and 67 % from *Anabaena flos-aquae* are ingested by the brown bullhead (Gunn, Qadri & Mortimer, 1977).

It is very common for bottom-living fish to graze on the sediments; in aquaria they can often be observed taking in the sediment and blowing it out again. Some of the organic material is filtered out in this process. The striped mullet (*Mugil cephalus*) derives a major part of its food from the micro-algae (epipelon) in the surface mud but, as Fig. 9.7 shows, the algal content of the food varies considerably with the habitat of the fish (Odum, 1970). A relationship between fish grazing on zooplankton whose food is phytoplankton and the density of the phytoplankton might be expected though there are few data in the

literature. In one experimental system using aquaria and the mosquito fish (*Gambusia affinis*) it was found that when fish depleted the rotifer, crustacean and larval insect populations in the water there was a massive development of phytoplankton (Hulburt, Zedler & Fairbanks, 1972).

In freshwaters, grazing studies have mainly involved the phytoplankton where the main grazing animals are the Rotifera and Crustacea. Both Nauwerck (1963) and Lund (pers. comm.) report that the rotifer *Asplanchna* selectively grazes certain plankton, e.g. *Staurastrum* is taken from a mixed population. Nauwerck (1963) found that Chrysophyta were more important food for crustacea than either diatoms or green algae. The effects of grazing on the composition and annual cycles of freshwater phytoplankton have received less attention than the study of grazing merely to ascertain the food source of the zooplankton but Bailey-Watts (1978) has shown that the return of *Daphnia* in large quantities to Loch Leven has paralleled the change in the phytoplankton from nannoplankton to a net plankton dominance. Other workers have shown that whilst phytoplankton is an important food for *Daphnia*, this organism cannot be maintained on phytoplankton alone, but requires organic detritus also (Saunders, 1972a). Taub & Dollar (1968) found that reproduction was particularly adversely affected when *Daphnia* was fed only on *Chlorella* or *Chlamydomonas*. Apparently *Daphnia* also has difficulty with large filamentous forms such as *Aphanizomenon* (Lefevre, 1950) and is unable to digest cellulose. Edmondson (1957) gives a good review of the subject.

Fig. 9.8 shows the annual cycle of phytoplankton and of zooplankton in Lake Erken: this hardly shows the usual text-book concept of zooplankton maxima succeeding the phytoplankton. Nauwerck made careful measurements of phytoplankton and zooplankton volumes in Lake Erken and during one year there were 5.5 times as many zooplankton as phytoplankton (extreme values were 1.3–30): to maintain this sort of relationship he considered that the phytoplankton would have to 'turn over' 15–25 times as fast as the zooplankton if the former were the only source of food for the animals. Such a rate of turnover is of course possible, since the predominantly unicellular phytoplankton reproduce fairly rapidly, but it does illustrate the point that comparisons of biomass need to be linked to rate studies to

Fig. 9.7. The estimated percentages of the material in the stomach contents of *Mugil cephalus* in seven different habitats. From Odum, 1970.

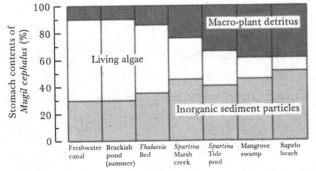

Fig. 9.8. *a*, phytoplankton volume (open circles, vol l⁻¹), proportion of net plankton (stippled area) and carbon assimilation (filled circles, μg day⁻¹ l⁻¹) in the 0–10 m depth in Lake Erken in 1957. *b*, percentage composition of total phytoplankton crop. *c*, zooplankton volume (open circles, vol l⁻¹) and proportion of carnivores (black areas) in the 0–20 m depth in Lake Erken in 1957. *d*, percentage composition of zooplankton and proportion of carnivores (black areas). From Nauwerck, 1963.

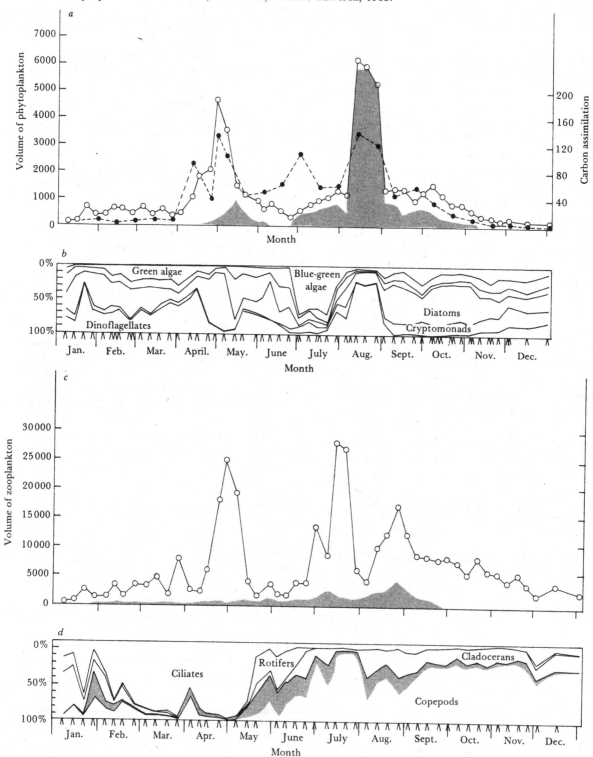

Fig. 9.9. The seasonal cycles of phytoplankton and zooplankton in Lake Lenore. *a*, phytoplankton; *b*, zooplankton. From Anderson *et al.*, 1955.

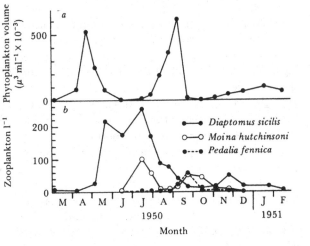

Fig. 9.10. An empty digestion cyst of *Asterocaelum algophilium* containing the remains of the centric diatoms which it had ingested. The cyst is surrounded by mucilage which also covers the spine-like projections. Photograph kindly supplied by Dr H. M. Canter.

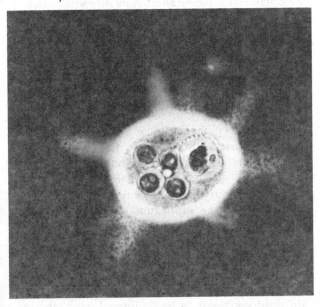

be really meaningful. In addition the utilisation of dissolved and particulate (non-living) matter must be investigated since under certain conditions this is used by grazing animals (especially in streams). Contrary to Nauwerck's experience, Davis (1958) found that the phytoplankton volume of western Lake Erie nearly always exceeded the zooplankton (on average it was 4.08 times greater).

A study showing an inverse relationship between phytoplankton and zooplankton is that of Anderson, Comita & Engstrom-Heg (1955) in Lake Lenore, which is a saline lake in the lower Grand Coulée, where the zooplankton flourishes between two peaks of phytoplankton (Fig. 9.9). They attributed the inverse relationship to grazing of the phytoplankton by the zooplankton. However Nauwerck (1963) found that in Lake Erken two zooplankton peaks (April–May and end of August) correlated with phytoplankton peaks but a large July peak occurred during a low phytoplankton period (Fig. 9.8). This graph shows also that sizes of organisms and taxonomic groups change throughout the year giving a quite different set of food–grazer relationship for each month. In a recent re-analysis of the Lake Erken data, Cushing (1976) suggests that the diatoms are controlled by rotifers in the autumn, throughout the year the green algae are controlled by copepods, the dinoflagellates are controlled by Cladocera in the autumn and even the other groups of algae are probably also controlled by some form of grazing.

In the tropical oceanic zone Tranter (1973) found that concurrent zooplankton and phytoplankton values were not well correlated, a two-week lag occurring between phytoplankton fluctuations and any subsequent zooplankton response.

Phytoplankton are grazed by organisms smaller than rotifers or copepods and isolated reports of protozoa ingesting algae have been made during the last hundred years. Recently Canter (1973) has described a protozoan (*Asterocaelum*, the digestion cyst of which is shown in Fig. 9.10) similar to the appropriately termed 'vampyrellids', grazing on the planktonic diatoms *Cyclotella pseudostelligera* and *Stephanodiscus rotula* (probably = *S. astraea*, but even this is an invalid name (Round, in press). It appears that such protozoa are much more widespread than previously suspected (compare the distribution of the chytridiaceous fungi as a result of the work of

Canter). Table 9.3 gives details of some infections in Windermere and Esthwaite Water between 1962 and 1966, and shows the dramatic effect of protozoan grazing. Fig. 9.11 illustrates how rapidly some planktonic species are depleted by such activity. The protozoan involved (*Pseudospora*) is a biflagellate cell in its free-swimming stage but becomes amoeboid within the colonies of palmelloid algae and moves from cell to cell ingesting the contents. It appears that different protozoa graze different algae; in one experiment the *Pseudospora* which grazes in *Gemellicystis* colonies was added to cultures of five other

algae but none were affected. As Canter & Lund (1969) point out, the larger phytoplankton are grazed by small protozoa and the small phytoplankton by the larger filter feeding rotifers and crustacea.

In rivers such as the Thames much of the phytoplankton is grazed by the filter feeding mussels (*Anodonta* and *Unio*) but large amounts of undigested algae also pass through their guts. This is a rather special case of phytoplankton grazing since the animals live in the benthos and presumably it is the turbulent movement of the water which carries the phytoplankton down to the bottom.

An intriguing relationship exists in the sea between bioluminescent dinoflagellates and grazing rates of copepods feeding on them: the highly bioluminescent species were always less grazed than the species exhibiting lower levels of bioluminescence (Esaias & Curl, 1972). A rather speculative early idea was that the movements of grazing copepods stimulate bioluminescence and the light attracts carnivores which then feed on the herbivores. Further modern studies of this system are required.

Another intimate association between algae and invertebrates is shown in the grazing on the surf zone diatoms *Chaetoceros armatus* and *Asterionella socialis* by razor clams burrowing in the sandy beaches of Washington State. No other algae are adapted to this unusual habitat (*see* p. 295) in the rough surf (Lewin & Mackas, 1972) so there is a very intimate relationship between the physical conditions and the organisms.

Most students must have observed the grazing of algae from epiphytic sites or artificial surfaces by tadpoles. These are voracious feeders and Dickman (1968*b*) found that they reduced the algal population to between $\frac{1}{50}$ and $\frac{1}{100}$ compared with control surfaces. Desmids were completely eaten but diatoms living close to the substratum were less affected. Chironomid larvae move onto *Typha* stems in spring in Alderfen Broad, Norfolk and by November the epiphyton is almost completely grazed; the larvae then move back into the sediment (Mason & Bryant, 1975). A similar habitat for grazers is provided by the dead vegetation in newly formed reservoirs: in some instances, for example in Lake Volta in Ghana this involves forest trees and the epiphyton of these is a rich grazing ground for both invertebrates and fishes such as *Tilapia* (Round, unpublished). Grazing

Fig. 9.11. Decreases in algal populations in relation to grazing by Protozoa during 1963. *a*. Continuous left-hand line *Eudorina elegans*; continuous right-hand line, *Gemellicystis imperfecta*; broken line, *Synura* sp., all in Esthwaite Water. *b*. Continuous line, left-hand side, *Gemellicystis imperfecta*, Windermere, south basin; continuous line, right-hand side, *Dictyosphaerium pulchellum* Wood; broken line, *Paulschulzia tenera* and *P. pseudovolvox*, Windermere, south basin; dotted line, as broken line, but in Windermere, north basin. A, Date when protozoans were first seen; B, 98%; C, < 1% and D, 88% of the algal colonies contained Protozoa. From Canter & Lund, 1968.

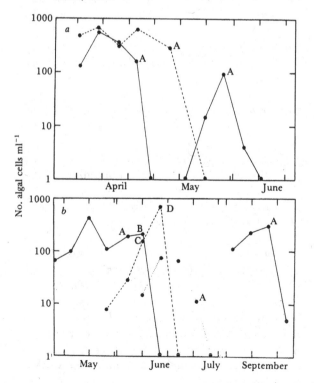

of the epilithon is very common but little studied; Calow (1973) showed that *Ancylus fluviatilis*, although eating all algae, prefers diatoms especially *Gomphonema* spp. He further showed that it is less selective when hungry. On the other hand, the snail *Limnaea* prefers green algae, a preference possibly related to its more active secretion of cellulase. In a marine environment, Nicotri (1977) found that gastropods selectively removed the chain-forming diatoms since they form an overstorey growing out from the substratum. However blue-green algae also growing outwards are not grazed. It is perhaps not surprising that the adpressed flora is untouched since the rasping of the radula may not penetrate to this layer.

Grazing of free-living benthic algae is extremely common and most samples of epipelic algal populations will also contain ciliates and amoebae, often packed with ingested algae. Little however is known of the rates or effects of this grazing. Brook (1952a) studied the protozoa on the surface of sand filters of a water works and found that *Chilodon* could ingest 30 cells of *Navicula dicephala* in 24 h and *Oxytrichia* could ingest 47 *N. dicephala* or 90 *Nitzschia palea* in 24 h. These seem to be large numbers but it is common to find such protozoa packed with diatoms; interestingly Brook did find that increased water movement reduced the feeding rate; is it more difficult to catch a moving alga? In a more natural habitat Goulder (1972) observed the ciliate *Loxodes magnus* feeding on *Scenedesmus* species in a small eutrophic pond and the ciliate consumed *Scenedesmus* cells at the hourly rate of 3.8–12.8 per ten ciliates. There were 65–1142 ciliates cm^{-2} at various times and these consumed what appears to be a very large crop of *Scenedesmus* (600–35090 cells cm^{-2} day^{-1}). But the crops of the alga were so large (5.20–18.52 × 10^6 cells cm^{-2}) that the grazed portion represented only 0.003–0.68 % of the population and Goulder concluded that this grazing was probably not significant as far as the algae are concerned. The converse may however not be true and several workers have suggested that the protozoan crops are related to the supply of algae. Provided algae are present in the quantity found at this site they are unlikely to be the controlling factor in regulating the animal population. On the other hand seasonal changes in the algal populations are common (*see* chapter 10) and these may be very important in regulation of the animals. In a similar habitat larval

stages of Lamprey (*Petromyzon marinus*) graze largely on algae and Moore & Beamish (1973) calculated that 0.6–0.7 × 10^5 algal cells were ingested per day, during winter, by large larvae but since there were some 3.3 × 10^5 cells m^{-2} on the sediment and the larvae graze over a much larger area the effect of removal on the algal population again may not be serious.

It might be expected that availability of algae would affect the growth rates of herbivore populations and, indeed, Edmondson (1962) showed that both rotifers (*Keratella cochlearis*) and female copepods have a higher rate of egg production when algal crops are large. Although rotifers are large zooplankton they seem to ingest small cells, e.g. Edmondson (1965) in a stimulating paper records that the μ-flagellate *Chrysochromulina* was very important in the diet of *Keratella* and *Kellicottia* and the reproductive rate of *Polyarthra* was strongly related to the ingestion of *Cryptomonas*. These rotifers are selective and *Chlorella* numbers were negatively correlated with the rotifer growth. Cushing (1976) in a reanalysis of Nauwerck's (1963) data stresses the importance of rotifers as the major grazers in the autumn in Lake Erken. Rotifers mainly graze the diatoms and *Peridinium* whilst cladocera control the spring growth of Chrysophyceae and copepoda strongly influence the Chlorophyta. It has been reported many times, e.g. by Porter (1973), that grazing by zooplankton decreases the small cells in the population allowing larger green algae, e.g. desmids, dinoflagellates and Chrysophyta to develop. A further positive effect of grazing on algal growth was reported by Porter (1976) who found that the gelatinous colonies of *Sphaerocystis* pass through the gut of *Daphnia* and during passage the colonies break up, absorb phosphorus and other nutrients from the gut fluid and after release from the animal have an enhanced growth, carbon fixation and division rate. The summer bloom of green algae in many lakes has never been adequately explained and Porter considers that an important factor for some algae is the nutrient regeneration by the zooplankton. Many algae with gelatinous sheaths seem to be protected from grazing by the sheaths. Grazing of the microbenthos by amphipods can affect the algal populations when the animals are present at high densities (Hargrave, 1970).

Faecal pellets of *Hyalella* were found to be

Table 9.3. *The occurrence of epidemics of Pseudospora sensu lato on planktonic algae in Windermere and Esthwaite Water from 1962 until 1966. From Canter & Lund, 1968*

Date	Lake	Alga	Cells (ml⁻¹)			Period (days)
			Before epidemic	After epidemic	Decrease (%)	
1962						
June–July	Windermere, S. basin	Coenococcus planctonicus	704	< 1	> 99.5	14
1963						
June	Windermere, S. basin	Gemellicystis imperfecta	208	1	99.5	13
May–June	Windermere, S. basin	Paulschulzia tenera	709	1	> 99.5	14
September	Windermere, S. basin	Dictyosphaerium pulchellum	300	5	98	7
June–July	Windermere, N. basin	Paulschulzia tenera	74	< 1	99	14
April–May	Esthwaite Water	Eudorina elegans	510	< 1	> 99.5	14
April–May	Esthwaite Water	Synura sp.	596	< 1	> 99.5	14
1965						
May–June	Windermere, S. basin	Gemellicystis imperfecta	44	4	93	14
June–July	Windermere, S. basin	Coenococcus planctonicus	548	1	> 99.5	14
July	Windermere, S. basin	Paulschulzia tenera	23	1	96	7
July	Windermere, N. basin	Coenococcus planctonicus	139	< 1	> 99.5	14
July	Windermere, N. basin	Sphaerocystis schroeteri	202	< 1	> 99.5	14
1966						
May	Windermere, S. basin	Gemellicystis imperfecta	334	< 1	> 99.5	14
June	Windermere, N. basin	Gemellicystis imperfecta	61	0.5	98	8

rapidly colonised by algae; the faecal flora of aquatic habitats is clearly one which requires further detailed study.

A conspicuous group which grazes on marine sediments is the Foraminifera. It requires a mixed diet of algae (diatoms are very commonly ingested) and bacteria for optimum growth and reproduction (Murray, 1973). In experimental studies the Foraminifera in general prefer to eat live rather than dead algae. An unusual feature recorded by Lee & Freudenthal (1964) is that the introduction of a foraminiferan (*Ammonia*) into a culture of *Dunaliella parva* stops the random motion of the alga and the cells actually swim towards the protozoan. The genus *Rosalina* clears paths along the algal coating on the glass surface of culture vessels (Myers, 1943), which must mean that the attached species are also grazed.

It is quite obvious from the outstanding work of Canter that grazing or parasitism, with attacks coming from all sides, is of very frequent occurrence in phytoplankton populations; Table 9.3 shows how frequently infections were observed in some English lakes and one reservoir (Farmoor). As can be seen from the data, *Asterionella* in Grasmere was infected three times during the year by chytrids, in each instance by a different species. A further example of the complexity of the algal–animal–fungal relationship is given by Canter (1974). She found that a bloom of the blue-green alga *Anabaena flos-aquae* in Windermere was being eaten by a ciliate, *Nassula*, which in turn was being captured by a suctorian and killed and later the suctorian itself succumbed to a biflagellate fungus. Only the identities of the alga and the ciliate are known with any certainty, showing again the amount of taxonomic work still to be done in this field.

Grazing on toxic algae*

The deleterious interactions between algae and animals have not been studied extensively. The most striking are those involving death of animals due to ingestion of algae or the toxins produced by the algae e.g. from the haptophycean, *Prymnesium parvum*, the cyanophytes, *Microcystis*, *Aphanizomenon*, etc. and the dinoflagellates *Gonyaulax* and *Gymnodinium*. There is little known of the ecological consequences of this kind of toxin production. Living cells

* See also comments in chapter 7, p. 307.

of *Microcystis aeruginosa* have no deleterious effect on *Daphnia longispina* or *Eucypris virens* but extracts of the pigment were toxic according to Stangenberg (1968*a*). Contrariwise Arnold (1971) reports some toxicity or at least inhibition of *Daphnia* by Cyanophyta and certainly their rates of ingestion, assimilation, survival and reproduction were lower on a blue-green algal diet than on green algae. Care needs to be taken with the assessment of feeding experiments when one reads that feeding of *Chlorella pyrenoidosa* to bacteria-free *Daphnia* was complicated by the finding that the algal culture medium was toxic to the *Daphnia* (Taub & Dollar, 1964). Whilst such an effect might be obtained by extracting any pigment system, it is not clear whether or not there is any ecological significance, though considerable amounts of pigment, etc. are released on the decay of water blooms. Blooms of blue-green algae have generally been studied in relation to poisoning of cattle caused by drinking water containing masses of cyanophycean cells. Not all *Microcystis* strains produce the toxins, and two different toxins have been isolated, one of which is a cyclic polypetide (Gorham, 1964). It is interesting that in the detailed work of Gorham and his colleagues, the *Microcystis* would fail, or grow poorly, when bacteria were eliminated from the cultures and they comment that 'the nature of the association with the alga looks suspiciously like symbiosis'.

Prymnesium kills fish and bivalve molluscs. It is a marine organism which can however tolerate great variations in salinity and it invades brackish waters, and even inland sites, where the chloride content varies from 250 to 10 000 mg l^{-1}. It has been especially serious in fish ponds in Israel and many fish deaths have now been noted also in East Anglian waters, whilst isolated records of its effects have come from many parts of the world. Apart from being photoautotrophic, it can also utilise organic sources of nitrogen and phosphorus, and it requires both thiamine and cobalamin. This flagellate is fairly adaptable and can grow in the dark in the presence of glycerol (Rahat & Jahn, 1965). *Prymnesium* toxin, unlike that produced by other algae, diffuses into the water and is apparently produced in the alga during darkness. Shilo (1967) reports a twentyfold increase in toxin in cells starved of phosphate. Little is known of the seasonal cycles of this alga which also produces a resting spore. The alga can fortunately be controlled

Table 9.4. *a, the preference for algal species shown by eleven benthic grazers in the Macrocystis beds, and b, the relative preference for the parts of a Macrocystis plant. From Leighton, 1971*

(a)

	Macrocystis	Egregia	Laminaria	Eisenia	Pterygophora	Cystoseira	Gigartina
Aplysia	×	×[a]	·		·	·	·
Astraea	×[a]	×	·	×	·	·	×
Lytechinus	×	·	×	·	·	·	×[a]
Norrisia	×	×[a]	×	·	·	·	·
Pugettia	×[a]	×	×	·	×	·	·
Haliotis corrugata	×	×[a]		·		·	·
Haliotis fulgens	×	×[a]	×	·	·	·	×
Haliotis rufescens	×[a]	×	×	·	·	·	·
Strongylocentrotus purpuratus	×[a]	·	×	·	·	·	×
Strongylocentrotus franciscanus	×[a]	·	×	·	·	·	×
Taliepus	×[a]	×	·	×	·	·	×

[a] Items most highly preferred.

(b)

Grazer species	Blades	Sporophylls	Stipes	Haptera	No. obs.
Aplysia	65.6	29.1	1.3	4.0	5
Astraea	42.9	41.7	14.7	0.7	5
Haliotis corrugata	38.6	47.4	6.9	7.1	5
Haliotis fulgens	55.8	27.4	15.9	0.9	5
Haliotis rufescens	31.3	50.1	15.0	3.6	5
Lytechinus	49.5	41.1	6.7	2.7	5
Norrisia	31.8	51.0	11.1	6.1	9
Pugettia	28.0	60.7	5.6	5.7	5
Strongylocentrotus franciscanus	32.7	51.4	10.3	5.6	8
Strongylocentrotus purpuratus	40.2	41.7	13.0	5.1	12
Taliepus	21.3	69.6	6.7	2.4	5

Groups of animals (2–10 individuals) were given equal quantities of blades, sporophylls, stipes and haptera. After feeding periods of 1–2 days, remaining fragments were blot-dried, sorted and weighed. Quantities consumed were then equated to relative values (%).

in lakes and ponds by the use of ammonium sulphate, either alone or with copper sulphate. The effective agent is free ammonia which readily enters the cell, and causes swelling and lysis. Sensitivity to ammonia is thought to be connected with the fact that the cell is under considerable stress in low salinity situations and entrance of ammonia into the cell increases this stress.

An interesting 'symbiotic' effect of bacteria reducing the toxicity of the blue-green alga *Anabaena flos-aquae* is reported by Carmichael & Gorham (1977).

Grazing on macroscopic algae

Grazing on subtidal populations has been studied in *Macrocystis* beds off south California; Leighton (1971) records a grazing front of sea urchins progressing through a bed at the rate of 33 ft per month. In three months, a bed about 300 × 600 ft had been completely destroyed leaving only the dead centres of the holdfasts and rock encrusted with *Lithothamnium*. The interior of *Macrocystis* holdfasts can also be grazed with the result that the plant is weakened and eventually breaks away. Leighton found 52 *Strongylocentrotus franciscanus* and 45 *S. purpuratus* individuals in a single holdfast weighing 62 pounds and the holdfast is of course a microcosm of many grazing organisms.

Selective grazing of benthic species grown on building blocks was shown to occur if sea urchins were excluded: the blocks were allowed to colonise first, and a succession developed with diatoms and then microscopic filamentous algae, followed by macroscopic growths of *Ulva* and *Gigartina*. If the blocks were then exposed to sea urchins they were grazed bare, but if placed in protective wire cages the *Ulva* and *Gigartina* continued growth and eventually *Egregia* also appeared.* Subsequent grazing by the more highly motile grazers removed the *Ulva* and although the *Gigartina* was chewed it was not completely destroyed. Blocks which were placed on sand had the *Ulva* and *Gigartina* cropped to a close turf (Leighton, 1971). There is in fact much indirect evidence that algal 'turfs' are maintained by grazing pressure and indeed Randall (1961) found that the turf species which normally grew to heights of only

---

* However caging of substrata in the subtidal can result in a total domination of the substratum by animals (Neushul *et al.*, 1976).

1–2 mm would reach up to 25–30 mm when protected from grazing. In one of the few experimental studies, Leighton followed this up by observing movement of animals into water which had passed over *Macrocystis* and showed that at least *Strongylocentrotus* moved into the channel supplied with water which had flowed over *Macrocystis* blades, suggesting that they were attracted to some chemical released from this alga. In a mixed kelp bed (*Macrocystis* and *Pterygophora*) the *Macrocystis* was grazed preferentially and only when this was denuded did the sea urchins attack the *Pterygophora*. Feeding various macroscopic algae to a range of grazing animals held in experimental tanks, so that each animal had the same opportunity, showed strong selectivity patterns: *Macrocystis* was preferred by seven of eleven animals. Of the animals, *Lytechinus* (sea urchin) had a strong preference for *Macrocystis* and *Gigartina* whilst *Pugettia* (crab), *Norrisia* and *Haliotis* (Mollusca) appeared to avoid *Gigartina* (Table 9.4; Leighton, 1971). Leighton also investigated the nutritional value of the various algae grazed by two sea urchins (Table 9.5) and showed that *Macrocystis* was utilised most efficiently and the coralline red alga *Bossiella* least. An interesting observation in these studies of *Macrocystis* is that grazing is much more intense in the beds growing on rock rather than those in sand: this is due to the preference of a rock habitat by many of the grazing animals.

Damage to *Laminaria* spp. and *Saccorhiza* by the gastropod *Patina pellucida* is common and undoubtedly results in considerable loss of photosynthetic material and eventually of whole plants. The young gastropods settle on the frond and in British populations migrate later on to the stipe and even on to the holdfast, yet in Norwegian populations the animal has never been found lower than the uppermost part of the stipe (Kain & Svendsen, 1969). At the moment there is no explanation for this variation in behaviour, but it does point to the dangers of relating conclusions drawn from data obtained at one site to other geographic locations. This work also showed that in deep water and in places where current flow was fairly fast the infestation by *Patina* was very reduced. Norton (1971) found 89 different animals on *Saccorhiza* from the Isle of Man and Lough Ine and also a decrease in the number of animals in the most wave exposed sites. Some species were frequent on the *Saccorhiza* but rare on surrounding substrata (e.g.

Table 9.5. *Summary of experiments on food values of different algae consumed by the sea urchins* Strongylocentrotus franciscanus *and* S. purpuratus. *From Leighton, 1971*

Food	Total consumption (ounces fresh weight)	Weight increase (ounces)	Ratio of consumption to increase	Mean gonadal index[a] at term
	S. franciscanus			
Macrocystis	6.40	0.33	19.4	0.124
Egregia	9.80	0.49	20.0	0.119
Pterygophora	4.87	0.12	40.5	0.146
Coralline alga (*Bossiella* sp.)	8.4	0.17	49.5	0.073
Starvation group	0.0	0.27	—	0.044
	S. purpuratus			
Macrocystis	8.9	0.47	18.8	0.188
Egregia	9.0	0.45	20.0	0.102
Pterygophora	6.9	0.18	38.4	0.128
Coralline alga (*Bossiella* sp.)	9.0	0.07	128.5	0.072
Starvation group	0.0	−0.13	—	0.015

[a] The gonadal index represents the ratio of gonad volume (ml) to total body wet weight (g).

Obelia geniculata, Nereis pelagica, Lineus longissimus, Gammarus locusta, Patina pellucida, Lacuna vincta, Gibbula cineraria and *Membranipora membranacea*) but only four species *Campanularia* sp., *Acmaea virginea, Antedon bifida* and *Psammechinus miliaris* were confined to *Saccorhiza* and absent from *Laminaria* spp. Grazing prunes beds of laminarians simply by damaging the stipes, which then break at the weakened point, e.g. *Nereocystis* was found to break only at the point at which sea urchins had grazed (Koehl & Wainwright, 1977).

There are very few data on variations in grazing intensity at different points in the subtidal region. One habitat which seems ideal for such studies is the coral reef: here grazing by fish during the day and by sea urchins at night severely reduces the flora of fleshy species on the reef itself (e.g. in Puerto Rico (Dahl, 1973)) and possibly also on the algal plain stretching seaward from the reef (*see also* p. 146). However further from the reef the algae appear to be relatively unaffected by grazing. Wiebe *et al.* (1975) report intensive grazing of *Calothrix crustacea* mats on the reef flat at Eniwetok. Grazing by the sea urchin *Strongylocentrotus dröbachiensis* off the coast of Nova Scotia seems to be intense in the 5–10 m depth range where it can eliminate the *Laminaria* forest: in the 10–30 m depth *Agarum cribosum* occurs and this is hardly grazed by the sea urchin (Table 9.6), which suggests that the animal is confined to the shallower water by its choice of food (Chapman, 1973). When the *Laminaria* is reduced by grazing, the *Agarum* can extend into the shallower water, where normally it does not grow due to competition from the *Laminaria* (Vadas, 1968, quoted in Chapman, 1973). The destruction of the *Laminaria* has led indirectly to a decrease in lobsters, which feed on the sea urchins (Breen & Mann, 1976). Similarly, Jones & Kain (1967) found that the sea urchin *Echinus esculentus* grazed below 6 m off the Isle of Man, and above this depth the *Laminaria hyperborea* forest was well developed. If a strip was cleared of urchins the *Laminaria* could then penetrate much lower and about five times as many *Laminaria* sporelings developed on the cleared compared with the uncleared shore. Kitching & Ebling (1961) also reported that at Lough Ine the sea urchin *Paracentrotus* destroyed the algae as fast as they grew. Paine & Vadas (1969)

Table 9.6. *Feeding preferences of* Strongylocentrotus dröbachiensis. *From Vadas, 1968, in Chapman, 1973*

	Weight eaten (g) in (h)				Total weight eaten (g) in four experiments
	36	48	60	72	
Nereocystis leutkeana	5.6	15.5	20.0	32.0	73.1
Laminaria saccharina	16.0	13.0	16.0	18.0	63
Costaria costata	7.5	19.7	3.0	2.0	32.2
Opuntiella californica	5.0	11.5	6.0	9.4	31.9
Laminaria groenlandica	5.0	5.6	13.4	3.0	27
Monostroma fuscum	9.4	3.8	10.2	2.5	25.9
Agarum fimbriatum	1.2	1.0	1.0	8.0	9.0
Agarum cribrosum	0.2	1.0	0.0	1.0	0.2

found that intertidal pools and the subtidal zone at sites on the coast of Washington were so severely grazed by sea urchins that only very small, fast-growing algae could survive. Removal of the *Strongylocentrotus* resulted in a sequence of recolonisation in which Paine & Vadas distinguished four stages. (1) An immediate increase in species, with usually 6–12 new species appearing. (2) Rapid establishment of a canopy by the large algae. (3) A succession until a single large alga achieved dominance, usually *Hedophyllum sessile* in the pools and *Laminaria complanatum* or *L. groenlandica* in the subtidal zone. (4) The establishment of an undergrowth of small, usually red algae. Of great interest in these experiments was the demonstration that grazing actually *maintains* species diversity by preventing a single alga from achieving dominance and this is a clear experimental proof of a theoretically predicted state. An even more complex state is reported by Dayton (1975*a*) who found that off Alaska the sea otter grazes the herbivores in the shallow water, down to 18–20 m, and this allows the algal populations to grow but at greater depths the ungrazed sea urchins destroy the algae except for *Agarum cribosum*. In addition, on some shores wave action can affect the predators of, for example, *Mytilis* (Lubchenco & Menge, 1978). Under extreme wave action starfish cannot attack *Mytilis* which can then outgrow the alga *Chondrus*. But on more wave-protected shores the removal of *Mytilis* by its predators allows *Chondrus* to invade and since it has a crustose holdfast which is resistant to grazing it can achieve dominance over other more delicate algae which are removed by *Littorina* grazing.

An interesting, positive advantage to algae of grazing is the reported pruning of *Egregia* populations by the limpet *Acmaea*: this keeps the alga to a reasonable size and it is not torn off the rocks during rough seas (Black, 1976). These examples illustrate the interaction between algal species and between algae and animals. Numerous comments in the literature indicate that grazing in the intertidal zone tends at times to be excessive and to denude areas of shore: this is in contrast to most terrestrial situations where a balance is more common. Indirect evidence of the effect of grazing can be seen from the vastly richer growth of many intertidal algae when the animals are removed experimentally or accidentally, e.g. after oil spills and treatment with detergent. Connel (1972) points out the interesting feature that in relatively quiet intertidal areas grazing is likely to be of prime importance in determining the dominance of the various algae but in harsh situations the animal may be removed and interspecific competition between the algae themselves determines dominance (*see* also above).

Experimentally cleared areas have to be sufficiently extensive or kept clear since even apparently fixed animals, e.g. *Patella*, rapidly migrate into the cleared zone and remove the algal colonists (Aitken, 1962). Experiments in which the grazing animals are cleared from shores have invariably shown that the vertical extent of algae can be extended. A classic investigation is that of Burrows & Lodge (1951) at Port St Mary, Isle of Man, where a 10 m strip was cleared of all limpets and large algae and after three years re-growth the common Fucoids were spread from the upper shore to low water mark with only slight increase in the proportion of *Fucus serratus* on the lower shore and of *F. vesiculosus* on the upper shore. This indicates that there is a competition factor operating between these two; *F. spiralis*, which is normally confined to the uppermost region of the shore, could clearly grow throughout the shore even in competition with the other species. Another notable observation was the high percentage of hybrids of *Fucus* present in the cleared strip: these presumably are always produced on any shore but only in the absence of competition from animals do they become established.

Most of the more detailed studies involving animals and macroalgae seem to have been undertaken using the large relatively easily observed sea urchins but the less easily estimated mobile fishes are also grazers. An example is the Angelfish (*Holocanthus bermudensis*) which feeds from December to March almost exclusively on *Caulerpa verticillata* but is omnivorous at other times (Menzel, 1959), so that grazing pressure of such a fish is not constant in its effect on the algae. In an interesting study, Vine (1974) found that the aggressive behaviour of one fish (*Pomacentrus lividus*) kept other fish away from its territory: the algae then flourished on the rocks whereas they were grazed off adjacent areas and the algae-covered rocks in turn were not easily colonised by corals or other sedentary invertebrates. Dawson *et al.* (1955) found that half the fish taken from a tropical atoll were herbivores: indeed these were better collectors of algae than the phycologists since more species were found in their guts than could be collected by the scientists. This is not surprising when one considers the range of foraging by the fish compared to man. Perhaps ecological surveys should utilise more intensively the natural collectors, especially for such resistant species as diatoms. The only problem is that this method disguises the actual habitats in which the grazed algae grow.

Grazing is responsible for the transfer of substances along food chains and in some instances the algae are suspected or known to produce toxins (*see* p. 413) which by this means can become concentrated, e.g. in fish. The transfer can be through intermediates, e.g. Lewin (1970) found a saccoglossan mollusc (*Oxynoe panamensis*) feeding on *Caulerpa* and producing an exudate which was toxic to fish. Recently Doty & Santos (1966) and Santos & Doty (1968) have isolated two new compounds from *Caulerpa*, caulerpin and caulerpicin which are physiologically active or even toxic to mice and rats. Since *Caulerpa* is grazed by herbivorous animals in the dense algal beds in the tropics it is likely that these compounds are transferred to man via fish. Direct transfer may also occur, since *Caulerpa racemosa* is widely eaten as a salad in the Pacific region. Earlier, Dawson *et al.* (1955) had suggested that the poisonous reef fishes may derive their poison from feeding on algae and noted the occurrence of the blue-green alga *Lyngbya* in the guts of the majority of such fish. It is interesting that recently the alga *Lyngbya* has been shown to cause a skin disease in humans (Moikeha, Chu & Berger, 1971).

Warmke & Almodóvar (1963) found distinct algal host preferences when studying mollusca in the Caribbean Sea. Gastropoda were the most common molluscs on the algae but they tended to be small and under 5 mm when adult. To illustrate this selectivity, 95% of the records of *Caecum nitidum* were from *Dictyota bartayresii*, whilst all the specimens of *Marginellopsis serrei* were found on *Dictyota divaricata* and none at all on *D. bartayresii* or *Padina gymnospora*. Clearly there is a multidisciplinary problem here to determine what factors cause these associations and what food sources the various algae supply. On *Caulerpa racemosa* in the Caribbean Sea a few gastropods were recorded whereas, off Fiji, opisthobranchs were the only common mollusc on *Caulerpa* (Burn, 1966). Alcala, Ortega & Doty (1972) carried the study further and showed that there was a positive correlation between algal species diversity and animal diversity in *Caulerpa* beds.

Grazing webs

The ratio of the yield in one trophic level to that in the one below is defined as the ecological efficiency (Slobodkin, 1961). Efficiencies of the order of 30–40% are reported for *Rhincocalanus* and *Calanus* feeding on diatoms (Mullin & Brooks, 1970). If yield is a constant fraction of production, this ratio may be termed a transfer coefficient (Cushing, 1973). In the Indian Ocean, Cushing found that this ratio for a 180 day period (i.e. the approximate period of the monsoon) ranged from 2% to 34% when the secondary production was considered relative to the primary production. Fig. 9.12 is a plot of the transfer coefficients from southwest monsoon and northeast monsoon periods in the Indian Ocean and the mean transfer coefficient is around 10%, which is near the value found from experimental work on *Daphnia* feeding by Slobodkin (1959): the highest transfer coefficient obtained with chickens is 40%. Since zooplankton spend time searching for food, probably have to live for periods on less than a maximum ration and devote parts of their body weight to reproduction (though so do chickens) the coefficient is readily reduced to 10%. Fig. 9.12 clearly shows the decline in the transfer coefficient from about 20% in areas of low production to about 5% in areas of high production. However the productivity of the major upwelling areas is greater than 1 g carbon m^{-2} day^{-1} and in the centres of the deep ocean is about 0.1–0.65 g carbon m^{-2} day^{-1}, a ten to twenty-fold difference; yet the difference in secondary production is reduced to 2.5–5.0, so some quantity is damped in the transfer from primary to secondary production. For the carbon transfer from phytoplankton to fish, Oglesby (1977a) found efficiencies varying over 2–3 orders of magnitude. It is generally assumed that a reduction in the standing crop of algae is synonymous with a reduction in food supply to the secondary consumers. However the rate of net fixation of carbon is also involved and there are several examples showing *increased* net primary productivity as the grazers increase up to a certain level after which it decreases as over-grazing sets in (Cooper, 1973); in other words, crop some of the algae and the rate of supply will increase. One example of a food web involving phytoplankton, zooplankton and fish is given in Fig. 9.13.

Nutrient enrichment can be a fairly steady event, e.g. in a freshwater spring, an oscillatory event, e.g. in lakes, or an intermittent event, e.g. in upwelling areas of the ocean, and thus in all sites, except the first, nutrient enrichment alone is a source of instability throughout the food web.

Another seasonally oscillating event is solar energy input and the combination of this and oscillating nutrient enrichment varies the primary production by as much as 50% even in a region such as the Indian Ocean.

Summary

Algae are involved in many symbiotic associations with animals from various phyla and these are detailed. It is now possible to assign the algal partners to distinct free-living genera and so the terminology 'zooxanthella', 'zoochlorella' has become superfluous. The degrees of interdependence of the partners and interchanges of metabolites between algae and animals are discussed. The ecological significance of symbiosis is obviously of considerable importance especially in the coral reef system but only recently have comprehensive studies been undertaken, involving carbon fixation rates, depth–photosynthesis profiles etc. A few less well-known symbioses are mentioned e.g. of *Noctiluca* and a flagellate alga, between *Nostoc* and midge larvae and between a brown alga and the Sea Pen. Saprophytism is much less developed amongst algae.

The intriguing symbiosis between opisthobranch molluscs and chloroplasts mainly from

Fig. 9.12. The transfer coefficient from primary to secondary production as a function of primary production over a 180-day period in the Indian Ocean. Open circles, SW monsoon; triangles, NE monsoon. Filled circles averages for both monsoons for intervals of primary production. From Cushing, 1973.

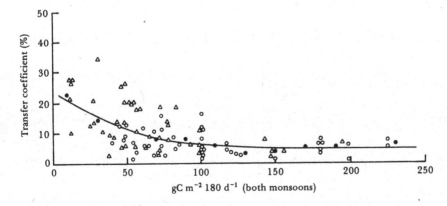

gC m^{-2} 180 d^{-1} (both monsoons)

Fig. 9.13. A tentative food chain in Saanich Inlet, June–July 1963. On the graph, the continuous line is the relative chlorophyll *a* concentration, the dotted line is the relative number of euphausiid eggs and the dashed line is the relative number of euphausiid furcilla. Start of young salmon feeding on furcilla → 1, Bloom of *Distephanos*; 2, *Chaetoceros socialis* and *C. debilis*; 3, μ-flagellates. From Parsons & LeBrasseur, 1970.

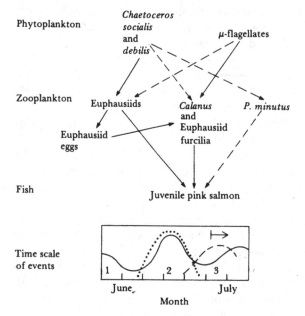

siphonaceous green algae has recently come into the forefront.

Algal–fungal and algal–bacterial symbioses are considered briefly.

Parasitism involves algae which are parasitic on other algae and a few which are parasitic on higher plants.

Viruses are being increasingly discovered within algal cells as further algal genera are studied by electron microscopy. Fungi are also very important parasites of algae and are responsible for many instances of population control. Protozoa and bacteria also attack algae though in the case of bacteria a few are also endosymbiotic.

Grazing on algae is not very often studied yet it is probably true to say that almost all algae both macroscopic and microscopic are grazed by some animals and in some situations, e.g. coastal upwelling systems, the algae are vital to the fish crops. Microscopic and macroscopic algal associations are controlled and dominance altered by the activity of grazers and grazing transfers substances, sometimes toxic ones, along food chains.

Toxic algae are known, some of which cause accumulation of poisons in shellfish which are then lethal to fish and man, while others are lethal to cattle.

10 Annual succession and growth

All plants have a cyclical development pattern with periods of growth alternating with periods of relative inactivity and, depending on the environmental conditions, there can be short bursts of activity (e.g. in deserts) or more subtle, damped activity as in tropical forests. Much macroscopic vegetation tends to be long lived and so the cycles are not so striking as they are amongst unicellular forms which have short life cycles. This is plainly seen when comparing studies of algae on rocky shores which in their relative permanence give little external evidence of annual cycles whereas the appearance and disappearance of species in the plankton is so striking that its study assumes greater significance. These features are very obvious when one considers the literature; it is very difficult, for example, to find data on the seasonal activity of any but a handful of rocky shore species, whilst almost every paper on plankton reveals some aspect of growth or succession.

This chapter can be divided into a consideration of gross annual changes in biomass, succession of individual species during the year and their respective occurrence each year and some of the factors responsible for the cycling.

Needless to say, to follow patterns of growth and succession the habitat must be sampled thoroughly and at frequent intervals. In some habitats it may be necessary to drift with the species assemblages, should they move from the sampling site, e.g. oceanic plankton. The assemblages may also move vertically in the water column during the period of observation, but this is rarely checked. In fact, normally in lakes the tendency is to sample the epilimnion, yet as Brook *et al.* (1971) have shown large populations of the flagellate *Mallomonas acaroides* and of Cyanophyta (*see* p. 348) can exist several metres below the thermocline and some species have diurnal migrations up and down, hence sampling for seasonal studies must take these into account. On the other hand some species are non-motile but nevertheless maintain a subsurface position (*see* for example Fig. 10.8 of *Asterionella* in Gull Lake).

Seasonal succession of species can only be detected by intensive sampling but rarely are samples taken more than once a week and often they are taken two-weekly or even monthly and such timing almost certainly misses many subtle changes in the populations. Doubling times are often in the range 12–60 h, the former common amongst small forms, especially naked flagellates, and the latter for the largest diatoms and dinoflagellates. Recent figures for the doubling times of the diatoms *Cyclotella cryptica*, *Thalassiosira fluviatilis* and *Coscinodiscus asteromphalus* are 5, 8 and 24 h respectively (Werner, 1970). These figures were obtained from cultures and it would be interesting to see if they are similar in nature. Banse (1976) reviewed the relationship between cell size and growth and confirmed that cell division rates are inversely and linearly related to the log of cell volume, i.e. in general smaller species have higher specific growth rates and shorter doubling times. He also found that respiration rate also falls with increases in size. In only four days, division of a cell at 12, 51 or 60 h intervals would yield 256, 64 or 2 cells. There is no doubt that populations of small species do appear and disappear very rapidly and the causes have never been adequately studied.

Seasonal changes in form of algae have rarely been investigated (*see* Hutchinson (1967) for some examples) and recently Gibson (1975) reports on such a feature (cyclomorphosis) in *Oscillatoria redekei* which has two peaks (spring and August) of filament length interspersed with shorter lengths.

Annual trends

The total production of algae in any habitat tends to follow the general pattern of variation of irradiance or temperature, though remarkably few studies show the overall annual pattern. One example, admittedly of only a portion of the phytoplankton, is that of desmids, temperature and radiation in Lake Windermere (Fig. 10.1) based on long-term means (Canter & Lund, 1966) which clearly shows the correlation. Another example (Fig. 10.2a) from the marine sphere and obtained by a different technique (^{14}C fixation) also shows a very clear correlation between production and temperature except in early spring (Steemann-Nielsen, 1964). Fig. 10.2b gives chlorophyll values for a similar latitude (Tett & Wallis, 1978) which clearly shows the build-up of a peak of chlorophyll in early spring corresponding to the production peak of the other graph. These examples are taken from work in a temperate climate with all the attendant seasonal complexities and the simplest situation is one in which the irradiance varies only slightly during the year: such a state occurs on the equator in Lake George where there is an irregular scatter of irradiance about an almost steady mean (Ganf & Viner, 1973; Ganf, 1974b). The studies on this lake by the IBP team are of immense value since they act as a 'control' situation against which other lakes can be set. The constancy does in fact result in a stability which one would predict from the ambient conditions and almost no seasonal succession is recorded (Fig. 10.3; Ganf, 1975). Relatively small overall seasonal changes have also been recorded in a coastal subtropical lake in South Africa, though there are peaks of growth of individual species (Hart & Hart, 1977).

The succession in Lake Ontario with peak biomass in the high light–high temperature period of the year (Fig. 7.25) is certainly not always recorded, for in many lakes there seems to be a mid-summer depression coming between a major spring peak and a lesser autumn peak, e.g. in Lake Mikolajskie (Fig. 10.4). However the two contrasting patterns could be related to the methods of analysis since the Ontario data are based on cell volume and the Mikolajskie on chlorophyll content (Szczepanski, 1966). Nevertheless the succession in Lake Mikolajskie is a pattern which can be found repeatedly both in freshwaters and in the sea; in the Mediterranean, for example, diatoms follow this pattern, which is probably related to nutrient depletion allied to a response to light-temperature regimes (Margalef, 1951). Using data from 1970–6 for Loch Creran, a sea loch off Scotland, Tett & Wallis (1978) report a distinct winter trough of chlorophyll (0.5 mg m^{-3}), a spring increase (12–37 mg m^{-3}) a lower summer–autumn level (1–7 mg m^{-3}) but no autumn bloom as in most lakes. After the 'spring' growth, there is often a period when a larger number of species are competing for nutrients, e.g. in Lake Windermere the average number of species (excluding the smaller algae) between 1 January and 31 May fluctuates between 5 and 10 but between 1 June and 31 October between (8) 12 and 23 (Lund, 1964). This feature may be coupled with niche diversity, since during the first period the waters are more or less well mixed providing a single spatial niche whereas once stratification occurs the epilimnion becomes structured and many niches are available.

Fig. 10.1. Monthly mean number of desmid cells per 10 ml (cross-hatched) for the period 1945–64 inclusive in the water of the South Basin of Windermere. Length of day (continuous line), mean amount of radiation per day (dotted line) for the last 13 yr and mean surface temperature (dashed line) for the last 18 yr are also plotted. From Canter & Lund, 1966.

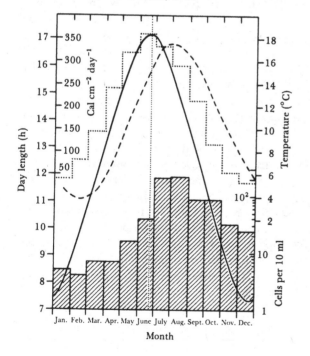

Fig. 10.2. *a*. The average monthly rate of gross phytoplankton production 1953–7, average surface temperature, 1921–30 at Halskov Rev in the Kattegat and average light intensity in Copenhagen during the year. From Steemann-Nielsen, 1964. *b*. A logarithmic plot of the combined chlorophyll data from the main basin of Loch Creran, 1–8 m depth zone, 1970–6. From Tett & Wallis, 1978.

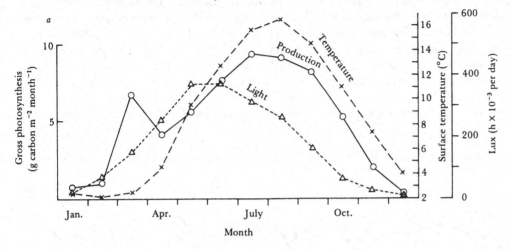

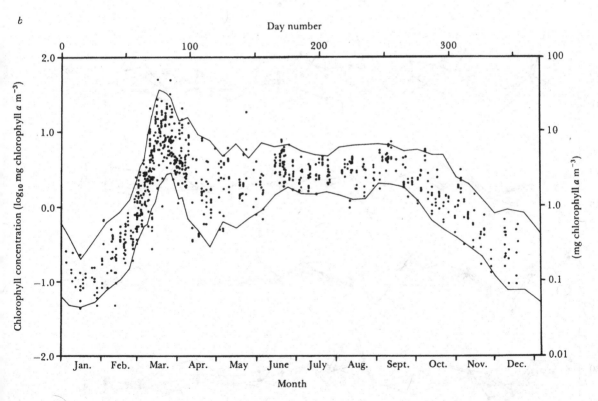

In lakes which are ice-covered for many months there tends to be a single mid-summer growing period as illustrated for the Finnish lake Pääjärvi (Fig. 7.22); this figure shows that at two different stations there is considerable fluctuation in carbon fixation during the short ice-free period and that although biomass is in general declining from June onward the amount of carbon fixation can be as high during this period as during that of maximum biomass (Ilmavirta & Kotimaa, 1974). The build-up to a single peak of growth will depend on the time of onset of growth and this will be correlated with latitude, and the result will be lower and later peaks at high latitude. Also, dependent upon the depth of circulation of the cells, the spring increase starts and ends later in deep lakes compared to shallow (Mortimer, 1969). In the seas also, Hart (1942) reported that the austral summer peak of phytoplankton growth in Antarctic waters moved from December in the northern through January to February in the southernmost waters. Although the fluctuations are more irregular in flowing waters

Fig. 10.3. Seasonal and horizontal variation of photosynthesis at 20 sites from Lake George, Uganda, in 1970, measured under laboratory conditions. From Ganf, 1976.

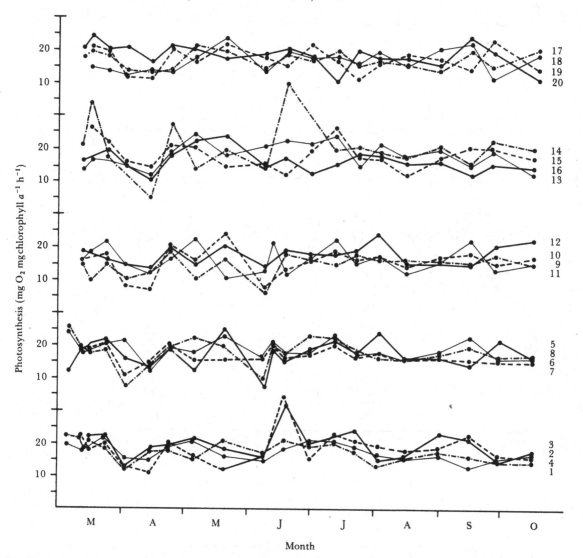

there is an approximate overall high cell count in summer and a low one in winter (e.g. in the Thames (Lack, 1971), Fig. 7.19).

A single summer peak of epipelic algal growth occurs at all depths in Marion Lake, British Columbia (Gruendling, 1971) with approximately 75% of the yearly production occurring between May and September (Fig. 10.5). By the use of multiple regression analysis, Gruendling concluded that temperature was the primary influence followed by total standing crop, then light, but there was no significant correlation with chemical factors. The latter feature is perhaps not unexpected in a habitat where nutrients are readily available from the sediment or water. An almost identical seasonal cycle, but peaking only in July to August, occurs also in the Finnish Pääjärvi (Kairesalo, 1977). Even in lakes which are not ice-covered for long periods there is a similar sequence e.g. in Windermere and Blelham Tarn (Round, 1961*b*) though here (Figs. 10.6, 10.7) the data were not obtained over a long enough period to correlate the cycles with light or temperature. The peak of diatom numbers at shallow stations, however, is more or less correlated with the spring diatom peak in the plankton though there can be local site variations. The single summer peak of biomass recorded for these epipelic algae (Fig. 10.7) is the result of integration of production at all depths; the deeper stations tend to peak production later in the year than the shallow stations.

A single summer peak of photosynthesis was found in a shallow temperate estuary (Williams, 1966) and this also seems to correlate well with temperature (Fig. 10.8).

Superimposed on the general seasonal cycle of growth there is a succession of different species throughout the year. For example in Lake Ontario (Munawar & Nauwerck, 1971) the winter plankton was dominated by cryptophycean flagellates (*Rhodomonas minuta* and *Cryptomonas erosa*), the spring by mixed Chlorophyta (*Scenedemus, Chlorella* and *Coelastrum*) and Bacillariophyta (*Melosira, Asterionella* and *Stephanodiscus*) the summer by Chlorophyta (*Pediastrum, Actinastrum, Chlorella, Ankistrodesmus, Oocystis, Scenedesmus*) and the autumn by Cyanophyta (*Chroccoccus, Oscillatoria*) plus some green algae and flagel-

Fig. 10.4. Seasonal changes in chlorophyll content of Lake Mikolajskie in 1961 (thin line) and 1963 (heavy line). From Szczepanski, 1966.

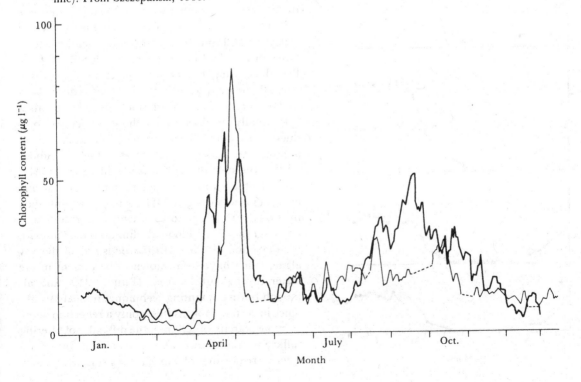

lates. Whenever detailed long-term records have been kept there appears to be a fairly precise repetitive pattern to the seasonal occurrence of species and the one quoted is fairly representative of temperate freshwaters. In the seas the succession is frequently from diatom-dominated spring assemblages to the dinoflagellate domination of summer (Conover, 1956).

Fig. 10.5. Weekly means of gross primary productivity of epipelic algae at five depths (*a–e*, 0.5, 1, 2, 3 and 4 m) in Marion Lake, British Columbia. From Gruendling, 1971.

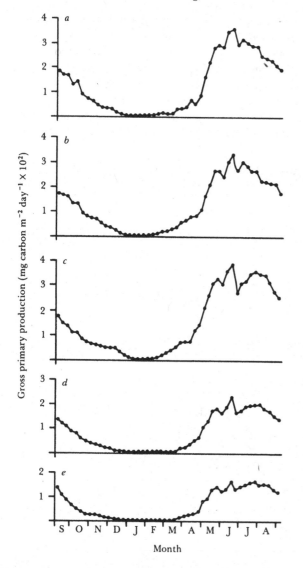

Annual cycles of selected species

Repeated annual appearances can be traced by long-term sampling of the individual habitats or by investigation of the algal remains deposited in the sediments (*see* Chapter 12). Studies of the former kind enable the recurrent growths to be correlated with the gross environmental changes and any new introductions related to subtle changes in the environment (*see*, for example, p. 25); the record from the sediment can be valuable in showing changing conditions within the basin and can be particularly valuable in detecting the onset and the course of pollution, etc.

Very few individual algae have been investigated in a detailed manner and even fewer have been investigated in different sites, however *Asterionella formosa* is a notable exception and one can compare its annual cycles from lakes in many parts of the world and it is clear that the species has its major growth in the spring with minor peaks in autumn. However, the start of growth each year varies slightly in each lake, e.g. it was almost invariably the beginning of April in Berlin, February to March in Abbot's Pool and Lake Windermere but variable in Lake Michigan but at some sites as late as May to June.

It is interesting that the commencement of the spring growth in the open Lake Windermere coincides approximately with that in the ice covered Lake Baikal. Lund (personal communication) attributes this to the long water retention time in Lake Baikal allowing the population to maintain itself beneath the ice whereas the population is diluted out in the winter in Lake Windermere. Also Lake Baikal is thermally stratified under the ice cover and production is confined to the upper 25 m (at 2.5 °C) as opposed to Lake Windermere which is about 4 °C at the start of the *Asterionella* increase. Shorter runs of observations suggest that the early spring onset of growth is a general feature and is controlled by growth rates responding to temperature and light and after the spring bloom declining coincident with the depletion of silica. *Asterionella* is one of the few algae which have been shown to be present in the water throughout the year (Lund, 1949) and although normal counting techniques indicate its absence in certain months this is only a reflection of the extreme scarcity of cells and the difficulty of filtering sufficient volume of water to catch the rare cells. There is really no problem of resting stages (and they

Fig. 10.6. The seasonal counts of diatom cells on samples from the littoral sediments of two stations in Windermere and two in Blelham Tarn. On the graphs for Windermere Station I and Blelham Tarn Station I the values for nitrate-nitrogen (dashed line) and silicate-silica concentrations (continuous line) in the water are also plotted. The scales are the same as those for cell numbers but divided by 100 to give silica in mg l⁻¹ and divided by 1000 to give nitrate in mg l⁻¹. Also plotted at the top of these graphs are the water-level changes for the two lakes. Above this is a combined graph of planktonic diatom cell numbers (continuous line) and histograms for monthly rainfall and sunlight hours. The period during which the lakes were stratified is shown thus. On the graphs for Windermere Station III and Blelham Station III are plotted the surface water temperatures in °C. From Round, 1957.

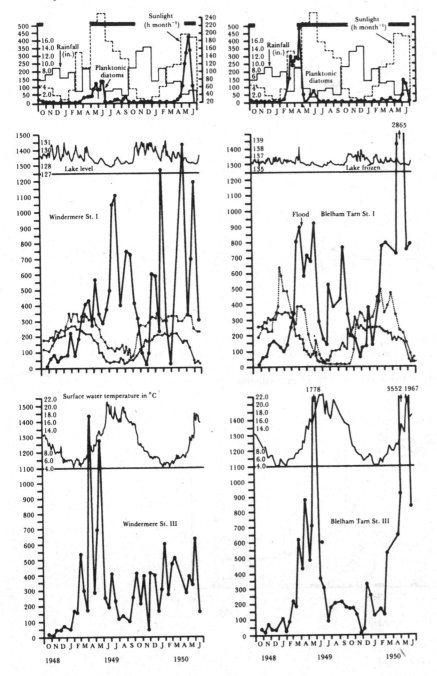

Fig. 10.7. The seasonal cell counts of, *a*, Cyanophyceae and, *b*, diatoms at the depth stations (1–6 m) summed for each sampling date, giving a measure of the annual biomass present in the epipelon of Windermere (continuous line) and Blelham Tarn (dashed line). From Round, 1961*b*.

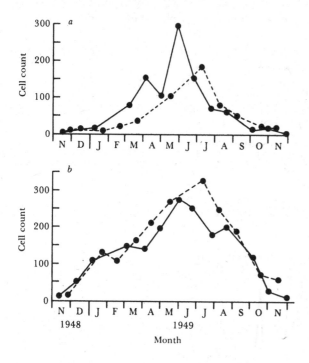

Fig. 10.8. The annual cycle of temperature and daily rates of gross photosynthesis (top pair of curves: expressed in m² and in m³) and respiration (lower pair of curves: expressed in m² and in m³) at a station on the coast of North Carolina. Continuous line, Temperature; broken line, photosynthesis and respiration, m⁻³; dotted line, photosynthesis and respiration, m⁻². From Williams, 1966.

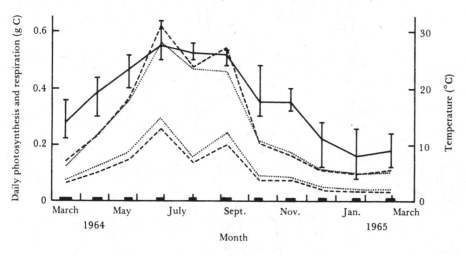

Table 10.1. *Rates of increase of blue-green algae in natural populations. From Reynolds & Walsby, 1975*

Species	Temperature (°C)	Mean doubling time (days)
Oscillatoria agardhii var. *isothrix* Skuja.	10–11	14.7
Oscillatoria redekei Van Goor	6	3.2
Anabaena circinalis Rabenh. ex Born. et Flah	11–1	7.0–9.9
Anabaena circinalis (stratified)	19–20	2.0
Anabaena flos-aquae Bréb. ex Born.	~26	3.9–5.7
Anabaena flos-aquae	~26	2.1
Microcystis aeruginosa Kütz. emend. Elenkin	15–20	4.4
Microcystis aeruginosa	~20	2.1
Aphanizomenon flos-aquae Ralfs ex Born. et Flah	15–20	2.1
Aphanizomenon flos-aquae	19–20	2.0
Aphanizomenon flos-aquae	23–26	1.1

Fig. 10.9. The time–depth distribution of *Asterionella formosa* in Gull Lake, Michigan. Vertical lines, periods of mixing of the water column and, between them, inverse stratification during ice cover (January to March) and direct stratification from late April until late October. From Moss, 1972*b*.

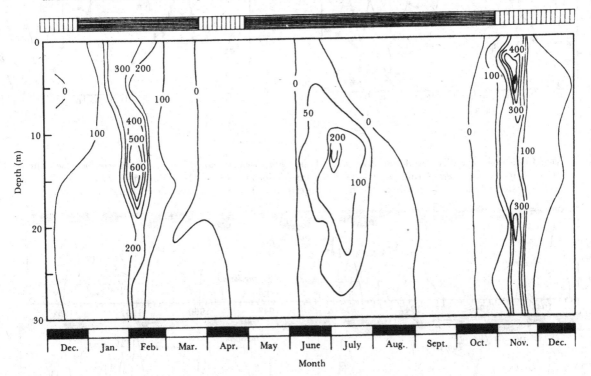

Fig. 10.10. *a*. The growth of *Asterionella formosa* and *Tabellaria flocculosa* var. *asterionelloides* in cultures suspended at 1 m (upper curves) and 6 m (lower curves) below the surface of Lake Windermere. Each exposure lasted one week or, occasionally, 4 or 8 days. Growth expressed as cell divisions (i.e. on a $\log_2$ basis) per week. *b*. The growth of *Asterionella formosa* and *Fragilaria* in the same circumstances as described above. From Lund, 1964.

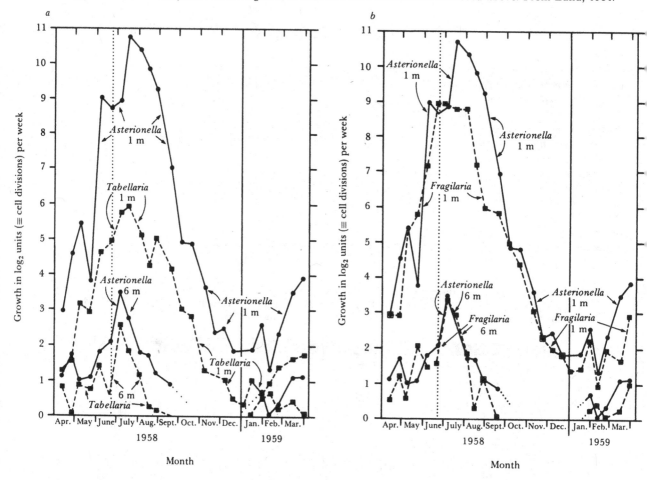

Fig. 10.11. The periodicity of *Asterionella formosa* Hass., *Fragilaria crotonensis* Kitton, *Tabellaria flocculosa* (Ehr) Grun., var. *asterionelloides* (Grun.) Knuds. and the fluctuations in the concentration of dissolved silica, in the 0–5 m water column of the northern basin of Windermere, 1945–60 inclusive. From Lund, 1964.

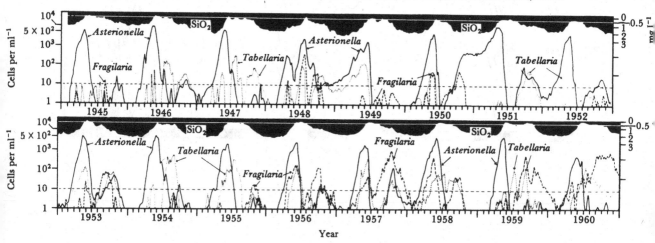

have never been reported) or need for re-inoculation, and the factors controlling the growth cycles are environmental. Moss (1972*b*) records the very interesting occurrence of *Asterionella* in mid-summer as a deep-lying population (Fig. 10.9) in Gull Lake, Michigan. Whether or not most algae are always present as isolated cells (spores) or deep-lying populations in the aquatic environment has not been determined but it is quite likely that at least all those which do not form spore stages are always present and even some of the groups which do form spores may be permanently present, e.g. desmids which are rarely recorded as forming spores in planktonic populations (Canter & Lund, 1966). Small populations of many algae probably grow in most waters with little chance of detection and only

when conditions become favourable do they bloom (*see* also p. 305); thus in several of the English Lake District lakes *Asterionella* is present but never forms large populations. Investigations of these populations and the related limnological conditions may lead to a definition of the conditions necessary for the occurrence of the species; such information is incredibly difficult to obtain for any species and it is certain that the precise environmental conditions for rapid growth of algae are known for only a small number of species (e.g., Table 10.1). One feature which seems definite is that increasing illumination in spring enables *Asterionella* to grow sufficiently fast to show a net increase in the water column and equally that this growth is ultimately stopped each year by silica depletion. But increasing illumination

Fig. 10.12. The relationship between the populations of *Asterionella formosa*, *Stephanodiscus rotula* and dissolved silica (continuous line) in the water column in Abbot's Pool. Stratification is indicated along the top of the figure and the overturn by the arrow. From Happey, 1970.

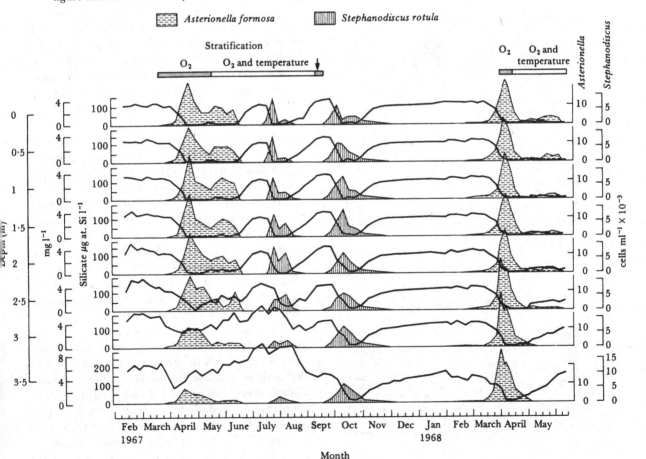

Month

does not enable the species to grow in other lakes, even when silica is in adequate supply. Temperature is a much less important factor since *Asterionella* commences its bloom at low temperatures in the English Lake District and in other regions it even forms large populations under ice e.g. Gull Lake,

Fig. 10.13. The abundance of total diatoms (open circles), *Skeletonema costatum* (filled triangles), an ebridian, *Phaeocystis poucheti* (as colonies ml, open triangles), and zooplankton (filled circles) biomass (*a, b*). Nutrients (*c, d*), temperature (*e*) and salinity (*e*) at 0 m in Narragansett Bay. From Smayda, 1973.

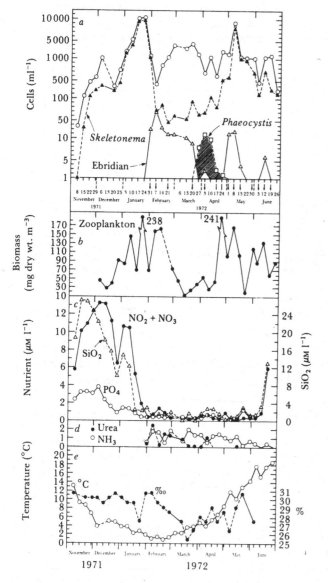

Michigan (Moss, 1972*b*) : it is not clear how a diatom with no flotation capabilities can maintain itself in the water column under ice but turbulence caused by the flow of water from basal springs may be effective in Gull Lake (Fig. 10.8). However, there are lakes in which *Asterionella* has extended its growth throughout the summer and this raises further problems one of which may involve adaptation of the population to high temperature. Equally in many lakes there is an autumn growth phase but this can hardly be controlled by the same conditions as the spring growth. It is interesting that the epidemics of chytrids which attack species such as *Asterionella* (*see* p. 402) do not appreciably affect the long-term sequence of annual cycles. Fig. 10.10*a* shows that under experimental conditions in bottles suspended in lakes *Asterionella* always outgrows *Tabellaria* so that it is not surprising that the former almost always precedes the latter or that, if they grow simultaneously, *Asterionella* forms larger populations (Fig. 10.10*b*). Competition with *Fragilaria crotonensis* is however of a different form since it can grow at the same rate as *Asterionella* (Fig. 10.10*b*) and Lund (1964) ascribes the lesser abundance in Windermere to more severe loss of cells during winter and hence a smaller inoculum at the commencement of growth. Although less obvious in flowing water, whenever a pronounced peak of diatoms develops, a decrease in silicate is recorded, e.g. in April–May in the River Thames (Fig. 7.19). Restriction of diatom growth by lack of dissolved silica is rarely recorded except at the time of the spring maxima (Fig. 10.11), probably due to the fact that this is usually the only time of the year when extremely large crops of diatoms are growing. Studies of the generally smaller autumn bloom of diatoms are less common though for at least one genus, *Stephanodiscus* (Fig. 10.12), in both Abbot's Pool (Happey, 1970) and Loch Leven (Bailey-Watts & Lund, 1973), silica limitation has been shown for this time of year.

A marine example showing a very similar growth pattern is *Skeletonema costatum* and Fig. 10.13 shows its cycle in Narragansett Bay, where nutrients declined as the *Skeletonema* population increased though there was an added complication in that zooplankton predation was important with peak numbers of zooplankton coincident with those of the diatom. Unlike *Asterionella*, the marine diatom was achieving peak growth early in the year when temperature was decreasing and light was low but increasing slightly. In fact, Pratt (1965) points out that the relationship between this diatom and light is actually inverse; a very unusual situation. However the general bloom of phytoplankton in Narragansett Bay is linked to light intensity and if this is abnormally low in any one year the bloom is delayed (Hitchcock & Smayda, 1977): the level of 40 ly day^{-1} seems to be critical for these temperate waters and confirms the figure proposed by Riley (1967). Another complication, especially in temperate to polar latitudes, is that the mixed upper layer of water is usually considerably deeper than the euphotic zone in winter to spring. Under these conditions cells can spend too long in the dark, lower, mixed water but in spring the depth of the euphotic zone increases and that of the mixed layer decreases as thermal stability increases (*see* also p. 313). The mixed layer is often deeper than the euphotic zone even during the spring bloom in the ocean but there is no reason why this should prevent growth, provided that the cells spend long enough in the euphotic zone to achieve sufficient photosynthesis for cell divisions to exceed losses. This important change in a physical factor can create suitable conditions for the spring bloom and attempts have been made to predict the timing from the physical data.

In Lake Windermere the *Asterionella* is succeeded by other planktonic diatoms and then by three desmid species *Cosmarium contractum*, *C. abbreviatum* and *Staurastrum lunatum*. These have succeeded one another each year in the 25 years of Lund's observations (Figs. 10.14, 10.15). The achievement of dominance by one or other desmid in Lake Windermere is related to growth rate; *C. contractum* grew more slowly than the other two, at all times of the year and at all depths, and hence reached its maximum in September to October, whilst the other two usually peaked in July to August. The general curves of abundance follow the temperature curve for the year (Fig. 10.1) and this and the change from the random distribution during isothermal conditions to concentration in the epilimnion in summer is shown in the time–depth plot in Fig. 10.15. The faster growth rate of *C. abbreviatum* and *S. lunatum* should, however, result in larger populations and Lund suggests that fungal parasitism may be a restraining factor since it can become most severe in summer when the algae are circulating in a smaller volume of water above the metalimnion and therefore closer to one another: this is an important factor, *see* p. 400. In addition, there is a larger number of other desmids present to provide fungal inocula. Since fungi parasitic on desmids show little host specificity, at times of high diversity of desmids in the plankton a large and variable complement of fungal parasites is ready to infect when desmid populations become dense. Yet even these factors do not wholly explain the differences in timing of the peaks and therefore, after 25 years' work, Lund considers that there are still unknown nutritional factors operating on the population. Variation in growth rates of communities has been known for a long time, e.g. Verduin (1952)

Fig. 10.14. The average monthly abundance of *a*, *Staurastrum lunatum*, *b*, *Cosmarium contractum*, and *c*, *C. abbreviatum* in Windermere, North Basin, 1945–69. From Lund, 1971.

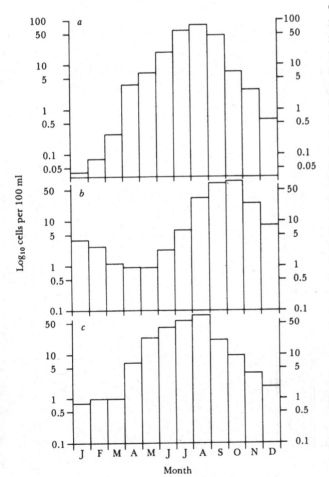

found that an *Asterionella–Cyclotella* community in Lake Erie grew at about three times the rate of a *Stephanodiscus* community. If the *Stephanodiscus* was a large-celled form the amount of silica needed would be many times that required for *Asterionella* and this alone may explain the rate difference. It is now

abundantly clear that species of algae have different specific growth rates (e.g. Table 10.1 gives some data for planktonic Cyanophyta) and that the interaction of these with light intensity and nutrient levels results in enormously varied growing periods (*see* p. 437); the competitive advantages of growth rates can

Fig. 10.15. Depth–time distribution of temperature (*a*) and the abundance of *Staurastrum lunatum* (*b*), *Cosmarium contractum*, (*c*) and *C. abbreviatum* (*d*) in Windermere, South Basin, 1953. From Lund, 1971.

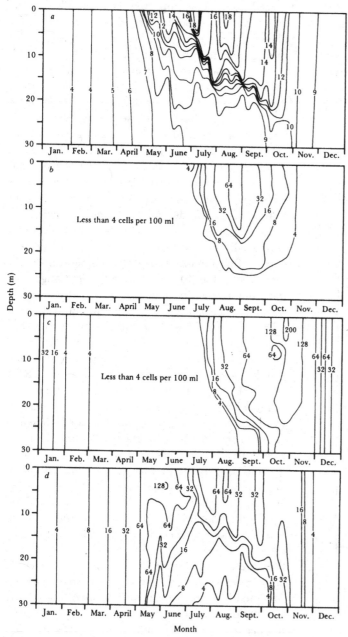

Month

be estimated and applied to natural populations (Eppley, Rogers & McCarthy, 1969). These workers suggest that *Emiliania* (*Coccolithus*) *huxleyi* would predominate over diatoms at 20 °C unless upwelling water increased the nitrate concentration and then diatoms would do well: this prediction has in fact been confirmed by observations off the coast of California.

Early workers were intrigued by the idea of populations building up on sediments, in the reeds, etc., and then being carried into the open water to form a bloom. This occurs, but on the whole is probably a relatively rare phenomenon. Perennation on the other hand is more common and as yet undocumented for genera such as *Mallomonas* and *Synura* which form copious cysts, although recently *Dinobryon* cysts have been observed to germinate (*see* p. 245). Two recently described examples of perennation are of the blue-green alga *Gloeotrichia echinulata* in which the akinetes (resting spores) are sedimented on to the bottom of Lake Washington (Roelofs & Oglesby, 1970) at the end of the summer growth. They germinate and small colonies can be found on the lake bottom in March prior to resuspension to form the next year's bloom. The other is *Anabaena circinalis* in Crose Mere, where the spores were lifted from the bottom of the lake by strong turbulent mixing during the winter, but without leading to growth in the water column at this time of the year (Reynolds, 1972). But by March conditions were presumably right and germination followed (Fig. 10.16). Resting type spores were then found free in the water column during the subsequent periods of decline of the *Anabaena* population: it is not clear whether or not the August *Anabaena* maximum is due to germination of the spores formed during the decline of the earlier maximum or merely

to re-growth of the filaments. The dominant dinoflagellate (*Peridinium cinctum*) in Lake Kinneret forms resting spores and these are known to sink to the sediments after the massive blooms in this lake. They then germinate in the next year, the initial increase in vegetative cells is recorded over sediments and the cells then move out into the main mass of water (Pollingher, personal communication).

The only well documented case of perennation by vegetative cells of diatoms is that of *Melosira italica* subsp. *subarctica* in lakes of the English Lake District (Lund, 1954, 1955). This heavy, filament forming diatom cannot maintain itself in the water column as the temperature increases (and viscosity decreases) and it sediments out, remaining on the sediments during the summer. Autumnal gales mix the water column and disturb the sediments thus bringing the cells of *Melosira* into suspension. Since this diatom can grow successfully at low light intensities it then builds up considerable populations over the winter period (Fig. 10.17). When ice forms, the importance of turbulence in maintaining planktonic populations is demonstrated; the sinking of the filaments under ice is due to the slowing down of turbulent mixing when the wind stress is cut off from the surface. Live cells of phytoplankton species have been recovered from lake sediments down to 40 cm in Lake Windermere and 13 cm in Esthwaite Water (Stockner & Lund, 1970) so storage of small inocula over fairly long periods of time is undoubtedly possible: their resuspension may be rare and associated with storm-induced turbulence. This feature may explain some of the intermittent records of planktonic species.

Long-term observations such as those described above are much more difficult to obtain at sea and the problem of perennation is a greater one, at least in the deeper ocean, where it is virtually impossible to raise sedimented material into the euphotic zone. I know of no studies attempting to solve this problem but presumably a very small number of live vegetative cells must be circulating somewhere in the mixed zone. Another possibility is that algae perennate attached to flocs of organic matter. Long-term observations made in the English Channel are illustrated in Fig. 10.18, the measurements here are based on ^{14}C fixation by the phytoplankton incubated *in situ* on deck. This is a common technique used at sea where stations cannot be occupied for the length of time required to perform incubations down

Fig. 10.16. Seasonal variations in the population of *Anabaena circinalis* (continuous line) and the number of free spores (dotted line) circulating in the water in Crose Mere. From Reynolds, 1972.

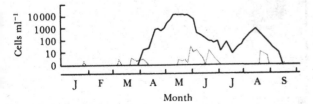

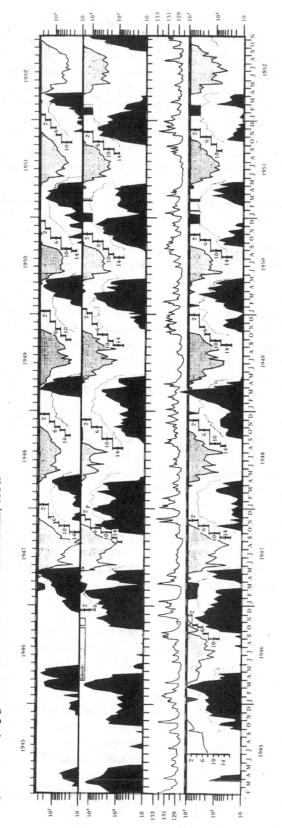

Fig. 10.17. *Melosira italica subarctica*, the seasonal cycle in Windermere (South Basin), Esthwaite Water and Blelham Tarn (top, middle and bottom graphs respectively) from February 1945 to May 1952. *Solid black*, number of cells per millilitre plotted on a log scale; *dotted areas*, direct thermal stratification expressed as the difference in °C between the surface and bottom temperature of the water column at the sampling stations; *chequered areas*, inverse stratification, expressed likewise, in both cases differences below 0.4 °C not shown; *dotted line*, surface temperature; *plain enclosed areas*, Windermere lake level based on 129 ft above O.D. P, epidemic of parasitism by *Zygorhizidium melosirae* Canter. From Lund, 1954.

a depth profile; they give a good measure of the potential production of samples. The precision of carbon fixation peaks in the spring was very distinct in every year except 1967 when it was delayed until summer. The steady increase from 1967 to 1971 is noteworthy but then in the succeeding two years the spring peak was low and large autumn peaks occurred reminiscent of the situation in temperate lakes. It is only by studying such long-term trends that real insight can be obtained into the overall factors controlling phytoplankton. Although based on rather different data, i.e. cell counts, the records of Bolin & Abbott (1963) confirm a general pattern of one major peak during the year in neritic phytoplankton but at this more southerly station the peak occurs in summer (Fig. 10.19). These data show great long-term fluctuations in abundance of certain genera, e.g. the increase in *Ceratium* in 1958–9 and the relative constancy of others, e.g. *Chaetoceros*.

The overall annual pattern of growth in algal populations is thus similar to that of the terrestrial vegetation with a short growing season in high latitudes, a long intermittently peaking sequence in temperate regions and an irregularly fluctuating non-seasonal pattern in the tropics. Obviously deviations can be instanced, especially where gross nutrient deficiencies occur, as in tropical oceans, but these are no more exceptional than the short growing seasons of certain tropical terrestrial areas. Having established a 'normal' annual succession for the various climatic zones any variations can be viewed with even greater interest and factors sought to explain lack of fit.

Obviously it is advantageous to know the exact recent history of species within a water mass before attempting to interpret annual cycles but such data are in fact very rarely available. The source of cells is generally assumed to be from relict inoculum remaining in the basin from previous growths. But sometimes inoculum may be derived from other waters within the drainage basin: I know of only one well-documented instance of this, the appearance of *Microcystis aeruginosa* and *M. flos-aquae* in Lake Kinneret carried in by flood waters from the River Jordan. If the *Microcystis* arrives early in the season it will spread out in the lake and bloom, but if it is late then the growth of *Peridinium cinctum* var. *westii* will out-compete the *Microcystis* in the uptake of nutrients (Serruya & Pollingher, 1971).

In the above discussion, the restriction of certain species to fairly precise seasons has already been demonstrated, but many hundreds more have equally short growing periods, detailed data about which are lacking or buried in the literature. Most examples discussed in the literature are of short growth periods in the spring for it is the so-called 'spring outburst' which is so striking and has been best documented, but summer, autumn and winter 'outbursts' also occur. One of these, the autumn *Asterionella* growth is as far as I know not yet analysed

Fig. 10.18. The seasonal variation in g carbon fixed under 1 m² per day in the English Channel at Station E1. From Boalch, Harbour & Butler, 1978.

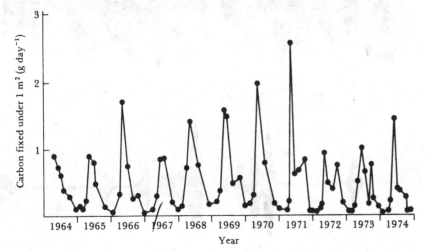

in detail, though Lund obviously has all the data and informs me that they will be published. I personally eagerly await their appearance since I know they will be based on many years' observations.

I have been impressed, whilst working on various communities, that only a very few species seem capable of growing over a whole annual cycle and a few like *Asterionella* have two growth periods. In fact from a consideration of the 'climatic' changes which occur in the water column during a year one would *a priori* expect either stability of the individual algal populations (e.g. in oligotrophic oceans or lakes on or within a few degrees of the equator) or of a succession of populations adjusted to the changing environment in all other regions. Stability is in fact a feature of the populations of individual species

in Lake George, one of the few lakes lying on the equator and investigated in some detail: all the species except *Anabaena* are present throughout the year and the annual variation in chlorophyll content is only slight (Ganf, 1974*a*, *c*). These changes will be compacted into a few months in the sub-polar regions but extend throughout the year in temperate climates. If one considers the changes which a temperate lake undergoes during an annual cycle then, as Hutchinson (1967) points out, the changes in 'climate' and water chemistry during an annual cycle in a lake are probably greater than those of soil chemistry and climate over several millennia and the latter changes have obviously induced great changes in the flora on the land, as evidenced by the replacement of species shown in pollen diagrams of the

Fig. 10.19. The monthly average volume of the phytoplankton standing crop (*a*) and the relative contribution (% of total, by counts) of *Chaetoceros* spp. (*b*) and *Ceratium* spp. (*c*) from 1954 to 1960 off California. From Bolin & Abbott, 1963.

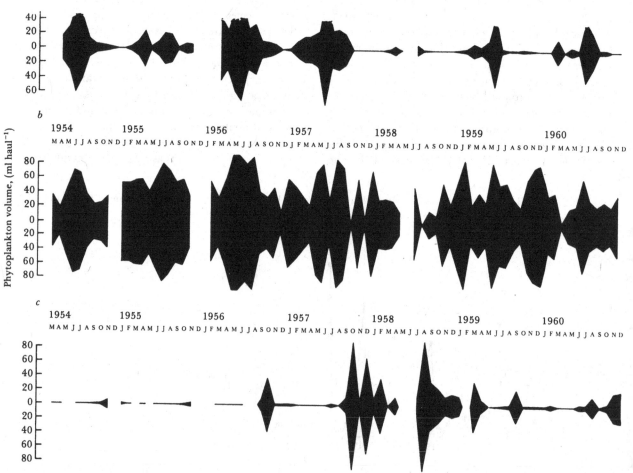

post-glacial period. There are of course seasonal chemical changes in the soil which have equally profound effects on the soil microflora but it was the long-term changes of which Hutchinson was thinking. Change is continuous but certain times in the year present 'shock' periods for the algae (Round, 1971*b*). Two of these periods are connected with the heating–cooling cycle, one in the spring when thermal stratification becomes established and the other in the autumn at the overturn, whilst two others are the mid-summer period of maximum light and stagnation and the mid-winter when temperature and light are at a minimum. For an alga to grow through these shock periods and thus straddle two sets of quite contrasted growing conditions assumes an extreme degree of eurytopy. On the other hand these 'shock' periods may be favourable times for some other algae, giving rise to the possibility of eight growing periods, though few species, in fact, seem to

Fig. 10.20. *a*. Hypothetical growth curves for freshwater populations, assuming fairly rapid death of population following termination of growth. *b*. Hypothetical growth curves with live cells remaining in the habitat for some time after division has ceased; arrows indicate times when the populations are subjected to 'shock' of some kind. *c*. Some patterns of epipelic diatom growths from two pools investigated by Round, 1971.

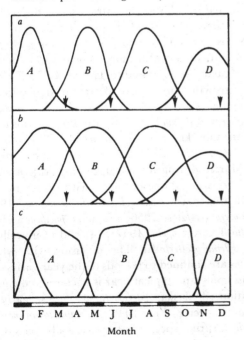

J F M A M J J A S O N D

Month

achieve maximum growth at the times of 'shock'. 'Shock' as such is probably quite great in the spring when the depth through which an alga circulates is reduced, the cells are maintained for a longer time in higher light and nutrients begin to change rapidly in the smaller volume of the epilimnion. An even more dramatic 'shock', since it may occur overnight, is the breakdown of thermal stratification which suddenly thrusts an alga into a colder, darker climate and a less 'stagnant' set of chemical conditions. Fig. 10.20 shows a hypothetical sequence of growths between 'shock' periods, *a*, and the sequence, *b*, if live cells remain in the habitat for some time after division has ceased. There are many examples of growth within this general four-cycle system including a mid-summer low if cell biomass is considered. Four periods have often been mentioned in the literature of temperate regions and designated by the light and temperature regimes: winter low light–low temperature, spring high light–low temperature, summer high light–high temperature and autumn high temperature–low light. The point has been made and is undoubtedly true that algae must be adapted to these conditions for growth in these periods but I believe equally important are the periods of maximal disturbance ('shock'). Fig. 10.20*c* shows how some epipelic diatom growths fit into this system. Some of these growths have been followed over a three-year period (Fig. 10.21) and maintain roughly the same timing each year. The species in Fig. 10.21 which extend over two or more growth periods are, with the exception of *Navicula hungarica* var. *capitata*, very 'difficult' taxa which may easily encompass slightly different morphological and physiological races. *A* in Fig. 10.20*c* is represented by *Stauroneis anceps*, *B*, by *Nitzschia acicularis*, *C*, by *Navicula pupula* var. *capitata* especially in the middle year and *D*, by *Navicula pupula*. Theoretically some species might occur in two periods, especially *B* and *C*, and possibly *Navicula oblonga* is an example. In rocky streams, Bursche (1962) found that *Navicula viridula* and *N. cryptocephala* grew in winter and early spring but *Cocconeis placentula* showed no periodicity. The first two species are motile and the last is attached and its attachment may explain these results, the cells being lost only very slowly from the substrata and therefore continual presence being recorded. This does not of course necessarily imply continual production. Similar restriction of growth

Fig. 10.21. The seasonal change in cell numbers of epipelic algae in two small pools (*a*, *b*) over three years. Algal counts plotted as $\log_{10}$ cell numbers. + indicates presence but no count. Abbreviations of algae as follows: St an, *Stauroneis anceps*; Na cr, *Navicula cryptocepha*; Na cu, *Navicula cuspidata*; Na pu, *Navicula pupula*; Na Pu ca, *Navicula pupula* var. *capitata*; Na hu ca, *Navicula hungarica* var. *capitata*; Na ob, *Navicula oblonga*; Na ra, *Navicula radiosa*; Ni ac, *Nitzschia acicularis*; Ni di, *Nitzschia dissipata*; Ni pa, *Nitzschia palea*; Ca am, *Caloneis amphisbaena*; Cy el, *Cymatopleura elliptica*; Cy so, *Cymatopleura solea*; Cl, *Closterium* spp.; Os, *Oscillatoria* spp.; Sy, *Synura* spp.; Ch, *Chlamydomonas* spp.; Cr, Cryptomonads; Eu, Euglenoids. From Round, 1972*c*.

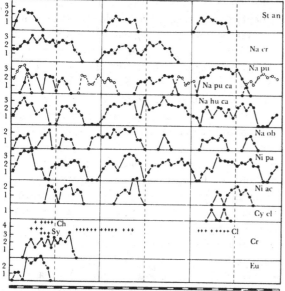

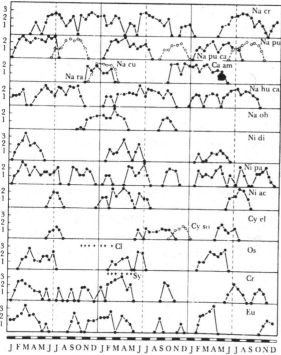

periods obviously occurs at sea and equally few species grow over a whole season at least in temperate zones. Margalef (1961) abstracted data from the literature, divided the year up into five periods and found that few planktonic forms occur in more than three periods. Table 10.2 from Pratt (1959) illustrates some of the patterns found in a shallow coastal zone and shows a greater tendency for a few species to extend throughout the year: the data only give presence or absence and if abundance was considered these species may nevertheless have distinct growing periods. In a bay such as Narragansett there may be more habitats available in which populations can maintain themselves, and the sampling is of different populations mixed into the water column. Some of the constant species might nevertheless prove to be metabolically more restricted, e.g. a study of the phytoplankton in Long Island Sound (Conover, 1956) certainly showed that post-bloom populations were in a state of senescence and could not be stimulated by addition of nutrients as can blooming populations. The period of upwelling along tropical coasts is not a time when a single assemblage is dominant, for here also a succession of species occurs, e.g. Smayda (1966) found that in the Gulf of Panama four growth phases could be recognised: (1) December–January, *Chaetoceros compressus–Skeletonema costatum* f. *tropicum* (presumably this is the diatom now known as *S. tropicum*); (2) February, *Rhizosolenia delicatula* (*Eucampia cornuta* in one year); (3) March, during intense upwelling *Nitzschia delicatissima* was abundant but with different co-dominants each year (*Launderia annulata*, 1955; *Chaetoceros costatus*, 1956; *Nitzschia pacifica* and *N. atlanticus* var. *pungens*, 1957); (4) end of upwelling, *Rhizosolenia stolterfothii*. The precision with which species are confined to periods of the year varies and occasionally an atypical year is encountered e.g. at a station off Alaska, *Asterionella japonica* is clearly an autumn species (Fig. 10.22), *Thalassiosira nordenskioldii* a spring species but occasionally extends to

Fig. 10.22. Daily average numbers of *Asterionella japonica* (*a*), *Thalassiosira nordenskioldii* (*c*), *Biddulphia aurita* (*b*) and *Chaetoceros debilis* (*d*) as cells l^{-1} in successive 7-day periods at Scotch Cap Light off Alaska, from 1926 to 1933. ■, cells in good condition; ▨, cells in poor condition. From Cupp, 1937.

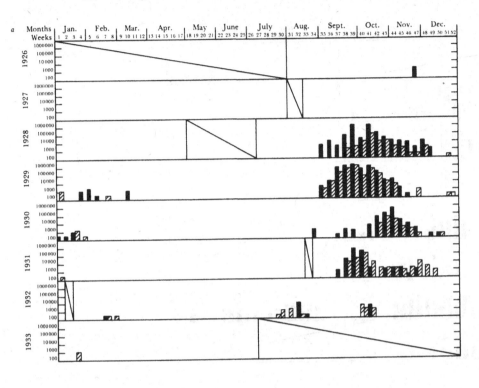

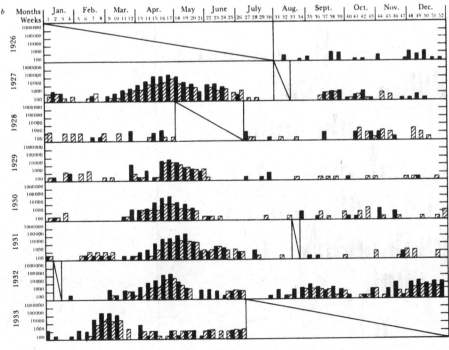

Fig. 10.22 (*cont.*)

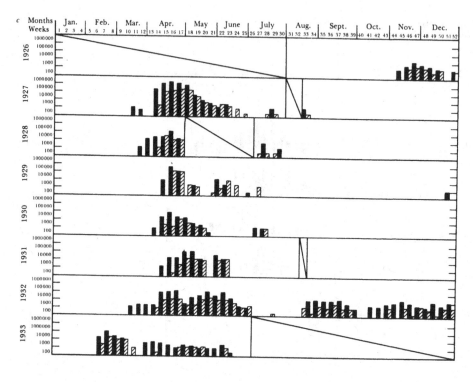

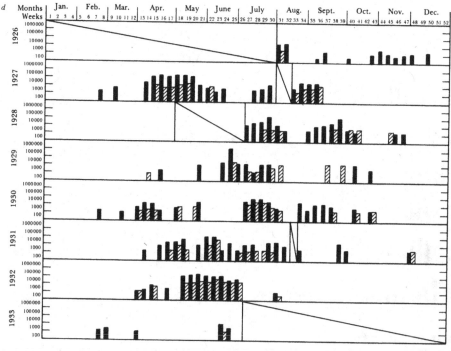

Table 10.2. *Examples of the occurrence of marine plankton in Narragansett Bay during the year. Data selected from Pratt, 1959*

Species	Month											
	J	F	M	A	M	J	J	A	S	O	N	D
Leptocylindrus danicus Cleve	×	×	×	×	×	×	×	×	×	×	×	×
Rhizosolenia setigera Brightwell	×	×	×	×	×	×	×	×	×	×	×	×
Skeletonema costatum (Greville) Cleve	×	×	×	×	×	×	×	×	×	×	×	×
Peridinium trochoideum (Stein) Lemmermann	×	×	×	×	×	×	×	×	×	×	×	×
Rhizosolenia faeroense Ostenfeld	×	×	×	×	.	.	.	.	.	.	.	.
Chaetoceros decipiens Cleve	.	×	×	×	×	.	.	×	×	×	×	.
Schröderella delicatula (Peragallo) Pavillard	.	.	×	×	×	.	.	.	.	.	.	.
Chaetoceros laciniosus Schütt	.	.	×	×	×	.	.	.	.	.	.	.
Peridinium triquetrum (Ehrenberg) Lebour	.	.	×	×	×	×	×	.	.	×	.	.
Dinophysis acuminata Claparede and Lachman	.	×	.	×	×	×	×	×	×	×	.	.
Rhizosolenia fragilissima Bergon	.	.	.	.	×	×	×	×	×	×	.	.
Ceratium lineatum (Ehrenberg) Cleve	.	.	.	.	×	×	×	×	×	×	.	.
Prorocentrum scutellum Schroeder	.	.	.	.	×	×	×	×	×	×	×	×
Chaetoceros simplex Ostenfeld	.	.	.	.	.	×	×	×	.	.	.	.
Prorocentrum gracile Schütt	.	.	.	.	.	×	×	×	×	×	×	.
Rhizosolenia delicatula Cleve	.	.	.	.	×	×	×	×	×	×	×	×
Chaetoceros teres Cleve	.	.	.	.	.	.	.	×	×	×	×	.
Pyrocystis (*Dissodinium*) *lunula* Schütt	.	.	.	.	.	.	.	.	×	×	.	.
Stephanopyxis sp.	.	.	.	.	.	.	.	.	×	×	×	×
Ditylum brightwellii (West) Grunow	×	.	.	.	.	.	.	.	×	×	×	×
Detonula cystifera (Cleve) Gran	×	×	×	×	.	.	.	.	.	.	.	×
Chaetoceros danicus Cleve	×	×	×	.	.	×	×	×	×	×	.	.
Chaetoceros curvisetus Cleve	×	×	.	.	.	.	.	×	×	×	×	×

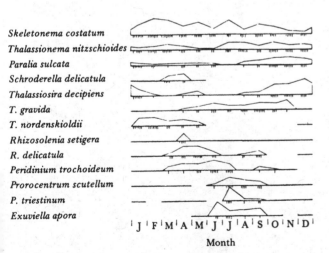

Skeletonema costatum
Thalassionema nitzschioides
Paralia sulcata
Schroderella delicatula
Thalassiosira decipiens
T. gravida
T. nordenskioldii
Rhizosolenia setigera
R. delicatula
Peridinium trochoideum
Prorocentrum scutellum
P. triestinum
Exuviella apora

J F M A M J J A S O N D

Month

Fig. 10.23. The seasonal occurrence, degree of dominance and frequency of year-to-year occurrence of the thirteen most important species of phytoplankton in Long Island Sound. All species had constituted 5% of the population during any one month for at least four of the eight years of sampling. The presence of a base line indicates presence of the species during the month in question. The small vertical lines indicate the number of years that the relative concentration exceeded 5% during any given month. The curves above the base line are average concentrations (% of total population) and the scale is 50% from one base line to the next. From Riley, 1967.

autumn, *Biddulphia aurita* is a spring species but also appears in small numbers during other months (especially in 1932 when *Thalassiosira* had its autumn peak) and finally *Chaetoceros debilis* is rather erratic but peaks occur in the summer months (Cupp, 1937). Scattered occurrences such as exhibited by *Chaetoceros* may indicate meroplanktonic tendencies

with resuspension of cells during periods of increased turbulence and certainly in the shallow Long Island Sound, *Skeletonema*, *Thalassionema* and *Paralia* (Fig. 10.23) tend to extend throughout the year (Riley, 1967). This study also shows the classic pattern of summer dinoflagellate blooms which is very apparent also from the plankton recorder data in Fig. 10.24

Fig. 10.24. The seasonal variation in phytoplankton colour *a*, visual estimate with three categories indicated by increasing stippling, *b*, abundance of *Ceratium* and, *c*, abundance of diatoms in the North Sea (Standard area C2) the three categories based on averaged monthly records. Data from plankton recorder studies (Reid, 1975*b*).

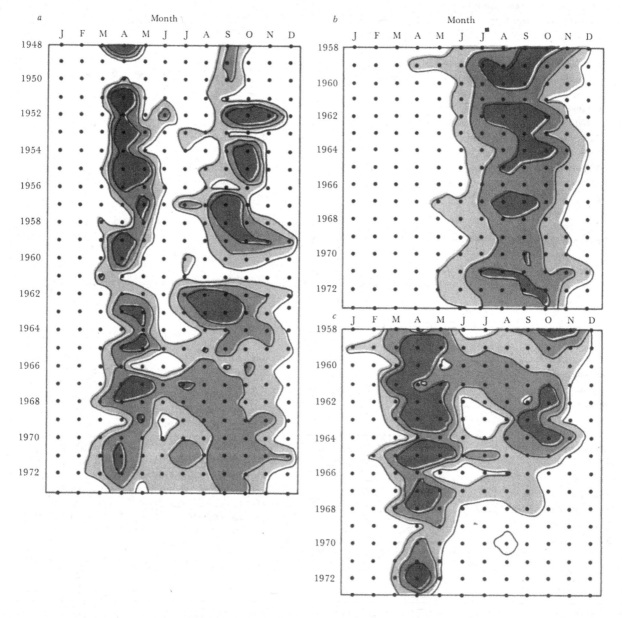

where the genus *Ceratium* alone is plotted and shows great stability (Reid, 1975*b*) and from other North Sea data (Fig. 10.25) obtained by Dodge & Hart-Jones (1974). The data in Fig. 10.24 show that the autumn diatom crop suddenly disappeared from the North Sea in 1966: Reid thought that this may have been due to an increase in eutrophication but if so it occurred at a time when overall diatom crops were beginning to increase at other stations (e.g. at E1 off Plymouth) (Fig. 10.18). Detailed comparison of the cell count data from these two regions will be very interesting.

The precision with which individual phytoplankton populations occur each year in lakes is illustrated in Figs. 10.11 and 10.12 of *Asterionella*, Fig. 10.17 of *Melosira*, Fig. 10.26 of the desmid *Staurastrum*

Fig. 10.25. The seasonal and vertical distribution in the water column of some common dinoflagellates in the North Sea. From Dodge & Hart-Jones, 1974.

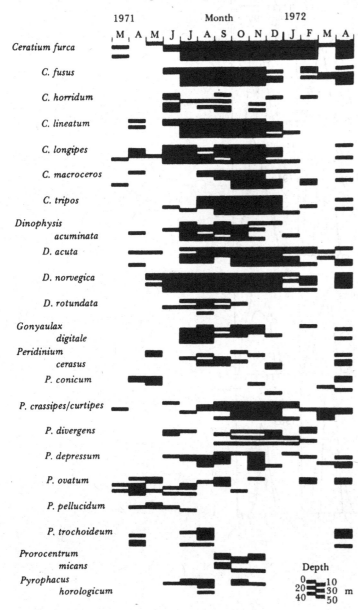

Fig. 10.26. The periodicity of *Staurastrum lunatum* as monthly averages in Windermere, 1945–69. From Lund, 1971.

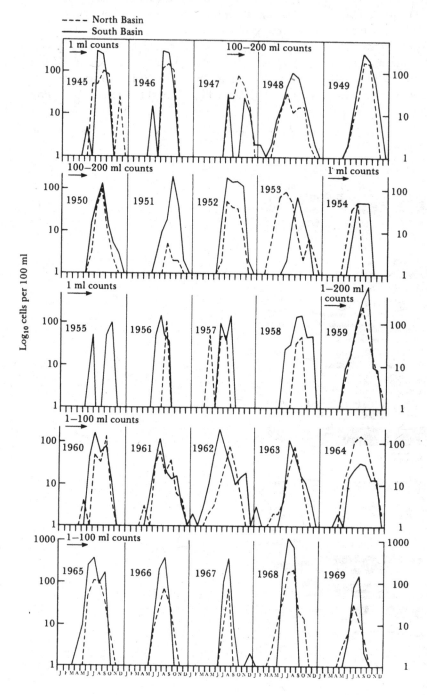

lunatum and in the data on Cyanophyta by Belcher & Storey (1968), who showed that over four years filamentous species only appeared in Rostherne Mere at the end of May, *Anabaena flos-aquae* and *Aphanizomenon flos-aquae* appearing first, followed by *Microcystis aeruginosa* which displaced the *Anabaena*.

The Cyanophyta then died out by the next January. Similar occurrences of Cyanophyta in the latter half of the year are very common e.g. *Aphanizomenon* has a very similar distribution in the Muggelsee (Bursche, 1953). Findenegg (1947) found *Dinobryon divergens* every spring from 1933 to 1938 in the Wörthersee

Fig. 10.27. *a*, monthly fresh-weight biomass of phytoplankton in Lake Kinneret (black area; dinoflagellates, mainly *Peridinium*: white area; other algae). *b*, *c*, *d*, the carbon, nitrogen and phosphorus levels in the biomass of Lake Kinneret. From Serruya & Berman, 1975.

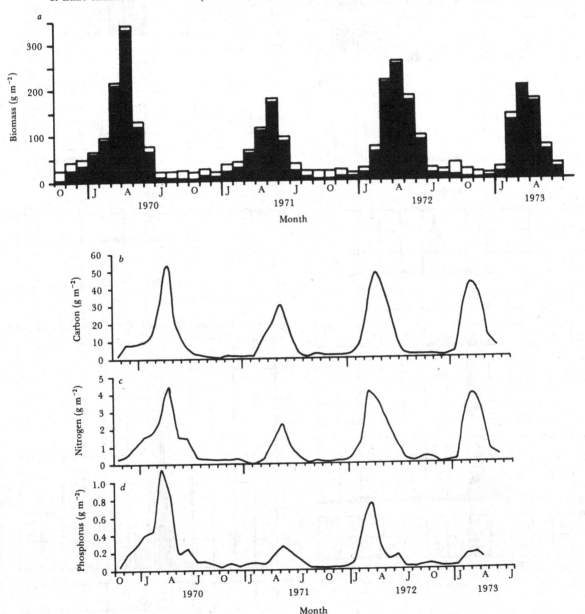

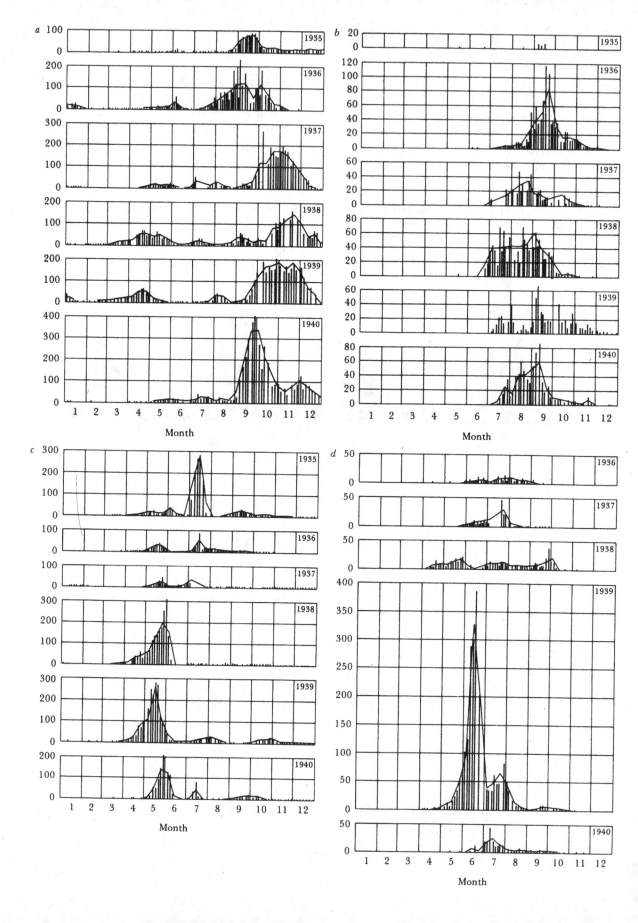

Table 10.3. *Seasonal cycle of nine common epiphyte species during a generation of* C. filum *sporophytes at Rhosneigr, Anglesey. Based on observations made from October 1963 to February 1965.* +, *vegetative plants;* *, *reproductive plants. From South & Burrows, 1967*

Species	A	M	J	J	A	S	O	N	D	J	F	M	A
Bolbocoleon piliferum	+	+	*	*	+	+	+	*	+	+	*	+	+
Acrochaete repens	+	+	+	+	+	+	+	*	*	*	*	*	*
Ectocarpus siliculosus	—	+	+	+	*	*	*	*	*	*	*	*	*
Litosiphon pusillus	—	+	*	*	*	*	*	*	—	—	—	—	—
Ceramium rubrum	—	+	+	+	+	+	*	+	+	*	*	*	+
Sphacelaria pennata	—	—	+	+	+	*	*	*	*	*	+	+	—
Polysiphonia violacea	—	—	+	+	*	*	*	*	*	*	—	—	—
Punctaria tenuissima	—	—	—	—	—	—	—	+	+	+	*	*	—
Melobesia minutula	—	—	—	—	—	—	—	+	+	*	*	—	—

and the cyanophycean *Gomphosphaeria lacustris* every autumn in the Millstättersee. Serruya & Berman (1975) record the occurrence of *Peridinium cinctum* f. *westii* in Lake Kinneret every spring (Fig. 10.27); this is an intriguing study and one of the few which has utilised weight of plankton rather than cell counts, although this was combined with microscopic examination. The alga grows in the period during which stratification is being built up and dies off when stratification sets in. In this study the carbon, nitrogen and phosphorus contents of the crop were also measured and Fig. 10.27 shows that, in particular, the phosphorus peak can occur a little before the carbon peak. Data from rivers are much more difficult to find but Fig. 10.28 illustrates the repeated occurrence of four algae during almost the same growing season over six years in a river in southern Sweden (Carlin, 1946).

Surprisingly few long-term observations have been reported for the macroscopic intertidal algae, although one such example is the growth of the upright thallus of *Nemalion helminthoides* between mid-May and mid-September on Anglesey coasts (Martin, 1969): this alga overwinters as a basal plate of cells. A study of seasonal growth of such algae may yield information on the environmental factors involved in reproduction, e.g. Fig. 10.29 (South & Burrows, 1967) shows the seasonal distribution of *Chorda filum* and the times of formation of the unilocular sporangia. From Spitzbergen to Portugal this alga grows actively only in spring and summer and the higher the latitude the later the sporulation commences. South & Burrows also record the seasonal distribution of epiphytes on *Chorda* (Table 10.3) and clearly there are algae such as *Litosiphon pusillus* which has a very restricted occurrence. Another example showing the occurrence of dormant periods for some coastal algae is given in Table 10.4.

Fig. 10.29. The percentage of *Chorda filum* sporophytes fruiting each month. 1, at Rhosneigr, Anglesey 1964–5; 2, Rhosneigr 1965; 3, Tvärminne, Gulf of Finland, 1964. From South & Burrows, 1967.

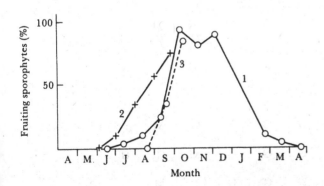

Fig. 10.28. The fluctuations of four planktonic species in a Swedish river over a period of six years. *a*, *Melosira*; *b*, *Aphanizomenon flos-aquae*; *c*, *Asterionella formosa*; *d*, *Cyclotella*. From Carlin, 1944.

Fig. 10.29. The percentage of *Chorda filum* sporophytes fruiting each month. 1, at Rhosneigr, Anglesey 1964–5; 2, Rhosneigr 1965; 3, Tvärminne, Gulf of Finland, 1964. From South & Burrows, 1967.

Table 10.4. *Check list and monthly distribution of benthic plants in Great Pond, Mass. T, trace present; D, dormant; G, growth indicated; M, period of maximum standing crop; ?, uncertain. From Conover, 1958*

	Month											
	Sept.	Oct.	Nov.	Dec.	Jan.	Feb.	Mar.	Apr.	May	June	July	Aug.
Xanthophyta												
Vaucheria thuretii Wor.	M	D	D	D	D	D	G	G	M	M	M	M
V. compacta	G	D	D	D	D	D	G	G	M	M	M	M
V. minuta Blum et Con.	G	D	D	D	D	D	G	G	M	M	M	M
V. coronata Nordst.	G	D	D	D	D	D	G	G	M	M	M	M
V. arcassonensis Dang.	G	D	D	D	D	D	G	G	M	M	M	M
V. intermedia Nordst.	G	D	D	D	D	D	G	G	M	M	M	M
Chlorophyta												
Ulothrix implexa Kütz.	—	—	—	—	—	—	G	G	G	—	—	—
Enteromorpha clathrata (Roth) J. Ag.	G	G	M	M	M	G	M	G	—	M	G	—
E. compressa (L.) Grev.	—	—	—	—	—	—	G	G	M	M	M	—
E. intestinalis (L.) Link	M	G	G	G	G	G	M	M	M	G	G	G
E. linza (L.) J. Ag.	M	M	G	G	G	G	M	G	M	G	M	G
E. plumosa Kütz.	M	G	G	M	—	—	—	G	M	G	G	G
E. torta (Mert.) Reinb.	—	—	—	—	—	—	M	G	G	M	G	G
Monostroma oxyspermum (Kütz.) Doty	—	G	M	G	G	—	G	G	G	M	—	—
Ulva lactura L.	G	G	G	M	G	G	M	M	M	M	G	G
Ulva lactuca var. *latissima* (L.) De Cand.	M	M	G	G	G	G	G	G	M	M	M	G
Ulva lactuca var. *rigida* (C. Ag.) Le Jolis	M	M	G	G	G	G	G	G	M	M	M	G
Chaetomorpha aerea (Dillw.) Kütz.	G	G	G	G	G	G	G	M	M	M	G	G
C. linum (Müll.) Kütz.	G	G	G	G	—	—	M	M	G	G	—	—
Cladophora gracilis (Griff. ex Harv.) Kütz. F. *tenuis* Farl.	M	M	M	G	—	—	G	G	G	M	G	G
C. refracta (Roth) Kütz.	G	M	G	—	—	G	G	M	M	G	—	—
C. rudolphiana (C. Ag.) Harv.	—	—	—	—	—	—	G	G	G	G	G	G
Rhizoclonium riparium (Roth) Harv.	—	—	—	—	—	—	G	M	M	M	G	—
Phaeophyta												
Ectocarpus confervoides (Roth) Le Jolis	M	M	?	?	G	G	?	G	M	?	?	—
E. siliculosus (Dillw.) Lyngb.	?	?	?	M	G	G	G	M	G	G	G	—
Giffordia mitchellae (Harv.) Hamel	G	M	M	G	?	G	G	?	M	G	?	G
Sphacelaria cirrosa (Roth) C. Ag.	M	M	G	—	—	—	G	G	M	M	G	—
Ralfsia verrucosa (Aresch.) J. Ag.	G	M	M	M	G	G	G	M	M	M	G	G

Table 10.4. (*cont.*)

	Month											
	Sept.	Oct.	Nov.	Dec.	Jan.	Feb.	Mar.	Apr.	May	June	July	Aug.
Leathesia difformis (L.) Aresch.	—	—	—	—	—	—	—	—	—	M	—	—
Acrothrix novaeangliae Taylor	—	—	—	—	—	—	—	—	M	G	—	—
Stilophora rhizoides (Ehrh.) J. Ag.	T	T	M	M	G	G	G	G	M	M	G	G
Desmarestia viridis (Müll.) Lamour.	—	—	—	—	—	—	M	M	G	G	—	—
Striaria attenuata (C. Ag.) Grev.	—	—	—	—	—	—	—	—	M	—	—	—
Petalonia fascia (Müll.) Kuntze	—	—	G	M	G	G	M	M	G	G	—	—
Punctaria latifolia Grev.	—	—	G	M	G	G	M	M	G	—	—	—
P. plantaginea (Roth) Grev.	—	—	G	G	G	G	M	M	G	G	—	—
Scytosiphon lomentaria (Lyngb.) J. Ag.	—	—	G	G	G	G	G	G	G	G	G	—
Dictyosiphon foeniculaceus var. *americanus* Coll.	—	—	—	—	—	—	M	M	M	G	—	—
Chorda filum (L.) Lamour.	—	—	—	—	—	—	M	M	G	—	—	—
Fucus spiralis L. (typical)	M	M	M	M	G	G	M	M	M	M	D	D
F. vesiculosus L.	M	M	M	M	G	G	M	M	M	M	D	D
Sargassum filipendula C. Ag.	M	G	D	D	D	D	D	M	M	M	M	M
Rhodophyta												
Bangia fuscopurpurea (Dillw.) Lyngb.	—	—	—	—	G	M	M	G	—	—	—	—
Porphyra leucosticta Thur.	—	—	—	G	G	M	M	G	G	—	—	—
Nemalion multifidum (Web. et Mohr.) J. Ag.	—	—	—	T	T	T	—	—	G	M	T	—
Gelidium crinale (Turn.) Lamour.	M	G	—	—	—	—	—	—	G	M	M	M
Corallina officinalis L.	M	G	D	D	D	D	D	G	M	M	M	M
Fosliella lejolisii (Rosen.) Howe	—	—	—	—	—	—	G	G	G	G	G	G
Agardhiella tenera (J. Ag.) Schm.	G	D	D	D	D	D	D	G	G	G	M	M
Hypnea musciformis (Wulf.) Lamour.	G	—	—	—	—	—	—	—	—	—	—	M
Gracilaria verrucosa (Huds.) Papenf.	M	M	G	G	G	G	G	G	M	M	M	M
Chondrus crispus (L.) Stackh.	M	G	G	D	D	D	D	G	M	M	M	M
Champia parvula (C. Ag.) Harv.	G	—	—	—	—	—	—	—	M	M	M	G

Table 10.4 (*cont.*)

	Month											
	Sept.	Oct.	Nov.	Dec.	Jan.	Feb.	Mar.	Apr.	May.	June	July	Aug.
Lomentaria baileyana (Harv.) Farl.	G	G	—	—	—	—	—	—	—	M	M	M
Callithamnion byssoides Arn.	—	—	M	G	G	G	M	G	G	G	—	—
Ceramium rubrum (Huds.) C. Ag.	M	M	M	D	D	D	M	G	G	G	G	G
C. rubrum var. *proliferum* Harv.	M	M	G	D	D	D	D	M	M	M	M	M
Griffithsia globulifera Harv.	G	G	—	—	—	—	—	—	—	—	—	G
Spyridia filamentosa (Wulf.) Harv.	M	D	D	D	D	D	G	G	G	M	M	M
Grinnellia americana (C. Ag.) Harv.	G	—	—	—	—	—	—	—	—	G	G	M
Dasya pedicellata (C. Ag.) C. Ag.	G	G	G	—	—	—	—	—	—	—	—	G
Bostrychia rivularis Harv.	—	—	—	—	—	—	—	—	—	—	M	M
Chondria baileyana (Mont.) Harv.	G	—	—	—	—	—	—	—	—	G	G	M
C. sedifolia Harv.	—	—	—	—	—	—	—	—	—	—	G	M
C. tenuissima (Good et Woodw.) C. Ag.	—	—	—	—	—	—	—	—	—	—	G	M
Polysiphonia harveyi var. *arietina* (Bailey) Harv.	G	G	G	G	—	—	—	—	M	M	M	G
P. novae-angliae Taylor	G	G	G	G	G	G	G	G	M	M	G	G
P. denudata (Dillw.) Kütz.	M	D	D	D	D	D	G	G	G	G	G	M.

The Chlorophyta and Phaeophyta appear to be growing over a greater part of the year than do the Xanthophyta and Rhodophyta.

Annual succession occurs in subtidal habitats but is not so obvious as in the plankton where some species *appear* to be totally eliminated at certain times. A recent study of *Antithamnion* around Newfoundland revealed that it over-wintered as fragments which would not normally be recorded as part of the flora but grew actively during summer. Clearly the existence of inoculum is just as much a problem amongst the attached as in the free-floating algae. Likewise, species which are present all the year round often have an annual rhythm of growth which is directly comparable to the appearance of unicellular species e.g. *Laminaria digitata* shows maximal growth rate during the first half of the year (Parke, 1948; Sundene, 1962a) and this is varied slightly according to latitude, e.g. maximal growth of the lamina occurs in February–April on the west coast of Norway and on the northern coast growth is later, March–May (Sundene, 1964). *L. hyperborea* has a similar growth pattern starting into growth in November and ceasing by July (Kain, 1971). The report of maximal assimilation surplus in summer in *L. hyperborea* (Lüning, 1971) is presumably a feature of growth slowing down and photosynthate being stored. Off the east coast of Canada these large seaweeds can grow at temperatures close to 0 °C (Mann, 1972d) but apparently only the large perennial species grow through the winter (Mann, 1973). *L. digitata* growing in Nova Scotia (Fig. 10.30)

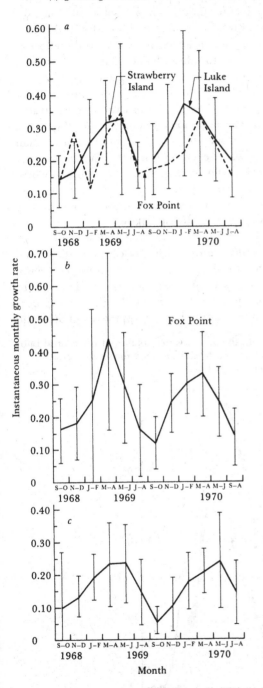

Fig. 10.30. Seasonal growth data for *Laminaria longicruris* (*a*), *L. digitata* (*b*) and *Agarum cribosum* (*c*) growing in Nova Scotia. From Mann, 1973.

behaves exactly like the populations on the European coast; *L. longicruris* and *Agarum cribosum* behave similarly along the north American coast and it appears that a similar growth strategy is adopted by most of the northern laminarians.

The growth of some of these large brown algae is truly prodigous, e.g. Scagel (1957) reports that *Nereocystis* can start as small plants a few inches high in April and become mature plants, 100 ft long by late June. By enclosing the blades of *Macrocystis* in plastic bags and measuring the uptake of ^{14}C into tip, mid and basal sections Towle & Pearse (1973) showed that different regions had different rates of carbon fixation and the tips were always the most active, incorporating about 0.315 mg carbon g^{-1} h^{-1}, and that blades in the centre of the canopy were most effective with a net carbon fixation of at least 6.8 g m^{-2} day^{-1}.

There have been very few studies of annual succession of intertidal or subtidal tropical algae.

In non-monsoonal tropical regions such as Hawaii the effects of storm waves may be more important than the slight seasonal changes in temperature and insolation. Doty (1971) provides evidence that storms influence the crops of frondose marine algae on reef flats of Hawaii. These are antecedent effects and are just one more example of the need to take into account happenings in the environment prior to the data of any particular study.

An important physical factor in a study of population dynamics is the retention time of water in a lake and of flow rate in a river. In very large lakes the whole water mass is replaced only over a long period of time, e.g. approximately 500 yr for Lake Baikal (Moskalenko, 1971 in Lund, 1972) whereas the eleven-mile long Lake Windermere is replaced in nine months. The effect of the renewal rate was shown by Dickman (1969) who placed wooden enclosures in a lake and showed that inside them, where flushing was prevented, the growth was much greater than outside, where only nannoplankton divided fast enough to overcome the flushing rate. Obviously the smaller lakes and ponds have more rapid replacement of the water and the flushing effect will add an important factor when considering population changes. Brook & Woodward (1956) found that the flora was most stable in lochs with a low rate of water replacement. Additionally however the rate of replacement varies throughout the year, e.g. in a monsoon area lakes may be flushed out over a period of days and then remain relatively static for long periods. Similarly in regions of low winter temperatures, freezing of the soil results in very low inflow during winter but high input during spring thawing.

Lack of inflow during the winter also means lack of supply of new nutrients. This may be of little importance in deep isothermal lakes where growth is slow at this time but in ice covered lakes turbulent mixing is reduced and although growth is also low some nutrient depletion may occur beneath the ice.

Temperate lakes on the other hand may have greatest inflow during winter months (e.g. 50% of the water in Lake Windermere is replaced between November and February, Lund, 1972) at a time when light and temperature are so low that cell division cannot proceed fast enough to replace cells lost by outflow. Rate of flow in rivers has even more drastic effects on the phytoplankton and many of the fluctuations on the annual graphs are simply the result of increased flow (e.g., Fig. 7.19).

Inflow waters tend to mix with those of the lake at a depth coinciding with temperature conditions, e.g. cold inflows will sink into the hypolimnion and warm will enter the epilimnion so not only must the amount of inflow be measured but also its distribution. In winter in an isothermal lake the effect of inflow will be to dilute the population, whereas in summer the inflow may have little effect if it sinks, but a great effect if it moves into the epilimnion. In summer the bulk of algal cells can be concentrated in a shallow epilimnion and if new water flows in at the surface it may dilute the population, or in small bodies of water even flush it out. Usually however any inflow water mixing with the epilimnion whilst diluting the population will also bring new supplies of nutrients into this zone and may overall stimulate the growth of the phytoplankton.

The dynamics of the populations must take into account the rate of loss of cells but this is one of the most difficult aspects to measure, especially since many sampling techniques are designed to measure only the live material (e.g. ^{14}C uptake, chlorophyll a estimations, trapping of phototactic

Fig. 10.31. Live cells of *Asterionella* in Windermere, North Basin in the 0–5 m water column at the central buoy (continuous line), on the surface sediments in Sawpit Bay at 3.5 m depth (dashed line) and in the surface sediment in Pull Wyke Bay at 0.25 m depth (dotted line). From Lund, 1949.

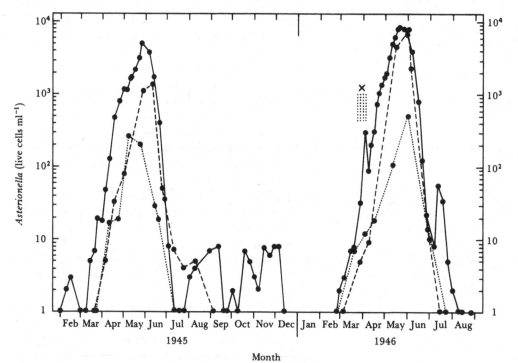

epipelic populations, etc). The lost cells also tend to disappear fairly rapidly from the habitat, usually being lost to the sediments, e.g. Fig. 10.31 shows the coincidence of *Asterionella* in the water mass and on the sediments. Losses from the epilimnion of a lake or mixed layer of the ocean, by sinking through the thermocline, have rarely been measured but this must be a continual process and Bailey-Watts (1976), for example, considered that a theoretical loss of 6.5% of the crop per day is possible in Loch Leven. In addition the losses of metabolically active cells by regular outflow from a lake can be appreciable. In an isothermal temperate lake in winter and in the temperate oceans the rate of algal cell division is governed by overall low light and temperature,

resulting in a very slow rate of division, and results in either a static population or a steadily declining population since losses from grazing (*see* p. 403) must also be added to losses by physical removal. Populations not subject to such losses ought to be able to increase, albeit slowly, during winter months. Such populations occur on solid substrata and on the sediments and it is interesting to note that there is often a steady increase in these populations during this time. Extreme intermittent events, e.g. gales and floods in winter, may however reduce these benthic populations, but this is not comparable to the continual 'loss effect' on the phytoplankton. In the temperate oceans light is limiting during winter but during spring and summer it is nutrient supply

Fig. 10.32. *a.* The relative penetration of photosynthetically active radiation in Lakes Kilotes (K), Esthwaite Water (ES), Windermere, North Basin (WN) and Ennerdale (Enn) (continuous lines), in relation to penetration defined by the minimal vertical extinction coefficient (dashed lines). *b.* The relative extent of the euphotic (unshaded) and mixed zones in four lakes during complete circulation. The mixed depth is represented both by the average lake depth (dashed lines) and the effective column depth (continuous lines). *c, d.* The daily variation of surface radiation at Windermere during a dull winter day (1–2 January 1970) and fine summer weather (18–19 June 1970). From Talling, 1971.

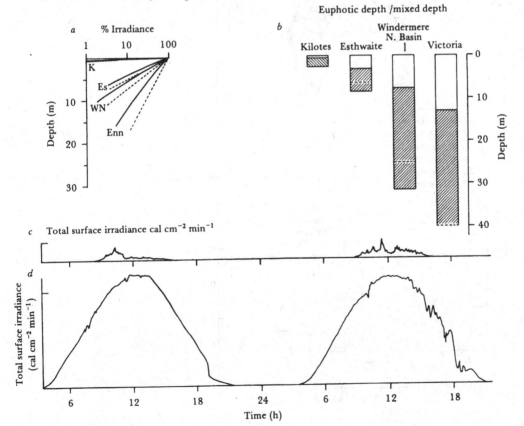

which determines the magnitude of production. There should be a northern and southern latitudinal belt of water which is roughly optimal for annual production where nutrients are kept in sufficient supply to cope with the utilisation of the incoming radiation but I know of no account of such a region.

The underwater light field affecting a circulating phytoplankton cell is very complex since there is a daily variation of irradiance on the surface, a variable penetration of the component wavelengths with depth and the cell itself may spend a varying amount of time in the 'light' or 'dark' zones as it undergoes vertical circulation. The daily energy flux on the water surface hides three sub-factors (Talling, 1971): (i), the fraction of photosynthetically available radiation in the total; (ii), the day length; and (iii) the variable irradiance during the day. Of these the photosynthetically available radiation is relatively fixed at about 0.5 of the total energy flux whereas day length and variation of the flux during the day are very variable (Fig. 10.32c, d shows this for a winter and summer day respectively).

Fig. 10.33. The percentage of green, blue, red and near ultra-violet (V) light penetrating into a series of English Lakes arranged in order of increasing ϵ_{min} (the minimum value of the vertical extinction coefficient) measured from June to September 1972. The three classes of lakes roughly correspond to the oligotrophic, mesotrophic and eutrophic categories. From Talling, 1971.

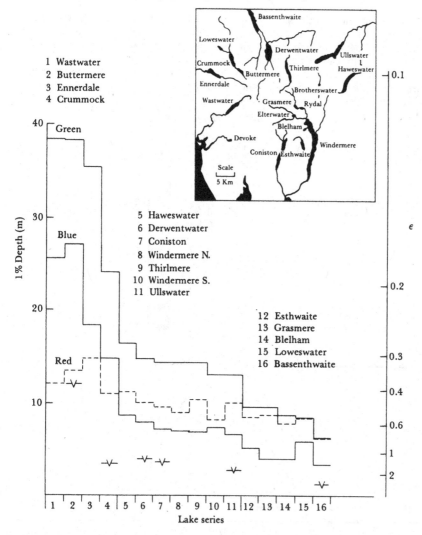

1 Wastwater
2 Buttermere
3 Ennerdale
4 Crummock

5 Haweswater
6 Derwentwater
7 Coniston
8 Windermere N.
9 Thirlmere
10 Windermere S.
11 Ullswater

12 Esthwaite
13 Grasmere
14 Blelham
15 Loweswater
16 Bassenthwaite

However there is *not* a linear relationship between the radiation flux and photosynthesis, a feature which is quite apparent when summer and winter fixation rates are compared. The winter rates are nowhere near as reduced as might be expected from the reduction in irradiance.

The penetration of light into the water varies greatly from water to water even in a small area such as the English Lake District, and is illustrated in Fig. 10.33, which also shows how the spectral value changes; in all but the four most oligotrophic lakes red light penetrates deeper than blue. This is the opposite of the theoretical expectations for clear waters and is due to the effect of dissolved matter in the water column.

The depth of the mixed zone will affect the length of time a cell spends in the most favourable light climate. In deep lakes during isothermal conditions and in shallow lakes with dense populations which 'self-shade' cells, the populations will tend to spend long periods below the euphotic zone and the 'column' photosynthesis is reduced. If the cells spend too long a period in the 'dark' zone, respiration losses will increase relative to gross photosynthesis: a so-called 'critical depth' or 'column compensation point' has been defined as the point at which photosynthesis and respiration balance and autotrophic growth ceases (*see* also p. 313). This is likely to be an important factor during winter mixed conditions in temperate lakes of moderate to considerable depth. Losses in deeper lakes during winter plus the greater depth of circulation will result in smaller populations in spring and the continuing deep circulation prior to the establishment of thermal stratification will result in a slow spring growth in such deep lakes. Fig. 10.34 shows the build-up in phytoplankton population in four Lake District basins in the months January–April with the euphotic, dark zones shown proportionately. At first sight it is surprising that the shallow Blelham Tarn has such low populations compared to Esthwaite Water but this is due to the operation of the rate of water replacement which is so much more rapid in the smaller Blelham Tarn (note the scale changes in the bathymetric profile data) and results in depletion of the winter populations by flushing out at a time when growth rate is low. The graphs clearly show the high winter biomass (chlorophyll *a*) levels in Esthwaite Water and the development of maxima

earlier than in the deeper basins, especially Windermere North basin. Some algae can, however, cope with the low winter irradiance and form quite large populations; an example is the filamentous diatom, *Melosira italica* subsp. *subarctica* (*see* p. 435 and Fig. 10.17).

As we have already seen, so long as nutrients are in ample supply, the time course of phytoplankton development in spring follows the increase in surface irradiance. However, diatoms in particular are susceptible to silica depletion in the water column. This has been studied in detail for *Asterionella formosa* by Lund (1949, 1950*a*, *b*) and Fig. 10.11 shows how, once the silica is exhausted, the spring population reaches a critical level at which further cell division is not possible. Less commonly, silica limitation of diatoms occurs in the autumn, e.g. *Stephanodiscus* is limited then in Loch Leven (Bailey-Watts & Lund, 1973). The marine *Skeletonema* can also be limited by lack of silica but more frequently nitrate is exhausted during the early growth, e.g. in Narragansett Bay (Pratt, 1966).

During summer the favourable light climate and reduction of the depth of circulation as a thermocline develops probably results in cells remaining in an advantageous position as far as light is concerned. At this time, rapidly changing nutrient conditions and subtleties of growth–grazing rates become the most important controlling factors. It is however surprising how few detailed studies of individual species have been made.

Even in tundra ponds where growth is confined to a very short summer period, nutrient deficiencies can be detected towards the end of summer (Kalff, 1971).

In the autumn the light climate again assumes prime importance: even before thermal stratification breaks down, the deepening of the epilimnion (mixed zone in the ocean) results in cells spending longer periods of time in the less well-illuminated zones whilst coincidentally the total energy flux is decreasing. At this time more and more cells are circulating below the euphotic zone. Fig. 10.35 shows the changes in chlorophyll *a* concentrations and temperatures in October and November, in Lake Windermere: by November, cells are circulating to approximately four times the euphotic depth.

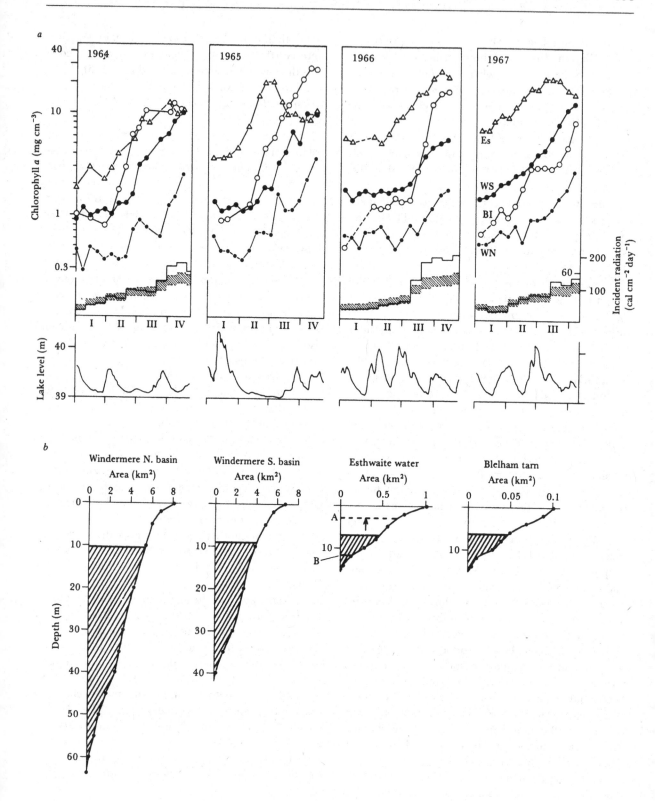

Summary

Cyclical growth is a feature common to all plants, and algae are no exception. The growth of microscopic forms however is often not measured on an annual basis but over periods of weeks and these may not be obviously synchronised with the aerial seasons. Overall annual trends of algal biomass, however, do tend to follow the pattern of irradiance or temperature and examples of these are given. Ice cover exerts a dramatic effect on annual cycles, whilst in tropical lakes the absence of pronounced climatic fluctuations results in lower amplitude algal variations.

Long-term observations over many growing seasons are often necessary before the details of the changes can be worked out and examples from lakes and from the ocean are given, illustrating also that techniques such as cell counting, volume estimates or ^{14}C fixation can all be used to trace the cyclical events.

The inability of most microscopic algae to grow over whole temperate annual climatic changes is fairly striking, whereas there is no such restriction in tropical situations. The annual onset of growth and later decline can often be fairly precisely predicted and this gives clues to the factors involved.

The annual cycles of growth of macroscopic algae have been less investigated but in all habitats

Fig. 10.35. The autumn overturn during 1966 in Windermere North basin showing (*a*) the depth–time distribution of temperature (°C) and (*b*) concentration of chlorophyll *a* (mg m^{-3}) in relation to (*c*) the bathymetric profile and the euphotic (1%) depth. Crosses denote estimates, based on temperature differences of the boundary of the upper mixed layer. z is the mean depth and z' the equivalent effective mixed depth. From Talling, 1971.

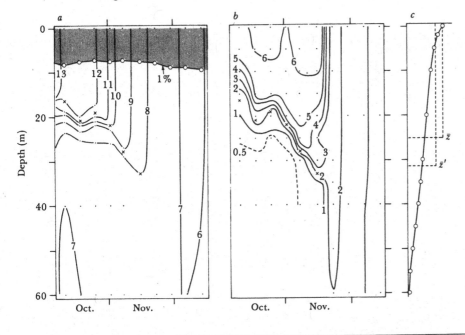

Fig. 10.34. *a*. Changes of phytoplankton density, as chlorophyll *a* content in Windermere North basin (WN), Windermere South basin (WS), Blelham Tarn (Bl) and Esthwaite Water (Es) between January and April in the years 1964–7. Broken lines indicate periods of ice cover. Also shown are the daily totals of surface incident radiation (10-day means, in cal cm^{-2} day^{-1}, continuous lines). Shaded portion is the radiation in logarithmic units relative to I_k values of 0.5 and 1.0 cal cm^{-2} h^{-1}, and daily records of Windermere lake level. *b*. Bathymetric profiles of four English lake basins, showing the limits of the euphotic zone in April, 1952, with deeper water shaded. For Esthwaite Water, the minimum (*A*) and maximum (*B*) recorded limits are also shown. There is a relatively greater volume of water below the euphotic zone in the deeper basins. From Talling, 1971.

there are distinct seasonal growth cycles although these have to be measured by actual growth features rather than presence or absence. The Fucoids on the shores are certainly not active all the year round.

Antecedent events are often important moderators of annual cycles, e.g. in the production of spores and cells for inocula, or by physical control of the size of previous populations. Water retention time in lakes also plays an important part in the build-up of populations as do also the mixing regime and the degree of light penetration.

11 Energy flow and nutrient cycling

Energy flow and productivity

In lakes and the sea, primary production of organic matter is performed almost entirely by algae and the biomass of other autotrophic plants contributes a negligible quantity on a global scale. All other 'production' in the aquatic habitats is only a transfer of organic matter (energy) from one trophic level to another. Fig. 11.1 illustrates the communities through which energy flows; this is a relatively simplified diagram and serves simply to illustrate the point that before the energetics of a system can be studied a spatial and functional analysis of assemblages is necessary to give an essential framework. Physically, the microscopic algae in the water affect

Fig. 11.1. The communities and habitats through which energy flows. Dashed lines indicate detrital pathways and continuous ones, grazing pathways.

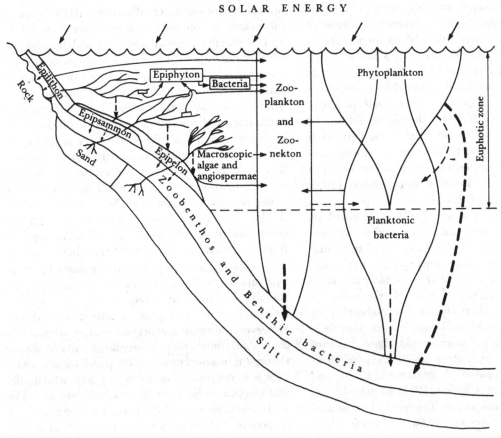

only the attenuation of energy but chemically they greatly affect the chemical components and the nutrient cycling especially of some of the minor elements.

In both marine and freshwater studies the gross primary production is essentially the same as the rate of photoassimilation, and the rate of storage of energy is proportional to the rate of synthesis of carbohydrate (Davies, 1967). It will be apparent from the many variables and complexities mentioned in other chapters that measurement of net photosynthesis is not the simple matter it at first seems and, to take but one factor, one must add the amount of organic matter lost by excretion. When the ^{14}C method is used various figures are quoted for this excretion; most have been obtained in laboratory experiments and they are therefore only approximations to the losses in nature. Ryther *et al.* (1970) found that surface phytoplankton from the Peru Current released 1–3% of the total carbon fixed (uncorrected for any dark uptake) whilst phytoplankton collected from the depth at which 1% incident light remained, lost 25% of the total. Thus down the depth profile the variation in secretion of extra-cellular products will lead to an underestimation of the organic production, a factor which cannot be calculated from isolated observations. At a less productive station they found 45% excretion which appears to confirm the findings of Fogg, Nalewajko & Wall (1965) that the release of dissolved organic matter (7–50% of total carbon fixed) is inversely proportional to the rate of organic production; it thus becomes a more important factor in oligotrophic waters. Detailed investigation rather than spot estimates of this extracellular production are rare but Wetzel *et al.* (1972) determined its contribution over an annual cycle and found that carbon release was 7.3 mg m^{-2} day^{-1} and an annual mean of 5.7% of phytoplankton primary production. The measurement of this extra-cellular material is complicated by the fact that it enters the pool of dissolved organic matter and some is re-absorbed by heterotrophic organisms. Though it is not possible to give exact figures for macroscopic algae, it seems likely to be in the region of 20–25% of gross photosynthesis (Khailiov & Burlakova, 1969), and Hatcher, Chapman & Mann (1977) estimated a loss of 35% of fixed carbon by *Laminaria* (but *see* also p. 117). Sieburth & Jensen (1969*a, b*) calculated that

Fucus vesiculosus exuded 33.7 mg carbon from 100 g thallus and *Laminaria* about 780 mg. Ryther *et al.* (1970) found that the loss of ^{14}C-labelled compounds was essentially similar at different light intensities though the rate of carbon fixation itself varied 10–20 fold; this aspect has yet to be satisfactorily explained.

Any consideration of energy flow must start with the influx of energy reaching the surface under which the algae live. The amount of photosynthetically usable energy reaching the ocean surface is about 8×10^{23} joules yr^{-1} (Fig. 11.2) and losses by absorption and reflection may reduce the amount available to 6×10^{23} joules; only a small fraction of this is used in photosynthesis (2% is often used in calculations and 1–2% is common under good agricultural practice over a *whole* season). Hence the maximal amount of fixation is estimated as 26×10^{10} tons of carbon in the ocean, assuming that 80% of the global fixation occurs here (Russell-Hunter, 1970). Comparing this figure with estimated carbon fixation (annually about $20–23 \times 10^9$ tons) Menzel (1974) suggests that less than 10% of the energy-set limit of production is actually achieved: the scope for increasing carbon fixation is therefore very limited and productivity could never expand beyond one order of magnitude. The actual usage by man of the energy fixed by plants either in the sea or on the land surface is minute, less than 1%, and so better utilisation of that currently fixed is likely to be more profitable than attempting to increase productivity, except in certain specific areas where crops can be improved or intensive aquaculture systems devised. The discrepancy between the estimated and calculated fixation is largely attributed to nutrient deficiency and probably can be overcome only in shallow-water regions or in lakes where nutrients can be added, although raising the pool of nutrients from deep water may be possible in some situations.

Fixation at any point on the earth's surface will vary according to the input of energy, population size, and nutrient supply but there are very few estimates of the variation in latitudinal fixation. However, the amount of organic matter produced in some algal communities is very large and can double the organic matter content of the population in a day. This is in contrast to many crop plants which only add increments between 10 and 20% per day. The likely standing crop of the oceans is given as 1–2 g carbon m^{-2} (tropical and temperate) and 2–10 g carbon m^{-2} for coastal or other fertile seas, except in winter when not more than 0.5–1 g carbon m^{-2} would be expected (Strickland, 1966). The organic production increases in proportion to the input of energy up to a certain energy level above which no further increase is possible. Ryther (1959) estimated the *theoretical maximum net photosynthesis* and this is shown in Fig. 11.3 with the compensation point at about 100 ly day^{-1} and assuming respiration as $\frac{1}{15}$ of photosynthesis. In mid-summer at latitude 45° an average input of 400 ly day^{-1} is reasonable and would enable the production of 18 g carbon m^{-2}

Fig. 11.2. A diagrammatic representation of energy flow from the sun to the water and through the organisms to the sediments. X^1–X^3 are points where energy is dissipated through reflection, water heating and excretion. Z^1–Z^5 are energy sinks.

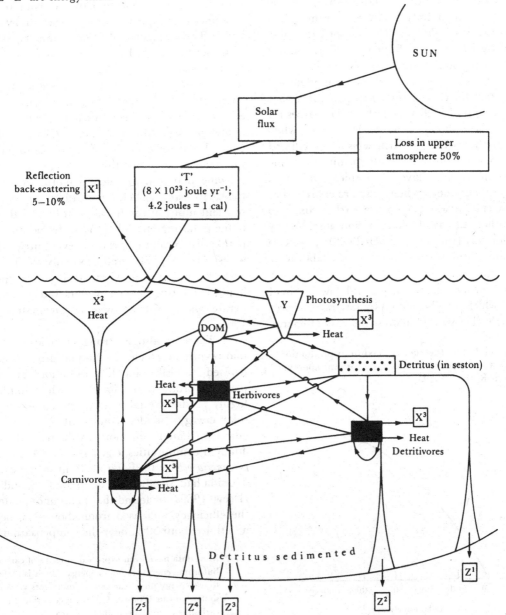

day^{-1}.* Laboratory models of sewage pond systems have achieved the remarkable fixation rate of 20 mg carbon m^{-3} day^{-1} (Uhlmann, 1978). Such a high figure has rarely been recorded but approximately half this value is reached in some very rich areas of sea giving an efficiency of 50%; this is, however, quite exceptional and efficiencies ranging from 1 to 5% are more normal. For example, in a marine diatom bloom, Ryther *et al.* (1958) estimated the efficiency on two successive days at 2.6% and 1.6% (3.2–5.3 g carbon m^{-2} day^{-1} and 5.1–6.2 g carbon m^{-2} day^{-1}), and in Lake Kinneret a range from 0.34% in August 1973 to 4.01% in April 1974 was recorded by Dubinsky & Berman (1976); the last figure was based on the integrated water column and calculated as the percentage of daily areal photosynthetic carbon fixation to total incident subsurface light. Talling *et al.* (1973) report similar values for rich Ethiopian lakes (0.51–3.34%) but when maximum photosynthetic efficiencies are calculated (carbon fixation to photosynthetically available radiation in a water layer 1 m thick) values of 10–12% are obtained. When these are expressed as quantum efficiencies (percentage carbon fixed to photosynthetically available radiation absorbed by algal chlorophyll only) they reach 25–30%, close to the maximum quantum efficiencies for laboratory cultures.

Lack of inoculum and general low levels of biomass (which may be caused by low rates of productivity) will obviously depress the rate of energy

fixation but these factors are rarely taken into account and in fact actual values are not often compared with theoretical estimates. One suspects that over the whole growing season the phytoplankton biomass is lower than the water could support, e.g. Ilmavirta (1975) found that biomass volume was responsible for 21%, 28% and 44% of the variation in production in three different years in Pääjärvi, a lake in Finland. The biomass volume of benthic communities may have less effect on energy fixation since a greater proportion of the available space (volume) is occupied by cells, at least in the shallow zones. The two other significant limiting factors to gross production in Pääjärvi were temperature and light intensity, whereas nutrients were not limiting. The recent analysis of data collected during the International Biological Programme study of lakes and reservoirs from the arctic to the tropics has been analysed (Brylinsky & Mann, 1973) and shows a general positive influence of energy input on phytoplankton production; this is reflected in Fig. 11.4*b* showing plankton production plotted against latitude. However, as Schindler (1978) points out, the nutrient input in the tropics can be 100 times that of the polar regions. The cold water oceanic phytoplankton community is rarely if ever nutrient limited whilst the temperate and especially the tropical is very obviously nutrient limited. Light, on the other hand, is limiting in polar waters and in some temperate waters during winter (Steemann-Nielsen, 1964).

There is also a strong correlation between photosynthetic efficiency (annual gross production divided by visible incident radiation) and phytoplankton production (Fig. 11.4*c*) which implies that highly productive lakes make more efficient use of solar energy. Shaded communities are also more efficient (though not more productive) than light-adapted communities, e.g. Drew (1974) found that *Laminaria ochroleuca* living at 55 m in the Straits of Messina had an efficiency of about 37% and earlier (Drew, 1969*a*) he found that in an underwater cave the efficiency increased from about 2% near the mouth to about 50% near the compensation point;

Fig. 11.3. The theoretical maximum production rate as a function of total daily radiant energy. From Ryther, 1959.

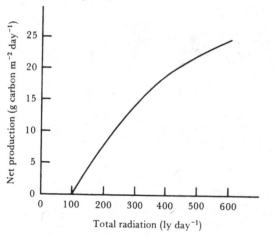

* Most data have been expressed in terms of carbon fixed but from general studies of energy flow it is preferable to use energy units; the conversion factors vary somewhat but Winberg (1971) suggests that 1 g of carbon is equivalent to about 10 kcal.

assimilation and assimilation number do not diminish on cloudy days as much as the light intensity (Odum, McConnell & Abbott, 1958) and thus community metabolism often varies less than expected in changing light conditions.

The energy entering the water does not circulate but 'diffuses' through a complex web of physico-chemical processes and from one organism to another with loss in the form of heat at each stage and ultimately a residual fraction of the initial energy input is deposited onto the sediment to contribute to the store of fossil fuel. The flow is thus a *one-way* system which is balanced by the influx of solar energy. Fig. 11.2 is a diagrammatic illustration of this, where the total incoming energy T at the surface is partially lost back to the atmosphere (x^1), part absorbed as heat (x^2) and part absorbed by plants and utilised in photosynthesis (y). The conversion (fixation) of energy at y is only 1–5 % of the absorbed energy and is then transmitted to herbivores, detritus and a small part to dissolved organic material (*DOM*). Both live and dead algae are consumed by herbivores *sensu lato* (animals, fungi, bacteria). Faeces and unconsumed particulate organic matter remain as detritus. The algal organic material eaten by the herbivores then passes into the carnivores or is deposited on to the sediments. On rocky shores and on salt marshes (Teal & Kanwisher, 1961) the main algal production probably passes not into animals but directly into the detritus pathway (Mann, 1972*a*). Some detritus is consumed by

detritivores but ultimately the detritus (z^{1-5}) from all plant and animal sources is incorporated in the sediment. The cycle of excreted organic products (x^3) is also added to this diagram plus the heat loss at each stage. The first law of thermodynamics applies to the system so all the 'packets' of energy x^{1-4} plus z^{1-5} plus the heat losses at each stage must equal the influx T at the water surface. To measure all these transfers and the rates of transference is a complicated matter and usually rather crude measures of composite pools are made. The one-way system expresses the overall transference of energy but it is important to remember that the system has many reversible stages which delay the progress of the energy down the profile to the sediments. The progress of this energy through the various organismal habitat pathways is shown in Fig. 11.1.

To determine the chemical energy at any level in the ecosystem the organisms can be harvested and the calorific value determined by complete combustion in a bomb calorimeter. This approach was used in an often quoted experimental study of a simplified ecosystem involving *Chlamydomonas* and *Daphnia* with 'predation' added, by the experimenter removing *Daphnia* from the system at four-day intervals (Slobodkin, 1959). More frequently indirect measurements are made and conversion factors are used instead of direct calorimetry.

According to Steemann-Nielsen (1974) about eight quanta are needed for the assimilation of one molecule of carbon dioxide. Only when illumination

Fig. 11.4. Relationship of phytoplankton production to chlorophyll *a* concentration (*a*), to latitude (*b*), and to photosynthetic efficiency (*c*). The graphs are based on data collected from 43 lakes and 12 reservoirs during the International Biological Programme. From Brylinsky & Mann, 1973.

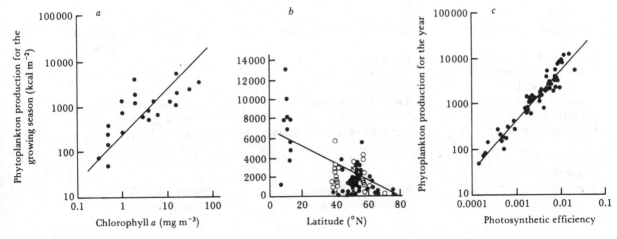

Table 11.1. *Chlorophyll* a *contents of various waters expressed in the authorities' own figures*

Community and reference	Location	Chlorophyll *a*
Phytoplankton		
Marine		
Tominaga (1971)	Antarctic	0.05–0.91 mg m^{-3}
Tominaga (1971)	Indian Ocean	0.02–0.60 mg m^{-3}
Strickland, Eppley & Mendiola (1969)	Peruvian coastal water	2 μg/l–10 μg l^{-1}
Strickland, Eppley & Mendiola (1969)	Open water off Peru	About 1 μg l^{-1}
El-Sayed (1971)	Antarctic, Weddell Sea	0.37 mg m^{-3} av.
Burkholder & Sieburth (1961)	Antarctic, Bransfield Straits	0.3–0.5 mg m^{-3}
	Antarctic, Gerlache Straits	10–25 mg m^{-3}
Strickland (1960)	Oligotrophic ocean (average)	0.05 mg m^3
	Fertile inshore blooms (average)	10–40 mg m^3
	Temperate ocean (average)	0.5 mg m^3 ($\times$ 5–10 for inshore during growing season)
Fenchel & Straarup (1971)	Kattegat	66–76 mg m^{-2}
Freshwater		
Talling *et al.* (1973)	Ethiopian Soda Lakes	2000 mg m^3 (179–325 mg m^{-2})
Dubinsky & Berman (0000)	Lake Kinneret	50–580 mg m^{-2} (1973–4)
Hickman (1973)	Abbot's Pool, England	18, 19, 26, 55 mg m^{-3} (means for 4 yr) 33, 35, 49, 108 mg m^{-2} (means for 4 yr)
Bindloss (1974)	Loch Leven, Scotland	Often near 430 mg m^{-2}
Ganf (1974*a, b*)	Lake George, Uganda	300–900 mg m^{-2} 650, 568, 632 for 3 yr at mid-lake station and 425–486 for lake as a whole
Epipelon		
subtidal		
Matheke & Horner (1974)	Marine shallow sea, Alaska	25–121 mg m^{-2} (Feb.–May) 138–321 mg m^{-2} (June–Aug.)
Bunt, Lee & Lee (1972)	Florida	20–200 mg m^{-2}
Eaton in Moss (1968)	River Niger, Africa	6.5–9.4 mg m^{-2}
Eaton in Moss (1968)	Rickford Spring, England	1.2–2.5 mg m^{-2}
Fenchel & Straarup (1971)	Two Danish beaches	1.5–3.9 and 10.2–19.9 mg m^{-2}
Boucher (1972)	French Atlantic coast	27–900 mg m^{-2}
Patmatmat (1968)	San Juan Island, Washington State	100 mg m^{-2}
Intertidal		
Moss (1968)	River Avon, England	22–124 mg m^{-2}
	Christchurch, England	Mean 42.5 mg m^{-2}

Table 11.1. (*cont.*)

Freshwater		
Moss (1968)	Blea Tarn, England	10.5–66.5 mg m^{-2}
Moss (1968)	Priddy Pool, England	2.1–3.8 mg m^{-2} (mean of 2 yr, max. 15)
	Abbot's Pool, England	7.6–7.3 mg m^{-2} (mean of 2 yr, max. 32)
Moss (1968)	Shearwater, England	12.3 (max. 50) mg m^{-2}
Epilithon		
Brock & Brock (1966)	Hot Springs	800 mg m^{-2}
McConnell & Sigler (1959)	Calcareous mountain river	300–1200 mg m^{-2}
Marker (1976*a*)	Calcareous lowland stream, England	20 mg m^{-2} (winter) 200–300 mg m^{-2} (spring) 40–100 mg m^{-2} (summer)
Gifford & Odum (1959)	Rocky shore (*Calothrix*)	350 mg m^{-2}
Epiphyton		
Brown in Moss (1968)	Priddy Pool, England (*Equisetum*)	110–303 mg m^{-2}
Edwards & Owens (1965)	River Ivel, England (*Hippuris*)	224 (Apr. 1960) mg m^{-2}
Epipsammon		
Moss & Round (1967)	Freshwater, Shearwater, England	86 mg m^{-2} (mean)
Under-ice flora		
Meguro, Ito & Fukushima (1967)	Alaska	24 mg m^{-2}

is low and therefore the rates of the photochemical reactions are limiting is there maximal usage of light energy and energies up to $10–20 \times 10^{15}$ quanta seem to be the optimum range. At higher levels the utilisation decreases and at the highest levels light inhibition sets in thus reducing still further the utilisation. There are however some reports (e.g. Burkholder, Repak & Sibert, 1965) of increasing carbon assimilation up to full tropical sunlight (13000 lux). In nature the utilisation at noon is reckoned to be about $\frac{1}{10}$ or less compared with the utilisation at low intensities.

The fixation by the algal populations is dependent upon their chlorophyll content and this varies from alga to alga and in the populations according to the biomass of the algae present and the conditions under which they are growing, but there is a general direct relationship between chlorophyll *a* and fixa-

tion (Fig. 11.4*a*). Biomass can be expressed, though it rarely is, as energy per unit area; chlorophyll content is more commonly used as a measure. Some representative figures are collected together in Table 11.1. Chlorophyll can be concentrated as a very thin veneer, e.g. on the surface of a salt marsh where Estrada *et al.* (1974) recorded 40–300 mg m^{-2} and in unshaded areas 200–500 mg m^{-2} (highest value recorded was 980 mg m^{-2}); these values appear to be high but several similar ones have been recorded in stable habitats, for example by Moss (1968). Another study, admittedly under experimental conditions in a laboratory, also showed that an attached stream community developed 140–1300 mg m^{-2} in low light and 480–2010 mg m^{-2} in high light (McIntire & Phinney, 1965). In oceanic, upwelling waters figures around 125 mg m^{-2} are recorded and during a bloom in the nutrient-rich Moriches Bay,

Ryther *et al.* (1958) recorded 200 mg m^{-2}. There is obviously an upper limit in water since absorption of light will prevent synthesis in deeper layers; Steemann-Nielsen (1962) considered that this was around 300 mg m^{-2} for open oceanic waters (at which level he considered that 46 g carbon m^{-2} would be fixed) and Talling *et al.* (1973) found actual values between 179 and 325 mg m^{-2} in two Ethiopian soda lakes where, due to self-shading, the euphotic zone was only 0.6 m deep. A striking example of the effects of self-shading is given by Ryther (1963) in a comparison of production in a 1.2 m deep sewage pond which fixed carbon at 1.9 g m^{-2} day^{-1} and the Sargasso Sea where 1.56 g m^{-2} day^{-1} is fixed by a population dispersed down the water column which is probably over 100 m deep. There is considerable annual variation in the algal biomass measured as cells or as chlorophyll in temperate waters: variations as much as 1000-fold have been reported but on theoretical grounds it can be expected that the variation would decrease as equatorial regions were approached. There have been few attempts to measure it but all recent work supports this proposition. Blackburn, Laurs, Owen & Zeitzschel (1970) found that the seasonal chlorophyll *a* content in Pacific waters off Central America varied by a factor of two. In the same area, Owen & Zeitzschel (1970) also found that carbon fixation only varied by a factor of two over the annual cycle. In freshwaters, Ganf (1974*a*, *c*) reported only a twofold annual fluctuation in Lake George. High biomass (chlorophyll) does not necessarily reflect high productivity, it can be merely the result of accumulation of net production and therefore it is necessary to discover how active the biomass is: expressed another way, what is its productivity? This is a rate term and is the amount of energy (dry weight) fixed by unit biomass, or in unit area per unit time. On the whole, most algal communities composed of unicells (or aggregations) tend to have high productivities (high photosynthesis:biomass ratio); since all the tissues are productive (i.e. there are no supporting or storage tissues) and since they have high surface area to volume ratio the exchange of nutrients is fairly rapid and their turnover times are rapid. In addition, the populations tend to die and decompose rapidly and do not accumulate in the water and so the energy flow through such systems is rapid. The turnover time of the cell carbon in algal

populations is another aspect which has been rarely computed, though Wetzel *et al.* (1972), found that in a temperate marl lake it took 0.30–2.25 days (mean 1.06 days from April–December) for the carbon in the phytoplankton to be replaced by primary production. The annual mean was 3.6 days and that for the more stable epiphyton was 8 days. A figure for the ocean is provided by McAllister *et al.* (1960) who found that the doubling time for the phytoplankton crop in the North Pacific at 50° N, 145 W was 4–5 days. However these workers found that in shipboard incubations it was lower at 1.75 days; such a result is not unexpected since attempts are made to optimise conditions in these experimental systems. Gross productivity in summer in lakes is often relatively low compared to that in spring since although the biomass may be high (the common green soup which some small lakes turn to in summer) compared to that in spring (when the waters look clean and empty) factors such as self-shading, nutrient depletion, inhibition, etc., reduce production in summer and the ratio of photosynthesis to biomass falls.

Values for the amount of carbon fixed vary greatly, but in the sea the upper estimate is usually about 200 g carbon m^{-2} yr^{-1} and the lower limit about 50 g carbon m^{-2} yr^{-1}. Taking a figure of 3.5×10^{14} m^2 for the open ocean, the total carbon fixation would be 70×10^9 tons per year at the upper rate of phytoplankton fixation and 17.5×10^9 tons at the lower limit (Russell-Hunter, 1970). Russell-Hunter further calculates that at the upper level such fixation would require 14×10^9 tons of nitrogen and 2×10^9 tons of phosphorus, amounts which are greatly in excess of the actual contents of the oceans. Such amounts can only be utilised by recycling the nutrient load which must therefore pass through the algae many times in any one year.

Calculations of fixation such as those just quoted are complicated by the fact that the different regions of the ocean vary greatly in their production. Table 11.2 gives a breakdown from Ryther (1969) which places the total closer to the lower limit of that given by Russell-Hunter. There are however many factors leading to underestimation and the final figure is likely to fall somewhere between the two extremes quoted above. One of the greatest problems is that most oceanic estimates are spot determinations and usually not coupled with data on depth of the

Table 11.2. *The productivity of various ocean areas. From Ryther, 1969*

Province	% fixation	Area (km^2)	Mean productivity (g carbon m^{-2} yr^{-1})	Total productivity (10^9 tons/carbon yr^{-1})
Open ocean	90	326×10^6	50	16.3
Coastal zone[a]	9.9	36×10^6	100	3.6
Upwelling areas	0.1	3.6×10^5	300	0.1

[a] Including coastal fishing zones.

Table 11.3 *Carbon fixation values of various aquatic and two terrestrial communities*

Community and reference	Location	Net production (g carbon m^{-2} yr^{-1} except where stated)
PHYTOPLANKTON		
Marine		
Antarctic		
Horne *et al.* (1969)	Inshore water	130 g carbon m^{-2} (for period Dec.–May)
Mandelli & Burkholder (1960)	Inshore water	0.21–1.5 g carbon m^{-2} day^{-1}
El-Sayed (1971)	Weddell Sea	1.56–3.62 g carbon m^{-2} day^{-1} (under ice probably 0.03 g max.; Strickland, 1966)
El-Sayed (1968)	Off Antarctic peninsula	Av. 0.89 g carbon m^{-2} day^{-1}
Indian Ocean		
Sournia (1965)	Madagascar (inshore)	15 g carbon m^{-3} yr^{-1} (little seasonal variation)
Jitts (1969)	110° E meridian, Southern Ocean	37 mg carbon m^{-2} h^{-1} (annual mean)
South Pacific		
Ryther *et al.* (1970)	Peru coastal current	10 g carbon m^{-2} av. for 5 days (calc. 350 g m^{-2} yr^{-1})
Strickland *et al.* (1969)	Peru coastal current	200–500 (highest, 4 g carbon m^{-2} day^{-1})
Pacific/Atlantic (unproductive)		
Strickland (1966)	Central gyres	50
North Pacific		
McAllister *et al.* (1960)	Sub-arctic water (50° N, 145° W)	205 mg carbon m^{-2} day^{-1}
North Pacific Coastal		
Brown & Parsons (1972)	British Columbia	120
Tropical Pacific		
Owen & Zeitzchel (1970)	Off Central America	75
Zeitzchel (1969)	Gulf of California	Av. 0.382 g carbon m^{-2} day^{-1}
Steven (1971)	Atlantic (near Barbados)	Av. 0.288 g carbon m^{-2} day^{-1}
Margalef (1971)	Black Sea	25–30
Minas (1970)	Mediterranean	78
Gulf of Mexico		
Margalef (1971)	Caribbean	40–120

Table 11.3. (*cont.*)

Community and reference	Location	Net production (g carbon m^{-2} yr^{-1} except where stated)
Caribbean		
Beers *et al.* (1968)	Off Jamaica	40
Bunt *et al.* (1972)		2.5–13.8 mg carbon m^{-2} h^{-1} (^{14}C); 0–32.6 mg carbon m^{-2} h^{-1} (O_2)
North Atlantic Coastal		
Platt (1971)	Nova Scotia	190
North Atlantic		
Menzel & Ryther (1960)	Northwest Sargasso Sea	72
Central Pacific		
Smayda (1966)	Gulf of Panama	255–280 g carbon m^{-2} gross
Equatorial waters (up-welling)		
Strickland (1966)	Pacific/Atlantic	209–246
North Pacific		
McAlister (1971)	Off Alaska	76 mg carbon m^{-2} day^{-1}
North Pacific		
Larrance (1971)	Mid-subarctic	50–100
Estuarine		
Williams (1966)	Carolina	Approx. 50
Temperate Atlantic		
Russell *et al.* (1971)	E^1 off Plymouth	127–212
	Mid-Channel	136–252 (400)
	South of mid-Channel	94–231
	Off Roscoff	152
Steele & Baird (1962)	Off Scotland	94
North Sea		
Steemann-Nielsen (1964)	Kattegat	51–82 (gross fixation)
	Baltic	35–40
North Atlantic		
Thórdandóttir (1973)	Off Iceland	172–373 mg carbon m^{-2} day^{-1}
Freshwater		
Arctic 'Taiga' lake		
Alexander & Barsdate (1971)	Alaska	90–175 μg carbon l^{-1} h^{-1} (11 g carbon m^{-2} yr^{-1})
Arctic ponds		
Miller & Reed (1975)	Alaska	0.6–0.9
North temperate		
Kalff (1972)	Near Montreal, Canada	78–82
Marion Lake		
Efford (1967)	British Columbia	8
Offshore Lake Victoria		
Talling (1965)	Uganda	Approx. 950 g carbon m^{-2} (gross)
Ultra-oligotrophic lakes		
Brown & Parsons (1972)	—	5
Ultra-rich soda lakes		
Talling *et al.* (1973)	Ethiopia	43–57 g carbon m^{-2} day^{-1}
Lawrence Lake		
Wetzel *et al.* (1972)	Michigan	43.41 (av. of 4 yr)

Table 11.3 (*cont.*)   Table 11.3. (*cont.*)

Community and reference	Location	Net production (g carbon m^{-2} yr^{-1} except where stated)
Lake Kinneret Data record	Israel	900–2400/4000 mg carbon m^{-2} day^{-1} during *Peridinium* bloom
Two lakes Hobbie (1964)	Alaska	6.6/7.5 (two years, Lake Schroder); 0.9 (one year, Lake Peters)
Lac Leman Serruya (1969)	Switzerland	1600 mg carbon m^{-2} day^{-1} (April max.)
Sylvan Lake Wetzel (1966)	Indiana (extreme eutrophic lake)	571
Loch Leven Bindloss (1974)	Scotland	785 (4-yr average; daily rate 2.2 g carbon m^{-2})
Two Arctic ponds Kalff (1967)	Alaska	0.38–0.85 (prob. lowest recorded annual rate)
Lake Tahoe Holm-Hansen *et al.* (1976)	California	94–240 mg carbon m^{-2} day^{-1}
Four Arctic lakes Frey & Stahl (1956)	Canadian Arctic	9–120 mg carbon m^{-2} day^{-1}
Oligotrophic lake Ilmavista & Kotimaa (1974)	Finland	9.5–13.9 (lake under ice for 5 months)
EPILITHON Subtidal		
Laminaria Mann (1972*b*)	Atlantic coast	1225–1750
Macrocystis Clendenning (1971)	California	400–820
Macrocystis Towle & Pearse (1973)	California	Approx. 2500[a]
Sargassum Wanders (1976*a*)	Curaçao	2550
Caulerpa Johnston (1969)	Canary Islands	365 net prod. of various communities up to 10.5 g carbon m^{-2} day^{-1}
Laminaria Bellamy (1968)	Southwest England	1225
Laminaria Johnston, Jones & Hunt (1977)	Scottish sea loch	120 plus
Fleshy reef algae Wanders (1976*b*)	Curaçao	2550

Table 11.3. (*cont.*)

Community and reference	Location	Net production (g carbon m^{-2} yr^{-1} except where stated)
Crustose corallines		
Littler (1973*c*)	Hawaii	5.7 g carbon m^{-2} day^{-1} or approx. 2080 g carbon m^{-2} yr^{-1}
Van den Hoek *et al.* (1975), quoting Wanders (1976*a*)	Curaçao	460
Coral reef lagoon		
Qasim, Bhattaihiri & Reddy (1972)	Indian Ocean	3.84 g carbon m^{-2} day^{-1} (160–1370 g carbon m^{-2} year for whole reef system)
Coral reef		
Qasim, Bhattaihiri & Reddy (1972)	Indian Ocean	4715
Smith (1973)	Eniwetok (Pacific)	569–2628
Taylor (1973)	Reefs in general	1500–3500
Wanders (1976*a*)	Curaçao	2330
Intertidal		
Mixed community		
Littler & Murray (1974*b*)	California	485 (135–1142)
Fucoids		
Mann (1972*b*)	Atlantic coast	640–840 (possibly > 1000)
Saltmarsh fucoids		
Brinkhuis (1977)	Atlantic coast	Approx. 315 (from photosynthesis data) or 155 (from standing crop data)
Freshwater		
Schindler *et al.* (1973)	Unproductive Canadian lakes Epipelon	5.18–5.19 for two lakes
Intertidal and shallow water		
Grøntved (1962)	Denmark	31–68 mg carbon m^{-2}
Marshall, Oviatt & Skauen (1971)	New England	81
Bunt, Lee & Lee (1972)	Caribbean	Mean 8.1 (max. 13.8) mg carbon m^{-2} h^{-1}
Boucher (1972)	French Atlantic coast	1 mg m^{-2} day^{-1} (winter) to 170 mg m^{-2} day^{-1} (spring)
Steele & Baird (1968)	Shallow sediments, Scotland	4–9
Mattheke & Horner (1974)	Alaskan coast	0.5 (Feb.–June), 1.9–20.7 (summer), mg carbon m^{-2} h^{-1}
Burkholder *et al.* (1965)	Long Island (sand)	Av. 4–5 g carbon m^{-2} day^{-1}
Mudflat		
Leach (1970)	Scotland	31
Estuary (benthos)		
Colign & Venekamp (1977)	Holland	57–209
Saltmarsh		
Gallaher & Daiber (1974)	Delaware	18–42 mg carbon m^{-2} h^{-1}
Sandflat		
Grøntved (1962)	Denmark	115–178 (recalculated by Patmatmat (1968))

Table 11.3. (*cont.*)

Community and reference	Location	Net production (g carbon m^{-2} yr^{-1} except where stated)
Sandflat		
Patmatmat (1968)	Washington, San Juan Island	Approx. same as recalculated Danish data
Freshwater		
Hargrave (1969)	British Columbia	40
Gruendling (1971)	Marion Lake	44
Wetzel *et al.* (1972)	Michigan	2
Hickman (1971)	Abbot's Pool, England	5.29 mg carbon h^{-1} m^{-2}
Miller & Reed (1975)	Arctic ponds (Alaska)	4.1–10.1
EPIPHYTON		
Wetzel *et al.* (1972)	Michigan	37
Hargrave (1969)	British Columbia (Marion Lake)	18
	Hot springs	
Iceland		
Sperling (1975)	—	Approx. 1800
		260–1130 mg carbon m^{-2} day^{-1}
TERRESTRIAL		
Forest		
Whittaker & Niering (1975)	—	700–1500 g carbon m^{-2} yr^{-1}
Woodland		
Whittaker & Niering (1975)	—	250–700 g carbon m^{-2} yr^{-1}

[a] An upper limit; probably not realistic for whole communities. Towle & Pearse (1973) give the range as 0.1–7.0 g carbon m^{-2}.

photic zone, mixed depth, layering etc. Table 11.3 gives some figures from various sites.

Daily fixation figures in regions of high production quoted by Strickland, Eppley & Mendiola (1969) are 4 g carbon m^{-2} day^{-1} and Ryther *et al.* (1970) have recorded 10 g carbon m^{-2} day^{-1} in hitherto poorly productive waters. Fig. 11.5 shows the range of variation in carbon fixation in the western Indian Ocean illustrating the tenfold range which can occur. These figures suggest that overall oceanic rates of production need to be revised upwards considering the extent of area in which production is relatively high. In addition, the finding of deep chlorophyll maxima and stratification in the euphotic zone (*see* p. 345) also indicates that rates probably need upgrading. Brown & Parsons (1972) consider that the theoretical maximum in natural communities is about 9 g carbon m^{-2} day^{-1} rather than the often quoted figure of 5 g. In fact, Hobson, Menzel & Barber (1973) report 9 g carbon m^{-2} day^{-1} in upwelling water off Peru and Talling *et al.* (1973) comment that yields of 10 g carbon m^{-2} day^{-1} are well-attested from experimental mass culture of algae and values of 4 g are probably common in very productive lakes. In an *Aphanizomenon* bloom a maximum of 11.1 g carbon m^{-2} day^{-1} was recorded though values around 7–8 are more common (Hammer *et al.*, 1975) and Table 11.4 gives the values for some subtropical marine macroscopic algal communities in the Canary Islands (Johnston, 1969) together with the production values of some individual species. The daily rates for individual species show a considerable variation and agree with the data of Littler & Murray (1974*a*, *b*) who observed a simiar variation off the Californian coast (2.16 – 32.40 mg carbon g dry wt). 4 g carbon m^{-2}

day^{-1} is about the value for land crops such as rice whilst other cereals may average 1–2 g, so the sea is not as unproductive as many think. Individual stands of vegetation can give higher values and some algal communities appear to exceed the theoretical maximum (*see* Table 11.3) e.g., the *Macrocystis* forests in which 30 g carbon m^{-2} day^{-1} has been recorded (Collier *et al.*, 1973), though Littler & Murray recorded only 1.5 g carbon m^{-2} day^{-1}; however for *Fucus* communities at Woods Hole 20 g carbon m^{-2} day^{-1} is recorded (Kanwisher, 1966). The fact that these values are higher than the theoretical maximum needs re-investigation but it is probably related to the techniques and conversion factors used,

and one can assume that such communities certainly seem to operate at a maximal level. Nevertheless the overall contribution from these benthic seaweeds has probably been underestimated and as Ryther (1963) commented their production may approach 10% of that of all the plankton yet they only occupy 0.1% of the area occupied by the plankton. Ryther speculated that in fact all the estimates may be too low and the coastal macrophytic production might even equal the oceanic; certainly since that date production in coastal areas has been shown to be greater than expected and a new detailed analysis is now required. Just how far some estimates of production fall below the theoretical maximum is

Fig. 11.5. The primary production in the Western Indian Ocean sampling stations, O. From Ryther *et al.* 1966.

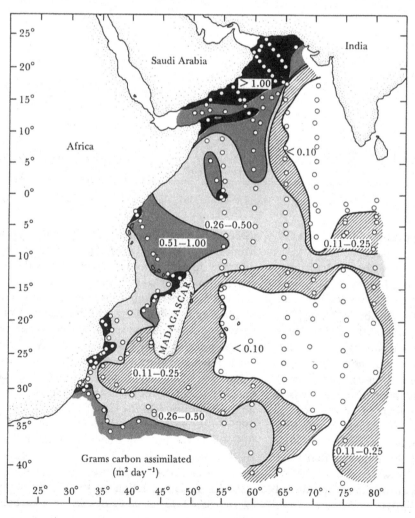

Table 11.4. *a, the production rates of some macroscopic algal communities in the Canary Islands, from oxygen measurements. b, the daily production rates of some of the commoner species in the communities in the Canary Islands. From Johnston, 1969*

(a)

Community type	Locality	Standing crop ($g\ m^{-2}$)	Net daily production ($g\ carbon\ m^{-2}\ day^{-1}$)
Cystoseira abies-marina	La Santa	1570	10.56
Padina–Halopteris–Jania	Jameos	1185	7.56
	Arrieta	390	2.81
	Jameos	313	2.04
Padina–Cystoseira fimbriata	Los Charcos	219	1.62
Caulerpa prolifera–Zostera marina	Arrecife (3 m)	250	1.09
Zostera marina	Arrieta	89	0.35
Caulerpa prolifera	Arrecife (35 m)	56	0.07

(b)

Species	Net daily production (dry wt as mg carbon g^{-1})
Dictyota dichotoma	21.36
Halopteris scoparia	8.28
Padina pavonia	7.56
Pocockiella variegata	7.28
Cystoseira abies-marina	6.41
Cystoseira fimbriata	5.53
Asparagopsis taxiformis	6.71
Galaxaura squalida	3.89
Caulerpa prolifera	4.52
Caulerpa mexicana	1.51
Codium bursa	1.72

shown by Steemann-Nielsen's (1953) estimate of gross fixation for the oceans as a whole of 0.15 g carbon m^{-2} day^{-1} and of 1.5×10^{10} tons carbon yr^{-1} for the whole aquatic system. These values are roughly a tenth of Riley's (1944) estimate (*see* p. 280) and are slightly less than Ryther's (1959), which is nevertheless indicative that the seas are more than twice as productive as the land.

Attempts to estimate the carbon fixation from light values and chlorophyll content have been made and as Fig. 11.3 shows there is a rough correlation. Ryther & Yentsch (1957) reckoned that carbon was fixed at 3.7 g h^{-1} g chl. a^{-1} at 2000 ft-c. Unfortunately this assimilation quotient (ratio)[*] is

not constant at light saturation (Curl & Small, 1965) and a considerable range (1–10) has been recorded which makes its use a little difficult; Cassie (1963) suggested from work on *Skeletonema* that a figure of 4.8 might be better. Steemann-Nielsen & Hansen (1961) quote various figures e.g. for tropical plankton (8), summer temperate arctic plankton (3–4) and winter temperate plankton (1), and Strickland, Eppley & Mendiola (1969) found values from 0.37 to 7.4 in Peruvian coastal waters. Light saturation values vary for the different groups of algae, e.g. Ryther (1956) quotes 500–750 ft-c for Chlorophyta,

[*] Also referred to as productivity index (Strickland, 1960; Jitts, 1965).

1000–2000 ft-c for diatoms and 2500–3000 ft-c for dinoflagellates. These are obviously only approximations and values will depend on many factors and especially important will be the 'sun' and 'shade' adaptation.

Another way of expressing the production is in terms of carbon *in excess* of respiratory needs and in the Atlantic Ocean, Riley (1951) found this to be about 14–37 mg cm^{-2} yr^{-1}. Studies of energy flow with data from ^{14}C fixation studies also use dark bottle controls, in which there is usually little fixation, although under certain conditions this can be equal to light fixation, e.g. in tropical oligotrophic oceans (Morris, Yentsch & Yentsch, 1971*a*), whilst Culver & Brunskill (1969) found that at 18–20 m depth in a rather exceptional meromictic lake (Fayetteville Green Lake) carbon fixation was 239 g m^{-2} yr^{-1} by bacteria and only 51 g m^{-2} yr^{-1} by the algae.

The relative amounts of plant and animal biomass vary considerably from the extreme of Lake George, Uganda where the algae form 95% of the biomass through to the reverse situation where the animals are in excess. The latter situation can only arise either where the turnover of the algae is rapid resulting in small algal biomass but large production or where allochthonous detritus is a major source of energy, e.g. in many streams.

Whilst it is important to improve techniques and obtain more precise data on the amounts of production in all types of water it is now perhaps even more important to concentrate some effort on disentangling the causes of the great variations encountered.

Nutrient cycles

The nutrients are as important to the algae as the solar energy but differ from the latter in the fundamental feature that they are not used up but are cycled through organism after organism and never totally lost to the biosphere. By far the greatest quantity is however locked up for a long time in the geochemical cycles involving pools in the atmosphere, in rocks, and in oceanic and lake sediments. Nutrient limitation is likely only in surface waters whilst light limitation *always* occurs in the deep waters. MacIsaac & Dugdale (1969) suggest that the crossover between these limiting factors occurs at a depth where the incoming radiation is reduced to 10% of the surface value.

Carbon cycle

The carbon cycle is illustrated diagrammatically in Fig. 11.6; essentially it involves a cycling of carbon from a small gaseous reservoir in the atmosphere to a large reservoir of aqueous carbon dioxide in the water, through the photosynthetic algae to the consuming animals and then from these to the detritus phase which is decomposed by the fungi and bacteria. This decomposition returns some of the carbon to the aqueous reservoir and from here a portion returns to the gaseous reservoir. Cycling is delayed at various points as respiration of all organisms returns carbon to the aqueous pool whilst a fraction is also side-tracked into the geological cycle by incorporation in the sediments, either directly or through the pool in the water. Sedimentary carbon returns to the atmospheric reservoir via the burning of fossil fuels and volcanic activity. In Fig. 11.6 there are three organismal pools (living pools) through which the carbon flows, algae, animals and fungi–bacteria; in many schemes these are labelled as producers or consumers, not entirely satisfactorily since each performs a series of different processes.

The influx of carbon is via equilibration with the atmosphere plus a small allochthonous input, mainly in freshwaters and coastal regions, though in streams the input from this source can be considerable. The non-living particulate carbon is located in three detrital pools: a small particulate inorganic pool (*PIC*, derived from calcium carbonate precipitation by algae, animals and physical mechanisms), a larger detrital pool of particulate dead and living material (*POC*) and a very large dissolved (plus colloidal) organic carbon pool (*DOC*). The remainder of the carbon is in the dissolved inorganic pool and in many waters this is the largest pool, the smallest being the particulate inorganic pool. Biologists are mainly concerned with the flow which is mediated by the organisms but one should not forget that there are also physico-chemical processes interacting with this cycle, e.g. temperature and pressure changes, complexing of calcium, magnesium, carbonate and biocarbonate ions, chemical precipitation and solution of carbonates, and other processes. At the air–water interface the carbon dioxide flux is slow and this can result in periods of imbalance due to carbon dioxide utilisation or production. But, as Schindler (1973) comments, the transference of carbon dioxide across the air–water interface has hardly been studied, although techniques involving calculating radon exchange rates have been developed and Schindler (1971) found that the carbon dioxide flux into a fertilised lake was 2.6 mg carbon dioxide-C m^{-2} day^{-1} whilst a nearby unfertilised lake lost 11.1 mg carbon dioxide-C m^{-2} day^{-1}.

The cycle is compounded of many sources, pathways and sinks and it is not very helpful to simplify the model except when necessary to make calculations from the partial data which are obtained in most studies. A good example of the complexities and limitations is provided by the study of Wetzel *et al.* (1972). The separation of the various pools in diagrams such as Fig. 11.1 inevitably gives the impression of discreteness; this is of course not true, for example, the bacteria are closely associated on and within the organismal fraction, on and within the detritus, and also free-living. Few data are available but Derenbach & Williams (1974) showed that in the English Channel, 1–50% of the photosynthetically produced *DOC* could be incorporated into bacteria. One should also clearly recognise that

there is a 'grazer food chain' and a 'detrital food chain'. At all stages there is both ingestion of carbon and egestion although the latter tends to be rather neglected. From the algal standpoint there is now a mass of evidence that organic molecules egested by the animals, bacteria and fungi can be taken up and used in algal metabolism. The paper by Antia, Berland, Bonin & Maestrini (1975) in which the growth of 26 planktonic species was tested on a range of inorganic and organic sources showed a remarkable number utilising organic compounds (*see* the discussion of nitrogen cycle in this chapter). However the uptake of organic compounds from the very minute concentrations in nature will involve considerable energy expenditure to overcome the concentration gradients and therefore their importance may be over-estimated by laboratory studies. Some may however be adsorbed on to the mucilage around the algae. The replenishment of the carbon dioxide in the aqueous pool from algal respiration is an almost unknown factor since few studies of respiration in natural habitats have been undertaken, but Wetzel *et al.* (1972) consider 30% of the

Fig. 11.6. Sources of carbon and flow through aquatic organisms. Dashed lines represent deposition pathways and solid lines transfer between pools.

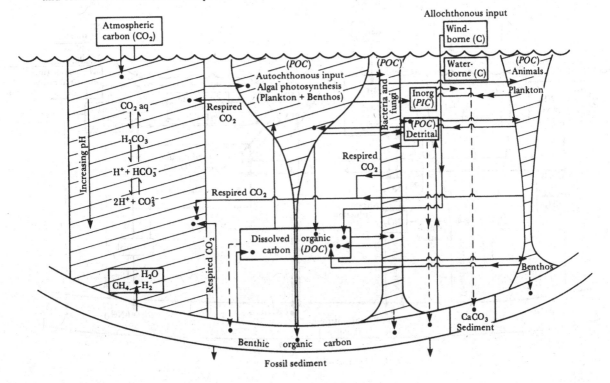

measured net primary production as a reasonable figure in their admittedly rather special lake.

The amount of algal production which passes through the grazing cycle varies according to the biomass of animals, amount of carbon fixed, size of the water body etc., but in the sea Menzel (1974) reckons that 90% of phytoplankton passes through the grazers; on the land the figure quoted is around 10%. However in coastal regions the situation is probably more like that of the land with much of the production passing direct into the detrital food chain, e.g. by erosion of fronds of Laminarians (Mann, 1972a). The total amount of organic detritus is high in many habitats and it can be used as a food source by the grazing animals when the algae are sparse, although there are divergent views on the value of detritus for many animals. The gross efficiency at each step in the food chain is often quoted as 10% but this figure cannot be taken very seriously since there are many examples of much higher efficiencies, even up to 80%. The considerable variation depends on cell densities, life cycle stages, etc. It is vitally important to obtain reliable figures if any attempt at construction of quantified flow diagrams is attempted. The contribution from the various organismal pools will vary greatly according to the type of water body, e.g. epiphytic populations in some systems contribute only about 1% of total fixation whilst in others they make a major contribution.

Nitrogen cycle

Nitrogen is utilised by algae in the form of nitrogen, ammonium ions, nitrite ions, nitrate ions and various organic compounds. Fig. 11.7 shows how nitrogen is cycled through algae and that in general nitrogen compounds only move through the biochemical cycle. Nitrogen cycles are complicated by the existence of nitrogen in nine oxidation states (-3 to $+5$) and the numerous processes involving this element (fixation, nitrification, denitrification). Inorganic nitrogen tends to follow the pathway, nitrate → nitrite → ammonium → amino acids → protein. Both nitrite and ammonium ions are present in aquatic systems together with nitrate ions and the recent experimental evidence that many marine planktonic algae can utilise all three (Antia, Berland, Bonin & Maestrini, 1975) and in addition organic

Fig. 11.7. Sources of nitrogen and flow through aquatic organisms.

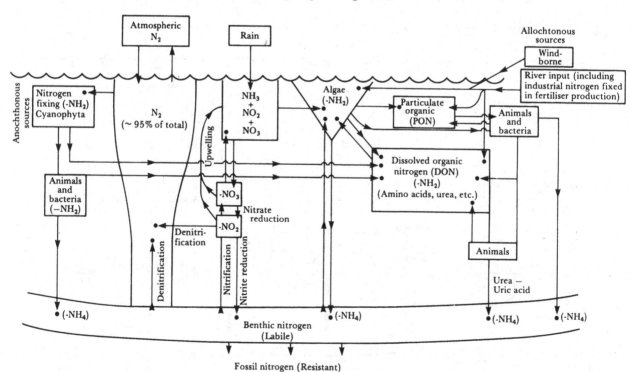

sources (urea, hypoxanthine and glycine were used to varying degrees) perhaps explains why nitrogen is only rarely limiting algal production in fresh and inshore marine waters,* though at times the phytoplankton can reduce the levels to extremely low values. Uptake of one form of nitrogen may be suppressed by the presence of another, e.g. when phytoplankton are supplied with both nitrate and ammonium ions, the rate of uptake of nitrate ions may be depressed (Bienfang, 1975).

Lund (personal communication) states that there does not seem to be any well attested example of a lake in which lack of nitrogen limits the growth of phytoplankton over a whole year, but there is evidence for nitrogen limitation over parts of the year and for specific plankton. For example it has been suggested that nitrate-nitrogen limits the growth of *Stephanodiscus hantzschii* in Lake Erken when it reaches values as low as 1 μg nitrate-nitrogen l^{-1} (Pechlaner, 1970). In tropical African lakes both Moss (1969a) and Viner (1973) considered that nitrate ions or nitrate and phosphate ions were the potential limiting nutrients whilst Holm-Hansen *et al.* (1976) regard the phytoplankton of Lake Tahoe to be limited by nitrogen. It has been calculated that the nitrate reserve in the oceans is sufficient to support the average phytoplankton production for 60–100 yr (Vaccaro, 1960) or, put in other terms, the annual usage of nitrogen by phytoplankton is only about 1 % of the reserves in the deep ocean (Dugdale & Goering, 1971). These estimates simply reflect the enormous stores in the deep water but in any one season the flux from the sediment may only support 30–40 % of the phytoplankton production, e.g. in the sea off Spanish Sahara (Rowe, Clifford & Smith, 1977). Another source of supply is the zooplankton, e.g. Smith & Whiteledge (1977) found the zooplankton in an upwelling system off northwest Africa supplied 44 % of the ammonia demand and 25 % of the total nitrogen required. Nevertheless, in the nutrient-poor surface layer of tropical oceans and during rapid algal growth in other oceans and especially in neritic regions, available nitrogen is frequently a limiting factor (Thomas, 1970; Ryther & Dunstan, 1971); this even applies in shallow bays such as Chesapeake Bay where one might expect a large input of nitrogen from the drainage basin

* Productive estuaries tend to be limited by nitrogen since the N:P ratio is often below 2:1.

(McCarthy *et al.*, 1977). Morris, Yentsch & Yentsch (1971b) concluded that, in subtropical waters, the phytoplankton population size is limited by nitrogen as it would be in a chemostat. Ammonia excretion by animals and its re-assimilation by phytoplankton has been shown many times and, in some oceanic regions, ammonia is often the only available inorganic form. In Lake Mendota assimilation of ammonium-nitrogen was always greater than that of nitrate-nitrogen and Brezonik (1972) is of the opinion that ammonia is generally more important to algae in aquatic systems than is nitrate. In the ocean nitrogen tends to be stripped from the water before phosphate and the effect of nitrogen starvation is to upset the enzyme balance of the cells, resulting in the synthesis of lipid rather than carbohydrate, whilst chlorophyll synthesis ceases, which leads to excess carotenoids in the cells.

Phosphorus cycle

The phosphorus cycle is illustrated in Fig. 11.8; the element is present in algal cells in relatively small amounts but it is even less abundant in the primary drainage sources and therefore is the element most likely to limit freshwater algal production; it certainly does so in many lakes (Schindler, 1977; Schindler, Fee & Ruszczynski, 1978), but not so often in the sea where the limiting value (often quoted as about 0.40–0.55 μg at. l^{-1}) is usually not reached before nitrate or silicate limitation occurs. In coastal waters Ryther & Dunstan (1971) reckon that there is about twice as much phosphate as can be used by the algae. In lakes, on the other hand, the reverse is true and in one of the rare compilations of amounts utilised by algae relative to the pool, Saijo & Sakamoto (1970) showed that in Lake Biwa some 11 200 tons of nitrogen and 1480 tons of phosphorus are consumed but the lake totals are estimated at 29 200 and 550 tons respectively. So only about $\frac{1}{3}$ of the nitrogen is required but three times the total phosphorus. If the inorganic phosphorus and nitrogen alone are considered then the pools in the water are 5200 tons of nitrogen and 200 tons of phosphorus and since it is likely that this form is utilised the nitrogen must turn over twice and the phosphorus seven times. Phosphorus is concentrated in marine phytoplankton by a factor of 10^7–10^8 over its concentration in the sea but, in spite of this, rapid growth of algae can deplete the phosphorus to a very low level. Phosphorus uptake by phytoplankton can continue even when all the nitrate in the seawater has been utilised (Corner & Davies, 1971). In fact most algal cells grown in water containing ample phosphate take up 'excess' quantities and store it as polyphosphates which can be utilised during periods of lower external phosphate concentration. It follows that absolute levels of phosphorus in cells have little ecological meaning. Another important feature of the phosphorus cycle is the very rapid turnover time, e.g. 0.9–7.5 min in summer in some lakes (Rigler, 1964) but extending to days in winter. The rapid uptake of added phosphate is well known but the cycling between particulate, low molecular weight compounds, colloidal phosphorus and phosphate is very rapid (fractions of a second) and part of the colloidal phosphorus enters an unavailable pool (Lean, 1973). Similarly in the ocean this turnover occurs in a matter of hours in summer and days in winter. Recent work (Harrison *et al.*, 1977) has shown that added ^{32}P is rapidly taken up by material passing through a 1 μm filter and so some of the rapid cycling is probably through bacteria and not algae. Organic sources of phosphorus can be used by some algae under experimental conditions but there are few data on their uptake or use in nature.

There have been many estimates of the excretion of phosphorus by animals, e.g. Pomeroy, Mathews & Min (1963) found that net zooplankton excreted almost the same amount per day as was contained in the organisms: slightly more than half was phosphate and the remainder insoluble organic compounds. The concentration of phosphate in the cells of algae is continually varying, e.g. in *Scenedesmus quadricauda* it decreases in the morning and afternoon and increases in the evening when cell division commences. The cells also excrete organic phosphorus during day-time (Overbeck, 1962*a*) and clearly all these cellular changes are superimposed on the general cycling of phosphorus and illustrate that time of day will influence attempts to estimate the various pools of this and other nutrients (*see also* diurnal effects, p. 329). *Scenedesmus* only absorbs phosphate ions, whilst pyrophosphate and organic phosphates are not taken up although the alga contains the enzymes to convert them (Overbeck, 1962*b*).

There are many reports that high phosphate concentrations in lakes prevent growth of some planktonic species (e.g. *Dinobryon*) and conversely Serruya & Pollingher have found that *Peridinium cinctum* f. *westii* in L. Kinneret forms very large blooms

Fig. 11.8. Diagrams of the phosphorus cycle.

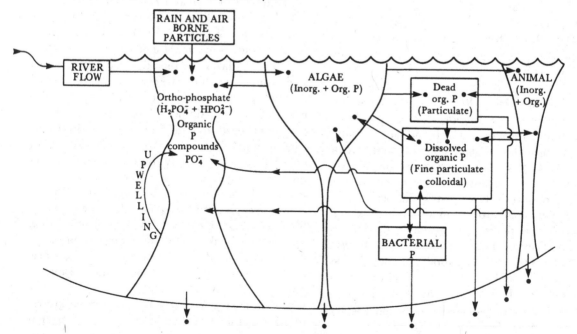

when phosphate concentration is low. Work by Lehman (1976) did not show high phosphate inhibition of three *Dinobryon* species and it appears that both these and the *Peridinium* have a competitive advantage in being able to take up phosphate from extremely low concentrations. Fogg (1973) comments that many species can absorb orthophosphate from phosphorus concentrations as low as $1\ \mu g\ l^{-1}$ and that when low phosphorus values are reached some algae are capable of producing powerful extracellular phosphatases which enable the cells to utilise the remaining inorganic and organic compounds. Fitzgerald (1972) found that algae grown in phosphorus-poor media can produce 25 times as much alkaline phosphatase as algae grown in phosphorus-rich media; these effects have also been seen in *Cladophora* growing in Lake Mendota. Not all algae are capable of induction of phosphatases, and those that are may have a competitive advantage (Kramer, Herbes & Allen, 1972).

Silicon cycle

Silicon is used by a number of algae (diatoms, silicoflagellates, some Chrysophycean flagellates) and the protozoan group Radiolaria (in marine ecosystems only). These algae are common in almost all habitats and silicon itself is very abundant and is present in waters in the form of orthosilicate. The supply of silicate is limited by the rate of rock and soil weathering, water flow, solubilisation of pre-existing populations and other factors and since it is used up very rapidly during periods of fast diatom growth it can easily become limiting; there are almost no data for use by the other algae except the report by Klaveness & Guillard (1975) that the chrysophyte *Synura* has comparable amounts of silica in its scales to diatoms and it can deplete the silica in the medium to very low levels. Silica can be depleted below the level of detection resulting in insufficient supply to support further cell divisions (*see* chapter 10). Yet in deep oceans it is present in amounts from 1000 to 4000 $\mu g\ l^{-1}$ and is the element showing the widest range of concentration (Armstrong, 1966) and in some areas, e.g. off northwest Africa, the rate of solution from the organisms in the upper 50 m was sufficient to supply the phytoplankton (Nielsen & Goering, 1977). Silicon cycles through organisms (Fig. 11.9) in a relatively simple way and

Fig. 11.9. Diagram of the silicon cycle.

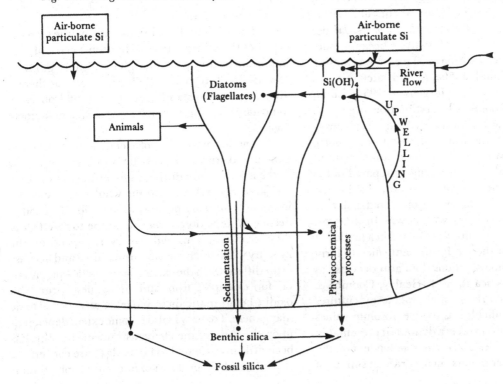

is only fixed into relatively immobile forms by the organisms mentioned above. The 'loss' from the system is often thought to be virtually complete in lakes but in the ocean the 'loss' is much less (though still enormous) since solution of much of the silica occurs during the long descent of the organisms to the sediments. In freshwater habitats few organisms other than diatoms use silica in quantity and hence changes in silica content are likely to follow diatom growth rather closely. In the ocean similar changes occur (*see* p. 432) but the other organisms utilising silica complicate the relationships. There have been remarkably few attempts to work out the silica budgets of lakes; one such by Dickson (1975) who estimated the flow of silica in Lough Neagh suggests that during an annual cycle 1.5 mg l^{-1} is deposited on the sediments, 0.5 mg l^{-1} is lost through the outflow and 4.2 mg l^{-1} is redissolved as the diatom crop sinks. He also estimated that some 65 % of the silica incorporated in the diatom crop is recycled during the summer period. Lough Neagh is of course an alkaline lake and such a high resolution rate must not be expected in lakes in general.

It was Pearsall (1932) who first pointed out that silica (at a concentration of approximately 500 μg SiO$_2$ l^{-1}) limited diatom growth in the English Lake District and soon after Weimann (1933) showed that blooms of *Stephanodiscus hantzschii* depleted silica. As methods of analysis became more refined it was concluded that 200 μg l^{-1} was a more realistic figure (Lund, 1950*a*, *b*) and later this was reduced to 60–80 μg l^{-1} (Jørgensen, 1957; Lund, 1965). Most methods for silica estimation attempt to determine the reactive silica only, and it is assumed that this is the form which the diatoms absorb, though Golterman (1967) showed that diatoms could use colloidal solutions of silica prepared in the laboratory and these were not estimated by the molybdate method. The extent of colloidal silicate or its uptake in nature is not known. In diatoms, cytoplasmic division continues at low silica levels but the cells die when there is insufficient silica to form the new walls. Photosynthesis does also decrease as the cells become silica starved (Healey, Coombs & Volcani, 1967). Work on silica uptake by cultures has shown how difficult it is to give absolute values for the silicon requirement of a diatom (it presumably differs for each species) since this is dependent on the interaction of other factors such as rate of supply of

the substance itself and the concentration of other substances in the water. Thus Hughes & Lund (1962) showed that in the presence of excess phosphorus, *Asterionella formosa* could deplete silica to the limits of detection and also that slowing the growth rate of this diatom by reducing the light level would increase the efficiency of uptake.

The amount of nutrient uptake by cells is obviously related to many factors but one which tends to be ignored is size, e.g. Reynolds (personal communication) showed that *Stephanodiscus astraea* of diam. 31.4 μm ($\pm$3.7) has an apparent uptake of 1620 μg of silica per 10^6 cells whilst large cells e.g. 43.3 μm ($\pm$4.2) can take up 2830 μg silica per 10^6 cells so that as far as diatoms are concerned knowing the number of cells in the population is little indication of the amount of silica likely to be removed or vice versa one cannot readily predict the size a population may achieve without knowing the cell size. Half saturation constants for silica uptake vary according to the source of the diatom. Thomas & Dodson (1975) showed that it was 0.33 μg at. silicon l^{-1} for *Leptocylindrus* isolated from high-nutrient water and 0.74 μg at. l^{-1} for a clone from low-nutrient water, the opposite from what one might expect.

Other nutrients

There is very little information on the seasonal cycles of the minor nutrients but some certainly are important as they pass through the algae, e.g. Lund, Jaworski & Butterwick (1975) consider that there is abundant circumstantial evidence that iron is an important limiting factor for freshwater planktonic algae.

The accumulation of elements such as manganese and iron and probably other heavy metals in the hypolimnion during summer stagnation (Fig. 1.7) and their release into the whole water column in the autumn may possibly account for the sudden decline of some organisms due to the toxic effect of these elements. This will, however, depend on the rapidity with which the salts are oxidised and because of the difference in the oxidation–reduction potentials of, for example, iron and manganese, iron will oxidise before manganese and leave an excess of toxic manganese. Toxicity is also to some extent dependent upon the ratio of the elements present, but this has been little investigated as far as algae are concerned.

The whole problem of nutrient supply, limitation, cycling and relationship to rates of primary production and standing crops needs very detailed studies and some of the effort put into measuring the movement of carbon by ^{14}C studies might profitably be directed to a re-assessment of the problems and investigation of the transfer of other nutrients amongst the various pools.

Summary

Algae are the major organisms involved in carbon fixation in the aquatic sphere taken as a whole. This fixed energy diffuses through the web of organisms in the water and into detritus, some of which is recycled and some incorporated into the sediments. The pathways are illustrated diagrammatically. Production and productivity values are discussed in relation to the input of solar energy and figures for maximum possible carbon fixation are discussed. Tables of chlorophyll contents of various algal associations and of amounts of carbon fixed have been collected together.

Nutrient cycles of carbon, nitrogen, phosphorus, and silicon are given with some brief comments.

12 Algal contribution to the sediments

Much of the organic and inorganic material incorporated into algae during their growth is recycled through the water after their death and becomes available for further uptake by organisms. But a proportion finds its way to the sediments and is preserved there in some form or other. The organic components of the sediment are the most difficult to study and relate to the algal populations, whilst the inorganic siliceous and calcareous plant remains are more easily correlated since they are usually recognisable by microscopic techniques. Unfortunately by no means all the siliceous and calcareous components of the sediment are derived from algal growth and it is necessary to distinguish those originating from other biogenic sources and, in the case of the inorganic components, those derived from inwash, aerial transport, inorganic precipitation, etc.

Lakes are temporary features of the landscape, gradually filling in and disappearing: infilling has been confused in many books and certainly in most elementary texts with 'eutrophication' yet it has little to do with this process. Unfortunately in some early definitions of eutrophic waters the amount of silt was taken into account and this misled other workers. All types of lake tend gradually to fill with material contributed both by the surrounding land and by the lake itself but there is no evidence that algal blooms are associated with the later stages of infilling (Moss, 1973c). Rapid infilling and the recognition of seral stages through marsh to a terrestrial flora, is found mainly in small lakes with hardly any outflow and study of this type of lake has also added to the general but erroneous impression of the importance of fairly rapid change from open water to land, an extremely long-term process in lakes of any size.

There are contradictory statements in the literature concerning the algal contribution to sediments, e.g. Mackereth (1965) considers that it is slight in the English Lake District lakes, but other workers report a much greater algal contribution. Each lake will differ depending on the nature of the drainage basin, renewal time, etc.

Silica deposition

The most important organisms contributing silica to the sediment are the diatoms, followed by the radiolarians, sponges and silico-flagellates, the last three in marine sediments, although a small quantity of sponge spicules and siliceous scales from Chrysophyta can also be detected in lake sediments. Varied figures have been quoted for the average silica content of oceanic sediments, e.g. 42.7% of which only some 8.5% is biogenic opal* (Wollast, 1974) whilst earlier calculations by El Wakeel & Riley (1961) gave an average of 61.5% of which less than half was biogenic. Another calculation suggests that between 70 and 90% of the suspended silica in the oceans is derived from diatoms and of this somewhere between 1 and 10% reaches the sediment (Lisitzin, 1971a). Fig. 12.1 shows examples of deep-sea sediments rich in diatoms after treatment in acid to remove the organic matter.

Lisitzin (1971a) attributes the difference between amorphous silica in suspension and in the sediments to solution of the diatoms, since silico-flagellates and radiolaria tend to sink unbroken to the bottom. The dissolution of silica needs careful re-investigation under a range of natural conditions but I suspect that some of the reports of the absence

* Biogenic refers to any substances formed by biological activity. Opal is the mineralogical form of silica in these organisms.

of diatom remains are due to fine comminution of the diatoms into unrecognisable fragments in their passage through the water column.

The silica is not depositied evenly over the ocean floor but in belts (Fig. 12.2) corresponding to the areas of high planktonic diatom production (Fig. 12.3). The sediments around Antarctica are extremely rich in silica and it has been calculated that more than three-quarters of the silica of the world's oceans accumulates here. According to Hays (1967) this is a relatively recent quaternary phenomenon and prior to that era red clay was accumulating. Other areas such as the central region of the Gulf of California have the highest absolute figures. An average accumulation of 50 g cm^{-2} 1000 yr^{-1} has been recorded here by Calvert (1966), and certainly the diatom counts (Round, 1967) confirm this very high rate of accumulation. Jousé, Kozlova & Muhina (1971) quote figures of 100–150 million

diatoms g^{-1} for some of the rich oceanic sediments (e.g. in the Bering and Okhotsk Sea) and it is probable that the figures exceed this in the centre of the Gulf of California. In contrast, under the two central gyres of the Pacific Ocean, no diatoms are recorded in the sediments and Johnson (1976) found low levels of biogenic silica in zones of extreme terrigenous sedimentation; this is specially noticeable in regions where mollusc remains are deposited (Milliman, 1977). It may be significant that these regions are marked on some sedimentary maps as deposits of red clay which, as mentioned on p. 489, could result in transference of silica from diatom fragments to the clay particles. Alternatively part of the red clay may be partially derived from minute diatom fragments.

Similar poor representation of diatoms is recorded in the grey-green or red-brown clay in glacial marine deposits around Antarctica (Hays, 1967;

Fig. 12.1. Photomicrograph (Round, unpublished) of diatoms in a sample from a core of sediment taken from the Gulf of California.

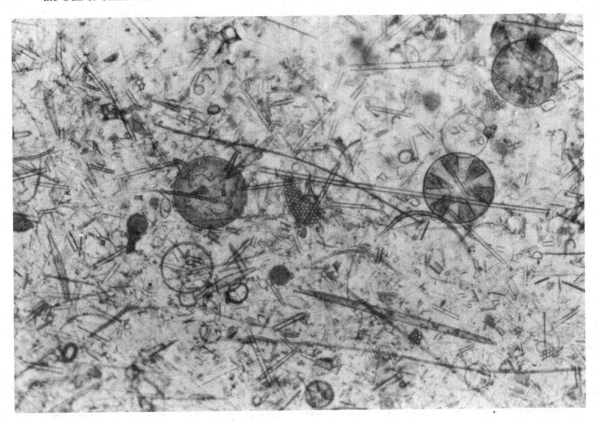

Du Saar & De Wolf, 1973); it is only north of the northern boundary of pack ice that the bottom deposits are rich in diatoms and this belt stretches northward (Fig. 12.4) to the region of the Antarctic polar front (Donahue, 1962). Deposition of silica derived from diatoms in coastal waters is slight and no good explanation of this phenomenon is available although it seems likely that the grinding movement of sediments in shallow water destroys the diatoms. Some shallow areas are rich in carbonates and, in them, diatoms are readily dissolved (e.g. Upper Florida Bay, DeFelice & Lynts, 1978). Where water movement is minimal, e.g. in the iron-forming sediments at Santorini in the Aegean Sea, the diatoms are preserved (Round, in press).

Much of the silica supplied to the oceans is derived from solution of siliceous minerals on the land and downwash via rivers. This almost certainly supplies the needs of the littoral diatom populations and probably also the neritic phytoplankton. It has been calculated that about 4.27×10^{14} g silica is added by rivers every year (Wollast, 1974) but this is nowhere near sufficient for the total calculated utilisation by siliceous organisms in the oceans (250×10^{14} g silica per year). This very large amount of silica which is cycled through the organisms in the photic zone comes from the pool of silica in the deep water; calculations of the amounts available from eddy diffusion and upwelling of deep water have shown these to be sufficient to supply this amount of silica to the trophogenic zone.* The amount of

* The term trophogenic (= biogenic) is used for the upper waters where in general a build-up of organisms is a major feature whilst the term tropholytic is sometimes used for the deeper waters where breakdown and solution of substances greatly exceeds any formative processes.

Fig. 12.2. Amorphous silica distribution in the surface layer of bottom sediments (% dry sediment). 1, < 1%; 2, 1–5%; 3, 5–10%; 4, < 10% (without subdivision); 5, 10–30%; 6, 30–50%; 7, 50–70%; 8, > 70%. From Lisitzin, 1971*a*.

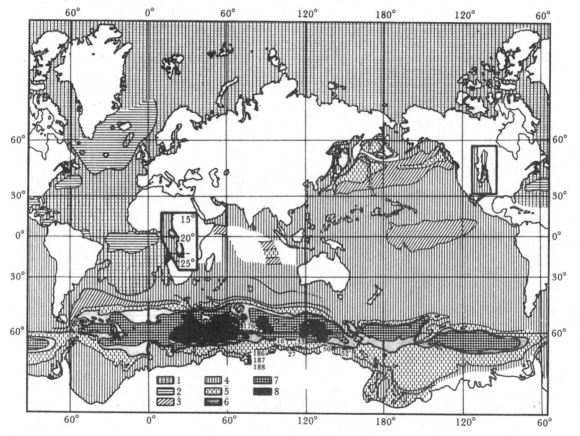

upward mixing of silica-rich water varies considerably from place to place in the ocean, e.g. the high productivities of the antarctic divergence are correlated with increased upwelling (Fig. 1.3). Study of some ocean sediments, such as in the centre of the Gulf of California or of Eocene–Miocene deposits of marine origin, reveal considerable numbers of diatoms in the material, giving the impression that ocean sediments are rich sinks for such organisms and recycling of silica in the water column is minimal. But in the case of the Gulf of California accumulation in the sediment occurs beneath the surface water which is richest in diatoms and suggests that a saturation factor is operating in the deep water preventing solution of the 'rain' of diatoms. However other sediments lying under moderately productive water, e.g. in the North Atlantic off the coast of Africa, yield only minute numbers of diatom frustules (Round, unpublished). Even the rich diatomaceous sediments are almost devoid of the delicate diatom genera (*Chaetoceros, Rhizosolenia, Bacteriastrum, Corethron,* etc.) whilst girdle bands of most centric species have also disappeared. This is partly brought about by fragmentation as the frustules pass through the guts of animals and largely by solution of the thin frustules. There have been few studies on solution of silica from diatoms and obviously it is necessary to determine this under natural as well as under experimental conditions. Golterman (1960) showed that dead *Stephanodiscus hantzschii* released about 20–30 % of its silicate after some weeks whilst Grill & Richards (1964) reported that 95 % of the silica of an unknown centric diatom dissolved in nine months; this suggests that one of the extremely

Fig. 12.3. The pattern of absolute masses, or the annual production of silica. Based on data on primary production by ^{14}C and oxygen as well as on the relationship between amorphous silica and organic carbon in suspended matter from different climatic zones expressed as amorphous silica (g m^{-2} yr^{-1}). 1, < 100; 2, 100–250; 3, 250–500; 4, > 500. From Lisitzin, 1971a.

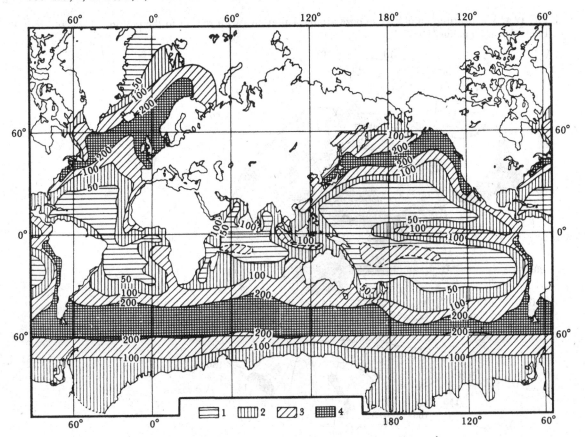

delicate species was used. There is evidence that the siliceous parts of the diatom cell have a thin coating or 'skin' of organic material and certainly material viewed in the scanning electron microscope often appears to have a fine organic layer. Logically, one would expect some organic material since the silica is laid down in organic vesicles, some of which probably attach to the valves and girdle bands. Hecky, Mopper, Kilham & Degens (1973) reported that carbohydrates occur associated with the siliceous material and suggested that they may retard dissolution though it seems unlikely that such material would be effective for any length of time. Solution occurs mainly in the upper 1000 m (Fig. 12.5) whilst below that level the percentage saturation of the water with silica, combined with the lower temperature, reduces the relative solution rate to 1/16 of that in the surface water at 25 °C. This temperature dependence of the process helps to explain the paucity of diatoms in tropical sediments and the richness in arctic and antarctic zones. However the richness in the warm water such as the

central Gulf of California (Round, 1967) must be due to rapid sinking and possibly also to high saturation values for the lower waters. Schrader (1971) has shown that diatoms become incorporated into faecal pellets in the surface waters and are then protected from solution and sink at much greater rates. Bacterial decay and secondary grazing then breaks up the pellets by the time they reach 500 m depth, but here the relative rate of solution of silica is some 3% compared to 16% in the surface water (Fig. 12.5). Even below 500 m depth it is likely that there is some aggregation of sedimenting material resulting in increased sinking rate since it would take diatoms such as *Thalassionema nitzschioides* 1900 days and a large *Coscinodiscus* species 45 days to sink 1500 m in the Gulf of California (Smayda, 1971*b*). I found such species at 3000 m and in sequence in the sediment which would hardly be possible if they took such a long time. In fact all the evidence of those who have worked on oceanic sediments indicates that accelerated sinking must occur via inversion currents, downwelling, aggregate formation and

Fig. 12.4. The distribution of diatom ooze around Antarctica with respect to the position of the antarctic Polar Front. From Donahue, 1967.

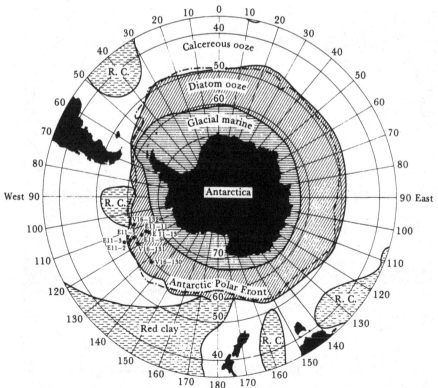

incorporation in faecal pellets. Since large cells sink faster than small (*see* p. 301) one would expect a sorting of diatoms resulting in the predominance of large forms in the sediment. Rapid sinking may explain the occurrence of belts of diatomaceous ooze in which the gigantic diatom *Ethmodiscus* is abundant since dead cells of this species sink at about 500 m d^{-1} and so only 5–12 days would be needed to sink to the depths (2780–5712 m) where the ooze is formed (Smayda, 1970). Smaller cells would take anything between one month and two years to sink to 5000 m. It is still not clear how the rich *Ethmodiscus* oozes are deposited in long thin belts across the oceans. *Ethmodiscus* ooze is apparently concentrated in the 10° band north and south of the equator but extends to 30 – 40° north and south (Mikkelsen, 1977). The ooze always occurs in depths of 4600–5000 m and often contains as much as 80% *Ethmodiscus*. Mikkelsen considers that this is due to differential solution

of species reaching this depth which also happens to be the carbonate compensation depth. The genus is resistant to solution under laboratory conditions.

Wollast (1974) suggests, from theoretical consideration of solution of opal debris, that the amount of silica which dissolves as a function of water depth is independent of the particle radius, which means *a priori* that small diatoms or thin fragments will disappear first. From his calculations, Wollast considers that this solution is almost entirely sufficient to meet the uptake requirements in the biogenic zone and so there is virtually a closed cycle of silica uptake and solution. Once incorporated into the sediments, solution of diatoms is further slowed down, though passage through animals and resuspension of the surface material still continues and results in further slight solution. The concentration of dissolved silicate increases in the interstitial water in the sediments, thus Hurd (1972a, b) found increases of 150 ± 20 μg at l^{-1} in the first 35 cm, but below this depth the concentration remained constant. There is an upward flux of dissolved silicate into the water which Wollast calculates to be about 3.2×10^{14} g yr^{-1} for the oceans (i.e. about three-quarters of that supplied by rivers). In addition, in some sediments, dissolved silicate also reacts with magnesium in the interstitial water to form sepiolite ($Mg_2Si_3O_6(OH)_4$) which is precipitated, or with clay minerals, to form even more siliceous clay. These factors combine to continue the solution process which diatoms and silicoflagellates undergo in sediments. Under some circumstances, e.g. anaerobic, stagnant conditions, the silica fraction of sediments is enriched in minor elements (Grant Gross, 1967) and study of the exchanges between diatom silica and the elements in the water column would be worthwhile. Sedimenting diatoms are reported to scavenge ions from the water.

The overall silica cycle, with reasonable values, is given in Fig. 12.6.

Data such as those presented above make it surprising to find any diatoms at all in sediments. Yet some are rich and show considerable variation in numbers of diatoms on a dry weight basis and appreciable changes in species composition down the cores (*see* p. 502). Below a certain depth in the sediment there can be little solution of silica otherwise there would be no remains of Pliocene–Miocene diatoms in deep sediments.

Fig. 12.5. Variation in % saturation and relative solution rates of biogenic opal with depth as a function of temperature and silicate concentration. From Hurd, 1972b.

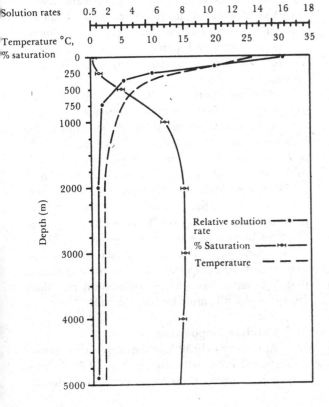

Solution of diatoms sinking through freshwater lakes has not to my knowledge been calculated, though a calculation by Parker, Conway & Yaguchi (1977a) suggested the very high figure of 90% applied to diatoms in Lake Michigan by dissolution and recycling. Similar values for Lake Michigan are also given by Parker, Conway & Yaguchi (1977b) and it would be interesting to have comparative data from other lakes. The distances to the sediments are however so small compared to those in the marine sphere that one can assume that virtually all the species arrive on the sediments or are lost from the system by outflow. Ferrante & Parker (1977) illustrate the incorporation of freshwater diatoms into faecal pellets and also show that most of the diatoms are crushed in the feeding and excreting process. Although the distances are much less than those involved in oceans, these workers also showed that the pellets decay after 6–14 days in deep lakes before they reach the sediments, since they calculate that the sinking rate of the pellets is 4.2 m day. In lakes and rivers the supply of silica is probably mainly from inflow and very little is actually required from the silica-rich sediments to maintain growth of diatom populations in the surface waters. Some

return of silica to the trophogenic zone undoubtedly occurs on the breakdown of thermal stratification but it is small compared to the amounts from inflow at this time of year. Studies on large, deep, tropical lakes may however show a silica cycle more like that of the oceans and recycling or lack of recycling of silica may stimulate or limit diatom growth in such waters. Deposits from the bottom of lakes often contain diatoms deposited during the late glacial period (i.e., about $10–15 \times 10^3$ yr ago) and these reveal very little solution of the frustules except in samples from the glacial clay.

The scarcity of diatom remains in varved glacial clays (Round, 1957b, 1967) deposited in freshwater basins may be related to chemical reactions such as those mentioned above occurring in marine clays, postulated by the author (Round, 1964). Somewhat more puzzling is the scarcity of diatoms in some peats when the surface pools and surface layer of peat usually support large populations. Underwater peat deposits on the other hand preserve diatoms well, e.g. the freshwater deposits underlying the marine sediments of Lough Ine in southwest Ireland (Buzer, 1975).

During the long history of the oceans, conditions have fluctuated and at some periods the silica of diatoms has been converted from opal to other forms, e.g. porcellanite, and such diagenesis has disturbed the fossil record of diatom assemblages in some strata.

The distribution of diatoms, radiolaria and silico-flagellates in sediments has been used to detect changes in climate, chemistry etc. An example of this is the use of the ratios of the silico-flagellates *Dictyocha* to *Distephanus* to determine cold and warm periods in the recent geological record. Schrader & Richert (1974) report that the percentage of *Distephanus* increases exponentially within the surface sea temperature range 16.5 °C to 8 °C whilst the percentage of *Dictyocha* increases within the range 19.5 °C to 25 °C. One complication in using such observations to interpret the geological record of these species is that *Distephanus* has a higher dissolution rate than *Dictyocha* and this must be corrected for.

Calcium deposition

Approximately 50% of the deep ocean bottom is rich in calcium carbonate (*see* Fig. 12.7) and the only major sources of this are biogenic formation by

Fig. 12.6. Tentative cycle for silica in the oceans. From Wollast, 1974.

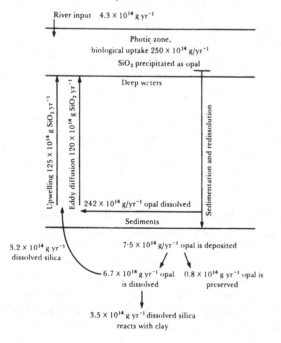

River input 4.3×10^{14} g yr^{-1}

Photic zone, biological uptake 250×10^{14} g/yr^{-1}

SiO$_2$ precipitated as opal

Deep waters

Upwelling 125×10^{14} g SiO$_2$ yr^{-1}

Eddy diffusion 120×10^{14} g SiO$_2$ yr^{-1}

Sedimentation and redissolution

242×10^{14} g/yr^{-1} opal dissolved

Sediments

3.2×10^{14} g yr^{-1} dissolved silica

7.5×10^{14} g/yr^{-1} opal is deposited

6.7×10^{14} g yr^{-1} opal is dissolved

0.8×10^{14} g yr^{-1} opal is preserved

3.5×10^{14} g yr^{-1} dissolved silica reacts with clay

protozoa of the Foraminifera group and algal flagellates belonging to the coccolithophorids, both of which provide calcite. It has been shown that most of the fine grained ($< 30\,\mu$) carbonate material in oceanic sediments is probably derived from coccoliths and about 50% of the finest grained material ($< 10\,\mu$) may also be derived from coccoliths. As with other planktonic remains the only conceivable way of getting such small particles as coccoliths to the deep sediments in a short enough time to prevent solution and mechanical destruction is by compaction in faecal pellets; Roth, Mullin & Berger (1975) found some coccoliths in faecal pellets in the East Tropical Pacific and in laboratory experiments they found large numbers in pellets. Honjo (1977) estimated a sinking rate of 150 m per day in faecal pellets but also showed some dissolution of coccoliths. At certain periods in post-Cretaceous time, material derived from coccoliths predominates over foraminiferan shells in the sediments. In coastal regions other groups of plants and animals also contribute calcium carbonate to the sediments.

Carbonate deposits are produced by algae in a variety of ways: the mucilage of algae traps the carbonate particles which themselves need not be of algal origin, crystals are formed on algal filaments or thalli and the remains of these are deposited, the larger calcified algae break down to form sand or silty calcareous sediments. Breakdown of some coralline red algae at first produces large 'stone-like' pieces of calcium carbonate which are then gradually broken down. On a coastal world-wide basis, Dawson (1966) considered that the Rhodophyta contain the majority of the carbonate depositing algae (some 590 species) and the green algae the next greatest number (89 species). In the open ocean it is only the coccolithophorids which make any real contribution and they are immensely important as their coccoliths form the bulk of many chalk deposits. The rate of accumulation of chalk has been variously estimated from 1.5 cm 1000 yr^{-1} to 50 cm 1000 yr^{-1} which as Bathurst (1971) put it 'would represent 500 μ yr or about 180 coccoliths a year piled one upon the other'. However, exactly as

Fig. 12.7. Map of calcium carbonate distribution in bottom sediments of the seas and oceans (% of dry sediment). 1, $< 1\%$; 2, 1–30%; 3, 30–70%; 4, $> 70\%$. From Lisitzin, 1971*b*.

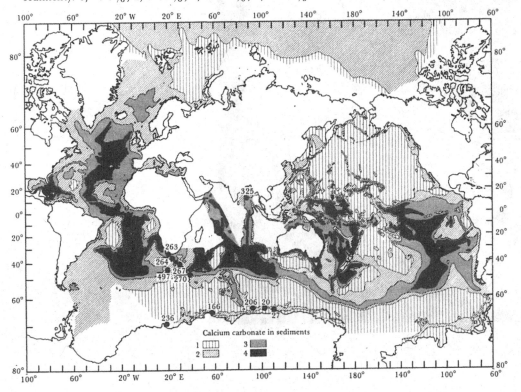

with diatoms it is only the coccoliths most resistant to solution which are found in sediments in any quantity and experiments have shown the variable resistance of species (Berger, 1973).

Calcareous microfossils were first found in chalk by Ehrenberg in 1836 and later, when the floor of the Atlantic was being investigated before the first transatlantic cable was laid, Huxley & Wallich (quoted without original reference in Black, 1965) found similar calcareous structures loose and called them coccoliths. They also noted that they could be built up into hollow spheres which they termed coccospheres. A few modern dinoflagellates form

cysts which have an outer calcite layer (e.g. (*Proto*) *Peridinium trochoideum*, Wall *et al.*, 1970) and similar structures have been recorded from Tertiary and Mesozoic strata. Calcareous bodies (Fig. 13.23) found in some deposits and classed as a fossil *Tetralithus* were regarded as rather enigmatic structures for many years but some have now been shown to form in the mucilage of the benthic stages of Prymnesiophyta (Green & Parke, 1975). Much more massive structures forming reefs and stromatolites existing right back to Precambrian times are known from geological studies (Wray, 1971). The dominant algae in the Palaeozoic reefs were en-

Fig. 12.8. Cross-section through a fossil stromatolite, probably of Upper Tertiary age. From the Etosha Pan, Southwest Africa. Photograph kindly supplied by Dr N. Wilczewski, from Martin & Wilczewski, 1972.

1 2 3 4 5 6 cm

crusting blue-green algae and red algae were rare; in the Mesozoic, Solenoporous red algae were important and throughout the Cenozoic, but especially in the Cretaceous, crustose red algae were abundant. Indeed the fossil record seems to indicate that algal activity in the production of massive deposits has been more active in the past than it is now. Fig. 12.8 shows a cross-section through an algal stromatolite of Tertiary age (*see* p. 186 for formation of modern stromatolites). Even in regions of the earth which are now subarctic, e.g. in Devonian strata in Canada, reefs are common (Jamieson, 1971).

The zones richest in calcium carbonate occur between 50° N and 50° S (*see* Fig. 12.7) and one might expect the greatest production of coccolithophorids here. However, the contribution from foraminifera has to be taken into account: it seems that in the centres of the major gyres the foraminiferal populations are low but increase considerably in the peripheral current areas. The centres of the gyres correspond with some rich carbonate sediments and therefore these might be expected to underlie regions of rich coccolithophorid production; Hasle (1959) found quite dense populations in the tropics. Other workers believe that the temperate zones are much richer and there is no doubt that coccolithophorids can live in very cold, even polar waters. In the tropics there is greater species diversity in the coccolithophorid flora, e.g. 35 species in the equatorial Pacific and only five in samples from subantarctic waters (of these four were rare and probably carried into the area). In subarctic and subantarctic sediments the coccolith content is down to less than 2% compared to 26% in the high productivity areas of the subtropics. In cold regions the carbonate compensation depth is shallower and hence more carbonate will be lost by solution here than in warm tropical waters.

In inshore tropical waters, calcium carbonate derived from siphonaceous genera of the Bryopsidophyceae can be deposited as aragonitic needles. In addition, abundant growth of phytoplankton in the shallow water over these inshore sediments has been shown to alter the carbonate balance to such an extent that aragonite crystals form in the water and precipitate, e.g. over the Bahama Banks. This is an indirect precipitating effect of algal growth: similar precipitation can be caused by purely physicochemical means and it is suggested that some 65–

75% of the calcium carbonate of the sediments around the Bahamas is produced in the latter manner, though other workers have suggested a much higher algal contribution (*see* p. 159). Stockman, Ginsburg & Shin (1967) estimate that *Penicillus* is the main producer of calcareous muds in the reef systems around Florida, and since this alga has a life span varying between 40 and 60 days and on average precipitates 0.52 g of aragonite per plant, which grow at a density of approximately 2 plants m^{-2}, the six to nine crops each year account for the massive deposition. Formation of colloidal calcium carbonate has been recorded in freshwaters and milky precipitates sometimes form in small lakes but nothing is known of the contribution to freshwater sediments.

A complicating factor when considering calcareous sedimentation is that in very deep water calcite begins to go into solution (Fig. 12.9) and aragonite, which is even more soluble, tends to disappear at depths around 2000–3000 m. This solubility is complicated by many factors, some physico-chemical and others biological, relating to the rate of supply. The depth where calcium carbonate content is reduced below 10% is often referred to as the compensation or critical depth: it is unfortunate that similar terminology is applied to the metabolism of algal populations, *see* p. 313.

In shallow marine waters, calcium carbonate is precipitated on and in the walls of numerous species of siphonaceous green algae (e.g. *Acetabularia, Dasycladus, Neomeris, Cymopolia, Halimeda*, etc.), on one brown alga (*Padina*), which often occurs as dense underwater swards in warm waters, and on and in coralline red algae (e.g. *Porolithon, Corallina, Lithothrix, Lithophyllum, Lithothamnium*, etc.). The calcium is formed as aragonite by the Chlorophyta, *Padina* and species of the Nemaliales and as calcite by other Rhodophyta. The remains of these algae are deposited in coastal waters, and in some areas fragments large enough to be identified occur on beaches, *Lithothamnium* fragments on European beaches and *Halimeda* fragments on tropical beaches. Underwater banks (ridges) are known especially from arctic seas, but also around Japan and even in the Bay of Naples; these sites need intensive study as do also the few sites where similar banks are reported in the tropics. They should not be confused with coral reefs which only develop in shallow water. Algal sediments mainly derived from *Halimeda* fragments are found in many coral atoll lagoons at intermediate depths between the shallow water coral debris sediment and the deeper finer sediment (often foraminiferal sand) (Stoddart, 1969). Sediments associated with coral reefs tend to be of biological origin with coralline algal remains contributing 7–51% and *Halimeda* from 1 to 43%, according to the compilation in Stoddart (1969). The green alga *Ostreobium* bores into the framework of recent 'cup reefs' in Bermuda and also into the reef-living gastropod *Dendropoma* which it infects very heavily, but it is also responsible for a secondary deposit of calcium carbonate in the form of radiating crystals around the boring filaments

(Schroeder, 1972). When the alga dies the space occupied by the coenocyte then also becomes filled with calcium carbonate. Algae can also bore into crystals of Iceland spar (calcite) and the algal filaments growing outwards become coated with fine carbonate crystals (micrite) (Kobluk, 1976).

A further source of carbonate in some regions is the calcareous epiphytes growing on sea grasses, e.g. Lund (1971) found that 7 g m^{-2} was formed by the calcareous epiflora on *Thallassia testudinum* which, when related to size of the *Thallassia* crop, gave an average contribution to the sediments of 40–180 g m^{-2}. Lund considered that this figure was on the low side and production from this source alone

Fig. 12.9. The solution of calcite balls relative to depth. The time exposure was three months. From Lisitzin, 1971*b*.

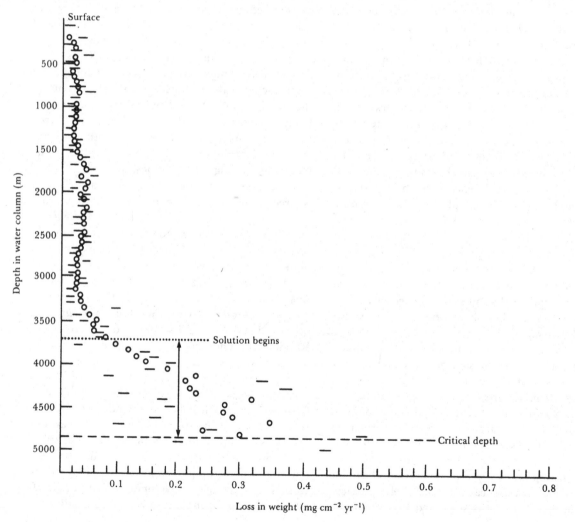

equals the average rate at which the local carbonate sediments accumulate.

In freshwaters the only large algae involved in the precipitation of calcium are the Charophyta, many of which are whitish-grey from a coating of crystalline calcium carbonate, in the form of calcite. Of wider distribution and probably of much greater significance in carbonate deposition in freshwaters are the blue-green algae, many of which aggregate crystals of calcium carbonate in the mucilage around the filaments. In genera such as *Rivularia* the process of calcite deposition can be followed as isolated crystals form in the mucilage and gradually more and more appear, until the mucilage between the threads becomes a hard mass of calcium carbonate and the colonies become converted into small 'stones', with live algal filaments only in the outer layer. Somewhat similar 'grains' of algal micrite were recorded on the surface of blue-green algal mats in lagoons along the Texas coast (Dalrymple, 1966): the grains contain 50% calcium carbonate which it was believed formed by precipitation from super-saturated seawater immediately above the algal mats. In some lakes, and quite commonly in streams, filamentous Cyanophyta form calcareous concretions on stones. Carbonate crystals are also deposited in the mucilage of *Chaetophora* colonies. These concretions can form a greyish-white sandy deposit along the lake shore and even extend out as a calcareous ooze into deeper water. Such ooze is reported to be the cause of the silting up of the section of Lake Constance known as the Untersee. Another site of calcite deposition by Cyanophyta is the region around hot springs where the blue-green algae are often dominant. A few freshwater diatoms, e.g. some *Nitzschia* species also precipitate calcium carbonate crystals on the outside of the valves and one desmid (*Oocardium*) forms tubes of calcium carbonate with the cells growing at the apices.

The deposition of lime within the mucilage and cell walls of algae is a specific property of the individual species and not a feature linked to the environment since species depositing lime and others completely free of such deposits are found growing side by side. The mass of calcareous material and algal filaments in some rivers and lakes forms the site also for the growth of Chironomid larvae which themselves live in tubes immersed between the algal mucilage-carbonate matrix. The 'algal pinnacles'

(small eroded hills) rising from the floor of the dry Lake Searles, California are thought to have been formed by precipitation of carbonate on algae around ancient submarine springs (Scholl, 1960; Reeves, 1968).

Beach rock

Sand grains compacted to form hard, usually flat, aggregates along sea shores is termed beach rock. The bonding of the sand grains is due to calcareous material (calcite) some of which is either directly or indirectly formed by algal growths. True beach rock only forms in tropical regions where water temperatures are above 68 °F for at least half the year (Russell & McIntire, 1965): formation in some areas begins along the water table on the beach and initially occurs in the zone of freshwater seepage and it is necessary for this water to be warm. Along the Gulf of Aqaba beach rock is actively forming and pebbles are firmly embedded in it to give a 'conglomerate' rock type. Here there is a very dense growth of coralline algae, Cyanophyta and diatoms which are matted together and fuse the sand grains and stones into pavement-like pieces of rock (Round, personal observation). There is certainly no trace of freshwater seepage in this area, which has no rainfall for most of the year. Later, other marine organisms also colonise the rock: not unexpectedly the first report of beach rock was made by Darwin during his stay in South America.

Organic remains

The organic remains of algae are preserved in soft deposits and also in lithified strata but until recently they have received little attention. In freshwaters the pigment degradation products have been extracted from lake sediments and Fig. 12.10 shows the fluctuation in these over a 14 000 yr period in an Indiana lake. The absorption spectra of diatoms and of sediment and surface sediment from Esthwaite Water show a very similar pattern (Fig. 12.11) to the spectra from diatoms (Gorham, 1960). The concentration of diatom pigments in the sediments increases with the general fertility of the lake, a feature noted also by Belcher & Fogg (1964), who suggest that the ratio of chlorophyll:epiphasic carotenoids is an inverse index to lake fertility and is in accord with known facts on lake history. On this basis Fig. 12.12 suggests that Ennerdale has become progressively

Fig. 12.10. *a*, *b*. Sedimentary chlorophyll decomposition products in a marl lake (Pretty Lake, Indiana), *a*, plotted as g dry wt and *b*, g organic wt; *c*, *d*. Carotenoid sedimentary production, *c*, plotted as g dry wt and *d* g organic wt. Over *b* the calculated production rate of carbon is plotted. From Wetzel, 1970.

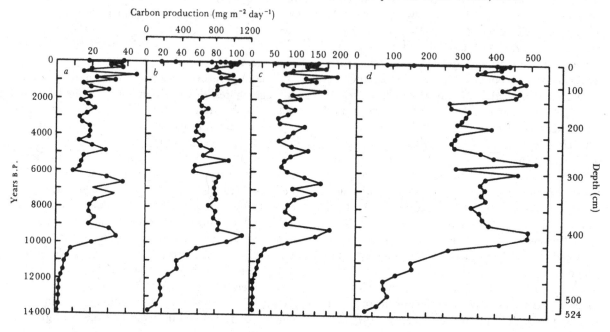

Fig. 12.11. Absorption spectra of acetone extracts of living and dead diatoms, of sedimenting debris and surface mud of Esthwaite Water. From Gorham, 1960.

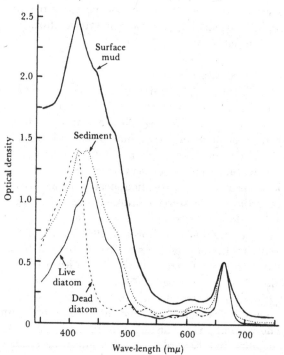

more oligotrophic, Esthwaite Water has remained at about the same trophic level and Lake Windermere has fluctuated much more than the other two.

Certain algal groups have distinctive pigments, e.g., myxoxanthophyll in blue-green algae, and even a particular carotenoid in the blue-green genus *Oscillatoria* (oscillaxanthin) and the occurrence of such pigments in sediments can be used to date precise algal groups (Griffiths, 1978).

The distribution of *n*-alkanes which are derived from lipids of plant cells is also being used to study the distribution of algal remains in sediments; *n*-heptadecane is the major hydrocarbon remnant of many green and blue-green algae and can be detected in sediments. Unfortunately many of these compounds and also of course the pigments, may be derived from the inwashed material and separation of the algal and angiosperm components is difficult. *n*-Alkanes, as well as isoprenoids, sterones, and porphyrins, have been detected in sediments as far back as Precambrian (*see* table in Schopf, 1971).

Summary

There are many parts of algae and also chemical compounds derived from algae which, when

Fig. 12.12. Variation of the chlorophyll/epiphasic carotenoid ratio in sediment cores from Esthwaite Water, Lake Windermere and Ennerdale Water. From Belcher & Fogg, 1964.

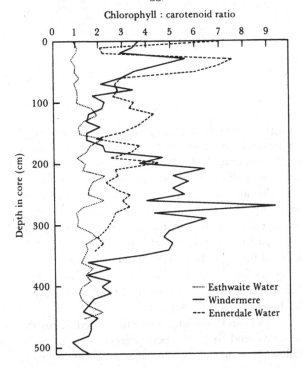

deposited on to aquatic sediments, are preserved as a fossil record. Siliceous algae – diatoms, silico-flagellates and chrysophycean cysts – are perhaps the most resistant remains, though the actual total quantities reaching the sediment are small due to dissolution of the delicate forms and comminution of others as they pass through animals. Some aspects of the rates of dissolution and accumulation are discussed. Diatoms are common only in certain sedimentary zones of the oceans but are generally abundant in lake sediments.

Calcareous deposits are formed from coccoliths in the open ocean and from fragments of calcified macrophytic algae in the inshore waters. The richest carbonate sediment zones occur in the tropics and subtropics. Solution of calcite is faster than that of silicon and this complicates the pattern of distribution. In freshwaters the Charophyta are the major algal precipitators of calcium carbonate though in some areas blue-green algae also play a large part.

13 Palaeoecology

Considerable amounts of data have now accumulated on the history of lake and ocean basins. Not all algal species contribute to the story, since only those with resistant skeletons or which produce chemically recognisable organic remains can be traced backwards through time in the sediments. This time span is very short, amounting to only some $10-15 \times 10^3$ yr for most lakes which have been formed by glacial action and from such habitats many of the data are derived. Deep tropical lakes have much longer histories than those of temperate regions but have been little explored whilst the oceans have been in existence for approximately 10^9 years but as yet there is only slight information on algal floras back beyond 20×10^6 yr, i.e. to approximately the middle of the Miocene. There seems little doubt, however, that algae existed even in the Precambrian in the form of Cyanophyta allied to the Chlorococcales (Schopf & Barghoorn, 1967: see Schopf, 1971, for a detailed review). From the Oligocene period the average temperature of the middle latitudes of the earth slowly declined with the most rapid fall in the Pliocene and then, in the Quaternary era which spans just over a million years, there was a series of temperature fluctuations with average highs around 10 °C and average lows below 0 °C. Climatologists think that there have been at least ten of these cycles, each with time scales of about 10^5 years. On this reckoning we are only at the beginning of a warm interglacial period though there is very good evidence that average temperatures have actually fallen slightly during the last $2-3 \times 10^3$ yr. Most of the data on algal remains in sediments are derived from the last period of warming though a few deposits from earlier interglacial periods have been studied and algae found within them. For most temperate regions the rapid period of warming started about 8×10^3 yr ago and was relatively rapid, peaking about 6×10^3

yr ago. Subsequent to this the sea level rose and changes of about 60 m have occurred with peak levels occurring $3-4 \times 10^3$ yr ago, since when there has been a slight decrease. Changes in sea level and also rises and falls in the land surface have resulted in many coastal regions being either exposed to seawater or to freshwater and the algae in the sediments reflect this, e.g. at Lough Ine in southwest Ireland there are thick freshwater sediments underlying marine clay and in the coastal lands around the Baltic Sea the freshwater sediments of lakes often lie on brackish–marine deposits.

In both freshwater and marine sediments the diatoms tend to be the best preserved algal fossils, followed in the sea by the coccoliths derived from species of the Prymnesiales. Smaller numbers of discoasters (see Fig. 12.23) derived from Prymnesiphyta and cysts (hystrichospheres, Fig. 13.20) of Dinophyta also occur. In freshwaters, organic walls of some Chlorophyta such as *Pediastrum*, *Botryococcus*, occasionally desmids, dinoflagellates and the akinetes of Cyanophyta are found. These walls are composed partly of cellulose and partly of other organic molecules, e.g. sporopollenin. In lake sediments a few cells in the surface layers are still viable and even down to depths of 35 cm, viable cells have been found but usually they are scarce below 5 cm (Stockner & Lund, 1970). These are live cells, sedimented out and existing under conditions where growth is not possible and they are vastly outnumbered by dead cells. Comparison of the algal 'communities' (sometimes termed 'thanatocoenoses' or dead assemblages) in sediments with those of living 'communities' (biocoenoses) generally reveals extreme depletion of species in the former, especially in marine basins. The most complete preservation of whole communities is that of diatoms in Quaternary lake sediments where a very high proportion, probably 90% or more, of the species which have grown in the lake are preserved, together with a few species washed in from the catchment area: obviously such an estimate can only be based on comparison of the present day flora with that in the surface sediments and one cannot be certain that similar proportions apply to the whole post and late glacial period. In oceanic sediments the proportion of species preserved is very much less (30–40 % at the most).* This is due to the fact that very many marine planktonic species are too delicate

* Estimated from my studies in the Gulf of California. Similar estimates (30 %) are given by Kozlova (1964) for littoral marine sediments but up to 80 % are mentioned for deep oceanic sites. Such a high figure I find hard to believe.

to survive the journey to the ocean floor in an easily recognisable form. Few marine cores have been taken in inshore waters and so the record of epipelic, epilithic and epiphytic species is almost unknown; in freshwaters, species of these communities can be abundant in cores.

Although there is, in general, an orderly sequence of new deposits lain upon old there are many subtleties of rates of accumulation, variation in the amount of inwashed material from the land (terrigenous), slipping of layers, etc., which can confuse the palaeoecologists' picture. Some of these show up quite clearly in seismic surveys of marine sediments but such techniques have rarely been used in lake basins. One such was made on Lac Leman

Fig. 13.1. Seismic profiles of sediments in Lac Leman (Lake Geneva) showing (*a*) a region of relatively undisturbed layered sediment and (*b*) one in which the layering is deformed. From Serruya *et al.*, 1966.

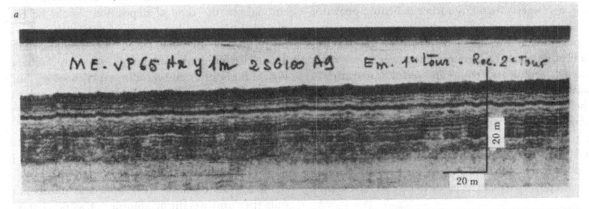

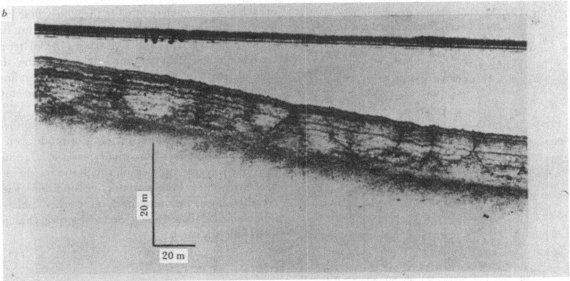

(L. Geneva) and showed details of variation in depth of the bands and in parts of the basin very complex patterns of sedimentation (Fig. 13.1; Serruya, Leenhardt & Lomrard (1966)).

Fossil algal assemblages are deposited on the bottom of every lake and ocean basin though not all sites are ideal for study, e.g. disturbance by turbidity currents can completely mix the material and destroy the sequential pattern of deposition. Reference to 'fossil' material in palaeolimnology and palaeo-oceanography simply implies dead organisms whatever their age: organisms deposited on to the sediments from last season's growth are merely the most recent piece of the fossil record! In order to study these assemblages it is necessary to take a core from the bottom sediments, preferably with the minimum of disturbance of recent deposits and in a region where an unbroken sequence exists. Usually it is necessary to employ rather delicate apparatus to sample the most recent sediments, which are in a liquid state and would be disturbed by the penetration of large coring devices. The lower, more compacted, sediments require powerful apparatus to push the core tube down into the sediment and in the most complex situations in the deep ocean it is necessary to use rigs of oil prospecting proportions, e.g. as on the *Glomar Challenger*. In some inland sites, old lake basins can be recognised from the topography of the land and are the easiest to sample since a peat borer can be used. In lakes it is necessary

to use piston cores in which the core tube is driven into the sediment by compressed air whilst the piston remains at the sediment surface thus preventing the sediment from rising in the core tube. The core is squeezed out of the tube by moving the core tube upwards over the piston, and it is then cut into convenient lengths from which subsamples can be removed for study. Obviously, precautions have to be taken to prevent contamination. Cores of this nature obtained from lakes are often complete back to the late glacial clay and hence a full record can be obtained. Cores obtained at sea are usually less complete and may have gaps in them owing to the drilling technique which has to be used and the variation in sediment type, compaction, etc.

Interpretation of the data from cores is not easy, both qualitative and quantitative studies are required and ideally the rates of sedimentation throughout the time of deposition should be estimated. Every possible means of dating the various layers must be used (e.g. ^{14}C, oxygen isotopes, radioactive iodine, palaeomagnetic surveys, archaeological finds, etc.) and a sediment chronology built up. Over the last decade palaeomagnetic reversal data has enabled oceanic cores to be dated with great accuracy and the subject will now advance rapidly: results can be correlated with the last abundant appearance of species which itself is a useful datum point (Burckle, Clarke & Shackleton, 1978). Most of the accumulated data are graphed as numbers of cells (per unit weight or volume) or as percentages of each taxon at each depth. On the whole these give a good indication of changes in the flora during the period of deposition, and Battarbee (1973a) terms the first method (cells per unit weight or volume) the absolute method. Battarbee added a further refinement by applying a correction factor which took into account the volume of each taxon (Fig. 13.2). Obviously this is important since a drop in numbers may be of no significance in terms of production if there is a change over from small species to large and conversely large number of small species do not necessarily imply highly productive conditions; such considerations are most important when palaeolimnological data are being used to interpret changes in trophic status.

Fig. 13.2. Absolute diatom quantities from a sediment core, Lough Neagh, Northern Ireland. *a*, total number of valves, in no. × 10⁶ (cc fresh sediment)$^{-1}$. *b*, total cell volume, in mm³ (cc fresh sediment)$^{-1}$. *c*, total cell volume, in mm³ cm^{-2} yr^{-1}. From Battarbee, 1973a.

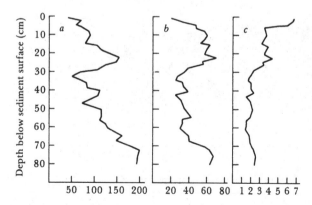

Coccoliths

The earliest records of coccoliths are from the Lower Lias of the Lower Jurassic which is approximately 190–195 × 10⁶ yr old; these records are from terrestrial strata and in the oceanic sediments the record goes back to the mid-Cretaceous.* The

coccolith-bearing flagellates (*see* Fig. 13.25*a*) of the present-day oceans are mainly a post-Tertiary development with just a few survivors from earlier

* For a detailed distribution of all algal groups during the geological time scale see *The Fossil Record* (The Geological Society, London, 1967).

Fig. 13.3. Electron transmission pictures of *Coccolithus huxleyi* (left column) and *Gephyrocapsa oceanica* (right column) from oceanic cores showing graded solution until the two genera become almost indistinguishable. From McIntyre & McIntyre, 1971.

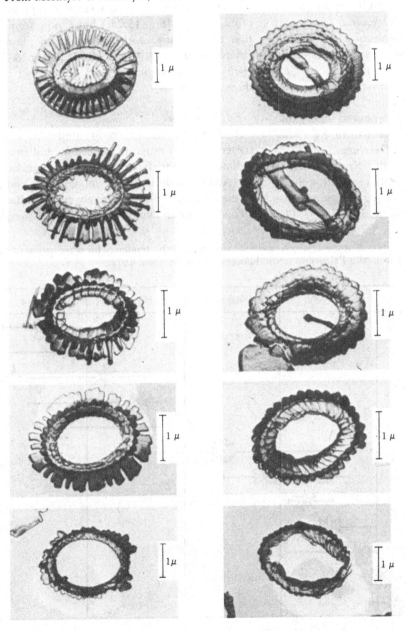

epochs. A striking feature of some coccoliths is their virtual constancy since the Cretaceous, e.g. *Braarudosphaera bigelowi* which has perhaps the largest geological range of any form and yet is apparently identical in form whether living or Cretaceous material is examined. However the common *Coccolithus pelagicus* of present-day plankton is not identical when viewed in polarised light with the numerous *C. pelagicus* types recorded from fossil deposits. Solution of coccoliths in deep sea deposits adds problems of identification (*see* Fig. 13.3) and interpretation of data. An interesting genus is *Deflandrius* which was abundant in the Cretaceous only but its coccolith structure is reminiscent of the organic scales of the modern *Chrysochromulina pringsheimii* (Fig. 13.25*b*). Any connection between *Deflandrius* and *Chrysochromulina* is purely hypothetical but it is now well established that coccoliths are deposited on organic scales of typical Prymnesciophycean form, e.g. in the motile stage (*Crystallolithus*) of *Coccolithus*.

Analysis of cores from the North Atlantic (north of 40° N) has shown that the coccolithophorid flora has migrated polewards during the period since the last glaciation (McIntyre & Bé, 1967). The majority of coccolithophorid species are tropical and subtropical and a few are stenothermal, e.g. *Umbellosphaera irregularis* (21–28 °C) and *Coccolithus pelagicus* (7–14 °C) and can therefore be a great help in determining past conditions (McIntyre & Bé, 1967). Gartner (1977) has divided the 1.65×10^6 yr of the Pleistocene into seven Coccolithophorid zones named after the dominant species.

Discoasters

For a long time these were an enigmatic group of microfossils (Fig. 13.25*c*) but it now seems likely that they belong to the coccoid and filamentous Prymnesiophyceae (Parke, 1971; Green & Parke, 1975). The geological record of some of these is shown in Fig. 13.4.

Diatoms

Diatoms are the most abundant and best preserved algal remains in oceanic and freshwater sediments, although preservation is not uniform, e.g. the Atlantic sediments are unfavourable compared to the Pacific, and the relatively enclosed Black Sea seems most unfavourable. Similarly across the

Fig. 13.4. The geological ranges of some *Discoaster* species showing also the tendency for a reduction of the number and width of rays from the Paleocene to the Pliocene. From Martini, 1971.

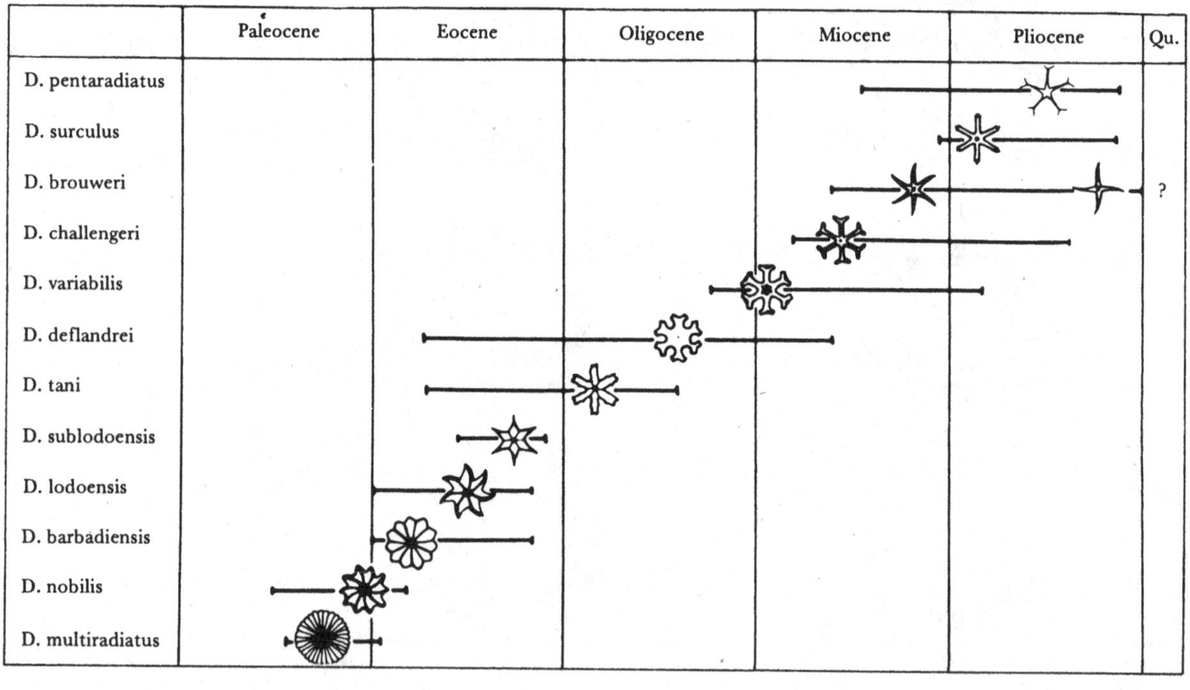

Pacific diatom preservation is better in the eastern than in the western sector (Mikkelsen, 1978). Fig. 13.5 shows the distribution of diatoms by numbers in the Pacific and Fig. 13.6 the zonation of the thanotocoenoses with the dominant species in each region. From Fig. 13.5 it is clear that each group is exclusive to a sedimentary region and indirectly to a water mass lying above. These water masses are expressions of the current systems and indirectly of the climatic zones. If cores are taken from these zones, the indicator diatoms within them will only change if the water mass above has been displaced and hence historical changes in water masses and of climate can be detected by such an analysis. However this is not quite as straightforward as it sounds, since recently lateral drift of species has been demonstrated

Fig. 13.5. Diatom numbers in the surface bottom sediment layer (million valves per g sediment). 1, devoid of diatoms; 2, 0.04–5; 3, 5–26; 4, 25–50; 5, 50–100; 6, > 100; 7, *Ethmodiscus* oozes; 8, sediments of Tertiary age; 9, stations investigated. From Jousé, Kozlova & Muhina, 1971.

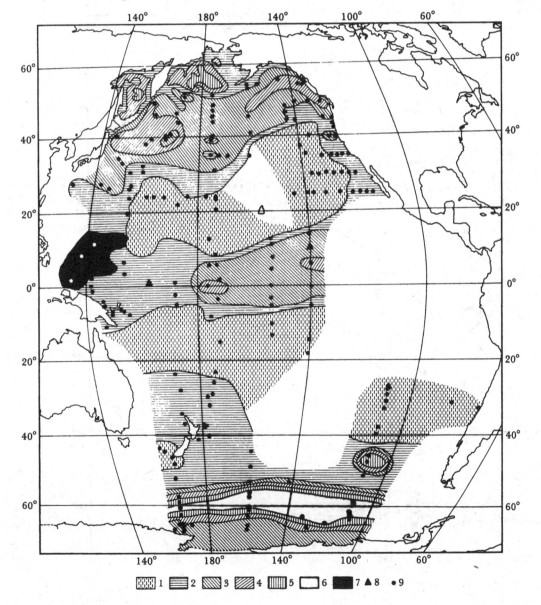

in the Argentine Basin where up to 500 000 antarctic-derived diatoms per gram of sediment have been found (Burckle & Stanton, 1975). These have been transported northwards in the antarctic bottom water (*see* p. 10) and provide a subtle means of detecting pathways of atlantic bottom water inflowing into the Argentine Basin (Fig. 13.7). If one compares the diatom distributions with the composite distribution maps of oceanic organisms including *Ceratium* spp. for each zone (Fig. 8.9) there are clear similarities of zonation. In one or two cores the record extends back to the Pliocene. All cores examined show great fluctuation in total number of diatom valves per gram (Fig. 13.8) and during the

Fig. 13.6. Diatom complexes in the surface sediment layer of the Pacific Ocean. I, arctoboreal; II, boreal; III, subtropical; IV, tropical; V, equatorial; VI, subantarctic; VII, antarctic; VIII, complexes of transitional type. Single finding of subtropical species, 1; tropical species, 2; subantarctic species, 3. 4, stations investigated. From Jousé, Kozlova & Muhina, 1971.

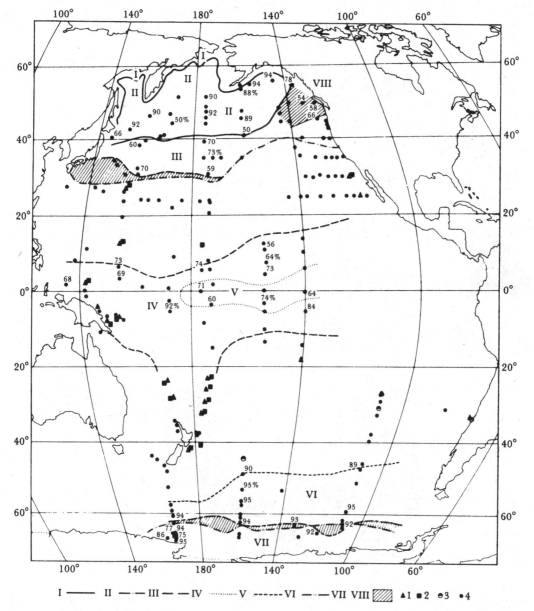

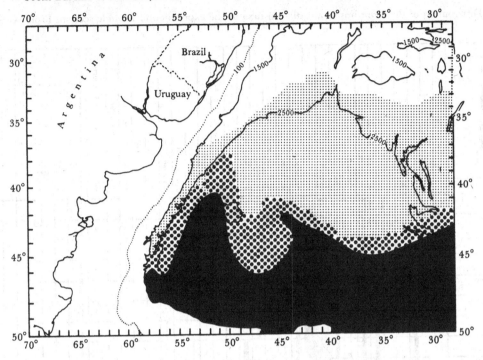

Fig. 13.7. The distribution of *Nitzschia kerguelensis* in surface sediment of the Argentine Basin. Fine dots, 0–10000 valve g⁻¹; heavy dots, 20000–200000 valve g⁻¹; black area, values in excess of 700000 valve g⁻¹. From Burckle & Stanton, 1975.

Fig. 13.8. Vertical distribution of some characteristic diatoms, at station V20–119, Pacific Ocean 47° 57′ N, 168° 47′ E, depth 2739 m. From Jousé, 1971.

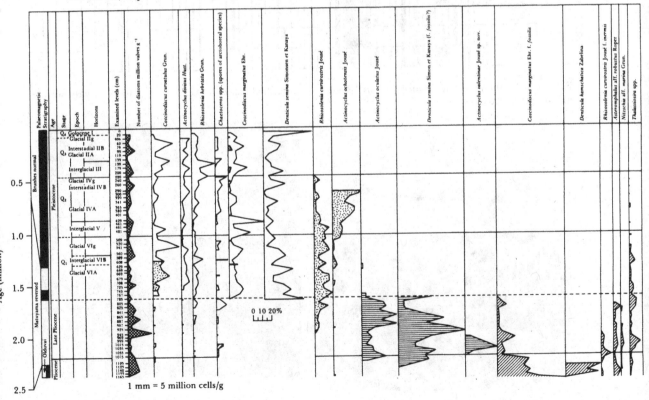

Fig. 13.9. Distribution of the more important diatom taxa in the experimental Mohole section. From Kanaya, 1971.

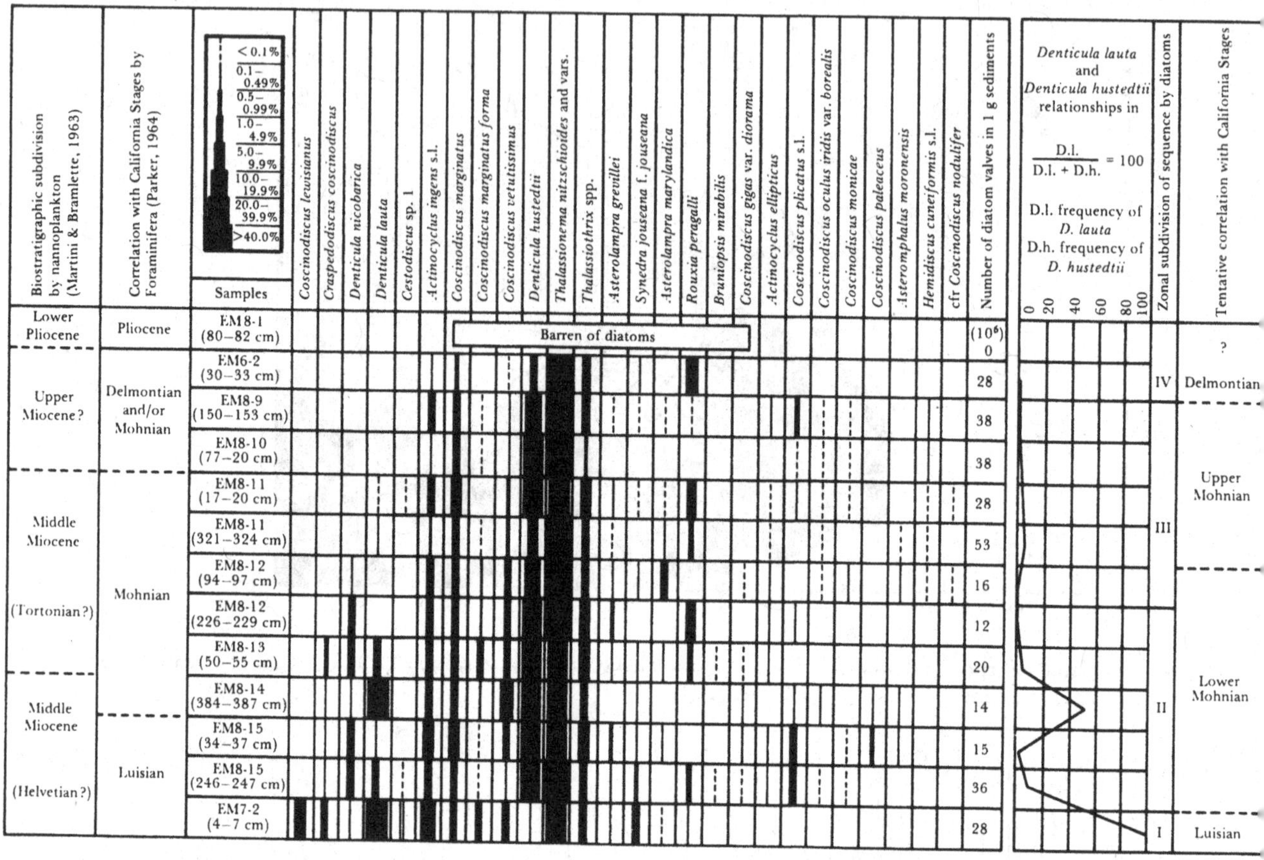

Legend (percentage frequency):
- < 0.1 %
- 0.1–0.49 %
- 0.5–0.99 %
- 1.0–4.9 %
- 5.0–9.9 %
- 10.0–19.9 %
- 20.0–39.9 %
- > 40.0 %

Taxa columns (left to right):
Coscinodiscus lewisianus, *Craspedodiscus coscinodiscus*, *Denticula nicobarica*, *Denticula lauta*, *Cestodiscus* sp. 1, *Actinocyclus ingens* s.l., *Coscinodiscus marginatus*, *Coscinodiscus marginatus forma*, *Coscinodiscus vetustissimus*, *Denticula hustedtii*, *Thalassionema nitzschioides* and vars., *Thalassiothrix* spp., *Asterolampra grevillei*, *Synedra jouseana* f. *jouseana*, *Asterolampra marylandica*, *Rouxia peragalli*, *Bruniopsis mirabilis*, *Coscinodiscus gigas* var. *diorama*, *Actinocyclus ellipticus*, *Coscinodiscus plicatus* s.l., *Coscinodiscus oculus iridis* var. *borealis*, *Coscinodiscus monicae*, *Coscinodiscus paleaceus*, *Asteromphalus moronensis*, *Hemidiscus cuneiformis* s.l., cfr *Coscinodiscus nodulifer*

Biostratigraphic subdivision by nannoplankton (Martini & Bramlette, 1963)	Correlation with California Stages by Foraminifera (Parker, 1964)	Samples	Number of diatom valves in 1 g sediments	Zonal subdivision of sequence by diatoms	Tentative correlation with California Stages
Lower Pliocene	Pliocene	EM8-1 (80–82 cm)	Barren of diatoms — (10⁶) 0		?
Upper Miocene?	Delmontian and/or Mohnian	EM6-2 (30–33 cm)	28	IV	Delmontian
		EM8-9 (150–153 cm)	38		
		EM8-10 (77–20 cm)	38		Upper Mohnian
Middle Miocene	Mohnian	EM8-11 (17–20 cm)	28		
		EM8-11 (321–324 cm)	53	III	
(Tortonian?)		EM8-12 (94–97 cm)	16		
		EM8-12 (226–229 cm)	12		
		EM8-13 (50–55 cm)	20		Lower Mohnian
Middle Miocene	Luisian	EM8-14 (384–387 cm)	14	II	
		EM8-15 (34–37 cm)	15		
(Helvetian?)		EM8-15 (246–247 cm)	36		
		EM7-2 (4–7 cm)	28	I	Luisian

Denticula lauta and *Denticula hustedtii* relationships in

$$\frac{D.l.}{D.l. + D.h.} = 100$$

D.l. frequency of *D. lauta*
D.h. frequency of *D. hustedtii*

Fig. 13.10. Scanning electron micrographs (Round, unpublished) of, *a*, a fossil and *b*, a modern valve of the diatom genus *Stephanopyxis*.

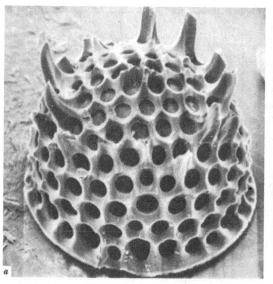
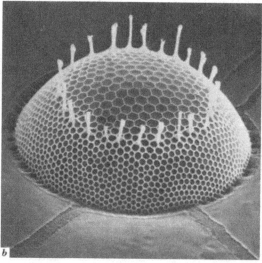

Pleistocene high numbers correspond with interglacial and low with glacial periods in cores taken from such northern stations. This figure shows at least five diatom epochs marked by definite assemblages. The Pliocene species are all characteristic of warm waters and those of the Pleistocene of colder water.

Tertiary deposits containing diatoms have been dredged up from the ocean floor in some localities where these strata are exposed. The dredged samples are valuable for diatom studies but stratigraphic data are best obtained from cores, e.g. the experimental Mohole drilling off Guadalupe where the sequence given in Fig. 13.9 was found. Such data can then be compared with samples taken from marine deposits which have been raised during geological time and occur now at terrestrial sites.

Groups of diatom species can be distinguished from various geological formations; these have been especially studied in California where characteristic species have been found in the Cretaceous Moreno shales (the oldest well authenticated material containing diatoms) in which some 136 of a total of 155 species seem to be restricted to this stratum (Wornardt, 1967). Diatom-containing material of similar age (Maastrichian) has also been studied from Russian deposits (Strelnikova, 1975) but these have only 18 species in common with the Californian

Fig. 13.11. Stratigraphic ranges of *Raphoneis* species. From Andrews, 1975.

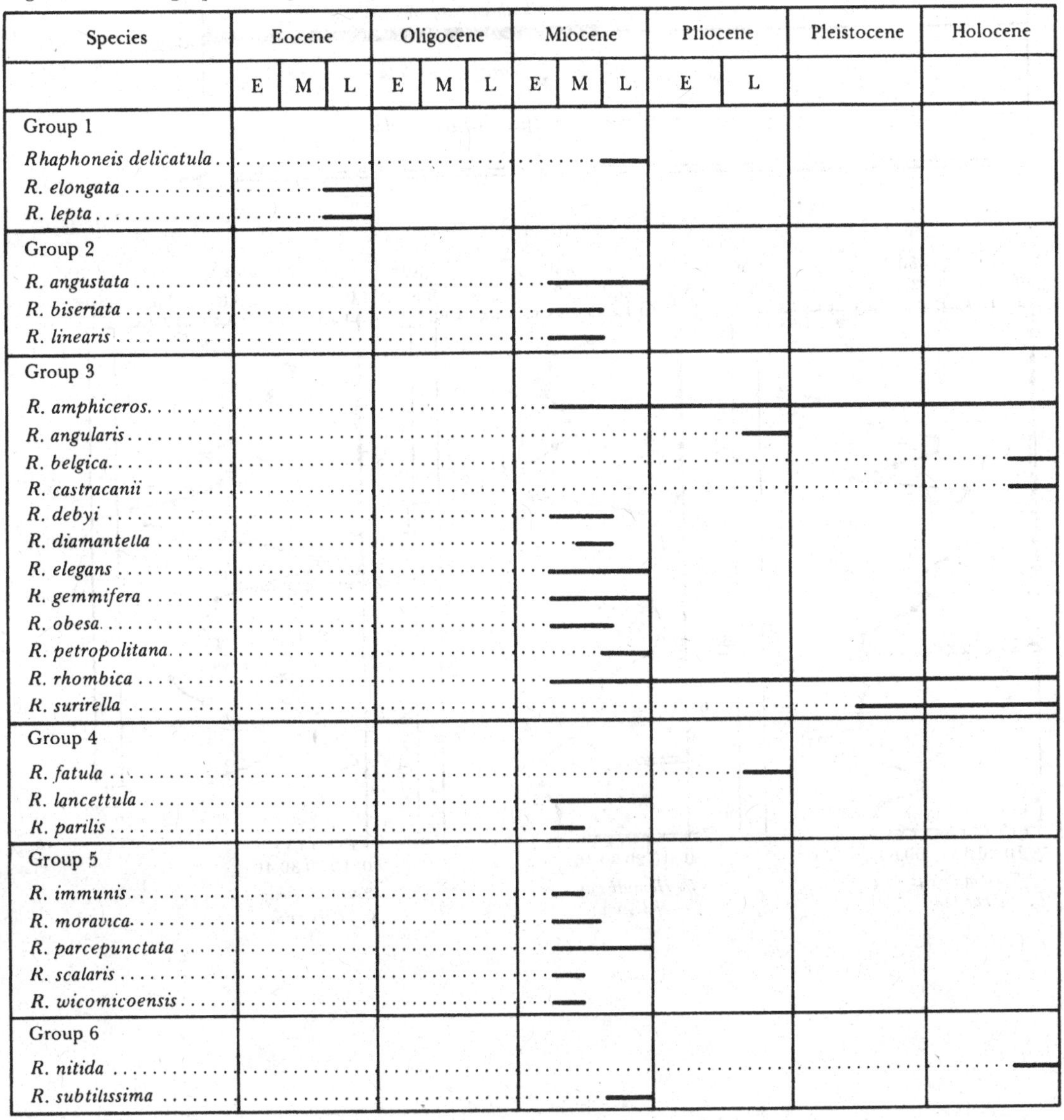

Species	Eocene			Oligocene			Miocene			Pliocene		Pleistocene	Holocene
	E	M	L	E	M	L	E	M	L	E	L		
Group 1													
Rhaphoneis delicatula													
R. elongata													
R. lepta													
Group 2													
R. angustata													
R. biseriata													
R. linearis													
Group 3													
R. amphiceros													
R. angularis													
R. belgica													
R. castracanii													
R. debyi													
R. diamantella													
R. elegans													
R. gemmifera													
R. obesa													
R. petropolitana													
R. rhombica													
R. surirella													
Group 4													
R. fatula													
R. lancettula													
R. parilis													
Group 5													
R. immunis													
R. moravica													
R. parcepunctata													
R. scalaris													
R. wicomicoensis													
Group 6													
R. nitida													
R. subtilissima													

Fig. 13.12. Two examples of the use of last abundant appearance data in dating cores. *a*, diatom datum levels for the equatorial Pacific. From Burkle, 1977. *b*, last abundant appearance of *Hemidiscus karstenii* in cores from the sub-antarctic (dotted line) and the oxygen isotope curve (continuous line). The hatched areas indicate the last abundant appearance. From Burkle, Clarke & Shackleton, 1978.

a

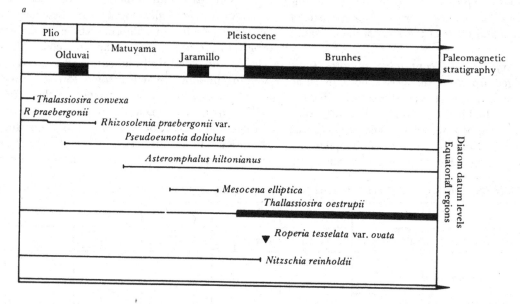

b

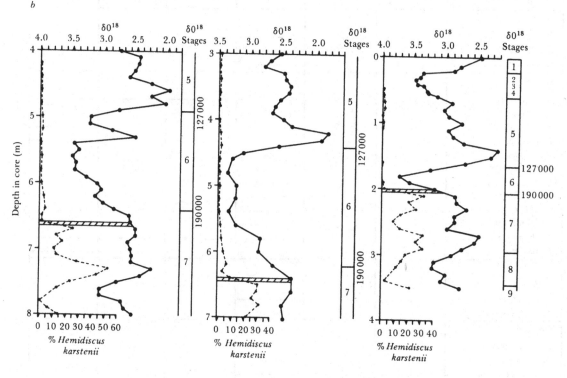

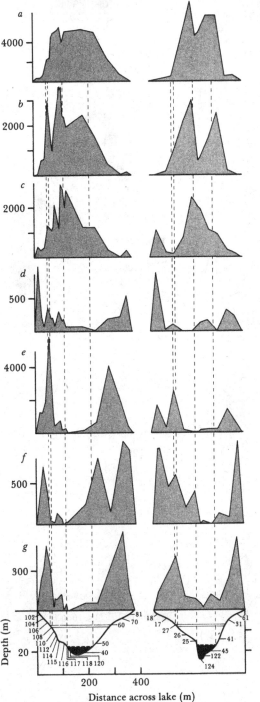

Fig. 13.13. The distribution of some diatoms in the surface sediments along two transects across Valkiajärvi. *a, Cyclotella kuetzingiana; b, Melosira distans* var. *alpigena; c, Tabellaria flocculosa; d, Anomoeoneis serians* var. *brachysira; e, Navicula; f, Frustulia rhomboides; g, Pinnularia interrupta.* From Meriläinen, 1971.

material (45 genera and 130 species in the data used for comparison) and this, as pointed out by Strelnikova, is only to be expected since the Russian sites belong to a Boreal paleobiogeographical region and the Californian to an Indo-Pacific paleobiogeographical region. According to Wornardt (1969) diatoms are unusual amongst fossilised groups in that so many living species extend a long way into the fossil record, e.g. about 15 % can be traced into the Eocene and even 6 % into the Upper Cretaceous. However it seems likely that in the more ancient sediments only the heavily silicified diatoms have survived (*see* Fig. 13.10 which is of a fossil and a modern valve of *Stephanopyxis*): an alternative explanation which is difficult to check is that the diatoms have gradually become more refined and less silicified. Few, if any, diatoms live in both marine and freshwater habitats and therefore they are extremely good indicators of salinity conditions under which the strata were deposited. The earliest diatoms are generally regarded as being marine and according to Lohman & Andrews (1968) the late Eocene is the generally accepted period when diatoms invaded freshwater, at least in the United States.

In an interesting study of a single genus, Simonsen & Kanaya (1961) found *Denticula lauta* in the middle Miocene, *D. hustedtii* more abundant in the upper Miocene and a mixture of these species plus *D. kamtschatica* in the Pliocene of California. There is also an apparent characteristic 'burst' of *Raphoneis* species (Fig. 13.11) in the mid–late Miocene period (Andrews, 1975). The word 'apparent' is used since, as Andrews points out, more Miocene sediments have been studied than any other and this may have led to a bias in the data. Clearly it is difficult to assess the geological conditions under which these fossil species lived but it is much easier to suggest the type of habitat and possibly community within the habitat – oceanic, neritic, benthic, epiphytic, etc. – by comparison with present day assemblages.

A few interesting studies have attempted to relate diatom cell size to environmental factors and then measure the cell size changes down cores. Burckle & McLaughlin (1977) have shown that the diameter of *Coscinodiscus nodulifer* valves is greater in sediments near the equator and in a core at latitude 3° 39′ N of the equator the size range of the *Coscinodiscus* varied at different depths, perhaps indi-

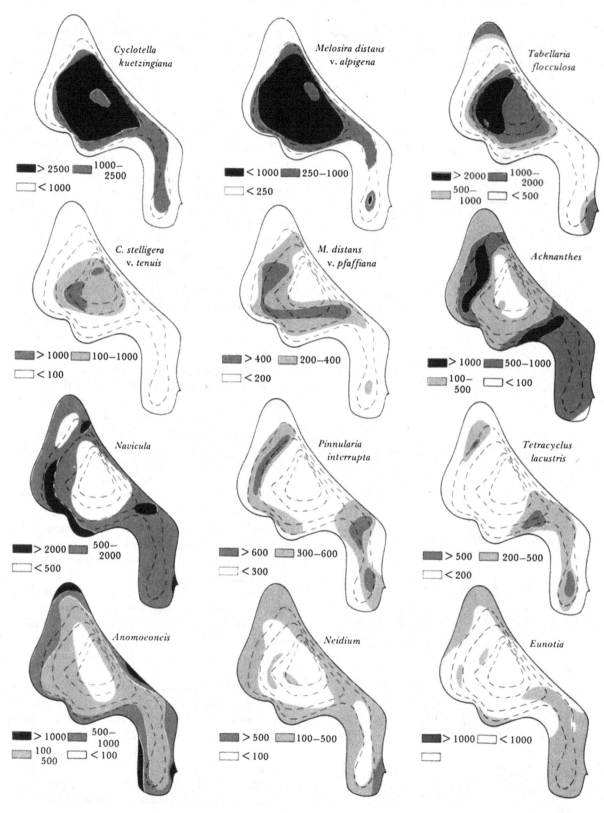

Fig. 13.14. The horizontal distribution of frustules of various diatom species in Valkiajärvi. From Meriläinen, 1971.

cative of fluctuating nutrient changes. Two further examples of the use of diatoms, in the form of the last abundant appearance datum, are given in Fig. 13.12 where it is particularly noteworthy how the *Hemidiscus* curve falls off around 190000 yr ago.

It is usually assumed that the deposition of diatoms onto the sediments is relatively uniform over a lake bottom except where the sediments are disturbed by waves along the margins and in obvious regions of disturbance by inflows or outflows. In large lakes one has to be aware of re-working of layers, slumping, etc. but in general cores taken at any reasonable site should contain a more or less undisturbed sequence (but *see* p. 499) though this always has to be very carefully checked usually by analysis of the pollen grain assemblages since these have received most attention and the sequences have been well established. However studies of some small Finnish meromictic lakes (Meriläinen, 1971) in sheltered basins and of Lake Sallie in Minnesota (Bradbury & Winter, 1976) revealed considerable lateral variation in diatom composition of the sediments. This is probably due to the rather precise and standardised water movements in these lakes, resulting in considerable 'sorting' of the diatoms and deposition in definite patterns in contrast to the more random mixing and deposition which occurs over many lakes. Figs. 13.13 and 13.14 show that some species are more abundant on the peripheral sediments and others in the central ones whilst adjacent to the shore line all species are relatively sparse. Species of *Cyclotella* and *Melosira* are planktonic and hence one would expect a greater concentration on the central sediments whilst the pennate genera *Anomoeneis*, *Navicula*, *Frustulia* and *Pinnularia* are benthic and therefore more likely to grow and be deposited in the shallower waters. The abundance of *Tabellaria flocculosa* which is basically an attached species in the central water is difficult to explain though its presence may be accounted for by the fact that when removed from its epiphytic site it can remain as a member of the pseudoplankton for some time. Some species are notably sparse in the monimolimnion and here the author reported corrosion of the valves, a feature which may have been related to the high iron concentration in this lake; corrosion is not however a constant feature of frustules in the monimolimnion of meromictic lakes (*see* p. 26). The whole problem of solution of diatom frustules requires further study. The author (Round, 1957*b*) suggested that clay deposits accelerate diatom solution and this has recently been reproposed to

Fig. 13.15. The distribution of *Thalassiosira gravida*, *a*, and *Thalassiothrix longissima*, *b*, in the surface sediments of the Sea of Okhotsk. Redrawn from Jousé, 1957.

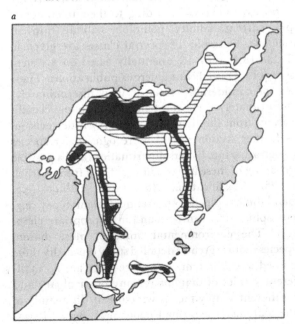

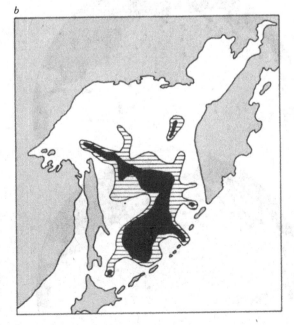

explain the distribution in some marine sediments. Striking patterns of diatom distribution in surface sediments can also be found in marine basins and they can to some extent be related to the living assemblages in the waters above, e.g. in Fig. 13.15 from the Sea of Okhotsk (Jousé, 1957). Over a wider area the study of samples from the surface layer of cores taken over the North Pacific shows that definite northern and southern sedimentation zones occur (Fig. 13.16) related to the pattern of production in the uppermost water mass (Kanaya & Koizumi, 1966). Duthie & Sreenivasa (1971) comment on the paucity of diatoms in some cores from the Laurentian Lakes: in Lake Ontario, four out of five cores had diatoms only in the upper metre (only the upper 3 cm in one core). Yet the fifth had diatoms preserved down to 2.77 m. Both this and a similar core from Lake Huron (Stoermer & Yang, 1969) were from deep

Fig. 13.16. The distribution of *Denticula seminae*, *a*, and *Coscinodiscus nodulifer*, *b*, in surface sediment samples from the north Pacific Ocean. Almost barren cores (black circle); cores in which species are absent (black square); cores in which fewer than 12 cells were present in a count of 200 samples, open triangles; cores in which more than 12 cells were present, filled triangles. Redrawn from Kanaya & Koizumi, 1966.

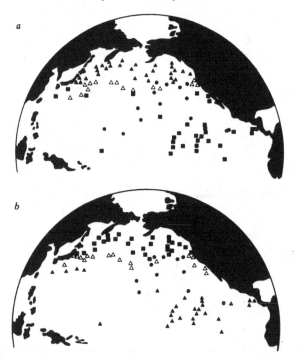

sites, suggesting either transport to such sites and little deposition elsewhere or solution of frustules in the shallower lying sediments. The major components of the diatom assemblages in these lakes are planktonic, as might be expected, though the composition is not quite the same as that in the present-day plankton; this may be a sorting effect or more likely the fact that there have been subtle changes in abundance of species in the post-glacial communities. One should not expect the assemblages, other than the most superficial, to have similar percentages of species. Duthie & Sreenivasa estimated the number of eutrophic–eurytopic–oligotrophic species in the core and found that the oligotrophic have decreased with time which indicates very tentatively a steady increase in eutrophication of the lake (Fig. 13.17). However comparison of data on diatom distributions in many other cores from lakes reveals no evidence for natural eutrophication over long periods of time: changes are always associated with fairly drastic shifts in climate, leaching of soil or change in land use. It may be possible to detect shifts between eutrophy and oligotrophy, e.g., in Lake Yoga-ko Hori (1969) found alternations between *Stephanodiscus carconensis* and *Melosira/Cyclotella* which were also accompanied by changes in nitrogen content of the core.

There has been a considerable usage of so-called spectra, in which individual diatoms are allocated to classes according to their tolerance of pH, current, salinity, pollution, habitat, nutrient richness. Examples of spectral classes are given in Table 13.1: they are essentially based on a system set up by Hustedt in numerous publications. These are of considerable value if used discriminately; unfortunately there is a tendency to follow classifications from the literature with very little revision. A further section should be recognised for species penetrating into hypersaline situations and a possible grouping of these is given in Fig. 13.18 from Ehrlich (1975). Classification into saprobic habitat (e.g. planktonic, epipelic, etc.) and nutrient types (e.g. eutrophic, etc.) can be found in appropriate chapters. The environmental limits of most diatom species require greater detailed study especially those classed as indifferent. I am sure there are no indifferent species of diatoms (or any other algae) and indifferent simply means widespread. In a study of the diatoms of two small ponds near Bristol it was

Table 13.1. *Spectra of physico-chemical tolerances used in palaeo-ecological studies of diatoms but equally applicable to other algae. Modified from Abbott & Van Landingham, 1972*

pH spectra
Acidobiontic – occur below pH 7.0 with optimum development below pH 5.5
Acidophilous – occur around pH 2.0 but with optimum development below pH 7.0
Indifferent – occur around pH 7.0
Alkaliphilous – occur at pH 2.0 but with optimum development above pH 7.0
Alkalibiontic – occur only in waters above pH 7.0

Current spectra
Limnobiontic – occur only in non-flowing water (lakes and ponds)
Limnophilous – have their optimum development in non-flowing waters
Indifferent – occur in non-flowing and flowing water
Rheophilous – have their optimum development in running water
Rheobiontic – occur only in flowing water (rivers)

Halobion (salt) spectra
Oligohabobious – freshwater, but occur in waters up to 5‰ salinity
(*a*) Halophobious – occur only in freshwater
(*b*) Indifferent – general freshwater forms
(*c*) Halophilous – widespread in freshwater but developing well in slightly brackish water
Mesohalobous – brackish water species in waters of salinity 5–20‰
Euhalobous – marine species developing in salinity of 20–40‰

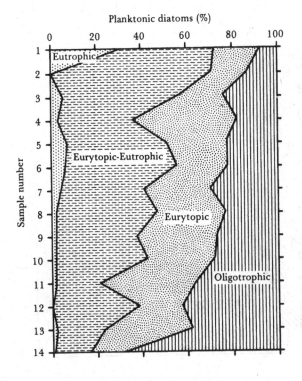

Fig. 13.17. Proportional representation of planktonic diatoms in Lake Ontario sediment according to their trophic requirements. From Duthie & Sreenivasa, 1971.

found that less than ten species appeared common to the two and closer examination of these revealed slight morphological differences which almost certainly indicates that separate entities are involved (Eaton, 1967).

Very little correlation has been attempted between fossil diatom floras in widely separated localities, but Wornardt (1967) has presented some data on a comparison of Miocene–Pliocene diatom floras in Java, Japan and California: of approximately 1200 species recognised from these regions some 900 appear to be geographically restricted to one or another area. A further 120 species are restricted to certain geological zones and are geographically widespread enabling comparison between strata of similar age from these distant points around the Pacific. Obviously with such large assemblages of species to study it will be a lengthy task to define the diatom floras of different strata and compare them

geographically. All these early fossiliferous deposits appear to be marine in origin. Extensive studies on marine cores collected from cruises such as those of the *Glomar Challenger* in the Pacific have yielded fragmented material from deep-sea deposits which are being correlated with those from terrestrial sites. A recent correlation of Pacific Ocean sites is given in Table 13.2 taken from McCollum (1975) which shows, as one would expect that the longer the time periods the more zones one can recognise. Unlike the situation in freshwaters, where about eight pollen zones are used for dating purposes, in these marine samples the zones are being named after indicator diatoms, and this may prove a burden as more sediments are worked.

Cores from many freshwaters reveal an assemblage of diatoms in the late glacial period associated with high calcium–magnesium values of the sediments and hence of the drainage waters; species such

Fig. 13.18. A simplified classification of diatoms according to their salt tolerance. From Ehrlich, 1975.

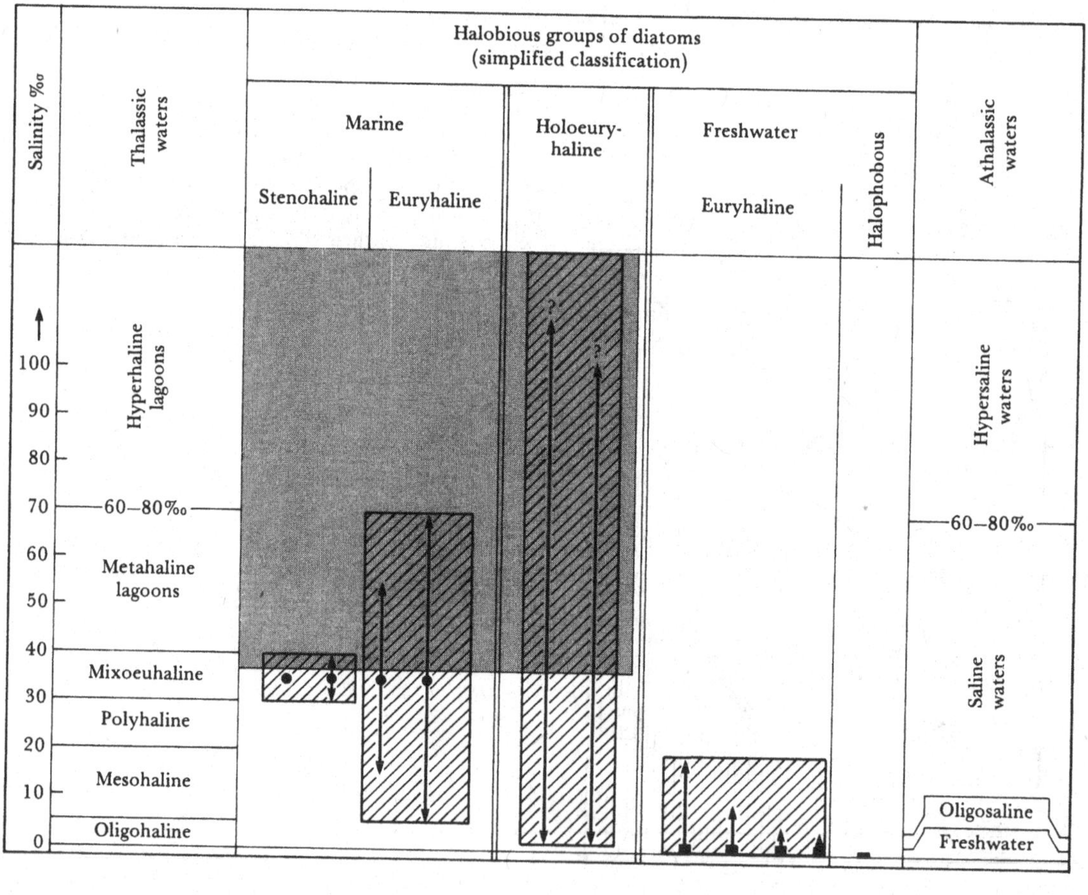

Table 13.2. *The correlation of various areas of the Pacific Ocean based on diatom stratigraphy from cores. From McCollum, 1975*

Southern ocean			North Pacific			Equatorial Pacific
McCollum (1975)	Donahue (1970b)	Jouse (1962)	Schrader (1973)	Koizumi (1973)	Donahue (1970a)	Burckle (1972)
C. lentiginosus	N. kerguelensis / B. californica	I / II	I / — — — — — / II		D. seminae / R. curvirostris	Pseudoeunotia doliolus
C. elliptipora A. ingenus	A. ingens	III	III			
R. barboi N. kerguelensis	R. barboi [Base undefined]	IV?	IV / V ? ? ? ? ? / VI	? ? ?	A. oculatus	Rhizosolenia praebergonii
C. Kolbei R. barboi			VII	D. seminae		
C. insignis			VIII -? ? ?	D. seminae / D. kamtschatica	?	? ? ?
N. interfrigidaria			IX			Nitzschia jouseana
N. pracinterfrigidaria			-? ? ?- / X	D. kamtschatica		
D. hustedtii			XI -? ? ?- / XII–XIV	D. hustedtii		Thalassiosira convexa
D. hustedtii D. lauta			XV–XVIII	D. hustedtii		
D. lauta D. antarctica			XIX -? ? ?	D. lauta		
D. antarctica C. lewisianus			XX / XXI–XXIV	D. lauta		
D. antarctica			XXV	[Base undefined]		
D. nicobarica			[Base undefined]			

as *Campylodiscus noricus* var. *hibernica*, *Cyclotella antiqua*, *Epithemia* spp. *Fragilaria construens*, *Gyrosigma attenuatum*, *Navicula oblonga* etc. were shown to be characteristic of such sediments and from their ecological requirements it was deduced by the author (Round, 1957*b*) that the early sediments were base-rich; later chemical measurements showed this to be so (Mackereth, 1965). Such conditions have been demonstrated since for other sites, e.g. Scotland (Pennington, Haworth, Bonny & Lishman, 1972) North Wales (Crabtree, 1969) Poland (Marciniak, 1973), etc. This early base-rich, late-glacial period is followed by the long post-glacial when indicators of base-poor conditions predominate and generally show a monotonous sequence until the recent period when man's activity has been of increasing importance in the environment.

The trend from a flora indicative of base-rich conditions to one indicative of base-poor conditions seems to be a fairly general feature of lakes formed in glacial basins and is related to the intermittent freezing and thawing cycles and general instability of the surface during late-glacial time, allowing a steady release of bases from the developing soils into the water. Once the leaching is reduced by changing climatic conditions and increasing vegetation cover the base-poor era sets in. In a study of the vegetation of interglacial lake basins in Denmark, Andersen (1966) found a similar succession from base-rich to base-poor conditions. However lakes are known which have undergone the opposite sequence, e.g. many north American lakes situated in base-rich glacial till; continual leaching throughout the lake history influences the flora towards the species requiring alkaline conditions, e.g. Figs. 13.19 and 13.20 show the distribution in Pickerel Lake, South Dakota (Haworth, 1972). The situation here is somewhat unpredicatable since distinct acidophilous species occur in the early diatom zones along with alkaliphilous species (*Fragilaria* and *Mastogloia* spp.)

which are the dominants. This imposition of acid-loving forms in an essentially base-rich environment could be explained by the fact that *Picea* (spruce) forest was the predominant vegetation at that time and under spruce an acid to neutral soil can be expected and drainage from this soil into shallow regions of the lake may produce a sufficiently acid environment to allow the development of acidophilous species. As Haworth writes 'the presence of *Picea* was a crucial factor in creating an environment suitable for acidophilous taxa' and this is one of the best examples I know of the subtle interaction between the diatom flora of a lake and the drainage

from a particular soil and vegetation type. Osborn & Moss (1977) and Phillips *et al.* (1978) have used cores to determine the enhanced growth of epiphytic species in recent time in the Norfolk Broads and to correlate this with the nutrient loadings and ultimate elimination of the macrophytic flora. There is little doubt that the study of algal remains in cores will yield valuable data about the onset of increasing eutrophication, pollution, etc.

The diagrams in Fig. 13.21 from Battarbee (1978) show how plotting diatom data incorporating percentage frequency and an input factor gives a more comprehensive picture of the changes in the

Fig. 13.19. Diatom stratigraphy of a sediment core from Pickerel Lake, South Dakota. Data selected from Haworth (1972) by Platt Bradbury (1975).

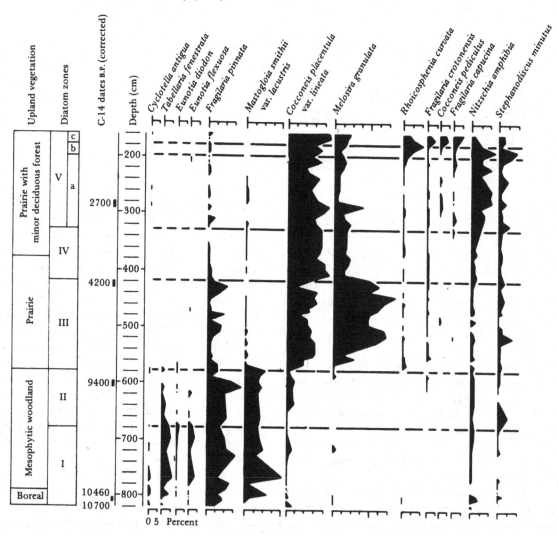

lake flora during the recent time period covered by this core (circa 4800 yr).

In the interesting case of Lake Lisan which was the Pleistocene precursor of the Dead Sea and extended north to Lake Kinneret and south to Hazeva, three diatom zones characterised by *Nitzschia vitrea*, *N. lembiformis* and *Rhopalodia gibberula* occur from the base upwards, though the surface sediments contain no diatoms since the salinity of the present day Dead Sea is too great for diatom growth (Begin *et al.*, 1974).

Dinoflagellate 'cysts'

In oceanic sediments the 'cysts' of dinoflagellates occur. These have been known for a long time and were previously called 'hystrichospheres' and their relationship to modern vegetative flagellates (mainly *Protoperidinium* (*Peridinium*) and *Gonyaulax* species) has been only recently confirmed (*see* review by Wall, 1970). Unfortunately the genus *Ceratium* is best known in the surface waters from a phytogeographical standpoint (*see* p. 3) and this is not

represented by a cyst stage. The cyst walls (Fig. 13.22) are organic and samples can be prepared for examination by the usual methods of pollen analysis. Whilst dinoflagellate cysts have been recorded from Mesozoic and Tertiary deposits and known for many years it is only during the last two decades that Quaternary material, including the surface oceanic sediments have been investigated. The results show that up to 25 000 cysts per gram of sediments can occur (Wall & Dale, 1973) though usually the numbers are much less than this. Some distinctive distributional patterns are also emerging and it is also apparent that the greatest concentration occurs in near-shore sediments and even within individual bays (Reid, 1972). This is related to the fact that unlike other sedimented algal remains the cysts are part of the life cycle of neritic dinoflagellates and under favourable conditions release motile cells which are then carried up into the plankton, i.e. they belong to dinoflagellates which are meroplanktonic. The cyst walls are resistant, probably containing a substance resembling sporopollenin found in terres-

Fig. 13.20. The pH spectrum of the diatom species occurring in a core from Pickerel Lake. From Haworth, 1972.

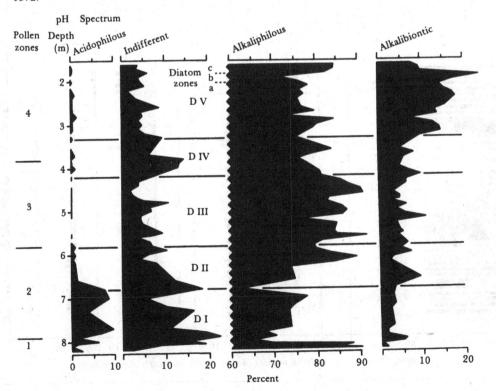

trial plants and some algal walls. If oceanic species form cysts it is unlikely that the products of germination could ever reach the surface again unless germination occured before sedimentation. In fact Wall (1971) implies that dinoflagellates do not form cysts in oceanic regions and that those found on the abyssal plains are merely the result of transport in surface, bottom or turbidity currents.

There is a small but rapidly increasing literature on the vertical distribution of dinoflagellates in Quaternary deposits, mainly based on the excellent work of Wall & Dale who have greatly extended our knowledge of the relationships between the various cysts and their parent dinoflagellates and also by their investigations of material mainly from oceanic

cores. They showed that the early Pleistocene dinoflagellate cyst assemblages corresponded to definite phases recognised from pollen and foraminiferal remains (Wall & Dale, 1968a). In one interesting work they were able to trace the change from freshwater dinoflagellates (*Tectatodinium psilatum* and *Spiniferites cruciformis*) which occurred in the almost freshwater New Euxinic stage of the Black Sea to haline species (Fig. 13.23; Table 13.3). The freshwater species grew at a time when glacial melt water was entering the basin and there was still no contact with the Mediterranean Sea through the Dardanelles. Then, about 7000 yr ago, saline water began to enter the Black Sea (small amounts may have entered earlier but without effect on the dinoflagellates)

Fig. 13.21. *a*, the relative frequency of occurrence of some diatom species in a core (SMii) from Lough Neagh, Northern Ireland and, *b*, the rate of input of some species to the same core over the same period of time. From Battarbee, 1978.

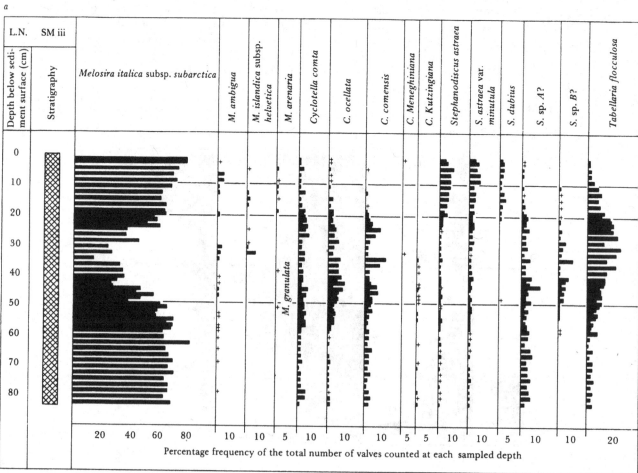

Percentage frequency of the total number of valves counted at each sampled depth

increasing the salinity to about 18‰ and allowing the cosmopolitan euryhaline dinoflagellates to colonise the waters. Many of the cysts deposited during the last 7000 yr belong to forms found in the Black Sea today, e.g. *Lingulodinium machaerophorum* which is the cyst of *Gonyaulax polyedra*. This cyst is also common in sediments around the British Isles but is most abundant off the west coast of Ireland and in the Irish Sea (Reid, 1972). In contrast *Nematosphaeropsis balcombiana* (the cyst of *Gonyaulax spinifer*) is more abundant in the north of Ireland, Scotland and northeast England and is correlated with the flow of North Atlantic drift water. Reid found 50 different species of cyst and it is clear that some of these have very definite distributional patterns related to water movements, tidal convergence, etc. (Reid, 1975a).

In an early study, Huber-Pestalozzi & Nipkow

(1922) found up to 150 dinoflagellate cysts per cubic millimetre in the dark varves laid down in autumn in the Zurichsee and by raising the temperature of samples they succeeded in germinating the spores. Both *Ceratium hirundinella* and *Peridinium cinctum* were recovered, the latter from deposits 16.5 yr old. In both marine (Wall & Dale, 1968) and freshwater (Pollingher, personal communication) encystation increases towards the end of the growth period. Excystation has been rarely studied and Wall & Dale (1968b) report that there is a nine-month period of dormancy in the case of *Gonyaulax digitale* off Woods Hole, which could not be broken by heat treatment during the first three months.

More rarely found in fossil deposits are the 'thecae' of vegetative dinoflagellates, and Fig. 13.24 shows an example of *Deflandrea* from the Cretaceous

Fig. 13.21 *b*

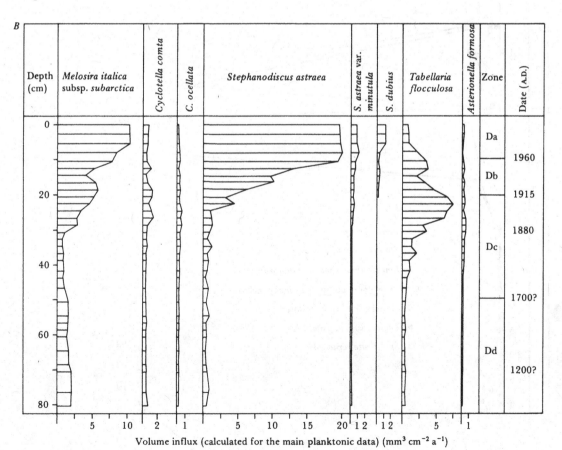

Volume influx (calculated for the main planktonic data) (mm³ cm⁻² a⁻¹)

Fig. 13.22. The dinoflagellate cyst *Spiniferites cruciformis* Wall & Dale. From Wall, Dale & Harada, 1973.

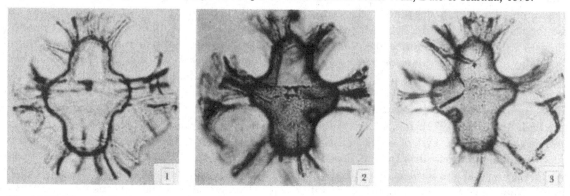

Fig. 13.23. The distribution of common species of dinoflagellate cysts in a core from the late quaternary of the Black Sea. From Wall & Dale, 1973.

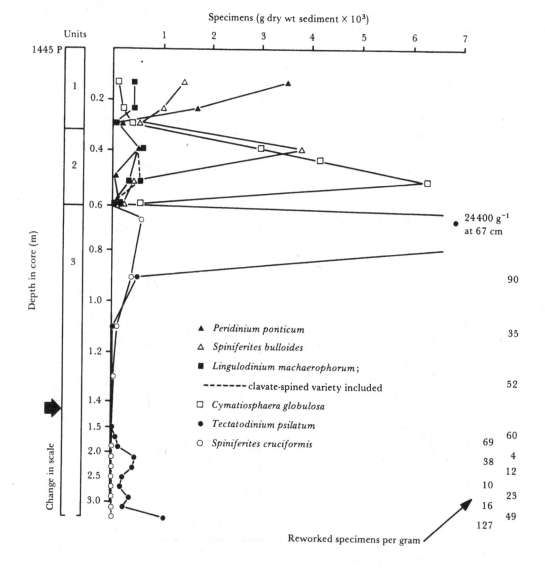

Table 13.3. *The distribution of dinoflagellate cysts down a core from the Black Sea. Details of sediment on left and stratigraphic zones of different workers on right. From Wall, Dale & Harada, 1973*

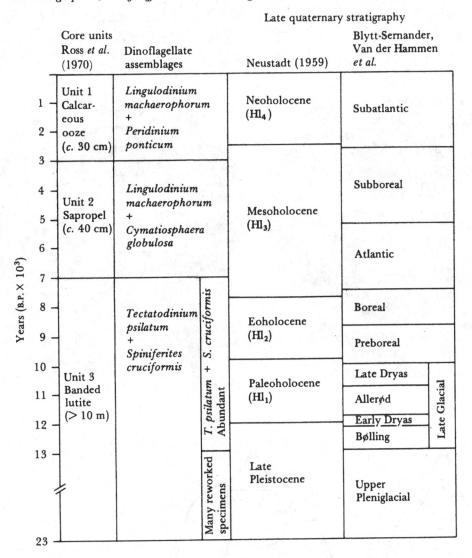

	Core units Ross *et al.* (1970)	Dinoflagellate assemblages		Late quaternary stratigraphy	
				Neustadt (1959)	Blytt-Sernander, Van der Hammen *et al.*
1	Unit 1 Calcareous ooze (*c.* 30 cm)	*Lingulodinium machaerophorum* + *Peridinium ponticum*		Neoholocene (Hl₄)	Subatlantic
2					
3					
4	Unit 2 Sapropel (*c.* 40 cm)	*Lingulodinium machaerophorum* + *Cymatiosphaera globulosa*		Mesoholocene (Hl₃)	Subboreal
5					
6					Atlantic
7					
8	Unit 3 Banded lutite (> 10 m)	*Tectatodinium psilatum* + *Spiniferites cruciformis*	*T. psilatum* + *S. cruciformis* Abundant	Eoholocene (Hl₂)	Boreal
9					Preboreal
10				Paleoholocene (Hl₁)	Late Dryas
11					Allerød
12					Early Dryas / Bølling
13			Many reworked specimens	Late Pleistocene	Upper Pleniglacial
23					

Years (B.P. × 10³)

Late Glacial

---

of Graham Land which clearly has affinities with modern species of *Protoperidinium* (*Peridinium*). Later workers (Eisenack & Fries, 1965; Evitt & Wall, 1968) showed that the cyst of some dinoflagellates forms as a capsule within the theca. The double structures are very like the fossil *Deflandrea* which is thus reinterpreted as a cyst. Presumably the capsule wall is imprinted with the plate markings of the vegetative cell. However, this is not always so, and my own observations with scanning electron microscopy of cysts of *Peridinium cinctum* show a smooth walled structure (Fig. 13.25*d*).

Gyrogonites (Charophyta)

The fossilised oogonia (Fig. 13.25*e*, *f*) of genera of the Charophyta (stoneworts) are found in deposits from the Devonian to the present. The forms known from the earliest deposits are rare and little studied and it is only in the middle Cretaceous that they become abundant. Fig. 13.26 shows the form of the fossilised oogonia (gyrogonite) during the long geological history.

Fig. 13.24. A fossil dinoflagellate *Deflandrea scheii* Manum. *A*, *A'*, ventral view; *B*, *B'* dorsal view. From Manum, 1962.

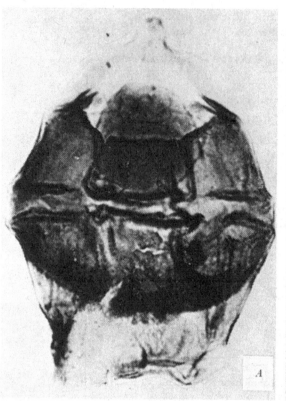
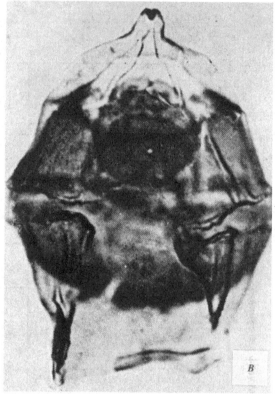
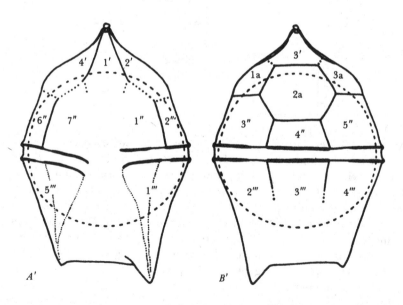

Fig. 13.25. *a*, Scanning electron micrograph of the coccolithophorid *Cyclococcolithus leptoporus*; *b*, transmission electron micrograph of the organic scales with and without spines of *Chrysochromulina pringsheimii*; *c*, light micrograph of the crystals formed in the mucilage of *Tetralithus*; *d*, scanning electron micrograph of the smooth walled cyst of *Peridinium cinctum* showing a few scales of the vegetative cell; *e*, *f*, scanning electron micrographs of fossilised oogonia (gyrogonites) of *Chara*. Photographs *a*, *b*, and *c* supplied by Dr G. Hasle, Dr M. Parke and Dr J. C. Green. *d*, *e*, *f* photographed by author.

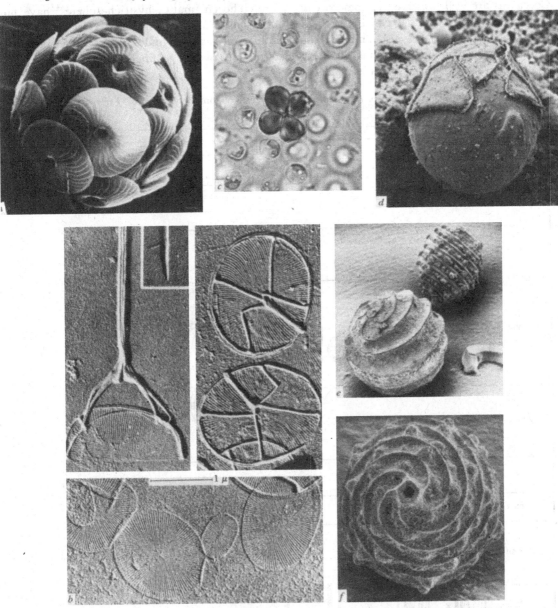

Summary

Remains of algae are preserved in rocks and in the unconsolidated sediments at the bottom of lakes and oceans. The kinds of algae which are preserved are listed and some of the problems discussed.

Coccoliths are well preserved in some ocean basins and in chalk deposits and can be used for dating sediments. Diatoms are also amongst the commonest preserved micro-organisms in both marine and freshwater basins and their use in palaeo-ecological studies has increased greatly in the last few years. In lakes there are many problems of sorting and winnowing of the cells as they are deposited. Changes in chemical composition of the drainage basin and water can be detected.

The enigmatic 'hystrichospheres' have now been shown to be dinoflagellate cysts and considerable work has been undertaken on the distribution of these in marine sediments.

Fig. 13.26. The general evolutionary trends of the Charophyte oogonia (gyrogonite). From Grambast, 1974).

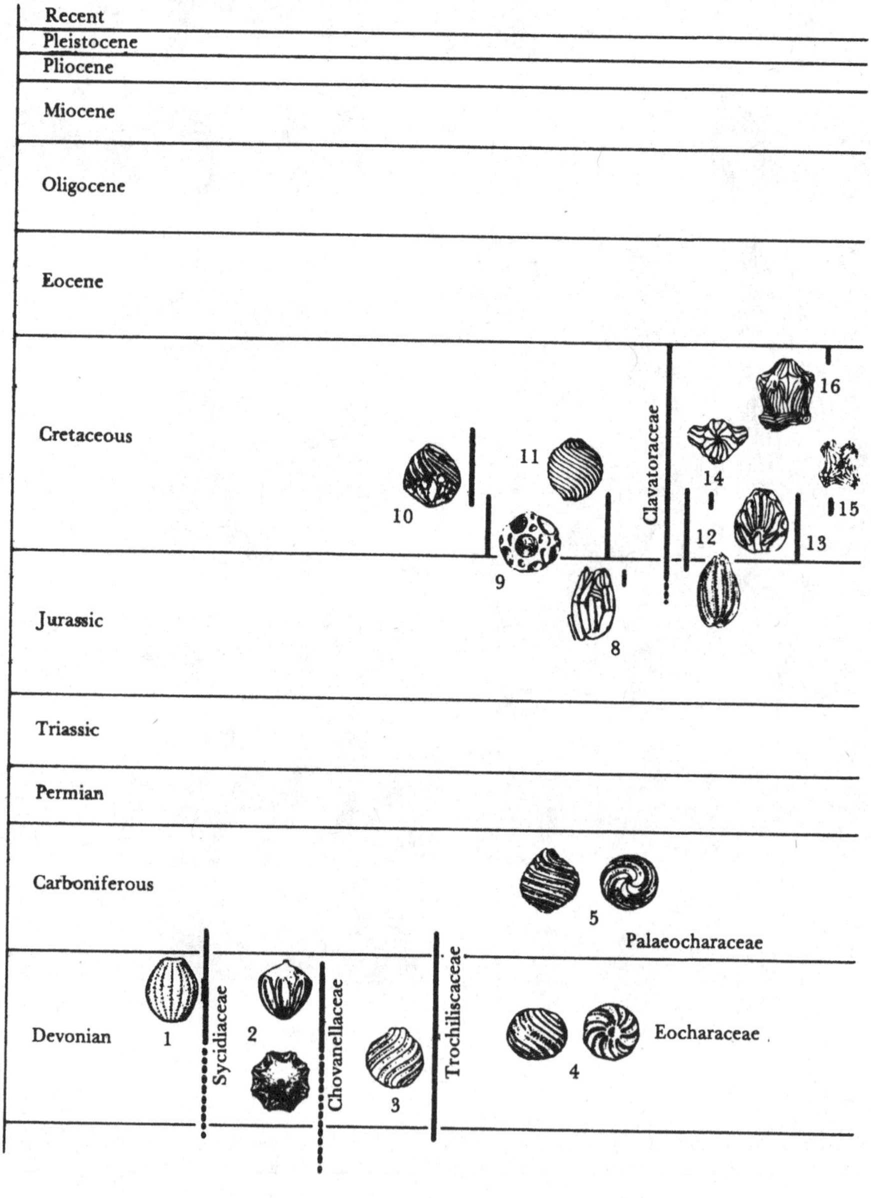

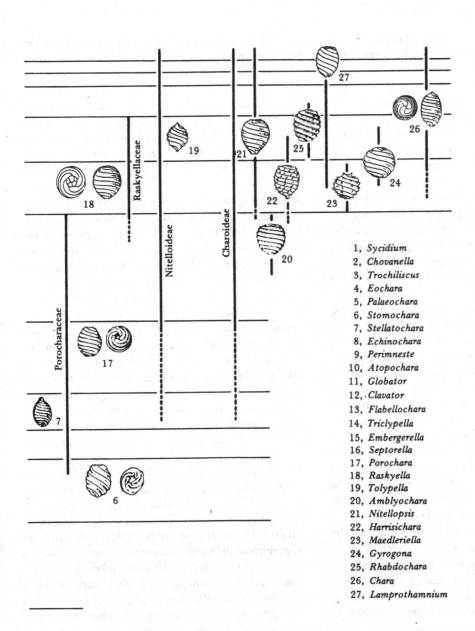

1, *Sycidium*
2, *Chovanella*
3, *Trochiliscus*
4, *Eochara*
5, *Palaeochara*
6, *Stomochara*
7, *Stellatochara*
8, *Echinochara*
9, *Perimneste*
10, *Atopochara*
11, *Globator*
12, *Clavator*
13, *Flabellochara*
14, *Triclypella*
15, *Embergerella*
16, *Septorella*
17, *Porochara*
18, *Raskyella*
19, *Tolypella*
20, *Amblyochara*
21, *Nitellopsis*
22, *Harrisichara*
23, *Maedleriella*
24, *Gyrogona*
25, *Rhabdochara*
26, *Chara*
27, *Lamprothamnium*

14 Eutrophication and pollution

Algae incorporate large quantities of inorganic material into their cells and on death this material is to a large extent deposited on to the sediments, either directly, or after passage through animals. Some material is also released into the water from decaying cells but it may not be sufficient to sustain further large populations and hence, wherever the rate of supply of nutrients from the inflow is low, the production of algae is low, e.g. in many tropical oceans. On the other hand production is high where nutrients, either inorganic or organic, are continuously supplied, e.g. in regions of upwelling, where nutrient-rich water enters a river or lake, or where nutrients are rapidly recycled. Looked at on a long-term basis the inevitable result of algal growth is a stripping of nutrients from the waters of the world and the abundance of sedimentary rocks of algal origin is further evidence of this. The continued algal growth over many millennia is thus directly related to the continual enrichment of the aquatic environment by water flow from the land, carrying with it all the products from the action of climatic factors on the land surface (wind, rain, freezing and thawing, oxidation, etc.). As shown in chapter 1 all these agencies contribute enormous amounts of nutrient material to the aquatic system. In addition many of the products of biological activity on the land surface also ultimately percolate into the waters. The effect can be demonstrated in such undisturbed areas as the Amazon basin where the different river systems carry quite different loads of sediment and the effect of the total outflow is detectable for hundreds of miles northwards into the Atlantic.

The input of material is a natural phenomenon and one which has been proceeding through geological time and, as far as one can deduce, has led to relatively stable ecosystems in which gradual changes occur almost imperceptibly over long periods. The long steady periods are broken by relatively sudden shifts, due mainly to climatic factors reaching critical values or to repeated occurrences of ice ages and interglacial periods in the Pleistocene. Considering only the most recent period there has been a state of relative equilibrium in the aquatic environment since the end of the last late-glacial period in temperate regions and probably for much longer in the subtropical and tropical regions. 'Ageing' of lakes is a natural process, whether the waters are eutrophic or oligotrophic, and there is little evidence to show that added nutrients speed up the process leading to filling in of the basin, which is largely due to inflow of sedimentary material off the land with some contribution from the algae and other aquatic organisms. It may be accompanied by massive algal growths, but these only supply the major components for the infilling material in eutrophic lakes. Another, almost certainly erroneous, concept is that lakes steadily get richer in nutrients (i.e. become eutrophic) over their history and again there is hardly any evidence for this from studies of algae in sediments though there must be a very gradual change in nutrient status over long periods; the term 'maturation' rather than eutrophication should perhaps be used for this process (Brooks, 1969; see Chapter 13). Very gradual changes are *not* what is understood by the term eutrophication; Beeton & Edmondson (1972) in an article entitled 'The eutrophication problem' state clearly that a steady *increase* (my italics) has no support. The long term (1850–1960) data in their paper demonstrate this stability in Lake Superior and the sudden increases in nutrient loading which have occurred in the other lakes (Fig. 14.1). The algal flora changes, but often rapidly when conditions suddenly change (e.g. from base-rich to base-poor, or when agricultural and human activity suddenly accelerates), and this is followed by long periods of stability (*see* for example the study of Lago di Montosori (Hutchinson, 1970). In lakes, rivers and coastal seas the annual succession of species is so varied (*see* p. 422) that increased nutrient supply is likely to have very different effects in different seasons. Many more detailed palaeolimnological and palaeo-oceanographic studies are needed to provide background data to the present situation.

The supply of nutrients from the land is as essential for algal growth as it is inevitable. In most natural, freshwater situations, the flow is of fairly concentrated solutions off the land into a realtively dilute medium, and as long as the dilution effect is operative, a fairly healthy and stable situation is maintained. However, in rivers such as the Rhine, Lower Thames, and Mississippi, inflows may be dilute compared to the river itself. The flow into saline inland basins and into the sea is of course one of dilute solutions into rich but again the subsequent dilution has a comparable stabilising effect. No two basins or areas of sea are alike and for reasons which are still only incompletely understood the nutrient levels of waters are extremely varied: they have been broadly classified into those rich in nutrients (eutrophic)* and those poor in nutrients (oligotrophic)* and these are in general states of equilibrium

* The terms eutrophic, oligotrophic, and the intermediate type mesotrophic were originally applied to waters distinguished on oxygen characteristics and then by the kinds of animal larvae found in the deeper water sediments. Eutrophic waters support *Chironomus plumosus*, mesotrophic *Sergentia coracina* and oligotrophic *Tanytarsus*

Fig. 14.1. The changes in the concentrations of the major ions in the Great Lakes from 1850 to 1960. *a*, Lake Ontario; *b*, Lake Erie; *c*, Lake Michigan; *d*, Lake Huron; *e*, Lake Superior. From Beeton & Edmondson, 1972.

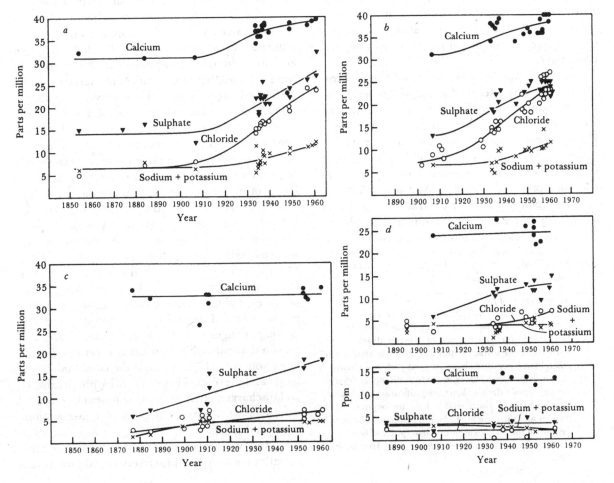

which are maintained for each particular water body over long periods of time (*see* also p. 363). In this chapter the term eutrophication is restricted to the phenomena of both *natural* and usually rapid increase in *nutrient* status, and man-made *artificial* increase; as such it is sometimes included under the general heading of pollution but this is not a good usage even though sometimes the effects are similar. In general, eutrophication involves the increase in growth promoting substances. In most agricultural areas flow of the excess fertilisers off the land (the main source of nitrate*) is causing a diffuse artificial eutrophication whilst individual point sources of addition of sewage (the main source of phosphate) and other organic wastes are the second major source of artificial eutrophication. Eutrophic conditions are often created in newly formed lakes (reservoirs) where initial leaching of the soils and decomposition of the vegetation produces high concentrations of nutrients and immense algal (and often angiosperm) growth which then tends to *decrease* as the nutrient supply equilibrates with that supplied by drainage. Examples are the large man-made African Lake Kariba, and Volta Lake.

There is nothing essentially bad about eutrophication; it may alter the balance of species but so does the natural varied nutrient loading of all waters. The extent of problems caused by algal growths is difficult to determine but one estimate suggests that all man-made water bodies and 50% of the United States surface water supplies have some alga-associated problems (Jewell & McCarty, 1968).

Pollution on the other hand is the addition of *harmful* substances to the environment, though one

* Drainage from good arable land in England contains an average of 10 mg l⁻¹ of nitrate-nitrogen and levels can become even higher (up to 100 mg l⁻¹) at certain times (Cooke & Williams, 1970).

Footnote continued from p. 527

lugens. In addition eutrophic lakes tend to be shallow lakes, turbid in summer due to high plankton production (usually green and blue-green algae) restricted to the surface waters and therefore resulting in a Secchi disc depth of about 1 m and with dense mats of rooted vegetation. Oxygen tends to disappear from the hypolimnia during summer stratification and under ice. Oligotrophic waters tend to have the opposite features and mesotrophic to be intermediate. The very first use of the terms was in the description of bog waters in north Germany (*see* comments in Hutchinson, 1969).

must admit that 'harmful' is difficult to define since it often merely involves excesses of otherwise harmless substances. Also many instances which are termed pollution by the news media may simply alter the balance of algal growth and in many instances actually increase the biomass; they will nevertheless be treated as pollution in this account. In fact, one view is that anything resulting in a change in the balance of species and the dominance of a few species can be regarded as pollution. Thus to the user of water, eutrophication and increased algal growth is often undesirable and therefore a form of 'pollution'. Pollutants, whilst generally harmful to species, may nevertheless stimulate some algae or provide conditions suitable for the growth of species which otherwise would be restricted.

Eutrophication

Eutrophication by animals is a fairly commonplace occurrence, e.g. by sea birds in high-lying rock pools, by ducks on ponds, by animals collecting around water holes. Such eutrophication leads to a reduction in algal diversity in the habitat, often accompanied by massive production of a few tolerant species. Eutrophic communities are characterised by high algal production and often by instability, so that decomposition and removal of decomposition products cannot proceed in a normal way. Excessive production stimulates respiration, increasing the utilisation of oxygen and rapidly leading to anaerobic conditions, the loss of oxygen resulting in accumulation of obnoxious decaying masses and death of animal populations. Eutrophication by man has been obvious since the Middle Ages when as a consequence of the growth of communities sewage was channelled into waters which could not dilute it sufficiently and a sudden but fairly localised burst of eutrophication occurred. The present-day demophoric explosion (i.e. the rapidly increasing growth of populations and technology) is resulting in the most striking changes which have ever taken place. It is certain that eutrophication has again erupted and is more widespread today, since the communities are larger and the sewage is even more highly channelled and discharged often without treatment into coastal waters or with varying degrees of treatment into rivers and lakes.

Inflow of river water often introduces large quantities of nutrients into reservoirs, almost always

Table 14.1. *The phosphate concentrations in the River Thames and River Lee, 1960–72. From Met. Wat. Bd. Rep. 1971–3*

	Phosphate (mg PO_4 l^{-1})			
	River Thames		River Lee	
	Laleham	Walton	New Gauge	Chingford Mill
1960	0.9	0.95	0.9	1.8
1961	1.0	1.1	0.95	1.9
1962	1.2	1.3	1.3	1.8
1963	1.2	1.3	1.6	2.6
1964	1.6	1.7	1.9	3.7
1965	2.0	2.0	2.2	4.1
1966	2.0	2.1	2.0	3.5
1967	2.1	2.2	1.9	3.4
1968	2.3	2.4	2.2	3.8
1969	2.7	2.7	2.2	4.5
1970	3.0	3.0	2.7	4.8
1971	2.7	2.7	2.4	4.7
1972	3.0	3.1	2.5	5.0

alters the natural balance of algal species and also increases the total production, which in itself may lead to extremely unfortunate effects such as deoxygenation, release of unpleasant chemicals when the algal population decays, clogging of filters, etc. It must not be assumed that increased chemical loading always leads to disastrous effects; thus from 1960 to 1972 the phosphate levels of the River Thames and Lee increased threefold (Table 14.1) but these increases did not increase the difficulties of the Metropolitan Water Board in managing their storage reservoirs and filtration works (Met. Water Bd. Rep. 1971–3), though it must be added that this authority continually improves its techniques and without their very careful and detailed basic scientific studies serious problems would have arisen. Much of the phosphate was removed, presumably cycling through the algae and other organisms and being deposited on the sediments so that the water flowing to the supply network was poorer in phosphate than the incoming water. Schindler, Frost & Schmidt (1973) found that 80% of the phosphorus added to a Canadian lake became locked in the sediments and did not return to the water even during periods of anoxia of the hypolimnion. This figure compares well with 90% of the phosphorus input which is trapped in Lake Kinneret in the Sea of Galilee: most of this enters as calcium phosphate in the River Jordan waters and is deposited in this form on the sediments. Likewise about 70–80% of the incoming nitrogen is also retained in Lake Kinneret (Serruya, 1975). Such high figures do not necessarily apply to other lakes though in very many the trapping of nutrients in the sediments is considerable. In a lake such as Lake Kinneret the low phosphorus load remaining in the water column allows very rich blooms of *Peridinium* to develop since this alga has a relatively low phosphorus requirement relative to carbon: increase in the phosphorus loading, however, modifies the population in favour of Cyanophyta and Chlorophyta. Systems for recycling or disposal of liquid waste on the land instead of disposal down water channels would have largely avoided the eutrophication problem and incidentally the fertiliser problem. A further source of eutrophication is detergents, a large proportion of which rapidly flow into the aquatic system and enhance the production of algae mainly by increasing the levels of phosphorus in the water. The detergent problem is being tackled by some countries and certain States of the USA now have regulations regarding phosphates in detergents. One of the suggested alternatives NTA (nitrilotriacetic acid) may however enhance eutrophication in coastal waters (Ryther & Dunstan, 1971).

Edmondson (1972 and earlier publications quoted there) gives the most detailed data obtained anywhere in the world on a large-scale study of the effects of sewage effluent over time in a densely populated area. In 1957 some 56% of total phosphorus entering Lake Washington came from effluent from secondary sewage treatment plants and improperly functioning septic tanks and this had increased to 75% by 1964, largely due to the increase in use of phosphorus-containing detergents. Diversion of this sewage effluent was started in 1963 and 28% was removed. The effect of this on algal populations and physico-chemical parameters can be seen in Figs. 14.2 and 14.3. The dramatic decrease in chlorophyll *a* and phosphorus was *not* accompanied by drops in nitrate or carbon dioxide levels. Fig. 14.3 shows the increase in the winter levels of phosphorus and nitrogen in Lake Washington between 1933 and 1963 owing to inflow of sewage, and the subsequent decreases in 1964–6 when diversions were in operation to carry the sewage effluent away from the lake. It shows also the almost total utilisation of the phosphate and nitrate during the spring and summer growth of phytoplankton and the decline in chlorophyll content of the surface waters as the concentration of nutrients was reduced. The graphs illustrate how deceptive measurements of nutrient loading can be if performed in spring and summer

when the nutrients are being depleted by the algal population. In streams the addition of phosphate via sewage (over 80% can enter from this source, Bolas & Lund, 1974) can result in massive increase in *Cladophora*, e.g. in the River Stour in Kent and elsewhere, though with improvement in the effluent flowing into the river this can be reversed and angiosperm growth encouraged. Borehole water in the same region supported *Tribonema* and addition

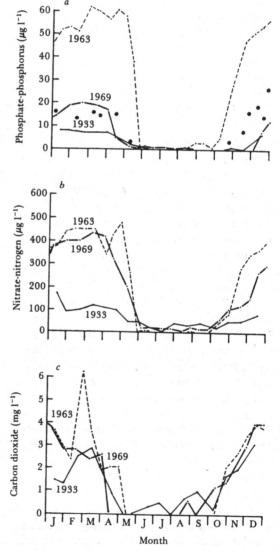

Fig. 14.3. Seasonal changes of, *a*, phosphate, *b*, nitrogen and, *c*, carbon dioxide in the surface water of Lake Washington during three years. Isolated dots are for 1950. From Edmondson, 1972.

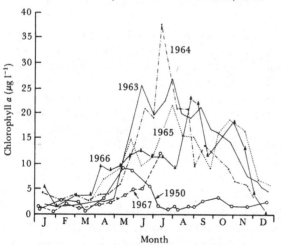

Fig. 14.2. The seasonal cycle of chlorophyll *a* content in the surface waters of Lake Washington in six different years. From Edmondson, 1972.

of phosphate did not greatly stimulate the *Tribonema* nor permit much growth of *Cladophora*. Zimmermann (1961) constructed artificial channels with stones over which river water was discharged at different flow rates and with differing additions of sewage and obtained much interesting information. With small amounts of sewage (5%) the summer population on the stones was dominated by *Chaetophora* and *Lyngbya* sp. and much calcium carbonate was precipitated. If the flow was increased *Tribonema* became dominant. With 10% sewage and slow flow *Closterium ehrenbergii*, *Nitzschia palea*, *Spirogyra*, *Fragilaria capucina* and *Tribonema* occurred, in medium flow only *Tribonema* grew, and in the fast channel *Phormidium autumnale*, *Cladophora*, *Tribonema* and *Stigeoclonium*

Fig. 14.4. The change in the seasonal variation in abundance of phytoplankton in Lake Erie between, *a*, 1935 and, *b*, 1962. From Davis, 1964.

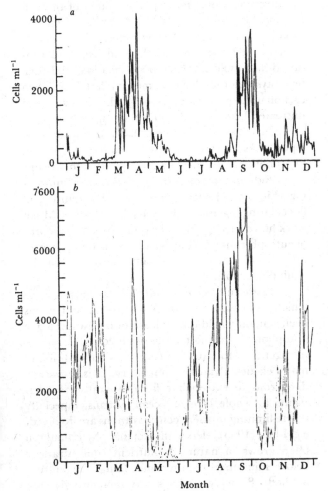

appeared. If even more sewage was added the filamentous growths were somewhat similar but they all became overgrown by diatoms and sewage fungus (a complex of a bacterium *Sphaerotilus* and a fungus *Leptomitus*) became more abundant. Some workers term this slimy mass a zoogloea.* Analyses of such data are difficult but the findings illustrate the complexities which must be faced in nature where more than two factors are varying.

Another effect of eutrophication is that the annual algal growths, instead of peaking in spring and autumn, which is very common in unenriched waters, tend to have additional peaks as one species after another grows intensively, an effect which can be illustrated by data from Lake Erie (Fig. 14.4). Extreme eutrophication need not disturb the normal seasonal sequence of algae, e.g. in the incredibly enriched Lake Onodaga there is still a diatom → Chlorococcales → Cyanophyta succession (Sze & Kingsbury, 1972). A similar merging of growth peaks can also be shown in the marine sphere, e.g. in the North Sea the distinct spring and autumn peaks seem to have merged in recent years (Fig. 10.24, Reid, 1975*b*) and though this is not necessarily an indication of eutrophication it is evidence of the effects of nutritional changes which have been recorded for the sea around the British Isles (Russell, Southward, Boalch & Butler, 1971) especially the return of high winter-phosphate levels accompanied by increasing annual phytoplankton production from 140 g m^{-2} in 1965 to 212 g m^{-2} in 1970 and from 1967 a steady increase in the size of the spring maximum in the English Channel (*see* Fig. 10.18). The changes are almost certainly linked but the phytoplankton reactions are not identical in the two areas – but neither are the hydrographic features. Eutrophication of coastal waters is very common but less well documented than that of freshwaters. In the Baltic Sea near Helsinki a four-year study of inshore and offshore sites showed that the annual carbon fixation was much greater in the eutrophic coastal water (80, 120, 174, 192 g carbon m^{-2} yr^{-1}) than in the offshore water (15, 28, 30 and 36 g carbon m^{-2} yr^{-1})

* A detailed account of the bacteria, fungi, protozoa and algae in 'sewage fungus' is given by Curtis & Curds (1972). In situations where this develops, algae are in fact only minor components since few seem to tolerate the excessive organic pollution required by sewage fungus.

and the maximum daily rate in the inshore was 1.0–1.2 g carbon m^{-2} compared with 0.07–0.28 g carbon m^{-2} in the Gulf of Bothnia (Bagge & Lehmusluoto, 1971). Recent increases in algae such as *Ulva* and *Enteromorpha* in the upper intertidal are recorded as man's activity has spread along tropical beaches, e.g. on Guam and along the shore at Waikiki, Hawaii (M. S. Doty, personal communication): on the shore at Waikiki *Ulva fasciata* and *U. reticulata* were not abundant in 1955, in the 1960s they began to grow richly in certain seasons and now they are prolific all the year round. Sewage is reported to increase the growth of the green alga *Dictyosphaerium* in Kaneohe Bay on the north coast of Oahu, smothering the corals (Johannes, 1977).

As any gardener knows the addition of nutrients, even in very small amounts, enhances the growth of plants except in soils which have been excessively fertilised, and aquatic systems behave in the same way, and even in regions where the population is low and the lake apparently almost undisturbed by man such changes can be detected. For example, Lund (1969) shows how the algal flora of Blelham Tarn, a small lake with a very sparsely populated drainage basin, has changed over the last 25 yr, owing mainly to increased use of fertiliser on the farms and the advent of piped water in the catchment area. Thus, in 1954 *Aphanizomenon flos-aquae* appeared in the plankton for the first time, followed by *Cyclotella pseudostelligera* in 1960, *Staurastrum chaetoceros* in 1962, *Fragilaria crotonensis* in 1964 and *Anabaena solitaria* in 1967. These species were present in neighbouring lakes for many years before appearing in Blelham Tarn and this shows that dispersal is not really a factor in the spread of a species in a small area. One of the few other detailed studies is that of Schindler *et al.* (1973) who artificially fertilised a 'virgin' lake in Ontario increasing the natural inputs of nitrogen and phosphorus by a factor of five. The 'virgin' flora of lakes of this type is given on p. 273 and the eutrophication resulted in a change to one dominated by Chlorophyta and Cyanophyta except during winter when, under ice, the flora reverted to a Chrysophyta–Cryptophyta type. The change must have been most dramatic with the appearance of *Staurastrum paradoxum*, *St. dejectum*, *St. boreale*, *Sphaerozosma granulatum*, *Spondylosium planum*, *Chroococcus limneticus*, *Mallomonas elongatum*, *M. caudata*, *M.* cf. *intermedia*, *M. pumilio* var. *canadensis* and *M. pseudo-coronata* in the first and second spring and summer followed by Cyanophyta, *Oscillatoria geminata*, *O.* cf. *amphigranulata*, *Pseudanabaena articulata* and *Lyngbya lauterbornii* in later summer. In the third year the conditions appear to be somewhat unstable and the desmids were not recorded but in unseasonally cool summer weather other green algae appeared (*Dictyosphaerium elegans*, *Oocystis submarina* var. *variabilis*, *Ankistrodesmus falcatus* var. *spirale*, *Chlorella* sp., *Scenedesmus* spp., *Gleotilopsis* and *Chlamydomonas* sp.). I quote these lists for comparison with the 'virgin' flora to emphasise the magnitude of the changes and also to illustrate that they differ in different years. Even more importantly they show that it would have been a brave phycologist who would have attempted prediction of such events or the fact that in such an area in the third year of fertilisation the chlorophyll *a* maximum would reach 296 μg l^{-1}, which is near the theoretical maximum per unit area of the epilimnion (Talling *et al.*, 1973). A similar effect was produced in a marl lake by Schelske, Hooper & Haertl (1962) who added chelated iron and produced a fourfold increase in ^{14}C fixation, and then ten days later added commercial fertilizer when carbon uptake increased to sixtyfold, whilst a second untreated lake showed no increase in fixation.

Increased use and especially overleading of septic tanks results in increased discharge of nitrogen, phosphorus and other algal nutrients into streams and ponds and it is desirable to find ways of improving efficiency of such systems, e.g. by addition of flocculating agents such as ferric chloride. Much work however remains to be done before this source of eutrophication is controlled but the dangers even in 'virgin' territory are admirably shown by Schindler's work.

There has been considerable discussion as to which elements stimulate algal growth and it has been generally held that at least for freshwater algae phosphorus is the most important followed closely by nitrogen; in laboratory experiments the addition of many elements can be stimulatory. It is, however, dangerous to extrapolate from laboratory experiments to whole lake or ocean systems, especially where limiting nutrient concentrations are involved, e.g. Lund (1965) refers to the fact that phytoplankton often grows in nature at nutrient concentrations below minimum requirements found by culture methods. Bioassay methods may therefore give the erroneous impression that addition of nutrients will be harmless to a natural system. Even enrichment experiments (Schindler, 1971*a*) in which the bottles are incubated in the natural environment may indicate that one nutrient is more important than another; Fig. 14.5, for example, illustrates the case of carbon. However carbon is rarely likely to be limiting in nature* and further experiments using larger volumes of water enclosed in plastic tubes failed to substantiate the bottle results; Schindler therefore concluded that 'there is some evidence that phosphorus is the "primary" limiting nutrient' even in his experimental lake where total available carbon is an order of magnitude lower than in most other lakes. Later Schindler *et al.* (1973) showed that although the eutrophication of a virgin lake decreased the dissolved carbon dioxide the flux from the

atmosphere accounted for 69–95 % of the inorganic and particulate carbon in the water.

A controversy has arisen, mainly in the USA, over the effect of carbon rather than phosphorus as a prime factor in eutrophication. In general there is very little evidence to support carbon as a factor except under isolated and exceptional circumstances. Most lakes are undoubtedly more productive of algal material as the phosphorus load increases which is really the crux of the matter (Schindler, Fee & Ruszczynski, 1978); standing crop of algae is highly correlated with phosphorus loading especially when mixing of the water column is taken into account (Oglesby, 1977*b*). Fig. 14.6 shows the relationship between annual primary production, phosphorus and surface chlorophyll *a* concentrations for the Laurentian lakes. A popular account of the controversy is given by Vallentyne (1974) and a detailed chemical review by Goldman, Porcella, Middlebrooks & Toerien (1972) who quote the approximate ratios in natural waters of $C:P \geqslant 580:1$ and $C:N \geqslant 27:1$ whilst in algae the ratios are $C:P \cong 72:1$ and $C:N \cong 6.31$ indicating that carbon would never be limiting assuming only 50 % carbon available for algal growth. In experiments involving fertilisation of a small Canadian lake (*see* Schindler *et al.*, 1973 and Schindler, 1971*b*) the added phosphorus and nitrogen was rapidly taken up by the seston (presumably mainly phytoplankton)† and although no carbon was added, the carbon content of the seston also increased proportionally indicating that there was no carbon limitation in the lake in spite of the fact that its total carbon dioxide content $(HCO_3^- + CO_{2(aq)} + CO_3^{2-})$ was only 0.025 mole l^{-1} (the world average for freshwaters is 0.99). Fertilisation of such a lake rapidly changed its nature from oligotrophic to eutrophic and the resulting decrease in carbon dioxide caused the pH to rise from 6.3–6.8 to 9.2–10.2.

Experimental eutrophication of freshwaters and even of narrow sea lochs has been attempted with interesting results. One of the earliest was the eutrophication of Loch Sween in Scotland. Also in

Fig. 14.5. Results of adding phosphate, bicarbonate and nitrate to bottles containing phytoplankton which were then suspended at various depths in a Canadian lake. Growth measured by ^{14}C technique. From Schindler, 1971*a*.

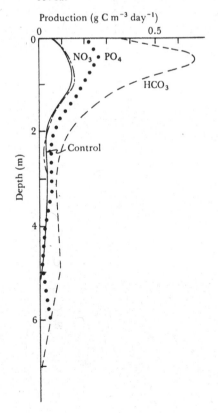

* Limitation certainly occurs during rapid growth or in dense blooms but it is never likely to inhibit overall annual production.
† Earlier Hays & Coffin (1951) had shown the rapid uptake of radioactive phosphorus added to a small lake in Nova Scotia.

Scotland, Brook (1958) and Brook & Holden (1957) fertilised a small oligotrophic lake with calcium superphosphate and succeeded in increasing its algal population – the attached algae on stones and stems increased strikingly. However this did not affect the desmid flora which is characteristic of the natural lake waters (pH 6.8–7.8) until cyanophycean and chlorophycean growths had increased the pH to 8–10 when the desmids finally disappeared. The plankton

had an initial burst of growth and then smaller growths in each successive year as the effect of fertilisation wore off.

Biological estimations of the degree of eutrophication of lakes are probably more informative than chemical determinations and in an interesting study Toerien, Hyman & Bruwer (1975) investigated the algal growth potential of 98 South African waters, ranking them according to the growth of

Fig. 14.6. *a*. The relationship between annual primary production and annual phosphorus loading and (*b*) between annual primary production and surface chlorophyll in the Great Lakes. Central, eastern and western basins refer to Lake Erie. From Vollenweider, Munawar & Stadelmann, 1974.

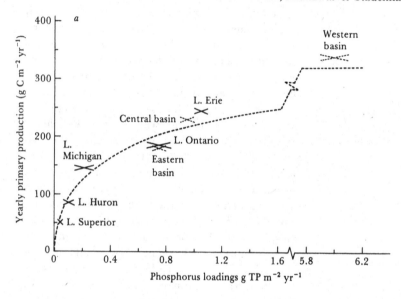

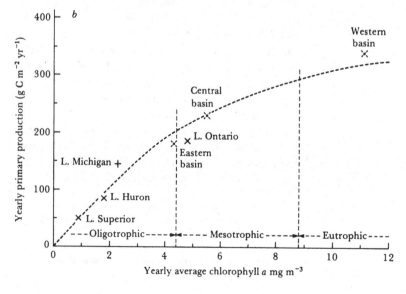

Selenastrum capricornutum. They found that lower algal growth potentials were in waters primarily phosphate limited and as the potential increased, nitrogen limitation became more important. Whilst growth of *Selenastrum* can give a measure of the available combined nitrogen in the water it is better to use a blue-green alga, e.g. *Anabaena flos-aquae*, to give a measure of the available phosphate (Walmsley & Ashton, 1977).

Carlson (1977) devised a trophic state index, based on a scale from 0 to 100 in which each 10 units implied a doubling over the previous and which can be used with measures such as Secchi disc readings, chlorophyll *a* contents, phosphate concentration, etc. Certainly there is a need to devise and test out simple ways of determining the trophic status or degree of eutrophication of natural waters using techniques which do not require refined methods.

Controlled artificial eutrophication has been attempted and experimental systems are in operation designed to increase the production of organic matter in aquatic systems. Such techniques have been used for years in Central European carp ponds and Malayan fish ponds. Similar techniques though with a different end in view and utilising the same principles are used for treatment of sewage effluent and for the nutrient stripping of waste water (e.g. irrigation water). Ideally the two aims can be combined with the nutrients stripped from the water, incorporated into algae which are then cropped by mollusca, crustacea or fish, and finally the effluent water returned to the natural environment without increased eutrophication. The system often involves piping sewage effluent into ponds, thus stimulating algal growth and providing extra food for the animals. Such aquaculture must however be performed under strict supervision since, as Ryther *et al.* (1972) succinctly comment 'the uncontrolled enrichment of natural ecosystems as a means of producing organisms that are useful and valuable to man is, at best, highly inefficient and unpredictable and at worst, catastrophic'. Growth of animals in sewage effluent is dependent upon converting the nutrients in the sewage to algae and their efficient cropping. A further complication however is that not all the nutrient is used and the animals excrete waste so that the process is only partially stripping the effluent of its nutrients. Ryther *et al.* propose a neat solution to this in that in a marine system a further

stage could be added in which a commercially valuable alga such as *Chondrus crispus* could be cropped in the residual liquor.

The effects of excessive eutrophication may be almost indistinguishable from those of pollution but as shown above for Lake Washington they may be reversible. However, long periods of excessive eutrophication may lead to diastrous, and according to some writers irreversible, changes in the environment, e.g. the deoxygenation of the lower layers of the hypolimnion. In Lake Erie, Burns & Ross (1972*a*, *b*) estimate that a layer of algae approximately 2–3 cm thick was deposited in the Central Basin after the algal bloom in July 1970, approximately 88% of the oxygen uptake in this layer was due to bacterial degradation of the algal sediment and 12% taken up in the reduction of metallic compounds, thus yielding an anoxic lower layer. Irreversible changes are probably rare except in the context of loss of species from a habitat and even in Lake Erie, Burns & Ross consider that since phosphate-deficient conditions (due to algal uptake) are often encountered, reduction of the phosphate loading in the inputs would soon return even this notorious basin to an acceptable state.

Lund (1972) has shown how difficult it is to predict the effect of increasing eutrophication when in one lake (Blelham Tarn) the increase in phosphate-phosphorus in the water column was accompanied by increasing crops of the diatom *Asterionella* whilst in an adjacent lake (Windermere, south Basin) the same alga decreased as phosphate-phosphorus increased (Fig. 14.7). The histograms also show that whilst *total* phosphorus has remained fairly static from 1945 to 1970 (in spite of the increasing population, tourism, etc.) the phosphate-phosphorus has risen considerably. The effect of supplying piped water and the operation of sewage plants is shown respectively in the increase and decrease in phosphate concentration in the lake waters. The effect of eutrophication on the hundreds of other organisms in the lakes has not been analysed and whilst the Blelham Tarn and Lake Windermere data do not suggest dramatic damage to the environment they do suggest that in view of the difficulties of prediction the safest course of action is to maintain systems in their natural state for as long as possible: this means concentrating effort on *preventing* excess nutrients from entering the aquatic environment rather than

Fig. 14.7. Left-hand columns, the mean weekly number of cells of *Asterionella formosa* for the years 1945–70; continuous line, the mean weekly concentration of phosphate phosphorus in the 0–5 m column of water in December and January, 1945–6 to 1970–1, with the exception of 1956–7. Right-hand graph, white columns, the mean weekly concentration of total phosphorus in 1945 (*A*), 1946 (*B*) and 1970 (*C*); black columns, the mean weekly concentration of phosphate phosphorus in December and January 1945–6 (*A*), 1946–7 (*B*) and 1970–1 (*C*). P; 1951, when piped water came to the main village in the drainage area of Blelham Tarn. SS, above lower left-hand columns, years of start of sewage plant and its satisfactory operation in the Blelham Tarn drainage basin; above upper left-hand columns, years of start of new sewage plant and its satisfactory operation near the shore of the south basin of Windermere. From Lund, 1972.

experimenting on *how much* the environment can take. This is even more evident when one considers the historical data and I can best quote Lund (1972) 'that it is difficult to forecast the detailed changes to be expected from eutrophication, particularly when only a limited amount of information is available' and 'that very erroneous conclusions may be obtained from facts obtained from short-term investigations'.

In a lake such as Buttermere in which phosphate-phosphorus is hardly detectable, addition of nutrients stimulated algal growth and, of the nutrients added, phosphate was the most effective (Lund, 1969). However, one other interesting feature was that *Asterionella formosa* could never be induced to grow when added as inoculum to the Buttermere water enclosed in plastic cylinders. If phosphate or soil extract was added *Asterionella* grew, with phosphate alone the attached diatoms outgrew the *Asterionella* but soil extract enabled the *Asterionella* to get the upper hand. *A. formosa* is absent from Buttermere and these experiments suggest that low levels of phosphate-phosphorus are not responsible for the failure to grow. Another experimental system devised by Stockner & Shortreed (1978) was to measure the growth of algae on the side of wooden troughs supplied with stream water on Vancouver Island. They added nitrogen, phosphorus and the two together and found that dissolved phosphorus was limiting but that the nitrogen to phosphorus ratio was a very important determinant of the composition of the diatom flora which developed.

An alga which is often associated with eutrophication is *Oscillatoria rubescens*, yet Gächter (1968) found that addition of phosphorus to waters containing this cyanophyte hardly affected its growth but did retard Chrysophyceae (an often reported result) and stimulated green algae and diatoms. It has often been suggested (e.g. Provasoli, 1969) that the requirement of sodium and/or potassium for blue-green algal growth is a factor in their rich growth in some eutrophic waters: more work is required on the ecological effects of sodium levels and blue-green algal populations to test this idea.

The effects of the addition of nutrients can be tested either in the laboratory or in actual lakes. An example of laboratory testing is that of Maloney, Miller & Shiroyana (1972) in which the chlorococ-

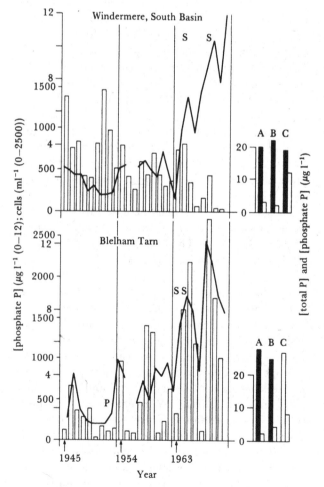

Fig. 14.8. The effect of the addition of various nutrients on the growth rate of *Selenastrum capricornutum* in, *a*, Woahink Lake, *b*, Waldo Lake, *c*, Upper Klamath Lake, *d*, Diamond Water, *e*, Odell Lake. From Maloney *et al.*, 1972.

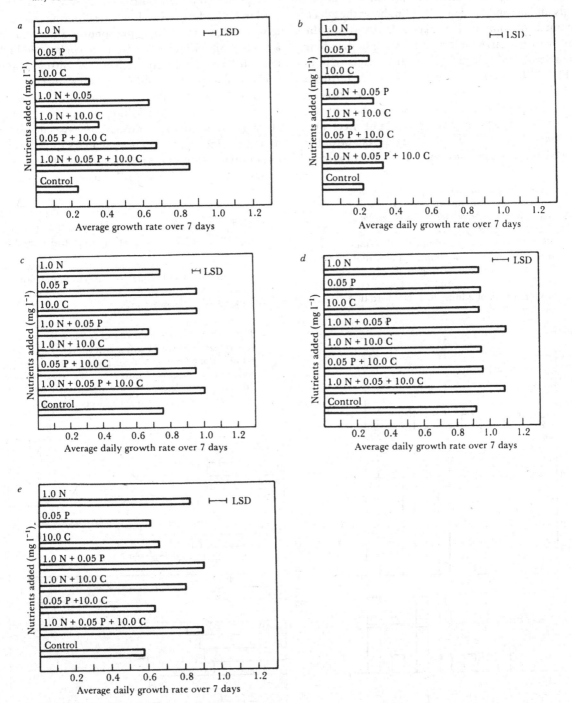

calean alga *Selenastrum capricornutum** was grown in the laboratory in water from nine Oregon lakes to which nitrogen, phosphorus and carbon were added singly and in combination. Fig. 14.8 shows the effects obtained for some of the lakes and illustrates the variety of responses: addition of phosphorus alone stimulated algal growth in four lake waters (e.g. in *a*, Fig. 14.8), in two waters addition of nitrogen slightly stimulated growth, e.g. Fig. 14.8*e*, three of the lakes supported high algal growth in the controls, Fig. 14.8*d* and additions had only a slight effect, and one lake, Fig. 14.8*b*, which is highly oligotrophic showed almost no response and at the other extreme

a highly eutrophic lake, Fig. 14.8*c*, also showed only slight response. Individual species also show a variety of responses to added nutrients, e.g. Fig. 14.9 from Jordan & Bender (1973) shows the differing response of seven algae and clearly phosphorus and combinations of this with nitrogen and the chelator EDTA tend to have the greatest effect; the nitrogen and chelator alone had little effect and for some algae the chelator blocks the effect of phosphorus. The latter effect is removed by the addition of nitrogen. Testing of waters by addition of nutrients, sewage etc. ought to be combined with chemical analyses since lack of response to additions may simply be due to the presence of an inhibitory substance, e.g. a heavy metal.

In some lakes the addition of 2 or 3 nutrients increased the growth rates slightly (e.g. in Lake Woahink, Fig. 14.8*a*) suggesting that the nitrogen

Fig. 14.9. The biomass responses of seven different algae (*a*, *Synedra nana*; *b*, *Fragilaria crotonensis*; *c*, *Synedra radians*; *d*, *Nitzschia* sp.; *e*, *Achnanthes* sp.; *f*, *Synechocystis aquatilus*; *g*, *Rhodomonas minuta*) to additions of various nutrients and combinations. *I*, the initial biomass; *C*, control; *N*, added nitrogen; *E*, added chelator EDTA; *P*, added phosphorus. Other columns are combinations of additives. From Jordan & Bender, 1973.

* This organism has now become the major bioassay alga for testing nutrient loading, toxicity, etc. in freshwaters in the United States, *see* for example Middlebrooks, Falkenborg & Maloney, 1976.

Fig. 14.10. ^{14}C uptake rates of *Selenastrum capricornutum* in water from, *a*, Triangle Lake, *b*, Cline's Pond and, *c*, Waldo Lake on four dates following additions of nutrients. SE, secondary effluent; LSD, least significant difference (95 % confidence interval). From Powers *et al.*, 1972.

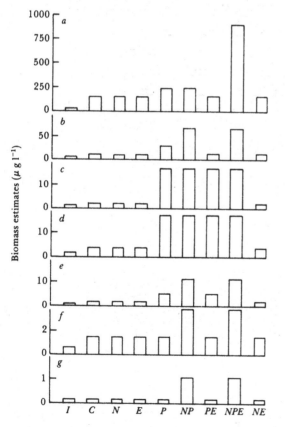

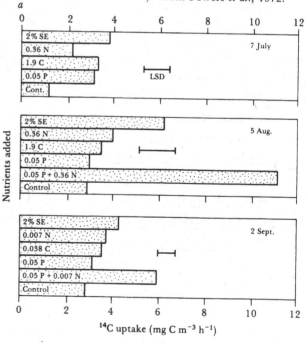

and carbon levels became limiting at high phosphorus concentration. Diamond Water and Upper Klamath lake (Fig. 14.8d) may be so eutrophic that much larger doses are needed to increase growth rates whilst Odell Lake appears to be nitrogen limited. When the lake water alone is used for the growth experiments (controls on Fig. 14.8) there is a linear relationship between dissolved phosphorus concentration and average daily growth rates of *S. capricornatum* but no such relationship between growth rates and nitrogen or carbon. The technique appears simple, but Murray, Scherfig & Dixon (1971) found that the recommended procedures of medium preparation actually removed ion and manganese and that addition of these and aeration with carbon dioxide enriched air, greatly stimulated growth. Similar lack of response to added nitrogen and phosphorus has been found by the author and G. Aykulu for the rich waters of the River Avon, using *Pandorina morum* as a test organism.

The intrinsic rate of growth of species tends to be lower in oligotrophic than in eutrophic lakes (Moss, 1973a) and this is viewed as an adaptation to low influx of potentially limiting nutrients in the oligotrophic lake (Moss, 1973b); uptake rates are also to a certain extent dependent on external concentration (see p. 336).

Lack of growth in oligotrophic water is perhaps not surprising since in such a 'poor' water, essential nutrients other than nitrogen and phosphorus may be limiting. A dangerous conclusion which might be drawn from some of the above data is that addition of nutrient-rich water is without effect on very poor or very rich waters. It would therefore be interesting to see whether addition of nitrogen, phosphorus or carbon to these lake waters would affect the growth rate of the indigenous algal populations and whether the effects were the same at different times of the year. Indeed, Powers *et al.* (1972), using slightly different techniques (e.g. *in situ* [14]C uptake by the light and dark bottles method) showed that in Waldo Lake the additions did have varied effects during four different months and in September overall productivity could be increased several fold (Fig. 14.10). Effect of nitrate or carbon was again insignificant but phosphorus and secondary sewage effluent increased ^{14}C uptake. Similar experiments on Triangle Lake (a nutrient-rich lake) showed significant

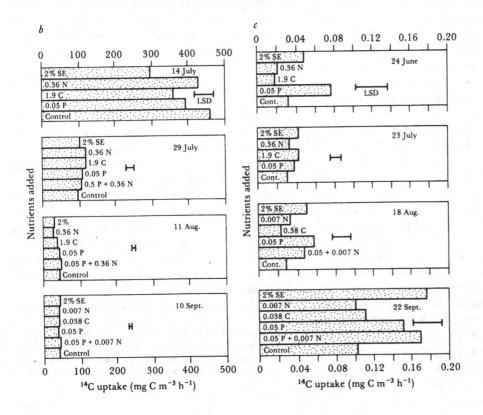

effects of secondary sewage effluent in all months and effects of phosphate and carbon in one month (July) and of nitrate plus phosphate in two months (Fig. 14.10). Whilst in another even more eutrophic lake (Clines' Pond) no effects were noticeable. As these workers point out, when dealing with natural phytoplankton populations, situations are frequently encountered in which more than one nutrient element appears to be limiting at the same time. Since the chemical composition, flora, etc. all change seasonally one would not expect reactions to eutrophication to be constant, indeed another study using a cultural approach with cultures of *Kirchneriella subsolitaria* revealed that winter water from the lakes studied, was phosphate limited but in summer, nitrate became limiting (Potash, 1956).

In the inshore marine habitat it appears that nitrogen is more often limiting than phosphate, e.g. waste waters fed into channels in the Pimlico River estuary, where flagellates were abundant, showed increases related to nitrogen content of the waste water (Carpenter, 1971). The usual concept of nutrient limitation is of only one limiting at any one time but there are well-known physiological and biochemical examples where rates of reaction depend on the concentrations of two reactants. Nevertheless most experiments tend to show that in freshwaters a single limitation, phosphate concentration, is the most important factor with nitrate second (e.g. Stadelmann, 1971), whilst carbon is only rarely important and then the effect is only of a 10–15% stimulation.

In general, deep lakes can withstand the imput of more nutrients than shallow lakes, without adverse effects. This fact was used by Vollenweider (1968) when he proposed permissible loading values related to mean depths. Mean depths of 5, 10 and 15 m can withstand, respectively, nitrogen concentrations of 1.0, 1.5 and 4.0 g m^{-2} yr^{-1} and phosphorus levels of 0.07, 0.10 and 0.25 g m^{-2} yr^{-1}. Doubling these loadings would lead to dangerous situations.

Another approach to the study of eutrophication is an adaptation of the large scale polythene bag experiments devised to estimate primary production by phytoplankton (*see* p. 339). Schelske & Stoermer (1972) used this technique in Lake Michigan and found that the natural phytoplankton populations and rates of carbon fixation increased when nitrate, phosphate and silicate were added to the plastic bags

(1000 and 4000 l capacities) suspended at 7.5 m depths. They found that the silica demand by diatoms is largely controlled by phosphorus supply and that continuous supplies of phosphorus will result in increased removal of silicate by diatoms (this had been shown earlier by Hughes & Lund, 1962). The normal diatom-dominated phytoplankton will be replaced by less desirable* Cyanophyta and green algae. In fact Stoermer (1967) analysed plankton data from Lake Michigan back to the turn of the century and found that diatoms were always the dominant component of the phytoplankton: even in 1962–3 they comprised 70% of the population on a cell count basis but by 1964 only one of 22 samples contained more than 50% diatoms. All the evidence pointed to an acceleration of the change in species composition and to the appearance of a set of more undesirable species.

A subtle, indirect approach to eutrophication has appeared from Lund's intriguing study of the algae and nutrient conditions within very large plastic tubes (enclosing 18 000 m^3 of water) in Blelham Tarn, a lake about which there is already an impressive amount of information. The tubes effectively isolate large cores of water from the lake as a whole and this means that whilst the lake water receives the normal input of nutrients from the land the water in the tubes receives only rainfall and a very small amount of splash. After 2.5 yr the flora in the tubes was qualitatively similar to that outside but the production was dramatically reduced, e.g. during 1971 the average chlorophyll *a* content of the 0–5 m water column in one tube was 3.7 μg l^{-1} whereas in the lake it was 25 μg l^{-1}. Only during winter when physical factors, rather than chemical, control production were the chlorophyll contents similar, 3.2 μg l^{-1} in the tube and 2.9 μg l^{-1} in the lake, lower in the latter due to loss from outflow (Lund, 1975). The result of cutting off normal nutrient sources to water in the tubes was to produce an oligotrophic situation in an otherwise eutrophic lake and is a clear proof of the relationship between eutrophication and nutrient loading.

Detection of eutrophication can be approached

---

* Cyanophyta form water blooms, are driven on to beaches, and their decay, affects amenities and water supplies. Diatoms on the other hand sink when their growth ceases and their decomposition occurs in deep water.

from several angles, e.g. agricultural records will indicate increasing input of nutrients from fertiliser, sewage discharge records will indicate changes in quality and quantity of inflow, long-term chemical records* of lakes, rivers and estuaries will indicate changes, algal analyses, particularly phytoplankton records, will indicate changes in cell numbers and species composition and fisheries statistics will reflect changing productivity. Analysis of organic remains in cores provides a longer record of changes than is usually available from other sources. Czeczuga (1965) records a fluctuating but increasing chlorophyll *a* content in the sediments of Mikolajskie lake from the late glacial to the present day which suggests a very steady enrichment of the basin. There are great problems in the use of sedimentary pigments to indicate changes, since in unproductive lakes the absorption spectra resemble those of woodland humus whereas in productive lakes the pigments are more akin to those of phytoplankton remains (Sanger & Gorham, 1972). These differences probably reflect the greater contribution of allochthonous material in the unproductive and autochthonous in the productive lake. Further subtleties can be deduced from the fact that algae decaying in lakes produce a high carotenoid value and hence a low chlorophyll:carotenoid ratio. Very few studies of pigments have been made but a study of pigments derived from blue-green algae in Esthwaite Water suggests an intensification of blue-green growth some 30 yr ago, a very slow increase in the previous 70 yr and few blue-green algae in the lake 100 yr ago (Griffiths, 1978). Combinations of the various methods have been used, for example by Lund (1969) and Edmondson (1972).

Detection of artificial eutrophication by changes in the flora is open to the criticism that there are often no base-line data and that natural changes are also occurring. However as shown from studies

on cores of sediment from lakes, 'natural' changes tend to be very slow and in fact most lakes are extremely stable during periods of stable climatic conditions. Controls in the sense that similar algal floras have been recorded over many years are available in well studied areas, e.g. the English Lake District where different forms of *Tabellaria flocculosa* occur (Fig. 7.16) in the different lakes (Knudsen, 1955) and these were recorded earlier in photomicrographs of West & West (1906). I have found similar stable species in Irish lakes sampled in 1953 and I suspect that if examination of the relevant historical records shows no drastic alterations in the lake basin it can be assumed that species complements have remained fairly stable.

Nutrient removal

The removal of excess nutrients from sewage effluent is feasible but involves modifications to most sewage plants which are designed to remove solids (primary process), followed by a secondary process which involves either activated sludge or trickling filters to remove nutrients (nitrates and phosphates, 30–50% removal). The original purpose of these processes was to reduce the biological oxygen demand to suitable levels to prevent de-oxygenation of the receiving waters. Removal of the nutrients left after secondary treatment is possible by a variety of processes, one of which involves growth and harvesting of algae from the effluent: others involve ion exchange, electrochemical methods, electrodialysis, reverse osmosis, distillation or chemical precipitation as tertiary processes (Rohlich & Uttormark, 1972). Costing of these is a problem, but Stephan & Schaffer (1970) give costs which work out at a fraction of the expenditure per capita per day on gas, electricity, telephones, food and tobacco. Removal of phosphorus by addition of ferric chloride in activated sludge plants in Switzerland adds 1–2% to the total cost: here the outflow of phosphate must not exceed 2 mg l^{-1} (Thomas, 1969). Figures in Rohlich & Uttormark (1972) indicate that municipal and industrial waste waters add between 36 and 88% phosphorus and 10–75% nitrogen to the budget of various important waters though the overall average is much less. Chemical precipitation (e.g., using aluminium sulphate) is the commonest method to remove these chemicals and some plants are working with tertiary treatment producing

* Care must be taken, however, since if the nutrient measured is one which is rapidly taken up and cycled through the population a long term *decrease* can be recorded in spite of increasing eutrophication, e.g. in Lake Michigan the average annual concentration of silica has decreased by at least 4 mg l^{-1} over the last 44 yr and this is purely due to increased crops of diatoms. Indeed Schelske & Stoermer (1971) suggest that data on long-term changes in silica may be a means of determining the rate of eutrophication of an oligotrophic lake.

nutrient stripped effluents. Cohen (1972) reports that over 95% of phosphorus can be removed at reasonable cost by precipitation and that pilot plant removal of nitrogen is also possible.

Forsberg & Hökervall (1972) have devised an algal measure of growth potential for the sewage effluent from a Swedish sewage treatment plant which uses mechanical, biological and chemical (aluminum sulphate precipitation) processes. They dilute the sewage effluent and then estimate the growth of algae in this medium. The algae used are *Ankistrodesmus falcatus*, *Euglena gracilis*, *Oocystis submarina* var. *variabilis*, *Scenedesmus quadricauda* and *Selenastrum capricornutum* and their growth is compared with that in a standard solution. In an eight-month trial of this technique, growth was shown to be variable and extremely low in July to August corresponding to a holiday period when phosphate in the effluent was extremely low. The low growth of the test species during this period may also be partly due to high aluminium content at a time when phosphate was low. An interesting side effect of these experiments was that mid-week growth was always higher than week-end! Forsberg (1972) showed that growth of *Selenastrum* was good in sewage effluent which had been biologically treated but was greatly reduced if chemical flocculation was also used.

Algal cultures have also been used to test the productivity of water as it passed from a river, through a reservoir and finally through a treatment plant. Fig. 14.11 shows that the growth of two test algae is greatly reduced during the passage, con-

Fig. 14.11. Histograms showing the growth of, *a*, *Synedra acus* and, *b*, *Haematococcus pluvialis* in water from the Wahnbach River, after passage through a pilot plant for controlling phosphorus and turbid materials, after passage through a pre-reservoir where settling can occur and in the main reservoir near the dam, i.e. after flow through the reservoir. Columns are chlorophyll *a* produced (μg l^{-1}). From Clasen & Bernhardt, 1974.

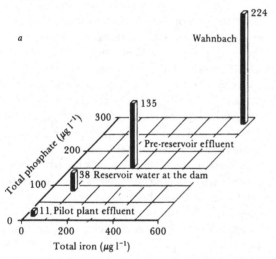

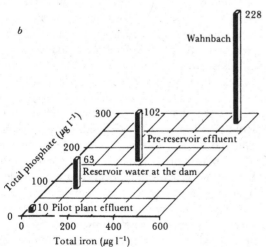

Fig. 14.12. Correlation between chlorophyll *a* and total iron concentrations at the end of growth tests using *Asterionella formosa* and water from the River Wahnbach, pilot plant and reservoirs. From Clasen & Bernhardt, 1974.

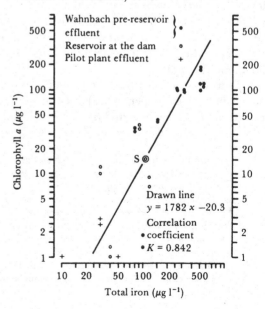

comitant with the reduction in phosphate and iron levels (Clasen & Bernhardt, 1974). In tests on water from highly eutrophic West Berlin waters, Bringmann & Kühn (1971) found that iron followed by silicon and nitrogen were the limiting factors for *Asterionella* and phosphorus was not limiting, as might be expected in such rich waters. Also using *Asterionella* Clasen & Bernhardt found a reasonable correlation between chlorophyll *a* and total iron concentration (Fig. 14.12). Similar relationships have been found in lakes (Fig. 14.13) e.g. in an analysis by Dillon & Rigler (1974) who used data from lakes on the Canadian Shield, and elsewhere suggested that the average summer chlorophyll *a* level could be computed from a single measurement of total phosphorus content at the spring overturn from the expression

$$\log_{10}[\text{Chl }a] = 1.449 \log_{10}[\text{P}] - 1.136 \quad (14.1)$$

This type of study needs testing through a greater range of eutrophic waters though Schindler, Fee & Ruszczynski (1978) have shown good agreement between phosphorus input, chlorophyte and phytoplankton levels.

Removal of nutrients from an actual lake has been attempted in Wisconsin, where Horseshoe Lake (a small body of water of 22 acres) was treated with slurried alum to give a final concentration of about 200 mg l^{-1}. This resulted in the precipitation of phosphate and a decrease in algal crops which has been troublesome in previous years. No obvious adverse results were recorded though no data are given on the actual populations (Peterson, Wall, Wirth & Born, 1973). Another technique used on Trummen in Sweden was to suction-dredge the upper 0.5 m of nutrient-rich sediment out of the lake (Gelin & Ripl, 1978). This resulted in a summer decrease of phosphorus content from 200 μg l^{-1} to < 10 μg l^{-1} and the phytoplankton changed from a blue-green net plankton to a diverse nannoplankton coupled with a decrease in light saturated photosynthesis from 10 g carbon m^{-3} day^{-1} to < 2 g carbon m^{-3} day^{-1}. Removal of hypolimnetic water in some lakes has improved the trophic status by deepening the epilimnion and has thus increased the depth through which phytoplankton can circulate and also remove the pool of rich hypolimnetic water. Hypolimnetic water can only be removed from lakes where there is a reasonable gradient on the outflow stream since it involves siphoning the water by pipe to a lower level on the outflow stream than the level of the hypolimnion of the lake. Many lakes on the continent of Europe and rivers in many countries are now being revitalised by various measures to improve the quality of water in the watershed and a noticeable improvement may be expected provided research and spending on improvements is carried out systematically.

Pollution

Aquatic environments can be polluted in many ways but always by agents which are not produced under natural conditions, e.g. by powdered wastes, toxic chemicals, hot effluents, radionuclides, etc. Unfortunately pollution is not a simple process and it is unlikely that simple solutions will always solve it; the detailed base-line data on natural populations often do not exist and much work has to be done to obtain them. Meanwhile the best solution is to avoid pollution wherever possible and to concentrate on

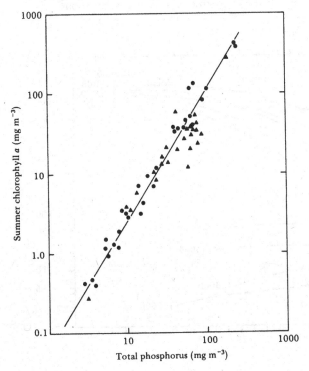

Fig. 14.13. Average summer chlorophyll concentration plotted against total phosphorus concentration at spring overturn. Data from various authorities and lakes. From Dillon & Rigler, 1974.

prevention rather than to attempt to define the pollution loading which can be tolerated, which is rather like experimenting on the degree to which one can tolerate cholera in the population. Pollution can destroy the natural habitats for algal growth, e.g. by blanketing the sediments and/or aquatic macrophytes with silt, coal mining waste, wood pulp waste, etc. Pollution is common in rivers in industrial regions and in extreme cases eliminates the algal flora. If the sediment load remains suspended for any time in the water it will reduce light penetration so much that algal growth is retarded or prevented. This has been demonstrated for some marine species (e.g. *Laminaria saccharina*, Burrows & Pybus, 1971) which can be cultured in water from turbid sites once the sediment has been removed. Temporary cessation of growth at low light intensity is not harmful to *L. saccharina* which can suspend growth for up to six months and then recommence when light becomes favourable (Burrows, 1961). A similar depression of carbon fixation has been shown in freshwaters where light is reduced by pulp-mill effluent (Stockner & Cliff, 1976). The species present were similar, but fixation was reduced from 338–369 mg

carbon m^{-2} day^{-1} to 29–24 mg carbon m^{-2} day^{-1}.

Vast quantities of chemicals are being added to the sea; to take just one example of dumping in a small area, Kayser (1971) reported that 180 tons of sulphuric acid and 250 tons of ferrous sulphate were being dumped northwest of Heligoland per day, 1200 tons of digested sewage sludge every other day and 2500–3500 tons of refinery waste diluted 1:1 at other points in the North Sea. This is the output from one region only. An example of the effect of this waste water on algal growth is given in Fig. 14.14.

Another example, taken from North (1971), quotes a figure of 800 million gallons of liquid waste discharging daily into the sea off Southern California. Most of this has only received primary treatment to remove the larger suspended solids. In spite of such massive additions to the sea, growth of algae in some areas is only slightly modified, e.g. Young & Barber (1973) found that even in York Bight where massive amounts of chemicals are dumped, phytoplankton growth is not consistently affected and the common alga *Skeletonema costatum* often grow well in tests on the water. Production of phytoplankton in fresh-

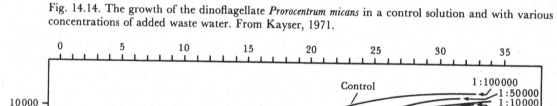

Fig. 14.14. The growth of the dinoflagellate *Prorocentrum micans* in a control solution and with various concentrations of added waste water. From Kayser, 1971.

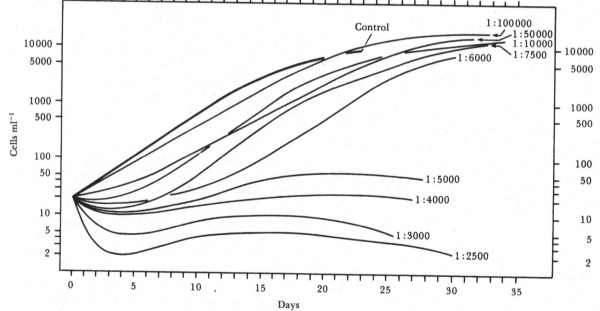

waters (polluted) can also be high, e.g. up to 350–450 g carbon m^{-2} yr^{-1} compared with only 100 g and 185 g in the eutrophic lakes, Lake Erken and Lake Esrom (Rodhe, 1969). In many regions biological arguments against pollution are less effective than aesthetic arguments.

Correlations of algal growth and pollution are not new, even though some publications give that impression; in 1911 Cotton *et al.* reported *Ulva latissima* (= *U. lactuca*) in polluted estuaries whilst at the same time Cotton showed its presence alone was not sufficient to indicate pollution. *Ulva* can use ammonia and under polluted conditions plants have a higher nitrogen and sulphur content than those growing in unpolluted regions (Wilkinson, 1964). *Ulva* is a convenient plant for pollution testing since discs can be cut from the thallus and incorporation of ^{14}C measured; many discs can be cut from a single thallus and thus the same genotype is involved in all the experiments (Burrows, 1971).

Heavy metals[*]

Algae growing in natural habitats react variously to addition of metals, e.g. copper solutions are frequently used to destroy algal populations in water reservoirs, but frequently this is followed by the growth of small coccoid species which are either resistant to the residual copper or more likely grow when the competition has been removed. Bartlett, Rabe & Funk (1974) give algicidal levels for copper, zinc and cadmium of 0.30, 0.70 and 0.65 mg l^{-1} respectively. Experiments in which lead nitrate was added to algal cultures have shown a decrease in carbon fixation with increasing concentration of lead: *Cosmarium* proved particularly sensitive, e.g. 50% reduction in fixation at 5 ppm whereas a diatom *Navicula pelliculosa* was initially stimulated and then was affected at approximately 28 ppm (Malanchuk & Gruendling, 1973). On the other hand an alga which is apparently a delicate flagellate was stimulated to increased photosynthesis by increasing concentrations at least up to 30 ppm.

Seaweeds seem to have little control over their uptake of heavy metals (Fuge & James, 1974) and this was clearly shown in a detailed study of the uptake of zinc and ^{65}Zn by *Laminaria digitata* (Bryan, 1969). He showed that there was a gradual net

[*] This is a very much studied topic and only a very brief treatment is possible here.

accumulation of zinc during growth of the plant and a linear relationship between amount of zinc in the frond and the concentration in the seawater. Very similar results were obtained by Hopkins & Kain (1971) for *L. hyperborea*. At the highest levels of zinc the growth of *Laminaria* was reduced. No effect on morphology could be found in four red algae grown in increasing concentrations of lead but there was merely a reduction in the overall growth (Stewart, 1977a). Burrows (1971) reports that concentrations of 0.25 ppm chromium caused a significant drop in the mean growth rate of *Laminaria saccharina* and concentrations greater than 1.0 ppm were lethal. The concentrations of some metals by *Fucus vesiculosus* and *Porphyra* collected on the shores of the Irish Sea are given in Table 14.2 (Preston *et al.*, 1972). Preston *et al.* were able to compare their data with results from 50 yr earlier and with the possible exception of cadmium, which had decreased, the concentrations had not changed appreciably.

Metals added in combination have varied effects on marine species, e.g. Cu and Zn acted synergistically on *Amphidinium carterii* and *Thalassiosira pseudonana* but antagonistically on *Phaeodactylum tricornutum*. On *Skeletonema costatum* the combination had a greater effect than the sum of the combined (Braek & Jensen, 1976).

As in other seaweeds (e.g. *Fucus vesiculosus*, Gutnecht, 1963) little of the absorbed metal could be washed out of *Laminaria* tissues and Bryan concluded that it was tenaciously bound. A similar relationship between concentration in the medium and in algae has been shown for freshwater species (*Chlorella vulgaris*, *Euglena viridis* and *Pediastrum tetras*) by Coleman, Coleman & Rice (1971). However a naked flagellate *Dunaliella tertiolecta* was affected by lead exactly as were algae possessing cell walls (Stewart, 1977b) and this same flagellate shows a remarkable tolerance to mercury (Davies, 1976). In *Asterionella formosa* there seems to be an active regulatory mechanism for excluding arsenic from the cells but no such mechanism for cadmium (Conway, 1978). The arsenic was apparently trapped in the organic layer outside the silica wall but the cadmium entered the cell.

Transference of heavy metals from algae to animals has been shown by Stewart & Schultz-Baldes (1976), who grew the brown alga *Egregia laevigata* in lead-enriched water and fed this to abalone (*Haliotis*

Table 14.2. *The concentrations of heavy metals in* Fucus vesiculosus *and* Porphyra. *From Preston* et al., *1972*

	Irish Sea, 1970	Concentration factors $\left(\dfrac{\mu g\ g^{-1}\ \text{dry material}}{\mu g\ g^{-1}\ \text{seawater}}\right)$
	Fucus (18 observations)	*Porphyra* (20 observations)
Zinc	2.0×10^4	1.0×10^4
Iron	2.4×10^4	3.6×10^4
Manganese	2.3×10^4	7.0×10^3
Copper	4.5×10^3	6.3×10^3
Nickel	2.8×10^3	1.1×10^3
Lead	2.4×10^3 (13 observations)	2.0×10^3 (15 observations)
Silver	5.0×10^3	1.9×10^3
Cadmium	2.7×10^3	6.6×10^3

spp.), which accumulated lead up to 21 $\mu g\ g^{-1}$ wet weight.

There is ample evidence that the response of algae to polluting chemicals varies according to degree of temperature or light stress. Experimental evidence of this was obtained by Jitts *et al.* (1964) when cultures of *Pavlova* (*Monochryris*) *lutheri* and *Thalassiosira nordenskiöldii* were accidentally poisoned with the result that growth only occurred in the tubes which happened to be in the optimum temperature and light regimes whilst elsewhere the combination of stress from three sources prevented growth. Antia & Cheng (1975) succeeded in adapting some algal species to high concentrations of borate and found that some could tolerate 50 mg l^{-1} but none 100 mg l^{-1}. A further important consideration is the length of time taken for an alga to adapt to increasing concentrations of nutrients, which may be 20–40 days (Stockner & Antia, 1976). Stockner & Antia very wisely point out the importance of distinguishing between 'shock' effects and long-term 'habitation' (Fig. 14.15).

Colloidal and other materials in natural waters can act as chelators and therefore reduce the toxicity of added metals, e.g. copper (Morris & Russell, 1973); such effects make tests under laboratory conditions somewhat unrealistic.

Some algae growing at low pH, e.g. *Eunotia exigua* (at pH 3.5–5.0) and *Pinnularia interrupta* f. *biceps* (pH 2.0–3.0) were found to be very resistant to zinc; levels > 10 mg zinc l^{-1} were tolerated (Besch, Ricard & Cantin, 1972).

Care must be taken to ensure that chemical analytical methods measure only the available amounts of an element. Many modern techniques merely measure totals on highly sophisticated apparatus. No one should assume that the analytical problems are solved, for not even the common methods for such ions as phosphate are free of criticism.

Radionuclides

Artificial radionuclides are absorbed by algae (*see* Robertson, 1972 for general discussion) and concentrated to excessive amounts in algal cells. Such radionuclides are derived mainly from atmospheric fallout from nuclear weapon testing, with additions from discharge by nuclear processing plants, nuclear power stations and nuclear powered ships. Many other uses for nuclear power have been suggested or are being developed and if these proceed the amount of radionuclides available to algae may well increase. Since many nuclear projects are planned along estuaries and shore lines the algae of the shore and inshore waters will be particularly vulnerable.

Adsorption on the surfaces of the algae occurs, e.g. iron can be adsorbed on to the surface of *Phaeodactylum tricornutum* (Davies, 1967) and [106]Ru on the surface of *Porphyra* (Jones, 1960). Work by

Bachmann & Odum (1960) showed that six marine benthic algae took up [65]Zn in the light but not in the dark. In more extensive experiments on *Chaetomorpha*, suspended in bottles at various depths, the uptake was proportional to gross oxygen production and accumulation was in proportion to net oxygen production. There has been disagreement about the mode of uptake of [65]Zn, e.g. Gutnecht (1961) thought that cation exchange is involved since uptake and loss was pH dependent, but it is probable that a whole complex of processes is involved. The uptake of ^{32}P by *Oedogonium* is increased considerably by increase in water flow over the species (Whitford & Schumacher, 1961), a feature which could have important consequences if radionuclides escaped into streams.

The fact that concentration of radionuclides occurs is merely a reflection of the fact that non-radioactive elements are also taken up in excess amounts by algae and this property can be utilised, as mentioned on p. 541, for removal of elements from aquatic systems.

Organic compounds

Organic pollution results in removal of excessive amounts of oxygen from the water as respiration leads to breakdown of the organic compounds. As mentioned above addition of organic compounds produces varying degrees of eutrophication and only when massive amounts are added does the material act as a pollutant. Algal growth in the same habitat will help to correct this oxidative removal of oxygen, e.g. Edwards & Owens (1965) found that bottom muds in the River Ivel consumed 2–4 g oxygen m^{-2} day^{-1} and that algal populations (mainly diatoms) on the sediments produced 0.28–2.8 g oxygen m^{-2} h^{-1}; in such a situation a balance would result. Production of oxygen by plankton will also contribute to re-aeration and the total photosynthetic oxygen production from algae is of great importance in

Fig. 14.15. Growth of *Skeletonema costatum* in various concentrations (as percentages) of pulp-mill effluent and in filtered seawater controls (C1, C2). Note the periods required for adaptation in the 20% and 30% experiments. From Stockner & Antia, 1976.

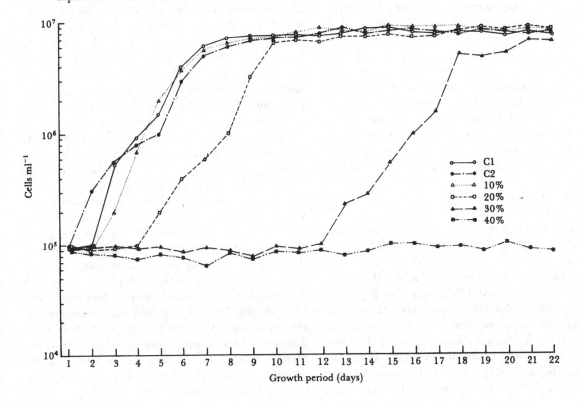

polluted waters. Knöpp (1960) working on the River Rhine found that from May to September photosynthetic oxygen formation was more important than atmospheric re-aeration. Gross rates between 3 and 35 g oxygen m^{-2} day have been recorded from various rivers and a considerable proportion of this is undoubtedly produced by algae. In many sites bubbles of oxygen can be seen covering the sediments, stones and aquatic plants, a direct result of algal photosynthesis.

Removal of organic compounds by algal growths is often possible, e.g. Maloney (1959) found that *Chlorococcum macrostigmatum* could use the sugars (glucose and mannose were present) in spent sulphite liquor both in the dark and in the light, with good growth in the latter. The experiments on sewage oxidation ponds in California have amply demonstrated this ability, the great complication is that the production of algal organic matter is colossal and the removal and utilisation of this is a major problem.

The differential effect of oil spillage on algae and animals was demonstrated when oil seeped from the tanker *Tampico Maru* off Baja California in 1957 and caused considerable damage to the benthic invertebrates whereupon the dwarf *Macrocystis* plants rapidly grew into a rich forest (North, Neushul & Clendenning, 1965). Preliminary studies have shown that outboard engines discharge oil and gas mixtures through the exhaust system into waters and these have been shown to have severe effects on ^{14}C uptake by planktonic algae. However the natural implications are quite unknown. On the other hand, concentrations of < 1 mg l^{-1} from oil pollution seems to have only mild inhibitory or stimulatory effects on marine phytoplankton (Prouse, Gordon & Keizer, 1976).

The concentration of insecticides in food webs is now a well-established fact and algae are no exception, e.g. filamentous algae in a laboratory stream have been shown to concentrate Dieldrin up to 30000 times the concentration in the water (Rose & McIntire, 1970). Several workers (Meeks & Peterle, 1967; Woodwell, 1967; Ware, Dee & Cahill, 1968) have shown that DDT is concentrated in algae, especially in *Cladophora*. There is a great range in toxicity of insecticides and this was shown by Hopkins & Kain (1971) who found that Atracine was toxic to *Laminaria hyperborea* in culture at 0.01 ppm, MCPA and 2–4 D at 1.0 ppm and Dalapon at 100 ppm. The effect of organic pollutants is complicated if certain compounds are incorporated into cells and others excluded, such as was found for estuarine diatoms by Thompson & Eglington (1976, 1979). These diatoms incorporated low molecular weight aliphatic hydrocarbons and excluded polycyclic hydrocarbons.

Thermal pollution

Pollution by heated effluents is becoming a common feature with the increasing release of warm water effluents from sources such as power stations sited on rivers, lakes and estuaries where cooling water is readily available. The term *heated streams* should be used for these artificially heated flowing waters to distinguish them from the natural warm *geothermal waters* (Castenholz & Wickstrom, 1975). If the increase in temperature is not too drastic the flora is not harmed. However, should the temperatures be maintained above 35 °C the usual flora of diatoms and Chlorophyta can be replaced by Oscillatoriaceae (Whitton, 1972) and the usual reduction in species diversity under stress is accompanied by excessive growth of the few tolerant species. When the source of thermal pollution is removed, the flora may quickly revert to its original state, e.g. Swale (1962) found that *Pithophora oedogonia*, *Chara braunii* and *Compsopogon* sp. occurred in the Reddish Canal when warm water was introduced but eighteen months after the removal of the effluent the species had disappeared. Results derived from the glass slide technique (Patrick, Crum & Coles, 1969) revealed a shift from diatoms to Cyanophyta when stream temperatures were raised to 35–40 °C. At 20–28 °C diatoms were dominant and at 30–35 °C green algae were most abundant. It is interesting that the sequence diatoms → green algae → Cyanophyta is also the common seasonal sequence in lakes as the waters warm during summer. Hickman & Klarer (1974) showed that heated effluent kept parts of a Canadian lake ice-free and enabled the diatom flora to extend its growing period throughout the year with subsequent production of higher crops, which may be a distinct advantage to secondary producers. Only during late summer and early autumn was there a change in the composition of the flora from diatoms to green algae, *Oedogonium/Spirogyra* (Hickman & Klarer, 1974). Hickman & Klarer wisely used the natural flora, epiphytic on *Scirpus* to examine the effects of the warm effluent.

Table 14.3. *The distribution of algae in American and British streams with pH below 3.0. Adapted from Hargreaves* et al., *1973*

Species	USA	England	Species	USA	England
Chlamydomonas applanata var. *acidophila*	·	+	*Navicula nivalis*	·	+
Chlamydomonas globosa	+	·	*Nitzschia subcapitellata*	·	+
Chlamydomonas spp.	+	+	*Nitzschia elliptica* var. *alexandrina*	·	+
Cryptomonas erosa	+	·	*Nitzschia palea*	·	+
Cryptomonas ovata	+	·	*Nitzschia ovalis*	·	+
Cryptomonas sp. (not above)	·	+	*Nitzschia* spp. (not above)	·	+
Chromulina ovalis	+	+	*Ochromonas* sp.	+	·
Characium sp.	·	+	*Penium jenneri*	+	·
Chlorogonium elongatum	+	·	*Phaeothamnium* sp.	+	·
Desmidium sp.	+	·	*Pinnularia acoricola*	·	+
Euglena mutabilis	+	+	*Pinnularia braunii*	+	·
Euglena sp.	+	·	*Pinnularia microstauron*	·	+
Eunotia exigua	+	+	*Pinnularia termitina*	+	·
Frustulia rhomboides	+	·	*Stichococcus bacillaris*	·	+
Gloeochrysis turfosa	·	+	*Synedra rumpens*	+	·
Hormidium rivulare	·	+	*Ulothrix zonata*	+	+
Lepocinclis ovum	+	+	*Ulothrix subtilis*	+	+
Microthamnion strictissimum	·	+	*Ulothrix tenerrima*	+	·
Mougeotia sp.	+	·	*Ulothrix* sp.	+	·
Navicula spp.	+	+	*Zygogonium ericetorum*	·	+

Water is abstracted from natural sources and passed through cooling towers in some electric power stations and if the temperature is not raised too dramatically the rate of photosynthesis of algae in the effluents can be increased, e.g. up to 16 °C, but at 23 °C the fixation was reduced to very low levels (Morgan & Stross, 1969). Experiments showed that no harm occurred during the cooler months of the year but in August the increase in temperature was such that photosynthesis was reduced from 99 to 51 mg carbon m^{-3} h^{-1}, but even this can vary according to the time of day, being higher in the morning. Fish and crustacea farming in warm effluents is proving possible and presumably enhanced algal growths are advantageous. There is still much experimental work to complete on such systems but wherever there is a supply of warm water it may be profitable to harness it for food production though it is necessary to check for heavy metal contamination from boilers, piping, etc.

Acid pollution

Drainage from mining sites and from some other industries can produce extremely acidic environments (pH below 3.0) and is also often contaminated with heavy metals. Nevertheless Harrison (1958) found the diatom *Frustulia rhomboides* var. *saxonica* in water from mine streams in South Africa at pH 2.3 and sulphate content of 475 mg l^{-1}, presumably protected by the mass of mucilage in which it was growing. High iron and sulphate concentrations have also been reported in other acidic waters; in some waters these are formed due to the activity of sulphur oxidising bacteria (*Thiobacillus*).

The most recent survey is that of Hargreaves, Lloyd & Whitton (1975) who investigated all the sites they could find in England with water below pH 3.0. Of the fifteen sites, thirteen were associated with coal mining, one with a barytes mine and one was caused by industrial effluent. Table 14.3 is compiled from their data and their search of the American literature; clearly this extreme habitat is species-poor and at the English sites the number of species varied from one to sixteen. Besch, Ricard & Cantin (1970) found that a few kilometres below an acid zone the diatom *Achnanthes microcephala* was colonising almost 100% of the surfaces.

A similar degree of acidity can be found in natural peat pools where *Sphagnum* removes cations and releases hydrogen ions into the water. These habitats usually have a pH between 3 and 5. The species most commonly recorded in such waters are flagellates of the genera *Chlamydomonas, Spermatozoopsis, Euglena, Lepocinclis, Trachelomonas, Peridinium, Cryptomonas* and *Gonyostomum*. Cassin (1974) found that *Chlamydomonas acidophila* could grow at pH 2.0, could be cultured in the dark on glucose and that hydrogen ions seem to be excluded from the cell by some mechanism. Presumably only algae which have the ability to exclude excess hydrogen ions are able to colonise such waters.

It is always necessary to check that extreme conditions recorded in nature can also support growth of algae, as opposed to simple existence until conditions improve. At least with some of these acid tolerant algae this has now been done and, in fact, some species have been found to grow at slightly lower pH values than recorded in nature, down to pH 1.3 for *Chlamydomonas applanata* var. *acidophila* and *Euglena mutabilis* (Hargreaves & Whitton, 1976).

Pollution testing

Testing of suspected pollutants for their effect on algae is carried out by adding known concentrations of pollutants to algal cultures and following growth rates, determining lethal concentrations, etc. Perhaps of even greater value is the testing of samples of water from variously contaminated sites to determine the effects of the suspected pollutants working within the complex of factors in the natural water. Even better is the study of algae themselves within the polluted habitat, either the indigenous flora or cultured populations immersed within it, e.g. on artificial substrata. It will be exceptional if doses, etc. determined under laboratory conditions have identical effects in nature (*see* also comments p. 535); thus Burrows (1971) found no consistent pattern of growth rate of test organisms when grown in waters taken from the northeast coast of England, where a gradient of pollution had been determined by Bellamy & Whittick (1968). As she suggests, this is not surprising, since there are tide-cycle variations in waters used for testing which are greater than the gradient of pollution. It is necessary therefore to know the rates of supply, rates of dispersal under differing tide patterns, seasonal variations of flow, salinity, temperature, etc. since all interact with the pollutant in nature.

Coastal pollution is very widespread and as Burrows (1971) suggests the large, attached algae of the intertidal are ideal test organisms since it is known that they are sensitive to changes in the composition of their growth media. Testing the effect on the natural ecosystem is complicated by all the features discussed in Chapter 3 and we still know too little of the natural cycles which undoubtedly occur; nevertheless valuable insight into the problem can be gained from a study of the effects on the total system and its subsequent recovery whenever pollution occurs, e.g. in the case of the *Torrey Canyon* oil pollution and subsequent detergent damage (*see* p. 417). Burrows used *Laminaria saccharina* and *L. hyperborea* as test organisms. Sporophytic plants are grown in the laboratory from single parent plants, using control seawater made up into Erd–Schreiber medium to ensure an adequate supply of nutrients. Growth rate is determined as mean percentage increase in surface area of the lamina. Alternatively the growth rate of young gametophytes from zoospores settled on slides can be used, though this technique has the disadvantage that different samples of seawater have selective effects on the zoospores, so that those developing already show some adaptation to the conditions under test (*see* p. 75 for use of this technique in nature). A suitable source of control seawater reveals certain problems and water had to be collected from the Irish sea, 16 km from the Isle of Man. Obviously coastal water would be unsuitable and even open oceanic water is now liable to pollution making the development of artificial media for growing large algae and especially oceanic phytoplankton an urgent task. Fig. 14.16*a* shows the growth rate of sets of *Laminaria saccharina* in water from various sites in Liverpool Bay; stations III, IV and V are close to the point of sewage sludge dumping on the day of collection of the water sample.

Culture of zoospores of *L. saccharina* to the fertile stage and subsequent production of young sporophytes require a minimum mean daily illuminance of 560 lux and 225 lux respectively (Burrows,

1971) and though this is relatively low such values and lower could easily be produced by heavy silting of the water. Fig. 14.16*b* shows that the total amount of light received, and not the day length, is important, 9 or 18 h being equally effective. The very low intensities simply result in a suspension of growth which, under experimental conditions, can last for as long as six months after which growth will start again if the cultures are transferred to a suitable light regime. Older sporophytes of *L. saccharina* could however grow with a mean daily illuminance of only 7.5 lux so that the effect of silt is likely to be most important at the establishment stage of the sporophyte. The breadth of the lamina of *L. saccharina* seems to be a good indicator of growth in older plants since plants grown under silted and polluted conditions were much narrower than those grown under non-polluted conditions (Burrows, 1971). It is important to determine the effects of pollution on plants which are abundant in the coastal regions and which, like *Laminaria*, have complicated life histories

since the various stages are liable to be affected to differing degrees by pollution. However, it is obvious that the complexities of growth make these plants very difficult to handle for routine pollution testing which is probably better performed on unicellular algae. This was done by Kayser (1971) using *Prorocentrum micans, Ceratium furca* and *Phaeocystis pouchetii* and Fig. 14.14 shows the effect on growth of *Prorocentrum micans* of various dilutions of waste water. This reveals the necessity for and degree of dilution required in batch culture; experiments using a turbidostat yielded very similar results. Axenic cultures of *Scenedesmus obliquus* were used by Braune (1971) in a careful study of river water; the photosynthesis of the cultures was measured by the oxygen production technique with platinum electrodes and Braune found, as is often the case, that polluted waters sampled below a town stimulated photosynthesis more than the pure waters from above the town.

In situ testing of the degree of eutrophication

Fig. 14.16. *a.* The growth rate of *Laminaria saccharina* in culture media made up with water from different sampling stations in Liverpool Bay. The percentage increase in surface area was measured over a 3-week period. Station III is at the Northwest Light Buoy where sewage sludge is deposited and water from Stations III, IV and V was brown from this dumping on the day of sampling.
b. The mean developmental stage reached by developing zoospores of *Laminaria saccharina* at time intervals under different mean daily illuminances, the values made up using daylengths of 9 h or 18 h and the appropriate light intensities. Stage 1, settled zoospore; Stage 2, developing gametophyte; Stage 3, fertile gametophyte; Stage 4, first sporophyte division; Stage 5, later sporophyte stages. From Burrows, 1971.

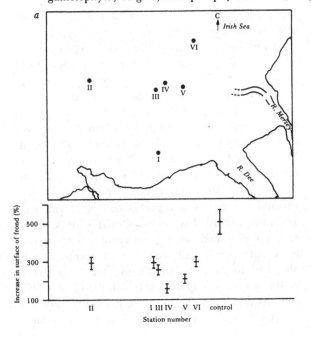

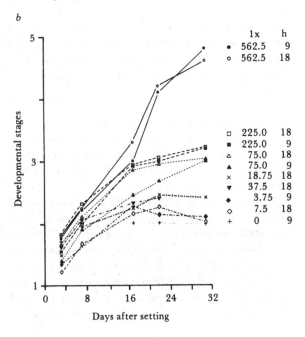

or pollution of rivers, estuaries and lakes has been achieved by analysing species composition of algae developed on glass slides held in position in a variety of frames. Biomass of algae on the slides is also sometimes determined. The method is highly selective of species capable of growing on such surfaces, e.g. the pennate diatoms *Cocconeis placentula*, *Gomphonema parvulum*, *Synedra ulna*, *Nitzschia* spp., *Navicula* spp., whereas the planktonic species which are often centric forms (*see* p. 267) and the ones likely to be stimulated by eutrophication are hardly detected by this technique. A discussion of the method and evidence against using this method to study algal floras is given on p. 126. Two species, *Nitzschia palea* and *Gomphonema parvulum* are tolerant of organic pollution and tend to be the dominant organisms in such situations (Weber & Raschke, 1970): they are however very widespread organisms and occur in many habitats which would not be regarded as polluted. The absence of other species is a surer indicator of polluted conditions.

Although glass slides are usually used since they can be observed directly under the microscope, Hohn (1964) tested many substrata and showed that styrofoam was the best for diatom growth.

A better *in situ* technique was developed by Braune (1966) who used *Scenedesmus obliquus* in 10 cm long glass tubes closed at either end by membrane filters and suspended in a river. Diffusion of ions was possible and in the polluted zone, growth was not only poor but the alga grew as single cells, *not* in the normal coenobial morphology. Similar experiments in the laboratory yielded good growth, illustrating yet again how important it is to do pollution testing in the field.

Indicator species and communities

Pollution detection based on an analysis of the species complement (plants and animals) has been used for many years, and Stein & Denison (1967) maintain that such biological indicators are better than chemical or physical features used alone, 'Chemical observations measure conditions whereas biological observations measure effects'. Those based on animals suffer from the drawback that the animals are often motile and move from the site. A similar objection can be made to the use of the phytoplankton which is often carried in currents and is usually regarded as less satisfactory than benthic

species (Fjerdingstad, 1971, who gives a detailed account and bibliography). Surprisingly, considering how often they are quoted, Fjerdingstad found diatoms to be rather unreliable, yet others, e.g. Zelinka & Marvan (1961) use them extensively. The classic scheme proposed by Kolkwitz & Marsson (1908) but actually derived from their publication of 1902, proposed five pollution zones containing characteristic species, which were termed polysaprobic, α-('stark') and β-('schwach') mesosaprobic, oligosaprobic and katharobic and they listed indicator species for all these zones except the last one which is in fact clean water. This system has been criticised by many workers, though practically all subsequent schemes involve similar divisions. In this decade, concern with pollution and ecology gives the impression that they have just been discovered – but if one reads Kolkwitz & Marsson all the basic ideas of relationships between flora (and fauna), chemistry of the water, purification ('selbstreinigung'), etc. were being worked out from 1870 onwards and by the turn of the century were well understood in some quarters. I recommend a careful reading of these papers by anyone starting work in this field, there is little point in merely confirming what was known a century ago! Kolkwitz & Marsson clearly recognised the almost complete absence of algae in the polysaprobic zone, except for *Arthrospira* (*Spirulina*) *jenneri* and *Euglena viridis*, the preponderance of Cyanophyta in the α-mesosaprobic zone, the rich diatom and green algal flora of the β-mesosaprobic zone and the occurrence of Peridiniales and Charales in any quantity only in the oligosaprobic zone. They recognised that the benthic flora was important and also that many of the species would have a widespread distribution in non-polluted waters since all waters have a lesser or greater loading of organic matter. Sixty years later, in the system of Zelinka & Marvan (1961) the pattern is the same but a few more species have been moved up into the α-mesosaprobic zone and using their algal indicators I would deduce that many lakes and rivers which are generally regarded as non-polluted waters would end up classified as α-mesosaprobic or even β-mesosaprobic! Equally unfortunate results can be obtained in some habitats using the Kolkwitz & Marsson system, e.g. Casper & Schultz (1960, 1962) showed that a site in Hamburg generally agreed to be polluted (polysaprobic) fell into the β-meso-

Table 14.4. *The saprobic zones with some common communities found within them. From Fjerdingstad, 1967*

Zone I   Coprozoic
> (a) Bacterium community; (b) *Bodo* community; (c) both communities

Zone II   α-polysaprobic
> (1) *Euglena* community; (2) Rhodo- and thio-bacterium community; (3) pure *Chlorobacterium* community

Zone III   β-polysaprobic
> (1) *Beggiatoa* community; (2) *Thiothrix nivea* community; (3) *Euglena* community

Zone IV   γ-polysaprobic
> (1) *Oscillatoria chlorina* community; (2) *Sphaerotilus natans* community

Zone V   α-mesosaprobic
> (a) *Ulothrix zonata* community; (b) *Oscillatoria benthonicum* community; (c) *Stigeoclonium tenue* community

Zone VI   β-mesosaprobic
> (a) *Cladophora fracta* community; (b) *Phormidium* community

Zone VII   γ-mesosaprobic
> (a) Rhodophyceae community (*Bactrachospermum moniliforme* or *Lemanea fluviatilis*); (b) Chlorophyceae community (*Cladophora glomerata* or *Ulothrix zonata* (clean-water type))

Zone VIII   Oligosaprobic
> (a) Chlorophyceae community (*Draparanaldia glomerata*); (b) pure *Meridion circulare* community; (c) Rhodophyceae community (*Lemanea annulata, Batrachospermum vagum* or *Hildenbrandia rivularis*); (d) *Vaucheria sessilis* community; (e) *Phormidium inundatum* community

Zone IX   Katharobic
> (a) Chlorophyceae community (*Chlorotylium cataractum* and *Draparnaldia plumosa*); (b) Rhodophyceae community (*Hildenbrandia rivularis*); (c) lime-encrusting algal communities (*Chamaesiphon polonius* and various *Calothrix* species)

(a), (b), (c), alternatives; (1), (2), (3), differences in degree.

saprobic with a tendency to oligosaprobic if one used the indicator species method. This I suspect is due to using species lists containing many forms which have a very wide range of tolerance. Obviously for this type of pollution index only species with narrow ranges are suitable and unfortunately the fewer the species listed the more difficult it becomes to find them. The method is however one of the most reliable and therefore careful re-evaluation of the lists should be made. Desmids which are usually regarded as characteristic of oligosaprobic waters can however be used to distinguish various trophic states but the situation is fairly complex and requires careful consideration of the whole desmid flora; Coesel (1975) gives a very valuable account.

Some studies of pollution (and eutrophication) are published in which long lists of genera are given as characteristic of certain states; such lists are virtually meaningless since few genera are characteristic of any one state of pollution.

A detailed discussion of the various classifications of polluted waters and incidentally the excessive terminology which has been developed can be found in Fjerdingstad (1971). Fjerdingstad comments that biological assessment is preferable, rapid and accurate in the hands of skilled biologists and the recent results of Lange-Bertalot (1978) show that this is an effective method when diatoms are utilised. How much more preferable if the efforts spent on measuring degrees of pollution were spent on developing methods to eliminate it!

Fjerdingstad (1950) in his now classic study of the River Mølleaa in Denmark designated 30 communities of micro-organisms in the zones of increasing organic pollution, in a system intended for use on waters receiving sewage plus some industrial waste. Such a large number of communities can lead to confusion in studies designed from a pollution angle rather than a plant sociological study and so by 1964 Fjerdingstad had reduced his system to nine zones of decreasing degree of pollution: coprozoic, a zone of undiluted faecal water, α, β, γ; polysaprobic, α, β, γ; mesosaprobic; oligosaprobic; and katharobic. In fact many algal ecologists would not consider those below the β-mesosaprobic zone as being particularly polluted. Table 14.4 gives a brief summary of Fjerdingstad's analysis of zones and communities (fullest details are given in his 1964 paper). Species have been designated as saprobiontic (occurring in the most heavily polluted waters), saprophilous (present in polluted but extending into other waters), saproxenous (found regularly in unpolluted water but may thrive even in the presence of pollutants) and saprophobous (organisms unable to thrive in polluted situations).

Numerous variations have been introduced since Kolkwitz & Marsson first proposed their system but many are terminological variants and the basic polysaprobic, mesosaprobic, oligosaprobic, katharobic classification remains. Attempts to evolve numerical systems have been devised but as Hawkes (1957) maintains 'biological communities are complex systems and cannot be reduced to exact numerical values to be entered into neat columns alongside the analytical (chemical) results'. No doubt very intensive and precise data obtained from refined sampling techniques will improve this approach and may enable certain regions to be submitted to numerical analysis but it is doubtful whether numerical values obtained from different sites can be compared. Indices such as that of Knöpp (1954) (see below) rely very heavily on correct identification, enumeration and classification of species and give rise to 'mathematical pseudo-exactness on uncertain basis' (Elster, 1960). A similar system which merely involves gross identification of groups was proposed by Dresscher & Mark (1976) who divided groups of unidentified species into Group A = Ciliates, indicating polysaprobic conditions, Group B = Euglenophyta, indicating α-mesosaprobic conditions, Group C = Chlorococcales + Diatoms, indicating β-mesosaprobic conditions, Group D = Peridinians + Chrysophyceae + Conjugate Green algae, indicating oligosaprobic conditions. The saprobic index is given by $\dfrac{C+3D-B-3A}{A+B+C+D}$ which varies from -3 to $+3$ giving six stages of saprobity. A common fallacy in studies using indicator organisms is that those indicative of certain degrees of pollution are restricted to such sites. They rarely are, and indeed may be very abundant in other sites due to the presence of other favourable factors. This does *not* mean that any site supporting these organisms is also polluted. This kind of evidence has been used quite unjustifiably to criticise systems such as those of Kolkwitz & Marsson and later schemes elaborated from their work. The problem arises because most studies use only the *presence* and not the *absence* of species, whereas in reality the absentees are as informative as the indicators but this approach would require more detailed studies approaching the problem of polluted waters from a study of clean waters. The system of using indicators has not failed as stated in Patrick (1973); it is the interpretors who have failed and I hope no one would expect to find any one species indicative of pollution in general.

An index which does not rely on identification of algae has been proposed by Weber (1973): he terms it an autrophic index (AI) and it is expressed as $\dfrac{\text{biomass (dry wt organic matter)}}{\text{chlorophyll } a}$.

This was found to be low at unpolluted and high at polluted stations. The high values are due to an excess of heterotrophic organisms.

The concept of small populations of many

$$\text{Relative load} = \frac{\Sigma\,(\alpha\text{-mesosaprobic} + \beta\text{-mesosaprobic})}{\Sigma\,(\text{oligosaprobic} + \alpha + \beta\text{-mesosaprobic} + \text{polysaprobic})} \qquad (14.2)$$

and Pantle & Buchs (1955)

$$\text{Saprobial index (S)} = \frac{\Sigma\,\text{Polysaprobic}\,\%\times 4 + \alpha\text{-mesosaprobic}\,\%\times 3 + \beta\text{-mesosaprobic}\times 2 + \text{oligosaprobic}\times 1)}{100}$$
$$\qquad (14.3)$$

species (i.e. high diversity) in unpolluted streams and of larger populations of fewer species (low diveristy) in polluted streams was developed by Patrick in 1949 and Fig. 14.17 shows the distribution of numbers of diatom cells and species collected on glass slides in a normal stream. Pollution caused a shift in this curve but it is admitted that certain types of pollu-

Fig. 14.17. Graphs of diatom species recorded on glass slides taken from a diatometer set, *a*, in an unpolluted river, *b*, in a river showing mild pollution and, *c*, in a river showing severe pollution. Plots are based on a truncated normal curve, and papers by Patrick give details. From Patrick, 1957.

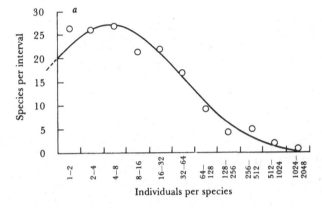

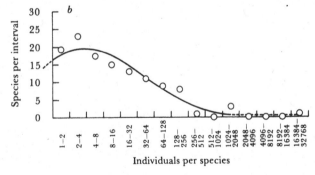

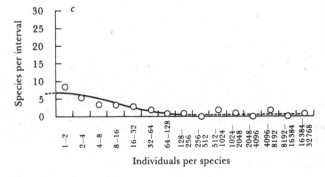

tion, e.g. those arising from low pH, give curves not so different from the unpolluted and make interpretation very difficult, apart from the difficulties of identifying and counting some 5000–8000 cells which is required by this method. It is not entirely satisfactory to base a system of monitoring pollution on a technique which requires expert taxonomic knowledge of a complex group of organisms such as the diatoms, particularly when there is as yet no complete freshwater or marine diatom flora. Similar changes in diversity are recorded in marine coastal sites, e.g. Murray & Littler (1974) found that on the Californian coast the number of macrophytic algae was reduced from 30 to 13 adjacent to a sewage outfall. Although there was also an 11.7% reduction in cover near the outfall the net primary production (485 g carbon m^{-2} yr^{-1}) was comparable to that in non-polluted areas. Near the sewage outfall the cover of blue-green algae was particularly well developed (Littler & Murray, 1975).

Diversity as such is a desirable feature within a community, since it spreads the risk and, assuming that there will always be stresses on the community, it prevents excessive damage, on the same principle that diversity in an industry is a defence against changes in supply and demand, etc. An interesting development from studies of diversity has been evolved by Cairns & Dickson (1971) and applied by some aquatic monitoring services. It is known as the Sequential Comparison Index (SCI) and simply compares two specimens at a time; if they are similar they are part of the same run, if not part of a new run. The observer need not know the name of any taxon. The greater the number of runs per number of species examined, the greater the biological diversity: thus healthy communities will have a much greater number of runs than polluted. Archibald (1972) tested this method using only diatoms in clean and polluted rivers and found that the moderately polluted station (Fig. 14.18) could be distinguished from the two clean ones but that the heavily polluted station had a similar form to the pure water station. This is not surprising since the cleaner the water the more *extreme* the environment becomes and low species diversity is expected. In this example the SCI does not decrease with deterioration of water quality, and Archibald concluded that species diversity in itself is not a reliable estimator of water quality. Not perhaps a justified conclusion, since clean water is not a good medium for algal growth and distilled water would be a pollutant! The enriching effect of mild pollution is so often overlooked: the tolerant species continue and the species enjoying the new nutrients also multiply. This is another example to support the contention that a good biologist can tell more of the water quality from a few samples than from complicated formulae; I would add there is certainly no need to count 5000 diatoms!

From the very simple (SCI) technique requiring only an awareness of differences in size, shape, colour, etc., Cairns *et al.* (1972, 1973) have begun to develop techniques for monitoring diatoms by laser holography. In this technique a Fourier transform hologram of a particular diatom is produced which serves as a matched spatial filter for the species. This is then used as a filter to detect diatoms of the same type when a sample is scanned. It is simply a refined matching process which is then linked up to electronic and computer apparatus to identify and count the kinds of diatoms automatically, a boon to those who cannot identify diatoms.

Fig. 14.18. The diversity of some diatom assemblages determined by the sequential comparison index (S.C.I.). *a*, *b*, clean water stations; *c*, moderately enriched; *d*, heavily polluted. From Archibald, 1972.

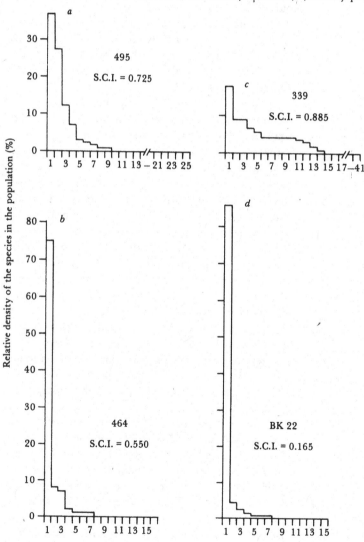

All the problems of distinguishing and sampling algal communities in unpolluted waters are equally attendant upon pollution studies. In most sites the phytoplankton is probably *not* the best community to sample, though methods for this are often the only ones known. The benthos is much better, since organisms associated with surfaces are present in all waters, whereas many flowing waters have little or no plankton. Unfortunately most lists of benthic species from polluted sites contain mixtures of species from epipelic, epilithic and epiphytic communities suggesting that very imprecise sampling methods have been applied. Few of the workers state how their samples were taken and how the species numbers were estimated. The species listed often provide a clue, e.g. Kolkwitz & Marsson (1908) and Zelinka & Marvan (1961) obviously studied the flora of benthic habitats although the methodology is not clear. The common method of hanging glass slides or other substrata in waters gives no real information about the flora of the various microhabitats and selects species which are primary colonisers of surfaces; later growths on the glass, etc. are confused by filtering effects, sloughing off, animal grazing, etc. As already mentioned the type of material used as a substratum for the algae affects the composition of the flora and Sládečková (1963) showed that 'Plexiglas' supported about twice as much organic material as other surfaces in unpolluted water but when pollution occurred this type of surface supported only half that of glass or porcelain. The method is discussed further on p. 126 where it is shown to be a very unsuitable and unreliable method for obtaining data on the productivity of a body of water. Obviously it will indicate some extremes but these will be obvious anyway. Comparison of the growths on slides placed at different sites is unlikely to yield useful information on the comparative pollution of the sites. Study of the natural flora will yield much more reliable information. Weber & Raschke (1970) found that in unpolluted zones of the Klamath River in Oregon the major component on slides was *Cocconeis placentula*, whilst in the polluted zone this species was subordinate to *Gomphonema parvulum* and *Nitzschia palea* though even here the *Cocconeis placentula* population was five times as dense as at the unpolluted site. Instead of waiting for these species to grow on slides, examination of the growth on natural or even permanent artificial substrata in the river would have given an instantaneous assessment. In fact a large number of sites can be rapidly assessed by such direct methods and there is absolutely no need to wait two weeks for an artificial population to grow on slides, a wasteful and time consuming process! Lange-Bertalot (1978) comments that all diatom populations can be grouped according to their sensitivity or resistance towards pollution. Species die out as pollution increases and this again emphasises the importance of information about the presence and the absence of species.

There are very few comprehensive studies of the tolerance of species of a genus to pollution or eutrophication though it is quite clear that this is an attribute of the particular species and not of the genus. Sládeček & Perman (1978) have provided such a list of 73 species of *Euglena* which will be very valuable to future workers.

Pollution is very common in estuaries where phytoplankton is often poorly developed leaving the microscopic and macroscopic benthic algae as the only indicators. Edwards (1972) considers the macrophytes good indicators with the green algae penetrating upstream higher than the brown and these higher than the red algae. In northern English estuaries he found only three algae confined to the polluted zone and these were *Fucus ceranoides*, *Vaucheria pilobaloides* and *Monostroma oxyspermum*. Species diversity of this attached community also shows a decrease upstream from the mouth, though this must surely be complicated by the lack of rocky sites upstream.

Algae as weeds

Excess growth of algae or growth in waters being utilised for special purposes results in algae which have to be treated as weeds, although this is not yet as great a problem as in land cultivation. The most serious offenders are the filamentous algae, in particular species of *Cladophora* but also *Oedogonium*, *Zygnema*, *Mougeotia* and *Spirogyra* (Whitton, 1971). The usually marine algae *Enteromorpha* has spread to many inland waters whilst the net-like *Hydrodictyon* also grows to stream-blocking proportions in some places. It is usually in streams that such algae grow to be a serious problem, though they are also a problem in some drainage channels and in rice fields. The algae tend to clog gates and weirs and generally slow down water flow with the resultant increase in silting. Their growth and decay tends to be more rapid than that of angiosperms and the changes in oxygen concentration can be lethal to fish, etc. Bolas & Lund (1974) show that a simple factor such as reduction of river flow stimulates the growth of *Cladophora*.

Exceptional growths of *Cladophora glomerata* have become a serious problem in some of the Laurentian Lakes: they clog harbours and water works, make bathing unsafe, form breeding sites for undesirable insects and, when decaying, release hydrogen sulphide. Similar problems are encountered in marine habitats where, in particular, species of *Sargassum* have become a problem along the Pacific coast of North America and may now become one around Britain (*see* p. 363).

The ecology of these loose-lying growths and their effect on other algal communities have never been adequately studied. Control has been attempted by the use of copper sulphate (0.1–0.5 mg l^{-1}), sodium arsenite (4 mg l^{-1}) and the organic compounds silvex (2,4,5 trichlorophenoxyproprionic acid), dichlorbenth and monuron (chlorphenoldimethylurea). Both paraquat and diquat have been used effectively but, interestingly, the Charophyta have been found to be resistant to these compounds. Some workers have however reported increases in growth of filamentous algae after the use of these to control angiosperms. Chlorine has been used to control planktonic algae (weeds) in cooling water and it has been shown that concentrations below that required to eliminate the phytoplankton will achieve great reduction in productivity and hence afford a degree of control (Carpenter *et al.*, 1972).

Summary

The steady natural cycling of chemicals through algae and into sediments is reiterated, followed by definitions of eutrophication and pollution which tend to be man-made events superimposed on the natural progression in aquatic systems. The consequences of these events are discussed and in particular the evidence that phosphorus is a key nutrient involved in freshwaters.

Detection of eutrophication from changes in the algal flora, changes in rates of carbon fixation and changes in annual cycles of growth is discussed. Artificial fertilisation studies of lakes have yielded fascinating insights into the process of eutrophication. Experimental bioassay of waters using algal growth as an indicator of fertility and determining changes in growth consequent on addition of nutrients is now well developed and some examples are quoted. Removal of nutrients to reverse eutrophication is possible.

Pollution is a more serious problem; some staggering figures for the amount of pollutants entering waters are given and it is quite surprising how algae have managed to tolerate such levels. Heavy metals, radionuclides, organic compounds, acids and heat all contribute to pollution. Pollution testing using macrophyte germlings or microscopic algae is well developed and some examples of techniques are quoted.

The use of indicator species or indicator communities goes back beyond the turn of the century and studies along these lines are still proceeding: used with discrimination they can be very valuable and certainly this is a field which would repay much more study. Classification of the various stages of pollution and various indices of pollution are discussed. The use of glass slides in such studies is well known and some of the various indices of pollution are discussed. A brief note on algae as weeds ends this chapter.

Conclusion

Concern with the environment tends to concentrate quite naturally on the land, yet large areas of land are desert or non-cultivatable mountain, swamp, etc. whilst the aquatic environment is much more extensive and is colonised throughout by plants and animals. Two factors are clearly of importance over the 70% of the earth's surface where algae grow – solar radiation and the supply of nutrients: the former is restrictive in polar to temperate zones and the latter in tropical and subtropical zones. Any change in the solar radiation reaching the surface of the planet will therefore change the pattern of algal growth, and, since the current systems of both the air and the water are linked to the input of solar radiation, these will be affected. Should the circulation pattern of the ocean change slightly due to increased solar radiation this may lead to increased nutrient supply from the deep water. The combined effects, increase in radiation plus increase in nutrients, could therefore drastically alter the primary production over 70% of the planet. The intensity of solar radiation at the water surface is linked to the conditions in the upper atmosphere and changes here caused by man's activities could have as great an effect on the system as does the addition of nutrients at the surface. Too little is known of the effects and consequences and hence prudence should dictate extreme care. The other factors which are of prime importance for algae are the supply of phosphorus especially to freshwater systems and of nitrate in the marine sphere. Increase of phosphorus from domestic and industrial sources has been the *single* most important factor changing the balance of production in freshwaters during this century. Soaps and detergents are available which are based on biodegradable substances and which do not contain phosphate and hopefully they may replace high phosphate types. Flow of nitrates, increased by additions from fertilisers, is becoming a problem in the marine coastal waters and must be tackled if production is to be controlled.

The mass of research aimed at determining levels of nutrient addition, pollution and eutrophication which is tolerable, is relatively pointless and should be replaced by studies of methods for *elimination* of the contaminants. It is better to attain pure water and in the unlikely event that this then proves unproductive, careful studies of additions can be contemplated.

The foregoing chapters may seem to encompass a mass of data, much of it disconnected, and in fact the amount of data is truly enormous even in this apparently neglected branch of ecology and only a small portion has actually been used. In preparing this account only a fraction of the works consulted have been quoted and works from many regions have hardly been consulted, for obvious linguistic reasons. Attempts at synthesis are remarkably scarce and whilst there are still many valuable data to be gathered it is essential that much more thought now be given to the design of experimental approaches to test hypotheses based on the synthesis of *sound* field data. Of course concepts of sound data vary, but it is illuminating that there seems to be a tentative return to qualitative studies on the grounds that some quantitative studies yield only a fraction of the available information and also that methods based on the identification of organisms yield many more data than those based on biochemical features, e.g. pigment and ^{14}C methods (Wood, 1971; P. G. Moore, 1974). Apparently some terrestrial ecologists are also turning back to qualitative surveys and certainly some of the most impressive data on algal ecology have been obtained from such, though with a sensible amount of quantification. But, if the sound field data are lacking this gap must be filled first, which often involves detailed knowledge of the systematics of algae. At every turn in algal ecology one is confronted by taxonomic problems and often no trained taxonomists to solve them. The problem exists in higher plant ecology, but in Europe at least, adequate floras exist. Few *modern* floras of any kind exist for algae,* and there is hardly a genus which does not require taxonomic revision. This, coupled with the fact that algal communities often contain frighteningly large numbers of species, often from several phyla, deters some students. Many ecological phycologists concentrate on a phylum or class with the result that whilst the studies may give a good account of, for example, the desmids of a lake or the diatoms of an oceanic mass of water, they are only fragmentary accounts and given an unbalanced impression, especially when read by someone unfamiliar with the field. The value and essentiality of taxonomic studies should need no stressing, though unfortunately it is often neglected by grant-aiding bodies and by students who imagine it a dusty, academic pursuit when, in reality, it is an exciting and wide ranging topic employing many advanced scientific techniques plus a strong element of detective work.

There is so much to do that it is saddening to see repeated studies of ^{14}C uptake down depth profiles, or of chlorophyll *a* values obtained without adding anything new to ecology. Unless some new aspect or approach is being adopted the question should be asked: what information will be gained from such studies which has not already been well documented? The same comment applies to zonation studies on rocky shores, the mere listing of plankton species from yet another lake or the study of the colonisation of glass slides.

On the other hand data collected over long periods of time are scarce and almost totally lacking from many habitats including the all-important

---

* For example, the only algal flora of note to appear in Britain for many decades, is the study of the British Rhodophyta in three volumes, the first one of which is now published (Dixon & Irvine, 1977). This aims at a comprehensive survey and anything less is still a hindrance to ecologists. I do not mean to imply that there is no more work to be done on the Rhodophyta – far from it – but at least a magnificent base-line study now exists and one can feel fairly sure that any question not resolvable from that text is a proper study for future research.

marine phytoplankton. The use of 'controls' in laboratory experiments is taken for granted but rarely can this methodology be applied to ecological studies. It is all the more important therefore to seek out the *simplified* situations which act as a form of control, e.g. the excellent work on Lake George with its virtual elimination of seasonal climatic changes, the work on chemically stable springs and the tideless seas. Such situations should be widely sought for they are not only simpler to work with but yield vital base-line data. Here the attempts at mathematical modelling are likely to be most successful.

Ecologists have suddenly been called upon to predict the consequences of the manipulation of populations and ecosystems, to construct models of ecosystems, and to assist in the planning of conservation measures, etc. To do this without an awareness of the intricacies of the interactions between the plant, animal, and physico-chemical environment is to court disaster. Only where long-term studies have been undertaken can any *reliable* prediction be attempted. This, I believe, is one of the strongest arguments for long-term approaches, rather than the haphazard tackling of odd problems. Not that the latter do not yield valuable data when properly attacked but they will rarely have the force of years of experience which are essential to advise on ecological problems. The recognition of basic principles and laws is essential to all science and the search for these in ecology is proving a long and arduous task. Nevertheless they exist and are equally as precise as those applying to other branches of science. Ecology is merely that much more difficult to study owing to the need to compound so many different biological and physico-chemical approaches.

In the present state of algal ecology it is still necessary to repeat discussions and report views on such fundamentals as terminology of phytoplankton, communities, shore zonation, etc. but fortunately these aspects are at last becoming clarified and the more important work of testing hypotheses derived from a study of the field data is accelerating.

Whilst it is obvious that there can be little understanding of ecology without the help of laboratory studies it is essential that the ecologist works from field studies backwards to laboratory testing. Only by gathering the data from nature, analysing them, setting up hypotheses and then designing experiments to test them will advances be made.

There is however an enormous pitfall here: the results of a laboratory experiment are *not* proof of an identical phenomenon in nature and their relevance must somehow be tested in the vastly more complex natural situation. Nowhere is this better seen than in the current methods used for testing pollutants on algal growth.

Modelling and general mathematical approaches have increased enormously in the last decades but many approaches are so simplified and based on incomplete data (sometimes completely unreliable data) that they have to be treated with caution. Margalef (1973) summed this up 'model building translates into mathematical terms only a small and biased part of what is known about the workings of the ecosystem'. Three of the most serious elements which are usually lacking are any assessment of mortality of algal populations (but *see* Jassby & Goldman, 1974, for one of the few discussions of this problem), any consideration of the behavioural pattern of individual species and the analysis of interaction between species or groups of species. For obvious reasons, efforts have been concentrated on the factors which affect the build-up of populations and not on the events as populations go into a decline. The problem is often that the ecologist is not competent to develop the techniques and the mathematician cannot judge the reliability of the field data whilst underestimating the complexity of the natural situation. I have no doubt that such approaches are essential and all the techniques at our disposal should be utilised, but so far the simplistic approaches have merely confirmed by complex mathematical manipulations what was obvious in the first place. Anyone doubting the difficulties and the problems of methodology still to be overcome should consult the series of papers in the *Journal of the Fisheries Research Board of Canada*, 1973, starting with the paper by Schindler on p. 1409. His comments are

apt; 'current approaches to systems modelling are based on erroneous assumptions...the belief that whole aquatic systems may be productively analysed without years of painstaking study of mechanisms is largely fictitious...the most important area in limnological research remains the construction and thorough testing of sound methodology, designed to illuminate ecosystem problems rather than describe them'. These papers give much interesting detail of methods and, in spite of the above extracts, do include some modelling techniques. Investigations designed from the beginning to lead to mathematical modelling are likely to be more productive than attempts to use incomplete data from isolated studies and a first essential is a good physico-chemical model to provide a basis for the energetic model. Such a carefully designed study is being undertaken on Lake Kinneret in Israel and will, I believe, act as a model for such attempts, for data from as long as ten years are being utilised!

What then for the future? There is an exciting one ahead, in which it is clear that ecology will play an increasingly important role in the world; an open one, where there are many more problems than there are ecologists, and a steadily clarifying one, as data are synthesised and the overall patterns emerge. Students sometimes find the complexities overwhelming but the solution is to isolate a fragment and study it intensively. This is not burying one's head in the sand but doing an absolutely essential task of completing one piece of the enormous jig-saw puzzle. At the same time one should maintain an awareness of the natural complexities and of the diverse approaches being used by ecologists in other fields – few can master more than one or a few techniques but all can appreciate that other approaches will interact, stimulate further work and ultimately explain the ecosystems.

Appendix

Algal ecologists have to contend with a vast range of algal groups, some of which are very well circumscribed and relatively easy to recognise, while others are more difficult to comprehend and identify. Central to all algal ecology is the correct naming of the species; an appreciation of algal classification is essential to the ecologist, therefore, and many excellent taxonomic/systematic studies have arisen from studies which were initially intended to be primarily ecological. As in all groups of organisms, there is continual taxonomic study and revision, even involving some well-known genera, e.g. the recent removal of the common *Rhodymenia palmata* to *Palmaria palmata* (Guirey, 1974*b*) and movement of *Coccolithus huxleyi* to *Emiliana huxleyi* (Hay *et al.*, 1967). Some ecological studies, especially of tropical, subtidal sites and even the much studied marine phytoplankton (in particular the deep populations of the tropics) are seriously hampered by the lack of systematic work. Comparative phytogeographic surveys are also sadly limited by the availability of comprehensive floras. Unfortunately algal systematics has been bedevilled by conflicting views on the grouping of classes, compounded by vague phylogenetic considerations. Only now are the exciting phylogenetic speculations of the last decade beginning to be tested by biochemical/genetical methods, but much remains to be done. An interesting account with emphasis on pigmentation and flagella structure is that of Casper (1974); this very detailed analysis cannot be incorporated here but should be consulted for the fine detail assembled.

The following survey is provided as a guide to the genera and groups mentioned in the text with some comment on the state of the art in each group. Neither fully comprehensive comment nor references can be given in this brief summary but some key references will be listed and the more modern proposals indicated.

The prokaryotic forms with algal-like gross morphology are clearly the most ancient of the genera and, although referred to as blue-green algae, it is clear that they are only half-algae with an equal admixture of bacterial characteristics. The term blue-green algae is so widespread it is difficult not to use it, but another common name should be found or the rather colloquial 'blue-greens' be used. It is equally unfortunate that some authors are using a terminology implying that these organisms are bacteria for this adds yet another misnomer. Bold (1973) coined the phylum name Cyanochloronta and this is certainly more appropriate than Cyanobacteria or Cyanophyta. Whilst not denying the admixture of bacterial characteristics shown by this phylum, the very interesting study of biochemical relationships of some selected organisms from various phyla by Schwarz & Dayhoff (1978) shows that the 'blue-greens' are quite distant from the bacteria when their ferredoxin sequences, *c*-type cytochromes and 5s ribosomal RNA sequences are considered.

One major algal group is involved plus a newly discovered small group which may prove to be as interesting as the blue-greens.

Cyanochloronta
(= Cyanophyta = Myxophyta)
Cyanophyceae (Christensen, 1979, typifies a new class name Nostocophyceae). These terms perhaps need changing since -phyceae implies algae and some less determinate class ending is needed. Five orders (some workers reduce them to three). Genera mentioned in the text are listed irrespective of whether they are synonyms of others.
Chroococcales. Single cells or colonies. *Agmenellum, Anacystis, Aphanocapsa, Aphanothecae, Chroococcus,*

Coccochloris, Coelosphaerium, Gloeocapsa, Gomphosphaeria, Holopedium, Merismopedia, Microcystis, Synechococcus.

Chamaesiphonales. Single cells or short filaments attached at the base. *Chamaesiphon, Chroococcidopsis, Chlorogloea, Clastidium, Entophysalis, Hydrococcus, Pseudoncobyrsa, Xenococcus.*

Pleurocapsales. Filaments, short branched or unbranched and often aggregating to form pseudoparenchyma. *Hormothonema, Hyella, Myxosarcina, Oncobyrsa, Pleurocapsa.*

Nostocales.* Filamentous, sometimes falsely branched, forming hormogonia. Heterocysts and akinetes present in some genera. *Anabaena, Anabaenopsis, Aphanizomenon, Arthrospira, Aulosira, Borgia, Calothrix, Cylindrospermum, Desmonema, Dichothrix, Fremeyella, Gloeotrichia, Homeothrix, Hydrocoleus, Hydrocoryne, Isocystis, Kyrtuthrix, Lyngbya, Mastigocladus, Microchaete, Microcoleus, Nodularia, Nostoc, Oscillatoria, Phormidium, Plectonema, Porphyrosiphon, Pseudanabaena, Raphidiopsis, Richelia, Rivularia, Spirulina, Scytonema, Schizothrix, Symploca, Tolypothrix, Trichodesmium.*

Stigonematales.† Filamentous, true branching forms, growing by cutting off pericentral cells. Reproduction by hormogonia, Heterocysts and akinetes occur. *Camptylonema, Fischerella, Hapalosiphon, Mastigocladus, Mastigocoleus, Stigonema.*

Widespread in soil, freshwater and marine habitats. The phylum is probably more widely distributed than any other algal group.

Prochlorophyta

A recently described phylum (Lewin, 1976, 1977) with one alga *Prochloron* of which a second species has now been found (Withers *et al.*, 1978). These algae are epizoic or symbiotic with Ascidians and they have chlorophyll *a*

and *b*, no bilin pigments and thylakoids in pairs or stacks. This is a new and controversial phylum but is included here since it is probably the beginning of an interesting new series of algae. See Antia (1977) for arguments against the setting up of this phylum and Chadefaud (1978) for comments on the origin of chlorophylls *a* and *b* several times during evolution and the comment that the Prochlorophyta are not on the direct line of evolution to the Chlorophyta.

All other algal forms belong within the Eukaryota, i.e. organisms with nuclear material enclosed within a membrane, possessing mitochondria, dictyosomes, plastids with thylakoids grouped within a surrounding membrane, etc. The segregation of these algae into phyla (divisions) is based on morphological, reproductive and physiological features but for most ecological purposes the morphological criteria are used. However it is increasingly obvious that there are genetic races with distinctive physiological characteristics and the study of these will become increasingly important in future ecological studies. In fact, I think it probably needs stressing that clones of most species isolated by differing environments (e.g. involving temperature, salinity, etc.) will prove to be genetically stable, physiological forms and a trinomial system of nomenclature may have to be devised. The phyla are listed below in what is believed to be a series of natural groupings which have phylogenetic significance. Throughout the lists which follow, the genera recorded by workers mentioned in the previous chapters are listed without consideration of synonymy etc., since this is merely a guide to the placing of the genera.

Rhodophyta

Traditionally one class Rhodophyceae (or Bangiophyceae according to Christensen (1979)) but two classes in Parke & Dixon (1976) Bangiophyceae, and Florideophyceae. This group is almost certainly closest to the Cyanochloronta in its possession of very simple plastids with single thylakoids and phycobilin pigments. The blue-greens and red algae are also reported to have a different unsaturated fatty acid biosynthetic pathway to that of the series of green algae (Erwin & Bloch, 1963). However, unlike the Cyanochloronta it has its major distribution in marine habitats.

---

* Bourelly (1970) placed the first three orders in a subclass Coccogonophycidae, since they reproduce by means of spores (nannocysts, endospores, exospores) and the other two orders in the subclass Hormogonophycidae since they reproduce by means of hormogonia or hormocysts. Bold & Wynne (1978) recognise only one order (Oscillatoriales) for the filamentous forms but the Stigonematales seem to me to exhibit more complex morphology – true branching, pleuriseriate thalli and pit connections between cells – and warrant an independent status.

† A recent paper (Cole & Conway, 1975) showed that the *Conchocelis* phase of *Porphyra* has characteristics of the Florideophyceae and indicates a return to subclass status for the two groups.

Bangiophyceae (*sensu* Parke & Dixon, 1976) (or subclass Bangiophycidae). Simple algae often with controversial life and reproductive cycles. Sexual reproduction of a simpler type than that in the other red algae. Cells uninucleate, chloroplasts usually single and pit connections rarely present (*see* footnote). Four orders.

Porphyridiales. Unicellular but also aggregating to form mucilaginous colonies. Only *Porphyridium* (*see* also note in Parke & Dixon (1976) regarding *Rhodella*) is common, forming purplish red coatings on soil and plant pots, but reported also in water varying from fresh to marine.

Goniotrichales. Small branched thalli often epiphytic on other red algae.

Compsopogonales. One genus *Compsopogon* which is a multicellular threadlike alga growing in tropical and subtropical rivers and streams.

Bangiales. Multicellular thalli, filamentous, uni- to multiseriate, (*Bangia*) disc-like, (*Erythrocladia*) branching, *Conchocelis*-stage; (*Erythrotrichia*) blade-like, (*Smithora* = *Porphyra naiadum*), *Porphyra*. The latter genus is the most common and widespread, generally found in the intertidal, though *Bangia* is equally dispersed but so small it is often overlooked and occurs in the less well investigated supratidal zone, and also rarely in freshwaters.

Florideophyceae (or Florideophycidae, subclass which becomes Nemaliophycidae in Christensen (1979)). Six orders. Most orders difficult to distinguish on vegetative features. Students will find the genera well described in the monumental work of Kylin (1956). European students will need to refer to Dixon & Irvine (1977 and two more volumes in preparation), elsewhere the specialist literature on individual genera must be consulted and many aspects are discussed in Irvine & Price (1978).

Nemalionales.* Contains the freshwater genera *Bactrachospermum* and *Lemanea* and the controversial cluster *Achrochaetium*, *Kylinia*, *Rhodochorton*, all placed in *Audouinella* by some workers. *Nemalion* is a branching worm-like alga, *Liagora* is loosely calcified and *Galaxaura* more so. *Bonnemaissonia–Trailliella*, *Asparagopsis–Falkenbergia* are linked in pairs as gametophyte and tetra-sporophytic phases which have been much studied (and even an order Bonnemaissoniales proposed, *see* Chihara & Yoshizaki (1972)). *Gelidum* is very widespread and used to produce agar (*see*

Santilices (1974) for review of this genus). Other genera are *Atractophora*, *Chaetangiun*, *Gelidiella*, *Helminthocladia*, *Naccaria*, *Pseudochantrania*, *Pterocladia*, *Pseudosciania*, *Sciania*, *Suhria*.

* **Cryptonemiales.** Both fleshy (*Dilsea*, *Dumontia*) and calcareous thalli are found in this order together with the very reduced, all but colourless, parasitic red algae (*Choreocolax*, *Harveyella*). One family contains the very familiar, widespread, calcified genera, *Archaeolithothamnion*, *Amphiroa*, *Corallina*, *Clathromorphum*, *Dermatolithon*, *Fosliella*, *Haliptylon*, *Jania*, *Leptophyllum*, *Leptophytum*, *Lithoporella*, *Lithophyllum*, *Lithothamnium*, *Lithothrix*, *Melobesia*, *Mesophyllum*, *Peysonnelia*, *Phymatolithon*, *Pseudolithophyllum*. (See reviews by Littler (1972) and Adey & McIntyre (1973).) Also in this order are *Acrosymphytum*, *Bossea*, *Calliarthron*, *Callophyllis*, *Choreonema*, *Constantinea*, *Corynomorpha*, *Crodelia*, *Dilsea*, *Dudresnaya*, *Endocladia*, *Erythrodermis*, *Erythrophyllum*, *Euthora*, *Gloiopeltis*, *Gloiosiphonia*, *Grateloupia*, *Halymenia*, *Hildenbrandia*, *Leptocladia*, *Neogoniolithon*, *Pachymenia*, *Porolithon*, *Porphyrodiscus*, *Prionites*, *Rhodopeltis*, *Schmitziella*, *Sporolithon*, *Tenarea*, *Thuretellopsis*.

Gigartinales. Mainly fleshy red algae, e.g. the two very common genera *Gigartina* and *Chondrus*. Also occurring in this order are *Agardhiella*, *Ahnfeltia*, *Calosiphonia*, *Calliblepharis*, *Catenella*, *Caulacanthus*, *Ceratocolax*, *Cruoria*, *Cruoriopsis*, *Curdiea*, *Cystoclonium*, *Eucheuma*, *Furcellaria*, *Gelidiopsis*, *Gracilaria*, *Gracilariopsis*, *Gymnogongrus*, *Halarchnion*, *Halymenia*, *Hypnea*, *Iridaea*, *Iridophycus*, *Melanthalia*, *Opuntiella*, *Petrocelis*, *Pikea*, *Phyllophora*, *Platoma*, *Plocamium*, *Polyides*, *Rhodoglossum*, *Rhodophysema*, *Rhodophyllis*, *Rissoella*, *Sarcodia*, *Schizymenia*, *Sebdenia*, *Solieria*, *Sphaerococcus*, *Stenogramma*.

Rhodymeniales. Some have hollow constricted thalli, e.g. *Champia*, *Lomentaria* and *Gastroclonium* or hollow unconstricted thalli, e.g. *Holosaccion*, others are flattened, e.g. *Rhodymenia*. Other genera are *Botryocladia*, *Chrysemenia*, *Chylocladia*, *Fauchea*, *Hymenocladia*, *Leptosomia*.

Palmariales. A new order erected by Guiry & Irvine in press; see also Parke & Dixon (1976). *Palmaria*.

Ceramiales. The largest group of the red algae but one of the easiest to recognise, e.g. of simple, branching filamentous nature (*Antithamnion*, *Callithamnion*, *Griffithsia*) or with a central axis of cells and bands of corticating filaments, e.g. *Ceramium*, *Dasya*, *Plumaria*, *Ptilota*, *Rhodomela* or blade/leaf-like but often with a distinct apical cell and very well-defined side branches, e.g. *Apoglossum*, *Caloglossa*, *Delesseria*, *Hypoglossum*, *Membranoptera*, *Murrayella*, *Myriogramme*, or with the pericentral cells as long as the axial cell,

---

* Dixon (1973) recognised a further phylum Gelidiales but this was not mentioned in Parke & Dixon (1976).

e.g. *Polysiphonia*. There are, however, other somewhat less obvious and difficult genera, e.g. the flat
leaf-like one *Polyneura* and the fleshy *Laurencia*. Also
classified here are *Acanthophora, Acrosorum, Aglao-
thamnion, Amansia, Antithamniella, Ballia, Börgeseniella,
Bornetia, Bostrychia, Botryoglossum, Bryocladia, Brongiartella, Bryothamnion, Caloglossa, Campylaeophora,
Centroceros, Chondria, Claudea, Compsothamnion,
Corynospora, Crouania, Cryptopleura, Dasyphylla, Dicty-
uris, Digenia, Dohrniella, Erythroglossum, Falkenber-
giella, Griffithsia, Grinnelia, Halurus, Herposiphonia,
Lophurella, Lophothamnia, Martensia, Microcladia,
Myriogramme, Nienburgia, Odonthalia, Pantoneura,
Platysiphonia, Plumariopsis, Pterosiphonia, Ptilotham-
nion, Sarcomenia, Seirospora, Spermothamnion, Spyridia,
Taenioma, Thuretia, Tolypiocladia, Vidalia, Wrangelia,
Yendonia*.

Cryptophyta

Cryptophyceae. The algae classified here also possess phycobilin pigments but located differently on
the thylakoids compared with the two previous
phyla and possessing many unusual ultrastructural
features suggesting that they are not really close to
the Rhodophyta. Biochemical studies are now
needed to establish the relationship of this group to
other phyla. A relatively small phylum of mainly
bi-flagellate organisms, widely distributed in fresh
and coastal waters. Often common on the surface
of sediments, e.g. on muddy beaches. Specific studies
of their ecology are almost non-existent. One order.
Cryptomonadales. All flagellates. *Hemiselmis,
Chroomonas, Cryptomonas, Rhodomonas*.

The remainder of the algae fall into (or between
in the case of the Euglenophyta) two distinct groups
but there is no classificatory niche for clusters of
phyla; the next group would be a sub-kingdom and
there is of course no intrinsic reason why this should

not be used. It is convenient however to use the terms
chlorophyte series for those algae in which chlorophyll
a and *b* are dominant pigments (resulting in a green
coloration), lutein is also present and starch is a
common storage product and a *chromophyte* (=
heterokont) series in which chlorophylls *a* and *c* are
present, together with a series of xanthophylls giving
the algae a general brown appearance. Within these
two series a number of distinct phyla are recognised
and it is fairly certain that these have very long
independent geological histories. The *chlorophyte*
series comprise three phyla.

Chlorophyta*

This phylum is common in all habitats except
the marine phytoplankton. Certain groups
and genera are almost confined to specific
habitats where they have probably been isolated for a considerable time, geologically.
Hence it is not difficult to subdivide the
phylum into classes. Four classes were recognised by the author (1971*c*, 1973*a*) and to these
Kornmann (1973) and Kornmann & Sahling
(1977) have added a fifth, the Codiolophyceae.
This class cuts across some of the traditional
boundaries, but I believe it is a valid grouping
since it is based on heteromorphy in the life
cycles. Van den Hoek (1958) tentatively raises
two other clusters of genera – the Cladophorophyceae and the Dasycladophyceae – to class
status and his interesting discussion should be
consulted. There is no doubt that there is still
much to achieve in the systematics of the
Chlorophyta, especially now that biochemical
studies of cytochromes, protein sequencing,

---

* Some workers (*see* e.g. van den Hoek, 1978) use this
designation for algae, bryophytes and vascular plants
and separate the algae off as a subphylum
Chlorophytina. For more detailed discussion of this and
several other points *see* Round (1971*c*), although since
that time more alterations have been made and I have
modified some aspects. The broad basis of classes
proposed then, is now becoming more widely accepted
and appears in many texts (e.g. Wartenberg (1972),
Kornmann & Sahling (1977), Trainor (1978)). Casper
(1974) also elevates some of the groupings I recognise as
classes, e.g. the Chlorophyceae, Oedogoniophyceae,
Bryopsidophyceae, Zygnemaphyceae (and he includes
here also the Prasinophyceae) become subphyla and in
a recent scheme of Margulis (1974) the latter two

groups become phyla (Siphonophyta and
Zygnemaphyta). On the other hand Bourelly (1972) in
the appendix in his 2nd edition also recognises the same
entities (four classes (Chlorophyceae, Prasinophyceae,
Zygophyceae and Charophyceae)) and splits the
Chlorophyceae into four subclasses Chlorophycidae
(containing the non-filamentous orders),
Ulotricophycidae (the filamentous, thalloid and
'chlorosarcenoid' orders), Oedogoniophycidae and
Bryopsidophycidae. Thus similar entities are being
recognised and the divergences of opinions concern only
their status (subclass, class and even phylum), except
for the first two subclasses of Bourelly which seem to me
to be very artificial.

DNA hybridisation etc., are being increasingly employed.

Chlorophyceae.* Contains the flagellate, coccoid (excluding desmids), simple filamentous and branched filamentous genera.

Volvocales. Flagellate unicells or colonies. *Brachiomonas, Carteria, Chlamydomonas, Chlainomonas, Chloromonas, Dunaliella, Eudorina, Gonium, Haematococcus, Pandorina, Platydorina, Pleodorina, Volvox, Volvulina, Spermatozoopsis, Sphaerellopsis, Stephanosphaera.*

Tetrasporales. Cells solitary or often in fours in mucilage colonies but reproducing by typical volvocalean flagellate states. *Emergococcus, Gemellicystis (= Pseudosphaerocystis), Gloeocystis, Hormotila, Kremastochloris, Nautocapsa, Nautococcopsis, Paulschultzia, Radiosphaera, Tetraspora.*

Chlorellales. An order segregated by Bold & Wynne (1978) to include the autosporic genera previously in the non-motile unicellular to colonial Chlorococcales. *Actinastrum, Ankistrodesmus, Botryococcus, Chlorella, Chodotella, Coccomyxa, Coelastrum, Coenococcus, Crucigeria, Dactylococcus, Dictyosphaerium, Elakatothrix, Eremosphaera, Franceia, Golenkenia, Kirchneriella, Lagerheimia, Monoraphidium, Muriella, Oocystis, Palmophyllum, Quadricoccus, Quadrigula, Raphidium, Scenedesmus, Scotiella, Selenastrum, Trochiscia, Trebouxia.*

Chlorococcales. Used in the sense of Bold & Wynne (1978) to include the unicellular and colonial genera which reproduce by means of zoospores, i.e. the zoosporic series of previous authors. How valid this subdivision into an autosporic and a zoosporic order will prove depends on detailed study of further genera, some of which, according to Bourelly (1972), have a tendency to be either zoosporic or autosporic. *Ankyra, Bracteococcus, Characium, Coelastrum, Chlorochytrium, Chlorococcum, Dictyochloris, Friedmania, Hydrodictyon, Myrmecia, Neochloris, Oocystis,*

Pediastrum, Phytoconis?, Radiococcus, Sphaerocystis (= Planktococcus), Spongiochloris, Trebouxia, Tetraedron.

Chlorosarcinales (*see* Groover & Bold, 1969). Erected by Herndon (1958) as the Chlorosphaerales for the genera, mainly soil algae, in which the vegetative growth occurs by the formation of packets of cells, the daughter cell walls being formed within those of the parents. *Chlorosarcina, Chlorosarcinopsis (= Chlorosphaera), Tetracystis.*

Ulvales. Genera growing as flat pseudoparenchymatous thalli. Marine. *Blidingia, Capsosiphon, Enteromorpha, Percusaria, Ulva.*

Microsporales. Filamentous algae with cell walls composed of H-pieces. Freshwater. *Microspora.*

Cylindrocapsales. A small, freshwater, filamentous, group characterised by cells with stellate chloroplasts. *Cylindrocapsa.*

Prasiolales (= **Schizogoniales**). This order is sometimes combined with the previous since the cells also have stellate chloroplasts but the morphology involves packets of cells or thalli (e.g. *Prasiococcus, Prasiola, Rosenvingiella*). More detailed studies are needed on these two orders.

Chaetophorales.† A mainly freshwater group (though some species are marine epiphytic and possibly epilithic) of branching filamentous genera, often with a basal attaching system of filaments and an upright free system. *Aphanochaete, Bulbocoleon, Chaetophora, Chaetosphaeridium, Draparnaldia, Endoderma, Fritschiella, Gongrosira, Microthamnion, Ochlochaete, Phaeophila, Pilinia, Pringsheimella, Pseudendoclonium, Pseudulvella, Protococcus (= Pleurococcus = Phytoconis), Stigeoclonium, Tellamia, Ulvella.* But also included here are the epiphytic species growing on trees which form packets of cells which more rarely elongate into short filaments (*Desmococcus, Apatococcus*).

Trentepohliales. Terrestrial algae growing on rocks or as epiphytes (and one or two even parasitic). A mainly tropical group requiring much further systematic study. *Trentepohlia, Cephaleuros, Phycopeltis, Stomatochroon.*

Coleochaetales. A small freshwater group of epiphytic genera with fairly advanced oogamous reproduction. *Coleochaete.*

Codiolophyceae. The separation of this group is the only major new departure since the classic work of Blackman & Tansley (1902) and is based on detailed life cycle studies of marine genera. The heteromorphic life cycle with, in this instance, a unicellular

---

* A more recent analysis (Stewart & Mattox, 1975) presents modern evidence based on ultrastructural and biochemical studies which ultimately may result in a complete reorganisation of the green algae. Anyone interested in this topic should consult this most stimulating review which tentatively places the Zygnematales, Coleochaetales and a new order Klebshormidiales in the Charophyceae. Certainly it has always appeared unnatural to have a complex group such as the Charales without any obvious simpler ancestral types, but this combination of orders cuts right across traditional concepts and, of course, the progenitors of the Charales may have been lost since they certainly grew in the Devonian or earlier geological period.

† The marine genera included in this group by classical workers are almost certainly out of place and will have to be re-located after further study.

sporophytic stage (*Codiolum* or *Chlorochytrium* stage) is surely an important evolutionary trend and although not all species of a group or genus may exhibit all stages this is hardly a reason for disregarding the concept in the systematics of the green algae. The filamentous gametophytes tend to grow in spring and summer and the unicellular sporophyte in autumn and winter. Four orders.

Ulotrichales. Simple unbranched filaments, often primarily attached but later forming flocculent masses. Only a very small number of genera of the classical Ulotrichales have been studied in detail and I think it would be wise to reserve the possibility that some genera will have to remain within the Chlorophyceae as a distinct order; if this is proved some further nomenclatural changes will be necessary. Some *Ulothrix* species are difficult to identify but are very commonly encountered by ecologists. *Chlorhormidium, Hormidium, Klebshormidium, Stichococcus.* *

Monostromatales. Flat pseudoparenchymatous thalli one cell thick (distinguishes them from Ulvales). *Monostroma, Gomontia*-stage.

Codiolales. Monosiphonous, unbranched filaments with net-like chloroplasts. *Urospora.*

Acrosiphoniales. Monosiphonous, branching filamentous thalli; growth is apical with cell extension only in the apical cells of the branches. Common as attached plants in marine habitats on rock, other plants, etc. *Acrosiphonia, Spongomorpha.*

Oedogoniophyceae (= Stephanokontae). A discrete, totally freshwater class with a highly specialised mode of cell division and complicated stages in the oogamous life cycle. They have a series of features unknown in any other group. One order.

Oedogoniales. *Oedogonium, Bulbochetae* and *Oedocladium.*

Zygnemaphyceae. Completely freshwater, filamentous (often fragmenting readily) or coccoid (desmids). Reproducing by conjugation and never forming flagellate stages. The filamentous species difficult to identify without the zygospores. Four orders.

Gonatozygales. *Gonatozygon.*

Mesotaeniales. Saccoderm desmids. *Ancyclonema, Cylindrocystis, Mesotaenium.*

Desmidiales. Placoderm desmids. *Amscottia, Closterium, Cosmarium, Desmidium, Euastrum, Gymnozygon, Hyalotheca, Micrasterias, Netrium, Oocardium, Pleurotaenium, Phymatodocis, Sphaerozosma, Sphondylosium, Staurastrum, Staurodesmus, Triplastrium, Xanthidium, Zygogonium.*

Zygnemales. Filamentous genera. *Mougeotia, Spirogira, Zygnema, Zygnemopsis, Zygogonium.* The first two orders are sometimes combined into a single order, viz. Mesotaeniales.

Bryopsidophyceae (= Siphonales *sensu lato* of earlier workers). This conglomeration (i.e. Siphonales) is rarely maintained by modern authors and further subdivision along the lines suggested by van den Hoek (1978) requires detailed study. Apart from the genera *Cladophora, Pithophora, Protosiphon* and *Rhizoclonium* which occur in freshwaters (and the epizootic *Basicladia*) this is a large class of marine algae, often calcified and mainly tropical in distribution. Cells are multinucleate and in many genera the organisation is siphonus (i.e. without cross walls in the thalli). They also differ from the other 'chlorophyte' series in possessing two extra pigments siphonxanthin and siphonein though these can be absent from certain genera. The seven orders still require further study from life cycle, ultrastructural and biochemical approaches to confirm the position of many genera, but a distinct pattern is now beginning to be accepted, especially as more workers accept the need for separation of clusters of genera showing distinctive features rather than following the early practice of lumping genera which had only a few characters in common.

Codiales (= **Bryopsidales, Caulerpales in part**). This order and the next is used either in a narrow sense or widened to combine them both (e.g. van den Hoek, 1978) Kornmann & Sahling (1977) working only with the Heligoland flora, use it to contain *Bryopsis, Derbesia* (which has also been placed in its own order Derbesiales) and *Codium.*

Caulerpales. See comments also in previous order. This order contains *Caulerpa, Chlorodesmis, Halimeda, Ostreobium,† Penicillus, Rhiphidodesmis, Rhipocephalus, Udotea.*

Dichotomosiphonales. This small order of filamentous non-septate algae contains only the freshwater genus *Dichostomosiphon.* Its position is tentative.

---

* In view of the doubt over the position of these genera relative to the Ulotrichales *sensu* Kornmann & Sahling (1977) it may be preferable to use the order Klebshormidiales of Stewart & Mattox (1975) for these genera but I am reluctant to do this without further knowledge of genera such as *Eugomontia, Gloeotila, Gloeotilopsis, Koliella.* Freshwater *Ulothrix* species may have to be transferred to another genus, possibly *Uronema.*

† *Ostreobium* requires re-investigation since Bourelly (1968) believes it may be in the Vaucheriales (Xanthophyta).

Dasycladales. There is less divergence of opinion over this distinctive order which is now recognised by most workers (e.g. Bold & Wynne (1978) and van den Hoek (1978) who even suggest class status). It has a long fossil record back to the Cretaceous. The genera are almost exclusively tropical, marine, genera with siphonous radial organisation, reproducing by the formation of operculate cysts. Calcium carbonate is regularly deposited on the thalli. *Dasycladus, Batophora, Bornetella, Cymopolia, Acetabularia, Neomeris.*

Protosiphonales. A small order has been proposed for the single rather simple genus *Protosiphon*, which grows on muddy banks, often above water (*see* Wartenberg, 1972). Unfortunately this publication has no references to authorities and the genus is placed in the Chlorococcales by other workers.

Cladophorales. Unbranched (*Chaetomorpha, Rhizoclonium*) or branched (*Cladophora*) algae with thick lamellate cell walls. Attached by basal cell. *Rhizoclonium* and *Cladophora* have both freshwater and marine species.

Siphonocladales. Characterised by segregative division of the thalli into multinucleate units. A tropical to semi-tropical marine group often growing in sand or over rocks. *Struvea, Valonia, Boodlea, Cladophoropsis, Siphonocladus, Anadyomene, Microdictyon, Chamaedoris.* In van den Hoek (1978) there is a tentative proposal to place the above two orders into a new class Cladophorophyceae.

Sphaeropleales. This small order contains the rather rare genus *Sphaeroplea* and is sometimes placed in the Cladophorales, e.g. by Bold & Wynne (1978) and sometimes in its own order (e.g. Bourelly, 1972).

Charophyta

A well defined phylum of freshwater (rarely brackish) macrophytic, whorled, oogamous algae having a long and well-authenticated fossil record. But see also footnote on p. 566. Structure and reproduction is amongst the most advanced in the algae.

Charophyceae. One living order.

Charales. *Chara, Lamprothamnium, Nitella, Nitellopsis, Tolypella.*

Prasinophyta

Although included in the Chlorophyceae by some workers, the genera have very distinctive features and the characters common with the Chlorophyceae are merely those common to green plants as a whole. The scaly ornamentation of the flagella and body of the motile cells is very distinctive. The group is largely known from marine habitats and the flagellate genera are very common in upper intertidal and supratidal rock pools. The 'green spheres' of oceanic plankton (*Halosphaera, Pterosperma*) belong here and require detailed ecological and distributional study.

Prasinophyceae. The following is the system of Parke & Green in Parke & Dixon, 1976.

Pedinomonadales. Contains the flagellate genera *Micromonas* and *Pedinomonas.*

Pyramimonadales. The main genus of flagellates, *Pyraminonas* occur here together with the coccoid *Halosphaera.*

Pterospermatales. Contains some flagellate and coccoid algae, *Nephroselmis,** Pachysphaera, Pterosperma.*

Prasinocladales. Flagellates (*Platymonas, Tetraselmis*) and the dendroid genus *Prasinocladus.* The allocation of genera to orders in this group is unlike that in other phyla in that motile and non-motile genera are intermingled.

Euglenophyta

This group of mainly flagellate algae have similar pigmentation to the 'chlorophyte' series but when natural material is examined can easily be distinguished by the presence of a red eyespot lying free in the cytoplasm (in contrast those 'chlorophytes' which possess eyespots have them embedded in the chloroplast). The euglenoids also have, unlike the 'chlorophytes' the two 'chromophyte' pigments, diadinoxanthin and diatoxanthin. The only other group with free eyespots is the Eustigmatophyta but these are rather rare flagellates compared with the Euglenophytes. The pigmented, carbon-fixing euglenoids are only a small branch of a complex series of flagellates, the majority of which are true protozoa without pigments and many with highly developed 'animal' organelles. Euglenoids are common in fresh and brackish water and frequently associated with nutrient-rich sites. They live primarily on the surface of freshwater sediments, often seen as green clouds of cells

* *Nephroselmis* is placed in the Pedinomonadales by Bourelly (1970).

lying over the sediments. Also found in temporary pools. A few species are marine.
Euglenophyceae. See Leedale (1967) for a comprehensive account. There are two pigmented orders.
Euglenales (= Euglenomonadales). Flagellates with two unequal flagella only one of which emerges from the subapical invagination. *Astasia, Euglena, Phacus, Lepocinclis, Trachelomonas.*
Eutreptiales. Flagellate with two equal emergent flagella. *Eutreptia.*

This phylum cannot be fitted into the 'chlorophyte' series since it has several ultrastructural and chemical features which are also found in the 'chromophyte' series, and hence it is placed in this survey between the two major clusters of phyla and adjacent to the Dinophyta which has some features akin to those of euglenoids and some to the 'chromophytes', e.g. chlorophyll *c*.

Dinophyta*

This is the phylum containing the dinoflagellates which are so common in the marine phytoplankton and in some lakes. The cells are very distinctive and hardly likely to be mistaken for those of any other group; a very small number of coccoid and filamentous genera are known. In addition to the pigmented species there are also many colourless and some parasitic forms which are most complex and one must conclude that the pigmented and unpigmented series have had very long divergent evolutionary histories. Two classes though some authorities merge these.
Desmophyceae.
Desmocapsales. Vegetative stage palmelloid.
Prorocentrales. Motile, theca of two watch-glass like halves. *Prorocentrum, Exuviaella.* These two are frequently recorded from the marine plankton but it has recently been suggested that *Exuviaella* be merged in *Prorocentrum* (Dodge & Bibby, 1973).
Dinophyceae. This group contains the majority of the genera known as dinoflagellates. Biflagellate, with one flagellum vibrating in a transverse furrow and one running down a longitudinal groove out into the water.
Dinamoebales. Free living, amoeboid forms.
Gloeodinales. Cells in palmelloid colonies. Motile cells, *Hemidinium*-like.

---

* The designation Pyrrophyta is sometimes used, but the prefix Dino- is to be preferred.

Pyrocystales. Cells non-motile, often lunate shaped. Common in tropical to subtropical, marine plankton. *Dissodinium, Pyrocystis.*
Dinophysiales. Cells with two very large plates and 16 small plates. *Dinophysis, Histioneis, Ornithocerus, Parahistioneis, Phalacroma.*
Gymnodiniales. Cells motile, of typical dinophycean form, but obvious thecal plates lacking though ultrastructural studies show homologous vesicles. *Amphidinium* (= *Endodinium*), *Gymnodinium, Gyrodinium, Oxyrrhis.*
Peridiniales. Motile, with a theca composed of obvious plates. *Cachonina, Ceratium, Glenodinium, Gonyaulax, Heteraulacus* (= *Goniodoma*), *Oxytoxum, Peridinium, Protoperidinium, Wolosynskia.*
Dinotrichales. Rare filamentous forms. There are several other classes and also orders of the two above classes which are colourless, and should not be included in the algae; a striking example is *Noctiluca* which by no stretch of the imagination could be an alga but is unfortunately put into some botanical texts.

The 'chromophyte' group comprises five phyla. Here again much work is still needed to determine the relationships and the degrees of separation between the phyla. Some authorities use slightly different criteria to group these algae, e.g. van den Hoek (1978) has a phylum Heterokontophyta (Chrysophyceae, Xanthophyceae, Bacillariophyceae, Phaeophyceae and Chloromonadophyceae) and keeps the Haptophyta (= Prymnesiophyta) as a separate group. The latter certainly have a series of unique characteristics but also similarities to the other brown pigmented algae. Whereas the 'chlorophyte' series do seem to have a considerable degree of coherence this is not so for the 'chromophyte' series which on biochemical and ultrastructural grounds show considerable heterogeneity and hence I prefer not to follow van den Hoek but to maintain the groups as individual phyla (*see* also Round, 1979).

Xanthophyta

This is the odd group of the 'chromophyte' series since the carotenoid pigments are not so abundant (fucoxanthin is absent) that they obscure the green pigment and hence these genera can be confused with those of the Chlorophyta. It is often necessary to make cultural studies of them to be sure of their identity, especially those which occur in soils. They are almost totally freshwater in distribution but apart from *Vaucheria** and *Tribonema* are rarely recorded and little is known with certainty about their ecological requirements. Determinations of cells in field material is often helped if it is remembered that the light-green coloured plastids are usually discoid, oil globules occur and careful observation of the cell walls often reveals a bipartite nature. The eyespot in the flagellate cells (as in Chlorophyta) is embedded in a chloroplast. Starch is absent so a simple iodine test may differentiate these algae from the chlorophytes. A valuable new publication on this group is provided by Ettl (1978).

Recently the phylum has been split into two with the recognition of fairly major ultrastructural and biochemical differences but as in many fields of taxonomy, as a result, much reinvestigation of earlier described taxa is needed in order to characterise fully the new group, Eustigmatophyta (see below), and to redefine the limits of the Xanthophyceae. The following division must be regarded as provisional but it has been brought into line with Silva's (1979) proposals.

Xanthophyceae. Six orders.

Heterochloridales (= Chloramoebales). Free living flagellates, flagella unequal.

Rhizochloridales. Amoeboid, sometimes forming net-like colonies. *Rhizochloris*.

Heterogloeales (= Heterocapsales in part). Coccoid, solitary or colonial.

Mischococcales (= Heterococcales in part). Zoospores with contractile vacuoles. Wall sometimes in two parts, e.g. *Ophiocytium*.

Tribonematales (= Heterotrichales). Filamentous, simple or branched. Cells with walls composed of two overlapping H-pieces. *Tribonema, Bumilleria*.

Vaucheriales (= Botrydiales and Heterosiphonales). Multinucleate, lacking crosswalls. Common on damp soil, mud flats and streams. *Botrydium, Vaucheria. Vaucheria* can only be safely identified to species when the reproductive struc-

---

* *Vaucheria* has numerous fine structural peculiarities and requires further study and possibly a revision of its status. It is also the only common genus of the group which can tolerate salt water. A few workers have elevated this and the related siphonaceous genera to the Vaucheriophyta (Maekawa, 1960) or Siphonophyta (Kimura, 1953) but this receives no support from the latest survey of Silva (1979).

tures are present and hence it is usually necessary to culture material.

A few genera (e.g. *Meringosphaera, Schilleriella*) of the Xanthophyta have recently been transferred to the Chrysophyta, see Leadbeater (1974). Chloromonadophyceae. Two genera of freshwater flagellates (*Vacuolaria and Gonyostomum*) are placed in this rather enigmatic class which is associated with the heterokont group, probably close to the Xanthophyceae (Heywood, 1977). The genera are not rare but few ecological observations have been recorded. Casper (1974) elevates this group to a phylum and indeed as a class must be eventually allied to an existing phylum or Casper's suggestion must be followed.

Eustigmatophyta†

Eustigmatophyceae. Members of this phylum are very rarely recorded in ecological works. They are mainly coccoid genera previously placed in the Xanthophyta but they are distinguished by possessing a single chloroplast and the pigment violoxanthin which is lacking in the Xanthophyta. The motile zoospores have a single flagellum and at the anterior end an eyespot free in the cytoplasm. For other details and a general review of this group and the Xanthophyta see Hibberd & Leedale (1972) *Pleurochloris, Ellipsoidion, Vischeria, Polyedriella* and *Pseudocharaciopsis* have so far been shown to belong here.

Nannochloris oculata is a common marine coccoid alga often used in experiments on marine phytoplankton and Antia, Bisalputra et al. (1975) have suggested it be transferred to the Eustigmatophyceae and a new taxon created.

Chrysophyta

Originally this phylum was, and still to some extent is, a miscellaneous collection of forms clustered around a central series of brown-pigmented genera of flagellates, coccoid and filamentous algae, some of which have characteristic siliceous scales and some have endogenous siliceous cysts. It has been clarified greatly by the removal of a whole series of genera into the Prymnesiophyta (= Haptophyta). The Chrysophyta contains mainly freshwater genera, principally growing in cold, clean waters. They are planktonic, epiphytic or epipelic (though these may be mainly

---

† Some recent phylogenetic discussions (e.g. Margulis, 1976) have elevated this group to a phylum.

sedimenting out of the overlying water). The ecological factors affecting the Chrysophyta are very inadequately known and autecological studies would repay the effort.

Chrysophyceae. Bourelly (1968) recognises three subclasses. (i) Acontochrysophycidae. No flagellate stages. Four orders.

Chrysosaccales (= **Chrysococcales**). Palmelloid genera.

Rhisochrysidales. Rhizopodial forms. *Rhizochrysis, Chrysarachnis.*

Stichogloeales. Solitary or colonial genera.

Phaeoplacales. Filamentous or thalloid genera. (ii) Heterochrysophycidae. Forms with flagellate stages having one or two unequal flagella. This is the largest and best known section. Two orders.

Chromulinales. Genera with a single flagellum. *Chromulina, Chrysococcus, Chrysoikos, Kephyrion, Chrysopyxis, Chrysocapsa;* the latter is a colonial genus forming packets of cells, *Hydrurus* (colonial, filamentous gel-like structure).

Ochromonadales. *Chrysamoeba,* rhizopodial. Flagellate, coccoidal or filamentous with motile stages possessing two flagella. *Ochromonas, Cyclonexis, Uroglena, Dinobryon, Epipyxis, Pseudokephyrion, Mallomonas, Olisthodiscus, Paraphysomonas, Synura, Chrysophaerella.* (iii) Isochrysophycidae. This subclass contains the genera with two equal flagella. One order.

Isochrysidales. Unicellular or colonial. The Isochrysidales has been transferred to the Haptophyta by Parke & Green (in Parke & Dixon, 1976) and these are now known as the Prymnesiophyta (Hibberd, 1976) but the freshwater species still require further investigation before the group is removed completely.

The over-riding impression from the literature is that, apart from the common flagellate genera *Dinobryon, Mallomonas* and *Synura,* the ecology of this phylum is a closed book.

The silicoflagellates are usually included in the Chrysophyta but according to Casper (1974) there is not enough evidence to justify this. They are marine only and are very common as fossils and used extensively for dating purposes. However, taxonomically, they are in a chaotic state (*see* Leoblich *et al.,* 1968 for an index of all the valid taxa). A single order Dictyochales is usually recognised. *Dictyocha, Distephanos, Ebria, Mesocena* are all reported in literature as living or fossil material, but see the above publication.

A further previously contentious group is the Choanoflagellates which have been placed in a subclass Craspedomonadophycidae by Bourelly with a single order Monosigales (= Craspedomonadales). However most genera are colourless and certainly should not be included in the algae (*see* Leadbeater, 1972). The Pedinellaceae (a section of the Choanoflagellates in some works) on the other hand do probably belong to the Chrysophyta since they possess chromatophores.

Prymnesiophyta (= Haptophyta)

This group, formally known as the Haptophyta, and renamed by Hibberd (1976) contains all the brown, biflagellate, algae with a haptonema and organic scales and those with organic scales on which calcareous deposits (coccoliths) occur. A few genera are coccoid or filamentous but produce scaly flagellate zoospores. Mainly found in marine habitats and very abundant in the nannoplankton of the oceans. The presence of a haptonema with a simpler internal organisation than that of the flagellum distinguishes this phylum from the Chrysophyta. One class.

Prymnesiophyceae (= Haptophyceae).

Isochrysidales.* Flagellate and filamentous forms. *Apistonema, Chrysotila, Emiliana, Gephyrocapsa, Hymenomonas.*

Coccosphaerales. This order contains the organisms commonly known as coccolithophorids. *Calyptrosphaera, Coccolithus, Crystallolithus, Rhabdosphaera, Syracosphaera.*

Prymnesiales. Flagellates and coccoid forms. *Chrysochromulina, Prymnesium, Phaeocystis.*

Pavlovales (*see* Green, 1976). Flagellates. *Pavlova* incl. *Monochrysis.* These orders and genera are based on the lists by Parke & Green in Parke & Dixon (1976).

Bacillariophyta (Diatoms)

This is the easiest of the phyla to distinguish, with all the taxa being unicells (sometimes joined to form filaments) and all with the same basic morphology. Silica is the wall material and hence the diatoms play an important role in the silicon cycle in nature. World-wide, marine, freshwater and on soils. This is another group in which considerable taxonomic revision is being undertaken. Although included as a

---

* But *see* Chrysophyta.

class of the Chrysophyta by some authors they clearly are so well defined and unlike any other group that they warrant phylum status, a conclusion reached also by Casper (1974). Bacillariophyceae (= Diatomophyceae). Two subgroups have long been recognised in the diatoms, a centric group which following Bourelly (1969) would be a subclass Centrophycidae or better Centrobacillariophycidae and would contain the centric diatoms of Hustedt and other workers and a second subclass (Pennatophycidae or better Pennatobacillariophycidae) to contain the pennate genera. The addition of -bacillario- here seems to me desirable in order to indicate clearly which phylum is involved. However some workers do not recognise *any* distinction and it must be admitted that the basic cell morphology is similar in both groups, though certain organelles are confined to one or the other and sexual reproduction has certain distinct features in each.

At the ordinal level some workers (e.g. Hendey, 1938) recognise only the single order Bacillariales, others two orders (Centrales and Pennales). The author and colleagues are in process of proposing a more comprehensive classification based mainly on ultrastructural features but taking into consideration the probable lines of phylogenetic development. Three subclasses will be recognised encompassing the centric, araphid and raphid genera and to bring the nomenclature in line with modern practice these are named after genera.*

(i) Eupodiscophycidae (the centric diatoms).

Coscinodiscales. Circular in outline and possessing only rimoportulae. Valves loculate with internal foramena. *Characotia, Coscinodiscus, Ethmodiscus, Gosleriella, Hemidiscus, Roperia.*

Melosirales. Circular in outline, cylindrical or hemicylindrical in valve view. Rimoportulae few or scattered. Valves loculate with complex closing plates both internally and externally. *Hyalodiscus, Melosira, Paralia, Podosira.*

Thalassiosirales. Circular in outline. Both rimoportulae and fultoportulae present. Areolae open to outside and closed internally by a poroid plate. *Cyclotella, Detonula, Planktoniella, Schroederella, Skeletonema, Stephanodiscus, Thalassiosira.*

Eupodiscales. *Eupodiscus.*

Astrolamprales. Valves sub-circular with radiating rays. *Astrolampra, Asteromphalus.*

Arachnodiscales. *Arachnodiscus.*

* These suggestions are tentative and work in progress may result in further substantial modifications.

Biddulphiales. Valves bipolar, sometimes extended into horns. *Biddulphia.* Further subdivision of this group is contemplated.

Leptocylindrales. *Corethrum, Leptocylindrus.*

Chaetocerales. Valves circular or bipolar, with numerous or two tubular extensions. *Attheya, Bacteriastrum, Chaetoceros.*

Rhizosoleniales. Valves oval with asymmetric horn. *Rhizosolenia.*

(ii) Fragilariophycidae. The araphid diatoms.

Fragilariales. Valves elongate with apical rimoportulae. *Amphicampa, Asterionella, Ceratoneis, Diatoma, Fragilaria, Meridion, Opephora, Raphoneis, Synedra.*

Tabellariales. Valves elongate, with apical areas of small pores. Plate-like septa on girdle bands. *Grammatophora, Tabellaria, Tetracyclus.*

Climacospheniales. Valves elongate, cuneate with apical girdle band septa or transverse girdle band septa. *Climacosphenia, Licmophora.*

Thalassionemales. Valves elongate, with complex spinous margins. *Thalassionema, Thalassiothrix.*

Ardissoniales. Valves elongate, without rimoportulae and with complex chambering. *Ardissonia* (= *Synedra* in part).

Pseudohimantidiales. Valves with numerous apical rimoportulae with external opening confluent. *Pseudohimantidum* (= *Sameioneis*).

(iii) Naviculophycidae. The raphid pennate diatoms.

Eunotiales. Valves with small apical raphe systems. Non-motile. *Actinella, Eunotia.*

Achnanthales. Heterovalvate with a raphe system on one valve only. *Achnanthes, Cocconeis, Eucocconeis.*

Naviculales. Valves with raphes more or less in central line. *Amphipleura, Amphiprora, Anomoeoneis, Berkeleya, Caloneis, Cymbella, Didymosphenia, Diploneis, Frustulia, Gomphonema, Gyrosigma, Mastogloia, Navicula, Neidium, Pinnularia, Rhoicosphenia, Trachyneis, Stauroneis, Stenoneis.* The *Didymosphenia, Cymbella, Gomphonema, Rhoicosphenia* group may in fact need relocating and several other groups may also be split off as further studies reveal the full structural details.

Epithemiales. Raphes eccentric, valves loculate. *Denticula, Epithemia, Rhopalodia.*

Nitzschiales. Raphe with bars beneath the slits, valves often needle-like. *Bacillaria, Cylindrotheca, Fragilariopsis, Hantzschia, Nitzschia* (marine *Denticula* spp. = *Denticulopsis*).

Surirellales. Valves evolved such that raphe slits are on margins of valve face and polar endings brought to be adjacent to one another. *Campylodiscus, Stenopterobia, Surirella.*

Phaeophyta

This is the classical brown algal phylum, almost entirely marine in distribution and containing the brown seaweeds ranging from microscopic, filamentous to massive thalloid structures. Their habitat is essentially the intertidal and subtidal rocky coast with a distinctly greater number of genera and species in temperate to polar waters. A few, however, tend to be pan-tropical (e.g. *Sargassum*).

Phaeophyceae. Christensen (1979) proposes Fucophyceae and typifies this name. The classification of certain genera has hardly changed over the last century but there is still confusion over a number of small orders and the following are the orders recognised by Wynne & Loiseaux (1976). See also the discussion in the volume by Irvine & Price (1978).

Ectocarpales. Heterotrichous, branching filaments. Isomorphic life cycle. *Acinetospora, Ectocarpus,* Feldmannia, Giffordia, Herponema, Laminariocolax, Mikrosyphar, Pilayella, Spongonema, Streblonema, Waerniella.*

Ralfsiales. Crustose thalli. *Elachista, Leathesia, Myrionema, Petroderma, Pseudolithoderma* (= *Lithoderma*), *Ralfsia.*

Chordariales. These algae have a heteromorphic life cycle and the sporophyte stage is a macroscopic plant. Vegetative morphology ranges from small discs of tissue to loosely aggregated pseudoparenchymatous structures *Chordaria, Elachista, Eudesme, Cladosiphon, Hecatonema, Liebmannia.*

Both the Chordariales and Ralfsiales are merged in the Ectocarpales by some workers. There is no doubt that great variations in life cycle and plasticity of form occurs and the final solution is not yet in sight.

Tilopteridales. Filamentous with polysiphonous development of the lower branches. *Haplospora, Tilopteris.*

Sphacelariales. Easily recognisable since the thalli have very large apical cells; these cut off cells at the base which then undergo longitudinal divisions and later transverse divisions. *Sphacelaria, Cladostephus, Halopteris.*

---

* A recent study by numerical taxonomic methods (Russell & Garbary, 1978) has shown that all the 43 species of the Ectocarpales recorded for the British Isles and allocated to 18 genera may belong in the genus *Ectocarpus*! A similar conclusion has been deduced from cultured studies (Ravanko, 1970*b*).

Dictyotales. Also with large apical cell but thallus expanding immediately into a flattened lamina (three cells thick in *Dictyota* but more in other genera). Plants either male, female or tetrasporic but with some variation in certain species. *Dictyota, Dictyopteris, Dilophus, Padinia, Taonia.*

Cutleriales. A small group of algae having a heteromorphic life cycle alternating between a small crustose stage (*Aglaozonia*) and a flattened laminate stage with a row of tufted hairs at the apices of the thalli (*Cutleria*). *Zanardinia* however is isomorphic.

Sporochnales. Much branched, pseudoparenchymatous thalli, each branch terminating in a tuft of hairs. Heteromorphic life cycle. *Sporochnus.*

Desmarestiales. Heteromorphic life cycle with a microscopic gametophyte and macroscopic sporophyte of branching pseudoparenchymatous structure with single filaments terminating each branch. A common genus in the subtidal of temperate and polar waters *Arthrocladia, Desmarestia, Himantothallus* (= *Phyllogigas* = *Phaeoglossum*).

Dictyosiphonales. Heteromorphic, as previous order but with much greater plasticity and thus not such a well-defined group. *Adenocystis, Asperococcus, Coilodesme, Isthmoplea, Litosiphon, Myriotrichia, Punctaria, Stictyosiphon, Striaria.*

Scytosiphonales. Separated from the previous order, though both these orders are fused with the Ectocarpales by some workers. Heteromorphic, with crustose and pseudoparenchymatous stages. Much work remains to be done to clarify this order *Colpomenia, Hydroclathrus, Petalonia, Scytosiphon.*

Laminariales. A distinctive order with a macroscopic sporophyte composed of lamina, stipe and holdfast. Gametophytes growing as few celled microscopic filaments. *Alaria, Agarum, Chorda, Costaria, Dictyoneura, Ecklonia, Egregia, Eisenia, Hedophyllum, Laminaria, Lessonia, Macrocystis, Nereocystis, Postelsia, Pterygophora, Sacchorhiza.*

Fucales. A unique group with a diploid, macroscopic thallus bearing sporangia in which meiosis occurs to produce gametes. The zygotes develop directly into the diploid plant. Common intertidal algae (*Ascophyllum, Bifurcaria, Fucus, Pelvetia*) or subtidal (*Cystoseira, Himanthalia, Sargassum,* floating also, *Halidrys*). Two further orders have been split off from the Fucales by some workers and these contain the genus *Durvillea* (Durvilleales) and the genus *Ascoseira* (Ascoseirales). More work is needed before these are confirmed. Included in the Fucales are all the genera in Table 8.4 but they are not listed again here.

References

Aàronson, S. 1971. The synthesis of extracellular macromolecules and membranes by a population of the phytoflagellate *Ochromonas danica. Limnol. Oceanogr.*, **16**, 1–9

Abbott, W. H. & Landingham, S. van 1972. Micropaleontology and paleoecology of Miocene non-marine diatoms from the Harper District, Malheur County, Oregon. *Nova Hedw.*, **23**, 847–906

Abdel Karim, A. G. & Saeed, O. M. 1978. Studies on the freshwater algae of the Sudan. III. Vertical distribution of *Melosira granulata* (Ehren.) Ralfs in the White Nile, with reference to certain environmental variables. *Hydrobiol.*, **57**, 73–9

Adey, W. H. 1968. The distribution of crustose corallines on the Icelandic coast. *Scientia Islandica*, Anniv. vol., 16–25

Adey, W. H. 1970. A revision of the Foslie Crustose Coralline Herbarium. *Kong. norske Vidensk. Selsk. skr.*, 46 pp.

Adey, W. H. 1971. The sublittoral distribution of crustose corallines on the Norwegian coast. *Sarsia*, **46**, 41–58

Adey, W. H. 1978. Coral reef morphogenesis: a multidimensional model. *Science*, **202**, 831–7

Adey, W. H. & Adey, P. J. 1973. Studies on the biosystematics and ecology of the epilithic crustose Corallinaceae of the British Isles. *Br. phyc. J.*, **8**, 343–408

Adey, W. H. & McIntyre, J. G. 1973. Crustose coralline algae; A re-evaluation in the geological Sciences. *Geol. Soc. Am. Bull.*, **84**, 883–904

Adey, W. H., Masaki, T. & Akioka, H. 1974. *Ezo epiyessoense* a new parasitic genus and species of Corallinaceae (Rhodophyta, Cryptonemiales). *Phycologia*, **13**, 329–44

Adey, W. H. & Vassar, J. M. 1978. Colonization, succession and growth rates of tropical crustose coralline algae (Rhodophyta, Cryptonemiales). *Phycologia*, **14**, 55–69

Admiraal, W. 1977*a*. Salinity tolerance of benthic estuarine diatoms as tested with a rapid polarographic measurement of photosynthesis. *Mar. Biol.*, **39**, 11–18

Admiraal, W. 1977*b*. Influence of various concentrations of orthophosphate on the division rate of an estuarine benthic diatom, *Navicula arenaria*, in culture. *Mar. Biol.*, **42**, 1–8

Aitken, J. J. 1962. Experiments with populations of the limpet *Patella vulgata* L. *Irish Nat. J.*, **14**, 12–15

Akagawa, H., Ikawa, T. & Nisizawa, K. 1972*a*. Initial pathway of dark $^{14}CO_2$ fixation in brown algae. *Bot. Mar.*, **15**, 119–25

Akagawa, H., Ikawa, T. & Nisizawa, K. 1972*b*. $^{14}CO_2$ – fixation in marine algae with special reference to the dark fixation in brown algae. *Bot. Mar.*, **15**, 126–33

Alcala, A. C., Ortega, E. & Doty, M. S. 1972. Marine, animal standing stocks in the Philippine *Caulerpa* communities. *Mar. Biol.*, **14**, 298–303

Aleem, A. A. 1973. Ecology of a kelp bed in southern California. *Bot. Mar.*, **16**, 83–95

Alexander, V. & Barsdate, R. J. 1971. Physical limnology, chemistry and plant productivity of a Taiga Lake. *Int. rev. ges. Hydrobiol.*, **56**, 825–72

Allen, H. L. 1971. Primary productivity, chemo-organotrophy and nutritional interactions of epiphytic algae and bacteria on macrophytes in the littoral of a lake. *Ecol. Monogr.*, **41**, 97–127

Allen, T. F. H. 1973. A microscopic pattern analysis of an epiphyllous tropical alga *Phycopeltis expansa* Jennings. *J. Ecol.*, **61**, 887–99

Allen, T. F. H. & Skagen, G. 1973. Multivariate geometry as an approach to algal community analysis. *Br. phycol. J.*, **8**, 267–87

Allender, B. M. 1977. Effects of immersion and temperature upon growth of the tropical brown alga *Padina japonica*, using a tide-simulation apparatus. *Mar. Biol.*, **40**, 95–9

Almodóvar, L. R. & Biebl, R. 1962. Osmotic resistance of mangrove algae around La Parguera, Puerto Rico. *Rev. Algol.*, **6**, 203–8

Almodóvar, L. R. & Págan, F. A. 1971. Notes on a mangrove lagoon and mangrove channels at La Parguera, Puerto Rico. *Nova Hedw.*, **21**, 241–53.

Almodóvar, L. R. & Rehm, A. 1971. Marine algal balls at La Parguera, Puerto Rico, *Nova Hedw.*, **21**, 255–8

Ålvik, G. 1934. Plankton-Algen Norwegischer Austernpollen, II. Licht und assimilation in verschiedenen Tiefen. *Berg. Mus., Årbok*, no. 10, 1–90

Amspoker, M. C. & McIntire, C. D. 1978. Distribution of intertidal diatoms associated with sediments in Yaquina Estuary, Oregon. *J. Phycol.*, **14**, 387–95

Andersen, S. T. 1966. Interglacial vegetational succession and lake development in Denmark. *Palaeobotanist*, **15**, 117

Anderson, D. M. & Wall, D. 1978. Potential importance of benthic cysts of *Gonyaulax tamarensis* and *G. excavata* in initiating toxic dinoflagellate blooms. *J. Phycol.*, **14**, 224–34

Anderson, G. C. 1969. Subsurface chlorophyll maximum in the northwest Pacific Ocean. *Limnol. Oceanogr.*, **14**, 386–91

Anderson, G. C. & Banse, K. 1963. Hydrography and phytoplankton production. In *Proc. Conf. on Primary Prod. Measures, Marine and Freshwater*, ed. M. S. Doty, 21 August–6 September, pp. 61–71. University of Hawaii, Publ. U.S. A.E.C. T.I.D./7633

Anderson, G. C., Comita, C. W. & Engstrom-Heg, V. 1955. A note on the phytoplankton-zooplankton relationships in two lakes in Washington. *Ecology*, **36**, 757–9

Anderson, G. C. & Munson, R. E. 1972. Primary productivity studies using merchant vessels in the north Pacific. In *Biological Oceanography of the northern north Pacific Ocean*, ed. A. Y. Takenouti, pp. 245–51. Tokyo.

Anderson, L. W. J. & Sweeney, B. M. 1977. Diel changes in sedimentation characteristics of *Ditylum brightwellii*. Changes in cellular lipid and effects of respiratory inhibitors and ion-transport modifiers. *Limnol. Oceanogr.*, **22**, 539–52

Anderson, L. W. J. & Sweeney, B. M. 1978. Role of inorganic ions in controlling sedimentation rate of a marine centric diatom *Ditylum Brightwelli*. *J. Phycol.*, **14**, 204–14

Anderson, O. R. 1976. Respiration and photosynthesis during resting cell formation in *Amphora coffeaeformis* (Ag.) Kütz. *Limnol. Oceanogr.*, **21**, 452–6

Anderson, R. R. 1969. The evaluation of species composition as a qualitative factor in primary productivity. *Chesapeake Sci.*, **10**, 307–12

Anderson, R. S. 1974. Diurnal primary production patterns in seven lakes and ponds in Alberta. *Oecologia, Berlin*, **14**, 1–17

Andrews, G. W. 1975. Taxonomy and stratigraphic occurrence of the marine diatom genus *Raphoneis*. In *3rd Symp. on recent and fossil marine diatoms, Kiel. Nova Hedw. Beih.*, **53**, 193–222

Antia, N. J. 1976. Effects of temperature on the darkness survival of marine micro-planktonic algae. *Microbial Ecol.*, **3**, 41–54

Antia, N. J. 1977. A critical appraisal of Lewin's Prochlorophyta. *Br. phycol. J.*, **12**, 271–6

Antia, N. J., Berland, B. R., Bonin, D. J. & Maestrini, S. Y. 1975. Comparative evaluation of certain organic and inorganic sources of nitrogen for phototrophic growth of marine microalgae. *J. mar. biol. Ass., U.K.*, **55**, 519–39

Antia, N. J., Bisalputra, T., Cheng, J. Y. & Kally, J. P. 1975. Pigment and cytological evidence for reclassification of *Nannochloris oculata* and *Monallantus salina* in the Eustigmatophyceae. *J. Phycol.*, **11**, 339–43

Antia, N. J. & Cheng, J. Y. 1975. Culture studies on the effects from borate pollution on the growth of marine phytoplankters. *J. Fish. Res. Bd. Can.*, **32**, 2487–92

Antia, N. J., McAllister, C. C., Parsons, T. R., Stephens, K. & Strickland, J. D. H. 1963. Further measurements of primary production using a large volume plastic sphere. *Limnol. Oceanogr.* **8**, 166–83

Archibald, R. E. M. 1972. Diversity in some South African diatom associations and its relation to water quality. *Wat. Res.*, **6**, 1229–38

Armstrong, F. A. J. 1966. Silicon. In *Chemical Oceanography*, Vol. 1, ed. J. P. Riley & G. Skirrow, pp. 409–82. Academic Press, London

Arnold, D. E. 1971. Ingestion, assimilation, survival and reproduction by *Daphnia pulex* fed seven species of blue-green algae. *Limnol. Oceanogr.*, **16**, 906–20

Ascione, R., Southwick, W. & Fresco, J. R., 1965. Laboratory culturing of a thermophilic alga at high temperature. *Science*, **153**, 732–85

Ashton, P. J. & Walmsley R. D. 1976. The aquatic fern *Azolla* and its *Anabaena* symbiont. *Endeavour*, **35**, 39–43

Asmund, B. 1968. Studies on Chrysophyceae from some ponds and lakes in Alaska. VI. Occurrence of *Synura* species. *Hydrobiol.*, **31**, 497–515

Atkinson, K. M. 1972. Birds as transporters of algae. *Br. phycol. J.*, **7**, 319–21

Austin, A. P. 1960. Observations on *Furcellaria fastigiata* (L) Lam. forma *aegagropila* Reinke in Danish waters, together with a note on other unattached algal forms. *Hydrobiol.*, **14**, 235–77

Ax, P. & Apelt, G. 1965. Die 'Zooxanthellen' von *Convoluta convoluta* (Turbellaria, Acoela) entstehen aus Diatomeen. Erster nachweis einer Endosymbiose zwischen Tieren und Kieselalgen. *Naturwiss.*, **52**, 444–6

Aykulu, G. 1978. A quantitative study of the phytoplankton of the River Avon, Bristol. *Br. phycol. J.*, **13**, 91–102

Azam, F. & Chisholm, S. W. 1976. Silicic acid uptake and incorporation by natural marine phytoplankton populations. *Limnol. Oceanogr.*, **21**, 427–35

Baalen C. van 1962. Studies on marine blue-green algae. *Bot. Mar.*, **4**, 129–39

Baalen, C. van 1967. Further observations on growth of single cells of coccoid blue green algae. *J. Phycol.*, **3**, 156–7

Baardseth, E. 1970. Synopsis of biological data on knotted wrack, *Ascophyllum nodosum* (L) Le Jolis. *FAO Fisheries Synopsis*, no. 38, rev. no. 1, 41 pp.

Baardseth, E. & Taasen, J. P. 1973. *Navicula dumontiae* sp. nov. an endophytic diatom inhabiting the mucilage of *Dumontia incrassata* (Rhodophyceae). *Norw. J. Bot.*, **20**, 79–87

Baatz, I. 1941. Die Bedeutung der Lichtqualität für Wachstum und Stoffproduktion planktonischer Meeresdiatomeen. *Planta*, **31**, 726–65

Bachmann, R. W. & Odum, E. P. 1960. Uptake of Zn^{65} and primary productivity in marine benthic algae. *Limnol. Oceanogr.*, **5**, 349–55

Bagge, P. & Lehmusluoto, P. O. 1971. Phytoplankton primary production in some Finnish coastal areas in relation to pollution. *Merentutkimuslait, Julk*, no. 235, 3–18

Bailey, D., Mazurak, A. P. & Rosowski, J. R. 1973. Aggregation of soil particles by algae. *J. Phycol.*, **9**, 99–101

Bailey-Watts, A. E. 1976. Planktonic diatoms and silica in Loch Leven, Kinross, Scotland: a one month silica budget. *Freshwater Biol.*, **6**, 203–13

Bailey-Watts, A. E. 1978. A nine-year study of the phytoplankton of the eutrophic and non-stratifying Loch Leven (Kinross, Scotland). *J. Ecol.*, **66**, 741–71

Bailey-Watts, A. E. & Lund, J. W. G. 1973. Observations on a diatom bloom in Loch Leven, Scotland. *Biol. J. Linn. Soc.*, **5**, 235–53

Bainbridge, R. 1957. The size, shape and density of marine phytoplankton concentrations. *Biol. Rev.*, **32**, 91–115

Baker, A. L. & Baker, K. K. 1976. Estimation of planktonic wind drift by transmissometry. *Limnol. Oceanogr.*, **21**, 447–52

Baker, A. L. & Brook, A. J. 1971. Optical density profiles as an aid to the study of microstratified phytoplankton populations in lakes. *Arch. Hydrobiol.*, **69**, 214–33

Baker, A. L., Brook, A. J. & Klemer, A. R. 1969. Some photosynthetic characteristics of a naturally occurring population of *Oscillatoria agardhii* Gomont. *Limnol. Oceanogr.*, **14**, 327–33

Baker, A. N. 1967. Algae from Lake Miers, a solar-heated Antarctic Lake. *N.Z. J. Bot.*, **5**, 453–68

Baker, S. M. 1909. On the causes of the zoning of brown seaweeds on the seashore. I. *New Phytol.*, **8**, 196–202

Baker, S. M. 1910. On the causes of the zoning of seaweeds on the seashore. II. *New Phytol.*, **9**, 54–67

Bakus, G. J. 1973. Some effects of turbulence and light on competition between two species of phytoplankton. *Invest. Pesq.*, **37**, 87–99

Balech, E. 1974. El genero 'Protoperidinium' Bergh, 1881 ('Peridinium' Ehrenberg 1831, partum.). *Revista Mus. Argent. Cienc. nat. Bernardino Invest. Cienc. Nat. Hidrobiol.*, **4**, 1–79

Ball, G. H. 1969. Organisms living on and in protozoa. In *Research in Protozoology*, ed. T. Chen, vol. 3, pp. 566–718. Pergamon Press

Ball, R. C. & Bahr, T. G. 1975. Intensive survey. Red Cedar River, Michigan. In *River Ecology*, ed. B. A. Whitton, pp. 431–60, Blackwell, Oxford

Banse, K. 1976. Rates of growth, respiration and photosynthesis of unicellular algae as related to cell size – a review. *J. Phycol.*, **12**, 135–40

Banse, K. 1977. Determining the carbon to chlorophyll ratio of natural phytoplankton. *Mar. Biol.*, **41**, 199–212

Barber, R. T. & Ryther, J. H. 1969. Organic chelators; factors affecting primary production in the Cromwell Current upwelling. *J. exp. mar. Biol. Ecol.*, **3**, 191–9

Barica, J. 1975. Collapses of algal blooms in prairie pothole-lakes: their mechanism and ecological impact. *Verh. Int. Ver. Limnol.*, **19**, 606–15

Barko, J. W., Murphy, P. G. & Wetzel, R. G. 1977. An investigation of primary production and ecosystem metabolism in a Lake Michigan dune pond. *Arch. Hydrobiol.*, **81**, 155–87

Barnes, D. J. & Taylor, D. L. 1973. *In situ* studies of calcification and photosynthetic carbon fixation in the coral *Montastrea annularis*. *Helgoländer wiss. Meeresunt.*, **24**, 284–91

Barnes, H. & Topinka, J. A. 1969. Effect of the nature of the substratum on the force required to detach a common littoral alga. *Am. Zool.*, **9**, 753–8

Barrales, H. L. & Lobban, C. S. 1975. The comparative ecology of *Macrocystis pyrifera*, with emphasis on the forests of Chubut, Argentina. *J. Ecol.*, **63**, 657–77

Bartlett, L., Rabe, F. W. & Funk, W. H. 1974. Effects of copper, zinc and cadmium on *Selenastrum capricornutum*. *Wat. Res.*, **8**, 179–85

Batham, E. J. 1956. Ecology of southern New Zealand, sheltered rocky shore. *Trans. Roy. Soc. New Zealand*, **84**, part 2, 447–65

Batham, E. J. 1958. Ecology of southern New Zealand exposed rocky shore at Little Papanui, Otago Peninsula. *Trans. Roy. Soc. New Zealand*, **85**, part 4, 647–61

Bathurst, R. G. C. 1966. Boring algae, micrite envelopes and lithification of molluscan biosparites. *Geol. J.*, **5**, 15–32

Bathurst, R. G. C. 1971. Carbonate sediments and their diagenesis. In *Dev. in Sedimentology*, 620 pp. Elsevier Publ. Co., Amsterdam, London & New York

Battarbee, R. W. 1972. Preliminary studies of Lough Neagh sediments. II. Diatom analysis from the uppermost sediment. In *Quaternary Plant Ecology. 14th Symp. Br. Ecol. Soc.*, ed. H. J. B. Birks & R. G. West

Battarbee, R. W. 1973a. A new method for the estimation of absolute microfossil number, with reference especially to diatoms. *Limnol. Oceanogr.* **18**, 647–53

Battarbee, R. W. 1973b. Preliminary studies of Lough Neagh sediments. II. Diatom analysis from the uppermost sediment. In *Quaternary Plant Ecology, 14th Symposium of the British Ecological Society*, ed. H. J. B. Birks & R. G. West. Oxford: Blackwell

Battarbee, R. W. 1978. Observations on the recent history of Lough Neagh and its drainage basin. *Phil. Trans. R. Soc. (Lond.)* **281B**, 303–45

Bauld, J. & Brock, T. D. 1974. Algal excretion and bacterial assimilation in hot spring algal mats. *J. Phycol.*, **10**, 101–6

Bayne, D. R. & Lawrence, J. M. 1972. Separating constituents of natural phytoplankton populations by continuous particle electrophoresis. *Limnol. Oceanogr.*, **17**, 481–9

Beers, J. R., Steven, D. M. & Lewis, J. B. 1968. Primary productivity in the Caribbean Sea off Jamaica and the tropical N. Atlantic off Barbados. *Bull. Mar. Sci.*, **18**, 86–104

Beeton, A. M. & Edmondson, W. T. 1972. The eutrophication problem. *J. Fish. Res. Bd. Can.*, **29**, 673–82

Beger, H. 1928. Atmosphytische Moosdiatomeen in den Alpen. *Schweiz. Festsch. Vjschr. naturf. Ges. Zürich*, **73**, 382–404

Begin, Z. B., Ehrlich, A. & Nathan, Y. 1974. Lake Lisan. The Pleistocene precursor of the Dead Sea. *Geol. Survey Israel*, Bull. 63, 30 pp.

Behre, K. 1956. Die Algenbesiedlung Seen um Bremen und Bremerhaven. *Veröff. Inst. Meersforsch. Bremerhaven*, **4**, 221–383

Behre, K. 1966. Zur Algensoziologie des Süsswassers (unter besonderer Berücksichtigung der Litoralalgen). *Arch. Hydrobiol.*, **62**, 125–64

Behre, K. & Schwabe, G. H. 1970. Auf Surtsey-Island in Sommer 1968 nachgewiesene nicht marine Algen. *Schr. Naturw. Ver. Schleswig-Holstein.*, pp. 31–100

Beklemishev, C. W., Petrikova, M. N. & Semina, G. I. 1961. On the cause of buoyancy of plankton diatoms. (Summary.) *Trudy Inst. Okeanol.*, **51**, 31–6

Belcher, J. H. & Fogg, G. E. 1964. Chlorophyll derivatives and carotenoids in the sediments of two English lakes. In *Recent researches in the fields of Hydrosphere, Atmosphere and Nuclear Geochemistry*. C. Marguen, Tokyo.

Belcher, J. H. & Storey, J. E. 1968. The phytoplankton of Rostherne and Mere Meres, Cheshire. *Naturalist*, April–June, no. 905

Bell, W. H., Lang, J. M. & Mitchell, R. 1974. Selective stimulation of marine bacteria by algal extracellular products. *Limnol. Oceanogr.*, **19**, 833–9

Bellamy, D. J., John, D. M. & Whittick, A. 1968. The 'Kelp forest ecosystem' as a 'phytometer' in the study of pollution of the inshore environment. *Underwater Ass. Rep.*, 79–82

Bellamy, D. J. & Whittick, A. 1968. Problems in the

assessment of the effects of pollution on marine inshore ecosystems dominated by attached macrophytes. *Field Studies*, **2**, 49–54

Belly, R. T., Tansey, M. R. & Brock, T. D. 1973. Algal excretion of ^{14}C-labelled compounds and microbial interactions in *Cyanidium caldarium* mats. *J. Phycol.*, **9**, 123–7

Ben, D. van der 1971. Les épiphytes des feulles de *Posidonia oceanica* Delile sur les côtes Francaises de la Mediterranée. *Mem. Inst. Sci. nat. Belg.*, no. 168, 101 pp.

Ben-Amotz, A. 1975. Adaptation of the unicellular alga *Dunaliella parva* to a saline environment. *J. Phycol.* **11**, 50–4

Bennett, A. G. 1920. On the occurrence of diatoms on the skin of whales. (With an appendix by E. W. Nelson.) *Proc. Roy. Soc., London*, **B, 91**, 352–7

Berger, W. H. 1973. Deep sea carbonates: evidence for a coccolith lysocline. *Deep Sea Res.*, **20**, 917–21

Bergquist, P. L. 1960a. Notes on the marine algal ecology of some exposed rocky shores of Northland, New Zealand. *Bot. Mar.*, **1**, 86–94

Bergquist, P. L. 1960b. A statistical approach to the ecology of *Hormosira banksii*. *Bot. Mar.*, **1**, 22–53

Berman, T., Hadas, O. & Kaplan, B. 1977. Uptake and respiration of organic compounds and heterotrophic growth in *Pediastrum duplex* (Meyer). *Freshwater Biol.*, **7**, 495–502

Berman, T. & Pollingher, U. 1974. Annual and seasonal variations of phytoplankton, chlorophyll and photosynthesis in Lake Kinneret. *Limnol. Oceanogr.*, **19**, 31–54

Berman, T. & Rodhe, W. 1971. Distribution and migration of *Peridinium* in Lake Kinneret. *Mitt. Int. Ver. Limnol.*, **19**, 266–70

Bernard, R. & Rampi, L. 1965. Horizontal microdistribution of marine phytoplankton of the Ligurian Sea. In *Proc. 5th Mar. Biol. Symp. Göteborg*, 13–24

Berthold, W. U. 1978. Ultrastrukturanalyse der endoplasmatischen Algen von *Amphistegina lessonii* d'Orbigny, Foraminifera (Protozoa) und ihre systematische Stellung. *Arch. Protistenk.*, **120**, 16–62

Besch, W. K., Ricard, M. & Cantin, R. 1970. Utilisation des diatomées benthiques comme indicateur de pollutions minieres dans le Bassin de la Muramichi N.W. *Ext. Tech. Rep. Fish, Res. Bd. Can.*, no. 202, 72 pp.

Besch, W. K., Ricard, M. & Cantin, R. 1972. Benthic diatoms as indicators of mining pollution in the Northwest Muramichi River system, New Brunswick, Canada. *Int. Rev. ges. Hydrobiol.*, **57**, 39–74

Beth, K. & Merola, A. 1960. Einige Experimente zum Epiphytismus in Zönosen Mariner Algen. *Delpinoa*, *N.S.*, **2**, 3–16

Beveridge, W. A. & Chapman, V. J. 1950. The zonation of marine algae at Piha, New Zealand, in relation to the tidal factor. *Pacif. Sci.*, **4**, 188–201

Beyer, R. J. 1965. The pattern of photosynthesis and respiration in laboratory microecosystems. In *Primary production in aquatic environments*, ed. C. P. Goldman, pp. 63–74. University of California Press, Berkeley

Bhakuni, D. S. & Silva, M. 1974. Biodynamic substances from marine flora. *Bot. Mar.*, **17**, 40–51

Bidwell, R. G. G. & Ghosh, N. R. 1963. Photosynthesis and metabolism of marine algae. V. Respiration and metabolism

of C^{14}, labelled glucose and organic acids supplied to *Fucus vesiculosus*. *Can. J. Bot.*, **41**, 155–64

Biebl, R. 1952. Resistenz der Meeresalgen gegen sichtbares Licht und gegen kurzwellige UV – Strahlen. *Protoplasma*, **41**, 353–77

Biebl, R. 1962. Temperaturresistenz tropische Meeresalgen, vergleichen mit jener von Algen in temperierten Meeresgebieten. *Bot. Mar.*, **4**, 241–54

Bienfang, P. K. 1975. Steady state analysis of nitrate – ammonium assimilation by phytoplankton. *Limnol. Oceanogr.*, **20**, 402–11

Bienfang, P. & Gundersen, K. 1977. Light effects on nutrient limited, oceanic primary production. *Mar. Biol.*, **43**, 187–99

Bienfang, P., Laws, E. & Johnson, W. 1977. Phytoplankton sinking rate determination: Technical and theoretical aspects, an improved methodology. *J. exp. mar. Biol. Ecol.*, **30**, 283–300

Bindloss, M. E. 1974. Primary productivity of phytoplankton in L. Leven, Kinross. *Proc. Roy. Soc. Edinb.* **B, 74**, 157–81

Bird, C. J. & McLachlan, J. 1974. Cold-hardiness of zygotes and embryos of *Fucus* (Phaeophyceae, Fucales). *Phycologia*, **13**, 215–25

Biswas, S. 1968. Hydrobiology of the Volta River and some of its tributaries before the formation of the Volta Lake. *Ghana J. Sci.*, **8**, 152–66

Black, M. 1933. The precipitation of calcium carbonate on the Great Bahama Bank. *Geol. Mag.*, **70**, 455–66

Black, M. 1965. Coccoliths. *Endeavour*, **24**, 131–7

Black, R. 1974. Some biological interactions affecting intertidal populations of the Kelp, *Egregia laevigata*. *Mar. Biol.*, **28**, 189–98

Black, R. 1976. The effects of grazing by the limpet, *Acmaea insessa*, on the Kelp, *Egregia laevigata* in the intertidal zone. *Ecology*, **57**, 265–77

Blackburn, M., Laurs, R. M., Owen, R. W. & Zeitschel, B. 1970. Seasonal and areal changes in standing stocks of phytoplankton, zooplankton and micronekton in the eastern tropical Pacific. *Mar. Biol.*, **7**, 14–31

Blackman, F. F. & Tansley, A. G. 1902. A revision of the classification of the green algae. *New Phytol.*, **1**, 17–24 pp.

Blasco, D. 1978. Observations on the diel migration of marine dinoflagellates off Baja California Coast. *Mar. Biol.*, **46**, 41–7

Blum, J. L. 1954. Evidence for a diurnal pulse in stream phytoplankton. *Science*, **119**, 732–34

Boaden, P. J. S., O'Connor, R. J. & Seed, R. 1975. The composition and zonation of a *Fucus serratus* community in Strangford Lough, Co. Down. *J. exp. mar. Biol. Ecol.*, **17**, 111–36

Boalch, G. T. 1961. Studies on *Ectocarpus* in culture. II. Growth and nutrition of a bacteria-free culture. *J. mar. biol. Ass.*, *U.K.*, **41**, 287–304

Boalch, G. T., Harbour, D. S. & Butler, E. I. 1978. Seasonal phytoplankton production in the western English Channel 1964–1974. *J. mar. biol. Ass.*, *U.K.*, **58**, 943–53

Bock, W. 1963. Diatomeen extrem trockener Standorte. *Nova Hedw.*, **5**, 199–254

Bodeanu, N. 1964. Contribution a l'etude quantitative du microphytobenthos du littoral Roumain de la mer Noire. *Rev. Roum. Biol. ser. Zool.*, **9**, 435–45

Bodeanu, N. 1968. Recherches sur la microphytobenthos du littoral roumain de la Mer Noire. *Rapp. P.-v. Réun. Commn, int. Explor. scient. Mer Mediterr.*, **19**, 205–7

Bodeanu, ·N. 1970. Contributions to the systematics and ecology of the benthic diatoms of the Romanian Black Sea littoral. *Rev. Roum. Biol. ser. Bot.*, **15**, 9–18

Bodeanu, N. 1971. Données qualitatives et quantitatives sur le microphytobenthos des fonds sablonneux et vaseau du littoral roumain de la Mer Noire. *Inst. Roman Cercet. Mar., Constanta*, **1**, 27–58

Bogorov, V. G. 1969. *The Pacific Ocean. 7. Biology of the Pacific Ocean*, translated from the Russian. Pt. 1. Plankton by V. G. Bogorov, iii. 411 pp. U.S. Naval Oceanographic Office Translation No. 435.

Böhm, E. L. & Goreau, T. F. 1973. Rates of turnover and net accretion of calcium and the role of calcium binding polysaccharides during calcification in the calcareous alga *Halimeda opuntia* (L). *Int. rev. ges. Hydrobiol.*, **58**, 723–40

Bolas, P. M. & Lund, J. W. G. 1974. Some factors affecting the growth of *Cladophora glomerata* in the Kentish Stour. *Water Treat. Exam.*, **23**, 25–51

Bold, H. C. 1970. Some aspects of the taxonomy of soil algae. *Ann. New York Acad. Sci.*, **175**, 607–16

Bold, H. C. 1973. *Morphology of Plants*, 3rd edition, 668 pp. Harper & Row, New York

Bold, H. C. & Wynne, M. J. 1978. *Introduction to the algae. Structure and reproduction*, 706 pp. Prentice Hall Inc., Englewood Cliffs, New Jersey

Boleyn, B. J. 1972. Studies on the suspension of the marine centric diatom *Ditylum brightwellii* (West) Grunow. *Int. rev. ges. Hydrobiol.*, **57**, 585–99

Bolin, R. L. & Abbott, D. P. 1963. Studies on the marine climate and phytoplankton of the central coastal area of California 1954–60. *Rep. Calif. Coop. Ocean. Fish. Invest.*, **9**, 23–45

Boltovskoy, E. 1963. The littoral foraminiferal biocoenoses of Puerto Deseado (Patagonia, Argentina). *Contr. Cushman Fdn. Foramin. Res.*, **14**, 58–70

Bond, G. & Scott, G. D. 1955. An examination of some symbiotic systems for nitrogen fixation. *Ann. Bot.*, **19**, 67–77

Boney, A. D. 1972. *In vitro* growth of the endophyte *Acrochaetium bonnemaisoniae* (Batt.) J. et G. Feldm. *Nova Hedw.*, **23**, 173–86

Boney, A. D. 1975. Mucilage sheaths of spores of red algae. *J. mar. biol. Ass., U.K.*, **55**, 511–18

Boney, A. D. & Corner, E. D. S. 1962. The effect of light on the growth of sporelings of the intertidal red alga *Plumaria elegans* (Bonnem.). *J. mar. biol. Ass., U.K.*, **42**, 65–92

Boney, A. D. & Corner, E. D. S. 1963. The effect of light on the growth of sporelings of the intertidal red algae *Antithamnion plumula* and *Brongiartella byssoides*. *J. mar. biol. Ass., U.K.*, **43**, 319–25

Booth, W. E. 1941. Algae as pioneers in plant succession and their importance in erosion control. *Ecology*, **22**, 38–46

Booth, W. H. 1946. The thermal death point of certain soil inhabiting algae. *Proc. Mont. Acad. Sci.*, **5/6**, 21–3

Børgesen, F. 1911. The algal vegetation of the lagoons in the West Indies. *Saertryk Biol. Arbejde. Eug. Warming.*, 41–56

Borowitzka, L. J. & Brown, A. D. 1974. The salt relations of marine and halophilic species of the unicellular green algae *Dunaliella*. The role of glycerol as a compatible solute. *Arch. Microbiol.*, **96**, 37–52

Borowitzka, M. A., Larkum, A. W. D. & Nockolds, C. E. 1974. A scanning electron microscope study of the structure and organisation of the calcium carbonate deposits of algae. *Phycologia*, **13**, 195–203

Boucher, D. 1972. Evaluation de la production primaire benthique en Baie de Concarneau. *C. r. hebd. Séanc. Acad. Sci., Paris*, **275**, (D), 1911–14

Boucher, D. 1975. Production primaire saisonniere du microphytobenthes des sables envases en Baie de Concarneau. Thesis submitted to L'Université de Bretagne Occidentale

Boudouresque, C. F. 1968. Contribution à l'étude du peuplement épiphyte de rhizomes de Posidonies (*Posidonia oceanica* Delile). *Recl. Trav. Stat. mar. Endoume.*, **59**, 45–64

Boudouresque, C. F. 1969. Étude qualitative et quantitative d'un peuplement algal à *Cystoseira mediterranea* dans la région de Banyuls-sur-Mer. (P.O.) *Vie et Milieu*, **B**, Oceanographie, **20**, 337–52

Boudouresque, C. F. 1971a. Recherches de bionomme analytique et expérimentale sur les peuplements benthiques sciaphile de Méditerranée occidentale (fraction algale): la sous-strate sciaphile des peuplements de grandes *Cystoseira* de mode battu. *Bull. Mus. Hist. nat. Marseille*, **31**, 141–51

Boudouresque, C. F. 1971b. Méthodes d'étude qualitative et quantitative du benthos (en particulier du phytobenthos). *Tethys*, **3**, 79–104

Bourelly, P. 1958. Algues microscopiques de quelques cuvettes supralittorales de la région de Dinard. *Verh. Int. Ver. Limnol.*, **13**, 683–6

Bourelly, P. 1968. *Les algues d'eau douce. Initiation à la systématique. Tome 11: Les algues jaunes et brunes.* 438 pp. Boubée et Cie, Paris.

Bourelly, P. 1970. *Les algues d'eau douce. Initiation à la systématique. Tome III. Les algues bleues et rouges, les Eugléniens, Peridiniens et Cryptomonadines*, 512 pp. N. Boubée et Cie., Paris

Bourelly, P. 1972. *Les algues d'eau douce. Initiation à la systématique. Tome I. Les algues vertes*, 572 pp. N. Boubée et Cie., Paris

Boyd, D. W., Korniker, L. S. & Rezak, R. 1963. Coralline algae microatolls near Cozumel Island, Mexico. *Contr. Geol. Dep. geol. Wyoming University*, **2**, (2), 105–8

Braarud, T. 1961. Cultivation of marine organisms as a means of understanding environmental influences on populations. ‘Oceanography’, *Amer. Ass. Adv. Sci.*, 271–98

Braarud, T. 1976. The natural history of the Hardangerfjord. 13. The ecology of taxonomic groups and species of phytoplankton related to their distribution in a fjord area. *Sarsia*, **60**, 41–62

Bradbury, J. P. & Winter, T. C. 1976. Areal distribution and stratigraphy of diatoms in the sediments of Lake Sallie, Minnesota. *Ecology*, **57**, 1005–14

Braek, G. S. & Jensen, A. 1976. Heavy metal tolerance of marine phytoplankton. III. Combined effects of copper and zinc ions on cultures of four common species. *J. exp. mar. Biol. Ecol.*, **25**, 37–50

Brand, F. 1925. Analyse der aerophilen Grünalgenauflüge,

insbesonderer der proto-pleurococcoiden Formen. *Arch. Protistenk*, **52**, 265–355

Braune, W. 1966. Experimentelle Ermittling der Zellvermehrung von *Scenedesmus obliquus* (Turp.) Kruger in Membranfilter-Kapseln unmittelbar im Fliessgewässer. *Verh. Int. Ver. Limnol.*, **16**, 830–6

Braune, W. 1971. Zur Ermittlung der potentiellen Productivität von Flusswasserproben im Algentest. *Int. Rev. ges. Hydrobiol.*, **56**, 795–810

Breen, P. A. & Mann, K. H. 1976. Changing lobster abundance and the destruction of Kelp beds by Sea Urchins. *Mar. Biol.*, **34**, 137–42

Brendemuhl, I. 1949. Über die Verbreitung der Erddiatomeen. *Arch. Mikrobiol.*, **14**, 407–69

Brezonik, P. L. 1972. Nitrogen: Sources and transformations in natural waters. In *Nutrients in natural Waters*, ed. H. E. Allen & J. R. Kramer, pp. 1–50. Wiley-Interscience

Briggs, J. C. 1974. *Marine Zoogeography*. McGraw-Hill Book Co. 475 pp.

Bringmann, G. & Kühn, R. 1971. Bestimmung der Begrenzungsfaktoren der Trophierung für die Kieselalge *Asterionella formosa* in West-Berliner Gewässer. *Gesundh. Ing.*, **92**, 176–83

Brinkhuis, B. H. 1976. The ecology of temperate salt-marsh Fucoids. I. Occurrence and distribution of *Ascophyllum nodosum* ecads. *Mar. Biol.*, **34**, 325–38

Brinkhuis, B. H. 1977. Comparisons of salt-marsh Fucoid production estimated from three different indices. *J. Phycol.*, **13**, 328–35

Brinkhuis, B. H. & Jones, R. F. 1976. The ecology of temperate saltmarsh fucoids. II. *In situ* growth of transplanted *Ascophyllum nodosum* ecads. *Mar. Biol.*, **34**, 339–48

Brinkhuis, B. H., Tempel, N. R. & Jones, R. F. 1976. Photosynthesis and respiration of exposed salt-marsh Fucoids. *Mar. Biol.*, **34**, 349–59

Bristol, B. M. 1919. On the retention of vitality by algae from old stored soils. *New Phyt.*, **18**, 92–107

Bristol, Roach, B. M. 1926. On the relation of certain soil algae to some soluble carbon compounds. *Ann. Bot.*, **40**, 149–201

Bristol, Roach, B. M. 1927a. On the algae of some normal English soils. *J. agric. Sci., Camb.* **17**, 583–8

Bristol, Roach, B. M. 1927b. On the carbon nutrition of some algae isolated from soil. *Ann. Bot.*, **41**, 509–17

Bristol, Roach, B. M. 1928. On the influence of light and glucose on the growth of a soil alga. *Ann. Bot.*, **42**, 317–45 South Orkney Islands. *Br. Antarct. Surv. Sci. Rep.*, no. 98

Broady, P. A. 1977. The terrestrial algae of Sidney Island, South Orkney Islands. *Br. Antarct. Surv. Sci. Rep.*, no. 98

Brock, T. D. 1967. Life at high temperatures. *Science*, **158**, 1012–19

Brock, T. D. 1969. Microbial growth under extreme environments. *Symp. Soc. Gen. Microbiol.*, **19**, 15–41

Brock, T. D. 1970. Photosynthesis by algal epiphytes of *Utricularia* in Everglades National Park. *Bull. Mar. Sci.*, **20**, 952–6

Brock, T. D. 1975. Effect of water potential on a *Microcoleus* (Cyanophyceae) from a desert crust. *J. Phycol.*, **11**, 316–20

Brock, T. D. & Brock, M. L. 1966. Temperature optima for algal development in Yellowstone and Iceland hot springs. *Nature, London*, **209**, 733–4

Brock, T. D. & Brock, M. L. 1971. Microbiological studies of thermal habitats of the central volcanic region, North Island, New Zealand. *N.Z. J. Mar. Freshw. Res.*, **5**, 233–58

Brody, M. & Emerson, R. 1959. The effect of wave length and intensity of light on the proportion of pigments in *Porphyridium cruentum*. *Am. J. Bot.*, **46**, 433–41

Broecker, W. S., Yuan-Hui, L. & Tsung-Hung, P. 1971. Carbon dioxide – mans unseen artifact. In *Impingement of man on the oceans*, ed. D. H. Wood, pp. 287–324. J. Wiley & Sons

Brook, A. J. 1952a. Some observations on the feeding of Protozoaon freshwater algae. *Hydrobiol.*, **4**, 281–93

Brook, A. J. 1952b. Occurrence of the terrestrial alga *Fritschiella* in Africa. *Nature, London*, **164**, 754

Brook, A. J., Baker, A. L. & Klemer, A. R. 1971. The use of turbidimetry in studies of the population dynamics of phytoplankton populations with special reference to *Oscillatoria agardhii* var. *isothrix*. *Mitt. Int. Ver. Limnol.*, **19**, 244–52

Brook, A. J. & Holden, A. V. 1957. Fertilization experiments in Scottish freshwater lochs. I. Loch Kinardorchy. *Freshwater Salmon Fish. Res.*, **17**, 30 pp.

Brook, A. J. & Rzòska, J. 1954. The influence of the Gebel Aulyia Dam on the development of Nile plankton. *J. anim. Ecol.*, **23**, 101–14

Brook, A. J. & Woodward, W. B. 1956. Some observations on the effects of water inflow and outflow on the plankton of small lakes. *J. anim. Ecol.*, **25**, 22–35

Brooks, J. L. 1969. Eutrophication and changes in the composition of the zooplankton. In *Eutrophication: Causes, consequences, correctives*, pp. 236–55. Nat. Acad. Sci., Washington D.C.

Brown, H. D. 1976. A comparison of the attached algal communities of a natural and an artificial substrate. *J. Phycol.*, **12**, 301–6

Brown, J. 1970. Studies on the biology of epiphytic algae. Ph.D. Thesis, University of Bristol

Brown, J. A. & Nielson, P. J. 1974. Transfer of photosynthetically produced carbohydrate from endosymbiotic chlorellae to *Paramesium bursaria*. *J. Protozool.*, **21**, 569–70

Brown, P. S. & Parsons, T. R. 1972. The effect of simulated upwelling on the maximization of primary productivity and the formation of phytodetritus. In *Detritus and its role in aquatic ecosystems. Proc. IBP–UNESCO Symp.*, pp. 169–83. Pallanza, Italy

Brown, R. M., Larson, D. A. & Bold, H. C. 1976. Air-borne algae. Their abundance and heterogeneity. *Science*, **143**, 583–5

Brown, S. D. & Austin, A. P. 1973. Spatial and temporal variation in periphyton and physico-chemical conditions in the littoral of a lake. *Arch. Hydrobiol.*, **71**, 183–232

Bryan, G. W. 1969. The absorption of zinc and other metals by the brown seaweed *Laminaria digitata*. *J. mar. biol. Ass.*, *U.K.*, **49**, 225–43

Brylinsky, M. 1977. Release of dissolved organic matter by some marine macrophytes. *Mar. Biol.*, **39**, 213–20

Brylinsky, M. & Mann, K. H. 1973. An analysis of factors governing productivity in lakes and reservoirs. *Limnol. Oceanogr.*, **18**, 1–14

Bunt, J. S. 1963. Diatoms of antarctic sea ice as agents of primary production. *Nature, London*, **199**, 1254–7

Bunt, J. S. 1967a. Microalgae of the Antarctic pack ice zone. *Proc. SCAR Symp. Antarctic Oceanogr.* Santiago, Chile

Bunt, J. S. 1967b. Some characteristics of microalgae isolated from Antarctic Sea Ice. *Antarctic Res. Ser.*, **11**, 1–14

Bunt, J. S. 1968b. Microalgae of the antarctic pack ice zone. In *Antarctic oceanogr.*, ed. R. J. Currie, pp. 198–218. Scott Polar Res. Int. Cambridge

Bunt, J. S., Cooksey, K. E., Heeb, M. A., Lee, O. C. & Taylor, B. F. 1970. Assay of algal nitrogen fixation in the marine subtropics by acetylene reduction. *Nature, London*, **227**, 1163–4

Bunt, J. S. & Lee, C. C. 1970. Seasonal primary production within antarctic sea ice at McMurdo Sound in 1967. *J. Mar. Res.*, **28**, 304–20

Bunt, J. S. & Lee, C. C. 1972. Data on the composition and dark survival of four sea-ice microalgae. *Limnol. Oceanogr.*, **17**, 458–61

Bunt, J. S., Lee, C. C. & Lee, E. 1972. Primary productivity and related data from tropical and subtropical marine sediments. *Mar. Biol.*, **16**, 28–36

Bunt, J. S., Owens, O. van H. & Hoch, G. 1966. Exploratory studies on the physiology and ecology of a psychrophilic marine diatom. *J. Phycol.*, **2**, 96–100

Bunt, J. S. & Wood, E. J. C. 1963. Microbiology and Antarctic sea-ice. *Nature, London*, **199**, 1254–5

Burkholder, P. R., Burkholder, L. M. & Almodovar, L. D. 1967. Carbon assimilation of marine flagellate blooms in neritic waters of Southern Puerto Rico. *Bull. Mar. Sci.*, **17**, 1–15

Burkholder, P. R. & Mandelli, E. F. 1965. Productivity of microalgae in antarctic sea ice. *Science*, **149**, 872–4

Burkholder, P. R., Repak, A. & Sibert, J. S. 1965. Studies on some Long Island Sound littoral communities of micro-organisms and their primary productivity. *Bull Torrey. Biol. Club*, **92**, 378–402

Burkholder, P. R. & Sieburth, J. M. 1961. Phytoplankton and chlorophyll in the Gerlache and Bransfield Straits of Antarctica. *Limnol. Oceanogr.*, **6**, 45–52

Burckle, L. H. 1972. Late Coenozoic planktonic diatom zones from the eastern equatorial Pacific. *Nova Hedw.*, **39**, 217–46

Burckle, L. H. 1977. Pliocene and pleistocene diatom datum levels from the equatorial Pacific. *Quart. Res.*, **7**, 330–40

Burckle, L. H., Clarke, D. B. & Shackleton, N. J. 1978. Isochronous last-abundant-appearance datum (LAAD) of the diatom *Hemidiscus karstenii* in the sub-Antarctic. *Geology*, **6**, 243–6

Burckle, L. H. & Stanton, D. 1975. Distribution of displaced Antarctic diatoms in the Argentine Basin. In *3rd Symposium on recent and Fossil Marine Diatoms*. Nova Hedw. 53, 283–92

Burckle, L. H. & McLaughlin, R. B. 1977. Size changes in the marine diatom *Coscinodiscus nodulifer* A. Schmidt in the equatorial Pacific. *Micropaleontology*, **23**, 216–22

Burn, R. 1966. The opisthobranchs of a Caulerpan microfauna from Fiji. *Proc. Malacol. Soc. London*, **37**, 45–65

Burns, R. L. & Mathieson, A. C. 1972. Ecological studies of economic red algae. II. Culture studies of *Chondrus crispus* Stackhouse and *Gigartina stellata* (Stackhouse) Batters. *J. exp. mar. Biol. Ecol.*, **8**, 1–6

Burns, N. M. & Ross, C. 1972a. Project Hypo. An intensive study of the Lake Erie Central Basin hypolimnion and related surface water phenomenon. *Canadian Centre for Inland Waters. Paper No. 6. EPA Technical Report TS–05–71–208–24*

Burns, N. M. & Ross, C. 1972b. Oxygen-nutrient relationships within the central basin of Lake Erie. In *Nutrients in natural Waters*, ed. H. E. Allen & J. R. Kramer, pp. 193–250. Wiley-Interscience

Burr, F. A. & West, J. A. 1970. Light and electron microscope observations on the vegetative and reproductive structures of *Bryopsis hypnoides*. *Phycologia*, **9**, 17–37

Burris, J. E. 1977. Photosynthesis, photorespiration and dark respiration in eight species of algae. *Mar. Biol.*, **39**, 371–9

Burrows, E. M. 1958. Sublittoral algal populations in Port Erin Bay, Isle of Man. *J. mar. biol. Ass., U.K.*, **37**, 687–703

Burrows, E. M. 1961. Experimental ecology with particular reference to the ecology of *Laminaria saccharina* (L.) Lamour. *Recent advances in botany*, 187–9

Burrows, E. M. 1971. Assessment of pollution effects by the use of algae. *Proc. Roy. Soc., London*, **B, 177**, 295–306

Burrows, E. M. & Lodge, S. M. 1950. A note on the inter-relationships of *Patella*, *Balanus* and *Fucus* on a semi-exposed coast. *Ann. Rep. Mar. Biol. Stat. Port Erin*, **62**, 30–4

Burrows, E. M. & Lodge, S. M. 1951. Autecology and the species problem in *Fucus*. *Jour. mar. biol. Ass., U.K.*, **30**, 161–76

Burrows, E. M. & Pybus, C. 1971. *Laminaria saccharina* and marine pollution in north east England. *Mar. Pollut. Bull.*, **2**, 53–6

Bursa, A. S. 1968. Epicenoses on *Nodularia spumigena* Martens in the Baltic Sea. *Acta Hydrobiol.*, **10**, 267–97

Bursche, E-M. 1953. Beitrag zum Phytoplanktonhaushalt des Müggelsees mit einer Betrachtung über die Verschiedenartigkeit von Netz-, Sieb- und Schopfplankton. *Zeitsch. f. Fisch. u. denen Hilfswissensch.* 11, 209–26

Bursche, E. M. 1962. Untersuchungen über die in der Polenz verherrshenden Bewuchsdiatomeen. *Arch. Hydrobiol.*, **58**, 474–89

Butcher, R. W. 1932. Studies in the ecology of rivers. II. The microflora of rivers with special reference to the algae on the river bed. *Ann. Bot.*, **46**, 813–61

Butcher, R. W. 1940. Studies in the ecology of rivers. IV. Observations on the growth and distribution of the sessile algae in the River Hull, Yorkshire. *J. Ecol.*, **28**, 210–23

Butcher, R. W. 1949. Problems of distribution of sessile algae in running water. *Verh. Int. Ver. Limnol.*, **10**, 98–103

Butcher, R. W., Longwell, J. & Pentelow, F. T. K. 1937. Survey of the River Tees. Part III. The non-tidal reaches. *Wat. Pollut. Res., Tech. pap.* no. 6. H.M.S.O., London

Buzer, J. S. 1975. A study of sediments from Lough Ine and Ballyally Lough, County Cork. S.W. Ireland. Ph.D. Thesis, University of Bristol

Cabioch, J. 1966. Contribution á l'étude morphologique, anatomique et systematique de deux Mélobésiées:

Lithothamnium calcareum et *Lithothamnium coralloides*. *Bot. Marina*, **9**, 33–53

Cadée, G. C. 1975. Primary production of the Guyana coast. *Neth. J. Sea Res.*, **9**, 128–43

Cairns, J. Jr. & Dickson, K. L. 1971. A simple method for the biological assessment of the effects of waste discharges on aquatic bottom dwelling organisms. *J. Wat. Pollut. Control. Fed.* **43**, 755–72

Cairns, J. Jr, Dickson, K. L. & Lanza, G. 1973. Rapid biological monitoring system for determining aquatic community structure in receiving systems. In *Biological methods for the Assessment of Water Quality*, ed. J. Cairns & K. L. Dickson, pp. 148–63. ASTM, STP 528

Cairns, J. Jr., Dickson, K. L. & Slocomb, J. 1977. The A.B.C.'s of diatom identification using laser holography. *Hydrobiol.* **54**, 7–16

Callame, B. & Debyser, J. 1954. Observations sur les mouvements des diatomées à la surface des sediments marins de la zone intercotidiale. *Vie et Milieu*, **5**, 242–49

Calow, P. 1973. The food of *Ancylus fluviatilis* (Müll.), a littoral stone dwelling, herbivore. *Oecologia, Berlin*, **13**, 113–33

Calvert, S. E. 1966. Accumulation of diatomaceous silica in the sediments of the Gulf of California. *Bull. geol. Soc. Am.*, **77**, 569–96

Cameron, R. E. 1960. Communities of soil algae occurring in the Sonoran desert in Arizonia. *J. Ariz. Acad. Sci.*, **1**, 85–8

Cameron, R. E. 1964. Terrestrial algae of southern Arizonia. *Trans. Am. Micros. Soc.*, **83**, 212–18

Cameron, R. E. & Blank, G. R. 1966. Desert Algae: Soil crusts and diaphanous substrata as algal habitats. *NASA Tech. Rep.* nos. 32–971, 1–45

Cameron, R. E. & Devaney, T. R. 1970. Antarctic soil algal crusts: Scanning electron and optical microscope study. *Trans. Am. Micros. Soc.*, **89**, 264–73

Cameron, R. E. & Fuller, W. A. Nitrogen fixation by some algae in Arizona soils. *Proc. Soil Sci. Soc. Am.*, **24**, 353–6

Canter, H. M. 1973. A new primitive protozoan devouring centric diatoms in the plankton. *J. Linn. Soc. London, Zool.*, **52**, 63–83

Canter, H. M. 1974. Parasites of algae. In *42nd Ann. Rep. Freshw. Biol. Ass.*, 44–6

Canter, H. M. 1979. Fungal and protozoan parasites and their importance in the ecology of the phytoplankton. *Ann. Rep. Freshwater Biol. Assc.* 43–50.

Canter, H. M. & Jaworski, G. H. M. 1978. The isolation, maintenance and host range studies of a chytrid *Rhizophidium planktonicum* Canter emend., parasitic on *Asterionella formosa* Hassall. *Ann. Bot.*, **42**, 967–79

Canter, H. M. & Lund, J. W. G. 1966. The periodicity of planktonic desmids in Windermere, England. *Verh. Int. Ver. Limnol.*, **16**, 163–72

Canter, H. M. & Lund, J. W. G. 1968. The importance of Protozoa in controlling the abundance of planktonic algae in lakes. *Proc. Linn. Soc. London*, **179**, 203–19

Canter, H. M. & Lund, J. W. G. 1969. The parasitism of planktonic desmids by fungi. *Öst. Bot. Zeitsch.*, **116**, 351–77

Capone, D. G. 1977. $N_2(C_2H_2)$ fixation by macroalgal epiphytes. In *Proc. 3rd Int. Coral Reef Symp. Miami. vol. I. Biology*, 337–42

Capone, D. G. & Taylor, B. F. 1977. Nitrogen fixation

(Acetylene reduction) in the phyllosphere of *Thalassia testudinum. Mar. Biol.*, **40**, 19–28

Capone, D. G., Taylor, D. L. & Taylor, B. F. 1977. Nitrogen fixation (acetylene reduction) associated with macroalgae in a coral-reef community in the Bahamas. *Mar. Biol.*, **40**, 29–32

Carlin, B. 1946. *Plankton i Motala Ström. 1935–40*. 49 pp. Norrköping

Carlson, R. E. 1977. A trophic state index for lakes. *Limnol. Oceanogr.*, **22**, 361–9

Carlucci, A. F. 1974. Production and utilization of dissolved vitamins by marine phytoplankton. In *Effects of the Ocean Environment on Microbial Activities*, ed. R. R. Colwell & R. Y. Morita, pp. 449–56

Carlucci, A. F. & Bowes, P. M. 1970*a*. Production of vitamin B_{12}, Thiamine and Biotin by phytoplankton. *J. Phycol.*, **6**, 354–7

Carlucci, A. F. & Bowes, P. M. 1970*b*. Vitamin production and utilization by phytoplankton in mixed cultures. *J. Phycol.*, **6**, 393–400

Carmichael, W. W. & Gorham, P. R. 1977. Factors influencing the toxicity and animal susceptibility of *Anabaena flos-aquae* (Cyanophyta) blooms. *J. Phycol.*, **13**, 97–101

Carpenter, E. J. 1970. Diatoms attached to floating *Sargassum* in the western Sargasso Sea. *Phycologia*, **9**, 269–74

Carpenter, E. J. 1971. Effects of phosphorus mining wastes on the growth of phytoplankton in the Pamlico River Estuary. *Chesapeake Sci.*, **12**, 85–94

Carpenter, E. J. & Cox, J. L. 1974. Production of pelagic *Sargassum* and a blue-green epiphyte in the western Sargasso Sea. *Limnol. Oceanogr.*, **19**, 429–36

Carpenter, E. J. & Guillard, R. R. L. 1971. Intraspecific differences in nitrate half saturation constants for three species of marine phytoplankton. *Ecology*, **52**, 183–5

Carpenter, E. J. & McCarthy, J. J. 1975. Nitrogen fixation and uptake of combined nitrogenous nutrients by *Oscillatoria* (*Trichodesmium thiebautii*) in the Western Sargasso Sea. *Limnol. Oceanogr.*, **20**, 389–401

Carpenter, E. J., Peck, B. B. & Anderson, S. J. 1972. Costing water chlorination and productivity of entrained phytoplankton. *Mar. Biol.*, **16**, 37–40

Carpenter, E. J. & Price, IV, C. C. 1976. Marine *Oscillatoria* (*Trichodesmium*). Explanation for aerobic nitrogen fixation without heterocysts. *Science*, **191**, 1278–80

Carpenter, E. J. & Price, IV, C. C. 1977. Nitrogen fixation, distribution and production of *Oscillatoria* (*Trichodesmium*) spp. in the western Sargasso and Caribbean Seas. *Limnol. Oceanogr.*, **22**, 60–72

Carpenter, E. J., Raalte, E. D. van & Valiela, I. 1978. Nitrogen fixation by algae in a Massachusetts salt marsh. *Limnol. Oceanogr.*, **23**, 318–27

Carpenter, E. J., Remsen, C. C. & Watson, S. W. 1972. Utilization of urea by some marine phytoplankton. *Limnol. Oceanogr.*, **17**, 265–9

Casper, S. J. 1974. *Grundzüge eines natürlichen Systems der Mikroorganismen*. 232 pp. VEB Gustav Fischer, Jena

Casper, S. H. & Schultz, H. 1960. Studien zur Wertung der Saprobiensysteme. *Int. Rev. ges. Hydrobiol.*, **45**, 535–65

Casper, S. H. & Schultz, H. 1962. Weitere Unterlagen zur

Prüfung der Saprobiensysteme. *Int. Rev. ges. Hydrobiol.*, **47**, 100–17

Cassie, R. M. 1963. Relationship between plant pigments and gross primary production in *Skeletonema costatum*. *Limnol. Oceanogr.*, **8**, 433–9

Cassin, P. E. 1974. Isolation, growth and physiology of acidophilic Chlamydomonads. *J. Phycol.*, **10**, 439–47

Castenholz, R. W. 1960. Seasonal changes in the attached algae of freshwater and saline lakes in the Lower Grand Coulee, Washington. *Limnol. Oceanogr.*, **5**, 1–28

Castenholz, R. W. 1963. An experimental study of the vertical distribution of littoral marine diatoms. *Limnol. Oceanogr.*, **8**, 450–62

Castenholz, R. W. 1969. The thermophilic cyanophytes of Iceland and the upper temperature limit. *J. Phycol.*, **5**, 360–68

Castenholz, R. W. 1977. The effect of sulphide on the blue-green algae of hot springs. II. Yellowstone National Park. *Microbial Ecol.*, **3**, 79–105

Castenholz, R. W. & Wickstrom, C. E. 1975. Thermal streams. In *River Ecology*, ed. B. A. Whitton. Blackwell, Oxford pp. 264–85

Cattaneo, A. & Kalff, J. 1978. Seasonal changes in the epiphyte community of natural and artificial macrophytes in Lake Memphremagog (Que. and VT). *Hydrobiol.*, **60**, 135–44

Cedercreutz, C. 1941. Beitrag zur Kenntnis der Felsenalgen in Finnland. *Mem. Soc. fauna flora Fennica*, **17**, 105–21

Cernichiari, E., Muscatine, L. & Smith, D. C. 1969. Maltose excretion by the symbiotic algae of *Hydra viridis*. *Proc. Roy. Soc., London*, **B**, **173**, 557–76

Chadefaud, M. 1978. Sur la notion de Prochlorophytes. *Rev. Algol. N.S.*, **13**, 203–6

Chalker, B. E. & Taylor, D. L. 1975. Light-enhanced calcification, and the role of oxidative phosphorylation in calcification of the coral *Acropora cervicornis*. *Proc. Roy. Soc., London*, **B**, **190**, 323–31

Chan, E. C. S. & McManus, E. A. 1969. Distribution, characterization and nutrition of marine microorganisms from the algae *Polysiphonia lanosa* and *Ascophyllum nodosum*. *Can. J. Microbiol.*, **15**, 409–420

Chandler, D. C. 1937. Fate of typical lake plankton in streams. *Ecol. Mongr.*, **7**, 445–79

Chapman, A. R. O. 1973. A critique of prevailing attitudes towards the control of seaweed zonation on the sea shore. *Bot. Mar.*, **16**, 80–2

Chapman, R. L. & Lang, N. J. 1973. Virus-like particles and nuclear inclusions in the red alga *Porphyridium purpureum* (Bory) Drew et Ross. *J. Phycol.*, **9**, 117–22

Chapman, V. J. 1942. Zonation of marine algae on the sea-shore. *Proc. Linn. Soc. London*, **154**, 239–53

Charters, A. C., Neushul, M. & Coon, D. 1973. The effect of water motion on algal spore adhesion. *Limnol. Oceanogr.*, **18**, 884–96

Chen, L. C. M., Edelstein, T. & McLachlan, J. 1974. The life history of *Gigartina stellata* (Stackh.) Batt. (Rhodophyceae, Gigartinales) in culture. *Phycologia*, **13**, 287–94

Cheney, D. P. & Babbel, G. R. 1978. Biosystematic studies of the red algal genus *Eucheuma*. I. Electrophoretic variation among Florida populations. *Mar. Biol.*, **47**, 251–64

Chihara, M. & Yoshizaki, M. 1970. Marine algal flora and communities along the coast of the Tsushima Islands. *Mem. Nat. Sci. Mus.*, **3**, 143–58

Chihara, M. & Yoshizaki, M. 1972. Bonnemaisoniaceae: Their gonimoblast development, life history and systematics. In *Contributions to the systematics of benthic marine algae of the North Pacific*, ed. I. A. Abbott & M. Kurogi, pp. 243–52. Sapporo, Japan

Chisholm, S. W., Azam, F. & Eppley, R. W. 1978. Silicic acid incorporation in marine diatoms on light:dark cycles. Use as an assay for phased cell division. *Limnol. Oceanogr.*, **23**, 518–29

Chock, J. S. & Mathieson, A. C. 1976. Ecological studies of the salt marsh ecad *scorpioides* (Hornemann) Hauch, of *Ascophyllum nodosum* (L.) Le Jolis. *J. exp. mar. Biol. Ecol.*, **23**, 171–90

Cholnoky, B. J. 1929. Symbiose zwischen Diatomeen. *Arch. Protistenk.*, **68**, 523–30

Cholnoky, B. J. 1968. *Die ökologie der Diatomeen in Binnengewässern*, 699 pp. *J. Cramer*

Christensen, T. 1979. Annotations to a textbook of phycology. *Bot. Tidsskr.*, **73**, 65–70

Christie, A. O. & Evans, L. V. 1962. Periodicity in the liberation of gametes and zoospores of *Enteromorpha intestinalis* Link. *Nature, London*, **193**, 193–4

Christie, A. O., Evans, L. V. & Shaw, M. 1970. Studies on the ship-fouling alga *Enteromorpha*. 2. The effect of certain enzymes on the adhesion of zoospores. *Ann. Bot.*, **34**, 467–82

Christie, A. O. & Shaw, M. 1968. Settlement experiments with zoospores of *Enteromorpha intestinalis* (L.) Link. *Brit. Phyc. Bull.*, **3**, 529–34

Chudyba, H. 1965. *Cladophora glomerata* and accompanying algae in the Skawa River. *Acta Hydrobiol.*, **7**, 93–126

Chudyba, H. 1968. *Cladophora glomerata* and concommitant algae in the River Skawa. Distribution and conditions of appearance. *Acta. Hydrobiol.*, **10**, 39–84

Churchill, A. C. & Moeller, H. W. 1972. Seasonal patterns of reproduction in New York populations of *Codium fragile* (Sur.) Hariot subsp. *tomentosoides* (van Goor) Silva. *J. Phycol.*, **8**, 147–52

Ciereszko, L. S. 1962. Chemistry of coelenterates. III. Occurrence of antimicrobial terpenoid compounds in the zooxanthellae of alcyonarians. *Trans. N.Y. Acad. Sci.*, ser. 2, **24**, 502–3

Clarke, E. de D. & Teichert, C. 1946. Algal structures in a west Australian salt lake. *Am. J. Sci.*, **244**, 271–6

Clarke, G. L. 1965. *Elements of Ecology*. 534 pp. Wiley

Clarke, W. D. & Neushul, M. 1967. Subtidal ecology of the Southern California coast. In *Pollution and Marine Ecology*, ed. T. A. Olsen & F. J. Burgess, pp. 29–42. Interscience

Clasby, R. C., Horner, R. & Alexander, V. 1973. An *in situ* method for measuring primary productivity of Arctic Sea ice algae. *J. Fish. Res. Bd. Can.*, **30**, 835–8

Clasen, J. & Bernhardt, H. 1974. The use of algal assays for determining the effect of iron and phosphorus compounds on the growth of various algal species. *Wat. Res.*, **8**, 31–46

Clauss, H. 1972. Der Einfluss von Rot – und Blaulicht auf die Photosynthese van *Acetabularia mediterranea* und auf die

Verteilung des assimilierten Kohlenstoffe. *Protoplasma*, **74**, 357–79

Clendenning, K. A. 1960. Physiology and biochemistry of giant kelp. *Prog. Rep. Univ. California Inst. Mar. Res.*, IMR ref. 60, 1–47

Clendenning, K. A. 1964. Photosynthesis and growth in *Macrocystis pyrifera*. In *Proc. 4th Int. Seaweed Symp.*, pp. 55–65. Pergamon Press

Clendenning, K. A. 1971. Organic productivity in kelp areas. In *The Biology of Giant Kelp Beds (Macrocystis) in California*, ed. W. J. North. *Nova Hedw. Beih.*, **32**, 259–63

Clesceri, L. S. 1979. The role of the surface sediments in lake metabolism. Rensselaer Fresh Water Institute. *Newsletter*, **9**, part 1, 1–4

Cloud, P. E. 1968. Atmospheric and Hydrospheric evolution on the primitive earth. *Science*, **160**, 729–36

Cloud, P. E., Holland, H. D., Commoner, B., Davidson, C. F., Fischer, A. G., Berkner, L. V. & Marshall, L. C. 1965. Contributions to Symposium on the evolution of the earth's atmosphere. *Proc. Nat. Acad. Sci.*, **53**, 1169–226

Coesel, P. F. M. 1975. The relevance of desmids in the biological typology and evaluation of fresh waters. *Hydrobiol. Bull.*, **9**, 93–101

Cohen, J. M. 1972. Nutrient removal from wastewater by physical-chemical processes. In *Nutrients in natural Waters*, ed. H. E. Allen & J. R. Kramer, pp. 353–89. Wiley-Interscience

Cohen, Y., Padan, E. & Shilo, M. 1975. Facultative anoxygenic photosynthesis in the Cyanobacterium *Oscillatoria limnetica*. *J. Bact.*, **123**, 855–61

Cohn, F. 1854. Untersuchungen über die Entwicklungs-geschichte der mikroscopishen Algen und Pilz. *Verh. Leop. Car.*, **24**, 103

Cole, K. 1964. Induced fluorescence in gametophytes of some *Laminariales*. *Can. J. Bot.*, **42**, 1173–81

Cole, K. & Conway, E. 1975. Phenetic implications of structural features of the perennating phase in the life history of *Porphyra* and *Bangia* (Bangiophyceae, Rhodophyta). *Phycologia*, **14**, 239–45

Colman, J. 1933. The nature of the intertidal zonation of plants and animals. *J. mar. biol. Ass.*, *U.K.*, **18**, 435–76

Coleman, J. S. & Stephenson, A. 1966. Aspects of the ecology of a 'tideless' shore. In *Some contemporary studies in Marine Science*, ed. H. Barnes, pp. 163–70. Allen & Unwin, London

Coleman, R. D., Coleman, R. L. & Rice, E. L. 1971. Zinc and cobalt bioconcentration and toxicity in selected algal species. *Bot. Gaz.*, **132**, 102–5

Coles, J. W. 1958. Nematodes parasitic on seaweeds of the genera *Ascophyllum* and *Fucus*. *J. mar. biol. Ass.*, *U.K.*, **37**, 145–58

Colijn, F. & Venekamp, L. 1977. Benthic primary production in the Ems-Dollard Estuary during 1975. (Summary.) *Hydrobiol. Bull.*, **11**, 16–17

Collier, A. 1958. Some biochemical aspects of red tides and related oceanographic problems. *Limnol. Oceanogr.*, **3**, 33–9

Collier, B. D., Cox, G. W., Johnson, A. W. & Miller, P. C. 1973. *Dynamic Ecology*. Prentice Hall, Inc., New Jersey

Colocoloff, M. & Colocoloff, G. 1973. Recherches sur la production primaire d'un fond sableux. 2. Méthodes. *Téthys*, **4**, 779–99

Colver, D. A. & Brunskill, G. J. 1969. Fayetteville Green Lake, New York. V. Studies of primary production and zooplankton in a meromictic marl lake. *Limnol. Oceanogr.*, **14**, 862–73

Comère, J. 1913. De l'action du milieu considéreé dans ses rapports avec la distribution générale des Algues d'eau douce. *Mem. Soc. Bot. France*, no. 25

Confer, J. L. 1972. Interrelations among plankton, attached algae, and the phosphorus cycle in artificial open systems. *Ecol. Monogr.*, **42**, 1–23

Connel, J. H. 1972. Community interactions on marine rocky intertidal shores. *Ann. Rev. Ecol. Systematics*, **3**, 169–92

Conover, J. T. 1958. Seasonal growth of benthic marine plants as related to environmental factors in an estuary. *Publ. Inst. Mar. Sci.*, *Univ. Texas*, **5**, 97–147

Conover, J. T. 1968. The importance of natural diffusion gradients and transport substances related to benthic marine plant metabolism. *Bot. Mar.*, **11**, 1–9

Conover, J. T. & Sieburth, J. N. 1964. Effect of *Sargassum* distribution and its epibiota and antibacterial activity. *Bot. Mar.*, **6**, 147–57

Conover, S. A. M. 1956. Oceanography of Long Island Sound, 1952–1954. IV. Phytoplankton. *Bull. Bing. Oceanogr. Collect.*, **15**, 62–112

Conway, H. L. 1978. Sorption of arsenic and cadmium and their effects on growth, micronutrient utilisation and photosynthetic pigment composition of *Asterionella formosa*. *J. Fish. Res. Bd. Can.*, **35**, 286–94

Conway, K. & Trainor, F. R. 1972. *Scenedesmus* morphology and flotation. *J. Phycol.*, **8**, 138–43

Cook, C. B. 1972. Benefit to symbiotic zoochlorellae from feeding by green *Hydra*. *Biol. Bull.*, **142**, 236–42

Cooke, G. W. & Williams, R. J. B. 1970. Losses of nitrogen and phosphorus from agricultural land. *Water Treat. Exam.*, **19**, 253–76

Cooke, W. J. 1975. The occurrence of an endozoic green alga in the marine mollusc, *Clinocardium nuttallii* (Conrad, 1837). *Phycologia*, **14**, 35–9

Cooksey, K. E. & Chansang, H. 1976. Isolation and physiological studies on three isolates of *Amphora*. (Bacillariophyceae). *J. Phycol.*, **12**, 455–60

Cooksey, K. E. & Cooksey, B. 1974. Calcium deficiency can induce the transition from oval to fusiform cells in cultures of *Phaeodactylum tricornutum* Bohlin. *J. Phycol.*, **10**, 89–90

Cooksey, K. E., Cooksey, B., Evans, P. M. & Hildebrand, E. L. 1975. Benthic diatoms as contributors to the carbon cycle in a mangrove community. In *Proc. 10th European Mar. Biol. Symp.*, *Ostend, Belgium*, **2**, 165–78

Coon, D., Neushul, M. & Charters, A. C. 1972. The settling behaviour of marine algal spores. In *Proc. 7th Int. Seaweed Symp.*, *Sapporo, Japan*, 237–42. University of Tokyo

Cooper, D. C. 1973. Enhancement of net primary productivity by herbivore grazing in aquatic laboratory microcosms. *Limnol. Oceanogr.*, **18**, 31–7

Corillion, R. 1957. Les Charophycées de France et d'Europe Occidentale. *Bull. Soc. Sci. Bretagne*, *H.S.*, **32**, 1–499

Corner, E. D. S. & Davies, A. G. 1971. Plankton as a factor in the nitrogen and phosphorus cycles in the sea. *Adv. mar. Biol.*, **9**, 101–204

Corner, E. D. S., Head, R. N. & Kilvington, C. C. 1972. On the nutrition and metabolism of zooplankton. VIII. The grazing of *Biddulphia* cells by *Calanus helgolandicus*. *J. mar. biol. Ass., U.K.*, **52**, 847–61

Cotton, A. D. 1911. On the growth of *Ulva latissima* in excessive quantity, with special reference to the *Ulva* nuisance in Belfast Lough. In *Royal Commission on Sewage Disposal. 7th Report II (Appendix IV)*, pp. 121–42. HMSO, London

Cox, E. J. 1975a. A reappraisal of the diatom genus *Amphipleura* Kützing using light and electron microscopy. *Br. phycol. J.*, **10**, 1–12

Cox, E. J. 1975b. Further studies on the genus *Berkeleya* Grev. *Br. phycol. J.*, **10**, 205–17

Cox, E. J. 1977. The distribution of tube dwelling diatom species in the Severn Estuary. *J. mar. biol. Ass., U.K.*, **57**, 19–27

Cox, E. J. 1977. The tube-dwelling diatom flora at two sites in the Severn estuary. *Bot Mar.*, **20**, 111–1119

Cox, E. R. & Hightower, J. 1972. Some corticolous algae of McMinn County, Tennessee, U.S.A. *J. Phycol.*, **8**, 203–5

Crabtree, K. 1969. Postglacial diatom zonation of linnic deposits in North Wales. *Mitt. Int. Ver. Limnol.*, **17**, 165–71

Craigie, J. S. 1969. Some salinity induced changes in growth, pigments and cyclo-hexanetetrol content of *Monochrysis lutheri*. *J. Fish. Res. Bd. Can.*, **26**, 2959–67

Craigie, J. S. & McLachlan, J. 1964. Excretion of coloured ultraviolet-absorbing substances by marine algae. *Can. J. Bot.*, **42**, 23–33

Cribb, A. B. 1954. Records of marine algae from south eastern Queensland. I. *Pap. Dep. Bot. Univ. Qd.*, 27–35

Cribb, A. B. 1956. The algal vegetation of Port Arthur, Tasmania. *Pap. Proc. Roy. Soc. Tasmania*, **88**, 1–44

Cribb, A. B. 1966. The algae of Heron Island, Great Barrier Reef, Australia, Part I. A general account. *Pap. Heron Island. Res. Sta.*, **1**, 1–23. University of Queensland

Cribb, A. B. 1969a. Algae on the Hawks-Bill turtle. *Queensland Nat.*, **19**, 108–9

Cribb, A. B. 1969b. The vegetation of north West Island. *Queensland Nat.*, **19**, 85–93

Cribb, A. B. 1969c Sea sawdust. *Queensland Nat.*, **19**, 115–17

Cribb, A. B. 1973. The algae of the Great Barrier Reefs. In *Biology and geology of coral reefs*, vol. 2, Biol. I, pp. 47–75. Academic Press

Crossett, R. N., Drew, E. A. & Larkum, A. W. D. 1965. Chromatic adaptation in benthic marine algae. *Nature, London*, **207**, 547–8

Crossland, C. J. & Barnes, D. J. 1976. Acetylene reduction by corals. *Limnol. Oceanogr.*, **21**, 153–5

Culkin, F. 1965. The major constituents of sea water. In *Chemical Oceanography*, ed. J. P. Riley & G. Skirrow, pp. 121–61. Academic Press, London

Cullimore, D. R. & McCann, A. E. 1977. The influence of four herbicides on the algal flora of a prairie soil. *Plant Soil*, **46**, 499–570

Cullimore, R. D. & Woodbine, M. 1963. A rhizosphere effect of the pea root on soil algae. *Nature, London*, **198**, 304–5

Culver, D. A. & Brunskill, G. J. 1969. Fayetteville Green Lake, New York. V. Studies of primary production and zooplankton in a meromictic marl lake. *Limnol. Oceanogr.*, **14**, 862–73

Cundell, W. M., Sleeter, T. D. & Mitchell, R. 1977. Microbial populations associated with the surface of the brown alga *Ascophyllum nodosum*. *Microbial Ecol.*, **4**, 81–91

Cupp, E. E. 1937. Seasonal distribution and occurrence of marine diatoms and dinoflagellates at Scotch Cap, Alaska. *Bull. Scripps Inst. Oceanogr.*, tech. ser., **4**, 71–100

Curl, H. C. & Small, L. F. 1965. Variations in photosynthetic assimilation ratios in natural marine phytoplankton communities. *Limnol. Oceanogr., Suppl.*, **10** (*Redfield Anniv. Vol.* R67–R73)

Curtis, E. J. C. & Curds, C. R. 1972. Sewage fungus in rivers in the United Kingdom: The slime community and its constituent organisms. *Wat. Res.*, **5**, 1147–59

Cushing, D. H. 1963. Studies on a *Calanus* patch. II. The estimation of algal reproductive rates. *J. mar. biol. Ass., U.K.*, **43**, 339–47

Cushing, D. H. 1971. A comparison of production in temperate seas and the upwelling areas. *Trans. Roy. Soc. S. Africa*, **40**, 17–33

Cushing, D. H. 1973. Production in the Indian Ocean and the transfer from the primary to the secondary level. In *The Biology of the Indian Ocean*, ed. B. Zeitschel, 475–86. Chapman & Hall

Cushing, D. H. 1976. Grazing in Lake Erken. *Limnol. Oceanogr.*, **21**, 349–56

Cushing, D. H. & Vúcetic, T. 1963. Studies on a *Calanus* patch. III. The quantity of food eaten by *Calanus finmarchicus*. *J. mar. biol. Ass., U.K.*, **43**, 349–71

Czeczuga, B. 1965. Quantitative changes in sedimentary chlorophyll in the bed sediment of the Mikalajki Lake during the post glacial period. *Schweiz. Z. Hydrol.*, **27**, 88–98

Daft, M. J., Begg, J. & Stewart, W. D. P. 1970. A virus of blue-green from freshwater habitats in Scotland. *New Phytol.*, **69**, 1029–38

Daft, M. J. & Stewart, W. D. P. 1971. Bacterial pathogens of freshwater blue-green algae. *New Phytol.*, **70**, 819–29

Dahl, A. L. 1971. Ecology and community structure of some tropical reef algae in Samoa. In *Proc. 7th Int. Seaweed Symp., Sapporo, Japan*, 36–9. University of Tokyo

Dahl, A. L. 1973. Benthic algal ecology in a deep reef and sand habitat off Puerto Rico. *Bot. Mar.*, **16**, 171–5

Daisley, K. W. 1969. Monthly survey of vitamin B_{12} concentrations in some waters of the English Lake District. *Limnol. Oceanogr.*, **14**, 224–8

Daly, M. A. & Mathieson, A. C. 1977. The effects of sand movement on intertidal seaweeds and selected invertebrates at Bound Rock, New Hampshire, U.S.A. *Mar. Biol.*, **43**, 45–55

Dalyrymple, D. W. 1966. Calcium carbonate deposition associated with blue green algal mats, Baffin Bay, Texas. *Inst. Marine Sci.*, **10**, 187–200. University of Texas

Damant, G. C. C. 1937. Storage of oxygen in the bladders of the seaweed *Ascophyllum nodosum* and their adaptation to hydrostatic pressure. *J. exp. Biol.*, **14**, 198–209

Darden, W. H. 1970. Hormonal control of sexuality in the genus *Volvox*. *Ann. New York Acad. Sci.*, **175**, 757–63

Darley, W. M., Dunn, E. L., Holmes, K. S. & Larew, H. G.

1976. A ^{14}C method for measuring epibenthic microalgal productivity in air. *J. exp. mar. Biol. Ecol.*, **25**, 207–17

Darnell, R. M. 1967. Organic detritus in relation to the estuarine ecosystem. In *Estuaries*, ed. G. H. Lauff. *Amer. Assoc. Adv. Sci.*, Washington D.C., no. 83, 376–82

Davies, A. G. 1976. An assessment of the basis of mercury tolerance in *Dunaliella tertiolecta*. *J. mar. biol. Ass., U.K.*, **56**, 39–57

Davies, C. C. 1967. Circulation of matter versus flow of energy in production studies. *Arch. Hydrobiol.*, **63**, 250–5

Davies, W. 1971. The phytopsammon of a sandy beach transect. *Am. Midl. Nat.*, **86**, 292–308

Davis, C. C. 1958. An approach to some problems of secondary production in the Western Lake Erie region. *Limnol. Oceanogr.*, **3**, 15–28

Davis, C. C. 1964. Evidence for the eutrophication of Lake Erie from phytoplankton records. *Limnol. Oceanogr.*, **9**, 275–83

Dawson, E. Y. 1966. *Marine Botany: an introduction*. 371 pp. Holt, Rinehart & Winston, New York

Dawson, E. Y., Aleem, A. A. & Halstead, B. W. 1955. Marine algae from Palmyra Island with special reference to the feeding habits and toxicology of reef fish. *Allan Hancock Found. Occ. Pap.*, **17**, p. 39

Dawson, E. Y. & Neushul, M. 1960. New records of marine algae from Anacapta Island, California. *Nova Hedw.*, **12**, 173–91

Dawson, E. Y., Neushul, M. & Wildmann, R. D. 1960. Seaweeds associated with kelp beds along southern California and north western Mexico. *Pac. Naturalist*, **1**, no. 14, 1–80

Dayton, P. K. 1971. Competition, disturbance and community organization: The provision and subsequent utilization of space in a rocky intertidal community. *Ecol. Monogr.*, **41**, 357–89

Dayton, P. K. 1973. Dispersion, dispersal and persistence of the annual intertidal alga, *Postelsia palmaeformis* Ruprecht. *Ecology*, **54**, 433–8

Dayton, P. K. 1975a. Experimental studies of algal canopy interactions in a Sea Otter dominated kelp community at Amchitka Island, Alaska. *Fish. Bull.*, **73**, 230–7

Dayton, P. K. 1975b. Experimental evolution of ecological dominance in a rocky intertidal community. *Ecol. Monogr.*, **45**, 137–59

Deacon, G. E. 1942. The Sargasso Sea. *Geogr. J.*, **99**, 16–28

Defant, A. 1961. *Physical oceanography*, vol. I. 729 pp. Pergamon Press, London

DeFelice, D. R. & Lynts, G. W. 1978. Benthic marine diatom associations: Upper Florida Bay (Florida) and associated sounds. *J. Phycol.*, **14**, 25–33

Degens, E. T., Okada, H., Honjo, S. & Hathaway, J. C. 1972. Microcrystalline sphalerite in resin globules suspended in Lake Kivu, East Africa. *Miner. Depos.*, **7**, 1–12.

Dellow, V. 1950. Intertidal ecology at Narrow Neck Reef, New Zealand. *Pacif. Sci.*, **4**, 355–74

Dellow, V. 1955. Marine algal ecology of the Hauroki Gulf, New Zealand. *Trans. Roy. Soc. New Zealand*, **83**, 1–91

Dellow, V. & Cassie, R. M. 1955. Littoral zonation in two caves in the Auckland district. *Trans. Roy. Soc. New Zealand*, **83**, 321–31

Denffer, D. 1948. Über einen Wachstumshemmstoff in alternden Diatomeenkulturen. *Biol. Zentralb.*, **67**, 7–13

Denman, K. L. 1976. Covariability of chlorophyll and temperature in the sea. *Deep Sea Res.*, **23**, 539–50

Derenbach, J. B. & Williams, P. J. Le B. 1974. Autotrophic and bacterial production: fractionation of plankton population by differential filtration of samples from the English Channel. *Mar. Biol.*, **25**, 263–9

Dickman, M. 1968a. Some indices of diversity. *Ecology*, **49**, 1191–3

Dickman, M. 1968b. The effect of grazing by tadpoles on the structure of a periphyton community. *Ecology*, **49**, 1188–90

Dickman, M. 1969. Some effects of lake renewal on phytoplankton productivity and species composition. *Limnol. Oceanogr.*, **14**, 660–6

Dickson, E. L. 1975. A silica budget for Lough Neagh, 1970–1972. *Freshwater Biol.*, **5**, 1–12

Dillard, G. E. 1969. The benthic algal communities of a North Carolina stream. *Nova Hedw.*, **17**, 9–29

Dillon, P. J. & Rigler, F. H. 1974. The phosphorus–chlorophyll relationship in lakes. *Limnol. Oceanogr.*, **19**, 767–73

Dinsdale, M. T. & Walsby, A. E. 1972. The interrelations of cell turgor pressure, gas-vaculoation and buoyancy in a blue-green alga. *J. exp. Bot.*, **23**, 561–70

Dixon, P. S. 1963. Changing patterns of distribution in marine algae. In *The biological significance of climatic changes in Britain*, ed. C. G. Johnson & L. P. Smith, pp. 109–15

Dixon, P. S. 1965. Perennation, vegetative propagation and algal life histories with special reference to *Asparagopsis* and other Rhodophyta. *Bot. Gothoburg.*, **3**, 67–74

Dixon, P. S. 1973. *Biology of the Rhodophyta*. 285 pp. Oliver & Boyd, Edinburgh

Dixon, P.S. & Irvine, L. M. 1977. *Seaweeds of the British Isles, vol. I. Rhodophyta. Part 1. Introduction, Nemaliales, Gigartinales*, 252 pp. British Museum Nat. Hist.

Dixon, P. S. & Richardson, W. N. 1970. Growth and reproduction in red algae in relation to light and dark cycles. *Ann. New York Acad. Sci.*, **175**, 764–77

Dodge, J. D. 1977. The early summer bloom of dinoflagellates in the North Sea, with special reference to 1971. *Mar. Biol.*, **40**, 327–36

Dodge, J. D. & Bibby, B. T. 1973. The Prorocentrales (Dinophyceae). A comparative account of the fine structure in the genera *Prorocentrum* and *Exuviaella*. *Bot. J. Linn. Soc.*, **67**, 175–87

Dodge, J. D. & Hart-Jones, B. 1974. The vertical and seasonal distribution of dinoflagellates in the North Sea. *Bot. Mar.*, **17**, 113–17

Doemel, W. N. & Brock, T. D. 1970. The upper temperature limit of *Cyanidium caldarium*. *Arch. Microbiol.*, **72**, 326–32

Dokulil, M. 1971. Atmung and Anaerobioseresistenz von Süsswasserlagen. *Int. Rev. ges. Hydrobiol.*, **56**, 751–68

Donahue, J. G. 1967. Diatoms as indicators of Pleistocene climatic fluctuations in the Pacific sector of the Southern Ocean. *Prog. Oceanogr.*, **4**, 133–40

Donahue, J. G. 1970a. Diatoms as quarternary biostratigraphic and paleoclimatic indicators in high latitudes of the Pacific Ocean. Ph.D. Thesis, University of Columbia, New York.

Donahue, J. G. 1970b. Pleistocene diatoms as climatic

indicators in North Pacific sediments. *Geol. Soc. Am. Mem.*, no. 126, 121–38

Dorgelo, J. 1976. Intertidal Fucoid zonation and desiccation. *Hydrobiol. Bull.*, **10**, 115–22

Doty, M. S. 1946. Critical tide factors that are correlated with the vertical distribution of marine algae and other organisms along the Pacific coast. *Ecology*, **27**, 315–27

Doty, M. S. 1957. Rocky intertidal surfaces. *Geol. Soc. Am. Mem.*, no. 67, 535–85

Doty, M. S. 1959. Phytoplankton photosynthetic periodicity as a function of latitude. *J. Mar. Biol. Ass. India*, **1**, 66–8

Doty, M. S. 1967. Pioneer intertidal populations and the related general vertical distribution of marine algae in Hawaii. *Blumea*, **15**, 95–105

Doty, M. S. 1971. Physical factors in the production of tropical benthic marine algae. In *Fertility of the Sea, Vol. 2*, ed. J. D. Costlow, pp. 99–121

Doty, M. S. 1973a. Inter-relationships between marine and terrestrial ecosystems in Polynesia. In *Proc. Symp. Nature conservation in the Pacific*, ed. A. B. Costin & R. H. Groves, pp. 241–52. Canberra, 1971

Doty, M. S. 1973b. Marine organisms – tropical algal ecology and conservation. In *Proc. Symp. Nature conservation in the Pacific*, ed. A. B. Costin & R. H. Groves, pp. 183–95. Canberra, 1971

Doty, M. S. & Archer, J. G. 1950. An experimental test of the tide factor hypothesis. *Am. J. Bot.*, **37**, 458–64

Doty, M. S., Gilbert, W. J. & Abbott, I. A. 1974. Hawaiian marine algae from seaward of the algal ridge. *Phycologia*, **13**, 345–57

Doty, M. S. & Morrison, J. P. E. 1954. Interrelationships of the organisms on Raroia aside from man. *Atoll Res. Bull.*, no. 35, 1–61

Doty, M. S. & Oguri, M. 1956. The island mass effect. *J. Cons. Intern. Explor. Mer.*, **22**, 33–7

Doty, M. S. & Oguri, M. 1957. Evidence for a photosynthetic daily periodicity. *Limnol. Oceanogr.*, **2**, 37–40

Doty, M. S. & Santos, G. A. 1966. Transfer of toxic algal substances in marine food chains. *Pacif. Sci.*, **24**, 351–8

Douglas, B. 1958. The ecology of the attached diatoms and other algae in a stony stream. *J. Ecol.*, **46**, 295–322

Doyle, R. W. & Poore, R. V. 1974. Nutrient competition and division synchrony in phytoplankton. *J. exp. mar. Biol. Ecol.*, **14**, 201–10

Drebes, G. 1974. *Marines phytoplankton. Eine Auswahl der Helgoländer Planktonalgen (Diatomeen, Peridineen)*. 186 pp. Georg Thieme, Stuttgart

Dresscher, G. N. & Mark, H. van der 1976. A simplified method for the biological assessment of the quality of fresh and slightly brackish water. *Hydrobiol.*, **48**, 199–201

Drew, E. A. 1969a. Photosynthesis and growth of attached marine algae down to 130 metres in the Mediterranean. In *Proc. 6th Int. Seaweed Symp.*, 157–9

Drew, E. A. 1969b. Uptake and metabolism of exogenously supplied sugars by brown algae. *New Phytol.*, **68**, 35–43

Drew, E. A. 1973. The biology and physiology of alga-invertebrate symbioses. III. *In situ* measurements of photosynthesis and calcification in some hermatypic corals. *J. exp. mar. Biol. Ecol.*, **13**, 165–79

Drew, E. A. 1974. An ecological study of *Laminaria ochroleuca* Pyl. growing below 50 m in the Straits of Messina. *J. exp. mar. Biol. Ecol.*, **15**, 11–20

Drew, E. A. & Smith, D. C. 1967. Studies in the physiology of lichens. VIII. Movement of glucose from alga to fungus during photosynthesis in the thallus of *Peltigera polydactyla*. *New Phytol.*, **66**, 389–400

Dring, M. J. 1967a. Effects of daylength on growth and reproduction of the *Conchocelis* phase of *Porphyra tenera*. *J. mar. biol., Ass., U.K.*, **47**, 501–10

Dring, M. J. 1967b. Phytochrome in the red alga, *Porphyra tenera*. *Nature, London*, **215**, 1411–12

Dring, M. J. 1970. Photoperiodic effects in micro-organisms. In *Photobiology of micro-organisms*, ed. P. Halldal, 345–68. J. Wiley & Sons

Dring, M. J. & Lüning, K. 1975. A photoperiodic response mediated by blue light in the brown alga *Scytosiphon lomentaria*. *Planta*, **125**, 25–32

Droop, M. R. 1953. On the ecology of flagellates from some brackish and freshwater rock pools in Finland. *Act. Bot. Fenn.*, **51**, 3–52

Droop, M. R. 1955. Carotenogenesis in *Haematococcus pluvialis Nature, London*, **175**, 42–3

Droop, M. R. 1958. Optimum, relative and actual ionic concentrations for growth of some euryhaline algae. *Verh. Int. Ver. Limnol.*, **13**, 722–30

Droop, M. R. 1963. Algae and invertebrates in symbiosis. In *Symbiotic Association*, ed. P. S. Nutman & B. Mosse, *13th Symp. Soc. Gen. Microbiol.*, pp. 171–99. Cambridge University Press

Droop, M. R. 1974. The nutrient status of algal cells in continuous culture. *J. mar. biol. Ass., U.K.*, **54**, 825–55

Droop, M. R. & Elson, K. G. R. 1966. Are pelagic diatoms free from bacteria. *Nature, London*, **211**, 1096–7

Druehl, L. D. 1968. Taxonomy and distribution of northeast Pacific species of *Laminaria. Can. J. Bot.*, **46**, 539–47

Druehl, L. D. 1970. The pattern of Laminariales distribution in the northwest Pacific. *Phycologia*, **9**, 237–47

Drum, R. W. & Webber, E. 1966. Diatoms from a Massachusetts salt marsh. *Bot. Mar.*, **9**, 70–7

Dube, M. A. & Ball, E. 1971. *Desmarestia* associated with the Sea Pen, *Ptilosarcus gurneyi* (Gray). *J. Phycol.*, **7**, 218–20

Dubinsky, Z. & Berman, T. 1976. Light utilization efficiencies of phytoplankton in Lake Kinneret. (Sea of Galilee). *Limnol. Oceanogr.*, **21**, 226–30

Ducker, S. 1958. A new species of *Basicladia* on Australian freshwater turtles. *Hydrobiologia*, **10**, 157–74

Dugdale, R. C. & Goering, J. J. 1971. A model of nutrient limited phytoplankton growth. In *Impingement of man on the oceans*, ed. D. W. Hood, pp. 589–600. Wiley-Interscience

Durbin, E. G. 1978. Aspects of the Biology of resting spores of *Thalassiosira nordenskioldii* and *Detonula confervacea*. *Mar. Biol.*, **45**, 31–7

Durbin, E. G., Krawiec, R. W. & Smayda, T. J. 1975. Seasonal studies on the relative importance of different size fractions of phytoplankton in Narragansett Bay (U.S.A.). *Mar. Biol.*, **32**, 271–87

Du Rietz, 1950. Phytogeographical mire excursion to the maritime birch forest zone and the maritime forest limit in the outermost archipelago of Stockholm. In *7th Int. Bot. Congr. Stockholm*, 1–11

Düringer, I. 1958. Über die Verteilung epiphytischer Algen auf den Blättern wasserbewohnender Angiospermen sowie systematisch entwicklungsgeschichtlicher Bemerkungen über einige grüne Algen. *Öst. Bot. Zeitsch.*, **105**, 1–43

Durrell, L. W. 1962. Algae of Death Valley. *Trans. Am. Micros. Soc.*, **81**, 267–73

Durrell, L. W. 1964. Algae in tropical soils. *Trans. Am. Micros. Soc.*, **83**, 79–85

Durrell, L. W. & Shields, D. N. 1961. Characteristics of soil algae relating to crust formation. *Trans. Am. Micros. Soc.*, **80**, 73–9

Du Saar, A. & De Wolf, H. 1973. Marine diatoms of sediment cores from Breid Bay and Brekilen, Antarctica, *Neth. J. Sea Res.*, **6**, 339–54

Duthie, H. C. & Sreenivasa, M. R. 1971. Evidence for the eutrophication of Lake Ontario from the sedimentary diatom succession. In *Proc. 14th Conf. Great Lakes Res. Int. Assoc.*, 1–13

Duvigneaud, P. & Symoens, J. J. 1949. Observations sur la strate algale des formations herbeuses du Sud du Congo Belge. *Lejeunia*, **13**, 67–98

Eaton, J. W. 1967. Studies on the ecology of epipelic diatoms. Ph.D. Thesis, University of Bristol

Eaton, J. W. & Moss, B. 1966. The estimation of numbers and pigment content in epipelic populations. *Limnol. Oceanogr.*, **11**, 584–95

Eaton, J. W. & Young, J. O. 1975. Studies on the symbiosis of *Phaenocora typhlops* (Vejdovsky) (Turbellaria; Neorhabdocoela) and *Chlorella vulgaris* var. *vulgaris* Fott & Novakova. *Arch. Hydrobiol.*, **75**, 50–75

Edelstein, T., Craigie, J. S. & McLachlan, J. 1969. Preliminary survey of the sublittoral flora of Halifax County. *J. Fish. Res. Bd. Can.*, **26**, 2703–13

Edelstein, T., Greenwell, M., Bird, C. J. & McLachlan, J. 1971–3. Investigation of the marine algae of Nova Scotia. X. Distribution of *Fucus serratus* L. and some other species of *Fucus* in the maritime provinces. *Proc. N.S. Inst. Sci.*, **27**, 33–42

Edgreen, R. A., Edgreen, M. A. & Tiffany, L. H. 1953. Some North American turtles and their epizoophytic algae. *Ecology*, **34**, 733–40

Edinburgh Oceanographic Laboratory 1973. Continuous plankton records: a plankton atlas of the North Atlantic and the North Sea. *Bull. Mar. Ecol.*, **7**, 1–174

Edmondson, W. T. 1957. Trophic relations of the zooplankton. *Trans. Am. Micros. Soc.*, **76**, 225–45.

Edmondson, W. T. 1962. Food supply and reproduction of zooplankton in relation to phytoplankton populations. *Rapp. Cons. Explor. Mer.* **153**, 137–41

Edmondson, W. T. 1965. Reproductive rate of planktonic rotifers as related to food and temperature in nature. *Ecol. Monogr.*, **35**, 61–111

Edmondson, W. T. 1972. Nutrients and phytoplankton in Lake Washington. *Limnol. Oceanog. Special Symposia.* I., 172–93

Edwards, P. 1972. Benthic algae in polluted estuaries. *Mar. Pollut. Bull.*, **3**, 55–60

Edwards, P. 1973. Life history studies of selected British *Ceramium* species. *J. Phycol.*, **9**, 181–4

Edwards, P. & Kapraun, D. F. 1973. Benthic marine algal ecology in the Port Aransas, Texas area. *Contr. mar. Sci.*, **17**, 15–57. University of Texas

Edwards, R. W. & Owens, M. 1965. The oxygen balance of streams. In *Ecology and the Industrial Society. 5th Symp. British Ecol. Soc.*, pp. 149–72

Efford, I. E. 1967. Temporal and spatial differences in phytoplankton productivity in Marion Lake, British Columbia. *J. Fish. Res. Bd. Can.*, **24**, 2283–307

Ehrenberg, C. G. 1854. *Zur Microgeologie*. Leopold Voss, Leipzig

Ehrlich, A. 1975. The diatoms from the surface sediments of the Bardawil Lagoon (Northern Sinai) – Paleoecological Significance. In *3rd Symp. recent and fossil marine diatoms. Kiel, Nova Hedw. Beih.*, **53**, 253–77

Eisenack, A. & Fries, M. 1965. *Peridinium limbatum* (Stokes) verglichen mit der tertiären *Deflandrea phosphoritica* Eisenach. *Geol. Fören. Förhandl. Stockholm.*, **87**, 239–48

Ellis, D. V. & Wilce, R. T. 1961. Arctic and subarctic examples of intertidal zonation. *Arctic*, **14**, 224–35

El-Sayed, S. Z. 1968. On the productivity of the southwest Atlantic Ocean and the waters west of the Antarctic Peninsula. In *Biology of the Antarctic Seas. III*, ed. G. A. Llano & W. L. Schmitt. *Antarct. Res. Ser.*, **11**, 15–47

El-Sayed, S. Z. 1971. Observations on phytoplankton bloom in the Weddell Sea. *Antarctic Res. Ser.*, **17**, 301–12

El-Sayed, S. Z. & Jitts, H. R. 1973. Phytoplankton production in the south eastern Indian Ocean. In *The Biology of the Indian Ocean*, ed. B. Zeitschel, pp. 127–42. Chapman & Hall

Elster, H. J. 1960. Lake Constance as an organism and alterations in its metabolism in the last decade. *Gas-u. Wasserfach*, **101**, 171–80

El Wakeel, S. K. & Riley, J. P. 1961. Chemical and mineralogical studies of deep-sea sediments. *Geochim. cosmochim. Acta*, **25**, 110–46

Emerson, S. E. & Zedler, J. B. 1978. Recolonization of intertidal algae: An experimental study. *Mar. Biol.*, **44**, 315–34

Ende, G. van den & Oorschot, R. van 1963. Weitere Beobachtungen über den Epiphytenbewuchs von *Himanthalia elongata* (L.) S. F. Gray. *Bot. Mar.*, **5**, 111–20

Ennis, G. L. 1977. Attached algae as indicators of water quality in phosphorus enriched Kootenay Lake, British Columbia. M.S. Thesis, University of British Columbia

Enright, J. T. 1969. Zooplankton grazing rates estimated under field conditions. *Ecology*, **50**, 1070–5

Eppley, R. W., Carlucci, A. F., Holm-Hansen, O., Keiper, D., McCarthy, J. J., Venrick, E. & Williams, M. M. 1971. Phytoplankton growth and composition in ship board cultures supplied with nitrate, ammonium or urea as the nitrogen source. *Limnol. Oceanogr.*, **16**, 741–51

Eppley, R. W., Holmes, R. W. & Paasche, E. 1967. Periodicity in cell division and physiological behaviour of *Ditylum brightwellii*, a marine planktonic diatom, during growth in light–dark cycles. *Arch. Microbiol.*, **56**, 305–23

Eppley, R. W., Holmes, R. W. & Strickland, J. D. H. 1967. Sinking rates of marine phytoplankton measured with a fluorometer. *J. exp. mar. Biol. Ecol.*, **1**, 191–208

Eppley, R. W., Holm-Hansen, O. & Strickland, J. D. H. 1968. Some observations on the vertical migration of Dinoflagellates. *J. Phycol.*, **4**, 333–40

Eppley, R. W. & Renger, E. H. 1974. Nitrogen assimilation of an oceanic diatom in nitrogen-limited continuous culture. *J. Phycol.*, **10**, 15–23

Eppley, R. W., Renger, E. H., Venrick, E. L. & Mullin, M. M. 1973. A study of plankton dynamics and nutrient cycling in the central gyre of the North Pacific Ocean. *Limnol. Oceanogr.*, **18**, 534–7

Eppley, R. W., Rogers, J. N. & McCarthy, J. J. 1969. Half-saturation constants for uptake of nitrate and ammonia by marine phytoplankton. *Limnol. Oceanogr.*, **14**, 912–20

Eppley, R. W. & Strickland, J. D. H. 1968. Kinetics of marine phytoplankton growth. *Adv. microbiol. sea.*, **I**, 23–62

Ercegović, A. 1934. Wellengang und Lithophytenzone an der Ostadriatischen Küste. *Acta Adriat.*, **1**, no. 3, 1–20

Ernst, J. 1959. Studien über die Seichtwasser Vegetation der Sorrentiner Küste. *Publ. Staz. Zool., Napoli*, **30**, 470–518

Erwin, J. & Bloch, K. 1963. Polyunsaturated fatty acids in some photosynthetic organisms. *Biochem. Z.*, **338**, 496–511

Esaias, W. E. & Curl, H. C. Jr. 1972. Effect of Dinoflagellate bioluminescence on copepod ingestion rates. *Limnol. Oceanogr.*, **17**, 901–6

Esmarch, F. 1910. Beitrag zur Cyanophyceenflora unser Kolonien. Jahrb. der Hamburgischen Wissenshaften Anstalten, **28**, 62–82

Estrada, M., Valiela, I. & Teal, J. M. 1974. Concentration and distribution of chlorophyll in fertilized plots in a Massachusetts salt marsh. *J. exp. mar. Biol. Ecol.*, **14**, 47–56

Ettl, W. 1978. Xanthophycean. In *Süsswasserflora von Mitteleuropa*, Bd. 3, Teil 1, **14**, 540 pp.

Evans, J. H. 1971. Biological applications of particle size analysis. *Proc. Soc. Analyt. Chem.*, 7/8, 260–4

Evans, J. H. & McGill, S. M. 1970. An investigation of the Coulter counter in 'biomass' determinations of natural freshwater phytoplankton populations. *Hydrobiol.*, **35**, 401–19

Evans, R. G. 1947. The intertidal ecology of selected localities in the Plymouth neighbourhood. *J. mar. biol. Ass., U.K.*, N.S. 27, **17**, 173–218

Evans, R. G. 1957. The intertidal ecology of some localities on the Atlantic coast of France. *J. Ecol.*, **45**, 245–71

Evitt, W. R. & Wall, D. 1968. Dinoflagellate studies. IV. Theca and cyst of recent freshwater *Peridinium limbatum* (Stokes) Lemmerman. *Geol. Sci.*, **12**, 15 pp. Stanford University.

Fager, E. W. 1963. Communities of organisms. In *The Sea*, vol. 2, ed. M. N. Hill, pp. 415–37. Interscience.

Fahey, E. M. 1953. The repopulation of intertidal transects. *Rhodora*, **55**, 102–8

Fairchild, E. & Sheridan, R. P. 1974. A physiological investigation of the hot spring diatom *Achnanthes exigua* Grun. J. Phycol., **10**, 1–4

Falkowski, P. G. 1975. Nitrate uptake in marine phytoplankton. Comparison of half-saturation constants from seven species. *Limnol. Oceanogr.*, **20**, 412–17

Falkowski, P. G. & Owens, T. G. 1978. Effects of light intensity on photosynthesis and dark respiration in six species of marine phytoplankton. *Mar. Biol.*, **45**, 289–95

Farnham, W. F. & Fletcher, R. L. 1976. The occurrence of a *Porphyrodiscus simulans* Batt. phase in the life-history of *Ahnfeltia plicata* (Huds.) Fries. *Br. phycol. J.*, **11**, 183–90

Fasham, M. J. R. & Pugh, R. R. 1976. Observations on the horizontal coherence of chlorophyll *a* and temperature. *Deep Sea Res.*, **23**, 527–38

Fauré-Fremiet, E. 1951. The tidal rhythm of the diatom *Hantzschia amphioxys*. *Biol. Bull.*, **100**, 173–7

Faust, M. A. & Correll, D. L. 1977. Autoradiographic study to detect metabolically active phytoplankton and bacteria in the Rhode River estuary. *Mar. Biol.*, **41**, 293–305

Fauvel, P. & Bohn, G. 1907. Le rhythme des marées chez les diatomeés littorales. *C. r. Séanc. Soc. Biol.*, **62**, 121–3

Fay, P. & Fogg, G. E. 1962. Studies on nitrogen fixation by blue-green algae. III. Growth and nitrogen fixation in *Chlorogloea fritschii* Mitra. *Arch. Mikrobiol.*, **42**, 310–21

Febvre, J. & Febvre-Chevalier, C. 1979. Ultrastructural study of zooxanthellae of three species of *Acantharia* (Protozoa, Actinopoda), with details of their taxonomic position in the Prymnesiales (Prymnesiophyceae, Hibberd, 1976). *J. mar. biol. Ass., U.K.*, **59**, 215–26

Fee, E. J. 1975. The importance of diurnal variation of photosynthesis vs. light curves to estimates of integral primary production. *Verh. Int. Ver. Limnol.*, **19**, 39–40

Feldmann, J. 1937. Recherches sur la végétation marine de la Méditeranée. *Rev. algol.* **10**, 1–339

Feldmann, J. 1955. La zonation des algues sur la côte atlantique du Maroc. *Bull. Soc. Sci. Nat., Maroc.*, **35**, 9–17

Feldmann, J. 1958. Origine et affinités du peuplement végétal benthique de l'Mediteranée. *Rapp. P.-v. Réun. Commn. int. Explor. scient. Mer Méditerr.*, **14**

Feldmann, J. & Lutova, M. J. 1963. Variations de la thermostabilité cellulaire des algues en fonction des changements de la temperature du milieu. *Cah. Biol. mar.*, **4**, 435–58

Fenchel, T. & Straarup, B. J. 1971. Vertical distribution of photosynthetic pigments and the penetration of light in marine sediments. *Oikos*, **22**, 172–82

Fenchel, T. & Riedl, R. J. 1970. The sulphide system: a new biotic community underneath the oxidised layer of marine sand bottoms. *Mar. Biol.*, **7**, 255–68

Fenical, W. 1975. Halogenation in the Rhodophyta. A review. *J. Phycol.*, **11**, 245–59

Ferguson-Wood, E. J. 1971. Phytoplankton distribution in the Caribbean region. In *UNESCO Symposium on investigations and resources of the Caribbean Sea and adjacent regions. WCNA*, pp. 399–410

Ferrante, J. G. & Parker, J. I. 1977. Transport of diatom frustules by copepod fecal pellets to the sediments of Lake Michigan. *Limnol. Oceanogr.*, **22**, 92–8

Fetzmann, E. 1961. Einige Algenvereine des Hochmoor-Komplexes Komosse. *Bot. Not.*, **114**, 185–217

Findenegg, I. 1947. Über die Lichtansprüche planktischer Süsswasseralgen. *Sitzungsb. Akad. Wiss. Wien. Math. naturw.*, Kl. Abt. I. **155**, 159–71

Findenegg, I. 1964. Types of planktonic primary production in the lakes of the Eastern Alps as found by the radioactive carbon method. *Verh. Int. Ver. Limnol.*, **15**, 352–9

Findenegg, I. 1971. Die Produktionsleistungen einer

planktische Algenarten in ihrem natürlichen Milieu. *Arch. Hydrobiol.*, **69**, 273–93

Fine, M. L. 1970. Faunal variation on *Sargassum*. *Mar. Biol.*, **7**, 112–22

Firth, R. J. & Hartley, B. 1971. A Pennine diatom site. *Microscopy*, **32**, 108–13

Fischer, E. 1928. Sur la distribution geographique de quelques organisms de rocher, le long des côtes de la Manche. *Trav. Lab. Mus. Hist. Natur. St. Servan.*, **2**, 1–16

Fitzgerald, G. P. 1967. Discussion. In *Environmental Requirements of Blue-green Algae*, pp. 97–102. Fed. Wat. Poll. Control Admin. N.W. Region. Corvallis, Oregon

Fitzgerald, G. P. 1972. Bioassay analysis of nutrient availability. In *Nutrients in natural Waters*, ed. H. E. Allen & J. R. Kramer, pp. 147–69. Wiley-Interscience

Fjerdingstad, E. 1950. The microflora of the River Mølleaa. With special reference to the relation of the benthal algae to pollution. *Folia Limnol., Scand.*, **5**, 1–123

Fjerdingstad, E. 1971. Microbial criteria of environment qualities. *Ann. Rev. Microbiol.*, **25**, 563–82

Fjerdingstad, E., Kemp, K., Fjerdingstad, E. & Vanggaard, L. 1974. Chemical analyses of red snow from East Greenland with remarks on *Chlamydomonas nivalis* (Bau) Wille. *Arch. Hydrobiol.*, **73**, 70–83

Flemer, D. A. 1970. Primary production in the Chesapeake Bay. *Chesapeake Sci.*, **11**, 117–29

Flensburg, T. 1967. Desmids and other benthic algae of Lake Kävsjön and Store Mosse, S.W. Sweden. *Act. Phytogeog. Suec.*, **51**, 1–132

Flensburg, T. & Sparling, J. H. 1973. The algal microflora of a string mire in relation to the chemical composition of water. *Canadian Journal of Botany*, **51**, 743–9

Fletcher, J. E. & Martin, R. D. 1948. Some effects of algae and molds in the rain crust of desert soils. *Ecology*, **29**, 95–100

Foerster, J. W. 1971. The ecology of an elfin forest in Puerto Rico, 14. The algae of Pico del Oeste. *J. Arnold Arbor.*, **52**, 86–109

Foerster, J. W. & Schlichting, H. E. 1965. Phycoperiphyton in an oligotrophic lake. *Trans. Am. Microsc. Soc.*, **84**, 485–502

Fogg, G. E. 1963. *Algal cultures and phytoplankton Ecology*, 126 pp. Athlone Press

Fogg, G. E. 1967. Observation on the snow algae of the South Orkney Islands. *Phil. Trans. Roy. Soc., Lond.*, ser. **B**, **252**, 279–87

Fogg, G. E. 1969. Survival of algae under adverse conditions. *Symp. Soc. exp. Biol.*, **23**, 123–42

Fogg, G. E. 1971. Extracellular products of algae in freshwater. *Arch. Hydrobiol. Beih. Ergebn. Limnol.*, **5**, 1–24

Fogg, G. E. 1973. Phosphorus in primary aquatic plants. *Wat. Res.*, **7**, 77–91

Fogg, G. E. 1977. Excretion of organic matter by phytoplankton. *Limnol. Oceanogr.*, **22**, 576–7

Fogg, G. E. & Horne, A. J. 1970. The physiology of Antarctic Freshwater Algae. In *Antarctic Ecology*, ed. M. W. Holdgate, pp. 632–8. Academic Press

Fogg, G. E. & Nalewajko, C. 1964. Glycollic acid as an extracellular product of phytoplankton. *Verh. Int. Ver. Limnol.*, **15**, 806–10

Fogg, G. E., Nalewajko, C. & Wall, W. D. 1965. Extracellular products of phytoplankton photosynthesis. *Proc. Roy. Soc., London*, **B**, **162**, 517–34

Fogg, G. E., Stewart, W. D. P., Fay, P. & Walsby, A. E. 1973. *The blue-green algae*. 459 pp. Academic Press, London

Fogg, G. E. & Walsby, A. E. 1971. Buoyancy regulation and the growth of planktonic blue-green algae. *Mitt. Int. Ver. Limnol.*, **19**, 182–8

Folger, D. W., Burkle, L. H. & Heezen, B. C. 1967. Opal phytoliths in a north Atlantic dustfall. *Science*, **155**, 1243–4

Foree, E. G., Jewell, W. J. & McCarty, P. L. 1971. The extent of nitrogen and phosphorus regeneration from decomposing algae. In *Proc. 5th Int. Conf. Adv. Wat. Pollut. Res.*, **2**

Forest, H. S. 1962. Analysis of the soil algal community. *Trans. Am. Micros. Soc.*, **81**, 189–98

Forest, H. S. 1965. The soil algal community. II. Soviet soil studies. *J. Phycol.*, **1**, 164–71

Forest, H. S., Willson, D. L. & England, R. B. 1959. Algal establishment on sterilised soil replaced in an Oklahoma prairie. *Ecology*, **40**, 475–7

Forsberg, C. 1965a. Nutritional studies of *Chara* in axenic cultures. *Physiologica Pl.*, **18**, 275–90

Forsberg, C. 1965b. Environmental conditions of Swedish charophytes. *Symb. bot. upsal.*, **18**, 67 pp.

Forsberg, C. & Hökervall, E. 1972. Algal growth potential (AGP-test) of sewage effluent. I. The treatment plant at Akeshov-Nockleby, Stockholm, February–September, 1971. *Vatten*, **28**, 17–26

Forsberg, C. G. 1972. Algal assay procedure. *J. Wat. Pollut. Control Fed., Washington*, **44**, 1623–8

Forsbergh, E. D. & Joseph, J. 1964. Biological production in the eastern Pacific ocean. *Bull. inter-Am. trop. Tuna Commn.*, **8**, 479–527

Forward, R. B. & Davenport, D. 1970. The circadian rhythm of a behavioural photoresponse in the dinoflagellate *Gyrodinium dorsum*. *Planta*, **92**, 259–66

Foslie, M. H. 1907. The Lithothamnia of the Percy Sladen Trust Expedition in HMS Sealark. *Trans. Linn. Soc., London*, zool. ser. 2, **12**, 93–100

Fott, B. & Nováková, M. 1969. A monograph of the genus *Chlorella*. The freshwater species. In *Studies in Phycology*, pp. 10–74. Prague

Føyn, B. 1955. Specific differences between northern and southern European populations of *Ulva lactuca*. *Publ. Staz. Zool., Napoli*, **27**, 261–70

Fournier, R. O. 1968. Observations of particulate organic carbon in the Mediterranean Sea and the relevance to the deep-living Coccolithophorid *Cyclococcolithus fragilis*. *Limnol. Oceanogr.*, **13**, 693–7

Fournier, R. O. 1970. Studies on pigmented microorganisms from aphotic marine environments. *Limnol. Oceanogr.*, **15**, 675–82

Francé, R. 1913. *Das Edaphon*. München

Fralick, R. A. & Mathieson, A. C. 1972. Winter fragmentation of *Codium fragile* (Suringar) Hariot, ssp. *tomentosoides* (van Goor) Silva (Chlorophyceae, Siphonales) in New England. *Phycologia*, **11**, 67–70

Franzisket, L. 1969. Riff Korallen können autotroph Leben. *Naturwiss.*, **56**, 144

Franzisket, L. 1970. The atrophy of hermatypic reef corals maintained in darkness and their subsequent regeneration in light. *Int. Rev. ges. Hydrobiol.*, **55**, 1–12

Fred, J. G. 1972. Compte rendu de plongée en S.P. 300 sur les fonds a *Laminaria rodrigueszii* Bornet de la pointe de Revellata (Corse). *Bull. Inst. Oceanogr., Monaco*, **71**, 42 pp.

Frémy, P. 1936. Les algues perforantes. *Mém. Soc. natn. Sci. Nat. math. Cherbourg*, **42**, 275–300

Frey, D. G. & Stahl, J. B. 1958. Measurements of primary production on Southampton Island in the Canadian Arctic. *Limnol. Oceanogr.*, **3**, 215–21

Friedmann, E. I. 1971. Light and scanning electron microscopy of the endolithic desert algal habitat. *Phycologia*, **10**, 411–28

Friedmann, I., Lipkin, Y. & Ocampo-Paus, R. 1967. Desert algae of the Negev (Israel). *Phycologia*, **6**, 185–200

Friedrich, H. 1969. *Marine biology: an introduction to its problems and results*. 474 pp. Sidgwick & Jackson

Fritsch, F. E. 1929. The encrusting algal communities of certain fast-flowing streams. *New Phytol.*, **28**, 165–96

Fritsch, F. E. & Salisbury, E. J. 1915. Further observations on the heath association of Hindhead Common. *New Phytol.*, **14**, 116–38

Fuge, R. & James, K. H. 1974. Trace metal concentrations in *Fucus* from the Bristol Channel. *Mar. Pollut. Bull.*, **5**, 9–12

Fukushima, H. 1961. Preliminary reports of the biological studies on coloured ocean ice. *Antarctic Rec.*, **11**, 164

Funk, G. 1955. Beiträge zur Kenntnis der Meeresalgen von Neapel. Zugleich mikrophotographischer Atlas. *Publ. Staz. Zool., Napoli*, **25**, 1–178

Gächter, R. 1968. Phosphorhaushalt und planktische Primärproduktion in Vierwaldstättersee (Horwer Bucht). *Schweiz. Z. Hydrol.*, **30**, 1–66

Gallagher, J. L. & Daiber, F. C. 1973. Diel rhythms in edaphic community metabolism in a Delaware salt marsh. *Ecology*, **54**, 1160–3

Gallagher, J. L. & Daiber, F. C. 1974. Primary production of the edaphic algal communities in a Delaware salt marsh. *Limnol. Oceanogr.*, **19**, 390–5

Ganf, G. G. 1974a. Phytoplankton biomass and distribution in a shallow eutrophic lake. (Lake George, Uganda.) *Oecologia, Berlin*, **16**, 9–29

Ganf, G. G. 1974b. Incident solar irradiance and underwater light penetration as factors controlling the chlorophyll *a* content of a shallow equatorial lake. (Lake George, Uganda.) *J. Ecol.*, **62**, 593–600

Ganf, G. G. 1974c. Diurnal mixing and the vertical distribution of phytoplankton in a shallow equatorial lake. (Lake George, Uganda.) *J. Ecol.*, **62**, 611–29

Ganf, G. G. 1975. Photosynthetic production and irradiance – photosynthesis relationships of the phytoplankton from a shallow equatorial lake. (Lake George, Uganda.) *Oecologia, Berlin*, **18**, 165–83

Ganf, G. G. & Horne, A. J. 1975. Diurnal stratification, photosynthesis and nitrogen fixation in a shallow, equatorial lake. (Lake George, Uganda.) *Freshwater Biol.*, **5**, 13–39

Ganf, G. G. & Viner, A. B. 1973. Ecological stability in a shallow equatorial Lake. (Lake George, Uganda.) *Proc. Roy. Soc., London*, **B**, **184**, 321–46

Gargas, E. 1970. Measurements of primary productivity, dark fixation and vertical distribution of the microbenthic algae in the Øresund. *Ophelia*, **8**, 231–53

Gargas, E. 1971. 'Sun-shade' adaptation in microbenthic algae from the Øresund. *Ophelia*, **9**, 107–12

Gartner, S. 1977. Calcareous nannofossil biostratigraphy and revised zonation of the Pleistocene. *Mar. Micropaleont.*, **2**, 1–25

Gause, G. F. & Witt, A. A. 1935. Behaviour of mixed populations and the problem of natural succession. *Amer. Nat.*, **69**, 596–609

Gauthier, M. J., Bernhard, P. & Aubert, M. 1978. Modification de la fonction antibiotique de deux diatomées marines. *Asterionella japonica* (Cleve) et *Chaetoceros lauderi* (Ralfs), par le dinoflagellé *Prorocentrum micans* (Ehrenberg). *J. exp. mar. Biol. Ecol.*, **33**, 37–50

Gebelein, C. D. 1969. Distribution, morphology and accretion rates of recent subtidal algal stromatolites, Bermuda. *J. sedim. Petrol.*, **39**, 49–69

Gebelein, C. D. & Hoffman, P. 1968. Intertidal stromatolites and associated facies from Cape Sable, Florida. *Geol. Soc. Am. Spec. Papers*, **121**, 109

Geissler, V. & Gerloff, J. 1965. Das Vorkommen von Diatomeen in menschlichen Organen und in der Luft. *Nova Hedw.*, **10**, 565–77

Geitler, L. 1948. Symbiosen zwischen Chrysomonaden und knospenden Bakterienartigen Organismen sowie Beobachtungen über Organisationseigentumlichkeiten der Chrysomonaden. *Öst. Bot. Zeitsch.*, **95**, 300–24

Gelin, C. & Ripl, W. 1978. Nutrient decrease and response of various phytoplankton size fractions following the restoration of Lake Trummen, Sweden. *Arch. Hydrobiol.*, **81**, 339–67

George, D. G. & Heaney, S. I. 1978. Factors influencing the spatial distribution of phytoplankton in a small productive lake. *J. Ecol.*, **66**, 133–55

Gerlach, S. A. 1954. Das Supralitoral der sandigen Meeresküsten als Lebensraum eine Mikrofauna. *Kieler Meeresforsch.*, **10**, 121–9

Gerletti, M. 1968. Dark bottle measurements in primary productivity studies. *Mem. Inst. Ital. Idrobiol.*, **23**, 197–208

Gerloff, G. C. & Skoog, F. 1954. Cell contents of nitrogen and phosphorus as a measure of their availability for growth of *Microcystis aeruginosa*. *Ecology*, **35**, 348–53

Gessner, F. 1955. *Hydrobotanik. I. Energiehaushalt*. 517 pp. VEB Deutsch Verlag der Wissenshaften, Berlin

Gessner, F. & Pannier, F. 1958. Influence of oxygen tension in respiration of phytoplankton. *Limnol. Oceanogr.*, **3**, 478–80

Gessner, F. & Simonsen, R. 1967. Marine diatoms in the Amazon? *Limnol. Oceanogr.*, **12**, 709–11

Ghelardi, R. J. 1971. 'Species' structure of the animal community that lives in *Macrocystis pyrifera* holdfasts. In *The Biology of Giant Kelp Beds (Macrocystis) in California*, ed. W. J. North. *Nova Hedw. Beih.*, **32**, 381–420

Gibb, D. C. 1938. The marine algal communities of Castletown Bay, Isle of Man. *J. Ecol.*, **26**, 96–117

Gibb, D. C. 1950. A survey of the commoner Fucoid algae on Scottish shores. *J. Ecol.*, **38**, 253–69

Gibb, D. C. 1957. The free-living forms of *Ascophyllum nodosum* (L.) Le Jol. *J. Ecol.*, **45**, 49–83

Gibson, C. E. 1975. Cyclomorphosis in natural populations of *Oscillatoria redekei* Van Goor. *Freshwater Biol.*, 5, 279–86

Gifford, C. E. & Odum, E. P. 1959. Chlorophyll *a* content of intertidal zones on a rocky sea shore. *Limnol. Oceanogr.*, 6, 83–5

Gilbert, J. S. & Allen, H. L. 1973. Chlorophyll and primary productivity of some green, freshwater sponges. *Int. Rev. ges. Hydrobiol.*, 58, 633–58

Gilet, R. 1954. Particularités de la zonation marine sur les côtes rocheuses s'etendent entre Nice et la frontiere Italienne. *Recl. Stat. mar. Endoume. Fasc.*, 12, 41–50

Gillbricht, R. 1952. Untersuchungen zur Productionsbiologie des Planktons in der Kieler Bucht. Teil I. *Kieler Meeresforsch.*, 8 (2), 173–91

Gilmartin, M. 1960. The ecological distribution of the deep water algae of Eniwetok Atoll. *Ecology*, 41, 210–21

Gilmartin, M. & Revelante, N. 1974. The 'island mass' effect and the phytoplankton and primary production of the Hawaiian Islands. *J. exp. mar. Biol. Ecol.*, 16, 181–204

Ginsburg, R. N. & Shroeder, J. H. 1973. Growth and submarine fossilization of algal cup reefs, Bermuda. *Sedimentology*, 20, 575–614

Gislen, T. 1930. Epibioses of the Gullmar Fjord. *Skr. Svenska Vetenskap. Akad.*, (4), 1–380

Glooschenko, W. A. & Blanton, J. O. 1977. Short term variability of chlorophyll *a* concentrations in Lake Ontario. *Hydrobiol.*, 53, 203–17

Glooschenko, W. A. & Curl, H. Jr 1971. Influence of nutrient enrichment on photosynthesis and assimilation ratios in natural North Pacific phytoplankton communities. *J. Fish. Res. Bd. Can.*, 28, 790–3

Godward, M. 1934. An investigation of the causal distribution of algal epiphytes. *Beih. Bot. Zbl.*, 52, 506–39

Goering, J. J., Nelson, D. M. & Carter, J. A. 1973. Silicic acid uptake by natural populations of marine phytoplankton. *Deep Sea Res.*, 20, 777–89

Goering, J. J. & Parker, P. L. 1972. Nitrogen fixation by epiphytes on sea grasses. *Limnol. Oceanogr.*, 17, 320–3

Goering, J. J., Wallen, D. D. & Nauman, R. M. 1970. Nitrogen uptake by phytoplankton in the discontinuity layer of the eastern subtropical Pacific Ocean. *Limnol. Oceanogr.*, 15, 789–96

Goetsch, W. & Scheuring, L. 1926. Parasitismus und Symbiose der Algengatting *Chlorella*. *Z. Morph. Ökol. Tiere*, 7, 220–53

Goff, L. J. 1976. The biology of *Harveyella mirabilis* (Cryptonemiales; Rhodoyphyceae). V. Host responses to parasite infection. *J. Phycol.*, 12, 313–28

Gold, K. & Pollingher, U. 1971. Occurrence of endosymbiotic bacteria in marine dinoflagellates. *J. Phycol.*, 7, 264–5

Goldman, C. R. 1964. Primary productivity studies in Antarctic Lakes. In *Biologie Antarctique*, 291–9

Goldman, C. R. 1967. Integration of field and laboratory experiments in productivity studies. In *Estuaries*, ed. G. H. Lauff. *Amer. Assoc. Adv. Sci.*, Washington D.C., 346–52

Goldman, C. R., Gerletti, M., Javornicky, P., Melchiorri-Santolini, V. & Amezaga, E. D. 1968. Primary productivity, bacteria, phytoplankton and zooplankton in Lake Maggiore: Correlations and relationships with ecological factors. *Mem. Inst. Ital. Idrobiol.*, 23, 49–127

Goldman, C. R., Mason, D. T. & Wood, B. J. B. 1963. Light injury and inhibition in Antarctic freshwater phytoplankton. *Limnol. Oceanogr.*, 8, 313–22

Goldman, J. C., Porcella, D. B., Middlebrooks, E. J. & Toerien, D. F. 1972. The effect of carbon on algal growth – its relationship to eutrophication. *Wat. Res.*, 6, 637–79

Golterman, H. L. 1960. Studies on the cycle of elements in fresh water. *Act. Bot. Neerl.*, 9, 1–58

Golterman, H. L. 1964. Mineralization of algae under sterile condition or by bacterial breakdown. *Verh. Int. Ver. Limnol.*, 15, 544–8

Golterman, H. L. 1967. Tetraethyl silicate as a 'molybdate unreactive' silicon source for diatom cultures. In *Proc. IBP Symp. Amsterdam-Nieuwersluis*, 56–62

Golterman, H. L. 1972. The role of phytoplankton in detritus formation. In *Detritus and its role in aquatic ecosystems, Proc. IBP–UNESCO Symp.*, pp. 89–103. Pallanza, Italy

Golubic, S. 1967. Algenvegetation der Felsen. *Die Binnengewässer*, 23, 183 pp.

Golubic, S. 1969. Distribution, taxonomy, and boring patterns of marine endolithic algae. *Am. Zool.*, 9, 747–51

Gooday, G. W. 1970. A physiological comparison of the symbiotic alga *Platymonas convolutae* and its free living relatives. *J. mar. biol. Ass., U.K.*, 50, 199–209

Goreau, T. F. 1963. Calcium carbonate deposition by coralline algae and corals in relation to their roles as reef-builders. *Ann. New York Acad. Sci.*, 109, 127–67

Goreau, T. F. & Goreau, N. I. 1959. The physiology of skeleton formation in corals. II. Calcium deposition by hermatypic corals under various conditions in the reef. *Biol. Bull. mar. biol. Lab., Woods Hole*, 117, 239–50

Goreau, T. F. & Goreau, N. I. 1960. The physiology of skeleton formation in corals. III. Calcification rate as a function of colony weight and total nitrogen content in the reef coral *Manicina areolata* (Linnaeus). *Biol. Bull. mar. biol. Lab., Woods Hole*, 118, 419–27

Goreau, T. F., Goreau, N. I. & Yonge, C. M. 1971. Reef corals: Autotrophs or heterotrophs? *Biol. Bull.*, 141, 247–60

Gorham, E. 1955. On the acidity and salinity of rain. *Geochimica et Cosmochimica Acta*, 7, 231–9

Gorham, E. 1960. Chlorophyll derivatives in surface muds from the English lakes. *Limnol. Oceanogr.*, 5, 29–33

Gorham, E. 1961. Factors influencing supply of major ions to inland waters, with special reference to the atmosphere. *Geol. Soc. Amer. Bull.*, 72, 795–840

Gorham, E., Lund, J. W. G., Sanger, J. E. & Dean, W. E. Jr 1974. Some relationships between algal standing crop, water chemistry and sediment chemistry in the English Lakes. *Limnol. Oceanogr.*, 19, 601–17

Gorham, P. R. 1964. Toxic algae. In *Algae and Man*, ed. D. F. Jackson, pp. 307–36. Plenum Press, New York

Gough, B. & Woelkerling, W. J. 1976. On the removal and quantification of algal Aufwuchs from macrophyte hosts. *Hydrobiol.*, 48, 205–7

Goulder, R. 1972. Grazing by the ciliated protozoan *Loxodes magnus* on the alga *Scenedesmus* in an eutrophic pond. *Oikos*, 23, 109–15

Graham, H. W. 1941. An oceanographic consideration of the dinoflagellate genus *Ceratium*. *Ecol. Monogr.*, **11**, 99–116

Grambast, L. J. 1974. Phylogeny of the Charophyta. *Taxon*, **23**, 463–81

Grant Gross, M. 1967. Concentrations of minor elements in diatomaceous sediments of a stagnant fjord. In *Estuaries*, ed. G. H. Lauff. *Amer. Assoc. Adv. Sci., Washington D.C.*, no. 83, 273–82

Grant, W. S. & Horner, R. A. 1976. Growth responses to salinity variation in four Arctic ice diatoms. *J. Phycol.*, **12**, 180–5

Green, J. C. 1976. Notes on the flagellar apparatus and taxonomy of *Pavlova mesolychnon* van der Veer, and on the status of *Pavlova* Butcher and related genera within the Haptophyceae. *J. mar. biol. Ass., U.K.*, **56**, 595–602

Green, J. C. & Parke, M. 1975. New observations upon members of the genus *Chrysotila* Anand, with remarks upon their relationship within the Haptophyceae. *J. mar. biol. Ass., U.K.*, **55**, 109–21

Greene, R. W. 1974. Saccoglossans and their chloroplast endosymbionts. In *Symbiosis in the Sea*, ed. W. B. Verberg, pp. 21–7. University of S. Carolina Press, Columbia

Greene, R. W. 1970. Symbiosis in saccoglossan opisthobranchs; translocation of photosynthetic products from chloroplast to host tissue. *Malacologia*, **10**, 369–80

Griffiths, M. 1978. Specific blue-green algal carotenoids in sediments of Esthwaite Water. *Limnol. Oceanogr.*, **23**, 777–84

Grill, E. V. & Richards, F. A. 1964. Nutrient regeneration from phytoplankton decomposing in sea water. *J. Mar. Res.*, **22**, 51–69

Gromov, B. V. 1968. Main trends in experimental work with algal cultures in the U.S.S.R. In *Algae, man and the environment*, ed. D. F. Jackson, pp. 249–78. Syracuse University Press.

Gromov, B. V. & Mamkaeva, K. A. 1969. Sensitivity of different *Scenedesmus* strains to the endoparasitic microorganism *Amoeboamphelidium*. *Phycologia*, **7**, 19–23

Grøntved, T. 1962. Preliminary report on the productivity of microbenthos and phytoplankton in the Danish Wadden Sea. *Medd. Dan. Fisk-Havunders.*, **3**, 347–78

Groover, R. D. & Bold, H. C. 1969. Phycological Studies. VII. The taxonomy and comparative physiology of the Chlorosarcinales and certain other edaphic algae. *Univ. Tex. Publs.*, no. 6907, 105 pp. Austin, Texas

Gross, F. & Zeuthen, E. 1948. The buoyancy of plankton diatoms: a problem of cell physiology. *Proc. Roy. Soc., London*, **B**, **135**, 382–9

Gruendling, G. K. 1971. Ecology of the epipelic algal communities in Marion Lake, British Columbia. *J. Phycol.*, **7**, 239–49

Guillard, R. R. L. & Cassie, V. 1963. Minimum cyanocobalamin requirements of some marine centric diatoms. *Limnol. Oceanogr.*, **8**, 161–5

Guillard, R. R. L. & Hellebust, J. A. 1971. Growth and the production of extracellular substance by two strains of *Phaeocystis pouchetii*. *J. Phycol.*, **7**, 336–8

Guillard, R. R. L. & Kilham, P. 1977. The ecology of marine plankton diatoms. In *The Biology of Diatoms*, ed. D. Werner, pp. 372–469. Blackwell, Oxford

Guillard, R. R. L., Kilham, P. & Jackson, T. A. 1973.

Kinetics of silicon limited growth in the marine diatom *Thalassiosira pseudonana* Hasle and Heimdal (= *Cyclotella nana* Hustedt). *J. Phycol.*, **9**, 233–7

Guiry, M. D. 1974*a*. The occurrence of the red algal parasite *Halosacciocolax lundii* Edelstein in Britain. *Br. phycol. J.*, **9**, 31–5

Guiry, M. D. 1974*b*. A preliminary consideration of the taxonomic position of *Palmaria palmata* (Linnaeus) Stackhouse = *Rhodymenia palmata* (Linnaeus) Greville. *J. mar. biol. Ass., U.K.*, **54**, 509–28

Guiry, M. D. & Irvine, D. E. G. in press. A critical assessment of infraordinal classification in the Rhodymeniales. In *Proc. 8th Int. Seaweed Symp., Bangor*

Gumtow, R. B. 1955. An investigation of the periphyton in a riffle of the West Callatin River, Montana. *Trans. Am. Micros. Soc.*, **74**, 278–92

Gunn, J. M., Qadri, S. V. & Mortimer, D. C. 1977. Filamentous algae as a food source for the brown bullhead (*Ictalurus nebulosus*). *J. Fish. Res. Bd. Can.*, **34**, 396–401

Gutnecht, J. 1961. Mechanism of radioactive zinc uptake by *Ulva lactuca*. *Limnol. Oceanogr.*, **6**, 426–31

Gutnecht, J. 1963. ^{65}Zn uptake by benthic marine algae. *Limnol. Oceanogr.*, **8**, 31–8

Hadfield, W. 1960. Rhizosphere effect on soil algae. *Nature, London*, **185**, 179

Haines, K. C. & Guillard, R. R. L. 1974. Growth of vitamin B_{12} – requiring marine diatoms in mixed laboratory cultures with vitamin B_{12} producing marine bacteria. *J. Phycol.*, **10**, 245–52

Halldal, P. 1968. Photosynthetic capacities and photosynthetic action spectra of endozooic algae of the massive coral *Favia*. *Biol. Bull.*, **134**, 411–24

Hallegraeff, G. M. 1976. Pigment diversity in freshwater phytoplankton. III. Summer phytoplankton of eight lakes with widely different trophic characteristics. *Hydrobiol. Bull.*, **10**, 87–95

Hallegraeff, G. M. 1977. Pigment diversity in freshwater phytoplankton. II. Summer-succession in three Dutch lakes with different trophic characteristics. *Int. Rev. ges. Hydrobiol.*, **62**, 19–39

Hamburger, B. 1958. Bakterien symbiose bei *Volvox aureas* Ehrenb. *Arch. Mikrobiol.*, **29**, 291–310

Hammen, van der, T., Maarleveld, G. C., Vogel, J. C. & Zagwijn, W. H. 1967. Stratigraphy, climatic succession and radiocarbon dating of the last glacial in the Netherlands. *Geol. Mijnb.*, **46**, 79–95

Hammer, V. T. 1964. The succession of 'bloom' species of blue-green algae and some causal factors. *Verh. Int. Ver. Limnol.*, **15**, 829–36

Hammer, V. T., Haynes, R. C., Heseltine, J. M. & Swanson, S. M. 1975. The saline lakes of Saskatchewan. *Verh. Int. Ver. Limnol.*, **19**, 589–98

Hanson, R. B. 1977. Pelagic *Sargassum* community metabolism: Carbon and Nitrogen. *J. exp. mar. Biol. Ecol.*, **29**, 107–18

Happey, C. M. 1970. The estimation of cell numbers of flagellate and coccoid Chlorophyta in natural populations. *Br. phycol. J.*, **5**, 71–8

Happey, C. M. & Moss, B. 1967. Some aspects of the biology of *Chrysococcus diaphanus* in Abbot's Pool, Somerset. *Br. phycol. Bull.*, **3**, 269–79

Happey-Wood, C. M. 1976. The occurrence and relative importance of nano-Chlorophyta in freshwater algal communities. *J. Ecol.*, **64**, 279–92

Haq, B. U. & Lohmann, G. P. 1976. Early cenozoic calcareous nannoplankton biogeography of the Atlantic Ocean. *Mar. Micropaleont.*, **1**, 119–94

Hardy, A. C. & Gunther, E. R. 1935. The plankton of the South Georgia whaling grounds and adjacent waters 1926–7. *Discovery Reports*, **11**, 1–456

Hargrave, B. T. 1969. Epibenthic algal production and community respiration in the sediments of Marion Lake. *J. Fish. Res. Bd. Can.*, **26**, 2003–26

Hargrave, B. T. 1970. The effect of a deposit-feeding amphipod on the metabolism of benthic microflora. *Limnol. Oceanogr.*, **15**, 21–30

Hargreaves, J. W., Lloyd, E. J. H. & Whitton, B. A. 1975. Chemistry and vegetation of highly acidic streams. *Freshwater Biol.*, **5**, 563–76

Hargreaves, J. W. & Whitton, B. A. 1976. Effect of pH on growth of acid stream algae. *Br. phycol. J.*, **11**, 215–23

Harlin, M. M. 1973. Transfer of products between epiphytic marine algae and host plants. *J. Phycol.*, **9**, 243–8

Harlin, M. M. 1975. Epiphyte – host relations in seagrass communities. *Aquatic Bot.*, **1**, 125–31

Harlin, M. M. & Craigie, J. S. 1975. The distribution of photosynthate in *Ascophyllum nodosum* as it relates to epiphytic *Polysiphonia lanosa*. *Br. phycol. J.*, **4**, 97–103

Harlin, M. M. & Lindbergh, J. M. 1977. Selection of substrata by seaweeds: Optimal surface relief. *Mar. Biol.*, **40**, 33–40

Harper, M. A. 1969. Movement and migration of diatoms on sand grains. *Br. phycol. J.*, **4**, 97–103

Harper, M. A. & Harper, J. F. 1967. Measurements of diatom adhesion and their relationship with movement. *Br. phycol. Bull.*, **3**, 195–207

Harris, D. O. 1970. An autoinhibitory substance produced by *Platydorina caudata*. *Plant Physiol.*, **45**, 210–14

Harris, D. O. 1971. Growth inhibitors produced by the green algae (Volvocaceae). *Arch. Mikrobiol.*, **76**, 47–50

Harris, E. 1959. The nitrogen cycle in Long Island Sound. *Bull. Bing. Oceanog. Collect.*, **17**, 31–65

Harris, G. P. 1973. The diel and annual cycles of net plankton photosynthesis in Lake Ontario. *J. Fish. Res. Bd. Can.*, **30**, 1779–87

Harris, G. P. & Piccinin, B. B. 1977. Photosynthesis by natural phytoplankton populations. *Arch. Hydrobiol.*, **80**, 405–57

Harrison, A. D. 1958. The effects of sulphuric acid pollution on the biology of streams in the Transvaal, South Africa. *Verh. Int. Ver. Limnol.*, **13**, 603–10

Harrison, P. G. 1978. Growth of *Ulva fenestrata* (Chlorophyta) microcosms rich in *Zostera marina* (Anthophyta) detritus. *J. Phycol.*, **14**, 100–3

Harrison, W. G. 1976. Nitrate metabolism of the red tide dinoflagellate *Gonyaulax polyedra* Stein. *J. exp. mar. Biol. Ecol.*, **21**, 199–209

Harrison, W. G., Azam, F., Renger, E. H. & Eppley, R. W. 1977. Some experiments on phosphate assimilation by coastal marine plankton. *Mar. Biol.*, **40**, 9–18

Hart, R. C. & Hart, R. 1977. The seasonal cycles of phytoplankton in subtropical Lake Siboya: A preliminary investigation. *Arch. Hydrobiol.*, **80**, 85–107

Hart, T. J. 1934. On the phytoplankton of the south west Atlantic and the Bellinghausen Sea, 1929–31. *Discovery Reports*, **8**, 1–28

Hart, T. J. 1935. On the diatoms of the skin film of whales, and their possible bearing on problems of whale movements. *Discovery Reports*, **10**, 247–82

Hart, T. J. 1942. Phytoplankton periodicity in Antarctic surface waters. *Discovery Reports*, **21**, 261–356

Hartley, W. R. & Weiss, C. M. 1970. Light intensity and the vertical distribution of algae in tertiary oxidation ponds. *Wat. Res.*, **4**, 751–63

Hartog, C. den 1955. A classification system for the epilithic algal communities of the Netherland coast. *Act. Bot. Neerl.*, **4**, 126–35

Hartog, C. den 1959. The epilithic algal communities occurring along the coast of the Netherlands. *Wentia*, no. 1, 241 pp.

Hartog, C. den 1968. The littoral environment of rocky shores as a border between the sea and the land and between sea and the freshwater. *Blumea*, **16**, 375–93

Hartog, C. den 1972. Substratum – Multicellular Plants. In *Marine Ecology. A comprehensive, integrated treatise on life in Oceans and Coastal Waters*, ed. O. Kinne, pp. 1272–89. Wiley-Interscience

Harvey, G. W. 1966. Microlayer collection from the sea surface. A new method and initial results. *Limnol. Oceanogr.*, **11**, 608–13

Hasle, G. R. 1954. More on phototactic diurnal migration in marine flagellates. *Nyt. Mag. Bot.*, **2**, 139–46

Hasle, G. R. 1959. A quantitative study of phytoplankton from the equatorial Pacific. *Deep Sea Res.*, **6**, 38–59

Hasle, G. R. 1968. *Navicula endophytica* sp. nov. A pennate diatom with an unusual mode of existence. *Br. phycol. Bull.*, **3**, 475–80

Hasle, G. R. 1969. An analysis of the phytoplankton of the Pacific Southern ocean. *Hvalråd. Skr.*, **52**, 168 pp.

Hasle, G. R. 1972. The distribution of *Nitzschia seriata* Cleve and allied species. *Nova Hedw., Beih.*, **39**, 171–90

Hastings, J. W. & Sweeney, B. M. 1958. A persistent diurnal rhythm of luminescence in *Gonyaulax polyedra*. *Biol. Bull.*, **115**, 440–58

Hatcher, B. G., Chapman, A. R. O. & Mann, K. H. 1977. An annual carbon budget for the kelp *Laminaria longicruris*. *Mar. Biol.*, **44**, 85–96

Hatton, H. 1938. Essais de bionomie explicative sur quelques espéces intercotidales d'algues et d'animaux. *Ann. Inst. Oceanogr. Monaco*, **17**, 241–348

Hattori, A. & Fujita, Y. 1959. Formation of phycobilin pigments in a blue-green alga, *Trichodesmium tenuis* as induced in illumination with coloured lights. *J. Biochem. Tokyo*, **46**, 521–4

Hawkes, W. A. 1957. Film accumulation and grazing activity in the sewage filters at Birmingham. *J. Inst. Sew. Purif.* (2) 88–112

Haworth, E. Y. 1969. The diatoms of a sediment core from Blea Tarn, Langdale. *J. Ecol.*, **57**, 429–39

Haworth, E. Y. 1972. Diatom succession in a core from

Pickerel Lake, Northeastern South Dakota. *Bull. geol. Soc. Am.*, **83**, 157–72

Hay, W. W., Mohler, H. P., Roth, P. H., Schmidt, R. R. & Boudreaux, J. R. 1967. Calcareous nannoplankton zonation of the Cenozoic of the Gulf Coast and Caribbean – Antillean area and transoceanic correlation. *Trans. Gulf-Coast Ass. Geol. Socs.*, **17**, 426–80

Hayes, F. R. & Coffin, C. C. 1951. Radioactive phosphorus and exchange of lake nutrients. *Endeavour*, **10**, 78–81

Hays, J. D. 1967. Quarternary sediments of the Antarctic Ocean. *Prog. Oceanography*, **4**, 117–31

Hayward, J. 1974. Studies on the growth of *Stichococcus bacillaris* Naeg. in culture. *J. mar. biol. Ass., U.K.*, **54**, 261–8

Healey, F. R., Coombs, J. & Volcani, B. E. 1967. Changes in pigment content of the diatom *Navicula pelliculosa* (Bréb). Hilse in silicon-starvation synchrony. *Arch. Mikrobiol.*, **59**, 131–42

Heaney, S. I. 1976. Temporal and spatial distribution of the dinoflagellate *Ceratium hirundinella* O. F. Müller within a small productive lake. *Freshwater Biol.*, **6**, 531–42

Hecky, R. E. & Kilham, P. 1973. Diatoms in alkaline, saline lakes: Ecology and geochemical implications. *Limnol. Oceanogr.*, **18**, 53–71

Hecky, R. E., Mopper, K., Kilham, P. & Degens, E. T. 1973. The amino-acid and sugar composition of diatom cell walls. *Mar. Biol.*, **19**, 323–31

Hedgepeth, J. W. 1963. Classification of marine environments. In *Treatise on Marine Ecology and Palaeoecology. Vol. 2*, ed. H. S. Ladd, *Geol. Soc. Amer. Mem.*, 67, pp. 93–100

Hellebust, J. A. 1965. Excretion of some organic compounds by marine phytoplankton. *Limnol. Oceanogr.*, **10**, 192–208

Hellebust, J. A. 1967. Excretion of organic compounds by cultured and natural populations of marine phytoplankton. In *Estuaries*, ed. G. H. Lauff. *Amer. Assoc. Adv. Sci., Washington D.C.*, no. 83, pp. 361–6

Hellebust, J. A. 1972. Light-Plants. In *Marine Ecology, vol. 1. Environmental factors*, ed. O. Kinne, pp. 125–58

Hendey, N. I. 1937. The plankton diatoms of the Southern Seas. *Discovery Reports*, **16**, 151–364

Hensen, V. 1887. Über die Bestimmung des Planktons oder des im Meere treibenden Materials an Pflanzen und Thieren. *Ber. Kommn. wiss. Unters. deutsch. Meere. Kiel.*, **12–14**, 1–107

Herbst, R. P. 1969. Ecological factors and the distribution of *Cladophora glomerata* in the Great Lakes. *Am. Midl. Nat.*, **82**, 90–8

Herdman, E. C. 1924. Notes on dinoflagellates and other organisms causing discolouration of the sand at Port Erin. *Proc. Trans. Lpool biol. Soc.*, **38**, 58–63

Herndon, W. R. 1958. Studies on Chlorosphaeracean algae from soil. *Am. J. Bot.*, **45**, 298–308

Hesse, R., Allee, W. C. & Schmidt, K. P. 1951. *Ecological Animal Ecology*, second edition, 715 pp. J. Wiley & Sons, New York

Heywood, D. 1977. Chloroplast structure in the chloromonadophycean alga *Vacuolaria virescens. J. Phycol.*, **13**, 68–72

Hibberd, D. J. 1976. The ultrastructure and taxonomy of the Chrysophyceae and Prymnesiophyceae (Haptophyceae): a survey with some new observations on the ultrastructure of the Chrysophyceae. *Bot. J. Linn. Soc.*, **72**, 55–80

Hibberd, D. J. & Leedale, G. F. 1972. Observations on the cytology and ultrastructure of the new algal class, Eustigmatophyceae. *Ann. Bot.*, **36**, 40–71

Hickman, M. 1969. Methods for determining the primary productivity of epipelic and epipsammic algal associations. *Limnol. Oceanogr.*, **14**, 936–41

Hickman, M. 1971. Standing crops and primary productivity of the epipelon of two small ponds in North Somerset, U.K. *Oecologia, Berlin*, **6**, 238–53

Hickman, M. 1973. The standing crop and primary productivity of the phytoplankton of Abbot's Pond, North Somerset. *J. Ecol.*, **61**, 269–87

Hickman, M. 1976. Phytoplankton population efficiency studies. *Int. Rev. ges. Hydrobiol.*, **61**, 279–95

Hickman, M. & Klarer, D. M. 1974. The growth of some epiphytic algae in a lake receiving thermal effluent. *Arch. Hydrobiol.*, **74**, 403–26

Hilliard, D. K. & Asmund, B. 1963. Studies on Chrysophyceae from some ponds and lakes in Alaska. II. *Hydrobiol.*, **22**, 331–97

Hinde, R. & Smith, D. C. 1972. Persistence of functional chloroplasts in *Elysia viridis* (Opisthobranchia, Saccoglossa). *Nature, London*, **239**, 30–1

Hirano, M. 1965. Freshwater algae in the Antarctic regions. In *Biogeography and Ecology in Antarctica*, ed. J. van Miegham & P. van Oye, pp. 127–93

Hitchcock, G. L. & Smayda, T. J. 1977. The importance of light in the initiation of the 1972–73 winter-spring diatom bloom in Narragansett Bay. *Limnol. Oceanogr.*, **22**, 126–31

Hobbie, J. E. 1964. Carbon 14 measurements of primary production in two arctic Alaskan lakes. *Verh. Int. Ver. Limnol.*, **15**, 360–4

Hobson, L. A., Menzel, D. W. & Barber, R. T. 1973. Primary productivity and sizes of pools of organic carbon in the mixed layer of the ocean. *Mar. Biol.*, **19**, 298–330

Hodgkiss, J. J. & Tai, Y. C. 1976. Studies on Plover Cove Reservoir, Hong Kong. III. A comparison of the species composition of diatom floras on different substrates and the effects of environmental factors on growth. *Freshwater Biol.*, **6**, 287–98

Hoek, C. van den 1958. The algal microvegetation in and on barnacle shells collected along the Dutch and French coasts. *Blumea*, **9**, 206–14

Hoek, C. van den 1967. Algal phytogeography of the European Atlantic coast. *Blumea*, **15**, 63–89

Hoek, C. van den 1969. Algal vegetation types along the open coasts of Curaçao, Netherlands Antilles. *Proc. K. ned. Akad. Wet.*, ser. C, **72**, 537–77

Hoek, C. van den 1975. Phytogeographic provinces along the coasts of the northern Atlantic Ocean. *Phycologia*, **14**, 317–30

Hoek, C. van den 1978. *Algen. Einführung in die Phykologie.* 481 pp. Georg Thieme, Stuttgart

Hoek, C. van den, Colijn, F., Cortel-Breeman, A. M. & Wanders, J. B. W. 1972. Algal vegetation-type along the shores of inner bays and lagoons of Curaçao and of the lagoon Lac (Bonaire) Netherlands Antilles. *Verh. K. ned. Akad. Wet., Afd. nat.*, **61** (2), 72 pp.

Hoek, C. van den, Cortel-Breeman, A. M. & Wanders, J. B. W. 1975. Algal zonation in the fringing coral reef of Curaçao, Netherlands Antilles, in relation to zonation of corals and gorgonians. *Aquatic Bot.*, **1**, 209–308

Hoffmann, C. 1942. Beiträge zur Vegetation des Farbstreifen-Sandwatts. *Kieler Meeresforsch.*, **4**, 85–108

Hoffman, L. R. 1978. Virus-like particles in *Hydrurus* (Chrysophyceae). *J. Phycol.*, **14**, 110–14

Hoham, R. W. 1973. Pleiomophism in the snow alga, *Raphidonema nivale* Lagerh. (Chlorophyta) and a revision of the genus *Raphidonema. Syesis*, **6**, 255–63

Hoham, R. W. 1974. *Chlainomonas kolii* (Hardy et Curl) comb. nov. (Chlorophyta, Volvocales) a revision of the snow alga *Trachelomenas kolii* Hardy et Curl. *J. Phycol.*, **10**, 392–6

Hoham, R. W. & Mullet, J. E. 1978. *Chloromonas nivalis* (Chod.) Hoh et Mull. comb. nov. and additional comments on the snow alga, *Scotiella. Phycologia*, **17**, 106–7

Hohn, M. H. 1964. Artificial substrate for benthic diatom-collection, analysis and interpretation. In *Organism-substrate relationships in streams*, ed. K. W. Cummins, C. A. Tyron Jr. & R. F. Hartman, University of Pittsburgh Spec. Publ. no. 4, Pymantuning Lab. Field Biol.

Hohn, M. H. & Hellerman, J. 1963. The taxonomy and structure of diatom populations from three eastern North American rivers using three sampling methods. *Trans. Am. Micros. Soc.*, **82**, 250–329

Holland, A. F., Zingmark, R. G. & Dean, J. M. 1974. Quantitative evidence concerning the stabilization of sediments by marine benthic diatoms. *Mar. Biol.*, **27**, 191–6

Holland, R. E. 1969. Seasonal fluctuation of Lake Michigan diatoms. *Limnol. Oceanogr.*, **14**, 423–36

Holland, R. E. & Claflin, L. W. 1975. Horizontal distribution of planktonic diatoms in Green Bay, mid-July, 1970. *Limnol. Oceanogr.*, **20**, 365–78

Holligan, P. M. & Harbour, D. G. 1977. The vertical distribution and succession of phytoplankton in the western English Channel in 1975 and 1976. *J. mar. biol. Ass.*, *U.K.*, **57**, 1075–93

Holm-Hansen, O. 1969. Environmental and nutritional requirements for algae. In *Proc. Eutroph.-Biostim. Assessment Workshop*, 19–21

Holm-Hansen, O. 1970. ATP levels in algal cells as influenced by environmental conditions. *Pl. Cell Physiol.*, **11**, 689–700

Holm-Hansen, O., Goldman, C. R., Richards, R. & Williams, P. M. 1976. Chemical and biological characteristics of a water column in Lake Tahoe. *Limnol. Oceanogr.*, **21**, 548–62

Holm-Hansen, O. & Paerl, H. W. 1972. The applicability of ATP determination for estimation of microbial biomass and metabolic activity. In *Detritus and its role in aquatic ecosystems. Proc. IBP-UNESCO Symp.*, pp. 149–68. Pallanza, Italy

Holt, C. & M. von 1968. Transfer of photosynthetic products from zooxanthellae to coelenterate hosts. *Comp. Biochem. Physiol.*, **24**, 73–81

Honjo, S. 1977. Biogeography and provincialism of living coccolithophorids in the Pacific Ocean. In *Oceanic Micropalaeontology*, ed. A. T. S. Ramsay, pp. 951–72. Academic Press, London

Hopkins, R. & Kain, J. M. 1971. The effect of marine pollutants on *Laminaria hyperborea. Mar. Pollut. Bull.*, **2**, 75–7

Horie, G. 1969. Asian lakes. In *Eutrophication; Causes, consequences, correctives*, pp. 98–123. Nat. Acad. Sci. Washington, D.C.

Horne, A. J. 1972. The ecology of nitrogen fixation on Signy Island, South Orkney Islands. *Brit. Antarct. Surv. Bull.*, **27**, 1–18

Horne, A. J. 1975. Algal nitrogen fixation in Californian streams: diel cycles and nocturnal fixation. *Freshwater Biol.*, **5**, 471–7

Horne, A. J. & Carmiggelt, C. J. W. 1975. Algal nitrogen fixation in Californian streams: seasonal cycles. *Freshwater Biol.*, **5**, 461–70

Horne, A. J., Dillard, J. E., Fujita, D. K. & Goldman, C. R. 1972. Nitrogen fixation in Clear Lake, California. II. Synoptic studies of the autumn *Anaebaena* bloom. *Limnol. Oceanogr.*, **17**, 693–703

Horne, A. J., Fogg, G. E. & Eagle, D. J. 1969. Studies *in situ* of the primary production of an area of inshore Antarctic Sea. *J. mar. biol. Ass.*, *U.K.*, **49**, 393–405

Horne, A. J. & Fogg, G. E. 1970. Nitrogen fixation in some English lakes. *Proc. Roy. Soc., London*, **B**, **175**, 351–66

Horne, A. J. & Goldman, C. R. 1972. Nitrogen fixation in Clear Lake, California. I. Seasonal variation and the role of heterocysts. *Limnol. Oceanogr.*, **17**, 678–92

Horne, A. J. & Viner, A. B. 1971. Nitrogen fixation and its significance in tropical Lake George, Uganda. *Nature, London*, **232**, 417–18

Horner, R. & Alexander, V. 1972. Algal populations in arctic sea ice. An investigation of heterotrophy. *Limnol. Oceanogr.*, **17**, 454–8

Hornsey, I. S. & Hide, D. 1974. The production of antimicrobial compounds by British Marine algae. I. Antibiotic-producing marine algae. *Br. phycol. J.*, **9**, 353–61

Hornsey, I. S. & Hide, D. 1976. The production of antimicrobial compounds by British Marine algae. II. Seasonal variation in production of antibiotics. *Br. phycol. J.*, **11**, 63–7

Horodyski, R. J., Bloes, E. R. & Von-der Haar, S. 1977. Laminated algal mats from a coastal lagoon, Laguna Mormona, Baja California, Mexico. *J. sedim. Petrol.*, **47**, 680–96

Howard, K. L. & Menzies, R. J. 1969. Distribution and production of *Sargassum* in the waters off the Carolina Coast. *Bot. Mar.*, **12**, 244–54

Howe, M. A. 1912. The building of 'coral' reefs. *Science, N.Y.*, **35**, 837–42

Hoyt, W. D. 1927. The periodic fruiting of *Dictyota* and its relation to the environment. *Am. J. Bot.*, **14**, 592–619

Hruby, T. & Norton, T. A. 1979. Algal colonization on rocky shores in the Firth of Clyde. *J. Ecol.*, **67**, 65–77

Hsiao, S. I. C. & Druehl, L. D. 1973. Environmental control of gametogenesis in *Laminaria saccharina*. IV. *In situ* development of gametophytes and young sporophytes. *J. Phycol.*, **9**, 160–4

Hubbs, C. L. 1952. Antitropical distribution of fishes and other organisms. In *Proc. 7th Pacif. Sci. Congress*, **3**, 324–9

Huber-Pestalozzi, G. & Nipkow, F. 1922. Experimentelle Untersuchungen über die Entwicklung von *Ceratium hirundinella* O.F.M. *Z. Bot.*, **14**, 337–71

Hughes, J. C. & Lund, J. W. G. 1962. The rate of growth of

Asterionella formosa Hass. in relation to its ecology. *Arch. Mikrobiol.*, **42**, 117–29

Hughes, M. K. & Whitton, B. A. 1972. Algae of Slapestone Sike, Upper Teesdale. *Vasculum*, **67**, 30–5

Hulburt, E. M. 1963. The diversity of phytoplanktonic populations in oceanic, coastal and estuarine regions. *J. Mar. Res.*, **21**, 81–93

Hulburt, E. M. 1968. Phytoplankton observations in the Western Caribbean Sea. *Bull. Mar. Sci.*, **18**, 388–99

Hulburt, E. M. & Corwin, N. 1969. Influence of the Amazon River outflow on the ecology of the western tropical Atlantic. III. The planktonic flora between the Amazon River and the Windward Islands. *J. Mar. Res.*, **27**, 55–72

Hulburt, E. M. & Corwin, N. 1970. Relation of the phytoplankton to turbulence and nutrient renewal in Casco Bay, Maine. *J. Fish. Res. Bd. Can.*, **27**, 2081–90

Hulburt, E. M. & Guillard, R. R. L. 1968. The relationship of the distribution of the diatom *Skeletonema tropicum* to temperature. *Ecology*, **49**, 337–9

Hulburt, E. M., Ryther, J. H. & Guillard, R. R. L. 1960. The phytoplankton of the Sargasso Sea off Bermuda. *J. Cons. perm. int. Explor. Mer.*, **15**, 115–28

Humm, N. J. 1964. Epiphytes of the sea grass, *Thalassia testudinum* in Florida. *Bull. mar. Sci. Gulf Carib.*, **14**, 306–41

Huntsman, S. A. & Barber, R. T. 1975. Modification of phytoplankton growth by excreted compounds in low density populations. *J. Phycol.*, **11**, 10–13

Huntsman, S. A. & Barber, R. T. 1977. Primary production off north west Africa; the relationship to wind and nutrient conditions. *Deep Sea Res.*, **24**, 25–33

Hurd, D. C. 1972a. Interactions of biogenic opal, sediment and seawater in the central equatorial Pacific. Ph.D. dissertation, University of Hawaii.

Hurd, D. C. 1972b. Factors affecting solution rate of biogenic opal in seawater. *Earth planet. Sci. Lett.*, **15**, 411–17

Hurlburt, S. H. 1971. The nonconcept of species diversity. A critique and alternative parameter. *Ecology*, **52**, 577–886

Hurlburt, S. H., Zedler, J. & Fairbanks, D. 1972. Ecosystem alteration by mosquito fish (*Gambusia affinis*) predation. *Science*, **175**, 639–41

Hustedt, F. 1921. Untersuchungen über die Natur der Harmatten-Trübe. *Dt. übersee. met. Beob.*, 1–3

Hustedt, F. 1958. Diatomeen aus der Antarktis und dem Sudatlantik. *Dtsch. Antarkt. Exped.*, 1938–9, **2**, 103–91

Hutchinson, G. E. 1957. *A treatise on limnology. Vol. I. Geography, Physics and Chemistry.* 1,015 pp. J. Wiley & Sons

Hutchinson, G. E. 1967. *A treatise on limnology. Vol. II. Introduction to lake biology and the limnoplankton.* 1,115 pp. J. Wiley & Sons

Hutchinson, G. E. 1969. Eutrophication, past and present. In *Eutrophication; causes, consequences, correctives.* pp. 17–26. Acad. Sci. Washington D.C.

Hutchinson, G. E. 1970. Ianula: An account of the history and development of the Lago di Monterosi, Latium, Italy. *Trans. Am. Phil. Soc.*, **60**, 1–178

Hutchinson, G. E. 1974. De rebus planctonicis. *Limnol. Oceanogr.*, **19**, 360–1

Hynes, H. B. N. 1970. The ecology of running waters. 553 pp. Liverpool University Press.

Ichimura, S., Nagasawa, S. & Tanaka, T. 1968. On the oxygen and chlorophyll maxima found in the metalimnion of a mesotrophic lake. *Bot. Mag. Tokyo.*, **81**, 1–10

Ilmavirta, V. 1975. Dynamics of phytoplanktonic production in the oligotrophic lake Pääjärvi, southern Finland. *Ann. Bot. Fennici*, **12**, 45–54

Ilmavirta, V., Jones, R. I. & Kairesalo, T. 1977. The structure and photosynthetic activity of pelagial and littoral plankton communities in Lake Pääjärvi, Southern Finland. *Ann. Bot. Fennici*, **14**, 7–16

Ilmavirta, K. & Kotimaa, A. L. 1974. Spatial and seasonal variations in phytoplanktonic primary production and biomass in the oligotrophic lake Pääjärvi, southern Finland. *Ann. Bot. Fennici*, **11**, 112–20

Irvine, D. E. G. & Price, J. H. 1978. Modern approaches to the taxonomy of red and brown algae. In *The Systematics Association.* Spec., vol. 10, 484 pp. Academic Press, London

Isaac, W. E. 1937. Studies of South African seaweed vegetation. I. West coast from Lamberts Bay to the Cape of Good Hope. *Trans. Roy. Soc., S. Africa*, **25**, 115–51

Isaac, W. E. 1949. Studies of South African seaweed vegetation. II. South Coast: Rooi Els to Gansbaai with special reference to Gansbaai. *Trans. Roy. Soc., S. Africa*, **32**, 125–60

Jaag, O. 1938. Kryptogamenflora des Rheinfalls und des Hochrheins von Stein bis Eglisau. *Mitt. naturf. Ges. Schaffhausen*, **14**, 1–158

Jaag, O. 1945. Untersuchungen über die Vegetation und Biologie der Algen des nackten Gesteins in der Alpen, im Jura und im schweizerischen Mittelland. *Beitr. Kryptog. Flora Schweiz*, **9**, 8–560

Jackson, J. E. & Castenholz, R. W. 1975. Fidelity of thermophilic blue-green algae to hot spring habitats. *Limnol. Oceanogr.*, **20**, 305–22

Jackson, D. F. & McFadden, J. 1954. Phytoplankton photosynthesis in Sanctuary Lake, Pymatuning Reservoir. *Ecology*, **35**, 1–4

Jacob, F. 1961. Zur Biologie von *Codium bursa* und seiner endophytischen Cyanophycees. *Arch. Protistenk.*, **105**, 345–406

Jamieson, E. R. 1971. Paleoecology of Devonian reefs in Western Canada. *Proc. natn. Am. Paleont. Conv.* vol. 2, 1300–40

Jansson, A. M. 1970. Production studies in the *Cladophora* belt. *Thalassia*, **6**, 143–55

Jassby, A. D. & Goldman, C. R., 1974. Loss rates from a lake phytoplankton community. *Limnol. Oceanogr.*, **19**, 618–17

Javorski, G. & Lund, J. W. G. 1970. Drought resistance and dispersal of *Asterionella formosa* Hass. *Nova Hedw. Beih.*, **31**, 37–48

Jefferies, R. L. 1972. Aspects of salt marsh ecology with particular reference to inorganic plant nutrition. In *The Estuarine Environment*, ed. R. S. K. Barnes & J. Green, pp. 61–85. Applied Science, London

Jeffrey, S. W. 1968. Pigment composition of Siphonales algae in the brain coral *Favia*. *Biol. Bull.*, **135**, 141–8

Jeffrey, S. W. 1974. Profiles of photosynthetic pigments in the ocean using thin-layer-chromatography. *Mar. Biol.*, **26**, 101–10

Jeffrey, S. W. 1976. The occurrence of chlorophyll c_1 and c_2 in algae. *J. Phycol.*, **12**, 349–54

Jeffrey, S. W. & Carpenter, S. M. 1974. Seasonal succession of phytoplankton at a coastal station off Sydney. *Aust. J. mar. Freshwat. Res.*, **25**, 361–9

Jeffrey, S. W. & Haxo, F. T. 1968. Photosynthetic pigments of symbiotic dinoflagellates (zooxanthellae) from corals and clams. *Biol. Bull.*, **135**, 149–65

Jeffrey, S. W. & Humphrey, G. F. 1975. New spectrophotometric equations for determining chlorophylls a, b, c_1 and c_2 in higher plants, algae and natural phytoplankton. *Biochem. Physiol. Pflanzen.*, **167**, 191–4

Jeffrey, S. W., Sielicki, M. & Haxo, F. T. 1975. Chloroplast pigment patterns in dinoflagellates. *J. Phycol.*, **11**, 374–84

Jenkin, P. M. 1937. Oxygen production by the diatom *Coscinodiscus excentricus* Ehr. in relation to submarine illumination in the English Channel. *J. mar. biol. Ass., U.K.*, **22**, 301–43

Jewell, W. J. & McCarty, D. L. 1968. Aerobic decomposition of algae and nutrient regeneration. *Tech. Rep. fed. Wat. Pollut. Control Admin.*, no. 91, 283 pp.

Jitts, H. R. 1965. The summer characteristics of primary productivity in the Tasman and Coral Seas. *Aust. J. mar. Freshw. Res.*, **16**, 151–62

Jitts, H. R. 1969. Seasonal variations in the Indian Ocean along 110° E. IV. Primary production. *Aust. J. mar. Freshwat. Res.*, **20**, 55–64

Jitts, H. R., McAllister, C. D., Stephens, K. & Strickland, J. D. H. 1964. The cell division rates of some marine phytoplankton as a function of light and temperature. *J. Fish. Res. Bd. Can.*, **21**, 139–57

Jitts, H. R., Morel, A. & Saijo, Y. 1976. The relation of oceanic primary production to available photosynthetic irradiance. *Aust. J. mar. Freshwater Res.*, **27**, 441–54

Johannes, R. E. 1972. Coral reefs and pollution. In *Marine pollution and sea life*, ed. M. Ruivo, pp. 364–75. Fishing News, West Byefleet

Johannes, R. E. 1978. Flux of zooplankton and benthic algal detritus. In *Coral reefs: research methods*, ed. D. R. Stoddart & R. E. Johannes, pp. 429–31.

Johannes, R. E., Ablens, I., D'Elia, C., Kinzie, R. A., Pomeroy, L. R., Scottile, W. & Wiebe, W. 1972. The metabolism of some coral reef communities. A team study of nutrient and energy flux at Eniwetok. *Bioscience*, **22**, 541–5

John, D. M. 1974. New records of *Ascophyllum nodosum* (L.) Le Jol. from the warmer parts of the Atlantic Ocean. *J. Phycol.*, **10**, 243–44

Johnson, P. W., Sieburth, J. McN., Sastry, A., Arnold, C. R. & Doty, M. S. 1971. *Leucothrix mucor* infestation of benthic Crustacea, fish eggs and tropical algae. *Limnol. Oceanogr.*, **16**, 962–9

Johnson, T. C. 1976. Controls on the preservation of biogenic opal in sediments of the eastern Tropical Pacific. *Science*, **192**, 887–90

Johnson, W. G., Gigon, A., Gulmon, S. L. & Mooney, H. A. 1974. Comparative photosynthetic capacities of intertidal algae under exposed and submerged conditions. *Ecology*, **55**, 450–3

Johnston, C. S. 1969. The ecological distribution and primary production of macrophytic marine algae in the Eastern Canaries. *Int. Rev. ges. Hydrobiol. Hydrogr.*, **54**, 473–90

Johnston, C. S., Jones, R. G. & Hunt, R. D. 1977. A seasonal carbon budget for a laminarian population in a Scottish sea-loch. *Helgoländer wiss. Meeresunt.*, **30**, 527–45

Jones, J. G. 1974. A method for observation and enumeration of epibenthic algae directly on the surface of stone. *Oecologia, Berlin*, **16**, 1–8

Jones, K. 1974. Nitrogen fixation in a salt marsh. *J. Ecol.*, **62**, 553–65

Jones, M. & Spencer, C. P. 1970. The phytoplankton of the Menai Straits. *J. Cons. int. Explor. Mer.*, **33**, 169–80

Jones, N. S. & Kain, S. M. 1967. Subtidal algal colonization following the removal of *Echinus*. *Helgoländer wiss. Meeresunt.*, **15**, 460–6

Jones, O. A. & Endean, R. 1976. *Biology and geology of coral reefs*. Vol. II. Biology I. 494 pp. Vol. III. Biology II, 435 pp. Academic Press, London.

Jones, R. I. & Ilmavirta, V. 1978. A diurnal study of the phytoplankton in the eutrophic lake Lovojärvi, Southern Finland. *Arch. Hydrobiol.*, **83**, 494–514

Jones, R. F. 1960. The accumulation of nitrosyl ruthenium by fine particles and marine organisms. *Limnol. Oceanogr.*, **5**, 312–25

Jones, W. E. & Babb, M. S. 1965. The motile period of swarmers of *Enteromorpha intestinalis* (L.) Link. *Br. phycol. Bull.*, **3**, 525–8

Jones, W. E. & Demetropoulos, A. 1968. Exposure to wave action; measurements of an important ecological parameter on rocky shores on Anglesey. *J. exp. Mar. Biol. Ecol.*, **2**, 46–63

Jones, W. E. & Dent, E. S. 1971. The effect of light on the growth of algal spores. In *4th European Mar. Biol. Symp., Cambridge*, ed. D. J. Crisp, pp. 363–74

Jónsson, S. 1970. Meeresalgen als Erstbesiedler der Vulkaninsel Surtsey. *Schr. Naturw. Ver. Schleswig-Holstein*, pp. 21–8

Jordan, A. J. & Vadas, R. L. 1972. Influence of environmental parameters on intraspecific variation in *Fucus vesiculosus*. *Mar. Biol.*, **14**, 248–52

Jordan, R. A. & Bender, R. E. 1973. Stimulation of phytoplankton growth by mixtures of phosphate, nitrate and organic chelators. *Wat. Res.*, **7**, 189–95

Jørgensen, E. G. 1956. Growth inhibition and substances formed by algae. *Physiologica Pl.*, **9**, 712–26

Jørgensen, E. G. 1957. Diatom periodicity and silicon assimilation. *Dansk. bot. Ark.*, **18**, 54 pp.

Jørgensen, E. G. & Steeman Nielsen, E. 1965. Adaptation in plankton algae. In *Primary production in aquatic environments*, ed. C. P. Goldman, pp. 39–46. University of California Press, Berkeley

Jost Casper, S. 1974. *Grundzüge eines natürlichen Systems der Mikroorganismen*. 232 pp. VEB Gustav Fischer, Jena

Joubert, J. J. & Rijkenberg, F. H. J. 1971. Parasitic green algae. *Ann. Rev. Phytopath.*, 45–64

Jousé, A. P. 1957. Diatoms in the surface layer of the Okhotsk Sea sediments. *Trudy Inst. Okeanol.*, **22**, 164–220

Jousé, A. P. 1971. Diatoms in Pleistocene sediments from the Northern Pacific Ocean. In *The Micropalaenotology of Oceans*, ed. B. M. Funnell & W. R. Riedel, pp. 407–21

Jousé, A. P., Kozlova, O. G. & Muhina, V. V. 1971. Distribution of diatoms in the surface layer of sediments from the Pacific Ocean. In *The Micropalaeontology of Oceans*, ed. B. M. Funnell & W. R. Riedel, pp. 263–9

Junge, C. E. & Werby, R. T. 1958. The concentration of chloride, sodium, potassium, calcium and sulfate in rain water over the United States. *J. Meteorol.*, **15**, 417–25

Kahn, N. & Swift, E. 1978. Positive buoyancy through ionic control in the non-motile marine dinoflagellate *Pyrocystis noctiluca* Murrey et Schuett. *Limnol. Oceanogr.*, **23**, 649–58

Kain, J. M. 1964. Aspects of the biology of *Laminaria hyperborea*. III. Survival and growth of gametophytes. *J. mar. biol. Ass.*, *U.K.*, **44**, 415–33

Kain, J. M. 1971. Synopsis of Biological data on *Laminaria hyperborea*. *FAO Fisheries Synopsis*, no. 87, 74 pp.

Kain, J. M. 1975a. The biology of *Laminaria hyperborea*. VII. Reproduction of the sporophyte. *J. mar. biol. Ass.*, *U.K.*, **55**, 567–82

Kain, J. M. 1975b. Algal recolonization of some cleared subtidal areas. *J. Ecol.*, **63**, 739–65

Kain, J. M. & Fogg, G. E. 1958. Studies on the growth of marine phytoplankton. I. *Asterionella japonica*. *J. mar. biol. Ass.*, *U.K.*, **37**, 397–413

Kain, J. M. & Svendsen, A. 1969. A note on the behaviour of *Patina pellucida* in Britain and Norway. *Sarsia*, **38**, 25–30

Kairesalo, T. 1976. Measurements of production of epilithophyton and littoral plankton in Lake Pääjärvi, southern Finland. *Ann. Bot. Fennici*, **13**, 114–18

Kairesalo, T. 1977. On the production ecology of epipelic algae and littoral plankton communities in Lake Pääjärvi, southern Finland. *Ann. Bot. Fennici*, **14**, 82–8

Kalff, J. 1967a. Phytoplankton dynamics in an arctic lake. *J. Fish. Res. Bd. Can.*, **24**, 1861–71

Kalff, J. 1967b. Phytoplankton abundance and primary production rates in two arctic ponds. *Ecology*, **48**, 558–65

Kalff, J. 1971. Nutrient limiting factors in an Arctic tundra pond. *Ecology*, **52**, 655–9

Kalff, J. 1972. Net plankton and nannoplankton production and biomass in a north temperate zone lake. *Limnol. Oceanogr.*, **17**, 712–20

Kamykowski, D. & Zentara, S. 1977. The diurnal vertical migration of motile phytoplankton through temperature gradients. *Limnol. Oceanogr.*, **22**, 148–51

Kanaya, T. 1971. Some aspects of Pre-Quaternary diatoms in the oceans. In *The Micropalaeontology of Oceans*, ed. B. M. Funnell & W. R. Riedel, pp. 545–65

Kanaya, T. & Koizumi, I. 1966. Interpretation of diatom thanatocoenoses from the North Pacific applied to a study of Core V20–130. (Studies of a Deep-sea Core V20–130. Part IV.) *Scient. Rep. Tohoku Imp. Univ., 2nd Ser. Geol.*, **37**, 89–130

Kann, E. 1941. Krustenstein in Seen. *Arch. Hydrobiol.*, **37**, 504–32

Kann, E. 1959. Die eulittorale Algenzone im Traunsee (Oberösterreich). *Arch. Hydrobiol.*, **55**, 129–92

Kann, E. 1973. Zur Systematik und Ökologie der Gattung *Chamaesiphon* (Cyanophyceae). 2. Ökologie. *Arch. Hydrobiol.*, suppl. 41, 243–82

Kanwisher, J. 1957. Freezing and drying in intertidal algae. *Biol. Bull.*, **113**, 275–85

Kanwisher, J. W. 1966. Photosynthesis and respiration in some seaweeds. In *Some contemporary Studies in Marine Science*, ed. H. Barnes, pp. 407–22. Allen & Unwin, London

Kanwisher, J. W. & Wainwright, S. A. 1967. Oxygen balance in some reef corals. *Biol. Bull. mar. biol. Lab., Woods Hole*, **133**, 378–90

Karakashian, S. J. 1963. Growth of *Paramesium bursaria* as influenced by the presence of algal symbionts. *Physiol. Zool.*, **36**, 52–68

Karayeva, N. I. & Makarova, J. V. 1973. Specific features and origin of the Caspian Sea diatom flora. *Mar. Biol.*, **21**, 269–75

Kawaguti, S. 1964. An electron microscope proof for a path of nutritive substances from zooxanthellae to the coral reef tissue. *Proc. Jap. Acad.*, **40**, 832–5

Kawaguti, S. 1966. Electron microscopy on the mantle of the giant clam with special reference to zooxanthellae and iridophores. *Biol. J. Okayama University*, **12**, 81–92

Kawaguti, S. & Yamasu, T. 1965. Electron microscopy on the symbiosis between an elysioid gastropod and chloroplasts from a green alga. *Biol. J. Okayama University*, **11**, 57–65

Kawarada, Y. & Sano, A. 1972. Distribution of chlorophyll *a* and phaeopigments in the northwestern north Pacific in relation to the hydrographic conditions. In *Biological Oceanography of the northern north Pacific Ocean*, ed. A. Y. Takenouti, pp. 125–38. Tokyo

Kayser, H. 1971. Pollution of the North Sea and rearing experiments on marine phytoflagellates as an indication of resultant toxicity. In *Advances in Water Pollution Research. Proc. 5th Int. Conf., San Francisco and Hawaii*, ed. S. H. Jenkins, vol. 2, pp. 111–12. Pergamon Press

Keating, K. I. 1978. Blue-green algal inhibition of diatom growth. Transition from mesotrophic to eutrophic community structure. *Science*, **199**, 971–3

Keeble, F. & Gamble, F. W. 1907. The origin and nature of the green cells of *Convoluta roscoffensis*. *Q. Jl. Micros. Sci.*, **51**, 167–219

Kellogg, W. W., Cadle, P. D., Allen, E. R., Lazrus, A. L. & Martell, E. A. 1972. The sulfur cycle: Mans contributions are compared to natural sources of sulfur compounds in the atmosphere and oceans. *Science*, **175**, 587–96

Kelly, M. G. 1968. The occurrence of dinoflagellate luminescence at Woods Hole. *Biol. Bull. mar. biol. Lab., Woods Hole*, **135**, 279–95

Kelly, M. G. & Katona, S. 1966. An endogenous diurnal rhythm of bioluminescence in a natural population of dinoflagellates. *Biol. Bull.*, **131**, 115–26

Kemp, C. I. 1974. Observations on the Biology and Life History of *Ulva reticulata* Forsk. M.Sc. Thesis, University of Hawaii

Kendall, C. G. St C. & Skipwith, P. A. d'E. 1968. Recent algal mats of a Persian Gulf Lagoon. *J. Sediment. Petrol.*, **38**, 1040–58

Kendall, C. G. St C. & Skipwith, P. A. d'E. 1969. Geomorphology of a recent shallow-water carbonate province: Khoral Bazam, Trucial Coast, South-west Persian Gulf. *Bull. geol. Soc. Am.*, **80**, 865–92

Kessler, E. 1974. Physiologische und biochemische Beiträge zur Taxonomie der Gattung *Chlorella*. IX. Salzresistenz als taxonomisches Merkmal. *Arch. Microbiol.*, **100**, 51–6

Khailiov, K. M. & Burlakova, Z. P. 1969. Release of dissolved organic matter by marine seaweeds and distribution of their total organic production to inshore communities. *Limnol. Oceanogr.*, **14**, 521–7

Khoja, T. & Whitton, B. A. 1971. Heterotrophic growth of blue-green algae. *Arch. Mikrobiol.*, **79**, 280–7

Kiefer, D. A., Holm-Hansen, O., Goldman, C. R., Richard, D. & Berman, T. 1972. Phytoplankton in Lake Tahoe: Deep-living populations. *Limnol. Oceanogr.*, **17**, 418–22

Kimura, Y. 1953. The system and the phylogenetic tree of plants. (In Japanese.) *J. Jap. Bot.*, **28**, 97–104

King, J. M. & Ward, C. H. 1977. Distribution of edaphic algae as related to land usage. *Phycologia*, **16**, 23–30

King, R. J. 1973. The distribution and zonation of intertidal organisms in Bass Strait. *Proc. Roy. Soc. Victoria*, **85**, part 2, 145–62

Kinne, O. 1971. *Marine Ecology. Vol. I.* Wiley-Interscience

Kitching, J. A. & Ebling, F. J. 1961. The ecology of Lough Ine XI. The control of algae by *Paracentrotus lividus* (Echinoidea). *J. Animal Ecol.*, **30**, 373–83

Kitching, J. A. & Ebling, F. J. 1967. Ecological studies at Lough Ine. *Adv. ecol. Res.*, **4**, 197–291

Kjellmann, F. R. 1883. The algae of the Arctic Sea. *Kung svenska Vetensk. Akad. Handl.*, **20**, no. 5, 350 pp.

Klarer, D. M. & Hickman, M. 1975. The effect of thermal effluent upon the standing crop of an epiphytic algal community. *Int. Revue ges. Hydrobiol.*, **60**, 17–62

Klasvik, B. 1974. Computerised analysis of stream algae. *Växtekologiska Studier*, **5**, 100 pp.

Klaveness, D. & Guillard, R. R. L. 1975. The requirement for silicon in *Synura petersonii* (Chrysophyceae). *J. Phycol.*, **11**, 349–55

Knaggs, F. W. 1967. *Rhodochorton purpureum* (Lightf.) Rosenvinge. Observations on the relationship between morphology and environment. III. *Nova Hedw.*, **12**, 521–8

Knapp, R. & Lieth, H. 1952. Über Ursachen des verstärkten Auftretens von erdbewohnenden Cyanophyceae. *Arch. Mikrobiol.*, **17**, 292–9

Knauss, J. A. 1965. Currents. 10. Equatorial Current Systems. In *The Sea, vol. 2*, ed. R. N. Hill, pp. 235–52. Interscience

Knight, M. & Parke, M. 1950. A biological study of *Fucus vesiculosus* L. and *F. serratus* L. *J. mar. biol. Ass. U.K.*, **29**, 439–514

Knoechel, R. & Kalff, J. 1975. Algal sedimentation: the cause of a diatom, blue-green succession. *Verh. Int. Ver. Limnol.*, **19**, 745–54

Knoechel, R. & Kalff, J. 1978. An *in situ* study of the productivity and population dynamics of five freshwater planktonic diatom species. *Limnol. Oceanogr.*, **23**, 195–218

Knöpp, H. 1954. A new method of displaying the results of biological examination of streams, illustrated by the quality of a longitudinal section of the Main. *Wasserwirtschaft*, **45**, 9–15

Knöpp, H. 1960. Untersuchungen über das Sauerstoff – Produktions Potential von Flussplankton. *Schweiz. Z. Hydrol.* **22**, 152–66

Knudsen, B. M. 1955. The distribution of *Tabellaria* in the English Lake District. *Proc. int. Ass. Theor. appl. Limnol.*, **12**, 216–18

Knudsen, B. M. 1957. Ecology of the epiphytic diatom *Tabellaria flocculosa* (Roth) Kütz. var. *flocculosa* in three English lakes. *J. Ecol.*, **45**, 93–112

Koblentz-Mishke, O. J., Bekasova, O. D., Vedernikov, V. I., Konovalov, B. V., Sapozhnikov, V. W. & Terskikh, V. A. 1972. Photosynthesis, pigments and underwater irradiation in the area of the Kurile-Kamchatka Trench during the summer of 1960. In *Biological Oceanography of the northern north Pacific Ocean*, ed. A. Y. Takennouti, pp. 263–74. Tokyo

Kobluk, D. R. 1976. Calcification of filaments of boring and cavity-dwelling algae, and the construction of micrite envelopes. In *Geobotany*, ed. R. C. Romans, pp. 195–207. Plenum Press, New York

Kochert, G. & Olson, L. W. 1970. Endosymbiotic bacteria in *Volvox carteri. Trans. Am. Micros. Soc.*, **89**, 475–8

Koehl, M. A. R. & Wainwright, S. A. 1977. Mechanical adaptations of a giant kelp. *Limnol. Oceanogr.*, **22**, 1067–71

Kohlmeyer, J. & Kohlmeyer, E. 1972. Is *Ascophyllum nodosum* lichenized? *Mar. Bot.*, **15**, 109–12

Kohn, A. J. & Helfritch, P. 1957. Primary organic productivity of a Hawaiian coral reef. *Limnol. Oceanogr.*, **2**, 241–51

Koisumi, I. 1973. The stratigraphic ranges of marine planktonic diatoms and diatom stratigraphy in Japan. *Mem. Geol. Soc. Japan*, **8**, 35–44

Kol, E. 1968. Kryobiologie. *Die Binnengewässer*, **24**, 216 pp.

Kolbe, R. W. 1955. Diatoms from equatorial Atlantic cores. *Rep. Swed. deep Sea Exped., Göteborg*, **9**, 1–50

Kolkwitz, R. & Marsson, M. 1908. Ökologie der pflanzliche Saprobien. *Ber. dt. bot. Ges.*, **26**, 505–19

Konopta, A., Brock, T. D. & Walsby, A. E. 1978. Buoyancy regulation by planktonic blue-green algae in Lake Mendota, Wisconsin. *Arch. Hydrobiol.*, **83**, 524–37

Koppen, J. D. 1975. A morphological and taxonomic consideration of *Tabellaria* (Bacillariophyceae) from the north central United States. *J. Phycol.*, **11**, 236–44

Kornas, J. & Medwecka-Kornas, A. 1949. Associations vegetales sous-marine dans le Gulfe de Gdansk. *Vegetatio*, **2**, 120–8

Kornas, J., Pancer, E. & Brzyski, B. 1960. Studies in sea-vegetation in the Bay of Gdansk off Pava. *Fragm. Flor. geobot.*, **6**, 1–92

Kornmann, P. 1973. Codiolophyceae, a new class of Chlorophyta. *Helgoländer wiss. Meeresunt.*, **25**, 1–13

Kornmann, P. & Sahling, P. H. 1977. Meeresalgen von Helgoland. Benthische Grün-, Braun- und Rotalgen. Sonderabdruck aus. *Helgoländer wiss. Meeresunt.*, **29**, 1–289

Kosswig, K. 1967. Der Sachwiesensee in den Ostalpen (Hochschwabgebiet) Zur Limnologie eines dystrophen Gipsgewässers. *Int. Rev. ges. Hydrobiol.*, **52**, 321–51

Koumans-Goedbloed, A. 1965. L'influence du broutage des patelles sur le peuplement algal. *Recl. Trav. Stat. mar. Endoume*, **55**, 221–35

Kowalczewski, A. & Lack, T. J. 1971. Primary production and respiration of the phytoplankton of the Rivers Thames and Kennet at Reading. *Freshwater Biol.*, **1**, 197–212

Kozlova, O. G. 1964. *Diatom algae from the Indian and Pacific Oceanic region of the Antarctic.* 176 pp. 'Nauka', Moscow

Kraft, G. T. 1972. Preliminary studies of Philippine *Eucheuma* species (Rhodophyta). Part I. Taxonomy and ecology of *Eucheuma arnoldii* Weber-van Bosse. *Pacific Sci.*, **26**, 318–34

Kramer, J. R., Herbes, S. E. & Allen, H. E. 1972. Phosphorus: Analysis of water, biomass and sediment. In *Nutrients in natural Waters*, ed. H. E. Allen & J. R. Kramer, pp. 51–100. Wiley-Interscience

Kremer, B. P. & Schmitz, K. 1973. CO₂-Fixierung und Stofftransport in benthischen marinen Algen. IV. Zur ¹⁴C-Assimilation einiger litoraler Braunalgen im submersen und emersen Zustand, *Pflanzenphysiol.*, **68**, 357–63

Krieger, W. 1927. Zur Biologie des Flussplanktons. Untersuchungen über das Potamoplankton des Havelgebiets. *Pflanzenforsch.* **10**, 1–66

Kriegstein, A. R., Castellucci, V. & Kandel, E. R. 1974. Metamorphosis of *Aplysia californica* in laboratory culture. *Proc. Nat. Acad. Sci.*, **71**, 3654–8

Kroes, H. W. 1971. Growth interactions between *Chlamydomonas globosa* Snow and *Chlorococcum ellipsoideum* Deason & Bold under different experimental conditions, with special attention to the role of pH. *Limnol. Oceanogr.*, **16**, 869–79

Kuentzel, L. A. 1969. Bacterial carbon dioxide and algal blooms. *J. Wat. Pollut. Control Fed., Washington*, **41**, 1737–47

Kullberg, R. G. 1971. Algal distribution in six thermal spring effluents. *Trans. Am. Micros. Soc.*, **90**, 412–34

Kurogi, M. 1972. Systematics of *Porphyra* in Japan. In *Contributions to the systematics of benthic marine algae of the North Pacific*, ed. I. A. Abbott & M. Kurogi, pp. 167–91. Sapporo, Japan

Kurz, W. G. W. & La Rue, T. A. 1971. Nitrogenase in *Anabaena flos-aquae* filaments lacking heterocysts. *Naturwiss.*, **8**, 417

Kylin, H. 1956. *Die Gattungen der Rhodophyceen.* 673 pp. C. W. K. Gleerups Förlag, Lund

Lack, T. J. 1971. Quantitative studies on the phytoplankton of the rivers Thames and Kennet at Reading. *Freshwater Biol.*, **1**, 213–24

Ladd, H. S. 1961. Reef building. *Science*, **134**, 703–15

Ladd, H. S. 1971. Existing reefs – geological aspects. In *Proc. natn. Am. Paleont. Conv.*, vol. 2, 1273–300

Ladd, H. S., Ingerson, E., Townsend, R. C., Russell, M. & Kirk Stephenson, H. 1953. Drilling on Eniwetok atoll, Marshall Islands. *Bull Am. Ass. Petrol. Geol.*, **37**, 2257–80

Lal, D. 1977. The oceanic microcosm of particles. *Science*, **198**, 997–1009

Lamb, I. M. & Zimmermann, M. H. 1964. Marine vegetation of Cape Ann, Massachusetts. *Rhodora*, **66**, 217–54

Land, L. S. 1971. Carbonate mud: Production by epibiont growth on *Thalassia testudinum*. *J. sedim. Petrol.*, **40**, 1361–3

Landingham, S. van 1964. Some physical and generic aspects of fluctuations in non-marine plankton diatom populations. *Bot. Rev.*, **30**, 437–78

Langangen, A. 1974. Ecology and distribution of Norwegian charophytes. *Norw. J. Bot.*, **21**, 31–52

Lange, W. 1971. Enhancement of algal growth in Cyanophyta-bacteria systems by carbonaceous compounds. *Can. J. Microbiol.*, **17**, 304–14

Lange-Bertalot, H. 1974. Das Phytoplankton im unteren Main unter dem Einfluss starker Abwasserbelastung. *Cour. Forsch-Inst. Senkenberg*, **12**, 1–88

Lange-Bertalot, H. 1978. Diatomeen-Differentialarten anstelle von Leit-formen: ein geeigneteres Kriterium der Gewässer belastung. *Arch. Hydrobiol. Suppl.*, **51**, 393–427

Larkum, A. W. D., Drew, E. A. & Crossett, R. N. 1967. The vertical distribution of attached marine algae in Malta. *J. Ecol.*, **55**, 361–71

Larrance, J. D. 1971. Primary production in the mid-subarctic Pacific region. 1966–68. *Fish. Bull.*, **69**, 593–613

Larson, D. W. 1978. Possible misestimates of lake primary productivity due to vertical migrations by dinoflagellates. *Arch. Hydrobiol.*, **81**, 296–303

Lasker, R. 1975. Field criteria for survival of anchovy larvae: The relationship between inshore chlorophyll maximum layers and successful first feeding. *Fish. Bull.*, **73**, 453–62

Latorella, A. H. & Vadas, R. L. 1973. Salinity adaptation by *Dunaliella tertiolecta*. I. Increases in carbonic anhydrase activity and evidence for a light-dependent Na⁺/H⁺ exchange. *J. Phycol.*, **9**, 273–7

Lawrence, J. M. & Dawes, C. J. 1969. Algal growth over the epidermis of Sea Urchin spines. *J. Phycol.*, **5**, 269

Lawson, G. W. 1957. Seasonal variation of intertidal zonation on the Gold Coast in relation to tidal factors. *J. Ecol.*, **45**, 831–60

Lawson, G. W. 1978. The distribution of seaweed floras in the tropical and subtropical Atlantic Ocean: a quantitative approach. *Bot. J. Linn. Soc.*, **76**, 177–93

Lawson, G. W. & John, D. M. 1977. The marine flora of the Cap Blanc peninsula: its distribution and affinities. *Bot. J. Linn. Soc.*, **75**, 99–118

Leach, J. H. 1970. Epibenthic algal production in an intertidal mud flat. *Limnol. Oceanogr.*, **15**, 514–21

Leadbeater, B. S. C. 1972. Fine-structural observations on some marine choanoflagellates from the coast of Norway. *J. mar. biol. Ass., U.K.*, **52**, 67–79

Leadbeater, B. S. C. 1974. Ultrastructural observations on nannoplankton collected from the coast of Jugoslavia and the Bay of Algiers. *J. mar. biol. Ass., U.K.*, **54**, 179–96

Lean, D. R. S. 1973. Movements of phosphorus between its biologically important forms in lake water. *J. Fish. Res. Bd. Can.*, **30**, 1525–36

Lebednik, P. A., Weinmann, F. C. & Norris, R. E. 1971. Spatial and seasonal distributions of marine algal communities at Amchitka Island, Alaska. *Bioscience*, **21**, 656–60

Lee, J. J., Crockett, L. J., Hagen, J. & Stone, R. J. 1974. The taxonomic identity and physiological ecology of *Chlamydomonas hedleyi* sp. nov., algal flagellate symbiont from the foraminifer *Archaias angulatus*. *Br. Phycol. J.*, **9**, 407–22

Lee, J. J. & Freudenthal, H. 1964. Neglected *Amoebas* in culture. *Nat. Hist. Mag. Am. Mus. nat. Hist.*, **54**–61

Lee, J. J., McEnery, M. E., Kennedy, E. M. & Rubin, H. 1975. A nutritional analysis of a sublittoral diatom assemblage epiphytic on *Enteromorpha* from a Long Island salt marsh. *J. Phycol.*, **11**, 14–49

Lee, J. J., McEnery, M. E., Shilo, M. & Reiss, Z. 1979. Isolation and cultivation of diatom symbionts from larger Foraminifera (Protozoa). *Nature, London*, **280**, 57–8

Lee, J. J. & Zucker, W. 1969. Algal flagellate symbiosis in the foraminifera *Archaias*. *J. Protozool.*, **16**, 71–80

Leedale, G. F. 1967. *Euglenoid flagellates*. 242 pp. Prentice Hall Inc., Englewood Cliffs, New Jersey

Leedale, G. F. 1969. Observations on endonuclear bacteria in Euglenoid flagellates. *Öst. Bot. Zeitsch.*, **116.**, 279–94

Lefevre, M. 1950. *Aphanizomenon gracile*, Lem. Cyanophyte dé favorable au zooplankton. *Ann. Stn. cent. Hydrobiol. appl.*, **3**, 205–8

Lehman, J. T. 1976. Ecological and nutritional studies on *Dinobryon* Ehrenb.: Seasonal periodicity and the phosphate toxicity problem. *Limnol. Oceanogr.*, **21**, 646–58

Lehman, J. T., Botkin, D. B. & Likens, G. E. 1975. The assumptions and rationales of a computer model of phytoplankton population dynamics. *Limnol. Oceanogr.*, **20**, 343–6

Leighton, D. L. 1971. Grazing activities of benthic invertebrates in southern California kelp beds. In *The Biology of Giant Kelp Beds (Macrocystis) in California*, ed. W. J. North. *Nova Hedw. Beih.*, **32**, 421–53

Levandowsky, M. 1972. Ecological niches of sympatric phytoplankton species. *Amer. Nat.*, **106**, 71–8

Levring, T. 1940. *Studien über die Algenvegetation von Blekinge, Südschweden.* 179 pp. Lund

Lewin, J. 1974. Blooms of surf-zone diatoms along the coast of the Olympic Peninsula, Washington. III. Changes in the species composition of the blooms since 1925. *Nova Hedw. Beih.*, **45**, 251–6

Lewin, J. & Hruby, T. 1973. Blooms of surf-zone diatoms along the coast of the Olympic Peninsula, Washington. II. A diel periodicity in buoyancy shown by the surf-zone diatom species, *Chaetoceres armatum* T. West. *Estuarine Coastal Mar. Sci.*, **1**, 101–5

Lewin, J. & Lewin, R. A. 1967. Culture and nutrition of some apochlorotic diatoms of the genus *Nitzschia*. *J. gen. Microbiol.*, **46**, 361–7

Lewin, J. & Mackas, D. 1972. Blooms of surf-zone diatoms along the coast of the Olympic Peninsula, Washington. I. Physiological investigations of *Chaetoceros armatum* and *Asterionella socialis* in laboratory cultures. *Mar. Biol.*, **16**, 171–81

Lewin, J. & Norris, R. E. 1970. Surf-zone diatoms of the coasts of Washington and New Zealand (*Chaetoceros armatum*, T. West and *Asterionella* spp.). *Phycologia*, **9**, 143–9

Lewin, R. A. 1966. Kultivo de zoochlorella apartigiata el spongo. *Sci. Rev. Int. Sci. Ass. Esperant.*, **17**, 33–6

Lewin, R. A. 1970. Toxin secretion and tail autotomy by irritated *Oxynoe panamensis* (Opisthobranchiata: Saccoglossa). *Pacif. Sci.*, **24**, 356–8

Lewin, R. A. 1972. Auxotrophy in marine littoral diatoms. In *Proc. 7th Int. Seaweed Symp.*, pp. 316–18. Sapporo, Japan. University of Tokyo

Lewin, R. A. 1975. A marine *Synechocystis* (Cyanophyta, Chroococcales) epizoic on ascidians. *Phycologia*, **14**, 153–60

Lewin, R. A. 1976. Prochlorophyta as a proposed new division of algae. *Nature, London*, **261**, 697–8

Lewin, R. A. 1977. *Prochloron*: type genus of the Prochlorophyta. *Phycologia*, **16**, 217

Lewis, J. B. 1974. The importance of light and food upon the early growth of the reef coral *Favia fragum* (Esper). *J. exp. mar. Biol. Ecol.*, **15**, 299–304

Lewis, J. R. 1961. The littoral zone on rocky shores – a biological or physical entity? *Oikos*, **12**, 280–301

Lewis, J. R. 1972. *The ecology of rocky shores*. English University Press, 323 pp.

Lewis, M. S. 1969. Sedimentary environments and unconsolidated sediments of the fringing coral reefs of Mahé, Seychelles. *Mar. Geol.*, **7**, 95–127

Lewis, W. M., Jr 1977. Net growth rate through time as an indicator of ecological similarity among phytoplankton species. *Ecology*, **58**, 149–57

Lewis, W. M. 1978. A compositional, phytogeographical and elementary structural analysis of the phytoplankton in a tropical lake. Lake Lanao, Philippines. *J. Ecol.*, **66**, 213–26

Lex, M., Silvester, W. B. & Stewart, W. D. P. 1972. Photorespiration and nitrogenase activity in the blue-green alga, *Anabaena cylindrica*. *Proc. Roy. Soc. London*, **B, 180**, 87–102

Liddle, L. B. 1971. Development of gametophyte and sporophyte populations of *Padina sanctae-crucis* Börg. in the field and laboratory. In *Proc. 7th Int. Seaweed Symp., Sapporo, Japan*, 80–2 University of Tokyo

Liddle, L. B. 1975. The effect of intertidal stress on *Padina sanctae-crucis* (Phaeophyta). *J. Phycol.*, **11**, 327–30

Lin, C. K. & Blum, J. L. 1977. Recent invasion of a red algae (*Bangia atropurpurea*) in Lake Michigan. *J. Fish. Res. Bd. Can.*, **34**, 2413–16

Linskens, H. F. 1963a. Oberflächenspannung an marinen Algen. *Proc. K. ned. Akad. Wet.*, sect. C., **66**, 205–17

Linskens, H. F. 1963b. Beitrag zur Frage der Beziehung zwischen Epiphyt und Basiphyt bei marinen algen. *Publ. Staz. Zool. Napoli*, **3**, 274–93

Linskens, H. F. 1966. Adhäsion von Fortpflanzungszellen benthontische Algen. *Planta*, **68**, 99–110

Lipkin, Y. 1972. Marine algal and sea-grass flora of the Suez Canal. *Israel J. Zool.*, **21**, 405–46

Lisitzin, A. P. 1971a. Distribution of siliceous microfossils in suspension and in bottom deposits. In *The Micropaleontology of Oceans*, ed. B. M. Funnel & W. R. Riedel, pp. 173–95

Lisitzin, A. P. 1971b. Distribution of carbonate microfossils in suspension and in bottom sediments. In *The Micropalaeontology of Oceans*, ed. B. M. Funnell & W. R. Riedel, pp. 197–218

Lisitzin, A. P. 1972. Sedimentation in the Worlds Oceans. *Soc. Econ. Paleo. Min. Spec. Pub.*, **17**, 218 pp.

Lisitzin, A. P. 1962. Distribution and composition of suspended materials in seas and oceans. (In Russian.) In *Recent sediments of seas and oceans*, USSR Acad. Sci. Comm. Sediments Div. Geol. Geogr. Sci., 175–231

Littler, M. M. 1971. Standing stock measurements of crustose coralline algae (Rhodophyta) and other saxicolous organisms. *J. exp. mar. Biol. Ecol.*, **6**, 91–9

Littler, M. M. 1973a. The distribution, abundance and communities of deep water Hawaiian crustose corallinaceae (Rhodophyta, Cryptonemiales). *Pacif. Sci.*, **27**, 281–9

Littler, M. M. 1973*b*. The population and community structure of Hawaiian fringing-reef crustose Corallinaceae (Rhodophyta, Cryptonemiales). *J. exp. mar. Biol. Ecol.*, **11**, 103–20

Littler, M. M. 1973*c*. The productivity of Hawaiian fringing-reef crustose Corallinaceae and an experimental evaluation of production methodology. *Limnol. Oceanogr.*, **18**, 946–52

Littler, M. M. 1972. The crustose Corallinaceae. *Oceanogr. Mar. Biol. Ann. Rev.*, **10**, 311–47

Littler, M. M. & Doty, M. S. 1975. Ecological components structuring the seaward edge of tropical Pacific reefs: The distribution, communities and productivity of *Porolithon*. *J. Ecol.*, **63**, 117–29

Littler, M. M. & Murray, S. N. 1974*a*. The primary productivity of marine macrophytes from a rocky intertidal community. *Mar. Biol.*, **27**, 131–5

Littler, M. M. & Murray, S. N. 1974*b*. Primary productivity of macrophytes. In *Biological features of intertidal communities near the U.S. Navy sewage outfall, Wilson Cove, San Clemente Island, California*, ed. S. N. Murray & M. M. Littler, pp. 67–85. U.S. Navy NUC. TP 396

Littler, M. M. & Murray, S. N. 1975. Impact of sewage on the distribution, abundance and community structure of rocky intertidal macro-organisms. *Mar. Biol.*, **30**, 277–91

Livingston, D. A. 1963. Chemical composition of rivers and lakes. *Prof. Pap. U.S. geol. Surv.*, no. 440, G1–G64

Loeblich, A. R., III, Loeblich, L. A., Tappan, H. & Loeblich, A. R., Jr 1968. Annotated index of fossil and recent silicoflagellates and ebridians with descriptions and illustrations of validly proposed taxa. *Geol. Soc. Am. Mem.*, no. 106, 319 pp.

Loeblich, A. R., III & Sherlev, J. L. 1979. Observations on the theca of the motile phase of free-living and symbiotic isolates of *Zooxanthella microadriaticum* (Freudenthal) comb. nov. *J. mar. biol. Ass., U.K.*, **59**, 195–205

Logan, B. W. 1961. *Cryptozoon* and associate stromatolites from the Recent, Shark Bay, Western Australia. *J. Geol.*, **69**, 517–33

Logan, B. W., Rezak, R. & Ginsburg, R. N. 1964. Classification and environmental significance of algal stromatolites. *J. Geol.*, **72**, 68–83

Lohman, K. E. & Andrews, G. W., 1968. Late Eocene non-marine diatoms from the Beaver Divide Area, Freemont County, Wyoming. *Prof. Pap. U.S. geol. Surv.*, **593**, 1–26

Lombard, E. H. & Capon, B. 1971. Observations on the tide pool ecology and behaviour of *Peridinium gregarium*. *J. Phycol.*, **7**, 188–94

Longhurst, A. R., Lorenzen, G. J. & Thomas, W. H. 1967. The role of pelagic crabs in the grazing of phytoplankton off Baja California. *Ecology*, **48**, 190–200

Lorenzen, C. J. 1971. Continuity in the distribution of surface chlorophyll. *J. Cons. perm. int. Explor. Mer.*, **34**, 18–23

Lovrić, A. Z. 1972. Signification de l'isolation geografique et ecologique dans la différentiation de la flore littorale de l'Adriatique du Nord. In *5th European Mar. Biol. Symp., Venice.* Battaglia, **B**, 53–9

Lowenstam, H. A. 1955. Aragonite needles secreted by algae and some sedimentary implications. *J. Sed. Petrol.*, **25**, 270–2

Lowenstam, H. A. & Epstein, S. 1957. On the origin of sedimentary aragonite needles of the Great Bahama Bank. *J. Geol.*, **65**, 364–75

Lubchenco, J. & Menge, B. A. 1978. Community development and persistence in a low rocky intertidal zone. *Ecol. Monogr.*, **48**, 67–94

Lucas, C. E. & MacNae, N. 1940. Phytoplankton in the North Sea, 1938–39. Part I. Diatoms. *Hull Bull. mar. Ecol.*, **2**, 19–46

Lucas, W. J. & Smith, F. A. 1973. The formation of alkaline and acid regions at the surface of *Chara corallina* cells. *J. exp. Bot.*, **24**, 1–14

Lund, J. W. G. 1942. The marginal algae of certain ponds with special reference to the bottom deposits. *J. Ecol.*, **30**, 245–83

Lund, J. W. G. 1945. Observations on soil algae. I. The ecology, size and taxonomy of British soil diatoms. *New Phytol.*, **44**, 196–219; **45**, 56–110

Lund, J. W. G. 1947. Observations on soil algae. II. Notes on groups other than diatoms. *New Phytol.*, **46**, 35–60

Lund, J. W. G. 1949. Studies on *Asterionella*. I. The origin and nature of the cells producing seasonal maxima. *J. Ecol.*, **37**, 389–419

Lund, J. W. G. 1950*a*. Studies on *Asterionella formosa* Hass. II. Nutrient depletion and the spring maxima. I. Observations on Windermere, Esthwaite Water and Blelham Tarn. *J. Ecol.*, **38**, 1–14

Lund, J. W. G. 1950*b*. Studies on *Asterionella formosa* Hass. II. Nutrient depletion and the spring maximum. II. Discussion. *J. Ecol.*, **38**, 15–35

Lund, J. W. G. 1954. The seasonal cycle of the plankton diatom, *Melosira italica* (Ehr.) Kütz, subsp. *subarctica*, O. Müll. *J. Ecol.*, **42**, 151–79

Lund, J. W. G. 1955. Further observations on the seasonal cycle of *Melosira italica* (Ehr.) Kütz. subsp. *subarctica*, O. Müll. *J. Ecol.*, **43**, 90–102

Lund, J. W. G. 1957. Fungal diseases of plankton algae. In *Biological aspects of the transmission of Disease*, by C. Horton-Smith. London

Lund, J. W. G. 1959. Buoyancy in relation to the ecology of the freshwater phytoplankton. *Brit. Phyc. Bull.*, no. 7, 1–17

Lund, J. W. G. 1964. Primary production and periodicity of phytoplankton. *Verh. Int. Ver. Limnol.*, **15**, 37–56

Lund, J. W. G. 1965. The ecology of the freshwater phytoplankton. *Biol. Rev.*, **40**, 231–93

Lund, J. W. G. 1967. Soil Algae. In *Soil Biology*, ed. A. Burges & F. Raw, pp. 129–48. Academic Press

Lund, J. W. G. 1969. Phytoplankton. In *Eutrophication: Causes, consequences, correctives*, pp. 306–30. Nat. Acad. Sci., Washington D.C.

Lund, J. W. G. 1971. The seasonal periodicity of three planktonic desmids in Windermere. *Mitt. Internat. Ver. Limnol.*, **19**, 3–25

Lund, J. W. G. 1972. Eutrophication. *Proc. Roy. Soc. London*, **B**, **180**, 371–82

Lund, J. W. G. 1975. The uses of large experimental tubes in lakes. In *The effects of storage on water quality*. *Wat. Res. Centre Symp.*, 12 pp.

Lund, J. W. G., Jaworski, G. H. M. & Butterwick, C. 1975. Algal bioassay of water from Blelham Tarn, English Lake

District and the growth of planktonic diatoms. *Arch. Hydrobiol. Suppl.*, **49**, 49–69

Lund, J. W. G. & Talling, J. F. 1957. Botanical limnological methods with special reference to the algae. *Bot. Rev.*, **23**, 489–583

Lüning, K. 1970. Tauschuntersuchungen zur Vertikalverteilung der sublitoralen Helgolände Algenvegetation. *Helgoländer wiss. Meeresunt.*, **21**, 271–91

Lüning, K. 1971. Seasonal growth of *Laminaria hyperborea* under recorded underwater light conditions near Helgoland. In *4th European Mar. Biol. Symp. Venice*, ed. D. J. Crisp, pp. 347–61

Lüning, K., Chapman, A. R. O. & Mann, K. H. 1978. Crossing experiments in the non-digitate complex of *Laminaria* from both sides of the Atlantic. *J. Phycol.*, **17**, 293–8

Lüning, K. & Neushul, M. 1978. Light and temperature demands for growth and reproduction of Laminarian gametophytes in southern and central California. *Mar. Biol.*, **45**, 297–309

Luther, G. 1976. Bewuchsuntersuchungen auf Natursteinsubstraten im Gezeitenbereich des Nordsylter Wattenmeeres: Algen. *Helgoländer wiss. Meeresunt.*, **28**, 318–51

Lylis, J. C. & Trainor, F. R. 1973. The heterotrophic capabilities of *Cyclotella meneghiniana*. *J. Phycol.*, **9**, 365–9

McAlice, B. J. 1970. Observations in the small-scale distribution of estuarine phytoplankton. *Mar. Biol.*, **7**, 100–11

McAllister, C. D., Parsons, T. R. & Strickland, J. D. H. 1960. Primary productivity at Station 'P' in the northeast Pacific Ocean. *J. cons. perm. int. Explor. Mer.*, **25**, 240–59

McAllister, C. D., Parsons, T. R., Stephens, B. & Strickland, J. D. H. 1961. Measurements of primary production in coastal sea water using a large volume plastic sphere. *Limnol. Oceanogr.*, **6**, 237–58

McAllister, H. A., Norton, T. A. & Conway, E. 1967. A preliminary list of sublittoral marine algae from West of Scotland. *Br. phycol. Bull.*, **3**, 175–84

McAlister, W. B. 1971. Oceanography in the vicinity of Amchitka Island, Alaska. *Bioscience*, **21**, 646–51

McCarthy, J. J. 1974. The uptake of urea by marine phytoplankton. *J. Phycol.*, **8**, 216–22

McCarthy, J. J., Rowland Taylor, W. & Loftus, M. E. 1978. Significance of nannoplankton in the Chesapeake Bay estuary and problems associated with the measurement of nannoplankton productivity. *Mar. Biol.*, **24**, 7–16

McCarthy, J. J., Rowland Taylor, W. & Taft, J. L. 1977. Nitrogenous nutrition of the plankton in the Chesapeake Bay. I. Nutrient availability and phytoplankton preferences. *Limnol. Oceanogr.*, **22**, 996–1011

MacCaull, W. A. & Platt, T. 1977. Diel variations in the photosynthetic parameters of coastal marine phytoplankton. *Limnol. Oceanogr.*, **22**, 723–31

McCollum, D. W. 1975. Diatom stratigraphy of the southern ocean. *Initial Rep. Deep Sea Drilling Project*, **27**, 515–71

McConnell, W. J. & Sigler, W. F. 1959. Chlorophyll and productivity in mountain rivers. *Limnol. Oceanogr.*, **4**, 335–51

McGowan, J. A. 1971. Oceanic biogeography of the Pacific. I. In *The Micropalaeontology of Oceans*, ed. B. M. Funnell & W. R. Riedel, pp. 3–74

McIntire, C. D. 1968. Structural characteristics of benthic algal communities in laboratory streams. *Ecology*, **49**, 520–37

McIntire, C. D. & Phinney, H. K. 1965. Laboratory studies of periphyton production and community metabolism in arctic environments. *Ecol. Monogr.*, **35**, 237–58

McIntyre, A. & Bé, A. W. H. 1967. Modern Coccolithophoridae of the Atlantic Ocean – I. Placoliths and Cyrtoliths. *Deep Sea Res.*, **14**, 561–97

McIntyre, A. & McIntyre, R. 1971. Coccolith concentrations and differential solution in oceanic sediments. In *The Micropalaeontology of Oceans*, ed. B. M. Funnell & W. R. Riedel, pp. 253–61

McIntyre, A. D., Munro, A. L. S. & Steele, J. H. 1970. Energy flow in a sand ecosystem. In *Marine Food Chains*, ed. J. H. Steele, pp. 19–31. Oliver & Boyd, Edinburgh

MacIsaac, J. & Dugdale, R. C. 1969. The kinetics of nitrate and ammonia uptake by natural populations of marine phytoplankton. *Deep Sea Res.*, **16**, 45–57

McKinley, K. R. 1977. Light mediated uptake of ^3H glucose in a small hard-water lake. *Ecology*, **58**, 1356–65

McLachlan, J. & Craigie, J. G. 1964. Algal inhibition by yellow ultra violet absorbing substances from *Fucus vesiculosus*. *Can. J. Bot.*, **42**, 287–92

McLachlan, J. & Edelstein, T. 1970–1. Investigations of the marine algae of Nova Scotia. IX. A preliminary survey of the flora of Bras d'Or Lake, Cape Breton Island. *Proc. Nova Scotian Inst. Sci.* **27**, 11–22

McLean, D. M. 1978. Land floras: The major late Phanerozoic atmospheric carbon dioxide/oxygen control. *Science*, **200**, 1060–1

McLeod, G. C. & Rhee, C. 1970. Physical measurements of our natural resources. In *Symp. Microbiological aspects of seawater pollution*, Kingston, Rhode Island, pp. 135–40

McRoy, C. P. & Goering, J. J. 1974. Nutrient transfer between the sea-grass *Zostera marina* and its epiphytes. *Nature, London*, **248**, 173–4

McRoy, C. P., Goering, J. J. & Shiels, W. E. 1972. Studies of primary production in the Eastern Bering Sea. In *Biological Oceanography of the northern north Pacific Ocean*, ed. A. Y. Takenouti, pp. 199–216. Tokyo

Machlis, L. 1973. The effect of bacteria on the growth and reproduction of *Oedogonium cardiacum*. *J. Phycol.*, **9**, 342–4

Mackereth, F. J. H. 1965. Chemical investigation of lake sediments and their interpretation. *Proc. Roy. Soc. London*, **B**, **161**, 295–309

Maekawa, F. 1960. A new attempt in phylogenetic classification of plant kingdom. *J. Fac. Sci. Univ. Tokyo Bot.*, **7**, 543–69

Maeda, O. & Ichimura, S. 1973. On the high density of a phytoplankton population found in a lake under ice. *Int. Revue. ges Hydrobiol.*, **58**, 673–85

Mague, T. H. & Holm-Hansen, O. 1975. Nitrogen fixation on a coral reef. *Phycologia*, **14**, 87–92

Mague, T. H., Mague, F. C. & Holm-Hansen, O. 1977. Physiology and chemical composition of nitrogen-fixing phytoplankton in the Central North Pacific Ocean. *Mar. Biol.*, **41**, 213–27

Mague, T. H., Weare, N. M. & Holm-Hansen, O. 1974.

Nitrogen fixation in the North Pacific Ocean. *Mar. Biol.*, **24**, 109–19

Maguire, B., Jr 1963. The passive dispersal of small aquatic organisms and their colonisation of isolated bodies of water. *Ecol. Mongr.*, **33**, 161–85

Maguire, B., Jr 1973. Niche response and the analytical potentials of its relationship to the habitat. *Amer. Nat.*, **107**, 213–46

Maguire; B., Jr & Neill, W. E. 1971. Species and individual productivity in phytoplankton communities. *Ecology*, **52**, 903–7

Main, S. P. & McIntire, C. D. 1974. The distribution of epiphytic diatoms in Yaquina estuary, Oregon. *Bot. Mar.*, **17**, 88–99

Malanchuk, J. L. & Gruendling, G. K. 1973. Toxicity of lead nitrate to algae. *Water, air and soil pollut.*, **2**, 181–90

Malinowski, K. C. & Ramus, J. 1973. Growth of the green alga *Codium fragile* in a Connecticut estuary. *J. Phycol.*, **9**, 102–10

Malone, T. C. 1971a. The relative importance of nannoplankton and net plankton as primary producers in tropical oceanic and neritic phytoplankton communities. *Limnol. Oceanogr.*, **16**, 633–9

Malone, T. C. 1971b. Diurnal rhythms in net plankton and nannoplankton assimilation ratios. *Mar. Biol.*, **10**, 285–9

Malone, T. C. 1977. Environmental regulation of phytoplankton productivity in the Lower Hudson estuary. *Est. & Coastal Mar. Sci.* **5**, pp. 157–71.

Malone, T. C., Garside, C., Anderson, R. & Roels, O. A. 1973. The possible occurrence of photosynthetic microorganisms in deep-sea sediments of the North Atlantic. *J. Phycol.*, **9**, 482–8

Maloney, T. E. 1959. Utilization of sugars in spent sulfite liquor by a green alga, *Chlorococcum macrostigmatum*. *Sewage ind. wastes*, 1395–400

Maloney, T. E., Miller, W. E. & Shiroyana, T. 1972. Algal responses to nutrient additions in natural waters. I. Laboratory assays. In *Nutrients and eutrophication: The limiting nutrient controversy*. Limnol. Oceanogr. Symp. I, 134–40

Mandelli, E. F. & Burkholder, P. R. 1966. Primary productivity in the Gerlache and Bransfield Straits of Antarctica. *J. Mar. Res.*, **24**, 15–27

Manea, V. & Skolka, H. 1961. Observatii asupra microfitobentosului marin, în dreptul grindului chituc. *Communle Acad. Rep. pop. Rom.*, **11**, 535–8

Manguin, E. 1960. Les Diatonées de la Terre Adélie Campagne du Commandant Charot, 1949–50. *Ann Sci. nat. Bot.*, **12**, 223–30

Mann, J. E. & Schlichting, H. E. 1967. Benthic algae of selected thermal springs in Yellowstone National Park. *Trans. Am. Micros. Soc.*, **86**, 2–9

Mann, K. H. 1972a. Introductory remarks. In *Detritus and its role in aquatic ecosystems. Proc. IBP–UNESCO Symp.*, pp. 13–16. Pallanza, Italy

Mann, K. H. 1972b. Macrophyte production and detritus food chains in coastal waters. In *Detritus and its role in aquatic ecosystems. Proc. IBP–UNESCO Symp.*, pp. 353–83. Pallanza, Italy

Mann, K. H. 1972c. Ecological energetics of the seaweed zone in a marine bay on the Atlantic Coast of Canada. I. Zonation and biomass of seaweeds. *Mar. Biol.*, **12**, 1–10

Mann, K. H. 1972d. Ecological energetics of the seaweed zone in a marine bay on the Atlantic coast of Canada. II. Productivity of the seaweeds. *Mar. Biol.*, **14**, 199–209

Mann, K. H. 1973. Seaweeds: their productivity and strategy for growth. *Science*, **182**, 975–81

Manum, S. 1962. Some new species of *Deflandrea* and their probable affinity with *Peridinium*. *Norsk Polarinstitut Årbok*, 55–67

Manzi, J. J., Stofan, P. E. & Dupuy, J. L. 1977. Spatial heterogeneity of phytoplankton populations in estuarine surface microlayers. *Mar. Biol.*, **41**, 29–38

Margalef, R. 1948. A new limnological method for the investigation of thin-layered epilithic communities. *Trans. Am. micr. Soc.*, **67**, 153–4

Margalef, R. 1949. A new limnological method for the investigation of thin-layered epilithic communities. *Hydrobiol.*, **1**, 215–16

Margalef, R. 1951. Ciclo anual del fitoplancton marino en la costa NE de la Peninsula Ibérica. *Publ. Inst. Biol. appl.*, *Barcelona*, **9**, 83–118

Margalef, R. 1960. Ideas for a synthetic approach to the ecology of running waters. *Int. Rev. ges Hydrobiol.*, **45**, 133–53

Margalef, R. 1961. Distributión ecológica y geográfica de las expecies del fitoplancton marino. *Invest. Pesq.*, **19**, 81–101

Margalef, R. 1963. On certain unifying principles in ecology. *Amer. Nat.*, **897**, 357–73

Margalef, R. 1969. Diversity and stability: A practical proposal and a model of interdependence. *Brookhaven Symp. Biol.*, **22**, 25–37

Margalef, R. 1971. The pelagic ecosystems of the Caribbean Sea. In *UNESCO Symposium on investigations and resources of the Caribbean Sea and adjacent regions*. WCNA, pp. 483–98

Margalef, R. 1973. Some critical remarks on the usual approaches to ecological modelling. *Invest. Pesq.*, **37** (3), 621–40

Margalef, R. & Estrada, M. 1971. Simple approaches to a pattern analysis of phytoplankton. *Invest. Pesq.*, **35**, 269–97

Margulis, L. 1974. The classification and evolution of prokaryotes and eukaryotes. In *Handbook of Genetics, vol. I*, ed. R. C. King. Plenum Press, 1–41

Margulis, L. 1976. The theme (mitotic cell division) and the variations (Protists): Implications for higher taxa. *Taxon*, **24**, 391–403

Marker, A. F. H. 1972. The use of acetone and methanol in the estimation of chlorophyll in the presence of phaeophytin. *Freshwater Biol.*, **2**, 361–85

Marker, A. F. H. 1976a. The benthic algae of some streams in southern England. I. Biomass of the epilithon in some small streams. *J. Ecol.*, **64**, 343–58

Marker, A. F. H. 1976b. The benthic algae of the English chalk-stream. II. The primary production of the epilithon in a small chalk-stream. *J. Ecol.*, **64**, 359–73

Markham, J. W. 1969. Vertical distribution of epiphytes on the stripe of *Nereocystis leutkeana* (Mertens) Postels & Ruprecht. *Syesis*, **2**, 227–40

Markham, J. W. 1973. Observations on the ecology of

Laminaria sinclairii on three northern Oregon beaches. *J. Phycol.*, **9**, 336–41

Marra, J. 1978. Phytoplankton photosynthetic response to vertical movement in a mixed layer. *Mar. Biol.*, **46**, 203–8

Marré, E. & Seruttaz, O. 1959. Sul meccanismo di adattamento a conditione osmotiche esteme in *Dunaliella salina*. II. Rapporto fra conzentrazione del mezzo esterno e composizione del succo cellulare. *Atti. Accad. naz. Lincei Rc.*, **26**, 272–8

Marsh, J. A. 1970. Primary productivity of reef-building calcareous red algae. *Ecology*, **51**, 255–63

Marshall, N. 1967. Some characteristics of the epibenthic environment of tidal shores. *Chesapeake Sci.*, **8**, 153–69

Marshall, N., Oviatt, C. A. & Skauen, D. M. 1971. Productivity of the benthic microflora of shoal estuarine environments in southern New England. *Int. Rev. ges Hydrobiol.*, **56**, 947–56

Marshall, P. 1931. Coral reefs – rough-water, calm-water types. *Rep. Gr. Barrier Reef Comm.*, **3**, 64–72

Marshall, S. M. & Orr, A. P. 1928. The photosynthesis of diatom cultures in the sea. *J. mar. biol. Ass., U.K.*, **15**, 321–60

Marshall, S. M. & Orr, A. P. 1930. A study of the spring diatom increase in Loch Striven. *J. mar. biol. Ass., U.K.*, **16**, 853–78

Martin, H. & Wilczewski, N. 1972. Algen-Stromatolithen aus der Etosha-Pfanne, Südwest-Afrikas. *Neues. Jb. Geol. Paläont. Abh.*, **12**, 720–6

Martin, J. H. 1970. Phytoplankton–Zooplankton relationships in Narragansett Bay. IV. The seasonal importance of grazing. *Limnol. Oceanogr.*, **15**, 413–18

Martin, M. T. 1969. A review of life-histories in the Nemaliales and some allied genera. *Br. phycol. J.* **4**, 145–56

Martini, E. 1971. The occurrence of Pre-Quaternary calcareous nannoplankton in the oceans. In *The Micropalaeontology of oceans*, ed. B. M. Funnell & W. R. Riedel, pp. 535–44. Cambridge University Press

Mason, C. F. & Bryant, R. J. 1975. Periphyton production and grazing by chironomids in Alderfen Broad, Norfolk. *Freshwater Biol.*, **5**, 271–7

Matheke, G. E. M. & Horner, R. 1974. Primary productivity of the benthic microalgae in the Chukchi Sea near Barrow, Alaska. *J. Fish. Res. Bd. Can.*, **31**, 1779–86

Mathieson, A. C. & Burns, R. L. 1971. Ecological Studies of economic red algae. I. Photosynthesis and respiration of *Chondrus crispus* (Stackhouse) and *Gigartina stellata* (Stackhouse). *J. exp. mar. Biol. Ecol.*, **1**, 197–206

Mathieson, A. C. & Norall, T. L. 1975. Physiological studies of subtidal red algae. *J. exp. mar. Biol. Ecol.*, **20**, 237–47

Mathis, B. J. 1972. Chlorophyll and the Margalef pigment ratio in a mountain lake. *Am. Midl. Nat.*, **88**, 232–4

Mattern, H. 1970. Beobachtungen über die Algenflora im Uferbereich des Bodensees. *Arch. Hydrobiol. Suppl.*, **37**, 1–163

May, V. 1976. Changing dominance of an algal species (*Caulerpa filiformis*) (Suhr.) Hering. *Telopea*, **1**, 136–8

Maynard, N. G. 1968. Significance of air-borne algae. *Z. allg. Mikrobiol.*, **8**, 225–6

Maxwell, W. G. H. 1968. *Atlas of the Great Barrier Reef*. Elsevier Publ. Co., Amsterdam

Maxwell, W. G. H. 1972. The Great Barrier Reef – Past, present and future. *Queensland Nat.* **20**, 65–78.

Meadows, P. S. & Anderson, J. G. 1968. Micro-organisms attached to marine sand grains. *J. mar. biol. Ass., U.K.*, **48**, 161–75

Meeks, R. L. & Peterle, T. J. 1967. R. F. Project 1794. Report No. COO-1358. Ohio State University, Columbus

Meguro, H. 1962. Plankton ice in the Arctic Ocean. *Antarctic Res. Tokyo*, **14**, 1192

Meguro, H., Ito, K. & Fukushima, H. 1967. Ice flora (bottom ecological conditions of their growth in sea ice in the Arctic Ocean. *Science*, **152**, 1089–90

Meguro, H., Ito, K. & Fukushima, H. 1967. Ice flora (bottom type). A mechanism of primary production in polar seas and the growth of diatoms on sea ice. *Arctic*, **20**, 114–33

Melack, J. M. & Kilham, P. 1974. Photosynthetic rates of phytoplankton in East African alkaline, saline lakes. *Limnol. Oceanogr.*, **19**, 743–55

Menzel, D. W. 1959. Utilization of algae for growth by the Angel fish, *Holocanthus bermudensis*. *J. Cons. perm. int. Explor. Mer.*, **22**, 308–12

Menzel, D. W. 1974. Primary productivity, dissolved and particulate organic matter and the sites of oxidation of organic matter. In *The Sea, Vol. 5, Marine chemistry*, ed. E. D. Goldberg, pp. 659–78. Interscience

Menzel, D. W. & Ryther, J. H. 1960. The annual cycle of primary production in the Sargasso Sea off Bermuda. *Deep Sea Res.*, **6**, 351–67

Menzel, D. W. & Ryther, J. H. 1961a. Nutrients limiting the production of phytoplankton in the Sargasso Sea, with special reference to iron. *Deep Sea Res.*, **7**, 271–81

Menzel, D. W. & Ryther, J. H. 1961b. Annual variations in primary production of the Sargasso Sea off Bermuda. *Deep Sea Res.* **7**, 282–8

Menzel, D. W. & Spaeth, J. P. 1962. Occurrence of vitamin B_{12} in the Sargasso Sea. *Limnol. Oceanogr.*, **7**, 151–4

Meriläinen, J. 1971. The recent sedimentation of diatom frustules in four meromictic lakes. *Ann. Bot. Fennici*, **8**, 160–76

Michanek, G. 1967. Quantitative sampling of benthic organisms by diving on the Swedish west coast. *Helgoländer wiss. meeresunt.*, **15**, 453–9

Middlebrooks, E. J., Falkenborg, D. H. & Maloney, T. E. 1976. *Biostimulation nutrient assessment*. 390 pp. Ann Arbor Sci.

Mikkelsen, N. 1977. On the origin of *Ethmodiscus* ooze. *Mar. Micropaleont*, **2**, 35–46

Mikkelsen, N. 1978. Preservation of diatoms in glacial to Holocene deep-sea sediments of the equatorial Pacific. *Geology*, **6**, 553–55

Millbank, J. W. 1975. Aspects of nitrogen metabolism. In *Progress and problems in Lichenology*, ed. Brown, D. H., Hawksworth, D. & Bailey, R. H. Academic Press, London

Miller, M. C. & Reed, J. P. 1975. Benthic metabolism of arctic coastal ponds, Barrow, Alaska. *Verh. Int. Ver. Limnol.*, **19**, 459–65

Miller, W. T., Montgomery, R. T. & Collier, A. W. 1977. A taxonomic survey of the diatoms associated with Florida Keys coral reefs. In *Proc. 3rd Int. Coral Reef. Symp. Miami, vol. I. Biology*, pp. 349–55

Milliger, L. E. & Schlichting, H. E. 1968. The passive

dispersal of viable algae and protozoa by an aquatic beetle. *Trans. Am. Micros. Soc.*, **87**, 443–8

Milliman, J. D. 1977. Effects of arid climate and upwelling upon the sedimentation regime off southern Spanish Sahara. *Deep Sea Res.*, **24**, 95–103

Minas, J. H. 1970. La distribution de l'oxygène en relation avec la production primaire en Méditerranée Nord-Occidentale. *Mar. Biol.*, **7**, 181–204

Mitra, A. K. 1951. The algal flora of certain Indian soils. *Ind. J. Agric. Sci.*, **21**, 357–73

Moe, R. L. & Silva, P. C. 1977. Antarctic marine flora: Uniquely devoid of kelps. *Science*, **196**, 1206–8

Moikeha, S. N., Chu, G. W. & Berger, L. R. 1971. Dermatitis-producing alga *Lyngbya majuscula* Gomont in Hawaii. I. Isolation and chemical characterisation of the toxic factor. II. Biological properties of the toxic factor. *J. Phycol.*, **7**, 4–7 and 8–13

Monty, C. 1967. Distribution and structure of some stromatolithic algal mats, eastern Andros Island, Bahamas. *Annls Soc. géol. Belg. Bull.*, **90**, 55–100

Monty, C. L. V. 1965. Recent algal stromatolites in the windward Lagoon, Andros Island, Bahamas. *Annls Soc. géol. Belg. Bull.*, **88**, 269–76

Moore, J. W. 1974a. The benthic algae of southern Baffin Island. I. Epipelic communities in rivers. *J. Phycol.*, **10**, 50–7

Moore, J. W. 1974b. Benthic algae of southern Baffin Island. II. The epipelic communities in temporary ponds. *J. Ecol.*, **62**, 809–19

Moore, J. W. 1974c. Benthic algae of southern Baffin Island. III. Epilithic and epiphytic communities. *J. Phycol.*, **10**, 456–62

Moore, J. W. 1976. Seasonal succession of algae in rivers. I. Examples from the Avon, a large slow-flowing river. *J. Phycol.*, **12**, 342–9

Moore, J. W. 1977. Seasonal succession of algae in a eutrophic stream in southern England. *Hydrobiol.*, **53**, 181–92

Moore, J. W. & Beamish, F. W. H. 1973. Food of larval sea lamprey (*Petromyzon marinus*) and American brook lamprey (*Lampetra lamottei*). *J. Fish. Res. Bd. Can.*, **30**, 7–15

Moore, P. G. 1974. The Kelp fauna of northeast Britain. III. Qualitative and quantitative ordinations and the utility of a multivariate approach. *J. exp. mar. Biol. Ecol.*, **16**, 277–300

Morgan, R. P., II & Stross, R. G. 1969. Destruction of phytoplankton in the cooling water supply of a stream electric station. *Chesapeake Sci.*, **10**, 165–71

Moriarty, D. J. W. 1973. The physiology of digestion of blue-green algae in the cichlid fish *Tilapia nilotica*. *J. Zool.*, **171**, 25–39

Moriarty, D. J. W. & Moriarty, C. M. 1973. The assimilation of carbon from phytoplankton by two herbivorous fishes: *Tilapia nilotica* and *Haplochromis nigripinnis*. *J. Zool.*, **171**, 41–55

Morris, I., Yentsch, C. M. & Yentsch, C. S. 1971a. Relationship between light, carbon dioxide fixation and dark carbon dioxide fixation by marine algae. *Limnol. Oceanogr.*, **16**, 854–8

Morris, I., Yentsch, C. M. & Yentsch, C. S. 1971b. The physiological state with respect to nitrogen of phytoplankton from low nutrient subtropical water as measured by the effect of ammonium ion on dark carbon dioxide fixation. *Limnol. Oceanogr.*, **16**, 859–68

Morris, O. P. & Russel, G. 1973. Effect of chelation on the toxicity of copper. *Mar. Pollut. Bull.*, **4**, 159–60

Mortimer, C. H. 1969. Physical factors with bearing on eutrophication in lakes in general and large lakes in particular. In *Eutrophication: Causes, consequences, correctives*, pp. 340–68. Nat. Acad. Sci., Washington D.C.

Mortimer, C. H. 1974. Lake hydrodynamics. *Mitt. Int. Ver. Limnol.*, **20**, 124–97

Moser, J. L. & Brock, T. D. 1971. Effect of wide temperature fluctuation on the blue-green algae of Bead Geyser, Yellowstone National Park. *Limnol. Oceanogr.*, **16**, 640–5

Moss, B. 1968. The chlorophyll *a* content of some benthic algal communities. *Arch. Hydrobiol.*, **65**, 51–62

Moss, B. 1969a. Limitation of algal growth in some Central African waters. *Limnol. Oceanogr.*, **14**, 591–601

Moss, B. 1969b. Vertical heterogeneity in the water column of Abbot's Pond. II. The influence of physical and chemical conditions on the spatial and temporal distribution of the phytoplankton and of a community of epipelic algae. *J. Ecol.*, **57**, 397–414

Moss, B. 1972a. The influence of environmental factors on the distribution of freshwater algae: An experimental study. I. Introduction and the influence of calcium concentration. *J. Ecol.*, **60**, 917–37

Moss, B. 1972b. Studies on Gull Lake, Michigan. I. Seasonal and depth distribution of phytoplankton. *Freshwater Biol.*, **2**, 289–307

Moss, B. 1973a. Diversity in freshwater phytoplankton. *Am. Midl. Nat.*, **90**, 341–55

Moss, B. 1973b. The influence of environmental factors on the distribution of freshwater algae – an experimental study. III. Effects of temperature, vitamin requirements and inorganic nitrogen compounds on growth. *J. Ecol.*, **61**, 179–92

Moss, B. 1973c. The influence of environmental factors on the distribution of freshwater algae: An experimental study. IV. Growth of test species in natural lake waters and conclusion. *J. Ecol.*, **61**, 193–211

Moss, B. 1977a. Factors controlling the seasonal incidence of *Pandorina morum* (Mull.) Bory (Chlorophyta, Volvocales) in a small pond. *Hydrobiol.*, **55**, 219–23

Moss, B. 1977b. Adaptations of epipelic and epipsammic freshwater algae. *Oecologia, Berlin*, **28**, 103–8

Moss, B. & Round, F. F. 1967. Observations on standing crops of epipelic and epipsammic algal communities in Shear Water, Wilts. *Br. phycol. Bull.*, **3**, 241–8

Moss, Betty 1974. Attachment and germination of the zygotes of *Pelvetia canaliculata*. *Phycologia*, **13**, 307–22

Moss, Betty 1975. Attachment of zygotes and germlings of *Ascophyllum nodosum* (L.) Le Jol (Phaeophyceae, Fucales). *Phycologia*, **14**, 75–80

Moss, Betty, Mercer, S. & Sheader, A. 1973. Factors affecting the *Himanthalia elongata* (L) S. F. Gray on the North-East coast of England. *Estuar. Cstl. Mar. Sci.*, **1**, 233–43

Mukai, H. 1971. The phytal animals on the thalli of *Sargassum serratifolium* in the Sargassum region, with reference to their seasonal fluctuations. *Mar. Biol.*, **8**, 170–82

Müller, D. 1962. Über jahres-und lunar periodische Erscheinungen bei einigen Braunalgen. *Bot. Mar.*, **4**, 140–55

Müller, D. G. 1964. Die Beteiligung eines Berühungzeit beim Festsetzen von Algenschwärmen auf dem Substrat. *Z. Bot.*, **52**, 193–8

Müller, D. G. 1976. Sexual isolation between a European and an American population of *Ectocarpus siliculosus* (Phaeophyta). *J. Phycol.*, **12**, 252–4

Müller-Haeckel, A. 1966. Diatomeendrift in Fliesgewässer. *Hydrobiol.*, **28**, 73–87

Müller-Haeckel, A. 1967. Tages-und Jahresperiodizität von *Ceratoneis arcus* Kütz. (Diatomeae). *Oikos.*, **18**, 351–6

Müller-Haeckel, A. 1970. Messung der Tagesperiodischen Neukolonisation von Algenzellen in Fliessgewässern. *Oikos*, suppl. **13**, 14–20

Müller-Haeckel, A. 1971*a*. Bewegungsverhalten einzellige Fliesswasser-Algen. *Natur. Mus.*, **101**, 167–72

Müller-Haeckel, A. 1971*b*. Circadiane Periodik der Kolonisationsaktivität driftender Algen. *Naturwiss.*, **5**, 273–4

Müller-Haeckel, A. 1973*a*. Experimente zum Bewegungsverhalten von einzelligen Fliesswasseralgen. *Hydrobiol.*, **41**, 221–46

Müller-Haeckel, A. 1973*b*. Different patterns of synchronization in diurnal and nocturnal drifting algae in the subarctic summer. *Aquilo. Ser. Zool.*, **14**, 23–8

Mullin, M. M. & Brooks, E. R. 1970. Growth and metabolism of two planktonic marine copepods as influenced by temperature and type of food. In *Marine Food Chains*, ed. J. H. Steele, pp. 74–95. Oliver & Boyd, Edinburgh

Munawar, M. & Nauwerck, A. 1971. The composition and horizontal distribution of phytoplankton in Lake Ontario during the year 1970. In *Proc. 14th Conf. Great Lakes Res. Int. Assoc.*, 69–78

Munda, I. M. & Lüning, K. 1977*a*. Combined effects of temperature and salinity on growth rates of germlings of three *Fucus* species from Iceland, Helgoland and the North Adriatic Sea. *Helgoländer wiss. Meeresunt.*, **29**, 307–10

Munda, I. M. & Lüning, K. 1977*b*. Growth performance of *Alaria esculenta* off Helgoland. *Helgoländer wiss. Meeresunt.*, **29**, 311–14

Munk, W. H. & Riley, G. A. 1952. Absorption of nutrients by aquatic plants. *J. Mar. Res.*, **11**, 215–40

Munro, A. L. S. & Brock, T. D. 1968. Distinction between bacterial and algal utilization of soluble substance in the sea. *J. gen. Microbiol.*, **51**, 35–42

Murray, J. & Hjort, J. 1912. *The Depths of the Ocean.* 821 pp. Macmillan & Co., London

Murray, J. W. 1963. Ecological experiments on Foraminiferids. *J. mar. biol. Ass., U.K.*, **43**, 631–42

Murray, J. W. 1973. *Distribution and ecology of living benthic Foraminiferids.* 274 pp. Heinemann Education

Murray, S., Scherfig, J. & Dixon, P. S. 1971. Evolution of algal assay procedures – PAAP, Batch test. *J. Wat. Pollut. Control Fed., Washington*, 1991–2003

Murray, S. N. & Littler, M. M. 1974. Analyses of standing stock and community structure of macro-organisms. Section 3. Primary Production of Macrophytes Section 5. In *Biological features of intertidal communities near the U.S. Navy sewage outfall, Wilson Cove, San Clemente Island, California*, ed. S. N. Murray & M. M. Littler, 85 pp. U.S. Navy NUC. TP 396

Muscatine, L. 1967. Glycerol excretion by symbiotic algae from corals and *Tridacna* and its control by the host. *Science*, **156**, 516–19

Muscatine, L. 1974. Endosymbiosis of cnidarians and algae. In *Coelenterate Biology: Reviews and new perspectives*, ed. L. Muscatine & H. M. Lenhoff, pp. 359–95. Academic Press, London

Muscatine, L. & Hand, C. 1958. Direct evidence for the transfer of materials from symbiotic algae to the tissues of a coelenterate. *Proc. Nat. Acad. Sci.*, **44**, 1259–65

Muscatine, L. & Lenhoff, H. M. 1965. Symbiosis of *Hydra* and algae. II. Effects of limited food and starvation on growth of symbiotic and aposymbiotic *Hydra. Biol. Bull.*, **129**, 316–28

Myers, E. H. 1943. Life activities of Foraminifera in relation to marine ecology. *Proc. Am. phil. Soc.*, **86**, 439–58

Nalewajko, C. 1966. Photosynthesis and excretion in various planktonic algae. *Limnol. Oceanogr.*, **11**, 1–10

Nalewajko, C. & Schindler, D. W. 1976. Primary production, extracellular release and heterotrophy in two lakes in the ELA, Northwestern Ontario. *J. Fish. Res. Bd. Can.*, **33**, 219–26

Naumann, E. 1917. Undersökningar öfver fytoplankton och under de pelagiska regionen forsiggående gyttje – och dybildningar inom vissa syo och mellasvenska ubergsvatten. *Kunn. svenska Vegtensk. Akad. Handl.*, **56**, 1–165

Naumann, E. 1925. Untersuchungen über einige sub-und elitorale Algenassosiation unserer Seen. *Ark. Bot.*, **19**, no. 16

Nauwerck, A. 1963. Die Beziehungen zwischen Zooplankton und Phytoplankton im See Erken. *Symb. bot. upsal.*, **17** (5), 163 pp.

Nelson, D. M. & Goering, J. J. 1977. Near-surface silica dissolution in the upwelling region off north-west Africa. *Deep Sea Res.*, **24**, 65–73

Neumann, A. C., Gebelein, C. D. & Scoffin, T. P. 1970. The composition, structure and erodability of subtidal mats, Abaco, Bahamas. *J. Sedim. Petrol.*, **40**, 274–97

Neumann, A. C. & Land, L. S. 1969. Algal production of lime mud deposition in the Bight of Abaco: a budget. *Geol. Soc. Am. Spec. Papers.*, **121**, 219

Neuscheler, W. 1967. Bewegung und Orientierung bei *Micrasterias denticulata* Bréb. im Licht. II. Photokinesis und phototaxis. *Z. Pflanzenphysiol.*, **57**, 151–67

Neushul, M. 1965. Diving observations of sub-tidal Antarctic marine vegetation. *Bot. Mar.*, **8**, 234–43

Neushul, M. 1967. Studies on the sublittoral marine vegetation in western Washington. *Ecology*, **48**, 83–96

Neushul, M. 1972*a*. Functional interpretation of benthic marine algal morphology. In *Contributions to the systematics of benthic marine algae of the North Pacific*, ed. I. A. Abbott & M. Kurogi, pp. 47–74. Sapporo, Japan

Neushul, M. 1972*b*. Underwater microscopy with an encased incident light dipping cone microscope. *J. Micros.*, **95**, 421–4

Neushul, M., Clarke, W. D. & Brown, D. W. 1967. Subtidal plant and animal communities of the southern Californian Islands. In *Proc. Symp. Biol. S. Calif. Islands*, ed. R. Philbrick, pp. 37–55. Santa Barbara Botanic Garden

Neushul, M. & Dahl, A. L. 1967. Composition and growth of subtidal parvosilvosa from Californian kelp forests. *Helgoländer wiss. Meeresunt.*, **15**, 480–8

Neushul, M., Foster, M. S., Coon, D. A., Woessner, J. W. & Harger, B. W. W. 1976. An *in situ* study of recruitment, growth and survival of subtidal marine algae. Technique and preliminary results. *J. Phycol.*, **12**, 397–408

Neushul, M. & Powell, J. H. 1964. An apparatus for experimental cultivation of benthic marine algae. *Ecology*, **45**, 893–4

Neustadt, M. I. 1959. Geschichte der Vegetation der UDSSR im Holözan. *Grana Palynologica*, **2**, 69–76

Nicholls, K. H. & Dillon, P. J. 1978. An evaluation of phosphorus – chlorophyll – phytoplankton relationships for lakes. *Int. Rev. ges. Hydrobiol.*, **63**, 141–54

Nicotri, R. E. 1977. Grazing effects of four marine intertidal herbivores on the microflora. *Ecology*, **58**, 1020–32

Nielsen, R. 1972. A study of the shell-boring marine algae around the Danish Island, Laesø. *Bot. Tidsskr.*, **67**, 245–69

Niemeck, R. A. & Mathieson, A. C. 1976. An ecological study of *Fucus spiralis* L. *J. exp. mar. Biol. Ecol.*, **24**, 33–48

Nienhuis, P. H. 1969. The significance of the substratum for intertidal algal growth on the artificial rocky shore of the Netherlands. *Int. Rev. ges. Hydrobiol.*, **54**, 207–15

Nienhuis, P. H. 1970. The benthic algal communities of flats and salt marshes in the Grevelingen, a sea arm in the south-western Netherlands. *Neth. J. Sea Res.*, **5**, 20–49

Nipkow, F. 1927. Über das Verhalten der Skelette planktischen Kieselalgen im geschichteten Tiefenschlamm des Zürich-und Baldeggersees. *Schweiz. Z. Hydrol.*, **4**, 71–120

Nival, P. 1965. Sur le cycle de *Dictyocha fibula* Ehr. dans les eaux de surface de la rade de Villefranche-sur-mer. *Cah. Biol. mar.*, **6**, 67–87

Nizamuddin, M. 1970. The phytogeography of the order Fucales (Phaeophyta). *Int. Rev. ges. Hydrobiol.*, **55**, 281–94

North, W. J. 1968. Effects of canopy cutting on kelp growth, comparison of experimentation with theory. In *Utilization of Kelp-bed resources in Southern California*, ed. W. J. North & C. L. Hubbs. *Fish. Bull. Calif.*, **139**, 223–54

North, W. J. 1971. Introduction and background. In *The biology of Giant Kelp Beds (Macrocystis) in California*, ed. W. J. North. *Nova Hedw. Beih.* **32**, 1–96

North, W. J. 1972. Observations on populations of *Macrocystis*. In *Contributions to the systematics of benthic marine algae of the North Pacific*, ed. I. A. Abbott & M. Kurogi, pp. 75–92. Sapporo, Japan

North, W. J., Neushul, M. & Clendenning, K. A. 1965. Successive biological changes observed in a marine cove exposed to a large spillage of oil ('Tampico' incident). In *Symp. Comm. Inst. explor. scient. mer. méd.*, Monaco, pp. 335–54

Northcraft, R. D. 1948. Marine algal colonization of the Monterey Peninsula, California. *Amer. J. Bot.*, **35**, 396–404

Norton, T. A. 1968. Underwater observations on the vertical distribution of algae at St. Mary's, Isles of Scilly. *Br. phycol. bull.*, **3**, 585–8

Norton, T. A. 1971. An ecological study of the fauna inhabiting the sublittoral marine alga *Saccorhiza polyschides* (Lightf.) Batt. *Hydrobiol.*, **37**, 215–31

Norton, T. A. 1978. The factors influencing the distribution of *Saccorhiza polyschides* in the region of Lough Ine. *J. mar. biol. Ass., U.K.*, **58**, 527–36

Norton, T. A. & Burrows, E. M. 1969. Studies on marine

algae of the British Isles. *Saccorhiza polyschides* (Lightf.) Batt. *Br. phycol. J.*, **4**, 19–53

Norton, T. A. & Parkes, H. M. 1972. The distribution and reproduction of *Pterosiphonia complanata*. *Br. phycol. J.*, **7**, 13–19

Norton, T. A. & Milburn, J. A. 1972. Direct observations on the sublittoral marine algae of Argyll, Scotland. *Hydrobiol.*, **40**, 55–68

Novichkova-Ivanova, L. N. 1972. Soil and aerial algae of polar deserts and arctic tundra. IBP Tundra Biome. In *Proc. 4th Int. Meeting Biol. Prod. Tundra*, ed. F. F. Wielgdasle & T. Rosswal

Nygaard, G. 1949. Hydrobiological studies of some Danish ponds and lakes. II. *Kon. danske Vidensk. Selsk. Biol. Skr.*, **7**, 293 pp.

Odintsova, S. V. 1941. Nitre formation in deserts. *C. r. Acad. Sci. URSS, N.S.*, **32**, 578–80

Odum, E. P. 1963. *Ecology.* 152 pp. Holt, Rinehart & Winston

Odum, E. P. 1971. *Fundamentals of Ecology.* 547 pp. Saunders, Philadelphia

Odum, H. T., McConnell, W. & Abbott, W. 1958. The chlorophyll *a* of communities. *Publ. Inst. Mar. Sci., Univ. Texas*, **5**, 66–96

Odum, H. T. & Odum, E. P. 1955. Trophic structure and productivity of a windward coral reef community on Eniwetok. *Ecol. Monogr.*, **25**, 291–320

Odum, W. E. 1968. The ecological significance of fine particle selection by the striped mullet *Mugil cephalus*. *Limnol. Oceanogr.*, **13**, 92–8

Odum, W. E. 1970. Utilization of the direct grazing and plant detritus food chains by the striped mullet *Mugil cephalus*. In *Marine Food Chains*, ed. J. H. Steele, pp. 222–40. Oliver & Boyd, Edinburgh

Ogawa, R. E. & Carr, J. F. 1969. The influence of nitrogen on heterocyst production in blue-green algae. *Limnol. Oceanogr.*, **14**, 342–57

Oglesby, R. T. 1977a. Relationships of fish yield to lake phytoplankton standing crop production and morphoedaphic factors. *J. Fish. Res. Bd. Can.*, **34**, 2271–9

Oglesby, R. T. 1977b. Phytoplankton summer standing crop and annual productivity as functions of phosphorus loading and various physical factors. *J. Fish Res. Board Can.* **34**, 2255–70

Ohwada, K. & Taga, N. 1973. Seasonal cycles of Vitamin B_{12}, Thiamine and Biotin in Lake Sagami. Patterns of their distribution and ecological significance. *Int. Revue ges. Hydrobiol.*, **58**, 851–71

Oppenheimer, C. H. & Vance, M. H. 1960. Attachment of bacteria to the surfaces of living and dead microorganisms in marine sediments. *Z. allg. Mikrobiol.*, **1**, 47–52

Osborne, P. L. & Moss, B. 1977. Palaeolimnology and trends in the phosphorus and iron budgets of an old man-made lake, Barton Broad, Norfolk. *Freshwater Biol.*, **7**, 213–33

Oschman, J. L. 1966. Development of the symbiosis of *Convoluta roscoffensis* (Graff) and *Platymonas* sp. *J. Phycol.*, **2**, 105–11

Ostrofsky, M. L. & Duthie, H. C. 1975. Primary productivity and phytoplankton of lakes on the Eastern Canadian Shield. *Verh. Int. Ver. Limnol.*, **19**, 732–8

Otsuki, A. & Hanya, T. 1972. Production of dissolved organic matter from dead green algal cells. Aerobic microbial decomposition. *Limnol. Oceanogr.*, **17**, 248–57

Overbeck, J. 1962a. Das Nannoplankton (μ-Algen) der Rügenschen Brackwässer als Hauptproduzent in Abhängigkeit von Salzgehalt. *Kieler Meeresforsch.*, **18**, 157–71

Overbeck, J. 1962b. Untersuchungen zum Phosphathaushalt von Grünalgen. I. Phosphathaushalt und Fortpflanzungsrhythmus von *Scenedesmus quadricauda* (Turp.) Bréb. an natürlichen Standort. *Arch. Hydrobiol.*, **58**, 162–209

Overbeck, J. 1962c. Untersuchungen zum Phosphathaushalt von Grünalgen. II. Die Verwertung von Pyrophosphat und organisch gebundene Phosphaten und ihre Beziehung zu den Phosphatasen von *Scenedesmus quadricauda* (Turp.) Bréb. *Arch. Hydrobiol.*, **58**, 218–308

Overbeck, J. & Babenzien, H. D. 1964. Bakterien und Phytoplankton eines Kleingewässers in Jahreszyklus. *Z. allg. Mikrobiol.*, **4**, 59–76

Owen, R. W. & Zeitzschel, B. 1970. Phytoplankton production: seasonal change in the oceanic eastern tropical Pacific. *Mar. Biol.*, **7**, 32–6

Paasche, E. 1964. A tracer study of the inorganic carbon uptake during coccolith formation and photosynthesis in the coccolithophorid *Coccolithus huxleyi*, *Physiologica Pl. Suppl.*, **3**, 1–82

Paerl, H. W. 1973. Detritus in Lake Tahoe: Structural modification by attached microflora. *Science*, **180**, 496–8

Paerl, H. W. & MacKenzie, L. A. 1977. A comparative study of the diurnal carbon fixation patterns of nannoplankton and net plankton. *Limnol. Oceanogr.*, **22**, 732–8

Pahwa, D. V. & Mehrotra, S. N. 1966. Observations on fluctuations in the abundance of plankton in relation to certain hydrological conditions of River Ganga. *Proc. natn. Acad. Sci. India*, **B**, **36**, 157–89

Paine, R. T. & Vadas, R. L. 1969. The effects of grazing by sea urchins, *Strongylocentrotus* spp. on benthic algal populations. *Limnol. Oceanogr.*, **14**, 710–19

Palmer, J. D. & Round, F. E. 1965. Persistent, vertical-migration rhythms in benthic microflora. I. The effect of light and temperature on the rhythmic behaviour of *Euglena obtusa. J. mar. biol. Ass., U.K.*, **45**, 567–82

Palmer, J. D. & Round, F. E. 1967. Persistent, vertical-migration rhythms in benthic microflora. VI. The tidal and diurnal nature of the rhythm in the diatom *Hantzschia virgata. Biol. Bull.*, **132**, 44–55

Pankow, H. 1961. Über die Ursachen der Fehlens von Epiphyton auf Zygnemales. *Arch. Protistenk.*, **105**, 417–44

Pankratova, Ye. M. & Vakhrushev, A. S. 1971. Field determination of the fixation of the atmospheric nitrogen by blue-green algae using N^{15}. *Soviet Soil Sci.*, **3**, 726–33

Pannier, F. 1957-8. El consumo de oxigéno de plantas acuaticas en relatión a distintas concentrationes de oxigéno. *Acta cient. venez.*, **8–9**. Reported in F. Gessner, *Hydrobotanik*, vol. 2

Pant, A. & Fogg, G. E. 1976. Uptake of glycollate by *Skeletonema costatum* in bacterized culture. *J. Exp. Mar. Biol. Ecol.*, **22**, 227–34

Pardy, R. L. 1974. Some factors affecting the growth and distribution of the algal endosymbiont of *Hydra viridis. Biol. Bull.*, **147**, 105–18

Pardy, R. L. & Muscatine, L. 1973. Recognition of symbiotic algae by *Hydra viridis*. A quantitative study of the uptake of living algae by aposymbiotic. *H. viridis. Biol. Bull.*, **145**, 565–79

Parke, M. 1948. Studies on British Laminariaceae I. Growth in *Laminaria saccharina* (L.) Lamour. *J. mar. biol. Ass., U.K.*, **27**, 651–709

Parke, M. 1971. The production of calcareous elements by benthic algae belonging to the class Haptophyceae (Chrysophyta). In *Proc. 2nd Plankton Conf., Rome*, ed. A. Farinacci, vol. 2, pp. 929–37

Parke, M. & Dixon, P. S. 1976. Check-list of British marine algae – third revision. *J. mar. biol. Ass., U.K.*, **56**, 527–94

Parke, M. & Moore, H. B. 1935. The biology of *Balanus balanoides*. II. Algal infection of the shell. *J. mar. biol. Ass., U.K.*, **20**

Parker, B. C. 1961. Facultative heterotrophy in certain soil algae from the ecological viewpoint. *Ecology*, **42**, 381–6

Parker, B. C. 1968. Rain as a source of vitamin B_{12}. *Nature, London*, **219**, 617–18

Parker, B. C. & Barsom, G. 1970. Biological and chemical significance of surface microlayers in aquatic ecosystems. *Bioscience*, **20**, 87–93

Parker, B. C. & Bold, H. C. 1961. Biotic relationships between soil algae and other microorganisms. *Am. J. Bot.*, **48**, 185–97

Parker, B. C., Bold, H. C. & Deason, T. R. 1960. Facultative heterotrophy in some chlorococcalean algae. *Science*, **133**, 761–3

Parker, B. C. & Hatcher, R. F. 1974. Enrichment of surface freshwater micro-layers with algae. *J. Phycol.*, **10**, 185–9

Parker, B. C., Samsel, G. L. & Prescott, G. W. 1973. Comparison of microhabitats of macroscopic subalpine stream algae. *Am. Midl. Nat.*, **90**, 143–53

Parker, B. C. & Wachtel, M. A. 1972. Seasonal distribution of cobalamins, biotin and niacin in rainwater. In *The structure and function of fresh-water microbial communities*, ed. J. Cairns, Jr, pp. 195–207. Virginia Polytechnic Institute and State University

Parker, B. C. & Wodehouse, E. B. 1972. Ecology and water quality criteria. In *The structure and function of fresh-water microbial communities*, ed. J. Cairns, Jr, pp. 111–34. Virginia Polytechnic Institute and State University

Parker, J. 1960. Seasonal changes in cold-hardness of *Fucus vesiculosus. Biol. Bull. mar. biol. Lab., Woods Hole*, **119**, 474–8

Parker, J. I., Conway, H. L. & Yaguchi, E. M. 1977a. Dissolution of diatom frustules and recycling of amorphous silicon in Lake Michigan. *J. Fish. Res. Bd. Can.*, **34**, 545–51

Parker, J. I., Conway, H. L. & Yaguchi, E. M. 1977b. Seasonal periodicity of diatoms, and silicon limitation in offshore Lake Michigan, 1975. *J. Fish. Res. Bd. Can.*, **34**, 552–8

Parr, A. E. 1939. Quantitative observations on the pelagic *Sargassum* vegetation of the Western North Atlantic. *Bull. Bing. Oceanogr. Collect.*, **6**, 1–94

Parsons, T. R. & LeBrasseur, R. J. 1968. A discussion of some critical indices of primary and secondary production for large scale ocean surveys. *Rep. Calif. Coop. Ocean. Fish. Invest.* **12**, 54–63

Parsons, T. R. & LeBrasseur, R. J. 1969. A discussion of some critical indices of primary and secondary production for large scale ocean surveys. *Rep. Calif. Coop. Ocean. Fish. Invest.*, **12**, 54–63

Parsons, T. R. & LeBrasseur, R. J. 1970. The availability of food to different trophic levels in the marine food chain. In *Marine Food Chains*, ed. J. H. Steele, pp. 325–43. Oliver & Boyd, Edinburgh

Parsons, T. R., LeBrasseur, R. J., Fulton, J. D. & Kennedy, O. D. 1969. Production studies in the Strait of Georgia. Part II. Secondary production under the Fraser River plume, February to May 1967. *J. exp. mar. Biol. Ecol.*, **3**, 39–50

Patmatmat, M. M. 1968. Ecology and metabolism of a benthic community on an intertidal sand flat. *Int. Rev. ges. Hydrobiol.*, **53**, 211–98

Patrick, R. 1949 A proposed biological measure of stream conditions, based on a survey of the Conestoga Basin, Lancaster County, Pennsylvania. *Prac. Acad. Nat. Sci. Philadelphia*, **101**, 277–341

Patrick, R. 1957. Diatoms as indicators of changes in environmental conditions. In *Biological Problems in Water Pollution. Trans. Sem. Biol. Probl. Wat. Pol. Cincinnati* 23–27 April 1956, ed. C. W. Tarzwell, pp. 71–83. U.S. Dept. Health Educ. Welfare.

Patrick, R. 1963. The structure of diatom communities under varying ecological conditions. *Ann. New York Acad. Sci.*, **108**, 359–64

Patrick, R. 1967. The effect of invasion rate, species pool and size of area on the structure of the diatom community. *Proc. Nat. Acad. Sci.*, **58**, 1335–43

Patrick, R. 1973. Use of algae, especially diatoms, in the assessment of water quality. In *Biological Methods for the Assessment of Water Quality*, ed. J. Cairns & K. L. Dickson, pp. 76–95. ASTM, STP 528

Patrick, R. 1957. Diatoms as indicators of changes in environmental conditions, pp. 71–83. In 'Biological problems in water pollution'. US Dept. Health, Educ. & Welfare. Publ. Health Serv. Cincinnati, Ohio.

Patrick, R., Crum, B. & Coles, J. 1969. Temperature and manganese as determining factors in the presence of diatoms or blue-green algal floras in streams. *Proc. natn. Acad. Sci., U.S.A.*, **64**, 472–8

Paul, D. J., Paul, J. M. & Nevé, R. A. 1978. Phytoplankton densities and growth of *Mytilus edulis* in an Alaskan artificial upwelling system. *J. Cons. perm. int. Explor. Mer.*, **38**, 100–4

Pearce, V. B. & Muscatine, L. 1971. Role of symbiotic algae (zooxanthellae) in coral calcification. *Biol. Bull.*, **141**, 350–63

Pearsall, W. H. 1932. Phytoplankton in the English Lakes. II. The composition of the phytoplankton in relation to dissolved substances. *J. Ecol.*, **20**, 241–62

Peary, J. A. & Castenholz, R. W. 1964. Temperature strains of a thermophyllic blue-green alga. *Nature, London,* **202,** 720–1

Pechlaner, R. 1967. Die Finstertaller Seen (Kühtai, Österreich). II. Das Phytoplankton. *Arch. Hydrobiol.*, **63**, 145–93

Pechlaner, R. 1970. The phytoplankton spring outburst and its conditions in Lake Erken (Sweden). *Limnol. Oceanogr.*, **15**, 113–30

Pechlaner, R. 1971. Factors that control the production rate

and biomass of phytoplankton in high mountain lakes. *Mitt. Int. Ver. Limnol.*, **19**, 125–45

Penhale, P. A. 1977. Macrophyte-epiphyte biomass and productivity in an eelgrass (*Zostera marina* L.) community. *J. exp. mar. Biol. Ecol.*, **26**, 211–24

Penhale, P. A. & Smith, W. O., Jr 1977. Excretion of dissolved organic carbon by eelgrass (*Zostera marina*) and its epiphytes. *Limnol. Oceanogr.*, **22**, 400–7

Pennington, W., Haworth, E. Y., Bonny, A. P. & Lishman, J. P. 1972. Lake sediments in northern Scotland. *Phil. Trans. Roy. Soc., London,* **B**, **264**, 191–294

Pérès, J. M. & Ricard, J. 1955. Biotopes et biocoenoses de la Méditerranée occidentale comparés à ceux de la Manche et de l'Atlantique nord-oriental. *Arch. Zool. exp. gén.*, **92**, 1–70

Pérès, J. M. & Ricard, J. 1957. Manuel de Biononie benthique de la Mer Méditerranée. *Recl. Trav. Stat. mar. Endoume. Bull.*, **14**, 5–122

Perkins, E. J. 1960. The diurnal rhythm of the littoral diatoms of the river Eden estuary. *J. Ecol.*, **48**, 725–8

Pérterfi, L. S. 1972. Cîteva aspecte ale cenologiei algelor dulciole. *Gradina Bot.*, 141–8

Peters, N. 1929. Über Orts und Geisselbewegung bei marinen Dinoflagellaten. *Arch. Protistenk.*, **67**, 291–321

Petersen, J. B. 1923. The fresh-water Cyanophyceae of Iceland. In *The Botany of Iceland Vol. II.* Nr. 7, ed. L. K. Rosenvinge & E. Warming

Petersen, J. B. 1928. The aerial algae of Iceland. In *The Botany of Iceland Vol. II*, ed. L. K. Rosenvinge & E. Warming. Pp. 327–447

Petersen, J. B. 1935. Studies on the biology and taxonomy of soil algae. *Dansk. Bot. Ark.*, **8**, 1–183

Peterson, J. O., Wall, J. P., Wirth, T. C. & Born, S. M. 1973. Nutrient inactivation by chemical precipitation at Horseshoe Lake, Wisconsin. *Tech. Bull. Dept. Nat. Res.*, no. 62, 20 pp. Madison, Wisconsin

Peterson, R. B., Friberg, E. E. & Burris, R. H. 1977. Diurnal variation in N_2 fixation and photosynthesis by aquatic blue-green algae. *Plant Physiol.*, **59**, 74–80

Petipa, T. S., Pavlova, E. V. & Mironov, G. N. 1970. The food web structure, utilization and transport of energy by trophic levels in the planktonic communities. In *Marine Food Chains*, ed. J. H. Steele, pp. 142–67. Oliver & Boyd, Edinburgh

Phillips, G. L., Eminson, D. & Moss, B. 1978. A mechanism to account for macrophyte decline in progressively eutrophicated freshwaters. *Aquatic Bot.*, **4**, 103–26

Pickett-Heaps, J. 1972. A possible virus infection in the green alga *Oedogonium. J. Phycol.*, **8**, 44–7

Pienaar, R. N. 1976. Virus-like particles in three species of phytoplankton from San Juan Island, Washington. *Phycologia*, **15**, 185–90

Pingree, R. D., Maddock, L. & Butler, E. J. 1977. The influence of biological activity and physical stability in determining the chemical distribution of inorganic phosphate, silicate and nitrate. *J. mar. biol. Ass., U.K.*, **57**, 1065–73

Plante-Cuny, M. R. 1969. Recherches sur la distribution qualitative et quantitative des diatomées benthiques de certains fonds meubles du Gulfe de Marseille. *Recl. Trav. Stat. mar. Endoume. Bull.*, **45**, 88–197

Plante-Cuny, M. R. 1973. Recherches sur la production primaire benthique en milieu marin tropical. I. Variations de la production primaire et des teneurs en pigments photosynthétique sur quelques fonds sableux. *Cah. Off. Rech. scient. tech. Outre – Mer.*, *Oceanogr.*, **11**, 317–48

Platt Bradbury, J. 1975. Diatom stratigraphy and human settlement in Minnesota. *Geol. Soc. Am. Spec. Pap.*, **171**, 74 pp.

Platt, T. 1971. The annual production of phytoplankton in St. Margarets Bay, Nova Scotia. *J. Cons. perm. int. Explor. Mer.*, **33**, 324–34

Platt, T. 1972. Local phytoplankton abundance and turbulence. *Deep Sea Res.*, **19**, 183–7

Platt, T., Dickie, L. M. & Trites, R. W. 1970. Spacial heterogeneity of phytoplankton in a near-shore environment. *J. Fish. Res. Bd. Can.*, **27**, 1453–73

Platt, T. & Subba Rao, D. V. 1973. Primary production of marine microphytes. In *Photosynthesis and productivity in different environments*, ed. J. P. Cooper, pp. 249–80. Cambridge University Press

Pohl, F. 1948. Tagesrhythmus im phototaktischen Verhalten der *Euglena gracilis*. *Z. Naturf.*, **3**b, 367–74

Pollingher, U. & Berman, T. 1976. Autoradiographic screening for potential heterotrophs in natural algal populations of Lake Kinneret. *Microbial. Ecol.*, **2**, 252–60

Pomeroy, L. R. 1959. Algal productivity in salt marshes of Georgia. *Limnol. Oceanogr.*, **4**, 386–98

Pomeroy, L. R. 1974. The oceans food web, a changing paradigm. *Bioscience*, **24**, 499–504

Pomeroy, L. R., Mathews, H. M. & Min, H. S. 1963. Excretion of phosphate and soluble organic phosphorus compounds by zooplankton. *Limnol. Oceanogr.*, **8**, 50–5

Poore, M. E. D. 1955. The use of phytosociological methods in ecological investigations. *J. Ecol.*, **43**, 226–69 and 606–51

Poore, M. E. D. 1962. The method of successive approximations in descriptive ecology. *Adv. ecol. Res.*, **1**, 35–68

Porter, K. G. 1972. A method for the *in situ* study of zooplankton grazing effects on algal species, composition and standing crop. *Limnol. Oceanogr.*, **17**, 913–17

Porter, K. G. 1973. Selective grazing and differential digestion of algae by zooplankton. *Nature, London*, **244**, 179–80

Porter, K. G. 1976. Enhancement of algal growth and productivity by grazing zooplankton. *Science*, **192**, 1332–3

Post, E. 1968. *Bostrychia kelanensis* Grun. *Hydrobiol.*, **31**, 81–150

Potash, M. 1956. A biological test for determining the potential productivity of water. *Ecology*, **37**, 631–9

Powell, H. T. 1957. Studies in the genus *Fucus* L. II. Distribution and ecology of forms of *Fucus distichus* (L)emend Powell in Ireland. *J. mar. biol. Ass.*, *U.K.*, **36**, 663–93

Powell, H. T. 1963. Speciation in the genus *Fucus* L. and related genera. In *Speciation in the sea. Syst. Assoc. Publ.*, **5**, 63–77

Powell, H. T. 1966. The occurrence of *Codium adhaerens* (Cabr.) C.Ag. in Scotland, and a note on *Codium amphibium* Moore & Harv. In *Some contemporary studies in Marine Science*, ed. H. Barnes, pp. 591–5. Allen & Unwin, London

Powell, H. T. 1972. The ecology of the macroalgae in sea lochs in Western Scotland. *Proc. Int. Seaweed Symp.*, **7**, 273

Powers, C. F., Schults, D. W., Malueg, K. W., Brice, R. M. & Schuldt, M. D. 1972. Algal responses to nutrient additions in natural waters. II. Field experiments. In *Nutrients and eutrophication: The limiting nutrient controversy*, Limnol. Oceanogr. Symp. 1, pp. 141–56

Prakash, A. 1971. Terrigenous organic matter and coastal phytoplankton fertility. In *Fertility of the sea, vol. 2*, ed. J. D. Costlow, pp. 351–68

Pratt, D. M. 1959. The phytoplankton in Narragansett Bay. *Limnol. Oceanogr.*, **4**, 425–40

Pratt, D. M. 1965. The winter–spring diatom flowering in Narragansett Bay. *Limnol. Oceanogr.*, **10**, 173–84

Pratt, D. M. 1966a. The Gulf Stream as a graded river. *Limnol. Oceanogr.*, **11**, 60–7

Pratt, D. M. 1966b. Competition between *Skeletonema costatum* and *Olisthodiscus luteus* in Narragansett Bay and in culture. *Limnol. Oceanogr.*, **11**, 447–55

Prentice, S. A. & Kain, J. M. 1976. Numerical analysis of subtidal communities on rocky shores. *Estuarine Coastal Mar. Sci.*, **4**, 65–70

Preston, A., Jefferies, D. F., Dutton, J. W. R., Harvey, B. P. & Steele, A. K. 1972. British coastal waters: The conservation of selected heavy metals in sea water, suspended matter and biological indicators – a pilot survey. *Environ. Pollut.*, **3**, 69–87

Proctor, V. W. 1957. Studies of algal antibiosis using *Haematococcus* and *Chlamydomonas*. *Limnol. Oceanogr.*, **2**, 125–39

Prouse, N. J., Gordon, D. C. & Keizer, P. D. 1976. Effects of low concentrations of oil accommodated in sea water on the growth of unialgal phytoplankton cultures. *J. Fish. Res. Bd. Can.*, **33**, 810–18

Provasoli, L. 1965. Organic regulation of phytoplankton fertility. In *The Sea, vol. 2*, ed. M. N. Hill, pp. 165–219. Interscience

Provasoli, L. 1969. Algal nutrition and eutrophication. In *Eutrophication: Causes, consequences, correctives*, pp. 574–93. Nat. Acad. Sci., Washington, D.C.

Provasoli, L. 1971. Nutritional relationships in marine organisms. In *Fertility of the Sea, vol. 2*, ed. J. D. Costlow, pp. 369–82

Provasoli, L., Yamasu, T. & Manton, I. 1968. Experiments on the resynthesis of symbiosis in *Convoluta roscoffensis* with different flagellate cultures. *J. mar. biol. Ass.*, *U.K.*, **48**, 465–79

Prowse, G. D. 1959. Relationship between epiphytic algal species and their macrophytic hosts. *Nature, London*, **183**, 1204–5

Purdy, E. G. & Kornicker, L. S. 1958. Algal disintegration of Bahamian limestone coasts. *J. Geol.*, **66**, 96–9

Qasim, S. Z., Bhattathiri, P. M. A. & Devassy, V. P. 1972. The influence of salinity on the rate of photosynthesis and abundance of some tropical phytoplankton. *Mar. Biol.*, **12**, 200–6

Qasim, S. Z., Bhattathiri, P. M. A. & Reddy, C. V. G. 1972. Primary production of an atoll in the Laccadives. *Int. Rev. ges. Hydrobiol.*, **57**, 207–25

Raalte, C. van, Stewart, W. C. Valiela, I. & Carpenter, E. J. 1974. A ^{14}C technique for measuring algal productivity in salt marsh muds. *Bot. Mar.*, **17**, 186–8

Rabinowitch, E. I. 1945. *Photosynthesis and related processes. Vol. 1*. 599 pp. Interscience

Rahat, M. & Jahn, T. L. 1965. Growth of *Prymnesium parvum* in the dark: Note on ichthyotoxin formation. *J. Protozool.*, **12**, 246–50

Ralph, R. D. 1977. The myxophyceae of the marshes of southern Delaware. *Chesapeake Sci.*, **18**, 208–21

Ramus, J., Beale, S. I. & Mauzerall, D. 1976. Correlation of changes in pigment content with photosynthetic capacity of seaweeds as a function of water depth. *Mar. Biol.*, **37**, 231–8

Ramus, J., Beale, S. I., Mauzerall, D. & Howerd, K. L. 1976. Changes in photosynthetic pigment concentration in seaweeds as a function of water depth. *Mar. Biol.*, **37**, 223–9

Ramus, J., Lemons, F. & Zimmerman, C. 1977. Adaptation of light-harvesting pigments to down welling light and the consequent photosynthetic performance of the eulittoral rock weeds *Ascophyllum nodosum* and *Fucus vesiculosus*. *Mar. Biol.*, **42**, 293–303

Randall, J. E. 1961. Overgrazing of algae by herbivorous marine fishes. *Ecology*, **42**, 812

Ranson, G. 1955. Observation sur l'agent essential de la dissolution du calcair sousmarin dan la zone côtiere des iles coralliennes de l'archipel des Tuamotu. *C. r. hebd. Séanc. Acad. Sci. Paris*, ser. **D**, **240**, 800–8

Raunkier, C. 1934. *The Life Forms of Plants and Statistical Plant Geography*. 632 pp. Clevedon Press, Oxford

Rautenberg, E. 1961. Zur Morphogie und Ökologie einiger epiphytischer und epi-endophytischer Algen. *Bot. Mar.*, **2**, 133–48

Rautiainen, H. & Ravanko, O. 1972. The epiphytic diatom flora of the benthic macrophyte communities on rocky shores in the south western archipelago of Finland. Seili Islands. *Nova Hedw.*, **23**, 827–42

Ravanko, O. 1970a. Observations on the taxonomy and ecology of the epilithic crustaceous brown algae in the SW Archipelago of Finland. *Bot. Not.*, **123**, 220–30

Ravanko, O. 1970b. Morphological, developmental and taxonomic studies in the *Ectocarpus* complex. (Phaeophyceae.) *Nova Hedw.*, **20**, 129–252

Ravanko, O. 1972. The physiognomy and structure of the benthic macrophyte communities on rocky shores in the south western archipelago of Finland. *Nova Hedw.*, **23**, 363–403

Reeves, C. C. 1968. *Introduction to paleolimnology*. 228 pp. Elsevier, Amsterdam

Reid, F. M. H., Stewart, E., Eppley, R. W. & Goodman, D. 1978. Spatial distribution of phytoplankton species in chlorophyll maximum layers off southern California. *Limnol. Oceanogr.*, **23**, 219–26

Reid, P. C. 1972. Dinoflagellate cyst distribution around the British Isles. *J. mar. biol. Assoc., U.K.*, **52**, 939–44

Reid, P. C. 1975a. A regional sub-division of dinoflagellate cysts around the British Isles. *New Phytol.*, **75**, 589–603

Reid, P. C. 1975b. Large scale changes in North Sea phytoplankton. *Nature, London*, **257**, 217–19

Reid, P. C. 1977. Continuous plankton records: Changes in the composition and abundance of the phytoplankton of the north eastern Atlantic Ocean and North Sea, 1958–74. *Mar. Biol.*, **40**, 337–9

Remane, A. 1940. Einführung in die zoologische Ökologie der Nord- und Ostsee. *Tierwelt N.- u. Ostee.*, **1a**, 1–238

Revelante, N. & Gilmartin, M. 1973. Some observations on the chlorophyll maximum and primary production in the eastern North Pacific. *Int. Rev. ges. Hydrobiol.*, **58**, 819–34

Reynolds, C. S. 1967. The breaking of the Shropshire meres: some recent investigations. *Shrops. Cons. Trust. Bull.*, no. 10, 9–16

Reynolds, C. S. 1972. Growth, gas-vacuolation and buoyancy in a natural population of a blue-green alga. *Freshwater Biol.*, **2**, 87–106

Reynolds, C. S. 1973a. The phytoplankton of Crose Mere, Shropshire. *Br. phycol. J.*, **8**, 153–62

Reynolds, C. S. 1973b. Phytoplankton periodicity of some North Shropshire meres. *Br. phycol. J.*, **8**, 301–20

Reynolds, C. S. & Rogers, D. A. 1976. Seasonal variations in the vertical distribution and buoyancy of *Microcystis aeruginosa* Kütz. emend. Elenkin in Rostherne Mere, England. *Hydrobiol.*, **48**, 17–23

Reynolds, C. S. & Walsby, A. E. 1975. Water-blooms. *Biol. Rev.*, **50**, 437–871

Reyssac, J. & Roux, M. 1972. Communautés phytoplanktoniques dans les eaux de Côte d'Ivoire. Groupes d'espèces associées. *Mar. Biol.*, **13**, 14–33

Richardson, D. H. S., Hill, D. T. & Smith, D. C. 1968. Lichen physiology. XI. The role of the alga in determining the pattern of carbohydrate movement between lichen symbionts. *New Phytol.*, **67**, 469–86

Richardson, J. L. 1968. Diatoms and lake typology in East and Central Africa. *Int. Rev. ges. Hydrobiol. Hydrogr.*, **53**, 299–338

Richman, S. & Rogers, J. N. 1969. The feeding of *Calanus helgolandicus* on synchronously growing populations of the marine diatom. *Ditylum brightwellii*. *Limnol. Oceanogr.*, **14**, 701–9

Ricketts, E. F., Calvin, J. & Hedgepeth, J. W. 1952. *Between Pacific tides*. 502 pp. Stanford University Press

Rider, D. E. & Wagner, R. H. 1972. The relationship of light, temperature and current to the seasonal distribution of *Batrachospermum* (Rhodophyta). *J. Phycol.*, **8**, 323–31

Rietz, E. G. du 1925. Die Hauptzüge der Vegetation der Insel Jungfurn. *Svensk. Bot. Tidskr.*, **19**, 323–46

Rietz, E. G. du 1932. Zur Vegetationsökologie der ostschwedischen Küstenfelsen. *Beih. Bot. Z.*, **49**, 61–112

Rigg, G. B. & Miller, R. C. 1949. Intertidal plant and animal zonation in the vicinity of Neah Bay, Washington. *Proc. Calif. Acad. Sci.*, **26**, 323–51

Rigler, F. H. 1964. The phosphorus fractions and turnover time of inorganic phosphorus in different types of lakes. *Limnol. Oceanogr.*, **9**, 511–18

Riley, G. A. 1944. The carbon metabolism and photosynthetic efficiency of the earth as a whole. *Amer. Scient.*, **32**, 132–4

Riley, G. A. 1951. Oxygen, phosphate and nitrate in the Atlantic Ocean. *Bull. Bing. Oceanog. Collect.*, **13**, 1–126

Riley, G. A. 1967. The plankton of estuaries. In *Estuaries*, ed. G. H. Lauff. *Amer. Assoc. Adv. Sci.*, Washington D.C., no. 83, pp. 316–26

Riley, G. A., Stommel, H. & Bumpus, D. F. 1949. Quantitative ecology of the plankton of the western North Atlantic. *Bull. Bing. Oceanog. Collect.*, **12**, 1–169

Ringer, Z. 1973. Phytoplankton of the southern Baltic Sea. *Polskie Arch. Hydrobiol.*, **20**, 371–8

Robertson, D. E. 1972. Influence of the physico-chemical forms of radionucleides and stable trace elements in seawater in relation to uptake by the marine biosphere. In *OECD Marine radioecology 2nd ENEA Seminar*, pp. 21–76

Robinson, D. G. & Preston, R. D. 1971. Studies on the fine structure of *Glaucocystis nostochinearum* Itzigs. II. Membrane morphology and taxonomy. *Br. phycol. J.*, **6**, 113–28

Robinson, G. A. & Waller, D. R. 1966. The distribution of *Rhizosolenia styliformis* Brightwell and its varieties. In *Some contemporary studies in Marine Science*, ed. H. Barnes, pp. 645–63. Allen & Unwin, London

Rodhe, W. 1958. The primary production in lakes, some results and restrictions of the ^{14}C method. *Rapp. Cons. Explor. Mer.*, **144**, 122–8

Rodhe, W. 1969. Crystallization of eutrophication concepts in northern Europe. In *Eutrophication: Causes, consequences, correctives*, pp. 50–64. Nat. Acad. Sci., Washington D.C.

Rodhe, W. 1974. Plankton, planktic, planktonic. *Limnol. Oceanogr.*, **19**, 360

Rodhe, W., Hobbie, J. E. & Wright, R. T. 1966. Phototrophy and heterotrophy in high mountain lakes. *Verh. int. Ver. Limnol.*, **12**, 117–22

Rodhe, W., Vollenweider, R. A. & Nauwerk, A. 1956. The primary production and standing crop of phytoplankton. In *Perspectives in Marine Biology*, ed. A. A. Buzzoti Traverso, 299–332

Roeder, D. R. 1977. Relationship between phytoplankton and periphyton communities in a Central Iowa Stream. *Hydrobiol.*, **56**, 145–51

Roelofs, T. D. & Oglesby, R. T. 1970. Ecological observations on the planktonic cyanophyte *Gleotrichia echinulata*. *Limnol. Oceanogr.*, **15**, 224–9

Roels, O. A., Gerard, R. D. & Bé, A. W. 1971. Fertilizing the sea by pumping nutrient-rich deep water to the surface. In *Fertility of the Sea, vol. 2*, ed. J. D. Costlow, pp. 401–15

Rohlich, G. A. & Uttormark, P. D. 1972. Waste water treatment and eutrophication. In *Nutrients and eutrophication: The limiting nutrient controversy*, Limnol. Oceanogr. Symp. I, pp. 231–45

Rojas de Mendiola, B. 1958. Breve estudio sobre la variation cualitativa annual del plancton superficial de la Bahia de Chimbote. *Bol. Cia. Adm. Guano*, **24**, 717

Rojas de Mendiola, B. 1971. Some observations on the feeding of the Peruvian anchoveta *Engraulis ringens* J. in two regions of the Peruvian coast. In *Fertility of the Sea, vol. 2*, ed. J. D. Costlow, pp. 417–40

Roll, H. 1938. Die Pflanzengesellschaften ostholsteinische Fliessgewässer. *Arch. Hydrobiol.*, **34**, 159–305

Rose, F. L. & McIntire, C. D. 1970. Accumulation of dieldrin by benthic algae in laboratory streams. *Hydrobiol.*, **35**, 481–93

Rosowski, J. R. & Willey, R. L. 1975. *Colacium libellae* sp. nov. (Euglenophyceae) a photosynthetic inhabitant of the larval damselfly rectum. *J. Phycol.*, **11**, 310–15

Ross, D. A., Degens, E. T. & Macilivaine, J. 1970. Black Sea: Recent sedimentary history. *Science*, **170**, 163–5

Roth, P. H., Mullin, M. M. & Berger, W. H. 1975. Coccolith sedimentation by faecal pellets: laboratory experiments and field observations. *Geol. Soc. Am. Bull.*, **86**, 1079–84

Rott, E. 1978. Chlorophyll *a* Konzentration und Zellvolumen als Parameter der Phytoplanktonbiomasse. *Ber. nat-med. Ver. Innsbruck*, **65**, 11–21

Round, F. E. 1953. An investigation of two benthic algal communities in Malham Tarn, Yorkshire. *J. Ecol.*, **41**, 174–97

Round, F. E. 1956. A note on some communities of the littoral of lakes. *Arch. Hydrobiol.*, **52**, 398–405

Round, F. E. 1957a. Studies on bottom-living algae in some lakes of the English Lake district. Part I. Some chemical features of the sediments related to algal productivities. *J. Ecol.*, **45**, 133–48

Round, F. E. 1957b. The late-glacial and post-glacial diatom succession in the Kentmere valley deposit. *New Phytol.*, **56**, 98–126

Round, F. E. 1957c. A note on some diatom communities in calcareous springs and streams. *Journal of the Linnean Soc. of London, Botany. LV*, pp. 662–8

Round, F. E. 1958a. The algal flora of Massom's slack, Freshfield, Lancashire. *Arch. Hydrobiol.*, **54**, 462–76

Round, F. E. 1958b. Algal aspects of lake typology. *Verh. Int. Ver. Limnol.*, **13**, 306–10

Round, F. E. 1958c. A note on the diatom flora of Harlech sand dunes. *J. Roy. Micros. Soc.*, **77**, 130–5

Round, F. E. 1959a. The algal flora of the Tornionjoki, Muonionjoki and Konkamaeno in north Finland. *Soc. Sci. Fenn. Commen. Biol.*, **21**, 1–34

Round, F. E. 1959b. A comparative survey of the epipelic diatom flora of some Irish loughs. *Proc. Roy. Ir. Acad.*, **B, 60**, 193–215

Round, F. E. 1960a. The epipelic algal fora of some Finnish lakes. *Arch. Hydrobiol.*, **57**, 161–78

Round, F. E. 1960b. The diatom flora of a salt marsh on the River Dee. *New Phytol.*, **59**, 332–48

Round, F. E. 1960c. A note on the diatom Flora of some springs in the Malham Tarn area of Yorkshire. *Archiv für Protistenkunde*, **104**, 515–26

Round, F. E. 1960d. Studies on bottom-living algae in some lakes of the English Lake District. Part IV. The seasonal cycle of the Bacillariophyceae. *J. Ecol.* **48**, 529–47

Round, F. E. 1961a. The diatoms of a core from Esthwaite Water. *New Phytol.*, **60**, 43–59

Round, F. E. 1961b. Studies on bottom-living algae in some lakes of the English Lake District. Part VI. The effect of depth on the epipelic algal community. *J. Ecol.*, **49**, 245–54

Round, F. E. 1964. The diatom sequences in lake deposits: some problems of interpretation. *Verh. Int. Ver. Limnol.*, **15**, 1012–20

Round, F. E. 1967. The phytoplankton of the Gulf of California. Part I. Its composition, distribution and contribution to the sediments. *J. exp. mar. Biol. Ecol.*, **1**, 76–97

Round, F. E. 1971a. Benthic marine diatoms. *Oceanogr. Mar. Biol. Ann. Rev.*, **9**, 83–139

Round, F. E. 1971b. The growth and succession of algal populations in freshwaters. *Mitt. Int. Ver. Limnol.*, **19**, 70–99

Round, F. E. 1971c. The taxonomy of the Chlorophyta. II. *Br. Phycol. J.* **6**, 235–64

Round, F. E. 1972a. Some observations on colonies and ultrastructure of the frustule of *Coenobiodiscus muriformis* and its transfer to *Planktoniella*. *J. Phycol.*, **8**, 222–31

Round, F. E. 1972b. *Stephanodiscus binderanus* (Kütz.) Krieger or *Melosira binderana* Kütz. (Bacillarophyta, Centrales). *Phycologia.*, **11**, 109–17

Round, F. E. 1972c. Patterns of seasonal succession of freshwater epipelic algae. *Br. phycol. J.*, **7**, 213–20

Round, F. E. 1973a. *The Biology of the Algae*, second edition. 278 pp. Arnold, London

Round, F. E. 1973b. On the diatom genera *Stephanopyxis* Ehr. and *Skeletonema* Grev and their classification in a revised system of the Centrales. *Bot. Mar.*, **16**, 148–54

Round, F. E. 1979. Botanical aspects of estuaries. In *Tidal Power and Estuary Management*, Colston Pap. no. 30. Wright Scientechnica, Bristol

Round, F. E. & Brook, A. J. 1959. The phytoplankton of some Irish Loughs and an assessment of their trophic status. *Proc. Roy. Ir. Acad.*, **60**, 168–91

Round, F. E. & Eaton, J. W. 1966. Persistent vertical-migration rhythms in benthic microflora. III. The rhythm of epipelic algae in a freshwater pond. *J. Ecol.*, **54**, 609–15

Round, F. E. & Happey, C. M. 1965. Persistent, vertical-migration rhythms in benthic microflora. IV. A diurnal rhythm of the epipelic diatom association in non-tidal flowing water. *Br. phycol. Bull.*, **2**, 463–71

Round, F. E. & Palmer, J. D. 1966. Persistent, vertical-migration rhythms in benthic microflora. II. Field and laboratory studies on diatoms from the banks of the River Avon. *J. mar. biol. Ass., U.K.*, **46**, 191–216

Rowe, G. T., Clifford, C. H. & Smith, K. L. 1977. Nutrient regeneration in sediments off Cap Blanc, Spanish Sahara. *Deep Sea Res.*, **24**, 57–63

Rübel, E. 1936. Plant communities of the world. In *Essays in Geobotany in honor of William Albert Setchel*, ed. T. H. Goodspeed, pp. 263–90

Russell, F. S., Southward, A. J., Boalch, G. T. & Butler, E. I. 1971. Changes in ecological conditions in the English Channel off Plymouth during the last-half century. *Nature, London*, **234**, 468–70

Russell, G. 1967a. The ecology of some free-living Ectocarpaceae. *Helgoländer wiss. Meeresunt.*, **15**, 155–62

Russell, G. 1967b. The genus *Ectocarpus* in Britain. II. The free-living forms. *J. mar. biol. Ass., U.K.*, **47**, 233–50

Russell, G. 1971. Marine algal reproduction in two British estuaries. *Vie et Milieu. Suppl.*, no. 22, 219–30

Russell, G. 1972. Phytosociological studies on a two-zone shore. I. Basic pattern. *J. Ecol.*, **60**, 539–45

Russell, G. 1973. The 'litus' line: a re-assessment. *Oikos*, **24**, 158–61

Russell, G. & Bolton, J. J. 1975. Euryhaline ecotypes of *Ectocarpus siliculosus* (Dillw.) Lyngb. *Estuarine Coastal Mar. Sci.*, **3**, 91–4

Russell, G. & Fielding, A. H. 1974. The competitive properties of marine algae in culture. *J. Ecol.*, **62**, 689–98

Russell, G. & Fletcher, R. L. 1975. A numerical taxonomic study of the British Phaeophyta. *J. mar. biol. Ass., U.K.*, **55**, 763–84

Russell, G. & Garbary, D. 1978. Generic circumscription in the family Ectocarpaceae (Phaeophyceae). *J. mar. biol. Ass., U.K.*, **58**, 517–25

Russell, P. T. & McIntire, W. G. 1965. Southern hemisphere beach rock. *Geog. Rev.*, **55**, 17–48

Russell-Hunter, W. D. 1970. *Aquatic productivity. An introduction to some basic aspects of biological oceanography and limnology*. 306 pp. Macmillan & Co., London

Ruttner, F. 1956. Einige beobachtungen über das Verhalten des Planktons in Seeabflüssen. *Öst. Bot. Zeitsch.*, **103**, 98–100

Ryther, J. H. 1954. Inhibitory effects of phytoplankton upon the feeding of *Daphnia magna* with reference to growth, reproduction and survival. *Ecology*, **35**, 522–37

Ryther, J. H. 1956. Photosynthesis in the ocean as a function of light intensity. *Limnol. Oceanogr.*, **1**, 61–70

Ryther, J. H. 1959. Potential productivity of the sea. *Science*, **130**, 602–8

Ryther, J. H. 1963. Geographic variations in productivity. In *The Sea*, vol. 2, ed. M. N. Hill, pp. 347–80. J. Wiley & Sons

Ryther, J. H. 1969. Photosynthesis and fish production in the sea. *Science*, **166**, 72–6

Ryther, J. H. & Dunstan, W. M. 1971. Nitrogen, phosphorus and eutrophication in the coastal marine environment. *Science*, **171**, 1008–13

Ryther, J. H., Dunstan, W. M., Tenore, K. R. & Huguenin, J. E. 1972. Controlled eutrophication – increasing food production from the sea by recycling human wastes. *Bioscience*, **22**, 144–52

Ryther, J. H., Hall, J. R., Pease, K., Bakun, A. & Jones, M. M. 1966. Primary organic production in relation to the chemistry and hydrography of the Western Indian Ocean. *Limnol. Oceanogr.*, **11**, 371–80

Ryther, J. H. & Hulburt, E. M. 1960. On winter mixing and the vertical distribution of phytoplankton. *Limnol. Oceanogr.*, **5**, 337–8

Ryther, J. H. & Menzel, D. W. 1959. Light-adaptation by marine phytoplankton. *Limnol. Oceanogr.*, **4**, 492–7

Ryther, J. H., Menzel, D. W. & Corwin, N. 1967. Influence of the Amazon River outflow on the ecology of the western tropical Atlantic. I. Hydrography and nutrient chemistry. *J. Mar. Res.*, **25**, 69–83

Ryther, J. H., Menzel, D. W., Hulbert, E. M., Lorenzen, C. J. & Corwin, N. 1970. The production and utilization of organic matter in the Peru coastal current. *Anton Brun Rep.*, no. 4, 403–12. A & M Press, Texas

Ryther, J. N. & Yentsch, C. S. 1957. The estimation of phytoplankton production in the ocean from chlorophyll and light data. *Limnol. Oceanogr.*, **2**, 281–6

Ryther, J. H. & Yentsch, C. S. 1958. Primary production of continental shelf waters off New York. *Limnol. Oceanogr.*, **3**, 327–35

Ryther, J. H., Yentsch, C. S., Hulburt, E. M. & Vaccaro, R. F. 1958. The dynamics of a diatom bloom. *Biol. Bull. mar. biol. Lab., Woods Hole*, **115**, 257–68

Rzóska, J., Brook, A. J. & Prowse, G. A. 1955. Seasonal plankton development on the White and Blue Nile near Khartoum. *Proc. int. Ass. theor. appl. Limnol.*, **12**, 327–34

Sagromsky, H. 1961. Durch Licht-Dunkel-Wechsel induzierter Rhythmus in der Entleerung der Tetrasporangien von *Nitophyllum punctatum. Publ. Staz. Zool. Napoli,* **32**, part 1, 29–40

Saijo, Y., Izuica, S. & Asaoka, O. 1969. Chlorophyll maxima in Kuroshio and adjacent area. *Mar. Biol.,* **4**, 190–6

Saijo, Y. & Sakamoto, M. 1970. Primary production and metabolism of lakes. I. In *Profile of Japanese Science and Scientists,* pp. 207–25. Kodansha, Tokyo

Saino, T. & Hattori, A. 1978. Diel variation in nitrogen fixation by a marine blue-green alga. *Trichodesmum thiebautii. Deep Sea Res.,* **25**, 1259–63

Salăgeanu, N. 1968. Considerations on the recognition of mineral salt requirements of plants on certain soils by means of the alga *Chlamydomonas reinhardi. Rev. Roum. Biol.* ser. Bot., **13**, 189–202

Samsel, G. L. & Parker, B. C. 1972. Limnological investigations in the area of Anvers Island, Antarctica. *Hydrobiol.,* **40**, 505–11

Sand-Jensen, K. 1977. Effect of epiphytes on eelgrass photosynthesis. *Aquatic Bot.,* **3**, 55–63

Sanger, J. E. & Gorham, E. 1972. Stratigraphy of fossil pigments as a guide to the postglacial history of Kirchner Marsh, Minnesota. *Limnol. Oceanogr.,* **17**, 840–54

Santilices, B. 1974. Gelidioid algae, a brief resume of the pertinent literature. *Marine Agronomy, U.S. Sea Grant Program, Hawaii, Tech. Rep. No. 1.* pp. 111

Santilices, B. 1977. Water movement and seasonal algal growth in Hawaii. *Mar. Biol.,* **43**, 225–35

Santos, G. A. & Doty, M. S. 1968. Chemical studies on three species of the marine algal genus *Caulerpa.* In *Drugs from the sea,* ed. H. D. Freudenthal, pp. 173–6

Sargent, M. C. & Austin, T. S. 1949. Organic productivity of an atoll. *Trans. Am. Geophys. Un.,* **30**, 245–9

Sassuchin, D. N., Kabanov, V. M. & Nieswestnova, K. S. 1927. Über die mikroskopische Pflanzen und Tierwelt der Sandflache des Okaufers bei Murom. *Russ. Gidrobiol. Z.,* **6**, 59–83

Saunders, G. W. 1972a. The transformation of artificial detritus in lake water. In *Detritus and its role in aquatic ecosystems. Proc. IBP–UNESCO Symp.,* pp. 261–88. Pallanza, Italy

Saunders, G. W. 1972b. Potential heterotrophy in a natural population of *Oscillatoria agardhii* var. *isothrix* Skuja. *Limnol. Oceanogr.,* **17**, 704–11

Saunders, M. J. & Eaton, J. W. 1976. A method for estimating the standing crop and nutrient content of the phytobenthes of stoney rivers. *Arch. Hydrobiol.,* **78**, 80–101

Scagel, R. F. 1947. An investigation on marine plants near Hardy Bay, B.C. *Rep. Prov. Dept. Fish., Victoria, B.C.,* no. 1, pp. 70

Scagel, R. F. 1957. Benthic algal productivity in the north Pacific with particular reference to the coast of British Columbia. In *Proc. 9th Pacif. Sci. Congress,* pp. 181–7

Scagel, R. F. 1963. Distribution of attached marine algae in relation to oceanographic conditions in the northern Pacific. In *Marine distribution,* ed. R. J. Dunbar, *Roy. Soc. Can.* Special Publ., no. 5, pp. 37–50

Schade, A. 1923. Die kyptogamischen Pflanzengesellschaften an den Felswänden der Sächsischen Schweiz. *Ber. dt. bot. Ges.,* **41**, 49–59

Schallgruber, F. 1944. Das Plankton des Donaustromes bei Wien in Qualitative und Quantitative Hinsicht. *Arch. Hydrobiol.,* **39**, 665–89

Schelske, C. L., Hooper, F. F. & Haertl, E. J. 1962. Responses of a marl lake to chelated iron and fertilizer. *Ecology,* **43**, 646–53

Schelske, C. L., Feldt, L. E., Santiago, M. A. & Stoermer, E. F. 1972. Nutrient enrichment and its effect on phytoplankton production and species composition in Lake Superior. In *Proc. 15th Conf. Great Lakes Res. Int. Assoc.,* Madison, Wisconsin

Schelske, C. L. & Stoermer, E. F. 1971. Eutrophication, silica and predicted changes in algal quality in Lake Michigan. *Science,* **173**, 423–4

Schelske, C. L. & Stoermer, E. F. 1972. Phosphorus, silica and eutrophication of Lake Michigan. In *Nutrients and eutrophication: The limiting nutrient controversy,* Limnol. Oceanogr. Symp. I, 157–71

Schiller, J. 1954. Über winterliche pflanzliche Bewohner des Wassers, Eises und des daraufliegenden Schneebreies. I. *Öst. Bot. Zeitsch.,* **101**, 236–84

Schindler, D. W. 1971a. Carbon, nitrogen and phosphorus and the eutrophication of freshwater lakes. *J. Phycol.,* **17** 321–9

Schindler, D. W. 1971b. Light, temperature and oxygen regimes of selected lakes in the Experimental Lakes Area, Northwestern Ontario. *J. Fish. Res. Bd. Can.,* **28**, 157–69

Schindler, D. W. 1973. Experimental approaches to limnology – an overview. *J. Fish. Res. Bd. Can.,* **30**, 1409–13

Schindler, D. W. 1977. Evolution of phosphorus limitation in lakes. *Science,* **195**, 260–2

Schindler, D. W. 1978. Factors regulating phytoplankton production and standing crop in the world's freshwaters. *Limnol. Oceanogr.,* **23**, 478–86

Schindler, D. W., Fee, E. J. & Ruszczynski, T. 1978. Phosphorus input and its consequences for phytoplankton standing crop and production in the experimental lakes area and in similar lakes. *J. Fish. Res. Bd. Can.,* **35**, 190–6

Schindler, D. W., Frost, V. E. & Schmidt, R. V. 1973. Production of epilithophyton in two lakes of the experimental lakes area, Northwestern Ontario. *J. Fish. Res. Bd. Can.,* **30**, 1511–24

Schindler, D. W. & Holmgren, S. K. 1971. Primary production and phytoplankton in the Experimental Lakes Area, northwestern Ontario and other low-carbonate waters and a liquid scintillation method for determining ^{14}C activity in photosynthesis. *J. Fish. Res. Bd. Can.,* **28**, 189–201

Schindler, D. W., Kling, H., Schmidt, R. V., Prokopowich, J., Frost, V. E., Reid, R. L. & Capel, M. 1973. Eutrophication of Lake 227 by addition of phosphate and nitrate: the second, third and fourth years of enrichment 1970, 1971, 1972. *J. Fish. Res. Bot. Can.,* **30**, 1415–40

Schlichting, H. E. 1964. Meteorological conditions affecting the dispersal of airborne algae and protozoa. *Lloydia.,* **27**, 64–78

Schlichting, H. E. 1969. The importance of airborne algae and protozoa. *J. Air Pollut. Cont. Ass.*, **19**, 946–51

Schlichting, H. E. 1975. Some subaerial algae from Ireland. *Br. phycol. J.*, **10**, 257–61

Schlichting, H. E., Jr & Gearheart, R. A. 1966. Some effects of sewage effluent upon phyco-periphyton in Lake Murray, Oklahoma. *Proc. Okla. Acad. Sci.*, **46**, 19–46

Schlichting, H. E. & Milliger, L. E. 1969. The dispersal of micro-organisms by a Hemipteran, *Lethoceres uhleri* (Montadon). *Trans. Am. Micros. Soc.*, **88**, 452–4

Schlichting, H. E. & Sides, S. L. 1969. The passive transport of aquatic micro-organisms by selected Hemiptera. *J. Ecol.*, **57**, 759–64

Schmidt, R. J., Gooch, Van, D., Loeblich, A. R., III & Woodland Hastings, J. 1978. Comparative study of luminescent and non-luminescent strains of *Gonyaulax excavata* (Pyrrhophyta). *J. Phycol.*, **14**, 5–9

Schmitz, K. & Lobban, C. S. 1976. A survey of translocation in Laminariales (Phaeophyceae). *Mar. Biol.*, **36**, 207–16

Schmitz, K., Lüning, K. & Willenbrink, J. 1972. CO_2-Fixierung und Stofftransport in benthischen mariner Algen. II. Zum Ferntransport ^{14}C-markierter Assimilate bei *Laminaria hyperborea* und *Laminaria saccharina*. *Z. Pflanzenphysiol.*, **67**, 418–29

Scholl, D. W. 1960. Pleistocene algal pinnacles at Searles Lake, California. *J. sedim. Petrol.*, **30**, 414–31

Schonbeck, M. & Norton, T. A. 1978. Factors controlling the upper limits of Fucoid algae on the shore. *J. exp. mar. Biol. Ecol.*, **31**, 303–13

Schöne, H. 1970. Untersuchungen zur ökologischen Bedeutung des Seegangs für das Plankton mit besonderer Beruchsichtigung mariner Kieselalgen. *Int. Rev. ges. Hydrobiol.*, **55**, 595–677

Schopf, J. W. 1971. Organically preserved Precambrian microorganisms. In *Proc. N. Am. Paleont. Conv. Chicago*, 1013–57

Schopf, J. W. & Barghoorn, E. G. 1967. Alga-like fossils from the early Precambrian of South Africa. *Science*, **156**, 508–12

Schorler, B. 1914. Die Algenvegetation an den Felswanden des Elbsandsteingebirgs. *Abh. Naturw. Ges. 'Isis', Dresden*

Schrader, H. J. 1971. Fecal pellets: role in sedimentation of pelagic diatoms. *Science*, **174**, 55–7

Schrader, H. J. 1973. Cenozoic diatoms from the north-east Pacific, Leg 18. In *Initial Report of the deep sea drilling project*, ed. L. D. Kulm & R. von Huene, vol. 18, pp. 673–797. U.S. Government Printing Office, Washington

Schrader, H. J. & Richert, P. 1974. Paleotemperature interpretation by means of percent amount of *Dictyocha/Distephanus* (Silicoflagellatae). Abstract in *Marine Plankton and sediments, 3rd Planktonic Conf., Kiel*

Schramm, W. 1968. Ökologisch-physiologische Untersuchungen zur Austrocknungs-und Temperaturresistenz an *Fucus vesiculosus*. L. der westlichen Ostsee. *Int. Rev. ges. Hydrobiol.*, **53**, 469–510

Schroeder, J. H. 1972. Calcified filaments of an endolithic alga in recent Bermuda reefs. *Jb. Geol. Paläont. Abh.*, 16–23

Schulenberger, E. 1978. The deep chlorophyll maximum and mesoscale environmental heterogeneity in the western half of the North Pacific central gyre. *Deep Sea Res.*, **25**, 1193–208

Schultz, M. E. 1971. Salinity-related polymorphism in the brackish water diatom *Cyclotella cryptica*. *Can. J. Bot.*, **49**, 1285–9

Schwabe, G. H. 1960. Zur autotrophen Vegetation in ariden Boden. Blaualgen und Lebensraum. IV. *Öst. Bot. Zeitsch.*, **107**, 281–309

Schwartz, R. M. & Dayhoff, M. 1978. Origins of prokaryotes, eukaryotes, mitochondria and chloroplasts. *Science*, **199**, 395–403

Schwartz, S. L. & Almodóvar, L. R. 1971. Heat tolerance of reef algae at La Parguera, Puerto Rico. *Nova Hedw.*, **21**, 231–9

Schwenke, H. 1959. Untersuchungen zur Temperatur-resistenz marine Algen der Westlicher Ostsee. I. Das Resistenzverhalten von Tiefen roten algen bei ökologisher und nicht ökologische Temperatur. *Kieler Meresforsch.*, **15**, 34–50

Scoffin, T. P. 1970. The trapping and binding of subtidal carbonate sediments by marine vegetation in Bimini Lagoon, Bahama. *J. sedim. Petrol.*, **40**, 249–73

Scott, B. D. & Jitts, H. R. 1977. Photosynthesis of phytoplankton and Zooxanthellae on a coral reef. *Mar. Biol.*, **41**, 307–15

Scott, C. T. & Hayward, H. R. 1955. Sodium and potassium regulation in *Ulva lactuca* and *Valonia macrophysa*. In *Electrolytes in Biological Systems*, ed. A. M. Schones, Am. Physiol. Soc., Washington, D.C., pp. 35–64

Scura, E. D. & Jerde, C. W. 1977. Acceptance of various species of phytoplankton as food by the larvae of the northern anchovy *Engraulis mordax* and the relative nutritional value of the dinoflagellates *Gymnodinium splendens* and *Gonyaulax polyedra*. *Fish. Bull.*, **75**

Sears, J. R. & Cooper, R. A. 1978. Descriptive ecology of offshore, deep water, benthic algae in the temperate western north Atlantic Ocean. *Mar. Biol.*, **44**, 309–34

Seliger, H. H., Carpenter, T. W., Loftus, M. & McElroy, W. D. 1970. Mechanism for the accumulation of high concentrations of dinoflagellates in a bioluminescent Bay. *Limnol. Oceanogr.*, **15**, 234–45

Semina, G. I. 1969. The size of phytoplankton cells along longitude 174° W in the Pacific Ocean. *Oceanology*, **9**, 391–8

Semina, G. I. 1967. Plankton. I. In *The Pacific ocean. Vol. 7. Biology of the Pacific Ocean*, ed. V. G. Bogorov, 262 pp.

Semina, H. J. 1972. The size of phytoplankton cells in the Pacific Ocean. *Int. Rev. ges. Hydrobiol.*, **57**, 177–205

Semina, H. J. & Tarkhova, I. A. 1972. Ecology of phytoplankton in the North Pacific Ocean. In *Biological Oceanography of the northern north Pacific Ocean*, ed. A. Y. Takenouti, pp. 117–24. Tokyo

Senft, W. H. 1978. Dependence of light-saturated rates of algal photosynthesis on intracellular concentrations of phosphorus. *Limnol. Oceanogr.*, **23**, 709–18

Sen Gupta, R. 1972. Photosynthetic production and its regulating factors in the Baltic Sea. *Mar. Biol.*, **17**, 82–92

Seoane-Camba, J. 1969. Sobre la zonación del sistena litoral y su nomenclatura. *Invest. Pesq.*, **33**, 261–7

Serruya, C. 1972. Metalimnic layer in Lake Kinneret, Israel. *Hydrobiol.*, **40**, 355–9

Serruya, C. 1975. Nitrogen and phosphorus balances and load-biomass relationship in Lake Kinneret (Israel). *Verh. Int. Ver. Limnol.*, **19**, 1357–69

Serruya, C. & Berman, T. 1975. Phosphorus, nitrogen and the growth of algae in Lake Kinneret. *J. Phycol.*, **11**, 155–62

Serruya, C., Edelstein, M., Pollingher, U. & Serruya, S. (1974). Lake Kinneret sediments: Nutrient composition of the pore water and mud water exchanges. *Limnol. Oceanogr.* **19**, 487–508

Serruya, C., Leenhardt, O. & Lomrard, A. 1966. Etudes géophysiques dans le lac Léman. Interprétation géologique. *Arch. Sci. Genève*, **19**, 179–96

Serruya, C. & Pollingher, U. 1971. An attempt at forecasting the *Peridinium* bloom in Lake Kinneret. (Lake Tiberias.) *Mitt. Int. Ver. Limnol.*, **19**, 277–91

Serruya, C., Serruya, S. & Pollingher, U. 1978. Wind, phosphorus release and division rate of *Peridinium* in Lake Kinneret. *Verh. Int. Ver. Limnol.*, **20**, 1096–102

Setchell, W. A. 1928. A botanical view of coral reefs, especially those of the Indo-Pacific region. In *Proc. 3rd Pacif. Sci. Congress*, pp. 1837–43

Setchell, W. A. 1930. Biotic cementation in coral reefs. *Proc. natn. Acad. Sci.*, **16**, 781–3

Sharp, J. H. 1969. Blue-green algae and carbonates – *Schizothrix calcicola* and algal stromatolites from Bermuda. *Limnol. Oceanogr.*, **14**, 568–78

Sharp, J. H. 1977. Excretion of organic matter by marine phytoplankton. Do healthy cells do it? *Limnol Oceanogr.*, **22**, 381–99

Sheath, R. G. & Hellebust, J. A. 1974. Glucose transport systems and growth characteristics of *Bracteococcus minor*. *J. Phycol.*, **10**, 34–41

Sheath, R. G., Hellebust, J. A. & Sawa Takasi 1975. The statospore of *Dinobryon divergens* Imhof: Formation and germination in a subarctic lake. *J. Phycol.*, **11**, 131–8

Shelford, V. E. & Eddy, S. 1929. Methods for the study of stream communities. *Ecology*, **10**, 382–91

Shepherd, S. A. & Womersley, H. B. S. 1970. The sublittoral ecology of West Island, South Australia. I. Environmental features and algal ecology. *Trans. Roy. Soc. S. Australia*, **94**, 105–37

Shepherd, S. A. & Womersley, H. B. S. 1971. Pearson Island Expedition 1969. The sub-tidal ecology of benthic algae. *Trans. Roy. Soc. S. Australia*, **95**, 155–67

Sheridan, R. P. 1976. Sun-shade ecotypes of a blue-green alga in a hot spring. *J. Phycol.*, **12**, 279–85

Shields, L. M. & Durrell, L. W. 1964. Algae in relation to soil fertility. *Bot. Rev.*, **30**, 92–128

Shields, L. M., Mitchell, C. & Drouet, F. 1957. Alga- and lichen-stabilised surface crusts as soil nitrogen sources. *Am. J. Bot.*, **44**, 489–98

Shilo, M. 1967. Formation and mode of action of algal toxins. *Bact. Rev.*, **31**, 180–93

Shtina, E. A. 1960. Methods for the calculation of algae as components of the soil microflora. *Soviet Soil Sci.*, **5**, 106–11

Sieburth, J. McN. 1964. Antibacterial substances produced by marine algae. *Dev. Indust. Microbiol.*, **5**, 124–34

Sieburth, J. McN. 1965. Role of algae in controlling bacterial populations in estuarine waters. In *Pollutions marines par les microorganisms et les produits petroliers*, pp. 217–33

Sieburth, J. McN. 1968. The influence of algal antibiosis on the ecology of marine organisms. In *Advances in microbiology of the Sea*, vol. I, pp. 63–94

Sieburth, J. McN. & Conover, J. T. 1965. *Sargassum* tannin, an antibiotic which retards fouling. *Nature, London*, **208**, 52–3

Sieburth, J. McN. & Jensen, A. 1968. Studies on algal substances in the sea. I. Gelb-stoff (humic material) in terrestrial and marine water. *J. exp. mar. Biol. Ecol.*, **2**, 174–89

Sieburth, J. McN. & Jensen, A. 1969a. Studies on algal substances in the sea. II. The formation of gelb-stoff (humic material) by exudates of Phaeophyta. *J. exp. mar. Biol.*, **3**, 275–89

Sieburth, J. McN. & Jensen, A. 1969b. Studies on algal substances in the sea. III. The production of extracellular organic matter by littoral marine algae. *J. exp. mar. Biol.*, **3**, 290–309

Sieburth, J. McN. & Pratt, D. M. 1962. Anticoliform activity of sea water associated with the termination of *Skeletonema costatum* blooms. *Trans. N.Y. Acad. Sci.*, ser. 2, **24**, 498–501

Sieburth, J. McN. & Thomas, C. D. 1973. Fouling on eelgrass. (*Zostera marina* L.) *J. Phycol.*, **9**, 146–50

Silva, M. W. R. N. de & Burrows, E. M. 1973. An experimental assessment of the status of the species *Enteromorpha intestinalis* (L.) Link and *Enteromorpha compressa* (L.) Grev. *J. mar. biol. Ass., U.K.*, **53**, 895–904

Silva, P. C. 1979. Review of the taxonomic history and nomenclature of the yellow-green algae. *Arch. Protistenk.*, **121**, 20–63

Silver, P. A. 1977. Comparison of attached diatom communities on natural and artificial substrates. *J. Phycol.*, **13**, 402–6

Simkiss, K. 1964. Possible effects of zooxanthellae on coral growth. *Experimentia*, **20**, 140

Simons, J. 1975a. *Vaucheria* species from estuarine areas in the Netherlands. *Neth. J. Sea. Res.*, **9**, 1–23

Simons, J. 1975b. Periodicity and distribution of brackish *Vaucheria* species from non-tidal coastal areas in the S.W. Netherlands. *Acta. Bot. Neerl.*, **24**, 89–110

Simonsen, R. 1968. Zur Küstenvegetation der Saro-Inseln im Roten Meer. '*Meteor*' *Forschungsergebnisse. Reihe*, **D**, **3**, 57–66

Simonsen, R. & Kanaya, T. 1961. Notes on the marine species of the diatom genus *Denticula* Kütz. *Int. Rev. ges. Hydrobiol.*, **46**, 498–513

Sjöstedt, L. T. 1928. Littoral and supralittoral studies on Scanian shores. *Lunds. Univ. Årsskr.*, **24**, 1–36

Skeen, J. M. 1975. An evaluation of chloroplast pigment characteristics as an indicator of successional status in terrestrial ecosystems. *Am. Midl. Nat.*, **94**, 370–84

Skuja, H. 1948. Taxonomie des Phytoplanktons einiger Seen in Uppland, Schweden. *Symb. bot. upsal.*, **9**, 1–400

Skuja, H. 1964. Grundzüge der Algenflora und Algenvegetation der Fjeldgegenden um Abisko in Schwechisch Lappland. *Nova Acta R. Soc. Scient. upsal.*, no. 4, **18**, 465 pp.

Sládeček, V. & Perman, J. 1978. Saprobic sequence within the genus *Euglena*. *Hydrobiol.*, **57**, 57–8

Sládečkova, A. 1963. Aquatic deuteromycetes as indicators of starch campaign pollution. Int. Rev. ges. Hydrobiol. *Hydrogr.* **48**, 35–42

Slobodkin, L. B. 1959. Energetics in *Daphnia pulex* populations. *Ecology*, **40**, 232–43

Slobodkin, L. B. 1961. *Growth and regulation of animal populations.* 184 pp. Holt, Rinehart & Winston, New York

Smayda, T. J. 1958. Biogeographical studies of marine phytoplankton. *Oikos*, **9**, 158–91

Smayda, T. J. 1966. A quantitative analysis of the phytoplankton of the Gulf of Panama. III. General ecological conditions and the phytoplankton dynamics at 8° 45′ N, 79° 23′ W from November 1954 to May 1957. *Bull. inter. Am. trop. Tuna Commn.*, **11**, 355–612

Smayda, T. J. 1970. The suspension and sinking of phytoplankton in the sea. *Oceanogr. Mar. Biol. Ann. Rev.*, **8**, 353–414

Smayda, T. J. 1971a. Further enrichment experiments using the marine centric diatom *Cyclotella nana* (clone 13-1) as an assay organism. In *Fertility of the Sea*, vol. 2, ed. J. D. Costlow, pp. 453–511

Smayda, T. J. 1971b. Normal and accelerated sinking of phytoplankton in the sea. *Mar. Geol.*, **11**, 105–22

Smayda, T. J. 1973. The growth of *Skeletonema costatum* during a winter-spring bloom in Narragansett Bay, Rhode Island. *Norw. J. Bot.*, **20**, 219–47

Smayda, T. J. 1974. Bioassay of the growth potential of the surface water of lower Narragansett Bay over an annual cycle using the diatom *Thalassiosira pseudonana* (oceanic clone 13-1). *Limnol. Oceanogr.*, **19**, 889–901

Smayda, T. J. 1975. Net phytoplankton and the greater than 20-micron phytoplankton size fraction in upwelling waters off Baja California. *Fish. Bull.*, **73**, 38–50

Smith, D. C. 1975. Symbiosis and the biology of lichenised fungi. In *Symbiosis Symp.*, *Soc. Exp. Biol.*, Vol. 29, ed. D. H. Jennings & D. L. Lee, pp. 373–405. Cambridge University Press

Smith, D. C. & Drew, E. A. 1965. Studies in the physiology of lichens. V. Translocation from the algal layer to the medulla in *Peltigera polydactyla*. *New Phytol.*, **64**, 195–200

Smith, D. C., Muscatine, L. & Lewis, D. 1969. Carbohydrate movement from autotrophs to heterotrophs in parasitic and mutualistic symbiosis. *Biol. Rev.*, **44**, 17–90

Smith, D. W. & Brock, T. D. 1973. The water relations of the alga *Cyanidium caldarium* in soil. *J. gen. Microbiol.*, **79**, 219–31

Smith, D. W., Fliermans, C. B. & Brock, T. D. 1972. Technique for measuring $^{14}CO_2$ uptake by soil microorganism *in situ*. *Appl. Microbiol.*, **23**, 595–600

Smith, G. M. 1947. On the reproduction of some Pacific coast species of *Ulva*. *Am. J. Bot.*, **34**, 80–7

Smith, J. E. 1968. ' *Torrey Canyon*' Pollution and marine life. 196 pp. Cambridge University Press

Smith, R. M. 1967. Sublittoral ecology of marine algae on the North Wales coast. *Helgoländer wiss. Meeresunt.*, **15**, 467–79

Smith, S. L. & Whitledge, T. E. 1977. The role of zooplankton in the regeneration of nitrogen in a coastal upwelling system off north west Africa. *Deep Sea Res.*, **24**, 49–56

Smith, S. V. 1973. Carbon dioxide dynamics: A record of organic carbon production, respiration and calcification in the Eniwetok reef flat community. *Limnol. Oceanogr.*, **18**, 106–20

Smith, S. V. & Marsh, J. A., Jr 1973. Organic carbon production on the windward reef flat of Eniwetok Atoll. *Limnol. Oceanogr.*, **18**, 953–61

Smith, W. O. J., Barber, R. T. & Huntsman, S. A. 1977. Primary production off the coast of north west Africa: excretion of dissolved organic matter and its heterotrophic uptake. *Deep Sea Res.*, **24**, 35–47

Sorokin, Ju I. 1971. On the role of bacteria in the productivity of tropical oceanic waters. *Int. Rev. ges. Hydrobiol.*, **56**, 1–48

Sorokin, Y. I. & Konovalova, I. W. 1973. Production and decomposition of organic matter in a bay of the Japan Sea during the winter diatom bloom. *Limnol. Oceanogr.*, **18**, 962–7

Sournia, A. 1965. Phytoplankton et productivité primaire dans une baie de Nossi-Bé (Madagascar). *C. r. hebd. Séanc. Acad. Sci., Paris*, **261**, 2245–8

Sournia, A. 1974. Circadian periodicities in natural populations of marine phytoplankton. *Adv. mar. Biol.*, **12**, 325–89

Sournia, A. 1976a. Primary production of sands in the Lagoon of an atoll and the role of Foraminiferan symbionts. *Mar. Biol.*, **37**, 29–32

Sournia, A. 1976b. Ecologie et productivité d'une Cyanophycée en milieu corallien: *Oscillatoria limosa* Agardh. *Phycologia*, **15**, 363–6

South, G. R. & Burrows, E. M. 1967. Studies on marine algae of the British Isles. 5. *Chorda filum* (L.) Stack. *Br. phycol. Bull.*, **3**, 379–402

Southward, A. J. 1958. The zonation of plants and animals on rocky sea shores. *Biol. Rev.*, **33**, 137–77

Sperling, J. A. 1975. Algal ecology of southern Icelandic hot springs. *Ecology*, **56**, 183–90

Sperling, J. A. & Grunewald, R. 1969. Batch culturing of thermophilic benthic algae and phosphorus uptake in a laboratory stream model. *Limnol. Oceanogr.*, **14**, 944–9

Stadelmann, P. 1971. Stickstoffkreislauf und Primärproduktion im mesotrophen Vierwaldstätersee (Horver Bucht) und im eutrophen Rotsee, mit besonderer Berücksichtigung des Nitrats als limitierenden Faktoren. *Schweiz. Z. Hydrol.*, **33**, 1–65

Stadelmann, P. & Munawar, M. 1974. Biomass parameters and primary production at a nearshore and a mid-lake station of Lake Ontario during IFYGL. In *Proc. 17th Conf. Great Lakes Res. Int. Assoc.*, 109–19

Stangenberg, M. 1968a. Toxic effect of *Microcystis aeruginosa* Kg. extracts on *Daphnia longispina* O. F. Müller and *Eucypris virens* Jurine. *Hydrobiol.*, **32**, 81–7

Stangenberg, M. 1968b. Bacteriostatic effects of some algae and *Lemna minor* extracts. *Hydrobiol.*, **32**, 88–96

Staley, J. T. 1971. Growth rates of algae determined *in situ* using an immersed microscope. *J. Phycol.*, **7**, 13–17

Stanley, D. W. 1976. Productivity of epipelic algae in tundra ponds and a lake near Barrow, Alaska. *Ecology*, **57**, 1015–24

Stanley, D. W. & Daly, R. J. 1976. Environmental control of

primary productivity in Alaskan tundra ponds. *Ecology*, **57**, 1025–33

Stark, L. M., Almodóvar, L. & Krauss, R. W. 1969. Factors affecting the rate of calcification in *Halimeda opuntia* (L.) Lam. and *Halimeda discoedea* Decaisne. *J. Phycol.*, **5**, 305–12

Starr, R. C. 1955. A comparative study of *Chlorococcum* Meneghini and other spherical, zoopore-producing genera of the Chloroccocales. *Sci. Ser.*, no. 20. University of Indiana

Starr, R. C. 1970. Control of differentiation in *Volvox*. In *29th Symp. developmental biology, Dev. Biol. Suppl.*, no. 4, pp. 59–100

Steele, R. L. 1965. Induction of sexuality in two centric diatoms. *Bioscience*, **15**, 298

Steele, J. H. & Baird, I. E. 1968. Production ecology of a sandy beach. *Limnol. Oceanogr.*, **13**, 14–25

Steele, J. H. & Yentsch, C. S. 1960. The vertical distribution of chlorophyll. *J. mar. biol. Ass.*, *U.K.*, **39**, 217–26

Steemann-Nielsen, E. 1952. The use of radio-active carbon (C^{14}) for measuring organic production in the sea. *J. cons. perm. int. Expl. Mer.*, **18**, 117–40

Steemann-Nielsen, E. 1953. On organic production in the oceans. *J. Cons. perm. int. Expl. Mer.*, **19**, 309–28

Steemann-Nielsen, E. 1962. On the maximum quantity of plankton chlorophyll per surface unit. *Int. Rev. ges. Hydrobiol.*, **47**, 333–8

Steemann-Nielsen, E. 1964. Investigations of the rate of primary production at two Danish Light ships in the transition area between the North Sea and the Baltic. *Medd. Dan. Fisk-Havunders. N.S.*, **4**, 31–77

Steemann-Nielsen, E. 1974. Light and Primary Production. In *Optical Aspects of Oceanography*, ed. N. G. Jerlov, pp. 331–88. Academic Press, London & New York

Steemann-Nielsen, E. & Hansen, V. K. 1961. Influence of surface illumination on plankton photosynthesis in Danish waters (56° N) throughout the year. *Physiol. Pl.*, **14**, 595–613

Steemann-Nielsen, E. & Jørgensen, E. G. 1968. The adaptation of plankton algae. I. General part. *Physiol. Plant.*, **21**, 401–13

Steemann-Nielsen, E. & Willemoes, M. 1971. How to measure the illumination rate when investigating the rate of photosynthesis of unicellular algae under various light conditions. *Int. Rev. ges. Hydrobiol.*, **56**, 541–56

Stein, J. E. & Denison, J. G. 1967. Limitations of indicator organisms. In *Pollution and marine ecology*, ed. T. A. Olsen & F. J. Burgess, pp. 323–35. Interscience

Stein, J. R. & Amundsen, C. C. 1967. Studies on snow algae and fungi from the Front Range of Colorado. *Can. J. Bot.*, **45**, 2033–45

Stein, J. R. & Brock, R. C. 1964. Red snow from Mt. Seymour, British Columbia. *Can. J. Bot.*, **42**, 1183–8

Steinbiss, H. H. & Schmitz, K. 1973. CO_2-Fixierung und Stofftransport in benthischen marinen Algen. V. Zur autoradiographischen Lokalisation der Assimilattransportbahnen im Thallus von *Laminaria hyperborea*. *Planta*, **112**, 253–63

Steneck, R. S. & Adey, W. H. 1976. The role of environment in control of morphology in *Lithophyllum congestum*, a Caribbean algal ridge builder. *Bot. Mar.*, **19**, 197–215

Stensland, G. L. 1976. Precipitation chemistry studies at Lake

George: Acid rains. *Rensselaer Freshw. Inst. Quart. Rep.*, no. 6, 1–3

Stephan, D. G. & Schaffer, R. B. 1970. Waste water treatment and renovation status of process development. *J. Wat. Pollut. Control Fed., Washington*, **42**, 399–410

Stephens, D. W. & Gillespie, D. M. 1976. Phytoplankton production in Great Salt Lake, Utah, and a laboratory study of algal response to enrichment. *Limnol. Oceanogr.*, **21**, 74–87

Stephens, G. C. & North, B. B. 1971. Extrusion of carbon accompanying uptake of amino acids by marine phytoplankton. *Limnol. Oceanogr.*, **16**, 752–7

Stephenson, T. A. & Stephenson, A. 1949. The universal features of zonation between tide marks on rocky coasts. *J. Ecol.*, **37**, 289–305

Stephenson, T. A. & Stephenson, A. 1972. *Life Between Tide Marks on Rocky Shores*. Pp. 425. W. H. Freeman & Co. San Francisco.

Stephenson, W. & Searles, R. B. 1960. Environmental studies on the ecology of intertidal environments at Heron Island. *Aust. J. mar. Freshwat. Res.*, **11**, 241–67

Steven, D. M. 1971. Primary productivity of the tropical western Atlantic Ocean near Barbados. *Mar. Biol.*, **10**, 261–4

Steven, D. M., Brooks, A. L. & Moore, E. A. 1970. Primary and secondary production in the tropical Atlantic. Final report. *Bermuda Biol. Sta.*, 124 pp.

Stewart, J. G. 1977a. Effects of lead on the growth of four species of red algae. *Phycologia*, **16**, 31–6

Stewart, J. G. 1977b. Relative sensitivity to lead of a naked green flagellate *Dunaliella tertiolecta*. *Water Air Soil Pollut.*, **8**, 243–7

Stewart, J. G. & Schultz-Baldes, M. 1976. Long-term lead accumulation in Abalone (*Haliotis* spp.) fed on lead-treated brown algae. (*Egregia laevigata*.) *Mar. Biol.*, **36**, 19–24

Stewart, K. D. & Mattox, K. R. 1975. Comparative cytology, evolution and classification of the green algae with some consideration of the origin of other organisms with chlorophylls *a* and *b*. *Bot. Rev.*, **41**, 104–35

Stewart, W. D. P. 1965. Nitrogen turnover in marine and brackish habitats. I. Nitrogen fixation. *Ann. Bot.*, **20**, 229–39

Stewart, W. D. P. 1967. Transfer of biologically fixed nitrogen in a sand dune slack region. *Nature, London*, **214**, 603–4

Stewart, W. D. P. 1970. Algal fixation of atmospheric nitrogen. *Plant and Soil*, **32**, 555–88

Stewart, W. D. P. 1971. Nitrogen fixation in the sea. In *Fertility of the Sea*, vol. 2, ed. J. D. Costlow, pp. 537–64. Gordon & Breach

Stewart, W. D. P., Mague, T., Fitzgerald, G. P. & Burris, R. H. 1971. Nitrogenase activity in Wisconsin lakes of differing degrees of eutrophication. *New Phytol.*, **70**, 497–501

Stewart, W. D. P. & Pearson, H. W. 1970. Effects of aerobic and anaerobic conditions on growth and metabolism of blue-green algae. *Proc. Roy. Soc. London*, **B**, **175**, 293–311

Stewart, W. D. P. & Pugh, G. F. J. 1963. Blue-green algae of a developing salt-marsh. *J. mar. Biol. Ass., U.K.*, **43**, 309–17

Steyaert, J. 1973a. Distribution of plankton diatoms along an African Antarctic transect. *Invest. Pesq.*, **37**, 295–328

Steyaert, J. 1973b. Difference in diatom abundance between

the two summer periods of 1965–1967 in Antarctic inshore waters (Breid Bay). *Invest. Pesq.*, **37**, 517–32

Stockman, K. W., Ginsburg, R. N. & Shin, E. A. 1967. The production of lime mud by algae in South Florida. *J. sedim. Petrol.*, **37**, 633–48

Stockner, J. G. & Antia, N. J. 1976. Phytoplankton adaptation to environmental stresses from toxicants, nutrients and pollutants – a warning. *J. Fish. Res. Bd. Can.*, **33**, 2089–96

Stockner, J. G. & Armstrong, F. A. J. 1971. Periphyton of the Experimental Lakes Area, northwestern Ontario. *J. Fish. Res. Bd. Can.*, **28**, 215–29

Stockner, J. G. & Cliff, D. D. 1976. Effects of pulp mill effluent on phytoplankton production in coastal marine waters off British Columbia. *J. Fish. Res. Bd. Can.*, **33**, 2433–62

Stockner, J. G. & Lund, J. W. G. 1970. Live algae in postglacial lake deposits. *Limnol. Oceanogr.*, **15**, 41–58

Stockner, J. G. & Shortreed, K. R. S. 1978. Enhancement of autotrophic production by nutrient addition in a coastal rainforest stream on Vancouver Island. *J. Fish. Res. Bd. Can.*, **35**, 28–34

Stoddart, D. R. 1969. Ecology and morphology of recent coral reefs. *Biol. Rev.*, **44**, 433–98

Stoermer, E. F. 1967. An historical comparison of offshore phytoplankton populations in Lake Michigan. In *Studies on the environment and eutrophication of Lake Michigan. Univ. Michigan Great Lakes Res. Div. Spec. Rep.*, ed. J. C. Ayres & D. C. Chandler, no. 30, pp. 47–77

Stoermer, E. F. & Yang, J. J. 1969. Plankton diatom assemblages in Lake Michigan. *Univ. Michigan Great Lakes Res. Div. Spec. Rep.*, no. 47, 268 pp.

Stokes, J. L. 1940. The influence of environmental factors upon the development of algae and other micro-organisms in soil. *Soil. Sci.*, **49**, 171–84

Stosch, H. A. von 1954. Die Oogamie von *Biddulphia mobiliensis* und die bisher bekannten Auxosporenbildungen bei den Centrales. *Com. 8th Congr. Int. Bot.*, sect. 17, 56–68

Stosch, H. A. von 1956. Die zentrischen Grunddiatomeen. Beiträge zur Floristik und Ökologie einer Pflanzengesellschaft der Nordsee. *Helgoländer wiss. Meeresunt.*, **5**, 273–91

Strain, H. M., Manning, W. M. & Harding, G. 1944. Xanthophylls and carotenes of diatoms, brown algae, dinoflagellates and sea anemones. *Biol. Bull. mar. biol. Lab., Woods Hole*, **86**, 169–91

Strelnikova, N. I. 1975. Diatoms of the Cretaceous Period. In *3rd Symp. on recent and fossil marine diatoms, Kiel, Nova Hedw. Beih.*, **53**, 311–21

Strickland, J. D. H. 1960. Measuring the production of marine phytoplankton. *Bull. Fish. Res. Bd. Can.*, no. 122, 172 pp.

Strickland, J. D. H. 1966. Production of organic matter in the primary stages of the marine food chain. In *Chemical Oceanography*, ed. J. P. Riley & G. Skirrow, vol. 1, pp. 477–610. Academic Press, London

Strickland, J. D. H. 1968. A comparison of profiles of nutrient and chlorophyll concentrations taken from discrete depths and by continuous recording. *Limnol. Oceanogr.*, **13**, 388–91

Strickland, J. D. H., Eppley, R. W. & Mendiola, B. R. de 1969. Phytoplankton populations, nutrients and photosynthesis in Peruvian coastal waters. *Bol. Inst. Mar. Peru*, **2**, 37–45

Strickland, J. D. H., Holm-Hansen, O., Eppley, R. W. & Linn, R. J. 1969. The use of a deep tank in plankton ecology. I. Studies of the growth and composition of phytoplankton crops at low nutrient levels. *Limnol. Oceanogr.*, **14**, 23–34

Strickland, J. D. H. & Parsons, T. R. 1968. A practical handbook of seawater analysis. *Bull. Fish. Res. Bd. Can.*, no. 167, 311 pp.

Strickland, J. D. H., Solorzano, L. & Eppley, R. W. 1970. The ecology of the plankton off La Jolla, California, in the period April through September 1967. Part I. General introduction, hydrography and chemistry. *Bull. Scripps Inst. Oceanogr.*, **17**, 1–22

Strickland, J. D. H. & Terhune, L. D. B. 1961. The study of *in situ* marine photosynthesis using a large plastic bag. *Limnol. Oceanogr.*, **6**, 93–6

Strøm, K. M. 1920. Norwegian mountain algae. *Skr. Norske Videnskaps. Math.-Natur.*, Kl, no. 6

Stull, E. A. 1975. The diversity of cell size in phytoplankton assemblages. *Verh. Int. Ver. Limnol.*, **19**, 630–4

Stull, E. A., Amezaga, E. de & Goldman, C. R. 1973. The contribution of individual species of algae to primary productivity of Castle Lake, California. *Verh. Int. Ver. Limnol.*, **18**, 1776–83

Subrahmanyan, R. 1960. Observations on the effect of the monsoons in the production of phytoplankton. *J. Ind. Bot. Soc.*, **39**, 78–89

Sud, G. C. 1969. Observations on zoochlorellae in fresh-water ciliates. *Res. Bull. Punjab Univ.*, *N.S.*, **20**, 429–41

Sukhanova, I. N. & Rodyakof, Y. A. 1973. Population composition and vertical distribution of *Pyrocystis pseudonoctiluca* (W. Thomson) in the western equatorial Pacific. In *Life activity of pelagic communities in the ocean tropics*, pp. 218–28. (Translated from Russian.) Israel Program. Sci. Transl.

Sullivan, M. J. 1975. Diatom communities from a Delaware salt marsh. *J. Phycol.*, **11**, 384–90

Sullivan, M. J. 1979. Epiphytic diatoms of three sea grass species in Mississippi Sound. *Bull. Mar. Sci.*, **29**, 459–464

Sullivan, M. J. & Daiber, F. C. 1975. Light, nitrogen and phosphorous limitation of edaphic algae in a Delaware salt marsh. *J. exp. mar. Biol. Ecol.*, **18**, 79–88

Sundene, O. 1953. The algal vegetation of Oslofjord. *Skr. Norsk. Videnskaps. Math.-Natur.*, Kl, no. 2, 244 pp.

Sundene, O. 1962a. Growth in the Sea of *Laminaria digitata* sporophytes from culture. *Nyt. Mag. Bot.*, **9**, 5–24

Sundene, O. 1962b. Reproduction and morphology in strains of *Antithamnion boreale* originating from Spitsbergen and Scandinavia. *Kon. norske Vidensk. Skr. Mat. Nat.*, Kl, **5**, 1–19

Sundene, O. 1962c. The implications of transplant and culture experiments on the growth and distribution of *Alaria esculenta*. *Nyt. Mag. Bot.*, **9**, 155–74

Sundene, O. 1964. The ecology of *Laminaria digitata* in Norway in view of transplant experiments. *Nyt. Mag. Bot.*, **11**, 83–107

Sundene, O. 1973. Growth and reproduction in *Ascophyllum nodosum* (Phaeophyceae). *Norw. J. Bot.*, **20**, 249–55

Svendsen, P. & Kain, J. M. 1971. The taxonomic status, distribution, and morphology of *Laminaria cucullata* sensu Jorde and Klavestad. *Sarsia*, **46**, 1–22

Swale, E. 1964. A study of the phytoplankton of a calcareous river. *J. Ecol.*, **52**, 433–46

Swale, E. 1969. Phytoplankton in two English rivers. *J. Ecol.*, **57**, 1–23

Swale, E. M. F. 1962. Notes on some algae from the Reddish Canal. *Br. phycol. Bull.*, **2**, 174–5

Swale, E. M. F. 1966. *Stephanosphaera* on acidic rock. *Br. phycol. Bull.*, **3**, 83–5

Swanson, C. D. & Bachmann, R. W. 1976. A model of algal exports in some Iowa streams. *Ecology*, **57**, 1076–80

Sweeney, B. M. 1971. Laboratory studies of a green *Noctiluca* from New Guinea. *J. Phycol.*, **7**, 53–8

Sweeney, B. M. 1976. *Pedinomonas noctilucae* (Prasinophyceae), the flagellate symbiotic in *Noctiluca* (Dinophyceae) in south east Asia. *J. Phycol.*, **12**, 460–6

Sweeney, B. M. & Hastings, J. W. 1958. Rhythmic cell division in populations of *Gonyaulax polyedra*. *J. Protozool.*, **5**, 217–24

Swift, D. G. & Guillard, R. R. L. 1978. Unexpected response to vitamin B_{12} of dominant centric diatoms from the spring bloom in the Gulf of Maine (Northeast Atlantic Ocean). *J. Phycol.*, **14**, 377–86

Swift, E., Biggley, W. H. & Seliger, H. H. 1973. Species of oceanic dinoflagellates in the genera *Dissodinium* and *Pyrocystis*: Interclonal and interspecific comparisons of the color and photon yield of bioluminescence. *J. Phycol.*, **9**, 420–6

Swift, E., Stuart, M. & Meunier, V. 1976. The *in situ* growth rates of some deep-living oceanic dinoflagellates. *Pyrocystis fusiformis* and *Pyrocystis noctiluca*. *Limnol. Oceanogr.*, **21**, 418–26

Szczepanski, A. 1966. Bemerkungen über die Primarproduction des Pelagials. *Verh. Int. Ver. Limnol.*, **16**, 364–71

Sze, P. & Kingsbury, J. M. 1972. Distribution of phytoplankton in polluted saline lake, Onondaga Lake, New York. *J. Phycol.*, **8**, 25–37

Szemes, G. 1967. Systematisches Verzeichnis der Pflanzenwelt der Donau mit einer zusammenfassenden Erläuterung. In *Limnobiologie der Donau*, ed. R. Liepolt. *Liefg.*, **3**, 70–131

Taasen, J. P. 1974. Remarks on the epiphytic diatom flora of *Dumontia incrassata* (Müll) Lamour (Rhodophyceae). *Sarsia*, **55**, 129–32

Talling, J. F. 1957. The growth of two plankton diatoms in mixed culture. *Physiol. Plant.*, **10**, 215–23

Talling, J. F. 1960. Comparative laboratory and field studies of photosynthesis by a marine planktonic diatom. *Limnol. Oceanogr.*, **5**, 62–77

Talling, J. F. 1965. The photosynthetic activity of phytoplankton in East African Lakes. *Int. Rev. ges. Hydrobiol.*, **50**, 1–32

Talling, J. F. 1966. Photosynthetic behaviour in stratified and unstratified lake populations of a planktonic diatom. *J. Ecol.*, **54**, 99–127

Talling, J. F. 1970. Generalized and specialized features of phytoplankton as a form of photosynthetic cover. In *Prediction and measurement of photosynthetic productivity. Proc. IPB-PP Tech. meeting*, pp. 431–45. Třeboň, Czechoslovakia

Talling, J. F. 1971. The underwater light climate as a controlling factor in the production ecology of freshwater phytoplankton. *Mitt. Int. Ver. Limnol.*, **19**, 214–42

Talling, J. F. 1976. The depletion of carbon dioxide from lake water by phytoplankton. *J. Ecol.*, **64**, 79–121

Talling, J. F., Wood, R. B., Prosser, M. V. & Baxter, R. D. 1973. The upper limit of photosynthetic productivity by phytoplankton: evidence from Ethiopian soda lakes. *Freshwater Biol.*, **3**, 53–76

Tanaka, N., Nakanishi, M. & Kadota, H. 1974. Nutritional interrelation between bacteria and phytoplankton in a pelagic ecosystem. In *Effects of the Ocean Environment on Microbial Activities*, ed. R. R. Colwell & R. Y. Morita, pp. 493–509

Taniguti, M. 1962. *Phytosociological study of marine algae in Japan*. 130 pp. Inoue & Co. Ltd, Tokyo

Taub, F. B. & Dollar, A. M. 1964. A *Chlorella-Daphnia* food chain study: The design of a compatible chemically defined culture medium. *Limnol. Oceanogr.*, **9**, 61–74

Taub, F. B. & Dollar, A. M. 1968. The nutritional inadequacy of *Chlorella* and *Chlamydomonas* as food for *Daphnia pulex*. *Limnol. Oceanogr.*, **13**, 607–17

Taylor, D. L. 1968. Chloroplasts as symbiotic organelles in the digestive gland of *Elysia viridis* (Gastropoda: Opisthobranchia). *J. mar. biol. Ass., U.K.*, **48**, 1–15

Taylor, D. L. 1970. Chloroplasts as symbiotic organelles. *Int. Rev. Cytol.*, **27**, 29–64

Taylor, D. L. 1971. On the symbiosis between *Amphidinium klebsii* (Dinophyceae) and *Amphiscolops langerhansi* (Turbellaria: Acoela). *J. mar. biol. Ass., U.K.*, **51**, 301–13

Taylor, D. L. 1972a. Symbiotic marine algae; taxonomy and biological fitness. In *Symbiosis in the Sea*, ed. W. B. Vernberg & F. J. Vernberg, pp. 245–66. University of S. Carolina Press, Columbia

Taylor, D. L. 1972b. Symbiotic pathways of carbon in coral reef ecosystems. Present status and future prospects. *Helgoländer wiss. Meeresunt.*, **24**, 276–83

Taylor, D. L. 1973. The cellular interactions of algal-invertebrate symbiosis. *Adv. mar. Biol.*, **11**, 1–56

Taylor, D. L. 1974. Nutrition of algal-invertebrate symbiosis. I. Utilization of soluble organic nutrients by symbiont-free hosts. *Proc. Roy. Soc. London*, **B, 180**, 357–68

Taylor, F. J. R. 1973. General features of dinoflagellate material collected by the 'Anton Bruun' during International Indian Ocean Expedition. In *The Biology of the Indian Ocean*, ed. B. Zeitzschel, pp. 155–9. Chapman & Hall

Taylor, J. E. 1969. Growth rate as a measure of primary productivity in benthic algae. *Chesapeake Sci.*, **10**, 299–300

Taylor, W. R. 1959. Associations algales des mangroves d'Amerique. *Coll. Int. Centre Nat. Rech. Scient.*, **81**, 143–52

Taylor, W. R. & Palmer, J. D. 1963. The relationship between light and photosynthesis in intertidal benthic diatoms. *Biol. Bull. mar. biol. Lab., Woods Hole*, **175**, 395

Taylor, W. R., Seliger, H. H., Fastie, W. G. & McElroy, W. D. 1966. Biological and physical observations on a phosphorescent bay in Falmouth Harbour, Jamaica, W.I. *J. Mar. Res.*, **24**, 28–43

Tchan, J. 1963. Study of soil algae. I. Fluorescence microscopy for the study of soil algae. *Proc. Linn. Soc., N.S.W.*, **77**, 265–69

Tchan, Y. T. 1952. Counting of soil algae by direct fluorescence microscopy. *Nature, London*, **170**, 328–9

Tchan, Y. T., Balaam, L. W., Hawkes, R. & Draette, F. 1961. Study of soil algae. IV. Estimation of the nutrient status of soil using an algal growth method with special reference to nitrogen and phosphorus. *Plant and Soil*, **14**, 147–58

Tchan, Y. T. & Beadle, N. C. W. 1955. Nitrogen economy in semi-arid plant communities. Pt. II. The non-symbiotic nitrogen fixing organisms. *Proc. Linn. Soc., N.S.W.*, **80**, 97–104

Tchan, Y. T. & Whitehouse, J. A. 1953. Study of soil algae. II. The variation of algal populations in sandy soils. *Proc. Linn. Soc., N.S.W.*, **78**, 160–70

Teal, J. M. & Kanwisher, J. 1961. Gas exchange in a Georgia salt marsh. *Limnol. Oceanogr.*, **6**, 388–99

Teiling, E. 1916. En Kaledonisk fytoplanktonformation. *Svensk Bot. Tidskr.*, **10**, 506–19

Tett, P. & Wallis, A. 1978. The general annual cycle of chlorophyll standing crop in Loch Creran. *J. Ecol.*, **66**, 227–39

Thayer, G. W. & Adams, M. 1975. Structural and functional aspects of a recently established *Zostera marina* community. In *Estuarine Research*, ed. L. E. Cronin, pp. 518–40. Academic Press, London

Thinh, L. V. & Griffiths, D. J. 1977. Studies on the relationship between the Ascidian *Diplosoma virens* and its associated microscopic algae. I. Photosynthetic characteristics of the algae. *Aust. J. mar. Freshwat. Res.*, **28**, 673–81

Thomas, E. A. 1969. The process of eutrophication in central European lakes. In *Eutrophication: Causes, consequences, correctives*, pp. 29–49. *Nat. Acad. Sci. Washington D.C.*

Thomas, W. H. 1966. Effects of temperature and illuminance on cell division rates of three species of tropical oceanic phytoplankton. *J. Phycol.*, **2**, 17–22

Thomas, W. H. 1969. Phytoplankton nutrient enrichment experiments off Baja California and in the eastern equatorial Pacific Ocean. *J. Fish. Res. Bd. Can.*, **26**, 1133–45

Thomas, W. H. 1970. On nitrogen deficiency in tropical Pacific oceanic phytoplankton: photosynthetic parameters in poor and rich water. *Limnol. Oceanogr.*, **15**, 380–5

Thomas, W. H. 1972. Observations on snow algae in California. *J. Phycol.*, **8**, 1–9

Thomas, W. H. & Dodson, A. N. 1975. On silicic acid limitation of diatoms in near-surface waters of the eastern tropical Pacific Ocean. *Deep sea Res.*, **22**, 671–7

Thompson, S. & Eglington, G. 1976. The presence of pollutant hydrocarbons in estuarine epipelic diatom populations. *Estuarine Coastal Mar. Sci.*, **4**, 417–25

Thompson, S. & Eglington, G. 1979. The presence of pollutant hydrocarbons in estuarine diatom populations. II. Diatom slimes. *Estuarine Coastal Mar. Sci.*, **8**, 75–86

Thórdardóttir, T. 1973. Successive measurements of primary production and composition of phytoplankton at two stations west of Iceland. *Norw. J. Bot.*, **20**, 257–70

Thunmark, S. 1945. Zur Soziologie des Süsswasser – planktons. Fine methodologisch – ökologische Studie. *Folia Limnol., Scand.*, **3**, 66 pp.

Tietjen, J. H. 1971. Pennate diatoms as ectocommensals of free living marine nematodes. *Oecologia, Berlin*, **8**, 135–8

Tilzer, M. M. 1973. Diurnal periodicity in the phytoplankton assemblage of a high mountain Lake. *Limnol. Oceanogr.*, **18**, 15–30

Tilzer, M. M., Paerl, H. W. & Goldman, C. R. 1977. Sustained viability of aphotic phytoplankton in Lake Tahoe (California-Nevada). *Limnol. Oceanogr.*, **22**, 84–91

Tippett, R. 1970. Artificial surfaces as a method of studying populations of benthic micro-algae in fresh water. *Br. phycol. J.*, **5**, 187–99

Titman, D. 1976. Ecological competition between algae: Experimental confirmation of resource based competition theory. *Science*, **192**, 463–5

Titman, D. & Kilham, P. 1976. Sinking in freshwater phytoplankton. Some ecological implications of cell nutrient status and physical mixing processes. *Limnol. Oceanogr.*, **21**, 409–17

Tittley, I. & Price, J. H. 1978. The benthic marine algae of the Eastern English Channel: a preliminary floristic and ecological account. *Bot. Mar.*, **21**, 499–512

Todd, D. J. 1971. New algal species record for Colorado. *J. Phycol.*, **7**, 266–7

Toerien, D. F., Hyman, K. L. & Bruwer, M. J. 1975. A preliminary trophic status classification of some South African impounds. *Water, S.A.* **1**, 15–23

Tominaga, H. 1971. Chlorophyll *a* and phaeophytin contents in the surface water of the Antarctic Ocean through the Indian Ocean. *Antarctic Rec.*, **42**, 124–34

Torrey, M. S. & Lee, G. F. 1976. Nitrogen fixation in Lake Mendota, Madison, Wisconsin. *Limnol. Oceanogr.*, **21**, 365–78

Toth, R. & Wilce, R. T. 1972. Virus-like particles in the marine alga *Chorda tomentosa* Lyngb. (Phaeophyceae.) *J. Phycol.*, **8**, 126–30

Tovey, D. J. & Moss, B. L. 1978. Attachment of the haptera of *Laminaria digitata* (Huds.) Lamour. *Phycologia*, **17**, 17–22

Towle, D. W. & Pearse, J. E. 1973. Production of the giant kelp, *Macrocystis* estimated by *in situ* incorporation of ^{14}C in polyethylene bags. *Limnol. Oceanogr.*, **18**, 155–9

Townsend, C. & Lawson, G. W. 1972. Preliminary results on factors causing zonation in *Entermorpha* using a tide simulating apparatus. *J. exp. mar. Biol. Ecol.*, **8**, 265–76

Tracey, J. I., Jr, Ladd, H. S. & Hoffmeister, J. E. 1948. Reefs of Bikini, Marshall Islands. *Bull. geol. Soc. Am.*, **59**, 861–78

Trainor, F. R. 1960. Temperature tolerance of algae in dry soil. *Phycol. News Bull.*, **15**, 3–4

Trainor, F. R. 1978. *Introductory phycology*. 525 pp. J. Wiley & Sons

Trainor, F. R. & McLean, R. J. 1964. A study of a new species of *Spongiochloris* introduced into sterile soil. *Am. J. Bot.*, **51**, 57–60

Trainor, F. R. & Shubert, L. E. 1974. *Scenedesmus* morphogenesis. Colony control in dilute media. *J. Phycol.*, **10**, 28–30

Tranter, D. J. 1973. Seasonal studies of a pelagic ecosystem (Meridian 110° E). In *The Biology of the Indian Ocean, Ecol. Studies, 3*, ed. B. Zeitzschel. Chapman & Hall

Travers, A. & Travers, M. 1968. Les siliolflagellés du Golfe de Marseille. *Mar. Biol.*, **1**, 285–8

Trench, M. F., Trench, R. K. & Muscatine, L. 1970. Utilization of photosynthetic products of symbiotic chloroplasts in mucus synthesis by *Placobranchus ianthobapsus* (Gould) Opisthobranchia, Saccoglossa. *Comp. Biochem. Physiol.*, **37**, 113–17

Trench, R. K. 1971*a*. The physiology and biochemistry of zooxanthellae symbiotic with marine coelenterates. I. The assimilation of photosynthetic products of zooxanthellae by two marine coelenterates. *Proc. Roy. Soc., London,* **B**, **177**, 225–35

Trench, R. K. 1971*b*. The physiology and biochemistry of zooxanthellae symbiotic with marine coelenterates. II. Liberation of fixed ^{14}C by zooxanthellae *in vitro*. *Proc. Roy. Soc., London,* **B**, **177**, 237–50

Trench, R. K. 1971*c*. Physiology and biochemistry of zooxanthellae symbiotic with marine coelenterates. III. The effect of homogenates of host tissues on the excretion of photosynthetic products *in vitro* by zooxanthellae from two marine coelenterates. *Proc. Roy. Soc., London,* **B**, **177**, 251–64

Trench, R. K. 1975. Of 'leaves that crawl'; functional chloroplasts in animal cells. In *Symbiosis*, ed. D. H. Jennings & D. L. Lee. *Symp. Soc. Exp. Biol.*, **29**, pp. 229–65 Cambridge University Press

Trench, R. K., Boyle, J. & Smith, D. C. 1973. The association between chloroplasts of *Codium fragile* and the mollusc *Elysia viridis*. I. Characteristics of isolated *Codium* chloroplasts. *Proc. Roy. Soc., London,* **B**, **184**, 51–61

Trench, R. K., Boyle, J. & Smith, D. C. 1974. The association between chloroplasts of *Codium fragile* and the mollusc *Elysia viridis*. III. Movement of photosynthetically fixed ^{14}C in tissue of intact, living *E. viridis* and in *Tridachia crispata*. *Proc. Roy. Soc., London,* **B**, **185**, 453–64

Trench, R. K., Green, R. W. & Bystrom, B. G. 1969. Chloroplasts as functional organelles in animal tissues. *J. Cell. Biol.*, **42**, 404–17

Trench, M. W., Trench, R. K. & Muscatine, L. 1970. Utilization of photosynthetic products of symbiotic chloroplasts in mucus synthesis by *Placobranchus ianthobapsus* (Gould), Opisthobranchia, Saccoglassa. *Comp. Biochem. Physiol.*, **37**, 113–17

Trench, R. K., Trench, M. F. & Muscatine, L. 1972. Symbiotic chloroplasts. Their photosynthetic products and contribution to mucus synthesis in two marine slugs. *Biol. Bull. mar. biol. Lab., Woods Hole*, **142**, 335–49

Trench, R. K. & Smith, D. C. 1970. Synthesis of pigment in symbiotic chloroplasts. *Nature, London*, **227**, 196–7

Treub, M. 1888. Notice sur la nouvelle flora de Krakatau. *Ann. Jard. bot. Buitenz.*, **7**, 221–3

Tsuda, R. T., Larson, H. K. & Lujan, R. J. 1972. Algal growth on beaks of live parrot-fishes. *Pacif. Sci.*, **26**, 20–3

Turner, J. B. & Friedmann, E. J. 1974. Fine structure of capitular filaments in the coenocytic green alga *Penicillus*. *J. Phycol.*, **10**, 125–34

Tyler, M. A. & Seliger, H. H. 1978. Annual subsurface transport of a red tide dinoflagellate to its bloom area. Water circulation patterns and organism distributions in the Chesapeake Bay. *Limnol. Oceanogr.*, **23**, 227–46

Uherkovich, G. 1969. Über die Quantitativen Verhältnisse des Phytosestons (Phytoplanktons) der Danau, Drau und Theiss. *Acta. bot. hung.*, **15**, 183–200

Uhlmann, D. 1978. The upper limit of phytoplankton production as a function of nutrient load, temperature, retention time of the water and euphotic zone depth. *Int. Rev. ges. Hydrobiol.*, **63**, 353–63

Ukeles, R. 1961. The effect of temperature on the growth and survival of several marine algal species. *Biol. Bull.*, **120**, 255–264

Ukeles, R. & Rose, W. E. 1976. Observations on organic carbon utilization by photosynthetic marine microalgae. *Mar. Biol.*, **37**, 11–28

Vaccaro, R. F. 1960. Inorganic nitrogen in sea water. In *Chemical Oceanography, vol. 1*, ed. J. P. Riley & G. Skirrow, pp. 365–408. Academic Press, London

Vaccaro, R. F. & Ryther, J. H. 1960. Marine phytoplankton and the distribution of nitrite in the sea. *J. Cons. perm. int. Explor. Mer.*, **25**, 260–71

Vadas, R. L. 1968. The ecology of *Agarum* and the kelp bed community. Ph.D. Thesis, University of Washington. 280 pp.

Valet, G. 1960. Algues épiphytes et endophytes du *Chaetomorpha area* (Dillw.) Kütz. *Naturalia monospel. Ser. Bot. Fasc.*, **17**, 89–101

Valkanov, A. 1968. Das Neuston. *Limnologica*, **6**, 381–408

Valkenburg, S. D. van & Norris, R. E. 1970. The growth and morphology of the silicoflagellate *Dictyocha fibula* Ehrenberg in culture. *J. Phycol.*, **6**, 48–54

Vallentyne, J. R. 1974. *The algal bowl. Lakes and Man. Misc. Spec. Publ., Dept. Envir.*, no. 22, 186 pp. Fisheries & Marine Service, Ottawa

Vandermeulen, J. H., Davis, N. D. & Muscatine, L. 1972. The effect of inhibitors of photosynthesis on zooxanthellae in corals and other marine invertebrates. *Mar. Biol.*, **16**, 185–91

Vandermeulen, J. H. & Muscatine, L. 1974. Influence of symbiotic algae on calcification in reef corals: critique and progress report. In *Symbiosis in the Sea*, ed. W. B. Vernberg & F. J. Vernberg, pp. 1–19. University of S. Carolina Press, Columbia

Velasquez, G. T. 1940. On the viability of algae obtained from the digestive tract of the Gizzard Shad, *Dorosoma cepedianum*. *Am. Midl. Nat.*, **22**, 376–412

Venrick, E. L. 1971. Recurrent groups of diatom species in the North Pacific. *Ecology*, **52**, 614–25

Venrick, E. L. 1972. Small-scale distribution of oceanic diatoms. *Fish. Bull. natl. oceanic atmos. Adm. U.S.*, **70**, 363–72

Venrick, E. L., McGowan, J. A. & Mantyla, A. W. 1973. Deep maxima of photosynthetic chlorophyll in the Pacific Ocean. *Fish. Bull. natl. oceanic atmos. Adm. U.S.*, **71**, 41–52

Verduin, J. 1952. Photosynthesis and growth rates of live diatom communities in western Lake Erie. *Ecology*, **33**, 163–9

Verduin, J. 1957. Daytime variations in phytoplankton photosynthesis. *Limnol. Oceanogr.*, **2**, 333–6

Verduin, J. 1959. Photosynthesis by aquatic communities in northwestern Ohio. *Ecology*, **40**, 377–83

Vesk, M. & Jeffrey, S. W. 1974. The effect of blue light on chloroplast number and thylakoid stacking in the marine diatom, *Stephanopyxis turris*. In *Proc. 8th Int. Electron Microscope Conf., Canberra*

Vielhaben, V. 1963. Zur Deutung des semilunaren Fortpflanzungszyklus von *Dictyota dichotoma. Z. Bot.*, **51**, 156–73

Villanueva, R., Jordan, R. & Burd, A. 1969. Informe sobre el estudio de comportamiento de cardumenes de anchoveta. *Infme. espec. Inst. Mar. Peru IM*, **45**, 18 pp.

Vine, P. J. 1974. Effects of algal grazing and aggressive behaviour of the fishes *Pomacentrus lividus* and *Acanthurus sohol* on coral reef ecology. *Mar. Biol.*, **24**, 131–6

Viner, A. B. 1973. Responses of a mixed phytoplankton population to nutrient enrichments of ammonia and phosphate and some associated ecological implications. *Proc. Roy. Soc., London*, **B**, **183**, 357–70

Visser, S. A. & Villeneuve, J. P. 1975. Similarities and differences in the chemical composition of waters from West, Central and East Africa. *Verh. Int. Ver. Limnol.*, **19**, 1416–28

Vollenweider, R. A. 1968. Scientific fundamentals of the eutrophication of lakes and flowing waters, with particular reference to nitrogen and phosphorus as factors in eutrophication. In *OECD Rep. DAS/CSC/68.27*, Paris

Vollenweider, R. A. 1974. A manual on methods for measuring Primary Production in aquatic environment. In *IBP Handbook*, no. 12. Blackwell, Oxford

Vollenweider, R. A., Munawar, M. & Stadelmann, P. 1974. A comparative review of phytoplankton and primary production in the Laurentian Great Lakes. *J. Fish. Res. Bd. Can.*, **31**, 739–62

Waern, M. 1952. Rocky-shore algae in the Öregrund Archipelago. *Act. Phytogeog. Suec.*, **30**, 1–298

Wainwright, S. A. 1965. Reef communities visited by the Israel South Red Sea Expedition, 1967. *Bull. Sea Fish. Res. Stn., Israel*, rep. 9, **38**, 40–53

Waite, T. D. & Mitchell, R. 1976. Some benevolent and antagonistic relationships between *Ulva lactuca* and its microflora. *Aquatic Bot.*, **2**,

Wall, D. 1970. Quaternary dinoflagellate micropaleontology 1959–1969. In *Proc. N. Am. Paleont. Conv., Chicago*, vol. 2, pp. 844–66

Wall, D. 1971. The lateral and vertical distribution of Dinoflagellates in quaternary sediments. In *Micropalaeontology of the Oceans*, ed. B. M. Funnell & W. R. Riedel, pp. 399–405. Cambridge University Press

Wall, D. & Dale, B. 1968a. Early Pleistocene dinoflagellates from the Royal Society Borehole at Ludham, Norfolk. *New Phytol.*, **67**, 315–26

Wall, D. & Dale, B. 1968b. Modern dinoflagellate cysts and evolution of the Peridiniales. *Micropalaeontology*, **14**, 265–304

Wall, D. & Dale, B. 1973. Paleosalinity relationships of dinoflagellates in the late Quaternary of the Black Sea – a summary. *Geosci. and Man*, 7, 95–102

Wall, D., Dale, B. & Harada, K. 1973. Descriptions of new dinoflagellates in the late Quaternary of the Black Sea. *Micropalaeontology*, **19**, 18–31

Wallace, J. B., Sherberger, S. R. & Sherberger, F. F. 1976. Use of the diatom *Terpsinoe musica* Ehrenb. (Biddulphiales: Biddulphiaceae) as case making material by *Nectopsyche* larva (Trichoptera: Leptoceridae). *Am. Midl. Nat.*, **95**, 236–9

Wallen, D. G. & Geen, G. H. 1971a. Light-quality and concentration of proteins, RNA, DNA and photosynthetic pigments in two species of marine algae. *Mar. Biol.*, **10**, 44–51

Wallen, D. G. & Geen, G. H. 1971b. The nature of the photosynthate in natural phytoplankton populations in relation to light quality. *Mar. Biol.*, **10**, 157–68

Wallen, D. G. & Geen, G. H. 1971c. Light quality in relation to growth, photosynthetic rates and carbon metabolism in two species of marine plankton algae. *Mar. Biol.*, **10**, 34–42

Walmsley, R. D. & Ashton, P. J. 1977. Algal growth potential and the succession of algae in a nitrogen limited impoundment. *S. Afr. J. Sci.*, **73**, 151–2

Walsby, A. E. 1969. The permeability of blue-green algal gas vacuole membranes to gas. *Proc. Roy. Soc., London*, **B**, **173**, 235–55

Walsby, A. E. 1971. The pressure relationships of gas vacuoles. *Proc. Roy. Soc., London*, **B**, **178**, 301–26

Walsby, A. E. 1978. The properties and buoyancy providing role of gas vacuoles in *Trichodesmium* Ehrenberg. *Br. phycol. J.*, **13**, 103–16

Walsby, A. E. & Xypolyta, A. 1977. The form resistance of chitan fibres attached to the cells of *Thalassiosira fluviatilis* Hustedt. *Br. phycol. J.*, **12**, 215–23

Walsh, J. J. 1969. Vertical distribution of Antarctic phytoplankton. II. A comparison of phytoplankton standing crops in the Southern Ocean with that of the Florida Strait. *Limnol. Oceanogr.*, **14**, 86–96

Walsh, J. J. 1971. Relative importance of habitat variables in predicting the distribution of phytoplankton at the ecotone of the Antarctic upwelling ecosystem. *Ecol. Monogr.*, **41**, 291–309

Walsh, J. J. & Dugdale, R. C. 1972. Nutrient submodels and simulation models of phytoplankton production in the sea. In *Nutrients in natural waters*, ed. H. E. Allen & J. R. Kramer, pp. 171–91. Wiley-Interscience

Walter, M. R. 1972. Stromatolites and the biostratigraphy of the Australian Pre-Cambrian and Cambrian. *Spec. Pap. Paleont.*, **11**, 1–190

Walter, M. R., Bauld, J. & Brock, T. D. 1972. Siliceous algal and bacterial stromatolites in hot spring and geyser effluents of Yellowstone National Park. *Science*, **178**, 402–5

Wanders, J. B. W. 1976a. The role of benthic algae in the shallow reef of Curaçao (Netherlands Antilles). I. Primary productivity in the coral reef. *Aquatic Bot.*, **2**, 235–70

Wanders, J. B. W. 1976b. The role of benthic algae in the shallow reef of Curaçao (Netherlands Antilles). II. Primary productivity of the *Sargassum* beds on the north-east coast submarine plateau. *Aquatic Bot.*, **2**, 327–35

Warmke, G. L. & Almodóvar, L. R. 1963. Some associations of marine molluscs and algae in Puerto Rico. *Malacologia*, **1** (2), 63–77

Watson, R. 1970. Distribution of epiphytic algae on palm fronds. In *A Tropical Rain Forest*, ed. H. T. Odum. Div. of Tech. Inf. US, AEC. Oak Ridge, Tennessee

Watt, W. D. 1969. Extracellular release of organic matter from two freshwater diatoms. *Ann. Bot.*, **33**, 427–37

Watt, W. D. 1971. Measuring the primary production rates of individual phytoplankton species in natural mixed populations. *Deep Sea Res.*, **18**, 329–40

Ware, G. W., Dee, M. K. & Cahill, W. P. 1968. Water flora as indicators of irrigation water contamination by DDT. *Bull. Envir. Contam. Toxicol.*, **3**, 333–8

Wartenberg, A. 1972. *Systematik der niederen Pflanzen.* 326 pp. Georg Thieme, Stuttgart

Webb, K. L. & Wiebe, W. J. 1978. The kinetics and possible significance of nitrate uptake by several algal-invertebrate symbioses. *Mar. Biol.*, **47**, 21–7

Webber, E. 1967. Blue-green algae from a Massachusetts salt marsh. *Bull. Torrey Bot. Club*, **94**, 99–100

Weber, C. A. 1907. Aufbau und Vegetation der Moore Norddeutschlands. *Bot. Jahrb. Beibl.*, **90**, 19–34

Weber, C. I. 1973. Recent developments in the measurement of the response of plankton and periphyton to changes in their environment. In *Bioasay techniques and environmental chemistry*, ed. G. E. Glass, pp. 119–38

Weber, C. J. & Raschke, R. L. 1970. *Use of a floating periphyton sampler for water pollution surveillance*, 22 pp. Fed. Wat. Poll. Control Admin. U.S. Dept. Int. Reprint

Wegmann, K. 1971. Osmotic regulation of photosynthetic glycerol production in *Dunaliella. Biochem. Biophys. Acta*, **234**, 317–23

Weiler, C. S. & Chisholm, S. W. 1976. Phased cell division in natural populations of marine dinoflagellates from shipboard cultures. *J. exp. mar. Biol. Ecol.*, **25**, 239–47

Weimann, R. 1933. Hydrobiologische und hydrographische Untersuchungen an zwei teichartigen Gewässern. *Botan. Centr. Beih.*, **51**, 397–476

Welch, E. B., Emery, R. M., Matsuda, R. I. and Dawson, W. A. 1972. The relation of periphytic and planktonic algal growth in an estuary to hydrographic factors. *Limnol. Oceanogr.*, **17**, 731–7

Wenrich, D. H. 1924. Studies on *Euglenamorpha Regneri* nov. gen., nov. sp. euglenoid flagellate found in tadpoles. *Biol. Bull.*, **47**, 149–74

Werner, D. 1970. Productivity studies on diatom cultures. *Helgoländer wiss. Meeresunt.*, **20**, 97–103

Werner, D. 1971. Der Entwicklungscyclus mit Sexualphase bei der marinen Diatomee *Coscinodiscus asteromphalus*. I. Kultur und synchronisation von Entwicklungstadien. *Arch. Mikrobiol.*, **80**, 43–9

West, J. A. 1972. Environmental control of sporulation of *Rhodochorton purpureum*. In *Contributions to the systematics of benthic marine algae of the North Pacific*, ed. I. A. Abbott & M. Kurogi, pp. 213–30. Sapporo, Japan

West, K. R. & Pitman, M. G. 1967. Ionic relations and ultrastructure in *Ulva lactuca. Aust. J. Biol. Sci.*, **20**, 901–14

West, W. & West, G. S. 1906. A comparative study of the plankton of some Irish loughs. *Proc. Roy. Ir. Acad.*, B, **32**, 77–116

Westlake, D. F., Casey, H., Ladle, M., Mann, R. H. & Marker, A. F. H. 1972. The chalk-stream ecosystem. In *Productivity problems of Freshwaters. IBP-UNESCO Symp.*, ed. Z. Kajak & Hillbricht-Ilkowska, pp. 15–35

Wetzel, R. G. 1966. Variations in productivity of Goose and hypereutrophic Sylvan Lakes, Indiana. *Invest. Indiana Lakes Streams*, **7**, 147–83

Wetzel, R. G. 1970. Recent and postglacial production rates of a marsh lake. *Limnol. Oceanogr.*, **15**, 491–503

Wetzel, R. G. 1975. Primary production. In *River Ecology*, ed. B. A. Whitton, pp. 230–47. Blackwell, Oxford

Wetzel, R. G. & Allen, H. L. 1971. Functions and interactions of dissolved organic matter and the littoral zone in lake metabolism. In *Proc. IBP-UNESCO Symp. Productivity Problems of Freshwaters*, ed. Z. Kayak & Hillbricht-Ilkowska, pp. 333–47

Wetzel, R. G., Rich, P. H., Miller, M. C. & Allen, H. L. 1972. Metabolism of dissolved and particulate detrital carbon in a temperate hard-water lake. In *Detritus and its role in aquatic ecosystems*. 185–243. Proc. IBP-UNESCO Symp. Pallanza, Italy

Wetzel, R. G. & Westlake, D. F. 1969. Perphyton. In *IBP Handbook*, no. 12, pp. 33–40. Blackwell, Oxford

White, A. W. 1974. Uptake of organic compounds by two facultatively heterotrophic marine centric diatoms. *J. Phycol.*, **10**, 433–8

White, E. B. & Boney, A. D. 1969. Experiments with some endophytic and endozoic *Achrochaetium* species. *J. exp. mar. Biol. Ecol.*, **3**, 246–74

White, E. B. & Boney, A. D. 1970. *In situ* and *in vitro* studies on some endophytic and endozoic *Achrochaetium* species. *Nova Hedw.*, **19**, 841–81

Whitford, L. A. 1960. The current effect and growth of freshwater algae. *Trans. Am. Micros. Soc.*, **79**, 302–9

Whitford, L. A. & Schumacher, G. J. 1961. Effect of current on mineral uptake and respiration by a freshwater alga. *Limnol. Oceanogr.*, **6**, 423–5

Whitford, L. A. & Schumacher, G. J. 1964. Effect of current on respiration and mineral uptake in *Spirogyra* and *Oedogonium. Ecology*, **45**, 168–70

Whittaker, R. H. 1973. *Handbook of vegetation science. Part V. Ordination and classification of communities*, 737 pp. W. Junk, The Hague

Whittick, A. 1977. The reproductive ecology of *Plumaria elegans* (Bonnem.) Schmitz (Ceramiaceae, Rhodophyta) at its northern limits in the western Atlantic. *J. exp. mar. Biol. Ecol.*, **29**, 223–30

Whittick, A. & Hooper, R. G. 1976. The distribution and reproductive biology of *Antithamnion cruciatum* (C.Ag.) Näg. at its northern limits in the Western Atlantic. *Br. phycol. J.*, **11**, 201

Whitton, B. A. 1971. Filamentous algae as weeds. In *3rd European Weed Res. Council Symp.*, *Oxford*, pp. 249–63

Whitton, B. A. 1972. Environmental limits of plants in flowing waters. In *Symp. zool. Soc., London*, no. 29, pp. 3–19

Whitton, B. A. 1974. Changes in the British freshwater algae. *Syst. Assoc. Spec. Vol.*, no. 6, pp. 115–61

Whitton, B. A. 1975. *River Ecology. Studies in Ecology*, vol. 2. 725 pp. Blackwell, Oxford

Whitton, B. A. & Dalpra, M. 1968. Floristic changes in the River Tees. *Hydrobiol.*, **32**, 545–50

Wiebe, P. H., Hulburt, E. M., Carpenter, E. J., Jahn, A. E., Knopp, G. P., III, Boyd, S. H., Ortner, P. B. & Cox, J. L. 1976. Gulf stream cold core rings: large scale interaction sites for open ocean plankton communities. *Deep Sea Res*, **23**, 695–710

Wiebe, W. J., Johannes, R. E. & Webb, K. L. 1975. Nitrogen fixation in a coral reef community. *Science*, **188**, 257–9

Wiebe, W. J. & Smith, D. F. 1977. Direct measurement of dissolved organic carbon release by phytoplankton and incorporation by microheterotrophs. *Mar. Biol.*, **42**, 213–23

Wieser, W. 1959. Zur Ökologie der Fauna mariner Algen mit besonderer Berücksichtigung des Mittelmeeres. *Int. Rev. ges. Hydrobiol.*, **44**, 137–80

Wilce, R. T. 1959. The marine algae of the Labrador Peninsula and Northwest Newfoundland (ecology and distribution). *Natl. Mus. Can. Bull.*, **158**, 1–103

Wilce, R. T. 1967. Heterotrophy in arctic sublittoral seaweeds: An hypothesis. *Bot. Mar.*, **10**, 185–97

Wilkinson, L. 1964. Nitrogen transformations in a polluted estuary. *Adv. Wat. Pollut. Res.*, **3**, 405–20

Wilkinson, M. 1974. Investigations on the autecology of *Eugomontia sacculata* Kornm, a shell-boring algae. *J. exp. mar. Biol. Ecol.*, **16**, 19–27

Wilkinson, M. 1975. The occurrence of shell-boring *Phaeophila* species in Britain. *Br. phycol. J.*, **10**, 235–40

Wilkinson, M. & Burrows, E. M. 1972a. The distribution of marine shell-boring algae. *J. mar. biol. Ass., U.K.*, **52**, 59–65

Wilkinson, M. & Burrows, E. M. 1972b. An experimental taxonomic study of the algae confused under the name *Gomontia polyrhiza*. *J. mar. biol. Ass., U.K.*, **52**, 49–57

Willen, T. 1961. The phytoplankton of Ösbysjön, Djursholm. I. Seasonal vertical distribution of species. *Oikos*, **12**, 36–69

Willenbrink, J., Rangoni-Kübbeler, M. & Tersky, B. 1975. Frond development and CO$_2$-Fixation in *Laminaria hyperborea*. *Planta*, **125**, 161–70

Willey, R. L., Bowen, W. R. & Durban, E. 1970. Symbiosis between *Euglena* and Damselfly nymphs is seasonal. *Science*, **170**, 80–1

Williams, L. G. 1962. Plankton population dynamics. *Hlth. Serv. Publs., Wash.*, no. 663, 90 pp.

Williams, L. G. 1964. Possible relationships between plankton-diatom species numbers and water-quality estimates. *Ecology*, **45**, 809–23

Williams, L. G. 1972. Plankton diatom species biomasses and the quality of American rivers and the Great Lakes. *Ecology*, **53**, 1038–50

Williams, L. G. & Scott, C. 1962. Principal diatoms of major waterways of the United States. *Limnol. Oceanogr.*, **7**, 365–79

Williams, P. J., LeB. & Askew, C. 1968. A method of measuring the mineralization by micro-organisms of organic compounds in sea-water. *Deep Sea Res.*, **15**, 365–75

Williams, R. B. 1964. Division rates of salt marsh diatoms in relation to salinity and cell size. *Ecology*, **45**, 877–80

Williams, R. B. 1965. Unusual motility of tube-dwelling pennate diatoms. *J. Phycol.*, **1**, 145–6

Williams, R. B. 1966. Annual phytoplanktonic production in a system of shallow temperate estuaries. In *Some contemporary studies in Marine Science*, ed. H. Barnes, pp. 699–716. Allen & Unwin, London

Willson, D. L. & Forest, H. S. 1957. An exploratory study of soil algae. *Ecology*, **38**, 309–13

Winberg, G. G. 1971. *Symbols, units and conversion factors in studies of fresh water productivity.* 23 pp. IBP Section (PM)

Windle Taylor, E. 1973. *Forty-fifth Report on the results of the bacteriological, chemical and biological examination of the London water for the years 1971–1973.* 188 pp. *Met. Wat. Bd.*

Wing, B. L. & Clendenning, K. A. 1971. Kelp surfaces and associated invertebrates. In *The Biology of Giant Kelp Beds (Macrocystis) in California*, ed. W. J. North. *Nova Hedw. Beih.*, **32**, 319–39

Winn, H. E. & Bardach, J. E. 1960. Some aspects of the comparative biology of parrot fishes at Bermuda. *Zoologica*, **45**, 29–34

Winner, R. W. 1969. Seasonal changes in biotic diversity and in Margalef's pigment ratio in a small pond. *Verh. int. Ver. Limnol.*, **17**, 503–10

Winner, R. W. 1972. An evaluation of certain indices of eutrophy and maturity in lakes. *Hydrobiol.*, **40**, 223–65

Withers, N., Vidaver, W. & Lewin, R. A. 1978. Pigment composition, photosynthesis and fine structure of a non-blue-green prokaryotic algal symbiont (*Prochloron* sp.) in a didemnid ascidian from Hawaiian waters. *Phycologia*, **17**, 167–71

Wollast, R. 1974. The silica problem. In *The Sea, vol. 5, Marine Chemistry*, ed. E. D. Goldberg, pp. 359–92. Interscience

Womersley, H. B. S. 1948. The marine algae of Kangaroo Island. II. The Pennington Bay Region. *Trans. Roy. Soc. S. Australia*, **72**, 143–67

Womersley, H. B. S. 1954. The species of *Macrocystis* with special reference to those on Southern Australian coasts. *Univ. Calif. Publs. Bot.*, **27**, 109–32

Womersley, H. B. S. 1959. The marine algae of Australia. *Bot. Rev.*, **25**, 545–614

Womersley, H. B. S. & Bailey, A. 1969. The marine algae of Solomon Islands and their place in Biotic Reefs. *Phil. Trans. Roy. Soc., London*, **B**, **255**, 433–42

Womersley, H. B. S. & Edmonds, S. J. 1952. Marine coastal zonation in southern Australia in relation to a general scheme of classification. *J. Ecol.*, **40**, 84–90

Womersley, H. B. S. & Edmonds, S. J. 1958. A general account of the intertidal ecology of south Australian coasts. *Aust. J. mar. Freshwat. Res.*, **9**, 217–60

Wood, E. J. F. 1965. *Marine microbial ecology.* Reinheld, New York 243 pp.

Wood, E. J. F. 1967. *Microbiology of oceans and estuaries.* Elsevier Publ. Co. 319 pp.

Wood, E. J. F. 1971. Phytoplankton study – an appraisal. *J. Cons. perm. int. Explor. Mer.*, **34**, 123–6

Woodwell, G. M. 1967. Toxic substances and ecological cycles. *Sci. Amer.*, **216**, 24–31

Woolacott, R. M. & North, W. J. 1971. Bryozoans of California and northern Mexico kelp beds. In *The Biology of Giant Kelp Beds (Macrocystis) in California*, ed. W. J. North. *Nova Hedw. Beih.*, **32**, 445–79

Wornardt, W. W. 1967. Miocene and pliocene marine diatoms from California. *Occ. Pap. Calif. Acad. Sci.*, no. 63, 108 pp.

Wornardt, W. W. Jr, 1969. Diatoms past, present, future. In *Proc. 1st Int. Conf. on Planktonic Microfossils, Geneva*, pp. 690–714

Wray, J. L. 1971. Algae in reefs through time. In *Proc. N. Am. Paleont. Conv., Chicago, vol. 2*, ed. E. L. Yochelson, pp. 1358–73

Wright, R. T. 1964. Dynamics of a phytoplankton community in an ice-covered lake. *Limnol. Oceanogr.*, **9**, 163–78

Wright, R. T. & Hobbie, J. E. 1966. Use of glucose and acetate by bacteria and algae in aquatic eco-systems. *Ecology*, **47**, 447–64

Wuthrich, M. & Matthey, W. 1978. Les Diatomées de la Toubiére du Cochan (Jura, suisse). II. Associations et distribution des espèces caractéristiques. *Schweiz. Z. Hydrol.*, **40**, 87–103

Wylie, P. A. & Schlichting, H. E. Jr. 1973. A floristic survey of corticolous subaerial algae in North Carolina. *J. Elisha Mitchell Scient. Soc.*, **89**, 179–83

Wynne, D. 1977. Alterations in activity of phosphatases during the *Peridinium* bloom in Lake Kinneret. *Physiologica Pl.*, **40**, 219–24

Wynne, M. J. & Loiseaux, S. 1976. Recent advances in life history studies of the Phaeophyta. *Phycologia*, **15**, 435–52

Yamauchi, K. 1973. Influence of salinity on the growth of sporelings of laver. *Bull. Jap. Soc. scient. Fish.*, **39**, 489–96

Yentsch, C. S. 1974. Some aspects of the environmental physiology of marine phytoplankton: A second look. *Oceanogr. Mar. Biol. Ann. Rev.*, **12**, 41–75

Yentsch, C. S., Backus, R. H. & Wing, A. 1964. Factors affecting the vertical distribution of bioluminescence in the euphotic zone. *Limnol. Oceanogr.*, **9**, 519–24

Yentsch, C. S. & Ryther, J. H. 1957. Short term variations in phytoplankton chlorophyll and their significance. *Limnol. Oceanogr.*, **2**, 140–2

Yentsch, C. S. & Scagel, R. F. 1958. Diurnal study of phytoplankton pigments. An *in situ* study in East Sound, Washington. *J. Mar. Res.*, **17**, 567–83

Yonge, C. M. 1940. The Biology of Reef-Building Corals. *Sci. Rep. Gt. Barrier Reef Expd.*, **1**, (13) 353–89

Yonge, C. M. 1968. Living corals. *Proc. Roy. Soc. London*, **B**, **169**, 329–44

Yonge, C. M. & Nicholls, A. G. 1931a. Studies on the physiology of corals. *IV Sci. Rep. Gt. Barrier Reef Expd.*, **1** (6), 135–76

Yonge, C. M. & Nicholls, A. G. 1931b. Studies on the physiology of corals. V. The effect of starvation in light and darkness on the relationship between corals and zooxanthellae. *Sci. Rep. Gt. Barrier Reef Expd.*, **1**, 178–213

Young, D. L. K. & Barber, R. T. 1973. Effects of waste dumping in New York Bight on the growth of natural populations of phytoplankton. *Environ. Pollut.*, **5**, 237–52

Young, J. O. & Eaton, J. W. 1975. Studies on the symbiosis of *Phaenocora typhlops* (Vejdovsky) (Turbellaria; Neo rhabdocoela) and *Chlorella vulgaris* var. *vulgaris*. Fott & Novakova (Chlorococcales). II. An experimental investigation in the survival value of the relationship to host and symbiont. *Arch. Hydrobiol.*, **75**, 225–39

Zaneveld, J. S. 1937. The littoral zonation of some Fucaceae in relation to dessication. *J. Ecol.*, **25**, 431–68

Zaneveld, J. S. & Barnes, W. D. 1965. Reproductive periodicities of some benthic algae in lower Chesapeake Bay. *Chesapeake Sci.*, **6**, 17–37

Zehnder, A. 1953. Beitrag zur Kenntnis von Mikroklima und Algenvegetation des nackten Gesteins in der Tropen. *Ber. Schweiz. Bot. Ges.*, **63**, 5–26

Zeitzschel, B. 1969. Primary productivity in the Gulf of California. *Mar. Biol.*, **3**, 201–7

Zeitzschel, B. 1970. The quantity, composition and distribution of suspended particulate matter in the Gulf of California. *Mar. Biol.*, **7**, 305–18

Zelinka, M. & Marvan, P. 1961. Zum Präzisierung der biologische Klassifikation der Reinheit fliessender Gewässer. *Arch. Hydrobiol.*, **57**, 389–407

Zimmermann, P. 1961. Experimentelle Untersuchungen über die ökologische Wirkung der Strömungsgeschwindigkeit auf die Lebengemeinschaften des fliessender Wassers. *Schweiz. Z. Hydrol.*, **23**, 1–81

Zimmermann, V. 1969. Ökologische und physiologische Untersuchungen an der planktonischen Blaualge *Oscillatoria rubescens* D.C. unter besonderer Berucksichtigung von Licht und Temperatur. *Schweiz. Z. Hydrol.*, **31**, 1–58

Zobell, C. A. 1977. 7. Substratum 7.1 Bacteria, fungi and blue-green algae. In *Marine Ecology. Environmental Factors*, vol. *1*, ed. O. Kinne, pp. 1251–76. Wiley-Interscience

Zobell, C. E. 1972. Bacteria, Fungi and Blue-green algae. In *Marine Ecology*, vol. I. *Environmental factors*, part 3, ed. O. Kinne, pp. 1251–70. Wiley-Interscience

INDEX

Numbers in italics refer to Tables and Figures; those in bold type refer to general discussions.